Medicinal and Aromatic Plants of the World

Volume 7

Series Editor
Ákos Máthé, Faculty of Agriculture and Food Science,
Széchenyi István University, Mosonmagyaróvár, Hungary

Medicinal and Aromatic Plants (MAPs) have been utilized in various forms since the earliest days of mankind. They have maintained their traditional basic curative role even in our modern societies. Apart from their traditional culinary and food industry uses, MAPs are intensively consumed as food supplements (food additives) and in animal husbandry, where feed additives are used to replace synthetic chemicals and production-increasing hormones. Importantly medicinal plants and their chemical ingredients can serve as starting and/or model materials for pharmaceutical research and medicine production. Current areas of utilization constitute powerful drivers for the exploitation of these natural resources. Today's demands, coupled with the already rather limited availability and potential exhaustion of these natural resources, make it necessary to take stock both of them and enrich our knowledge regarding research and development, production, trade and utilization, and especially from the viewpoint of sustainability. The series Medicinal and Aromatic Plants of the World is aimed to look carefully at our present knowledge of this vast interdisciplinary domain, on a global scale. In the era of global climatic change, the series is expected to make an important contribution to the better knowledge and understanding of MAPs.

Budapest, Prof. Dr. Ákos Máthé.

More information about this series at http://www.springer.com/series/11192

Ákos Máthé • Arnaldo Bandoni
Editors

Medicinal and Aromatic Plants of South America Vol. 2

Argentina, Chile and Uruguay

Springer

Editors
Ákos Máthé
Faculty of Agriculture and Food Science
Széchenyi István University
Mosonmagyaróvár, Hungary

Arnaldo Bandoni
Facultad de Farmacia y Bioquímica,
Farmacognosia
Universidad de Buenos Aires
Buenos Aires, Argentina

ISSN 2352-6831 ISSN 2352-684X (electronic)
Medicinal and Aromatic Plants of the World
ISBN 978-3-030-62817-8 ISBN 978-3-030-62818-5 (eBook)
https://doi.org/10.1007/978-3-030-62818-5

This Springer imprint is published by the registered company Springer Nature Switzerland AG
The registered company address is: Gewerbestrasse 11, 6330 Cham, Switzerland

Preface

This book, as the seventh volume of the series Medicinal and Aromatic Plants of the World (MAPW), focuses on the medicinal and aromatic plants (MAPs) of the so-called Southern Cone (SC), the southernmost geographic and cultural region of South America. South of the Tropic of Capricorn, this region of the huge South American continent comprises Argentina, Chile and Uruguay. Frequently, also Paraguay is considered in this context, though due to certain deliberations, in our volume, we focus on the three aforementioned countries.

The Southern Cone, this remote, Southern edge of the South American continent, is home to numerous medicinal and aromatic plants that have been scientifically less explored.

Due to this fact and also to the adventurous history of the area, much of the traditional knowledge of original peoples has been lost or remained less known to the world. This relates also to the traditional knowledge and use of medicinal and aromatic plants. Scientific information on the frequently unique, endemic species is scarce or limited.

The aim of the collective of authors, who are specialists working in the relevant fields of medicinal and aromatic plants, is to explore/collect/summarize and evaluate/validate the still available information on these resources.

With the selection of 41 occasionally less-known species, the editors' aim was to complement the knowledge on species already published in similar monographs, e.g. MAPs of Brazil (vol. 5 of this series). Due to the vast diversity and large number of endemic species, this task has been easy to comply with.

By dealing with less-known and/or less-utilized species or – on the contrary – on already overexploited species, the volume desires to call attention to the need to study the huge wealth of MAP biodiversity in this area, with the ultimate aim of either sustainable utilization or management.

The volume also aims to contribute to the written chronicle of traditional knowledge of original peoples of the area, which – in view of the deep-cutting political and cultural changes of the past – has been losing its deep-rooted and social values. In this respect, this volume can be considered a gap-filling work.

Finally, the editors express special thanks to Professors Dr. Robert Verpoorte and Dr. Eduardo Dellacassa, members of the Scientific Advisory Board of this volume, for their support and critically reading relevant parts of the manuscript.

Mosonmagyaróvár, Hungary Ákos Máthé

Buenos Aires, Argentina Arnaldo L. Bandoni

Contents

Part I
Introduction

Medicinal and Aromatic Plants in the Southern Cone

Ákos Máthé and Arnaldo Bandoni

"Yo marcharé por los caminos en procura de hierbas
En elección de plantas textiles y aromáticas
Que luego estrujaré, ayudadora, sobre la greda.

Cuando el alfarero ponga el vaso en las manos de Dios
Tendrá también el olor vegetal de las selvas.

Y Dios dirá con plácida sorpresa:
"¡Qué brillantes son y qué bien huelen,
Mis tierras de América!"

Juana de Ibarbourou – Poemas
Ed. Colección Austral
Buenos.Aires. 1946

"I will walk the roads in search of herbs
Choosing fibre plants, and aromatic,
That I, the helper, will squeeze over the clay.

When the potter puts the vessel in God's hands
It will still have the leafy jungle smell.

And God, with calm surprise, will say:
"How bright they are, how sweet they smell,
My lands of America!"

Juana de Ibarbourou – Poems
Ed. Colección Austral
Buenos.Aires. 1946

Abstract The Southern Cone (SC) is a geographic and cultural region composed of the southernmost areas of South America. Although traditional use of plants (also as medicines) has a long history here despite the uniqueness and richness of the floras have not been completely explored and utilized.

Á. Máthé (✉)
Faculty of Agriculture and Food Sciences, Széchenyi István University, Mosonmagyaróvár, Hungary
e-mail: mathe.akos@sze.hu; acatius2005@gmail.com

A. Bandoni (✉)
Facultad de Farmacia y Bioquímica, Cátedra de Farmacognosia, Universidad de Buenos Aires, C.A. de Buenos Aires, Argentina
e-mail: albandoni@speedy.com.ar

Á. Máthé, A. Bandoni (eds.), *Medicinal and Aromatic Plants of South America Vol. 2*, Medicinal and Aromatic Plants of the World 7,
https://doi.org/10.1007/978-3-030-62818-5_1

The present volume deals with the medicinal and aromatic plants of Argentina, Chile and Uruguay, the three countries that due to geographical, historical and population aspects, are usually "fused" into one region and as a culture. For certain, mainly practical reasons, South of Brazil and Paraguay are not subjects of this volume.

The countries of the Southern Cone are in the Neotropical region. Remarkably, however, the biodiversity in Chile is sufficiently distinguishable from the relevant biodiversities of Argentina and Uruguay, which difference is rooted in the isolation caused by the Andes Mountains in the East, the Pacific Ocean in the West and the Andean Plateau (Puna) in the North of its territory.

The history of the Southern Cone can be broken down into four main eras: pre-Columbian era, colonial era, era of independence from Spain and era of the nineteenth century to the present day. As the indigenous population in these countries has been diminished by the historic events, as a consequence, the contribution of indigenous people to the dominant culture, including the knowledge of medicinal and aromatic plants is rather modest.

The natural flora of Southern Cone comprises a total of *ca.* 12,000 species, distributed in several eco-regions. Certain areas, like the Subantarctic forest, are rich in endemic or exclusive families, while there is a smaller number of endemic species in relation to other areas. In Chile, as a result of geographic isolation, a large degree of endemism is observed with some 1043 species.

Medicinal and Aromatic Plants are a rich source and potential providers of not only herbal medicines, but plant materials also for other applications, different branches of industries. Therefore, as a positive trend, the exploration and utilization of medicinal and aromatic plant resources is expanding. In view of the numerous quality influencing factors, like the uncertain botanical identity or diverse or frequently unknown origin of plant materials, this seems to be paramount task, though the mapping of vegetation has already started and as a precondition of good quality medicinal and aromatic plant produce. A pertinent regulatory framework for the appropriate use and control of these native natural resources already exists in the region. It is to be hoped that all these efforts will lead to the sustainable utilization of the rich MAP resources for the benefit of the populations of this peculiar region of the world.

Keywords Medicinal and aromatic plants · Southern Cone · Argentina · Chile · Uruguay

1 Introduction

The present volume deals with the Medicinal and Aromatic Plants (MAPs) of the so called Southern Cone, a geographic and cultural region of South America that is composed of the southernmost areas of this huge continent, South of the Tropic Capricorn. Traditionally, the Southern Cone (SC) covers Argentina, Chile and Uruguay. It should be noted that some authors include also Paraguay and Southern Brazil in the same region, though for practical reasons this aspect, herewith, will be omitted (Fig. 1).

The question might arise: Why Argentina, Uruguay and Chile in a single volume of this series? The answer is simple: These three countries share various geographical, historical and population aspects that fuse them as a region and as a culture.

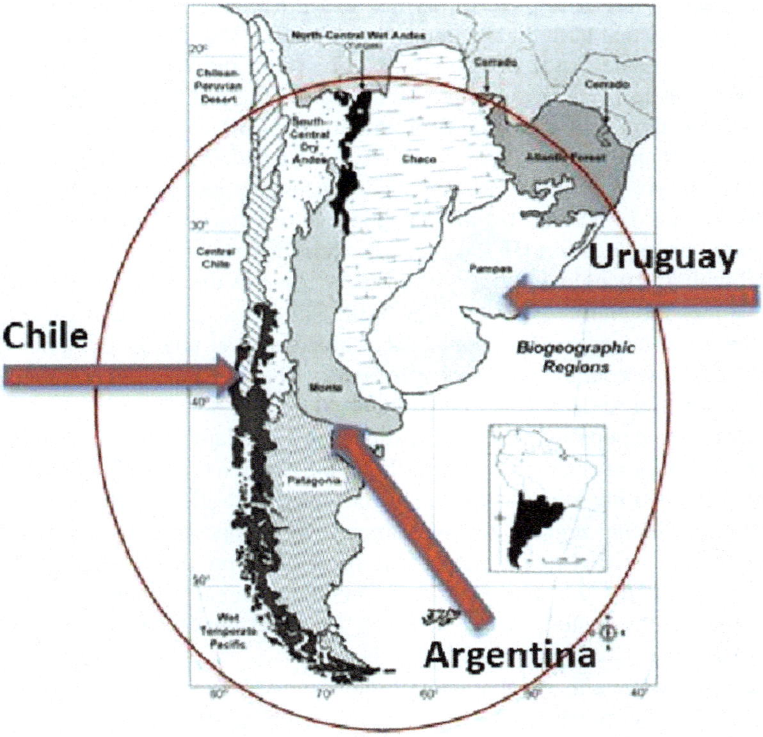

Fig. 1 Biogeographic Regions of the Southern Cone. (After: Zuloaga and Belgrano 2015)

2 Geographic Characteristics of the Southern Cone

The Southern Cone is rich in geographic variations. This variation contributes to the continent's large number of biomes.

2.1 Argentina

The Argentine Republic occupies a major part of the territory of the Southern Cone, with 2,791,810 square kilometers in the continental area (the maritime area extends to approximately 1 million ha) (IGN 2020). It borders, to the West, the Republic of Chile (the largest border, is longer than 5300 km). To the North, Argentina borders Bolivia and Paraguay, and to the East, Southern Brazil, Uruguay and the Atlantic Ocean, where it has more than 4700 km of coastline, on its territorial sea.

Argentina's great latitudinal extension (more than 3500 km), is home to a great diversity of climates and soils. In the North, stand out the last foothills of the Amazon, with tropical to subtropical climate. In the Northeast, the climate and therefore the flora and agricultural production are influenced by the presence of large wetlands and extensive rivers, such as Paraná and Uruguay. The Center of the country is an agro-industrial area par excellence, encompassing the Pampean region. The Southern area includes mainly Patagonia (more than 1 million square kilometers), large areas of arid steppes, meadows and deserts, not yet exploited and in general with a very cold and rigorous climate. Towards the West, the highest peaks of the Andes mountains serve as limits with Chile, and it has the highest peak in America: the Aconcagua (6962 m). The major classical phytogeographic regions identified in this country are shown in Fig. 2: Chaco, Amazonian, Antarctic and Subantarctic, and Andean-Patagonic dominions.

The Argentine population density is very high in the large cities (in the North and Center of the country) and very low in the Southern area (*ca.* 2 inhabitants/km^2).

Argentina is a net exporter of grains and meat, with global relevance. Its most important producing regions are the Pampa and Mesopotamia (in the Northeast). Argentina has a varied range of regional agricultural products, both for domestic market and exports: the most important production areas include the viticulture, citrus, the "yerba mate", and to a much lesser extent the medicinal and aromatic plants.

In 2000, it was estimated that natural forests covered a total area of 34,6 million hectares (12.7% of the country's total area) (SENASA 2014).

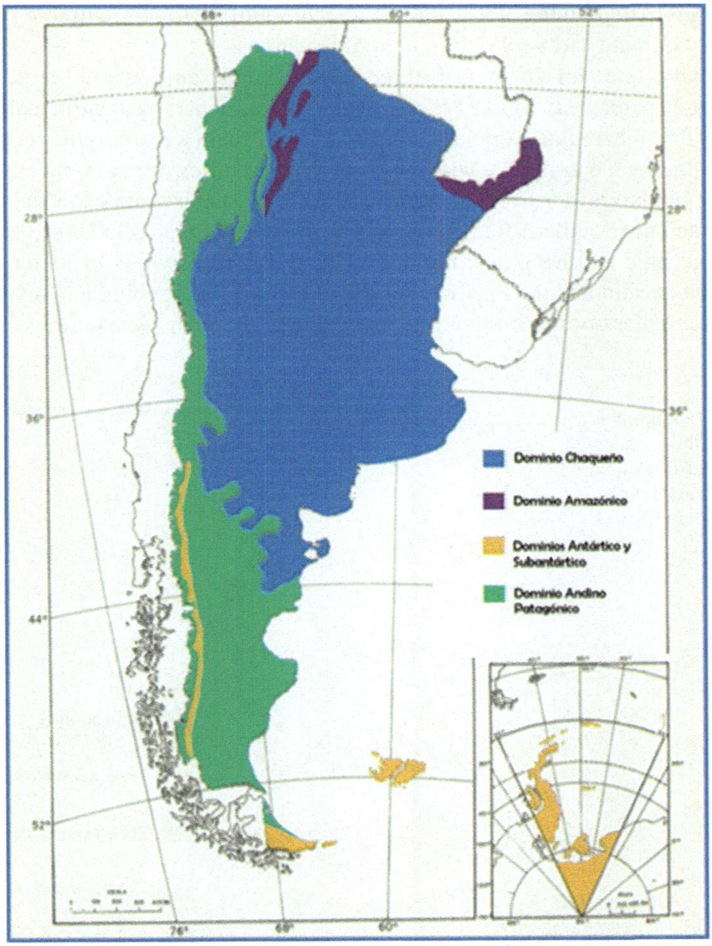

Fig. 2 Major phytogeographic regions of Argentina, adapted from Cabrera (1971)

2.2 *Chile*

This country has the particularity of having a very vast latitudinal extension (more than 4700 km), but a very narrow longitudinal amplitude (with maximums of 350 km). This continental area (756,626 square kilometers) is also limited by the Pacific Ocean to the West, by the Andes Mountains to the East, by desert regions to the North, on its border with Peru, and by the Antarctic continent to the South (Andrade 1999). It is a territory that, in its entirety, has great heights on its Eastern side (the Andes), and decreases to the West, to the Pacific coast.

These geographical limitations have a strong influence on the climate and soils of their territory, and therefore on their floristic biodiversity. Between the extreme temperatures (high in the North and low in the South), it has a Central region with a temperate climate and very rich in natural resources.

Basically, one can distinguish three large climatic and edaphic regions in the continental sector (Fig. 3): the North, covered in large part by the Puna area, desert and very dry, where the average rainfall is 1.1 mm/year (Xeromorphic region). As it goes South, the Central area, with a temperate to Mediterranean climate, where the highest population density and agro-industrial production appear (Mesomorphic region); and the Southern area, very wet and cold (more than 4000 mm/year of rain) (Hydromorphic and patagonic regions). Due to the narrowness of its territory, the Humboldt current and the Pacific winds regulate the climate in much of Chile's territory. Regardless of these three regions, a fourth, Andean (Andean region), which

Fig. 3 Biogeographic zones of Chile. (Adapted from https://download.rincondelvago.com/biogeografia-chilena)

is defined according to its height, can be described by a particular vegetation of cacti, snouted plants and peatlands according to latitudes (Andrade 1999).

It is noteworthy to mention the importance of products derived from Chilean native flora in the export statistics, something that is not common in the SC. Boldo, algae, quillay, as well as *Sphangum magellanicus* moss make a significant contribution not only to Chilean foreign trade but also to the international statistics of these commodities.

Chile has some 2 million hectares of forest plantations (2019). The most commonly cultivated species are *Pinus* spp. and *Eucalyptus* spp. The most commercialized native forest species are *Nothofagus* spp., but exports of these woods have declined sharply in recent years, due to the reduced reforestation (INFOR 2020).

2.3 Uruguay

The Oriental Republic of Uruguay has a territory of 176,215 square kilometers, bordered to the North by Brazil, to the East by the Atlantic Ocean, to the South by the Río de la Plata and to the West by the Uruguay River, which separates it from Argentina (IGM 2020).

Uruguay's territory consists almost exclusively of a Pampean plain (at its highest point within the country it is 500 meters high). The Pampean plain has three phytogeographic zones (Fig. 4): the Central zone; the Eastern zone, ranging from the Atlantic coast to the center of the country, and the West coast of the Uruguay River.

The climate here is basically temperate, with an annual average temperature of 17 °C accompanied usually by high humidity (annual average above 70%) and rainfall between 1200 and 1600 mm/year (INIA 2011).

Agricultural and livestock production occupies about 14,5 million ha (*ca*. 96% of the territory). This reflects the importance of these activities for the country's economy (MGAP 2015). These figures also indicate the limited area of wildland and explain that there is currently a strong tendency in the country to preserve their native flora (Soutullo et al. 2013).

In Uruguay, the large scale (industrial) production of citrus deserves to be highlighted: it occupies about 17,000 ha land area and produces some 315,000 tons of fruits (INIA 2020).

In the forest areas of Uruguay, like in Chile and Argentina, the dominant species are *Eucalyptus* spp. and *Pinus* spp. These are utilized mostly by the paper industry (2600 million tons of wood-pulp, in 2019) (MGAP 2020).

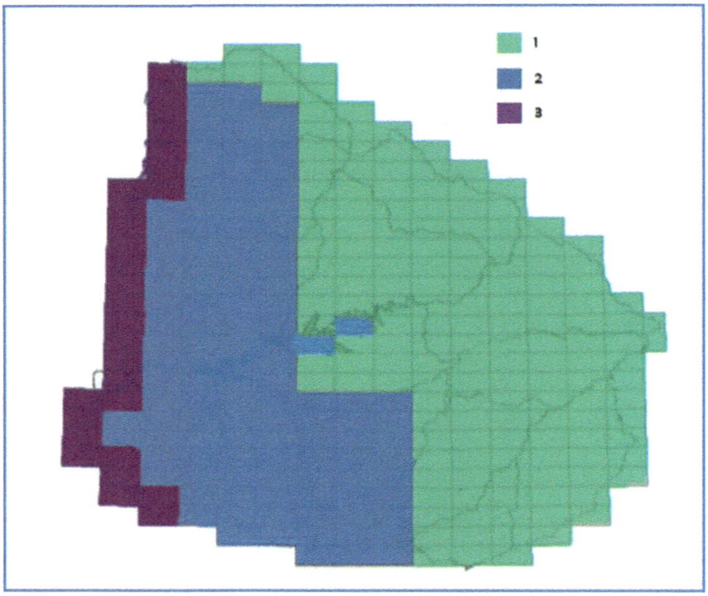

Fig. 4 Biozones of woody species from Uruguay: East (1), Central (2) and West coast (3). (Adapted from Brazeiro et al. 2012)

3 Phytogeographic Features and Native (Endemic) Floras of the Southern Cone

From a phytogeographic point of view, the countries of the Southern Cone are in the Neotropical region. Remarkably, however, the biodiversity in Chile is sufficiently distinguishable from the relevant biodiversities of Argentina and Uruguay. This difference is rooted in the isolation caused by the Andes Mountains in the East, the Pacific Ocean in the West and the Andean Plateau (Puna) in the North of its territory.

Although some species are common to the floras on both sides of the Andes, there are also several differentiating elements in these two regions of the Southern Cone. In particular, it is the historical and cultural elements that also weigh when it comes to analyzing their ethnobotanical potential, as a source of phytomedicines.

The systematic distributional analysis of the floras of SC carried out lately by Zuloaga and Belgrano (2015) has resulted in a checklist of the vascular plants of the Southern Cone. The analysis (that beyond the target area of our review, covers also the territories of Southern Brazil: Paraná, Santa Catarina and Rio Grande do Sul states, as well as Paraguay) presents updated information of 19,787 taxa distributed in 18,139 species and in 2679 genera and 318 families. 7787 species are endemic to the region.

According to Ulloa Ulloa et al. (2017), as of the middle of 2010ies, there had been no previous attempts to catalog the plant diversity of the Americas in its entirety.

They published a plant checklist compiled from specimen-based data from floras and checklists that cover all sectors of the Americas that they consider verifiable and which can be readily updated with input from taxonomic specialists. They state that keeping these listings up to date is challenging, given the constant stream of discoveries and taxonomic and nomenclatural changes.

Ulloa Ulloa et al. (2017) also attempted to explore floristic similarities among 12 geographical areas in the Americas, including the Southern Cone. The degree of floristic similarity among the geographic areas is represented as a nonmetric multidimensional scaling (NMDS). Distance and placement are indicative of similarity among areas (Fig. 5).

Only the temperate part of Southern Cone has the most in common with Brazil and Bolivia (Fig. 5, Ulloa Ulloa et al. 2017). After this, the authors explain the permanent update of the list of new species in America. To demonstrate the validity of these findings, some authors (Ulloa Ulloa et al. 2017) stated that in the past 25 years, between 439 and 1022 (average 744) species have been described each year and only *ca.* 10 years after the publication of Brazil.

Pitman and Jørgensen (2002) estimated that as many as 10 to 20% of species could remain undescribed in tropical American biodiversity hotspots. Compared to the number of new species described between 1990 and 2014, by 2050 the exploratory work would yield about 152,000 species for the Americas.

A comprehensive analysis of the vascular plants of the SC with a special focus on Argentina carried out by Zuloaga et al. (2019) summarizes 11 years of research after the publication of their original study. The study offers floristic information for the Southern Cone and each country.

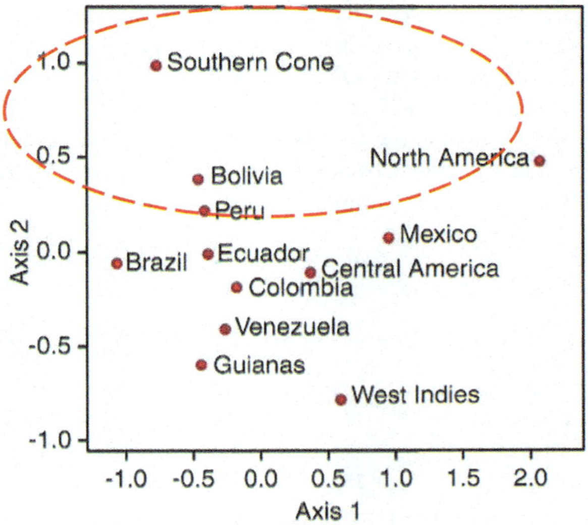

Fig. 5 Floristic similarities among geographical areas of the Americas. (Source: Ulloa Ulloa et al. 2017)
Note: Axes 1 and 2 show the floristic similarities among regions according to an NMDS ordination based on Sørensen's distances

The following brief compilation is meant to offer an insight into the floras of the individual countries. Owing to the extreme diversity of (edaphic as well as climatic, biotic) factors to be considered, any floristic comparison is hardly realistic.

3.1 *Flora of Argentina*

Although the revision (updating) of the Flora of Argentina is currently still under elaboration, reliable figures are available from the previously cited publication by Zuloaga and Belgrano (2015). For the Flora of Argentina, there is a current record of 10,944 taxa distributed in 10,006 species, 2090 genera and 283 families, with a total of 1749 endemic species (Table 1) representing 17.48% of the country's flora.

3.2 *Flora of Chile*

The flora of Chile is composed of *ca.* 813 genera. Analyzing the distribution of each genus, 7 floristic elements were identified by Moreira-Muñoz 2011) (Table 2).

As regards the **medicinal plant flora** of this extremely diverse country, an overview of the flora of Central Chile by San Martín (1983) can offer good orientation: in this, 131 species are mentioned as species used in popular medicine. They were found from 10 sites in Central Chile and include Pteridophytes (2 families, 2 species) and Angiosperms (52 families and 122 species of Dicotyledoneae; 3 families and 7 species of Monocotyledoneae). Sixty-six of these are native and the rest introduced species.

According to Niemeyer (1995) the geographical isolation of Chile provided by the Andean range, the Pacific Ocean, and the Atacama Desert, has produced a unique flora of *ca.* 5000 species, with a high degree of endemism (*ca.* 50%).

The extant traditional pharmacopoeia consists of nearly 300 native species, of which only some 90 have been studied chemically. Regarding their origin, the number of native and exotic plant species was equal. Preference in usage was found to be evenly shared between both types of plants. Remarkably, however, cultivated plants are not as commonly used as wild plants. This may reflect the impact of the

Table 1 Argentina, summary of the taxa recorded for the country

Group	Families	Genera	Species	% Species	Endemics	% Endemics
Pteridophyta	34	97	390	3,90	16	4,10
Gymnospermae	5	12	26	0,26	1	3,85
Dicotyledoneae	194	1519	7162	71,58	1383	19,31
Monocotyledoneae	50	462	2428	24,27	349	14,37
Total	**283**	**2090**	**10,006**		**1749**	

Source: Zuloaga and Belgrano (2015)

Table 2 Floristic elements for the vascular plant genera (Moreira-Muñoz 2011)

Floristic elements	No. of Genera	%
1. Pantropical	88	11
2. Australasiatic	59	7
3. Neotropical	216	27
4. Antitropical	152	19
5. South-temperate	81	10
6. Endemic	83	10
7. Cosmopolitan	134	16
Total	**813**	**100**

Table 3 Summary of Chilean flora (From Moreira-Muñoz 2011)

	Orders	Families	Genera	Native species
Ferns and fern allies	18	24	50	121
Gymnosperms	2	4	9	16
Monocots	6	30	163	880
Dicots	33	121	591	3316
Total	59	179	813	4333

local flora as a readily available natural resource. Based on the number of species used, families of greater significance are: Compositae (I7), Rosaceae (12), and Labiatae (9) (Moreira-Muñoz 2011) (Table 3).

Localities, where medicinal plants were more intensively used are Talca, Curepto, Constitución, Corel and San Clemente; all, except Corel, are urban centers (with competent health centers and established ambulatory commerce of medicinal plants). According to this study (Moreira-Muñoz 2011) species like *Senecio pycnanthus*, *Quinchamalium majus*, *Calceolaria thyrsiflora* and *Myosclzilos oblonga* are often subject to massive collection by the inhabitants and as thus, they are facing the danger of extinction.

3.3 *Flora of Uruguay*

The **flora of Uruguay** numbers some 2500 species distributed among 150 native and foreign families. Approximately 80% of Uruguay's territory is prairie, with grasses predominating. Uruguay is primarily a grass-growing land, with a vegetation that is essentially a continuation of the Argentine Pampas (Brussa et al. 2007).

Uruguay has traditionally been considered as a part of the Pampean Province, a region characterized by extensive grasslands or grass-dominated steppes. Forested areas are much smaller than the pampas grasslands but contain a mix of hardwoods and softwoods. *Eucalyptus* spp. were imported from Australia (Haretche et al. 2012), as in all of the SC regions.

A study by Brussa et al. (2007) seems to suggest that the number of tree and shrub species in the flora might be greater than that expected for the Pampean region. Trees and shrubs are characteristic elements of forests and shrublands, respectively. They represent 12% of all vascular plants in Uruguay, which adds up to approximately 2750 species. In Uruguay, the surface occupied by native forests is 835,389 ha (MGAP 2019). This represents 4.77% of the continental territory and 136 species of trees and 150 species of shrubs. The main types of forests present in the country are gallery forests, highlander forests, palms, open forests and coastal forests (MGAP 2019). Some of the most common species are: *Scutia buxifolia, Allophylus edulis, Eugenia uniflora, Celtis tala, Prosopis affinis, Lithraea molleoides, Vachellia caven* and *Sebastiana* spp. Regarding the shrub-land character of the country, the floristic studies deal mostly, with the areas occupied by such woody vegetation. The data compilation in Table 4 is an attempt to compare the richness of trees in the Uruguay flora with the neighboring regions (Haretche et al. 2012).

The comparison could offer further support for the notion. There are differences between Uruguay and the Province of Buenos Aires floras, and this is consistent with the demarcation of an Uruguayan District and the Pampean Province, as proposed by Chebataroff (1942) (Fig. 6) and Morrone (1999), although critically it should be mentioned this conclusion is not mentioned in the later publication by Morrone (2001).

The **medicinal plants' flora** of Uruguay can be estimated by the species cited in two of the most relevant and recognized local references devoted to the use of medicinal plants in the country (Arrillaga de Maffei 1969; Alonso Paz et al. 2007), as 95 and 30 native species, respectively.

3.4 Endemism in the Southern Cone

According to Zuloaga and Belgrano (2015), the analysis of the distribution of endemisms through the different biogeographic regions of the Southern Cone allows for the conclusion that the number of endemic or exclusive families and genera is high in the Subantarctic forests, while there is a smaller number of endemic species in other areas.

These authors point out that Mediterranean Chile and deserts of Patagonia and Monte have a high number of endemic families, genera and species, when compared with the totality of the analyzed flora. As such, the indices are decreasing

Table 4 Richness of trees at the level of families, genera species and tree density in Uruguay and neighboring regions (Haretche et al. 2012)

	Families	Genera	Species	Species /1000 km^2
Buenos Aires	30	50	62	0,20
Entre Ríos	35	84	106	1,34
Uruguay	46	105	151	0,86
Rio Grande do Sul (Brazil)	79	238	510	1,81

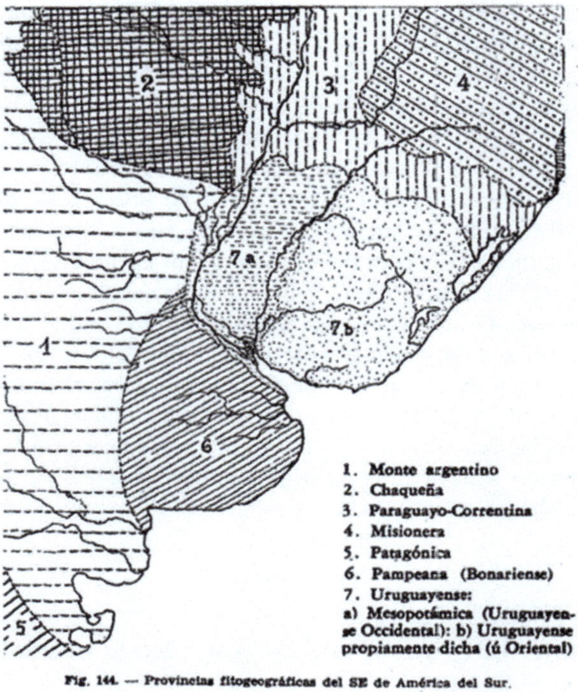

Fig. 6 Phytogeographic provinces of Southeastern South America. (Chebataroff 1942)

with the latitude increasing towards Patagonia, together with the general decrease in species richness of the vascular flora. The Chilean-Peruvian desert and adjacent high Andean and Puna areas in Northern Argentina do not have endemic families and the number of endemic genera is smaller here, while the number of endemic species is outstanding. Remarkably, the amount of endemism diminishes from West to East, with low numbers in the Chaco and Pampa, probably due to the extensive agronomic exploitation. It starts growing again, in terms of the number of endemic species, in the Atlantic Forest from of Eastern Paraguay, Northeastern Argentina and Southern Brazil.

The territories of Uruguay and Argentina are broadly similar in their border areas for their orography and phytogeographic expression (Haretche et al. 2012). Regarding **Argentina**, about 20% of the species identified in its territory are considered endemic (Zuloaga 2013). Barboza et al. (2009) and Rondina et al. (2003) report about 1600 medicinal species, many of which are also valid for **Uruguayan** medicinal flora. In this country, (Cracco et al. 2007; Soutullo et al. 2013) the flora was estimated as *ca.* 2500–2700 vascular species, with 95% of knowledge and 20% of threatened species. Haretche et al. (2012) estimate 313 native species for Uruguay: this list includes species such as *Ilex paraguariensis, Schinus molle, Aspidosperma quebracho blanco, Erythrina crista-galli* (Argentinean and

Uruguayan national flower), *Geoffroea decorticans*, *Prosopis nigra, Heimia salicifolia, Eugenia uniflora, Phytolacca dioica,* and several others, that are actually typical species from all over the Rio de La Plata and adjacent regions, including the entire Uruguayan territory. In **Chile**, as a result of geographical isolation, a large degree of endemism is observed with some 1043 species (Urban 1934; Mellado Campos et al. 1997; Chile 2014, 2019). 624 species of its flora are used for medicinal purposes by native peoples (Chile 2016). The flora of its islands (e.g.: the Chilean Juan Fernández Archipelago,), like the majority of the world isolated islands, with significant distances from the continents, is home to a large number of endemic species (Vargas et al. 2011).

The **endemism** of some plant species in territories of the Southern Cone is caused by the different phytogeographic subregions it covers. Its extent has been influenced by human intervention through the various population expansions and their needs for food, medicine, utensil manufacturing, and even playful, cosmetic, religious and social customs, etc. The obligatory nomadism suffered by some civilizations, and the crossing of different original cultures, such as the Incas, Guaraníes, Mapuches, Calchaquíes, and Collas, among others, have also contributed to its dissemination.

4 Inheritance of Ethnobotanical Knowledge in the Light of Cultural Roots

The history of the Southern Cone can be broken down into four main eras: pre-Columbian era, colonial era, era of independence from Spain and Portugal and era of the nineteenth century to the present day.

4.1 Pre-Columbian Era

From the ethnographic point of view, in the pre-Columbian times, these territories represented the southern boundary of two of the great native American civilizations: the Inca and the Tupi-Guarani. Their influences on the original peoples were partly limited by the great distances and weather conditions. Also, the belligerence of some local tribes, such as Mapuches and the ancient Pampas tribes (e.g. Tehuelches and Ranqueles), had put an end to Inca expansion only a few years before the arrival of the Spaniards.

Added to this, it must be mentioned that no writing existed in these cultures. In the case of Guarani, certain colonial-era backgrounds and the knowledge of their language by the Jesuit missionaries have made it possible to generate perpetuating documentations (Melià 2006; Wilde 2013). Among the Incas, the use of so-called "quipus" may have partially resolved this shortcoming (Urton and Bresine 2005;

Urton and Chu 2018), but the ability to "read" their complex knotted and colored strings has been lost. Most of these cultures were transmitted and survived by orality and not by writing or through the Spanish sixteenth century chroniclers.

4.2 *Colonial Era*

Spanish colonization (the Treaty of Tordesillas of 1494 attempted to impose limits on the Portuguese entering these regions, despite the several attempts to occupy the territories of the present Uruguay and a part of northeastern Argentina), (1. Picture), established the *Virreinatos* in the South American territories, and subsequently, quasi as a consequence of the legal system, the *Audiencias*, during the sixteenth and seventeenth centuries.

1. Picture – Sanson, Nicolas, 1600–1667. Norman B. Leventhal Map Center Collection. Available in https://collections.leventhalmap.org/search/commonwealth:x633f861p

The Virreinato of the Río de La Plata (created in 1776) was a segmentation of the Virreinato of Peru, due to its enormous territorial extension. It established to rule all these Southern lands. To this end, it divided the region into *Capitanías Generales*; Chile was located between them. Chile already existed from 1541 and always had problems with its Southern boundaries, a region that it could not dominate due to the strong resistance of the Mapuches, called Araucanos by Spaniards, during the so-called Guerra de Arauco. This region continued to depend on both the Virreinato of Peru, until 1798, and Buenos Aires. The latter included also the territories of the present-day Uruguay. The numerous impregnable border areas (that always meant the Andes Mountains in this region) made much of Chile's present-day territories politically separated. Even from a legal point of view, the Real Audiencia de Concepción (1565) was formed and then called Chile (created in 1609). Thereby, it was separated from the Real Audiencia of Buenos Aires (1661).

These territorial unions and divisions were the first political and cultural links between the territories on both sides of the Andes, although they were separated by economic and strategic factors.

The collision with the Western culture, at the time of Spanish colonization from the sixteenth century onwards was a major shock to the possibility of inheriting the extensive knowledge that the Pre-Hispanic populations collected from the surrounding plant world. Still worse, the colonizers tried to erase any indication of those who considered "the savagery" of the inhabitants of these soils. Even doubting, in the early years, whether they were human beings or not, considering them mere beasts (Dumont 2003; León Guerrero and Gervás 2013). This removed much of the remains that could be found from the original cultures.

4.2.1 Beneficial Role of Jesuits in Preserving Written Traditions

An exception to above policy was the work of the Jesuit settlements, who knew how to amalgamate, even in part, both the European and the American traditions. The strategy they used for this was to maintain, and still employ, the Guarani language in their communication with the indigenous people (they made use of the time they used to study Latin to get to know the regional languages) (Martín and Valverde 1995). The result of this practice was the fusion of the two cultures for which a written documentary acquis was generated. This allowed much of the Guarani traditions to be preserved and, in turn, these cultures were prepared to adopt European knowledge. This is how many texts of that time included words of the indigenous, written based on phonetics, and many of which have survived to this day: "chipi" (*Petiveria alliacea*), "iviraró guasú" (*Aspidosperma quebracho blanco),* "karaguatá" (*Bromelia balansae*), "karandai" (*Trithrinax campestris*), "mburukuyá" (*Passiflora* spp.), and so on.

4.2.2 Transculturation

Much of our present knowledge of these ancient civilizations has been strongly influenced by European beliefs, traditions, sciences and arts from the 16th centuries onwards. South American toponymy is a clear example of this transculturation, singularly in the terms Argentina, Uruguay and Chile, reminiscent of the search for precious metals, or Guarani and Aymara or Mapuche words. The influence of the Quechua language (of the Inca Empire) on South American Spanish is also remarkable. This phenomenon was not so disruptive in Bolivia and Paraguay, possibly due to the logical isolation that the orography and the Amazon present in these countries provoked.

4.3 Independence from Spain

In the early nineteenth century, the American revolutions that demanded independence from Spain began. The historical feat of General San Martín crossing the Andes was one of the main milestones of this epic enterprise, which ended in the strengthening of local governments, with the independence of Argentina (which at that time reached the territory of today's República Oriental del Uruguay), Chile and Peru.

In Argentina, in the middle of the same century, the so-called "Conquista del Desierto" was launched to prevent the attack of the indigenous people in the Pampean region and take territorial possession in Patagonia, East of the Andes. This military campaign achieved to consolidate the national territory, but at the cost of the extermination of most of the original peoples, with their oral traditions. Something similar happened in Chile, with the so-called "Pacificación de la Araucaria", almost simultaneously. The difference was that they did not exterminate the Mapuches in the South of the country. In addition, in both Northern Chile and Argentina, the "colla culture" had remained but was totally modified by the Inca (pre-Columbian) and Christian (colonial era) influences.

4.4 Nineteenth and Twentieth Centuries

A fourth historical period in the history of Southern Cone is observed during the late nineteenth and early twentieth centuries. The famines, wars and industrialization in Europe forced large social groups to emigrate from their countries and seek new horizons with better futures. South America was an ideal port for these goals. Spanish, Italian, German and so many other communities populated this region. They brought their culture and traditions and populated many of the territories that still followed deserts, abandoned or disintegrated areas. As nations they were already well delimited, the countries that received most of this immigration were Argentina, Chile and Uruguay (in addition to Brazil).

To-date, in Chile it is estimated on the basis of the national census of 2002 that 4.6% of the population is indigenous: in Argentina it is 1.5% (national census 2004) and in Uruguay, 0.4% (national census 2006), excluding African or black ancestry. (In the neighboring countries, such as Bolivia and Paraguay, these figures are historically far greater) (Paraguay 2018; IGWIA 2019).

Regarding the **inheritance of ethnobotanical knowledge**, both the orality and the phonetic transcription of some languages such as Guarani or Mapuche, represent a fragile connection that has allowed these peoples to get to know the traditions of their native ancestors. It was crossed by the Europeanizing culture but, nevertheless, it has been able to survive in isolated niches, reluctant to communicate, because of the ancestral arbitrariness with which they were treated. It is a challenge for today's scientists to rescue this cultural universe that is gradually lost in a globalized and dehumanizing world.

For these reasons, the work of a handful of authors, who have recognized this situation and tried to collect and compile the remaining ethnobotanical knowledge, is invaluable. The pioneering works of chroniclers, expeditions and naturalists were published in view of these objectives. Some important examples are: de Rosales (1674), Ruiz and Pavon (1798), Molina (1788, 1810), Gay (1845/1854), Murillo (1889), Sánchez Labrador (1948), Paucke (2010), Parodi (1881), Domínguez (1928), Dobrizhoffer (1967), Azara (1805), Hieronymus (1882), Acosta (1979), Girault (1984), de Mösbach (1992), Pastor Arenas (1981), Martínez Crovetto (1981), Bertoni (1922–1927), Scarpa (2009) and Duke (2009). Recently, Álvarez (2019) reviewed the ethnobotanical data published in the last years for Argentina.

In this context, the re-editions of the manuscript called *Herbario de las Misiones* and attributed to the P.J. Pedro de Montenegro (1945) is worth highlighting, especially in the virtual absence of written records. This document that was originally dated at *ca.* 1710 is considered as a compilation. The manuscript has been repeatedly published and it represents for the Southern Cone region a document of great value, like the Badiano Codex, for Nahuatl culture: the first written indigenous ethnopharmacological document of these regions. Several authors (Arata 1898; Martín and Valverde 1995) attribute enormous historic importance to this document, although – like in the case of several botanical books written before the Linnean classification- even despite its plant illustrations-it is difficult to identify the species cited. Nonetheless, the scientific significance of these records is unquestionable.

Recently (Otazú Melgarejo 2014) reported the existence of a similar work, written entirely in the Guarani language and without drawings. It is preserved in the Wellcome Library, in London. Remarkably, the author of this manuscript seems to be a layman (it is known that these manuals were written by indigenous people under the orders of some Jesuits, Dominicans or some other religious orders).

5 Role of Ethnobotany in the Search for Value Added MAP Products

As a tool for the modern recognition of the medicinal virtues of plants, the science of **Ethnobotany** has served the search for promising indigenous species, for their inclusion in therapeutics. This has made it possible to narrow down phytochemical and pharmacological screenings to species that are by tradition considered either suitable for medicinal use or dangerous because of their toxicity. In the latter case, their use depends on the detection of the bioactivity, the correct formulation, purification and dosage of their products.

Herbal traditions – in folk medicine – have never been completely forgotten. Moreover, in modern times, where the Occident has recognized the value of natural products (although often by senselessly mystifying them), these medicinal resources have reemerged. There is a tendency to return to the past and reconnect with secular values, customs and tools and supported by scientific research. This is the cause for the rebirth of ethnomedicine, now globalized and therefore often denatured and offered without the proper quality or safety safeguards.

However, this strategy is not so useful in the exploration and use of promising aromatic species. There are several causes of this limitation. First, in the Southern Cone region, there is very little background to demonstrate a traditional use of flora based exclusively on the aromatic properties of species. And this is due to the ignorance of our primeval inhabitants of suitable isolation and preparation techniques for aromatic fractions. The forms used were the "vapors" or incenses produced by burning the vegetable, the impregnation of its parts in animal fats or the preparation of infusions or water-based cooking. They preferred the smell of nature, when they passed through a mountain, stepped on grass, smoked or adorned the house (see the epigraph of this chapter). The flowers were used to beautifie the women's in hats. The smell was like an alternative to interpreting the human mind (psychosomatic studies) or symptomatology of pathologies (medicinal use). In their cosmetics, apart from soaps made from vegetable ash and some saponin-rich plants, they used extensively dyes and hair treatments. They had smooth skin, and long, bushy hair, of which women were proud. But not much else.

These practices were based on two basic aspects: the **medicinal properties of the plants and their use in culinary or food traditions**. With this last objective, the first chroniclers of American colonizers recorded great virtues attributed to certain plants, such as vanilla, cocoa, and countless fruits – until then – unknown in Europe. In regions of the Southern Cone, the key and disruptive component was the so-called **"Paraguay's yerba" or "yerba mate"**: the leaves of *Ilex paraguariensis*. In the texts recounting the trade in these territories, the transport of "yerba mate" is repeatedly cited from the Jesuit Missions (settled in present-day Paraguay, southern Brazil, Northeastern Argentina and Uruguay) to Chile and Peru, then sent to Spain (Haenke 1789/1794). This aromatic species from the sixteenth century constitutes an advantageous source of income in the South American market even today. The marketing tools used were initially not related to the aromatic virtues of the plant. The high esteem of local cultures was used by the Jesuits to change their habits of getting drunk with spirits made by fermenting certain plants: the benefit of "yerba mate" was that people changed their habit of stalking, drinking spirits. But the stimulating effect of its consumption gave incentives also for the dissemination of its use and marketing. Bioactivity covered up its relevant aromatic properties.

To date, it is already generally recognized that it is necessary to use the current extractive and analytical techniques to give a fair value to species that remain, maybe, unnoticed or unknown in the global market, but could have other values, e.g.: for suppliers of aromatic products. **Marcela** (*Achyrocline satureioides*) is a clear example of this uniqueness. As a medicinal plant, it has already been included in the Brazilian Pharmacopoeia IV. edition and was also included in the First Supplement of the VIIth Argentine Pharmacopoeia (2018). Recently, however, an exhaustive analysis of its chemical composition and properties has identified this species also as a valuable **source of aromatic substances** suitable for use in both the fragrance, tobacco and food industry (Retta 2012; Retta et al. 2012; Pires et al. 2019).

Several texts have been published in the region with the purpose of disseminating and promoting the use and management of these aromatic plant resources and their sustainability (FIA 1999/2009; Bandoni 2003, 2005; Davies 2005; Niemeyer and Teillier 2007; Elechosa 2009). However, unlike in the case of exotic species (citrus, oregano, mint, lavandins, coriander, cumin, etc.), the marketing of native aromatic plants and derived products relies almost entirely on wild extraction, or extractivism, which by being often informal, threatens its sustainability (Martínez et al. 2006). The cultivation of these species is very limited and sporadic, which is the result of the exclusively local demand supported by extractivism, and an unstandardized quality.

The natural floras of Southern Cone are potential providers of not only herbal medicines, but plant materials also for other applications, different branches of industries, like the cosmetic, food industry, veterinary applications, tobacco industry, synthesis chemistry, etc. In view of the imposing diversity of vegetation in this area, and most importantly, the large number of endemic and probably also not yet sufficiently explored plant species, the search for both new MAPs and new uses remain to be an important activity in Research and Development, in these countries.

6 Contribution of Southern Cone to Medicinal and Aromatic Plant Research

If information is sought about developing countries, a lot of material runs through non-traditional channels, as they are of little value at the global level. As a result, publications of special interest to these countries are valued with an exceptionally low impact factor, even if their intrinsic value for the region is transcendent. This is quite common also in the agricultural sciences, where domestication of an endemic species or the management of its cultivation, is of a modest significance for other regions. But still, we can assess the global influence of this information.

With such a focus, and also in view of the previously asserted limitations, the contributions of the three South Cone countries to MAP-research, in the period 2018 and 2019, can be assessed on the basis of the SCOPUS database of special literature (Table 5).

As presented in Table 5, and according to the selected model, Argentina was evidently the most cited country in this particular scientific database during the

Table 5 SCOPUS records obtained by a search (2018 to 2019) and compared with the population of each country

	Argentina	Chile	Uruguay
Inhabitants aprox. (millions, 2018)	44,5	18,7	3,4
SCOPUS: 2018 to 2019. in **all fields**			
Second filter used			
Medicinal plant	**1193**	784	159
Aromatic plant	**302**	126	42
Essential oil	**1148**	588	153
As affiliation country of authors	34,920	**35,699**	4373
As % of inhabitants	0,1	**0,2**	0,1
Values per inhabitants			
Medicinal plants	27	42	**47**
Aromatic plants	7	7	**12**
Essential oils	26	31	**45**
By authors affiliation country	785	**1909**	1286

period and subjects evaluated. But, in a comparison of the author's affiliation of scientific documents, irrespective of the subject, Chile has a discrete primacy. It is even more so, if the population of each country is weighted, Chile is, by far, the most representative. However, if the keywords "medicinal plant" or "aromatic plant" or "essential oil" are used as limit per inhabitants, Uruguay markedly prevails in all of the categories. What does this tell us? Three conclusions could be assumed:

- Argentina would appear as the globally most perceptible area of this region,
- Chile appears with the best research/population ratio (probably due to a better social appraisal of the research value?), and when related to the number of inhabitants,
- Uruguay is the country that has worked most on these issues, which could suggest greater social importance given to the knowledge of these resources.

One could ask: Are these preconceptions, realities or… premonitions? In order to know, if it is a preconception, the answer should be approached over a longer period, 20 years (Table 6) and by adding some other keywords. With these new sets of data, and with only minor differences, it is possible to confirm the former conclusions. Due to the demonstrated weak difference between Argentina and Chile, values for "Affiliation country of authors" presented in Table 5 and Table 6, the order is reversed in absolute values, but the prevalence is maintained in percentages by population. Other changes appear with the use of new filters: "native flora", "threatened plant", "herbal medicine" and "phytotherapy". Chile emerges as the certain leader for "threatened plant" filter, both in absolute values and concerning the number of inhabitants. This figure can be easily explained by the remarkable endemism recognized for its flora. Finally, it is curious how Uruguay prevails in its production of publications referring to "herbal medicine" in relation to its population, even if

Table 6 Values obtained through SCOPUS (1980–2019 period) and compared with the population of each country

	Argentina	Chile	Uruguay
Inhabitants aprox. (millions, 2018)	44,5	18,7	3,4
SCOPUS: 1980 to 2019. In: **Article title, abstract, keywords, authors**			
Second filter used (underlines)			
Medicinal plant	**377**	186	42
Aromatic plant	**51**	11	3
Essential oil	**235**	113	29
Native flora	**47**	36	5
Threatened plant	2	**7**	0
Herbal medicine	**54**	26	8
Phytotherapy	**77**	38	2
As affiliation country of authors	**252,692**	176,163	22,581
As % of inhabitants	0,6	**0,9**	0,7
Values by inhabitants (underlined)			
Medicinal plant	8,5	9,9	**12,4**
Aromatic plant	**1,1**	0,6	0,9
Essential oil	5,3	6,0	**8,5**
Native flora	1,1	**1,9**	1,5
Threatened plant	0,0	**0,4**	0,0
Herbal medicine	1,2	1,4	**2,4**
Phytotherapy	1,7	**2,0**	0,6
Authors affiliation country	5678	**9420**	6641

this terminology is not used in the legislation of this country (but rather phytotherapeutical medicines). The rest of the relations stays approximately constant.

In view of the above-mentioned, it appears that the previous hypotheses were not preconceptions, but instead a reality. Time will tell.

It is also possible to make some other conclusions based on the SCOPUS data. If the documents included in this period are analyzed by country (Fig. 7), new deductions arise. The area more intensively studied in "medicinal plants" is obviously the pharmacological features, followed by agricultural issues.

However, when data are analyzed by the term "aromatic plants" (Uruguay did not show, but only three references): the main subject remains "agricultural" features, as a first option, followed by "pharmacological" and "medical" reports. Importantly, the "chemical" studies and within that the chemical/phytochemical study of medicinal plants, remarkably, barely counts Fig. 8.

Finally, when the search is focused on "essential oils" (Fig. 9), there seems to be a remarkably diverse, and almost chaotic situation regarding, the predominance of subjects according to the countries. In the case of each country, the agricultural aspects seem to dominate among the scientific publications, whereas the repercussions of chemical aspects are far from being so harmonic. In fact, these are curiously and absolutely, different for the three countries.

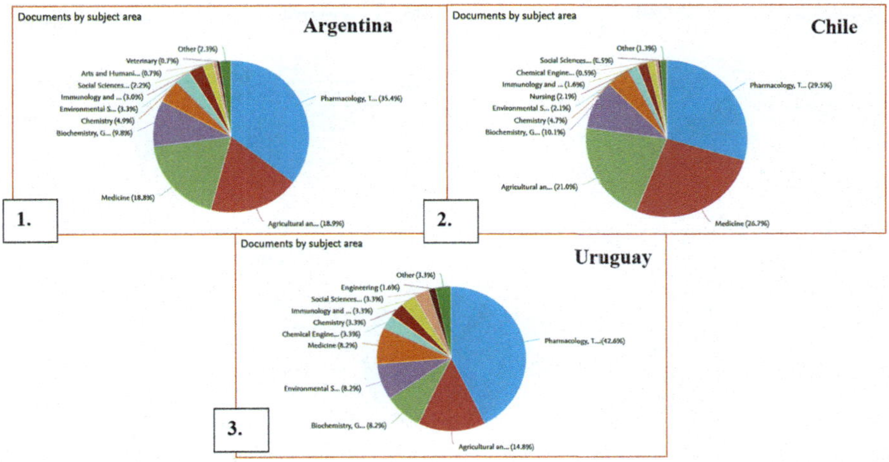

Fig. 7 Main objective of MAP related scientific publications in SCOPUS (1980–2019)

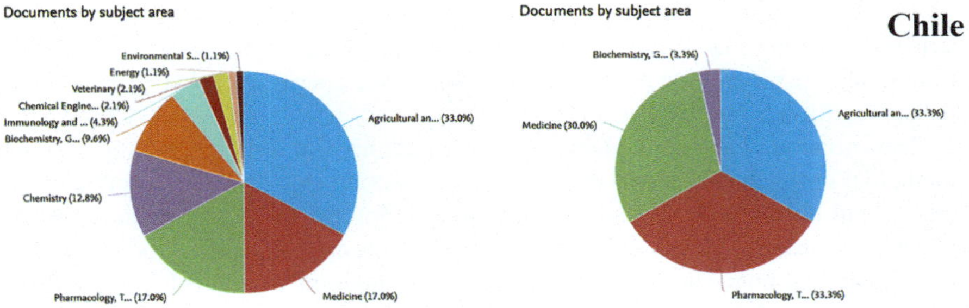

Fig. 8 Main objective of "aromatic plants" related scientific publications in SCOPUS (1980–2019)

7 Resources of MAPs in the Southern Cone Countries

7.1 Argentina

In a contribution to the Medicinal Plant Conservation Newsletter of the International Union for Conservation of Nature (IUCN), Villamil (2004) states that despite the potential important economic activity that it could create, cultivation of medicinal plants has not been well developed in Argentina. Remarkably, in one of the technical publications on this topic, only three species are mentioned as native (Milano, 1964, cited in Villamil 2004).

It can be seen in Argentine statistics that in some cases the demand for products made from aromatic plant derivatives (extracts or essential oils) far exceeds that of the phytotherapeutics, even if it regards the same plant species. In the Argentine beverage market, flavored waters, as an example, occupy an estimated export

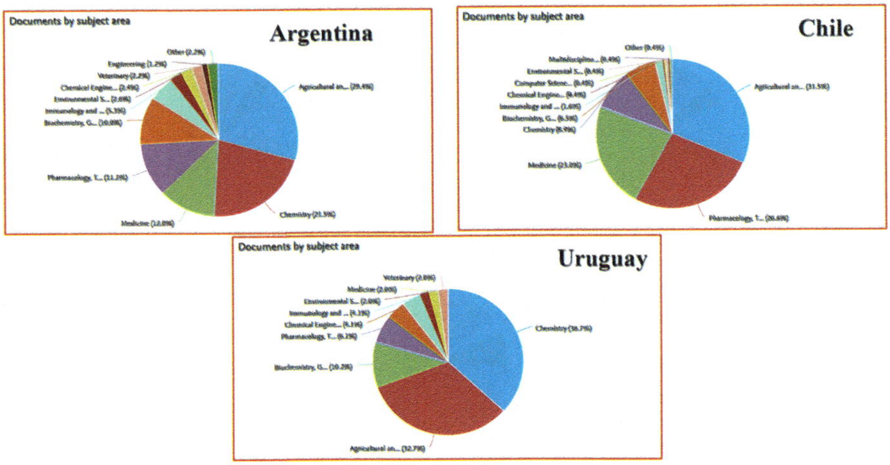

Fig. 9 Main objective of documents in the database SCOPUS resulting from the search on Essential Oils (1980–2019)

market of about US$ 5 million (2017) according to the Ministry of Agroindustry (2018) and the production of bitter drinks, fernet and aperitifs reached 2018 to more than 80 million liters (INDEC 2019).

The situation of aromatic plant crops in Argentina, similarly to the case of native medicinal crops, has minimal economic or marketing significance as compared to other local agroindustrial produce. Argentina is known to be an agricultural producer of global relevance and its national pharmaceutical industry is a leader in Latin America; however, there are almost no agropharmaceutical (pharmacoergasy, according to Tschirch 1909) endeavors involving crop areas to produce medicines or aromatic products. Even in regional economies, such as the Northeast, Central, Northwestern and sub-Andean region of the country, it does not have the due implication that is seen in other regions of the world. However, and exactly in these geographical regions, the production and processing of MAP species could become a valid alternative for smallholding areas, or where it is imperative to detect a diversification of primary agricultural production. Despite this, there is an important local market for native medicinal and aromatic plants, supplied almost exclusively by the collection of wild materials, with the consequent threat of sustainability of these species.

7.1.1 Cultivation of Medicinal and Aromatic Plants

Argentina has a restricted number of aromatic species that are exported in large quantities: chamomile flowers (produced mainly in the East central part of the country, more than 2 millions US$ in 2017), "yerba mate" (Northeast Argentina, approximately 79 million US$ in 2017), and Northwest lemon essential oil, mainly from the province of Tucumán, with the export of more than US$ 190 million in 2017 and around US$ 10 million in 2017 of other citrus essential oils (Ministerio de Agroindustria 2018). These volumes are even remarkable in the global market, and their constancy in quality and quantity are recognized. Other aromatic exports, according to the latest published statistics, are coriander seeds: US$ 2.5 million, oregano and mustard US$ 1.1 million each. Items of lower value: hop cones, anise, oregano and cumin seeds, "cedrón", essential oils of citronella, and other oils not detailed into annual national statistics.

So far, native species have not been detailed. If evaluating the vernacular species, one species emerges as emblematic, owing to its transcendence to international markets: "cedrón" or lemon verbena (*Aloysia citrodora*). It is an indigenous species with international demand. Its success is not shared with any other native species. For this reason, it is valuable to appraise the entire commercial structure of the production of the native aromatic species in the country, and the related problems. Argentinean export volumes of "cedrón" show rather strong fluctuations (from 0 to 10 tn in the last 5 years, being 1.5 tn in 2017) which is even also valid for the imports (from 0 to 120 tn in the same period, 102 tn 2017). Something similar trend can be observed in the neighboring countries such as Chile, with export volumes of approximately 65 tn from 1998 to 2000 (FIA 1999/2009), or Paraguay, which exports for approximately US$ 2.5 million in 2011 (MAG/ALADI 2011). These figures denote a lack of permanence of local supply, and even deal with deserved demand. Viewed from the point of view of the local market, the causes of this commercial inconsistency can be explained by three factors: (a) absence of significant agricultural production; b) variation in quality, and (c) insufficient profit margins.

Note how these variables arise in this species in a very special way, but also how they could relate to other species, justifying the situation of their markets.

In the absence of a critical mass of "cedrón" crops, the domestic supply is small and is therefore subject to the fluctuations of the lowest costs traded on the international market.

The supply dimension is another valuable trading tool. But in turn, it is a weakness for small producers. However, a small growing area is competitive in the diversification of production, a strategy that would allow independence from the oscillations of the single-product international market.

Consequently, the two useful strategies for bridging these constraints by small producers are: permanent or transitional association (cooperatives, export networks or export consortiums) and crop diversification.

In all of the above points, there is another underlying factor, which may not be given the importance it deserves when regarding only the domestic market, but which -in a way- seems to be characteristic of many regions in Latin America: this factor is the lack of continuity, and the opportunistic / sporadic offer. There may be no notion of the disadvantages generated by this inconsistency in supply, but it is important to be aware that it is a factor that gravitates negatively to both producers and buyers. The sudden and sporadic entry of new bidders can lead to abrupt changes in prices and demands. On the contrary, strong temporary demand -due to climate changes, economic or political problems among others- encourages the income of non-traditional bidders, who take advantage of the opportunity.

The other limiting factor identified in the national offer of aromatics is quality. This element is often related to increased costs: higher technology, better seeds, labor-intensive, better quality in packaging, transportation, etc. But it should be borne in mind that, while this is the case at the first stage, in the medium term the improvement of supply by quality broadly justifies these investments. Of course, an early-stage venture is not always able to finance these higher costs, and the permanent or transitional partnership again appears as a solution, as a mechanism to apportion initial investment costs in quality.

Another aspect to consider is the laxity in the quality requirements that the local market manages, which benefits the low-cost product and impairs the offer of superior quality. In this regard, the producers themselves should promote stricter standards and quality differentiation criteria, such as defense mechanisms against indiscriminate offers.

7.1.2 Wildcrafting of Medicinal and Aromatic Plants

There are large regions in the country that are suppliers of crude drugs gathered by extractivism: e.g.: the situated in the Northeast (provinces of Misiones and Corrientes mainly) and Northwest Argentina (essentially the provinces of Jujuy, Salta and Tucumán).

In the province of Córdoba, possibly the most important supplier region, it has been found that from 48 medicinal or aromatic species collected, 10% are threatened (Martínez 2005; Martínez et al. 2006). It is expected that something similar is happening in the Northeastern and Northwestern parts of the country. This warns of the urgent need to study the indigenous flora, and above all to promote its possible management. There is only one Argentine endemic aromatic and medicinal native species in the CITES, Appendix II: *Bulnesia sarmientoi* (UNEP-WCMC 2011).

Some private estimates consider that there are about 50 aromatic and medicinal species in the local markets, but the number of recognized native plants amounts to more than 1000. Information on the already available native potentially useful species, as sources of uncommon applications, as antimicrobials, pediculicide, mosquito repellent, veterinary or fine chemicals synthesis among others, has been published, e.g.: Demo et al. 2005; Bandoni 2005; Gillij et al. 2008; Toloza et al. 2010; Martínez and Luján 2011; Echeverry and Rossini 2012.

7.1.3 Organic Production of Medicinal and Aromatic Plants

Argentina, like most Latin American countries, has great areas of untouched land, where it is very easy and economically profitable to promote organic cultivation. In fact, there is an offer of aromatics of organic quality in the country, which has even been exported, logically in very small volumes.

Except for citrus essential oils, domestic production of other organic essential oils is very low and distributed throughout the country. There are numerous existing distilleries with approximately 500 to 3000 liters of capacity, many of them located in the main aromatic producing areas: mainly in the provinces of the Northeast and Center, and the province of Buenos Aires, although there are also in the Northwest and the Andean-Patagonia region.

7.2 Chile

The geographical isolation of Chile provided by the Andean range, the Pacific Ocean and the Atacama Desert, has produced a unique flora of *ca.* 5000 species, with a high degree of endemism (*ca.* 50%) according to Niemeyer (1995).

7.2.1 Wildcrafting

At present, Chilean export of medicinal and aromatic plants, in 2018 (Chile Aduana 2019) was dominated by the following products: rosehips (plant material: 6000 tn, 19,5 million US$, oil: 153 tn, 8,5 million US$), boldo (plant material: 2300 tn, 1,8 million US$), quillay (plant material: 160 tn, 1,1 million US$, extract: 619 tn, 11,1 million US$), *Hypericum perforatum* (376 tn, 1,7 million US$) and algae (58,750 tn, 101,5 million US$, comprise also food algae).

The preeminence of boldo and quillay, as two native species, in the Chilean economy is distinguishable. Therefore, it is important to analyze why the commercial success achieved with these plants, has made an aromatic or medicinal plant, iconic for Chile. Boldo (*Peumus boldus*) is a leathery leaf, which is obtained by pruning from a tree that it is not threatened by extinction. The relatively easy post-harvest procedures of producing marketable produce could be an intuitively comparative advantage. Also, the diffusion of its medicinal properties by different researchers, contributed to its knowledge in the world. Originally, scientific investigation into boldo began in France, where it was introduced in 1869, with the isolation of its essential oil and boldine (Verne 1874). Speisky and Cassels (1994) renewed information about this plant. A chapter of this book gives more details and updates on it.

Several authors have collected ethnobotanical information about the traditional medicinal uses of Chilean MAPs. Zin (1930) mentions hundreds of species, between exotic and natives. de Mösbach (1992) analyzed the ethnobotany of the countless plants, mainly from the Southern region. Other more contemporary authors (Montes and Wilkomirsky 1987; Hoffmann et al. 1992; Muñoz et al. 2001) screened this information with more emphasis on the chemical and pharmacological targets.

7.2.2 Production of MAPs in Chile

In the case of boldo, the increased global demand has accelerated the threat to its natural populations; and that is why, since mid-century past, the Chilean government is implementing policies to control its extraction and promote its cultivation (Chile-Instituto Forestal 2009; Benedetti and Barros 2011). Due to the slow growth of this arboreous species, cultivation is not yet economical.

This limitation does not exist with arbustival species, favoring their cultivation to not only ensure their offer, but also to homogenize and guarantee their quality. But even so, the cultivation of medicinal and native aromatic plants in Chile, as in all of the SC region, is rather limited and reduced mostly to cover the local demands.

However, the cultivation of other native medicinal and aromatic species (FIA 1999/2009; Niemeyer and Teillier 2007) has been touted for more than 30 years.

7.3 Uruguay

7.3.1 Wildcrafting of MAPs in Uruguay

The flora of Uruguay numbers 2500 species distributed among 150 native and foreign plant families. Approximately 80% of Uruguay's territory is prairie, with grasses predominating. Uruguay is primarily a grass-growing land, with a vegetation that is essentially a continuation of the Argentine Pampas. Forest areas are relatively small. Trees grow in bunches (Brussa et al. 2007).

Some commonly wild-growing and traditionally used MAPs in Uruguay include *Achyrocline satureioides* (Lam.) DC., *Aloysia citrodora* Paláu, *Baccharis trimera* (Less.) DC., *Chamomilla recutita* (L.) Rauschert, *Ilex paraguariensis* A. St.-Hil., *Malva parviflora* L., *Mentha × piperita* L., *Phyllanthus niruri* L. and *Tilia* spp. as examples (Arrillaga de Maffei 1969; Alonso Paz et al. 2007).

7.3.2 Production of MAPs in Uruguay

The production of aromatic or medicinal native plants does not figure in national statistics, as occasionally it happens in Argentina. Only some rural cooperative groups are engaged in this area of crop production, for a very focused local market of herbalists, cosmetics or infusions, and where the supply of native species is combined with other conventional ones to be able to sustain such a narrow market.

Soutullo et al. (2013) published a list of 687 plant species with the priority of being preserved, because of the threat to their populations; and they include numerous aromatic and medicinal species. Subsequently (Berretta et al. 2007; Rivas 2010; Uruguay 2016b) they have counted 623 species within the country's native plant genetic resources, including 116 medicinal and 44 aromatic resources.

According to a report on a local website (Montevideo Portal 2019), the total volume of yerba mate consumed in Uruguay is estimated at 34,8 million kg/year, approximately US$ 143 million, nevertheless the main supplier of this product is Brazil.

In view of the aforestations – with species of *Eucalyptus* covering approximately 1 million ha – the territory dedicated to intensive non-traditional productions, such as medicinal or aromatic plants is rather modest, only a small percentage (less than 2% according to DIEA 2019). Among these, conventional species predominate (*Citrus* -about 13,900 ha in 2018, exporting 108 tn of essential oils, mint, oregano, rosemary, thyme, basil, etc.). Since the legalization of psychoactive and non-psychoactive plants of *Cannabis* in the country in 2013 by the law 19,172 (IRCCA 2019), this crop also emerges as a medicinal agroindustrial and research alternative (*ca.* 1000 ha in 2019).

8 Peculiar Plant Stories from the Southern Cone

The long-lasting history of individual plants is frequently accompanied by remarkable and inexplicable events. The long and adventure-rich history of the Southern Cone also has not been an exception. Some peculiarities related to characteristic native Southern Cone species are meant to offer insight into this rich tradition of selected species, bridging thus past with their future.

8.1 Pepper Tree (Schinus molle)

Schinus molle, the plant named "pepper tree", is known for its bright pink fruits, often sold as "pink peppercorns". Remarkably, however, *S. molle* is unrelated to the true Asiatic pepper (*Piper nigrum*) or the American allspice (Jamaica's pepper, *Pimenta dioica*).

The traditional use of pepper tree by the ancestral South American cultures has been cited by European chroniclers for the medicinal properties attributed to its leaves, fruits, and resin, since the early sixteenth century. The most popular names used in the region are: "**molle**" or "molli" from Quechua, or "aguaribay" according

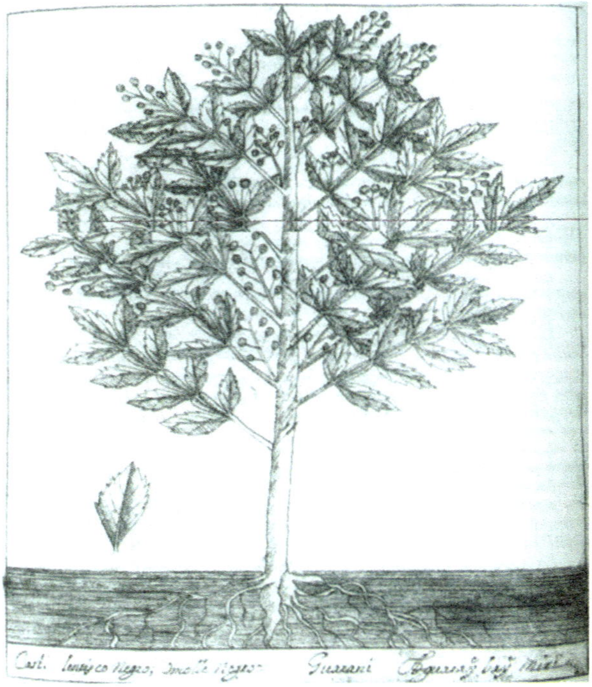

Fig. 10 This is one of the drawings included in "Materia Médica Misionera", attributed to the P.J. Pedro de Montenegro (1945) and therein named as "aguaribay mini". Today, it was identified as *Schinus molle* L., a typical tree from the American Southern Cone

to Guarani, among others. Given the great morphological similarity between the two species *S. molle* and *S. areira* (syn. *S. molle* var. *areira*) there has always been much confusion in the special literature about the correct botanical identity of the species cited.

Figure 10 shows one of the laminae included in the text of de Montenegro (1945), named as "aguaribay mini", and today identified as *Schinus molle* L. Remarkably, even though several authors cite this name, the species of global distribution is more probably *S. areira*; *S. molle* is a species growing in Southern Brazil, Paraguay, Uruguay and Northeastern Argentina that is initially considered as an endemic plant.

Countless works have been published on its bioactivities, its aromatic characteristics, its virtues as an ornamental plant and the ease of its cultivation in most varied climates and soils (Viturro et al. 2010).

Its fruits are now known in the world market as "red pepper". Also, its essential oil, with an aromatic profile similar to that of Asian pepper (*Piper nigrum*), is offered for the flavors and fragrance industry. In the subtropical and tropical areas, it is a species widely cultivated as an ornamental.

8.2 *Boldo* (Peumus boldus), *Quillay* (Quillaja saponaria) *and Rosa Mosqueta* (Rosa eglanteria)

There are two native trees and one exotic species that characterize Chile in the international market for medicinal and aromatic plants: **boldo** (*Peumus boldus*), **quillay** (*Quillaja saponaria*), and "**rosa mosqueta**" (*Rosa eglanteria*).

Boldo and quillay were botanically identified, for the first time, by S.J. Molina (1788/1810). He described the benefits of the **boldo** (Molina 1788) as: "su fruta…tiene tan dura la cáscara, que sirve para hacer rosarios. Los labradores llaman Boldo a esta última especie, y se valen de su corteza para sahumar las vasijas en que envasan sus vinos. Los frutos de todos los demás Peumos se comen cocidos en agua tibia...sus avellanas abundan de un aceite, que tal vez será bueno para la comida…" [its fruit... has such a hard shell, so that it is used for making rosaries. The farmers call this latter species Boldo, and they use their bark as incense to aromatize the vessels in which they pack their wines. The fruits of all other Peumos are eaten cooked in warm water... Boldo fruits contain an oil, which will perhaps be good for food..."]. Until Ruiz and Pavón (1798) there had been no mention of its medicinal uses. Regarding **quillay**, Molina (1810) wrote: "…aquello que lo hace precioso a los chilenos es su corteza, la cual, machacada y desleída en agua, produce -como el más perfecto jabón- una gran espuma, saca bien las manchas, desgrasa las lanas y limpia óptimamente toda suerte de tejidos y telas…Se encuentran en Chile dos suertes de árboles de corteza saponaria, poco diferentes en el nombre y follaje, esto es, el quillay, que he descrito, y el cúllay, ambos así llamados por los verbos *quillcan* y *cúllcan*, que significan lavar, con ciertas modificaciones…" ["... what makes it precious for Chileans is its bark, which, crushed and dissolved in water, produces – like the most perfect soap – a great foam, removes the stains well, degreases the wools and optimally cleanses all kinds of fabrics... In Chile, there are two sorts of trees yielding saponary bark. They are slightly different in their names and foliage. One of them, the Quillay, which has been described, and the Cúllay: both are called so, by the verbs *quillcan* and *cúllcan*, which mean washing, with certain modifications..."]. In these books, as in some others, recognized among the foremost texts that described the medicinal Chilean flora (besides to de Rosales (1674), who quotes quillay but not boldo), several Chilean native medicinal and aromatic plants were described. However, no other species has a current commercial relevance just like boldo and quillay. These, together with "rosa mosqueta" hips and oil, as well as algae, are presently the only important export items of value from the Chilean flora.

8.3 *Chamomile,* Chamomilla recutita

The Southern Cone, more specifically Argentina, is considered as one of the main global producers of chamomile flowers, with an export volume of *ca.* 1500 to 6000 tn/year (variable according to the year) obtained from about 20,000 ha

(Cameroni 2010). Local consumption is around 200 tons, thus, the export objective of this production is evident. In this context, it could be useful to look into the causes of this Argentine prevalence in the global market, especially as this species is not native to this country.

The introduction of chamomile to Argentina started at the beginning of the nineteenth century, possibly as the result of German pharmaceutical companies promoting these initiatives. Originally, the northwest area of the province of Buenos Aires was chosen for the production, specifically the city of Pehuajó, due to the favorable quality of its soils and climate. This area was originally populated by Spanish and Italian immigrants. Today, this city has the flower of chamomile in its shield, as a symbol of its importance in the region.

Most of the exports go to Germany and to Italy. The largest local producers and exporters have always been Germans or Italians. They established and consolidated the business and increased it according to the existing demands. As the production increased, the technology improvements included the introduction of mechanical harvesting and drying systems with automatic ovens.

Argentine chamomile production can be evaluated not only in terms of quantities, but also in quality. Argentine chamomile is highly valued for its aroma, and for this reason, it is usually used in mixtures with other qualities. Although its bisabolol content is not high, it corresponds to the high bisabololoxide B chemotype, and can be normalized by its content in essential oil yield, its content of flavonoids (apigenin glycosides) and polyphenols, and by the absence of the sesquiterpene lactone anthecotulide, a highly irritating compound of the skin. This lactone has never been found in significant quantity in samples of *Matricaria recutita*. Certain reported cases of contamination might have been caused by a different botanical species, the so called stinking chamomille (*Anthemis cotula*). Today, it is also known that the bisaboloxide chemotypes, like the one produced by Argentina, have a marked antiphlogistic effect (Kobayashi et al. 1986; Tomić et al. 2014).

8.4 "Cedrón", Aloysia citrodora

An emblematic case for Argentina is *Aloysia citrodora* Paláu, lemon verbena or "cedrón". Several authors have placed its center of genetic diversity to the Northern Argentine provinces of Salta and Jujuy, where the Inca culture arrived (Siedo 2006; Elechosa et al. 2017). It is generally accepted that the kallawayas, connoisseurs of herbal medicine, as itinerant doctors of the Incas, who had been active not only in Bolivia, Peru and Ecuador, but also reached the north of Argentina and Chile, must have been the mentors of the medicinal use of this plant, what they called "wari pankara" or "chake saya saya" (Girault 1984). It was gradually also known to the Guarani (Bertoni 1922/27), and when the Spaniards came to the coasts of South America, it had already been regarded as a typical species of the indigenous traditional medicine. The first samples of this species that arrived in Europe (eighteenth century) were transported by the expeditions of Bougainville (the French physician and botanist Philibert Commerçon of this expedition took it from Montevideo) and by Ruiz and Pavón (the French botanist Joseph Dombey, who participated in this trip, sent it from Lima). From the plant individuals grown in the Royal Garden of Madrid and the Royal Garden of Paris, its medicinal and aromatic properties spread throughout the Occident. Today, it is cultivated even in some Asian and African countries and is already a product of global trade of aromatic and medicinal herbs.

The above described dispersion of some species and their medicinal applications also justify the inclusion of these species, native to territories of the SC, in the volume of this series dedicated to Brazil (Albuquerque et al. 2018). Similar stories could be told not only about "cedrón" but also about *Valeriana carnosa* and other species that have an uncertain origin in the region, e.g.: *Baccharis* spp., *Achyrocline satureioides, Tabebuia* spp., *Lantana camara,* etc.

It is also most likely that our volume contains species that are widely known as medicinal plants both in Brazil and in Paraguay, Bolivia or Peru. This verifies the saying plants do not know territorial or political boundaries.

"Cedrón", as discussed above, is an example of a native species that, without too much local motivation, has acquired an already consolidated international market. In the case of chamomile, the opposite has occurred. It is an exotic species to Argentina, whose local production has been consolidated as a result of close collaboration with the main international stakeholders. Two strategies that have led to MAP-production in Argentina: a) the diffusion of the ethnobotanical tradition (in "cedrón", started with the first European chroniclers in the eighteenth century), b) through a specific demand (the need for Germany to strengthen its offer of chamomile, from twentieth century).

9 Legal Status of MAPs in Countries of Southern Cone

From the viewpoint of modern societies, the phenomenon of transculturation is evident. With the contemporary habit of preferring the natural and valuing traditions, the use of medicinal plants has been re-popularized. In some rural areas, they remain as the only therapeutic resource. In large cities, however, the use of medicinal plants has become a practice that imitates other cultural contexts, implying a potential misinterpretation or overestimation of criteria that originally used to be well-founded values. These circumstances require the official control of plant materials. The objective of World Health Organization (WHO), similarly to many of the developed countries, is to implement legislation in line with these requirements of the society.

9.1 Argentina

In Argentina, the activity of herbalists or locals engaged in the dispensing of herbal medicines is regulated by the laws 16,463 (medicines law) and 17,565 (law of exercise of pharmaceutical activity) and their regulatory decrees. Formerly, Herbal medicines were regulated only by the Resolution of the Ministry of Health No.144 of 1998, using as a model a work that was proposed to the national health authorities at the time (Bandoni et al. 1993). This Resolution was subsequently amended by (ANMAT 1999–2015) Dispositions No. 2671, 2672 and 2673 of 1999, which regulate the requirement of establishments dedicated to these medicinal products, the respective Good Manufacturing and Control Practices, and the standards for its registration. The Disposition No.1788/2000 regulated a list of plants considered to be dangerous and therefore not permitted in Phytotherapy; and Disp. 4980/2005 regulated the advertising of these drugs. Finally, from 2013 Resolutions No.1817/2013, 5418/2015, 5482/2015 were published, where, inter alia, these products are being designated as herbal medicines, in line with European legislation.

Unfortunately, mainly under the influence of US norms and local industries, parallel other rules were drafted for the registration and dispensing of dietary supplements based on plant raw materials (ANMAT, Disp. No. 1637/2001). These were lacking a sufficient scientific basis. Since a dietary supplement as such is meant to correct a certain deficiency in the diet, no one has been reported having a diet reduced in ginseng saponins, or glycyrrhizin. In any case, there is a need for a drug that prevents, cures or contributes to the homeostasis of the individual. This practice has made the industry more prone to market these products and not herbal medicines (in 2012, according to ANMAT, 33% of dietary supplements marketed in Argentina contained herbs as the main ingredient), because a tolerant legislation for supplements obviously results in an advantage for the industry, but a danger to society. This is how many medicinal plants are currently marketed in this form, without knowing the potential undesirable effects, contraindications or risks of frequent use. Moreover, this situation has caused the addition of synthetic drugs to be detected in numerous cases in dietary supplements, some dangerous to health, without declaring it (Calahan et al. 2016; O'Malley Olsen et al. 2019). A strategy to correct these anomalies may be a regulation that requires dietary supplement processors to have the same requirements as for herbal drug development. In many cases, transculturation has erased some elementary knowledge, such as the correct identification of the species, harvest time, the part used, the way of use, the appropriate dosage, the dangers it hides its use, etc. As of 2018, less than 90 herbal medicines have been registered in Argentina.

The Farmacopea Argentina (2013/2018) VII. edition lists 4 monographs of native medicinal plants ("marcela" -*Achyrocline satureioides*-, "cedrón" -*Aloysia citrodora*-, "yerba mate" -*Ilex paraguariensis*- and "pasionaria" -*Passiflora caerulea*-), although, in the previous editions, up to 91 plant drugs had been registered. In the previous Edition VI of 1976, only 9 natives remained (Bandoni 2011): "peperina" (*Minthostachys mollis* now *M. verticillata*), "sarandí blanco" (*Phyllanthus sellowianus*), "poleo" (*Lippia physsicalyx* and *L. turbinata*), "cola de caballo" (*Equisetum giganteum*), "ambay" (*Cecropia adenopus*, now *C. pachystachya*), "carqueja" (*Baccharis articulata* and *B. crispa*), "yerba del pollo" (*Alternathera pungens*), "cedrón" (*Aloysia triphylla*, now *A. citrodora)* and "quebracho blanco" *(Aspidosperma quebracho blanco*). The elaboration of the pharmacopoeial monograph of a certain species requires an enormous effort, as it should evaluate not only the quality, the suitability of the material for medicinal use, but also the variability of plants and known or potential adulterants, as detected by the cutting edge phytochemical technologies.

As for the registration of varieties of native medicinal plants, so far there are three species listed by the National Seed Institute (INASE 2019): "incayuyo" or *Lippia integrifolia* (Grey.) Hieron. accession TAWA INTA, "peperina" or *Minthostachys mollis* (Kunth) Griseb. accession CHAMPAQUI FCA and 15 cultivars of "yerba mate" or *Ilex paraguariensis* A. St.-Hil., the last with an important agronomic development (approximately 165,000 ha grown in 2016, according to the National Institute of Yerba Mate -INYM) (INASE 2019).

9.2 Chile

In Chile, the most cited works on medicinal plants, after the colonial era, began with the book of Zin (1930) and followed with Montes and Wilkomirsky (1974, 1978, 1987), Hoffmann et al. (1992), de Mösbach (1992) and Muñoz et al. (2001), as it was cited before. These antecedents, added to the importance that herbal medicine continues to be used by the local population, has been assumed by the national health authorities. Thus, they recognized these traditions in various projects and regulations to validate and normalize their use.

In 1935, Decreto No. 2 was published in Chile, which for the first time regulates the dispensing of herbal medicines in "yerberías" or by street vendors. Concurrently with the development of herbal medicine in Argentina, in Chile, the first initiatives emerged to normalize complementary medicine, at the end of the twentieth century. The Chilean Ministry of Health proposed, in 1992, the creation of the Unit of Traditional Medicine and Other Alternative Medical Practices. One of its objectives was to revive the use of traditional herbal medicine, and as a first goal, a diagnosis of the herbal medicine of Chile was written (Mellado Campos et al., 1996). This work, inspired by the monumental work published in Mexico (Zolla 1994), provided the first foundations for all future normative work. For this purpose, a bibliographic review was carried out starting with 1941 (date of the edition of the Chilean Pharmacopoeia III, last edition by then). With this ethnobotanical, chemical and pharmacological information, the national medicinal flora was evaluated, dividing the territory into three sections: North, Central and South. Interestingly, it was in the Southern region that the greatest variety of species was detected: 296 out of a total of 464, prove Mapuche influence. Another interesting fact to learn from this work is that, by classifying the registered species, it was observed that only 65 species had enough scientific information to support their medicinal use; and of these only 8 species were native: *Aloysia citrodora, Drimys winteri, Equisetum giganteum, Haplopappus baylahuen, Lomatia hirsuta, Oyholobium glandulosum, Peumus boldus* and *Quillaja saponaria.*

In the year 2005, Decreto No. 42 was issued. This regulates the exercise of alternative (complementary) medical practices, as auxiliary professions of health, and the conditions of the premises in which they are performed. Based on this framework regulation, Acupuncture (Decreto No. 123/2008), Homeopathy (Decreto No. 19/2010) and Naturopathy (Decreto No. 5/2013) have been evaluated and recognized as auxiliary health professions. In this same sense, the relevance of recognizing Floral Therapies and Masotherapy was also studied.

In recent years the regulations have been amended and supplemented according to current requirements. For example, the Regulations of the National Pharmaceutical Control System (D.S. No. 1876/95) were amended, in 1998, by D.S. No. 855/98, creating the Product Pharmacists Supplementary class, a decree that was subsequently repealed and replaced by D.S. No. 286 of 2001, modifying the D.S. 1876/95 and created the phytopharmaceutical category. Next, D.S. 1876 was replaced by D.S. No. 3/2010, which was effective from December 26, 2011 (approves the Regulations of the National System of Control of Pharmaceuticals for Human Use), which includes definitions of phytopharmaceuticals and requirements for their regulation, with the Institute of Public Health being who should apply them. Finally, D.S. 3/2010, ratified by The Technical Standard of the Ministry of Health 128/2012, regulates the registration of the establishments that process these plant drugs; D.S. 133/2012 approves a list of traditional herbal medicines (43 native species out of 103 species!): *Luma chequen, Haplopappus.spp. [but not italian or cursive fonts!], Peumus boldus, Cuscuta chilensis, Centaurium cachanlahuen, Drimys winteri, Aloysia citrodora, Acaena splendens, Paspalum vaginatum, Fuchsia magellanica, Durvillea antarctica, Flaveria bidentis, Otholobium glandulosum, Porlieria chilensis, Verbascum thapsus, Calceolaria thyrsiflora, Senecio fistulosus, Lampaya medicinalis, Laretia acaulis, Aristotelia chilensis, Buddleja globosa, Ecballium elaterium, Schinus areira, Gunnera tinctoria, Solanum ligustrinum* (now *S. crispum*), *Chenopodium chilense* (now *Dysphania chilensis*), *Bauhinia forficata, Fabiana imbricata, Ephedra chilensis, Senna stipulacea, Quillaja saponaria, Muehlenbeckia hastulata, Quinchamalium chilense, Tristerix tetrandrus, Lomatia hirsuta, Acantholippia deserticola, Rumex conglomeratus, Margyricarpus pinnatus, Libertia sessiliflora, Pseudognaphalium viravira, Geum chiloensis, Polypodium feuillei* and *Ribes cucullatum.*

To date, three official editions of the Chilean Pharmacopoeia have been published: I. in 1882 (212 species cited), II. in 1905 (150 monographs of plants and 10 native products) and III. in 1941. In the latter, 124 botanical species, but almost no native species have been recorded. The University of Valparaiso presented a IV. edition, in 2016. This is, however, not yet official.

9.3 Uruguay

The state of herbal medicines in Uruguay shows a very similar picture to Argentina. The reason for this is that the two countries are members of MERCOSUR, the Southern Common Market. The South American trade block was established by the Treaty of Asunci6n in 1991, and the Protocol of Ouro Preto, in 1994. Politics and regulations are being unified between member countries.

Although the specific issue of herbal medicines has not made much progress within MERCOSUR, meetings with experts from these countries have allowed several common criteria to be homogenized, although not yet fully internalized. There is also a commitment to draft a MERCOSUR Pharmacopoeia (MERCOSUR 2014), although it is still in initial stages.

As for the background, the most widespread written records on the local medicinal flora, if we disregard the aforementioned works of European travelers and chroniclers, are those of González et al. (1936, 1939), followed by the works of Arrillaga de Maffei (1969) and late Alonso Paz et al. (2007).

The current Uruguayan legislation on herbal medicines is covered by Decreto 403/2016 (Uruguay 2016a), which includes – in a single rule – all the requirements that relate to this subject: from the enabling of establishments to the list of species prohibited as phytotherapeutical medicines (not called herbal medicines). Uruguay does not have a national pharmacopoeia but uses those of MERCOSUR member states (Brazil and Argentina) or USP and EuPH, BPh, RFE, Pharmacopoeia International, etc.

10 MAPs Export Trade in the Southern Cone

Statistics that show the level of supply of the product category medicinal and aromatic plants (MAP) are scarce. To obtain comparable estimates, we have analyzed the most important MAPs items on the basis of the European Commission's Customs Tariff Codes or Harmonized Commodity Description and Coding System. Values published by the International Trade Center (ITC 2020) for the period of the last 5 years were used.

MAP exports are nearly imperceptible, in value, as compared to other products exported by the three Southern Cone countries analyzed.

In the case of Argentina (year 2019), total exports of item sorted by values $US is ranked 43 (data not shown): lemon essential oil. However, in the global ranking of exporters of this product, Argentina is the major player. The second MAPs derivate product - in values - for this country is "mate", the first derivative of native MAP (number 85 in this ranking!).

For Chile, the first product derived from MAP appears ranked number 56 of this ranking: quillay extracts, and the second is algae (rank 67). However, again the primacy of these countries arises in the global market: Chile is considered as the first exporter of some algae derivatives and quillay extracts, both derived from native species.

In the case of Uruguay, Argentina's situation is repeated. The first item (109, in order of US$ values) of export of some MAP's derivative is the lemon essential oil and the next is the "mate" (located 172 in this ranking).

These figures show that each country in the region has a very different profile, depending on the biodiversity, the edaphoclimatic conditions they possess and the industrial development they have achieved. But they also demonstrate the important contribution they give to ethnobotanical heritage.

These international statistics show the modest of trade significance of MAPs in the economies of these countries, but they also highlight their global importance as the main traders for locally sourced raw materials and international importance. It is necessary to deepen the search with more specific data, to find the real components of are the regional contributions to the international market. The scale of production becomes so small in these cases that they escape most statistics, both international and national. However, some significant values can still be monitored (Table 7), if the searched is carried out using 4 or more digits customs codes (ITC 2020).

So far, the visible, the statistically ponderable. However, a lower level assessment revealing the influence these products on regional microeconomics is till lacking. These production volumes tacitly involve an important mass of labor, often coming from areas with mini-funds or cooperative groups. At this lower level, it weighs much more on the social and cultural value of production than its economic significance. This scale of supply is basically based on traditions, on ethnobotany that is kept alive in each country.

11 Conclusions

The Southern Cone, this remote, Southern edge of the huge South American continent is still harboring numerous medicinal and aromatic plants that have been scientifically less explored and as such, are less known to the world. Regarding the large number of native species in the Southern Cone, scientific information is frequently scarce or limited. This volume can endeavor to discuss only selected examples of mostly local relevance. In this attempt the authors' aim was to avoid possible overlaps with species listed in similar monographs on Medicinal and Aromatic Plants of Argentina, Uruguay or Chile or other any related publications. As a rule, the most commonly known or the most commonly used species in traditional medicine were preferred.

The countries of the Southern Cone have several aspects in common, which fact has been recognized and investigated by several researchers. As an example, Ulloa Ulloa et al. (2017) also attempted to explore floristic similarities among 12 geographical areas in the Americas, including the Southern Cone.

The sometimes remarkable huge differences in topographic features, have also brought about farther differences. As such, the geographical isolation of Chile

Table 7 Comparison of main export items for products derived from MAPs. (Argentina (AR), Chile (CL) and Uruguay (UY). n.e.s.: not elesewhre specified; nd: not data

Country	Custom tariff code	Products	Quantity (tons) 2015	2016	2017	2018	2019	Main product included
AR	0903	"Mate"	35,708	27,149	31,030	41,115	39,961	"Mate"
CL	0903		1	7	6	3	4	
UY	0903		107	136	125	143	208	
AR	0909	Seeds of anise, badian, fennel, coriander, cumin or caraway; juniper berries	7686	9255	2993	8457	5644	Coriander + etc.
CL	0909		nd	nd	nd	nd	nd	
UY	0909		nd	nd	nd	232	nd	
AR	0910	Ginger, saffron, turmeric "curcuma", thyme, bay leaves, curry and other spices (excluding pepper of the genus *Piper*, the fruit of the genus *Capsicum* or the genus *Pimenta*, vanilla, cinnamon, cinnamon tree flowers, cloves [wholefruit], clove stems, nutmeg, mace, cardamoms, seeds of anise, badian, fennel, coriander, cumin and caraway, and juniper berries)	328	362	157	113	235	Coriander + cumin + etc.
CL	0910		61	74	59	88	85	Turmeric + etc.
UY	0910		1	5	2	1	2	
AR	1210	Hop cones, fresh or dried, whether or not ground, powdered or in the form of pellets; lupulin	41	11	21	nd	nd	Hop
CL	1210		1	0	1	1	1	
UY	1210		nd	nd	nd	nd	nd	
AR	121190	Plants and parts of plants, incl. seeds and fruits, of a kind used primarily in perfumery, medicaments or for insecticidal, fungicidal or similar purposes, fresh or dried, whether or not cut, crushed or powdered	1104	1403	865	998	712	Oregano + etc.
CL	1211		8925	8988	9748	9933	9256	Boldo + rosehip + chamomile + *Hypericum* + etc.
UY	1211		nd	85	38	3	46	
AR	121229	Seaweeds and other algae, fresh, chilled, frozen or dried, whether or not ground, unfit for human consumption	101	201	168	171	nd	Algae

(continued)

Table 7 (continued)

Country	Custom tariff code	Products	Quantity (tons)					Main product included
			2015	2016	2017	2018	2019	
CL	1212	Locust beans, seaweeds and other algae, sugar beet and sugar cane, fresh, chilled, frozen or dried, whether or not ground; fruit stones and kernels and other vegetable products, incl. unroasted chicory roots of the variety *Cichorium intybus sativum*, of a kind used primarily for human consumption, n.e.s.	68,575	76,447	81,436	58,618	60,001	Algae
UY	1212		nd	18	nd	nd	10	
AR	1302	Vegetable saps and extracts; pectic substances, pectinates and pectates; agar-agar and other mucilage and thickeners derived from vegetable products, whether or not modified,	312	388	496	541	769	Hop extract + etc.
CL	13021910	Vegetable saps and extracts (exc. licorice, hop, opium): quillay extract	10,866	12,792	12,178	11,052	11,620	Quillay extract
UY	1302	Vegetable saps and extracts; pectic substances, pectinates and pectates; agar-agar and other mucilage and thickeners derived from vegetable products, whether or not modified,	31	11	17	22	9	
AR	1404	Vegetable products, n.e.s.	nd	nd	nd	nd	nd	
CL	14049010	Vegetable products, n.e.s: quillay barks	nd	nd	nd	nd	131,043	Quillay bark
UY	1404	Vegetable products, n.e.s.	nd	nd	nd	nd	nd	
AR	3301	Essential oils, whether or not terpeneless, incl. concretes and absolutes; resinoids; extracted oleoresins; concentrates of essential oils in fats, fixed oils, waxes or the like, obtained by enfleurage or maceration; terpenic by-products of the deterpenation of essential oils; aqueous distillates and aqueous solutions of essential oils	7072	6799	6717	7951	7091	Lemon oil + etc.
CL	3301		742	301	134	270	569	Peppermint oil + etc.
UY	3301		120	119	140	108	148	Orange oil + lemon oil + etc.
AR	38061000	rosin	8337	11,945	13,004	14,542	16,959	rosin

provided by the Andean range, the Pacific Ocean and the Atacama Desert, has produced a unique flora of *ca.* 5000 species, with a high degree of endemism (*ca.* 50%), according to Niemeyer (1995). These and similar differences in the medicinal and aromatic plant genetic resources have been recognized and utilized by local populations for centuries. Remarkably, the traditional knowledge has only partially survived, which is the result of the rich and peculiar sociological development of the area.

As a result, trade figures in medicinal and aromatic plant commerce do not yet reflect the significance of MAPs, although the present states of Argentina, Chile and Uruguay are inheritors of this huge, rich historic and biological heritage. More and more, however, these countries are becoming aware of the importance of these traditions, so that the exploration and utilization of medicinal and aromatic plant resources are encouraging. In view of the numerous quality influencing factors, like the uncertain botanical identity or frequently unknown origin of plant materials, diverse methods and harvest time/post-harvest treatments, etc., the tasks are manifold. There is also an increasing amount of knowledge about the diversity, quality (including safety, efficacy, control), marketing and regulatory aspects of the botanical resources of the Southern Cone. This will, surely, lead to the sustainable utilization of MAP resources for the benefit of new successor populations of this peculiar region of the world.

Acknowledgments To Dr. Robert Verpoorte, Dr. Eduardo Dellacassa and Dr. Bruce K. Cassels, for critically reading, valuable comments and your willingness to correct English. To Vanina Tavernise, for the design of figures.

References

Acosta, Joseph de (1590) Historia Natural y Moral de las Indias. Ed. Fondo de Cultura Económica, México (Reedition from 1979), 444 pp

Albuquerque UP, Patil U, Máthé Á (2018) Medicinal and Aromatic Plants of South America – Brazil, vol 5. Springer, Netherland, 486 pp

Alonso Paz E, Bassagoda MJ, Ferreira F (2007) Yuyos. Uso racional de las plantas medicinales. Ed. Fin de Siglo, Montevideo, Uruguay, 222 pp

Álvarez MA (2019) Pharmacological properties of native plants from Argentina. Ed. Springer, Cham, Switzerland, 255 pp

Andrade B (1999) El marco natural de Chile: su diversidad de paisajes. Estudios Geográficos 60(234):7–24

ANMAT (1999–2015) Administración Nacional de Medicamentos, Alimentos y Tecnología Médica, Argentina. Disposiciones 2671/1999, 2672/1999, 2673/1999, 1788/2000, 1637/2001, 2819/2004, 4980/2005; and Resolutions 1817/2013, 5418/2015, 5482/2015

Arata PN (1898) Botánica Médica Americana. Los Herbarios de Las Misiones del Paraguay. La Biblioteca 2(7):1–192

Arrillaga de Maffei BR (1969) Plantas Medicinales. Ed. Nuestra Tierra, Montevideo, Uruguay, 60 pp

Azara F (1805) Viajes por la América Meridional. Tomo I. Ed. Elefante Blanco (reedition 1998), Buenos Aires, Argentina, 239 pp

Bandoni AL (2005) Especies aromáticas argentinas como fuente de terpenoides de interés. In: Fernández Barrero A (ed) Plantas Iberoamericanas como Fuente de Terpenoides útiles en Química Fina. Ed. Universidad de Granada – CYTED, Granada, Spain, pp 7–33
Bandoni AL (2011) Evaluación farmacopeica de la calidad de drogas vegetales y productos relacionados. Estado actual en las farmacopeas argentina y brasilera. Dominguezia 27(2):35–56
Bandoni AL (ed) (2003) Los Recursos Vegetales Aromáticos en Latinoamérica, 2nd edn. Ed. Universidad Nacional de La Plata – CYTED), La Plata, Argentina, 410 pp
Bandoni AL, Debenedetti SL, Nadinic EL, Sánchez PL, Coussio JD (1993) Lineamientos Generales para la Implementación de una Reglamentación sobre Medicamentos herbarios. Unpublished text. Prize Luis de Prado 1993, Academia Nacional de Farmacia y Bioquímica, Argentina, 82 pp
Barboza GE, Cantero JJ, Núñez C, Pacciaroni A, Ariza Espinar L (2009) Medicinal plants: a general review and a phytochemical and ethnopharmacological screening of the native Argentine Flora. Kurtziana 34(1/2):7–365
Benedetti S, Barros S (2011) Boldo (*Peumus boldus* Mol.) Rescate de un Patrimonio Forestal Chileno. Manejo Sustentable y Valorización de sus Productos. Ed. Instituto Forestal, Santiago, Chile, 236 pp
Berretta A, Condón F, Rivas M (2007) Segundo Informe País sobre el estado de los Recursos Fitogenéticos para la Alimentación y la Agricultura, Uruguay, p 86
Bertoni MS (1922/27) La Civilización Guaraní. Ed. Ministerio de Agricultura y Ganadería del Paraguay. Three volumes. Asunción, Paraguay, 531 + 546 + 240 pp
Brazeiro A, Panario D, Soutullo A (2012) Clasificación y Delimitación de las Eco-Regiones de Uruguay. Montevideo, Uruguay, 40 pp
Brussa C, Grela I, González G (2007) Flora Arbórea Del Uruguay: con énfasis en las Especies de Rivera y Tacuarembó. Montevideo (Uruguay). Ed. COFUSA, Montevideo, Uruguay, 544 pp
Cabrera A (1971) Fitogeografía de la República Argentina. Bol Soc Argentina de Botánica 14(1/2):1–50
Calahan J, Howard D, Almalki AJ, Gupta MP, Calderón AI (2016) Chemical adulterants in herbal medicinal products: a review. Planta Medica 82(6):505–515
Cameroni MG (2010) Ficha Técnica: Manzanilla. Ed. Ministerio de Agricultura, Ganadería y Pesca de la Nación. Argentina, 22 pp
Chebataroff J (1942) La Vegetación del Uruguay y sus Relaciones Fitogeográficas con la del Resto de la América del Sur. Revista Geográfica (IPGH) 2(4/5/6):49–90
Chile (2014) Quinto Informe Nacional de Biodiversidad de Chile. Ed. Ministerio de Medio Ambiente, Chile, 141 pp
Chile (2016) Informe País. Estado del Medio Ambiente en Chile. 1999–2015. Ed. Universidad de Chile, 604 pp
Chile (2019) Inventario Nacional de Especies de Chile. http://especies.mma.gob.cl/CNMWeb/Web/WebCiudadana/pagina.aspx?id=89&pagId=85 (last access, July 2019)
Chile Aduana (2019). https://www.aduana.cl/aduana/site/edic/base/port/estadisticas.html (last access, August 2019)
Chile-Instituto Forestal (2009) Sistematización Silvícola, Tecnológica y Comercial de Boldo (*Peumus boldus* Mol.) en Chile, 218 pp
Cracco M, García L, González E, Rodríguez L, Quintillán A (2007) Importancia Global de la Biodiversidad de Uruguay. Proyecto Fortalecimiento de las Capacidades para la Implementación del Sistema Nacional de Áreas Protegidas (URU/05/001), 37 pp
Davies P (ed) (2005) Estudios en Domesticación y Cultivo de Especies Medicinales y Aromáticas Nativas. Ed. INIA, Montevideo, Uruguay, 264 pp
Montenegro P de -attributed at- (1945) Materia Médica Misionera. Reedition. Ed. Imprenta de la Biblioteca Nacional, Buenos Aires, Argentina, 479 pp
de Mösbach EW (1992) Botánica Indígena de Chile. Ed. Fundación Andes y Ed. A. Bello, Santiago, Chile, 140 pp

Rosales D de (1674) Historia General del Reyno de Chile. New edition by Mackenna BV (1877), Ed. Imprenta del Mercurio, Valparaíso, Chile, 506 pp
Demo M, Oliva del M, López ML, Zunino MP, Zygadlo JA (2005) Antimicrobial activity of essential oils obtained from aromatic plants of Argentina. Pharm Biol 43(2):129–134
DIEA (2019) Anuario Estadístico Agropecuario 2019. Oficina de Estadísticas Agropecuarias, Ministerio de Ganadería, Agricultura y Pesca, Uruguay, 256 pp
Dobrizhoffer M (1967) Historia de los abipones. 3 vol. Ed. Universidad Nacional del Nordeste, Resistencia, Argentina, 569 pp
Domínguez JA (1928) Contribuciones a la Materia Médica Argentina. Ed. Peuser, Buenos Aires, Argentina, 433 pp
Duke JA (2009) Duke's handbook of medicinal plants of Latin America. Ed. CRC, Boca Raton, USA, 908 + VIII pp
Dumont J (2003) El Amanecer de los Derechos del Hombre. 3° ed. Universidad Autónoma de Guadalajara, México, 279 pp
Echeverry R, Rossini C (eds) (2012) Productos naturales contra parásitos externos del ganado bovino y ovino, tales como mosca de los cuernos y garrapatas. Ed. Universidad de Magallanes, Chile, 166 pp
Elechosa MA (2009) Manual de recolección sustentable de plantas aromáticas nativas de la región central y noroeste de la Argentina. Ed. INTA, Buenos Aires, Argentina, 48 pp
Elechosa MA, Di Leo Lira P, Juárez MA, Viturro CI, Heit CI, Molina AC, Martínez AJ, López S, Molina AN, van Baren CM, Bandoni AL (2017) Essential oil chemotypes of *Aloysia citrodora* (Verbenaceae) in Northwestern Argentina. Biochem Syst Ecol 74:19–29
Farmacopea Argentina (2013/2018) VII edition and First Supplement (2018). Ed. ANMAT, Buenos Aires, Argentina, 2745 pp + 393 pp
FIA (1999/2009) Fundación para la Innovación Agraria, Chile. Plantas Medicinales y Aromáticas en Chile: situación actual y perspectivas. Ed. Fundación para la Innovación Agraria, Ministerio de Agricultura, Santiago, Chile, 24 pp. Also: (2009). Resultados y Lecciones en Especies Vegetales Nativas Productoras de Aromas, 28 pp; (2003) Plantas medicinales y aromáticas evaluadas en Chile, 316 pp
Gay C (1845/1854) Historia Física y Política de Chile. París, France. 8 volumes
Gillij YG, Gleiser RM, Zygadlo JA (2008) Mosquito repellent activity of essential oils of aromatic plants growing in Argentina. Bioresource Technol 99:2507–2515
Girault L (1984) Kallawaya: Guérisseurs itinérants des Andes. Ed. ORSTOM, París, France. 669 pp
González M, Coppetti V, Lombardo A (1936) Plantae Diaphoricae Florae Uruguayensis. Anales de la Universidad (R.O. del Uruguay) (137):119–377
González M, Lombardo A, Vallarino AJ (1939) Plantas de la Medicina vulgar del Uruguay. Montevideo, Uruguay
Haenke RG (1789–1794) Descripción del Reyno de Chile. Versión of Edwards A (1942). Ed. Nascimento, Santiago, Chile, 280 pp
Haretche F, Mai P, Brazeiro A (2012) Woody flora of Uruguay: inventory and implication within the Pampean region. Acta Botanica Brasilica 26(3):537–552
Hieronymus J (1882) Plantas Diafóricas – Flora Argentina. Reedición. Ed. Atlántida, Buenos Aires, Argentina, 421 pp
Hoffmann A, Farga C, Lastra J, Veghazi E (1992) Plantas Medicinales de uso común en Chile. Ed. Fundación Claudio Gay, Santiago, Chile, 274 pp
IGM (2020) Instituto Geográfico Militar, Uruguay. www.igm.gub.uy (last access, May 2020)
IGN (2020) Instituto Geográfico Nacional. Argentina. www.ign.gob.ar (last access, June 2020)
IGWIA (2019) Grupo Internacional de Trabajo sobre Asuntos Indígenas. El Mundo Indígena 2019. Ed. IGWIA, Copenhague, Denmark, 736 pp
INASE (2019) Instituto Nacional de Semillas, Argentina. Resoluciones N° 22 de 2006 and 318 de 2008. https://gestion.inase.gov.ar/consultaGestion/gestiones (last access, July 2019)
INDEC (2019) Instituto Nacional de Estadísticas y Censos, Argentina. https://www.indec.gob.ar/indec/web/Nivel4-Tema-3-6-18 (last access, July 2019)

INFOR (2020) Instituto Forestal de Chile. www.infor.cl (last access, June 2020)
INIA (2011) Instituto Nacional de Investigaciones Agropecuarias, Montevideo. Uruguay. Serie Técnica 193, 40 pp
INIA (2020) Instituto Nacional de Investigaciones Agropecuarias, Montevideo, Uruguay. Producción citrícola. Antecedentes. http://www.inia.uy (last access, June 2020)
IRCCA (2019) Instituto de Control y Regulación de *Cannabis*. Montevideo, Uruguay. https://www.ircca.gub.uy/ (last access, September 2019)
ITC (2020) International Trade Center. http://www.intracen.org/itc/market-info-tools/ (last access, May 2020)
Kobayashi Y, Suzuki A, Kobayashi A, Kasai A, Ogata Y, Kumada Y, Takahashi R, Ogino F (1986) Suppression of Sensory Irritation by Chamomile Essential Oil and Its Active Component – Bisabololoxide A. Acta Horticult 749:163–174
León Guerrero MM, Gervás JMA (2013) La Controversia de Valladolid, 1550–1551. El concepto de igualdad del «otro». Boletín Americanista 1(76):135–154
MAG/ALADI (2011) Proyecto Fortalecimiento de las Capacidades Técnicas de Funcionarios del MAG en Cadenas Productivas como parte del Proceso de Fortalecimiento de la Competitividad del Paraguay. Ed. Ministerio de Agricultura y Ganadería del Paraguay, Asunción, Paraguay. 34 pp
Martín CM, Valverde JL (1995) La Farmacia en la América Colonial. El arte de preparar medicamentos. Ed. Universidad de Granada, Granada, Spain, 633 pp
Martínez Crovetto R (1981) Plantas utilizadas en Medicina en el NO de Corrientes. Ed. Fundación Miguel Lillo, Tucumán, Argentina, 139 pp
Martínez GJ (2005) Recolección y Comercialización de Plantas Medicinales en el Departamento Santa María, Provincia de Córdoba, Argentina. Acta Farm Bonaerense 24(4):575–584
Martínez GJ, Luján MC (2011) Medicinal plants used for traditional veterinary in the Sierras de Córdoba (Argentina): An ethnobotanical comparison with human medicinal uses. J Ethnobiol Ethnomed 7:23
Martínez GJ, Planchuelo AM, Fuentes E, Ojeda E (2006) A numeric index to establish conservation priorities for medicinal plants in the Paravachasca Valley, Córdoba, Argentina. Biodivers Conserv 15:2457–2475
Meliá B (2006) Escritos guaraníes como fuentes documentales de la historia paraguaya. Lecture of incorporation as a Member of Number, at the Paraguayan Academy of History. April 15, 2004
Mellado Campos V, Medina E, San Martín Saldías C (1997) Herbolaria Médica de Chile. Ed. Ministerio de Salud, Santiago, Chile, 273 pp
MERCOSUR (2014) Resolución RE- 22-2014-GMC. Estructura y Organización de la Farmacopea MERCOSUR
MGAP (2015) Ministerio de Ganadería, Agricultura y Pesca, Uruguay. Estadísticas Agropecuarias DIEA, Montevideo, Uruguay, 42 pp
MGAP (2019) Ministerio de Ganadería, Agricultura y Pesca. Dirección General Forestal. Resultados del Inventario Nacional Forestal del Bosque Nativo, Montevideo, Uruguay, 43 pp
MGAP (2020) Ministerio de Ganadería, Agricultura y Pesca, Uruguay. Estadísticas Forestales 2020, 66 pp
Ministerio de Agroindustria (2018) Industria de Alimentos y Bebidas. Informe de actividad. Febrero 2018. Presidencia de la Nación, Argentina, 51 pp
Molina JI (1788) Compendio de la Historia Geográfica Natural y Civil del Reyno de Chile. First part. Madrid, Spain, 417 pp
Molina JI (1810) Ensayo sobre la Historia Natural de Chile. New edition from 1986. Ed. Maule, Santiago, Chile, 385 pp
Montes M, Wilkomirsky T (1974) Quelques aspects de la medication vegetale populaire dans la region de Concepción (Chile). Planta Medica 25:183–192
Montes M, Wilkomirsky T (1978) Plantas Chilenas en Medicina Popular- Ciencia y Folklore. Ed. Univ. de Concepción, Chile, 135 pp
Montes M, Wilkomirsky T (1987) Medicina Tradicional Chilena. Ed. Univ. de Concepción, Chile, 205 pp

Montevideo Portal (2019) https://www.montevideo.com.uy/Negocios-y-Tendencias/Los-uruguayos-consumimos-10-kilos-de-yerba-mate-per-capita-por-ano-uc728131 (Last access, October 2020)
Moreira-Muñoz A (2011) Plant Geography of Chile. Ed. Springer. Dordrecht, Netherland. 343 pp
Morrone JJ (1999) Presentación Preliminar de un Nuevo Esquema Biogeográfico de América del Sur. Biogeographica 75(1):1–16
Morrone JJ (2001) Biogeografía de América Latina y el Caribe. Ed. CYTED-ORCYT-UNESCO-Sociedad Entomológica Aragonesa (SEA). Zaragoza, Spain, 148 pp
Muñoz O, Montes M, Wilkomirsky T (2001) Plantas Medicinales de uso en Chile. Ed. Universitaria, Santiago, Chile, 330 pp
Murillo A (1889) Plantes Médicinales du Chili. Ed. Exposition Universelle de Paris, Section Chilienne, Paris, France, 234 pp
Niemeyer HM, Teillier S (2007) Aromas de la Flora Nativa de Chile. Ed. Universidad de Chile, Santiago, Chile, 446 pp
Niemeyer HM (1995) Biologically active compounds from chilean medicinal plants. In: Arnason JT, Mata R, Tomeo JT (eds) Phytochemistry of medicinal plants. Ed. Springer, New York, USA, pp 137–159
O'Malley Olsen O, O'Donnell J, Mattson CL, Schier JG, Wilson N (2019) Unintentional drug overdose deaths with Kratom detected. Morb Mortal Wkly Rep 68(14):326–327
Otazú Melgarejo A (2014) Contribución a la medicina natural: Pohã Ñana, un Manuscrito inédito en Guaraní (Paraguay, S. XVIII). Corpus Archivos virtuales de la alteridad americana 4(2):1–15. http://journals.openedition.org/corpusarchivos/1301 (last access July, 2019)
Paraguay (2018) Población Indígena. Ed. Dirección General de Estadísticas y Censos, Asunción, Paraguay, 53 pp
Parodi D (1881) Ensayo de Botánica Médica Argentina Comparada. Thesis. Facultad de Ciencias Médicas de la Universidad de Buenos Aires. Ed. Coni, Buenos Aires, Argentina, 103 pp
Pastor Arenas (1981) Etnobotánica Lengua-Maskoy. Ed. CONICET, Buenos Aires, Argentina, 358 pp
Paucke F (2010) Hacia allá y para acá. Ed. Ministerio de Innovación y Cultura de la Provincia de Santa Fe., Argentina, 706 pp
Pires VP, Almeida RN, Wagner VM, Lucas AM, Vargas RMF, Cassel E (2019) Extraction process of the *Achyrocline satureioides* (Lam.) DC. essential oil by steam distillation: modeling, aromatic potential and fractionation. J Ess Oil Res 31(4):286–296
Pitman NCA, Jørgensen PM (2002) Estimating the size of the world's threatened flora. Science 298:989
Retta D, Dellacassa E, Villamil J, Suárez SA, Bandoni AL (2012) Marcela, a promising medicinal and aromatic plant from Latin America: a review. Ind Crop Prod 38:27–38
Retta DS (2012) Parámetros fitoquímicos para la determinación de calidad de "marcela" *Achyrocline satureioides* (Lam.) DC. (Asteráceas). Thesis, Universidad de Buenos Aires, Argentina, 243 pp
Rivas M (2010) Valorización y Conservación de la Biodiversidad en Uruguay. In: García F et al.: Intensificación Agrícola: oportunidades y amenazas para un país productivo y natural. Ed. Universidad de la República, Montevideo, Uruguay, pp 89–109
Rondina RVD, Bandoni AL, Coussio JD (2003) Plantas Silvestres Argentinas con reconocidas Propiedades Medicinales o Tóxicas (database in CD). 3° edition. ISBN: 987-43-6073-9
Rondina RVD, Bandoni AL, Coussio JF (2010) Argentine medicinal plants with potential antifungal activity. Dominguezia 26(1):31–39
Ruiz H, Pavon J (1798) Flora Peruvianae e Chilense. Ed. Gabriel de Sancha. Madrid, Spain. Volume I, 455 pp
San Martín J (1983) Medicinal plants in Central Chile. Econ Bot 37:216–227
Sánchez Labrador P (1948) La Medicina en el Paraguay Natural 1771–1776. Ed. Universidad Nacional de Tucumán, Argentina, 474 pp

Scarpa GF (2009) Etnobotánica Médica de los Indígenas Chorote y su Comparación con la de los criollos del Chaco Semiárido (Argentina). Darwiniana 47(1):92–107
SENASA (2014) Servicio Nacional de Sanidad y Calidad Agroalimentaria. Bosques Argentinos, Actividad Forestal y Economías Regionales. http://www.senasa.gob.ar/senasa-comunica/infografias/bosques-argentinos-actividad-forestal-y-economias-regionales (last access, June 2020)
Siedo SJ (2006) Systematics of *Aloysia* (Verbenaceae). Thesis. The University of Texas at Austin, USA, 309 pp
Soutullo A, Clavijo C, Martínez-Lanfranco JA (eds) (2013) Especies prioritarias para la conservación en Uruguay. Vertebrados, moluscos continentales y plantas vasculares. Ed. SNAP/DINAMA/MVOTMA y DICYT/MEC, Montevideo, Uruguay, 222 pp
Speisky H, Cassels BK (1994) Boldo and Boldine: an emerging case of natural drug development. Pharmacol Res 29(1):1–12
Toloza AC, Zygadlo J, Biurrun F, Rotman A, Picollo M (2010) Bioactivity of Argentinean essential oils against permethrin-resistant head lice, *Pediculus humanus capitis*. J Insect Sci 10:185. Available online: insectscience.org/10.185
Tomić M, Popović V, Petrović S, Stepanović-Petrović R, Micov A, Pavlović-Drobac M, Couladis M (2014) Antihyperalgesic and Antiedematous Activities of Bisabolol-Oxides-Rich Matricaria Oil in a Rat Model of Inflammation. Phytother Res 28:759–766
Tschirch, A (1909) Handbuch der Pharmakognosie. Verlag von Chr. Herm. Tauchnitz, Leipzig, Germany. vol I, 286 pp
Ulloa Ulloa C, Acevedo-Rodríguez P, Beck S, Belgrano MJ, Bernal R, Berry PE et al (2017) An integrated assessment of the vascular plant species of the Americas. Science 358(6370):1614–1617
UNEP-WCMC (2011) Review of *Bulnesia sarmientoi* from Paraguay. Ed. UNEP-WCMC, Cambridge, UK, 8 pp
Urban O (1934) Botánica de las Plantas endémicas de Chile. Ed. Concepción, Concepción, Chile, 292 pp
Urton G, Bresine CJ (2005) Khipu accounting in Ancient Peru. Science 309:1065–1067
Urton G, Chu A (2018) The Invention of Taxation in the Inka Empire. Latin American Antiquity:1–16. https://doi.org/10.1017/laq.2018.64
Uruguay (2016a) Decreto 403. Aprobación del Marco Regulatorio para Hierbas Medicinales, Especialidades Vegetales y Medicamentos Fitoterápicos. Poder Ejecutivo, Ministerio de Salud Pública. Montevideo, Uruguay
Uruguay (2016b) Estrategia Nacional para la Conservación y Uso Sostenible de la Biodiversidad Biológica de Uruguay. Ed. Ministerio de Vivienda, Ordenamiento Territorial y Medio Ambiente, Montevideo, Uruguay, p 59
Vargas R, Reif A, Faúndez MJ (2011) The forests of Robinson Crusoe Island, Chile: an endemism hotspot in danger. Bosque 32(2):155–164
Verne C (1874) Étude sur le boldo. Ed. J. B. Baillière et fils, Paris, France, 52 pp
Villamil CB (2004) Conservation of medicinal plants in the southern cone of South America. Med Plants Conserv (IUCN) 9(10):12–14
Viturro C, Bandoni A, Dellacassa E, Atti Serafini L, Elder H (2010) Problemática *Schinus* en Latinoamérica. In: Dellacassa E (ed) Normalización de productos naturales obtenidos de especies de la Flora Aromática Latinoamericana. Ed. EdiPUCRS, Porto Alegre, Brazil. pp 205–280
Wilde G (2013) Fuentes Indígenas en la Sudamérica Colonial y Republicana. Escritura, Poder y Memoria. Corpus. Archivos virtuales de la alteridad americana 3:N°1. 1er. semestre 2013
Zin J (1930) La Salud por medio de las plantas medicinales. Ed. Salesiana, Chile, 702 pp
Zolla C (1994) Biblioteca de la Medicina Tradicional Mexicana. Ed. Instituto Nacional Indigenista, México D.F., 12 volumes
Zuloaga FO, Belgrano MJ (2015) The Catalogue of Vascular Plants of the Southern Cone and the Flora of Argentina: their contribution to the World Flora. Rodriguésia 66(4):989–1024
Zuloaga FO, Belgrano MJ, Zanotti CA (2019) Actualización del Catálogo de Las Plantas Vasculares del Cono Sur. Darwiniana (nueva serie) 7(2):208–278
Zuloaga FO (2013) Ciencia para conocer y proteger la Flora Argentina. Ed. CONICET. Buenos Aires, Argentina, 1 pp

Part II
Selected Medicinal and Aromatic Plants

Adesmia boronioides Hook. *f.*

Silvia Beatriz González

Adesmia boronioides Hook. *f.* (Photo by the author)

Abstract *Adesmia boronioides* Hook. *f.* (Leguminosae), known as "paramela", is an aromatic and medicinal plant widely used in the Patagonia region (South of Argentina). It has been used traditionally as medicinal and also for its aromatic and

S. B. González (✉)
LIPAM (Laboratorio de Investigación de Plantas Aromáticas y Medicinales), Facultad de Ciencias Naturales y Ciencias de la Salud, Universidad Nacional de la Patagonia San Juan Bosco sede Esquel, Esquel (9200), Chubut, Argentina
e-mail: quim-esq@unpata.edu.ar; silviaesquel@speedy.com.ar

Á. Máthé, A. Bandoni (eds.), *Medicinal and Aromatic Plants of South America Vol. 2*, Medicinal and Aromatic Plants of the World 7,
https://doi.org/10.1007/978-3-030-62818-5_2

ornamental qualities. This species has received an increasing interest by its fragrant odor and the potential use of its essential oil in perfumery. However, the oil composition of "paramela" wild populations from Patagonia showed a wide chemical variability, so experimental crops from selected plants was introduced in the Andean region in order to select the best plant material for essential oil production.

Keywords "Paramela" · Leguminosae · Patagonia · Aromatic · Esquelenone

1 Introduction

Adesmia boronioides Hook. *f.* ("paramela") is a very well-known aromatic and medicinal species growing wild in the Patagonia region of Argentina and Chile. It is used in rural areas mainly as a medicinal infusion, and in urban areas both as ornament and aromatic ingredient for the preparation of an alcoholic beverage. This native plant of the Patagonian region is almost exclusively found in its wild state. However, in the rural communities *A. boronioides* is cultivated in family orchards. Recently its essential oil has been introduced as a valuable raw material for perfumery, placing the species in an incipient domestication process.

2 Taxonomic Characteristics

Adesmia boronioides Hook. *f.*, Leguminosae (Fabaceae), is popularly known as "paramela", "yagneu", "lonckotrevo", "té pampa", "té silvestre", "yerba Carmelita" and "éter" (González 2002; Barboza et al. 2009). The *Adesmia* DC genus includes almost 240 species and can been found only in South America (Burkart 1967). The different species reported are mainly distributed in the center of Chile and in the South and West regions of Argentina (Burkart 1967; Ulibarri and Burkart 2000).

In particular, the genus is represented by more than 100 species for Argentina, making it the most numerous genus of the Leguminosae (subfamily Papilionoideae) in the country, while 55 species are present in the Patagonian region (Burkart 1984; Ulibarri and Burkart 2000; Ulibarri and Simpson 2010).

3 Crude Drug Used

The aerial parts of *A. boronioides* have been used in traditional medicine to treat rheumatic pains, colds, and digestive disorders; as well as a decongestant for the respiratory tract (Montes and Peltz 1963; Montes and Wilkomirsky 1987; Simirgiotis et al. 2012). The aerial parts have been also used popularly in preparing incenses to cleanse the houses. Occasionally, the dried leaves and stems are sold in herb-shops and as ornamental (Green and Ferreyra 2011).

4 Major Chemical Constituents and Bioactive Compounds

The leaves and stems of *A boronioides* from Patagonia, Argentina yielded variable amounts of essential oil according to the season latitude and its ecological conditions (González et al. 2004). The main components of its essential oil are two bisnorsesquiterpenes (esquel-6-en-9-one and esquel-7-en-9-one, named from Esquel, city of the Argentinean Patagonia, where "paramela" grows) as well as α-copaen-11-ol, δ-cadinene, 10-epi-γ-eudesmol, 4α-hydroxydihydroagarofuran, 1-epi-cubenol and α-pinene. The structural elucidation of the main components was performed through NMR (mono and bidimensional), vibrational circular dichroism (VCD) and its absolute configuration was finally reassigned (González et al. 2002; Cerda-García-Rojas et al. 2015).

The very pleasant sweet-woody, licorice-spicy odor and strong fixation properties of the oil was suggested to be used in the fragrance industry (Bandoni 2000; González et al. 2002).

Moreover, it was also demonstrated that the essential oil composition varies according to latitude and its ecological conditions, as soil, weather, altitude and living organisms associated (González et al. 2016; González and Martin 2019).

5 Morphological Description

A. boronioides ("paramela") is a perennial bush. It has a medium size (0.4–2.0 m), highly ramified with glandular branches, fragrant and very resinous and sticky to the touch. The leaves 3–6 cm, shortly petiolate, 10–20 leaflets, leaf rachis with erect, brief hair; obovate, fleshy, glabrous, toothed, shiny, 4–6 mm leaflets with crateriform glands especially at the edge; short, amplexicaul, glabrous, glandular stipules. The flowers are 7–10 mm, colorful, yellow, perfumed, with campanulate chalice, pubescent, glandulous, with short teeth, serice-pubescent in their interior; glabrous vexillum (banner), glabrous wings and keel shorter than vexil. Ovary with some marginal hair. Narrow, pubescent, glandular, 35-articulate loments; semicircular, dehiscent 4.5–6 mm trusses. Roots are axonomorph (González et al. 2019).

A. boronioides stands out as the only species of this genus highly glandulous-resinous (Burkart 1967). Its anatomy was initially studied by Nájera et al. (2000). Subsequently, in an evaluation of leaves and stems of diverse origins in their Patagonian distribution, very similar microscopic characters were found. They all have cyclocytic stomata on the leaf's epidermis. The secretory pores are located, in greater number, on the abaxial surface, although some pores can be observed on the adaxial surface. On some samples, secretion pores were found at the end of the leaves (González et al. 2014).

6 Geographical Distribution

The *A. boronioides* distribution ranges from Mendoza to Tierra del Fuego provinces, in Argentina, including Neuquén, Río Negro, Chubut, and Santa Cruz, and the XI and XII Regions of Chile (http://www2.darwin.edu.ar/Proyectos/FloraArgentina/fa.htm; Burkart 1967; Ulibarri and Burkart 2000). The species grows wild up to 2200 m.a.s.l.

It inhabits sunny areas, shrubs, river sides, roads and ravines, mainly in the Patagonian steppe, shrubland areas and steppe-forest transition zones (https://www.sib.gov.ar/ficha/PLANTAE*adesmia*boronioides). It has also been found in the Atlantic littoral, Santa Cruz Province, at Rio Gallegos area (González et al. 2014). It grows in shrubland vegetation near Los Molles, province of Mendoza (Argentina).

7 Ecological Requirements

A. boronioides is native of the Patagonian region inhabiting low irrigation sites being of slow growth. It is found almost exclusively in its natural state, being its culture of interest (Barthélémy et al. 2008; Contardi et al. 2016a, b).

Reproduction studies made from *A. boronioides* seeds allowed the development of propagation protocols and plants production in greenhouses (González et al. 2009; Sánchez and Riat 2012; Mazzoni et al. 2014, 2016). As of overharvesting and excessive commercial exploitation of wild populations, its domestication has gained on importance. Since 2015, an experimental culture in the Andean region of the Argentinean Patagonia was developed in order to evaluate the productivity and quality of the cultivated plant's essential oil by comparison with the wild populations (González et al. 2018).

8 Traditional Use (Part(s) Used) and Common Knowledge

Aerial parts, with or without flowers (Photo 1), are used as medicine by the Patagonian rural population, mostly as digestive, antirheumatic, diaphoretic and antiemetic (Campos et al. 1997; González et al. 2004, 2005a, b; Estomba et al. 2006). Traditionally, the plant material is collected in autumn-winter, dried and kept in dark places to use it as home medicine by rural communities during the whole year (Richeri et al. 2013a, b). The people distinguish it by its "perfumed" character,

Photo 1 Flowers and leaves of *A. boronioides*. (Photos by the author)

and it is described as a plant possessing "magical soul", "sweet smell and bitter taste" (Molares and Ladio 2009; Ladio and Molares 2014).

A. boronioides is also valued by "heating the body" when prepared in the form of steam inhalations and baths (González 2005). Steam inhalation is a common practice for cold and cough discomfort in Neuquén, Río Negro and Chubut Provinces communities (Igon et al. 2006; Eyssartier et al. 2011a, b; Richeri et al. 2013a, b). Richeri (2016) also records the use of the plant as incense to perfume and cleanse the houses of evil spirits. According to Ochoa (2015), the species is also used in ointment as a sedative for rheumatic problems and to heal wounds.

The infusion of "paramela" is widely used as digestive (Igon et al. 2006), being also included in the "mate", a traditional drink of the region, mainly composed of *Ilex paraguariensis* dried leaves (Weigandt et al. 2004). It is usually found marketed in tourist shops and crafts businesses where is sold in San Carlos de Bariloche (South Argentina) as medicinal plant mainly for its digestive properties (Cuassolo et al. 2010) and to wash the hair in order to kill lice and as hair vitalizer (Martínez-Crovetto 1980; Conticello et al. 1997; Igon et al. 2006). Some studies have also reported its use as aphrodisiac (Muñoz et al. 2001; Igon et al. 2006; Barboza et al. 2009). In other Argentine regions of Province of Río Negro, people emphasize the use of the plant's decoction to ease flu, stomach pains, diarrhea, and fever (Eyssartier et al. 2009, 2011a, b, 2013).

9 Modern Medicine Based on Its Traditional Medicine Uses

The properties of infusions from aerial parts of *A. boronioides* were investigated showing anticancer potential due to their antiproliferative activity on human cancer cell lines (Gastaldi et al. 2018). Antioxidant activity, most likely due to the presence of phenolic compounds and flavonoids has also been reported (Gastaldi et al. 2016a; Silva Sofrás et al. 2016). Resinous exudate obtained from the aerial parts of A. *boronioides* were evaluated by its anti-phytopathogenic effects (Montenegro et al. 2019). The plant is also a scent agent in the fragrance industry (González et al. 2002). Moreover, the essential oil showed *in vitro* trypanocidal effect even with IC_{50} value lower than that of benzimidazole, used as reference drug (Villagra et al. 2008).

The dermal irritability test performed on albino rabbit skin using *A. boronioides* essential oil evaluating erythema and edema formation, showed the oil safety use under the test conditions according to the methodology of Draize (González 2002).

The use of *A. boronioides* to joint and muscle pains in traditional medicine could be linked to their anti-inflammatory activity tested *in vitro* (González et al. 2003).

In addition, the toxicity values found for an infusion of *A. boronioides* tested using *A. salina* suggests not risk of acute toxicity to humans (Mongelli et al. 1995; Pérez and Lazo 2010; Gastaldi et al. 2016b, 2018).

Actually, *A. boronioides* is commercialized in three forms: (a) as an ingredient in alcoholic beverages, (b) as a medicinal herb, mainly in infusions and decoctions, being the one of the most common native medicinal plant commercialized in the region (Cuassolo et al. 2010), and (c) its essential oil is used more and more as a raw material for perfumery.

More recently the species has been incorporated into the Argentine Food Code (Código Alimentario Argentino 2019) as a food additive for use as a flavoring, and an oil/water emulsion was developed for the encapsulation of the essential oil (Martinez et al. 2019).

10 Conclusions

A. boronioides has a long history of use in Patagonian communities as an aromatic and medicinal plant, making it one of the most commercialized native species in the urban centers of the region. Over the past years, there has been a notable increase in the demand for this plant because of its outstanding aromatic quality, increasing the pressure of harvesting on wild populations. As consequence, a few experiences of propagation have been developed whose results suggest that it would be feasible to produce large-scale seedlings to establish future crops as a productive alternative for the region.

Acknowledgments Thanks are due to Secretaría de Ciencia y Técnica, Universidad Nacional de la Patagonia San Juan Bosco, Argentina.

References

Bandoni AL (ed) (2000) Los Recursos Vegetales Aromáticos en Latinoamérica, su aprovechamiento industrial para la producción de aromas y sabores. Ed. Universidad Nacional de La Plata. La Plata, Argentina, 410 pp

Barboza GE, Cantero JJ, Núñez C, Pacciaroni A, Ariza Espinar L (2009) Medicinal plants: a general review and a phytochemical and ethnopharmacological screening of the native Argentine Flora. Kurtziana 34(1–2):7–365

Barthélémy D, Brion C, Puntieri J (2008) Plantas de la Patagonia. Ed. Vázquez Mazzini. Buenos Aires, Argentina, 238 pp

Burkart AE (1967) Sinopsis del género sudamericano de Leguminosas *Adesmia* D.C. Darwiniana 14:463–468

Burkart AE (1984) *Adesmia*. In: Correa MN (ed) Flora Patagonica, Leguminosae. Col. Cient. INTA 4b:92–161

Campos VM, Cárdenas EM, San Martín Saldías C (1997) Herbolaria médica de Chile. Diagnóstico de su estado actual y perspectivas futuras para la medicina oficial chilena. Ed. Ministerio de Salud, Santiago, Chile, 273 pp

Cerda-García-Rojas CM, Bucio A, González SB, García Gutiérrez H, Joseph-Nathan P (2015) Absolute configuration of esquelane derivatives from *Adesmia boronioides* by vibrational circular dichroism. Tetrahedron: Asymmetry 26:136–140

Código Alimentario Argentino (2019) Artículo 1398, ítem 174. Boletín Oficial de la República Argentina del 23 de noviembre de 2019

Contardi LT, Sofrás FM, González SB (2016a) Paramela, una especie nativa de la Patagonia: análisis de frutos y semillas. Revista Patagonia Forestal XXI(1), Jun 2016

Contardi LT, Silva Sofrás FM, González SB (2016b) Reproducción de *Adesmia boronioides,* especie aromática y medicinal actualmente cosechada de poblaciones naturales en Patagonia. Revista Naturalia Patagónica 9:50–60

Conticello L, Gandullo R, Bustamante A, Tartaglia C (1997) El uso de plantas medicinales por la comunidad Mapuche de San Martín de los Andes, provincia de Neuquén (Argentina). Parodiana 10(1–2):165–180

Cuassolo F, Ladio AH, Ezcurra C (2010) Aspectos de la comercialización y control de calidad de las plantas medicinales más vendidas en una comunidad urbana del NO de la Patagonia Argentina. BLACPMA 9(3):166–176

Estomba D, Ladio AH, Lozada M (2006) Medicinal wild plant knowledge and gathering patterns in a Mapuche community from North-Western Patagonia. J Ethnopharmacol 103:109–119

Eyssartier C, Ladio AH, Lozada M (2009) Medicinal use of plant species cultivated in a semi-rural community of the patagonian steppe. BLACPMA 8(2):67–76

Eyssartier C, Ladio AH, Lozada M (2011a) Horticultural and gathering practices complement each other: a case study in a rural population of Northwestern Patagonia. Ecol Food Nutr 50:429–451

Eyssartier C, Ladio AH, Lozada M (2011b) Traditional horticultural knowledge change in a rural population of the Patagonian steppe. J Arid Environ 75:78–86

Eyssartier C, Ladio AH, Lozada M (2013) Traditional horticultural and gathering practices in two semi-rural populations of Northwestern Patagonia. J Arid Environ 97:18–25

Gastaldi B, Assef Y, Sofrás F, Catalán C, González SB (2016a) Actividad antioxidante y toxicidad frente a *Artemia salina* de infusiones de nuevo especies aromáticas del noroeste de la Patagonia. Dominguezia 32(2):39

Gastaldi B, Assef Y, van Baren C, Di Leo Lira P, Retta D, Bandoni A, González SB (2016b) Antioxidant activity in teas, tinctures and essential oils of native species from Patagonia Argentina. Rev Cub Pl Med 21(1):51–62

Gastaldi B, Marino G, Assef Y, Sofrás F, Catalán C, González SB (2018) Nutraceutical properties of herbal infusions from six native plants of Argentine Patagonia. Plant Foods Human Nutr 73(3):180–188

González SB (2002) Estudio de *Adesmia boronioides* Hook. *f.*, una especie aromática y medicinal nativa de la Patagonia. Thesis. Facultad de Farmacia y Bioquímica, Universidad de Buenos Aires, Argentina, 219 pp

González SB (2005) *Adesmia boronioides* Hook. *f.*: una especie aromática y medicinal nativa de la Patagonia. Naturalia Patagónica 2(1):85–91

González SB, Martin G (2019) Aspectos que se entrelazan: interacciones ecológicas, coevolución y propiedades emergentes, en el estudio de la relación planta-insecto. Análisis con una especie aromática y medicinal nativa de la Patagonia *Adesmia boronioides* Hook. *f.* Naturalia Patagónica 15:1–43

González SB, Bandoni AL, van Baren C, Di Leo Lira P, Cerda-García-Rojas CM, Joseph-Nathan P (2002) Structure, Conformation and Absolute Configuration of Novel Bisnorsesquiterpene from the *Adesmia boronioides* Essential Oil. Tetrahedron 58:3065–3071

González SB, Houghton P, Hoult J (2003) The activity against leukocyte eicosanoid generation of essential oil and polar fractions of *Adesmia boronioides* Hook. f. Phytother Res 17:290–293

González SB, Bandoni AL, van Baren C, Di Leo Lira P, Cerda-García-Rojas CM, Joseph-Nathan P (2004) The Essential Oil of the Aerial Parts of *Adesmia boronioides* Hook. *f.* J Essent Oil Res 16:513–516

González SB, Gallardo C, Guerra P (2005a) Etnobotánica de plantas medicinales en comunidades rurales de la Provincia del Chubut, Patagonia Argentina. Libro de Resúmenes V Reunión de la Sociedad Latinoamericana de Fitoquímica, Montevideo, Uruguay. November 28 – December 2, 102

González SB, Gallardo C, Recalde J, Guerra P (2005b) Talleres etnobotánicos en comunidades del NO del Chubut. Actas de las XXX Jornadas Argentinas de Botánica, Rosario, Argentina, November 6–10, 83

González SB, Pasquini N, Contardi L (2009) Calidad de semillas y producción de plantas de *Adesmia boronioides* Hook. *f.*, especie aromática y medicinal nativa de la Patagonia. Proyecto de Ciencia y Técnica de la UNPSJB sede Esquel. CAFCN N° 374/08, Resol. C.S. N° 101/08

González SB, Guerra, P, Gastaldi B, van Baren C, Di Leo Lira P, Bandoni A (2014) Avances en el estudio de plantas aromáticas nativas y naturalizadas en la Patagonia Argentina. Variabilidad y constancia en la composición de sus aceites esenciales. Lilloa 52 (Suplemento): IV Jornadas Nacionales de Plantas Aromáticas Nativas y sus Aceites Esenciales, pp 37–42

González SB, Gastaldi B, Silva Sofrás F, van Baren C, Di Leo Lira P, Retta D, Bandoni A (2016) Diversidad química en una especie aromática y medicinal endémica de Patagonia. Dominguezia 32(2):89–90

González SB, Silva Sofrás F, Gastaldi B, Mazzoni A, Sánchez G, Di Leo Lira P, Retta D, van Baren C, Bandoni A (2018) Estudios fitoquímicos y botánicos comparativos de una planta aromática y medicinal endémica de la patagonia argentina llevada a cultivo. Actas del XI Congreso Internacional de Fitoterapia Oviedo, Spain. April 13–15:85–86

González SB, Ladio AH, Gastaldi B, Silva Sofrás FM, Mazzoni A, Sánchez G (2019) Paramela (*Adesmia boronioides* Hook. f.): from popular uses to commercialization. In: Martínez JL, Muñoz-Acevedo A, Rai M (eds) Ethnobotany application of medicinal plants. Ed. Taylor & Francis Group, Boca Raton, USA, pp 89–104

Green L, Ferreyra M (2011) Flores de la Estepa patagónica. Ed. Vázquez Mazzini, Buenos Aires, Argentina, 288 pp

Igon P, Ladio AH, Lozada M (2006) Plantas Medicinales utilizadas en las comunidades de Villa Traful y Cuyín Manzano. UNCOMA. Ed. Imaginaria, Argentina, 67 pp

Ladio AH, Molares S (2014) El paisaje patagónico y su gente. Capítulo 9. In: Raffaele E, de Torres Curth M, Morales CL, Kitzberger T (eds.): Ecología e historia natural de la Patagonia Andina: Un cuarto de siglo de investigación en biogeografía, ecología y conservación. Ed. Fund. Azara, Buenos Aires, Argentina, 256 pp

Martinez Y, Retta D, Gastaldi B, González SB (2019) Encapsulamiento del aceite esencial de *Adesmia boronioides* Hook. *f* en una emulsión aceite/agua. Caracterización. XIX Encuentro de superficies y materiales nanoestructurados – NANO 2019, June 5–7, Buenos Aires

Martínez-Crovetto R (1980) Apuntes sobre la vegetación de los alrededores del Lago Cholila. Publicación Técnica N° 1, Facultad de Ciencias Agrarias, Corrientes, Argentina, pp 1–22

Mazzoni A, Sánchez G, Riat M, Hagiwara J (2014) Propagación de Adesmia boronioides Hook. f., planta nativa medicinal de la Patagonia Argentina. Libro de resúmenes y poster presentado al XXXVII Congreso Argentino de Horticultura (ASAHO). Mendoza, Argentina. September 23–26

Mazzoni A, Sánchez G, Riat M, Hagiwara J (2016) Propagación de *Adesmia boronioides* Hook. f., planta nativa medicinal de la Patagonia Argentina. Presentación oral: Experiencias de Domesticación. V Jornadas Nacionales de plantas aromáticas nativas y sus aceites esenciales y I° Jornadas Nacionales de plantas medicinales nativas. Esquel. Dominguezia 32(2):70

Molares S, Ladio AH (2009) Ethnobotanical review of the Medicinal Mapuche Flora: use patterns on a regional scale. J Ethnopharmacol 122:251–260

Mongelli E, Coussio J, Ciccia G (1995) Estudio de toxicidad aguda de plantas medicinales argentinas mediante el bioensayo de *Artemia salina* Lech. Dominguezia 12:35–42

Montenegro I, Valenzuela M, Zamorano N, Santander R, Báez C, Madrid A (2019) Activity of *Adesmia boronioides* resinous exudate against phytopathogenic bacteria. Nat Prod Res. https://doi.org/10.1080/14786419.2019.1648465

Montes AL, Peltz L (1963) Esencias de plantas aromáticas del Parque Nacional Nahuel Huapi y sus aledaños. 2. *Adesmia boronioides* Hooker, o paramela. An Soc Cient Argent 175:91–101

Montes M, Wilkomirsky T (1987) Medicina Tradicional Chilena. Ed. Universidad de Concepción. Concepción, Chile, 205 pp

Muñoz O, Montes M, Wilkomirsky T (2001) Plantas medicinales de uso en Chile. Química y Farmacología. Ed. Universitaria. Santiago, Chile, 205 pp

Nájera MT, Spegazzini ED, Castro MT, Guerra P, González SB (2000) Anatomía de *Adesmia boronioides* Hook. *f.* (Leguminosae - Papilionoideae). Acta Farmacéutica Bonaerense 19(4):245–250

Ochoa J (2015) Uso de plantas silvestres con órganos de almacenamiento subterráneos comestibles en Patagonia: perspectivas etnoecológicas. Thesis. Universidad Nacional del Comahue. S.C. de Bariloche (Río Negro), Argentina, 250 pp

Pérez O, Lazo F (2010) Ensayo de *Artemia*: útil herramienta de trabajo para ecotoxicólogos y químicos de productos naturales. Rev Protección Veg 22(1):34–43

Richeri M (2016) Evaluación de los reservorios etnobotánicos en las poblaciones rurales del centro norte de la Provincia del Chubut. Thesis. Universidad Nacional del Comahue. S.C. de Bariloche (Río Negro), Argentina, 170 pp

Richeri M, Cardoso MB, Ladio AH (2013a) Soluciones locales y flexibilidad en el conocimiento ecológico tradicional frente a procesos de cambio ambiental: estudios de caso en Patagonia. Ecología Austral 23:184–193

Richeri M, Ladio AH, Beeskow AM (2013b) Conocimiento tradicional y autosuficiencia: la herbolaria rural en la Meseta Central del Chubut (Argentina). BLACPMA 12(1):44–58

Sánchez G, Riat M (2012) Estudio de *Adesmia boronioides* Hook. *f.* "paramela" (Fabaceae). Un acercamiento para su reproducción y manejo en viveros. Informe técnico. Carrera Tecnicatura en Viveros. UNRN. Sede Andina. San Carlos de Bariloche, Río Negro, Argentina

Silva Sofrás F, Gastaldi B, González SB (2016) Compuestos fenólicos de *Adesmia boronioides* Hook *f.*: Análisis de su variabilidad en dos estaciones diferentes. Dominguezia 32(2):65–66

Simirgiotis M, Silva M, Becerra J, Schmeda-Hirschmann G (2012) Direct characterization of phenolic antioxidants in infusions from four Mapuche medicinal plants by liquid chromatography with diode array detection (HPLC-DAD) and electrospray ionization tandem mass spectrometry (HPLC-ESI–MS). Food Chem 131(1):318–327

Ulibarri E, Burkart A (2000) Sinopsis de las especies de *Adesmia* (Leguminosae-Adesmiae) de la Argentina. Darwiniana 38(1–2):59–126

Ulibarri EA, Simpson BB (2010) Una nueva especie de *Adesmia* (Leguminosae, Papilionoideae) para la Argentina. Darwiniana 48(2):204–207

Villagra S, González S, González Rojas S, Garriz N, Longarzo A, Errecalde J, Zaidenberg A (2008) Inhibición *in vitro* de *Trypanosoma cruzi* por aceites esenciales. VIII Congreso Argentino de Protozoología y Enfermedades Parasitarias, November 2–5, 2008, Rosario, Argentina. Revista Médica de Rosario 74 (Supl)

Weigandt M, Ladio AH, Lozada M (2004) Plantas medicinales utilizadas en la comunidad Mapuche Curruhuinca. Secretaría de Extensión Universitaria. Ed. UNCo. Imaginaria. Bariloche, Argentina, 73 pp

Aloysia gratissima (Gillies & Hook. ex Hook.) Tronc. var. *gratissima*

Oscar Ariel Risso, Susana Amalia Suárez, and Ariana Cristina Posadaz

A. gratissima var. *gratissima*. (Photo: Risso OA)

O. A. Risso (✉)
AER INTA Concarán, Instituto Nacional de Tecnología Agropecuaria, Centro Regional La Pampa-San Luis, San Luis, Argentina
e-mail: risso.oscar@inta.gob.ar

S. A. Suárez
Departamento de Ciencias Naturales, Morfología Vegetal. Facultad de Ciencias Exactas, Físico-Químicas y Naturales, Universidad Nacional de Rio Cuarto, Córdoba, Argentina

A. C. Posadaz
Departamento de Aromáticas y Jardinería. Facultad de Turismo y Urbanismo, Universidad. Nacional de San Luis., San Luis, Argentina

Á. Máthé, A. Bandoni (eds.), *Medicinal and Aromatic Plants of South America Vol. 2*, Medicinal and Aromatic Plants of the World 7,
https://doi.org/10.1007/978-3-030-62818-5_3

Abstract In Argentina and the neighboring countries, the aboveground part of *Aloysia gratissima* (Gillies & Hook. ex Hook.) Tronc. "whitebrush" is used for its aromatic and medicinal properties. The plant is a shrub reaching 3 meters height that is distributed in both South and North America. In Argentina, it is mainly used by the so called "yerba mate" industries, in compound products. As in the case of other native aromatic and medicinal species, the demand is mainly covered by the collection of wild plants, which leads to the possible overexploitation, deterioration of both resources and the heterogeneity of the commercialized product.

Keywords "Usillo" · 1,8-Cineol · Spathulenol · Cadinol · Antispasmodic

1 Introduction

Aloysia gratissima is an aromatic shrub, a close relative of *Aloysia citrodora,* the "lemon verbena". Since pre-hispanic times, it is widely used by traditional popular medicine and is consumed in the traditional South American infusion: "mate" (Martínez 2011; Galli et al. 2015). In the last decades, the demand of the "yerba mate" industry has increased considerably. Numerous works endorse the properties of its secondary metabolites due to which it is consumed (Bucciarelli and Skliar 2007; Berardi 2010; Consolini et al. 2011). Phenotypes of varying chemical composition have been described according to the regions of collection. To date, collection is still the main source of raw material supply (Ricciardi et al. 2006; Trovati et al. 2009; Benovit et al. 2015). It appears possible to select chemotypes of stable chemical composition for use in a variety of products.

2 Taxonomic Characteristics

The *Aloysia gratissima* complex (Verbenaceae) is a group of specific and infraspecific taxa, the limits of which are controversial. Various authors (Troncoso 1964; Botta 1979; Siedo 2006; O'Leary et al. 2012, 2016) have assigned different significance to individual characters in delimiting taxa in the complex, resulting in a variable number of species and varieties recognized over time. Currently, *A. gratissima* (Gillies & Hook. ex Hook.) Tronc. has been delimited in two varieties: *gratissima* and *sellowii,* depending on the qualitative and quantitative characteristics of their leaves. *A. gratissima* var. *gratissima* leaves of smaller, blade length 9.9–19.0 × blade width 3.6–7.3 mm and leaves of *A. gratissima* var. *sellowii* of larger, blade length 20.7–32.4 × blade width 7.7–13.0 mm. Furthermore, the variety "*sellowii*" is distributed only in South America, being frequent in Northern Argentina, in the provinces of Misiones, Corrientes and Salta, and Southern Brazil. It has a more restricted distribution than *A. gratissima* var. *gratissima* (Moroni et al. 2016). Due to the difficulty of the delimitation of species within the complex, numerous synonyms have been assigned to the taxon by different authors, even being included in other genera. Most used synonymies: *Aloysia chacoensis* Moldenke, *Aloysia gratissima* (Gillies & Hook. ex Hook.) Tronc. var. *paraguariensis*, *Aloysia ligustrina* (Lag.) Small var. *paraguariensis*, *Aloysia lycioides* Cham., *Lippia ligustrina* (Lag.) Kuntze var. *paraguariensis*, *Lippia lycioides* (Cham.) Steud., *Verbena gratissima* Gillies & Hook. ex Hook., *Lippia ligustrina* (Lag.) Kuntze var. *lasiodonta*, *Lippia gratissima* (Gillies & Hook.) L.D. Benson, *Aloysia gratissima* (Gillies & Hook. ex Hook.) Tronc. var. *revoluta*, *Aloysia lycioides* Cham. var. *revoluta*, *Aloysia famatinensis* Ravenna.

Nearby species are *Aloysia citrodora* Paláu ("cedrón", lemon verbena) and *Aloysia polystachya* (Griseb.) Moldenke ("té de burro" – donkey tea, "burrito" – little donkey). *A. citrodora* has larger leaves, elliptic-lanceolate sheet 3.5–7.5 cm long. × 1–1.5 cm wide, while *A. polystachya* is a lesser and lax-bearing bush with a tendency to layer naturally, its inflorescences are speciform clusters of 0.5–1.2 (−2) cm long, dense, sessile, solitary or branched of 0.5–1.2 (−2) cm long. White, tiny flowers, arranged in whorls of 4, on the spine; Very broadly obovate bracts, hugging the calyx (Barboza et al. 2006; Elechosa 2009; Múlgura et al. 2012).

3 Crude Drug Used

As a vegetable drug, aerial parts are consumed: leaves, stems and flowers, in infusions (Del Vitto et al. 1997; Barboza et al. 2006; Martínez 2011; Zeni 2011), particularly in "mate", a traditional infusion in the southern region of South America. Hence, the herb industry is an important source of demand. It is mixed with *Ilex paraguariensis* "yerba mate" to produce herb blends (Photo 1). In the central region of Argentina, annually one million kg/year is collected for use in this industry (Galli et al. 2015).

4 Major Chemical Constituents and Bioactive Compounds

The essential oil yield, obtained by hydrodistillation, ranges between 0.15–1.2%, in the different populations studied and varies according to the phenophase of the plant: higher yields were observed when the harvest was carried out in autumn and summer (Ricciardi et al. 2000, 2006; Alonso and Desmarchelier 2015; Risso et al. 2018). It was also found that the maximum production of essential oil is obtained when the plant is grows in full sun (Dos Santos 2007; Pinto et al. 2007). The composition of its essential oil is complex and diverse, with sesquiterpene compounds prevailing in proportion over oxygenated terpenes (Ricciardi et al. 2000). In addition, the composition and percentage of the constituents are markedly modified by the season of the year in which the vegetable is harvested (Bailac et al. 1999;

Photo 1 "Yerba mate", with blend herbs. (Photo: Risso OA)

Duschatzky et al. 2004; Ricciardi et al. 2006; Dambolena et al. 2010), by the part of the plant analyzed and by the place of collection (Table 1). In the northeast of the province of San Luis, in materials collected in autumn, the predominance of sesquiterpenes (S) was detected in the populations located above the valley, while in the mountain areas, the compositions with monoterpenes and sesquiterpenes prevail (M + S). The presence of both profiles in the same population added to the higher performance in essential oils of the M + S profile is indicative of the existence of chemotypes (Risso 2018). Among the main components are 1,8-cineole, in the group of monoterpenes, and cadinol (isomer not identified) and spathulenol for sesquiterpenes. Flavonoids, such as quercetin and hesperidin, were found in decoctions and methanolic extracts of the leaves (Berardi 2010). The flavonoids apigenin 5-hydroxy-7,4′-dimethylether, genkwanin and luteolin 7,3′,4′-trimethylether; the *ent*-kaurane hoffmanniaketone and rutin were isolated from ethanol extracts (Da Silva et al. 2006). As for the aqueous extract, ferulic acid is reported as its main constituent (Zeni 2011; Zeni et al. 2014).

5 Morphological Description

Aloysia gratissima is an aromatic shrub that grows up to a height of 3 meters. It has glabrous stems, gray-white and striated bark. Opposite leaves, sometimes solitary, or more than two per knot; petiole of (0.7-) 2.3 (−6.7) mm, elliptical to lanceolate sheet, of 16.5 × 6.3 mm, obtuse or acute apex, conspicuous lateral venation or not, partially serrated-serrated margin, sometimes revolute (Barboza et al. 2006; Moroni et al. 2016). The flowers, tetramers, are arranged in clusters that resemble spikes up to 10 cm long, lonely or arranged in terminal panicles (Photo 2). The calyx is hairy, with a white corolla, scented, briefly pedicelled with the limbo divided into 4 uneven lobes. It has ovate-elliptical floral bracts, occasionally linear. It blooms numerous times from early spring to late autumn. The fruit is dry, schizocarpic, separated at maturity in two clusters of 1.5 mm, glabrous, each with one seed (Davies 2004; Del Vitto et al. 2011).

Table 1 Qualitative and quantitative variability of the main chemical components of the essential oil of *Aloysia gratissima* according to the part of the plant and the place of collection

Part of the plant	Chemical composition	Harvest location	References
Leaves	Sabinene (30%), β-pinene (8%)	Minas, Uruguay	Soler et al. (1986a)
	trans-pinocamphone (10.9%), *trans*-pinocarvyl acetate (9.3%), β-caryophyllene (6%), germacrene D (4.6%)	Sao Paulo, Brazil	Sartoratto and Augusto (2003)
	Pinocarvyl acetate (17.6%), pinocamphone (16.3%) and guaiol (11.5%)	Lavras, Brazil	Dos Santos et al. (2013)
	1,8-cineole (13.7%), germacrene D (13.4%), β-caryophyllene (12.7%), β-pinene (11.7%)	Guabiruba, Brazil	Santos et al. (2015)
	1,8-cineole (18.5%), sabinene (9.5%), guaiol (6.8%), bicyclogermacrene (5.1%)	Santa Maria, Brazil	Benovit et al. (2015)
	Isopinocamphone (25.4%), limonene (15.1%), guaiol (12.7%)	Sao Carlos, Brazil	Trovati et al. (2009)
	β-Pinene (14.1%), isopinocamphone (18.4%), pinocarvyl acetate (13.5%)	Goiânia, Brazil	Franco et al., (2007)
	Cadinol (isomer not identified) (32–33%), caryophyllene oxide (11–8.6%), β-caryophyllene (4.3–3.3%) and *trans*-verbenol (5.8–2.8%)	San Luis, Argentina	Bailac et al. (1999)
	1,8-cineole (45.5%), sabinene (8.3%), carvacryl acetate (8.2%), spathulenol (8.7%)	La Rioja, Argentina	Dambolena et al. (2010)
Flowers	Pulegone (65.8%)	Córdoba, Argentina	Zygadlo et al. (1995)
	Globulol, sabinene, β-caryophyllene, caryophyllene oxide	Uruguay	Soler et al. (1986b)
	Guaiol (19.5%), germacrene B (10.5%), bulnesol (10%), *β*-caryophyllene (8.9%)	Lavras, Brazil	Dos Santos et al. (2013)
Aerial parts	Pinocamphone (13.5–16.3%), β-pinene (10.5–12%), pinocarvyl acetate (7.3–8.3%) guaiol (6.6–8.7%), bulnesol (3.7–4.1%)	Campinas, Brazil	Silva et al. (2007)
	1,8-cineole (17.6%), guaiol (10.3%), germacrene D (6.2%), bicyclogermacrene (5.6%), β-caryophyllene (5%), δ-elemene (5%), sabinene (4.3%)	Rio de Janeiro, Brazil	García et al. (2018)
	Cadinol (isomer not identified) (17.4%), caryophyllene oxide (15.8%), limonene oxide (5.3%), chrysantenyl acetate (5.6%), β-caryophyllene (4.8%)	San Luis, Argentina	Duschatzky et al. (2004)
	γ-Elemene (20%), globulol (19%), spathulenol (13%)	La Rioja, Argentina	Juliani et al. (2004)

(continued)

Table 1 (continued)

Part of the plant	Chemical composition	Harvest location	References
	β–elemene (tr - 35.7%), viridiflorol (0.9–33.6%), β-caryophyllene (1.8–28%), α-thujone (6.8–17.5%), 10-*epi*-cubebol (0.1–13.4%), bicyclogermacrene (3.8–12.8%), germacrene D (1.9–10.1%)	Corrientes, Argentina	Ricciardi et al. (2006)
	Spathulenol, 1,8-cineole	San Luis, Argentina	Risso et al. (2018)
	β-Thujone (36.1%), α-thujone (32.2%), 1,8-cineol (10.7%), sabinene (6.2%)	Northwestern Argentina	Gálvez et al. (2018)
	Kunzeaol (16.3%), β-caryophyllene (8.5%), cubebol (5.1%), viridiflorol (3.1%), germacrene D-4-ol (3.3%)	Cochabamba, Bolivia	Arze et al. (2013)
	1,8-cineole (22.1%), germacrene D (19.6%), bicyclogermacrene (7.9%)	Uruguay	Davies (2004)

Photo 2 *Aloysia gratissima* inflorescences. (Photos: AO Risso and A Posadaz)

6 Geographical Distribution

Aloysia is an American genus. Its area extends from the Southern United States of America to Central Chile and Argentina. In particular, *A. gratissima* var. *gratissima* has a bicentric distribution (Fig. 1). In North America, it is found in the Southern United States and Northern and Central Mexico. In South America, in Bolivia, Paraguay, the Southern point of Brazil, Uruguay, Northern and Central Argentina and Central Chile.

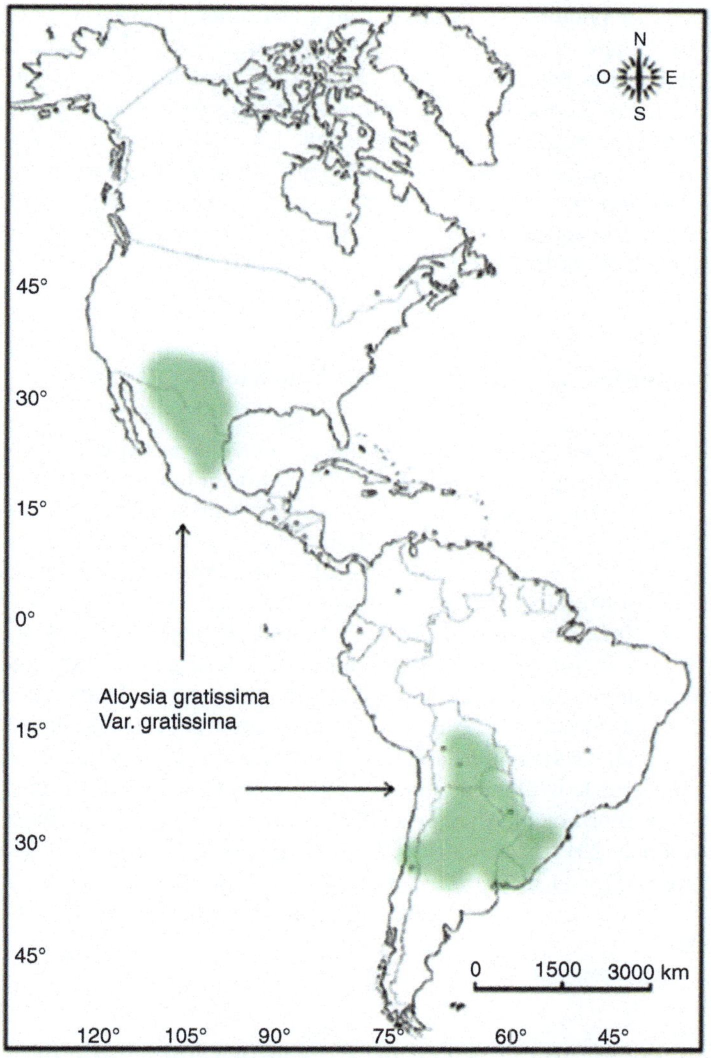

Fig. 1 Geographic distribution *A. gratissima* var. *gratissima*. (Adapted from Moroni et al. 2016)

7 Ecological Requirements

A. gratissima grows in xerophilous environments and dry forests, in Argentina. It is also found in temperate or temperate-warm climate in Paraguay and Uruguay (Álvarez 2019). It also grows in rocky and shallow soils of mountains and clayey and deep plain soils.

The seeds are photoblastic positive. Under controlled cultivation conditions, the total dry phytomass is superior in plants grown in full sun. The thickness of the adaxial cuticle, the palisade parenchyma and the total thickness of the leaf are greater in full sun. Furthermore, under this condition a greater number of glandular trichomes develop on the abaxial surface of the leaves (Dos Santos 2007).

Some recent successes and achievements are the study of the agronomic behavior of the species in Uruguay (Davies 2004), study of ecophysiology; in Brazil (Dos Santos 2007), studies of chemical and morphological variability, and Argentina (Ricciardi et al. 2006; Risso 2018).

8 Traditional Use and Common Knowledge

It is a species with numerous vernacular names, which in part alludes to its wide use throughout its geographical distribution. In Argentina it is known as "usillo", "palo amarillo" (yellow stick), "niño rupá" (baby's crib), "azahar del campo", among others (Del Vitto et al. 1997; Botta 1979; Barboza et al. 2006; Ricciardi et al. 2006; Martínez 2011; Galli et al. 2015; Moroni et al. 2016); in Brazil it is known as "mimo do Brasil" (treat from Brazil), "brazilian-lavender", "erva santa" (Dos Santos 2007; Franco et al. 2007; Trovati et al. 2009; Zeni et al. 2014), in Uruguay "cedrón del monte" (Davies 2004). Its leaves and stems are cited mainly for their digestive use and to relieve upset stomach, while their infusion flowers are mainly used to relieve bronchial conditions. It is also used to soothe menstrual cramps, against hemorrhoids and to relieve symptoms associated with nervous system disorders. In addition, infusions made with the aerial parts are used to treat varicose veins, dizziness and against hemorrhoids (Alice et al. 1991; Del Vitto et al. 1997; Barboza et al. 2006; Martínez 2011; Zeni 2011; Zeni et al. 2014; Nuñes Gonçalves 2016; Bernasconi Salazar et al. 2017). It is also a species considered ornamental and honey due to its numerous blooms of pleasant aroma (Cardoso 2005; Sérsic et al. 2006).

9 Modern Medicine Based on Its Traditional Medicine Uses

Studies carried out by Berardi (2010) endorse its traditional use as intestinal antispasmodics, since they reduce the spasmogenic effects of acetylcholine. The effects on acetylcholine can be explained by a non-competitive antagonism of the Ca^{+2} influx (Consolini et al. 2011). In addition, tests performed on mice show that orally administered aqueous extract show gastroprotective activity, exhibited significant antiulcer activity (Bucciarelli and Skliar 2007). Zeni (2011), in studies with the aqueous extract (AE), validated its potential use in the prevention or treatment of diseases that involve glutamatergic and serotoninergic systems. The main component of the AE, ferulic acid, was also able to exert an antidepressant effect in mice.

In addition, studies done with infusion on rats show that there is no toxicity due to acute exposure at a dose of 2000 mg/kg (Zeni 2011; Rihl et al. 2017). On the other hand, Ricco et al. (2010) in genotoxicity tests using the alkaline kite technique (Singh et al. 1988), revealed the absence of genotoxicity and a marked antioxidant power in extracts prepared in infusion and cooking (dose 0.05 mg/ml and 0.5 mg/ml). As regards toxic compounds, the presence of germacrene has been documented, whose structure, by oxidation, gives rise to a series of lactone-type compounds that may produce allergic or cytotoxic reactions. It also highlights the presence of ketones, such as thujone (Ricciardi et al. 2000) and pinocamphone (Trovati et al. 2009; Benovit et al. 2015), recognized as nerve toxins. Thus, seasonal variations of the major constituents modify the pharmacological applications of the species (Ricciardi et al. 2000, 2006). On the other hand, Benovit et al. (2015) have reported the anesthetic effects of the essential oil and its bioactive fractions on a species of "catfish" used in aquaculture (*Rhamdia quelen*) as an alternative to the synthetic compounds traditionally used. The data obtained for crude extracts and methanolic fractions of *A. gratissima* showed a significant reduction in the intensity of inflammations (Vandresen et al. 2010). Alcoholic, hydroalcoholic and decoction extracts showed selective antibacterial activity (inhibition/inactivation) on bacterial inoculums of *Staphylococcus aureus*, *Rhodococcus equi*, *Enterococcus faecalis*, *Salmonella enteritidis*, *Escherichia coli* and *Pasteurella multocida*. Two extracts showed greater antibacterial activity for *Rh. equi* and *P. multocida*, both bacteria related to cases of pneumonia and other respiratory conditions in humans and animals. In addition, the alcoholic extract of fresh plants showed greater activity, compared to decoctions and extracts of dried plants (Souza and Wiest 2007). Studies carried out by Freires et al. (2015) with essential oils and bioactive fractions on *Streptococcus mutans* and *Candida albicans* highlight the promising antimicrobial activity of this plant and suggest avenues for future translational research on the treatment of dental caries and oral candidiasis. It was observed that the essential oil has antimicrobial properties, especially the fraction corresponding to the inflorescences, against *Candida albicans* and *Streptococcus pneumoniae*, which supports its use in folk medicine in pneumonia and bronchial conditions (Dos Santos et al. 2013). A study done on *Allium cepa* cells shows that the infusion and the essential oil have antiproliferative effects, therefore possessing potential activity against tumor cells (Hister et al. 2009). It was also observed that it has activity against the Junín virus (responsible for hemorrhage fever) and herpes simplex virus type 1 (HSV-1) (García et al. 2003). The antiparasitic activity of the essential oil against *Leishmania amazonensis* has also been reported, with guaiol being the constituent with the highest activity (García et al. 2018). Further, it has antifungal properties against *Ascosphera apis* (Dellacassa et al. 2003), and against *Aspergillus* and *Fusarium* species (Gálvez et al. 2018).

10 Conclusions

A. gratissima is a native plant species of Argentina with a wide geographic distribution. It is used in folk medicine by numerous communities. Scientific studies endorse many of the known properties observed by these communities and explore new attributes. The low genotoxicity of the species strengthens this potential as a medicinal plant (Ricco et al. 2010; Zeni 2011). However, the presence of potentially toxic components has been detected in its essential oils (thujone and pinocamphone) in some wild populations, or in certain phenological states, which should be considered at the time of use or in domestication programs (Ricciardi et al. 2006; Trovati et al. 2009; Benovit et al. 2015). Importantly, the boom in the use of this species by "yerba mate" industries emphasizes the need to study the ways and means of its possible domestication including the selection-breeding of this species. Cultivation and conservation, in the wild, would yield characterized materials for the industries, strengthening a more sustainable and secure marketing chain for consumers.

References

Alice CB, Vargas VMF, Silva GAAB, Siqueira NCS, Schapoval EES, Gleye J, Henriques JAP, Henriques AT (1991) Screening of plants used in south Brazilian folk medicine. J Ethnopharmacol 35:165–171

Alonso J, Desmarchelier C (2015) Plantas Medicinales Autóctonas de la Argentina. Bases Científicas para su Aplicación en Atención Primaria de la Salud. 1° ed. Ed. Corpus, Ciudad Autónoma de Buenos Aires, Argentina. 748 pp

Álvarez MA (2019) Pharmacological properties of native plants from Argentina. Ed. Springer Nature, Switzerland, 268 pp

Arze JBL, Collin G, Garneau FX, Jean FI, Gagnon H (2013) Essential oils from Bolivia. XI. *Verbenaceae*: *Aloysia gratissima* (Gillies & Hook.) Tronc. and Boraginaceae: *Cordia chacoensis* Chodat. J Essent Oil Bear Plants 16(4):545–550

Bailac PN, Dellacassa E, Duschatzky CB, Firpo N, Ponzi MI (1999) Composición del aceite esencial y actividad antimicrobiana de *Aloysia gratissima* de San Luis - Argentina. Anales Asociación Química Argentina 87:149–153

Barboza EG, Cantero JJ, Núñez CO, Espinar LA (eds.) (2006) Flora medicinal de la provincia de Córdoba (Argentina). Pteridófitas y Antófitas silvestres o naturalizadas. Ed. Museo Botánico de Córdoba. Argentina, 1252 pp

Benovit SC, Silva LL, Salbego J, Loro VL, Mallmann CA, Baldisserotto B, Flores EMM, Heinzmann BM (2015) Anesthetic activity and bio-guided fractionation of the essential oil of *Aloysia gratissima* (Gillies & Hook.) Tronc. In silver catfish *Rhamdia quelen*. Anais da Academia Brasileira de Ciências 87(3):1675–1689

Berardi A (2010) Etnofarmacología gastrointestinal de plantas medicinales argentinas del género *Aloysia*, familia Verbenaceae: mecanismos de acción y relación con los principios activos. Thesis. Universidad Nacional de La Plata, La Plata, Argentina, 71 pp

Bernasconi Salazar JR, Karlin M, Arnulphi AS, Karlin UO, Accietto RH, Cora A (2017) Plantas del centro de la República Argentina: guía de campo, 1° ed. Ed. Ecoval. Córdoba, Argentina, 353 pp

Botta SM (1979) Las especies argentinas del género *Aloysia* (Verbenaceae). Darwin 22:67–108

Bucciarelli A, Skliar MI (2007) Medicinal plants from Argentina with gastro-protective activity. Ars Pharmaceutica 48(4):361–369

Cardoso JCW (2005) Niveis de luz e homeopatia sobre caracteres morfofisiológicos e óleo essencial e atividade fungitóxica do óleo essencial de *Aloysia gratissima* (Gilles & Hook.) Tronc. Thesis. Universidade Federal de Lavras, Mina Gerais, Brazil, 100 pp

Consolini AE, Berardi A, Rosella MA, Volonté MG (2011) Antispasmodic effects of *Aloysia polystachya* and *A. gratissima* tinctures and extracts are due to non-competitive inhibition of intestinal contractility induced by acetylcholine and calcium. Braz J Pharmacognosy 21(5):889–900

Da Silva CC, Vandresen F, Oliveira CMA, Kato L, Tanaka CMA, Heleno D, Ferreira HD (2006) Chemical composition of *Aloysia gratissima* (Gill. ex Hook.) Tronc. (Verbenaceae). Biochem Syst Ecol 34:593–595

Dambolena JS, Zunino MP, Lucini EI, Zygadlo JA, Banchio E, Biurrun F, Rotman A, Ahumada O (2010) Aromatic plants of Northwest Argentina. Constituents of the essential oils of aerial parts of seven Verbenaceae: *Lantana* and *Aloysia*. J Essent Oil Res 22:289–293

Davies P (2004) Estudios de domesticación y cultivo de especies medicinales y aromáticas nativas. Ed. Instituto Nacional de Investigación Agropecuaria-INIA, Montevideo, Uruguay. 261 pp

Del Vitto LA, Pettenati EM, Petenatti ME (1997) Recursos herbolarios de San Luis (República Argentina) primera parte: plantas nativas. Multequina 6:49–66

Del Vitto, LA, Petenatti EM, Petenatti ME (2011) Flores y plantas de la reserva natural Mogote Bayo y costa de la Sierra de Comechingones, Merlo, San Luis, Argentina. 1° edición. Ed. Fundación Espacios Verdes, Buenos Aires, Argentina, 75 pp

Dellacassa AD, Baillac PN, Ponzi MI (2003) *In vitro* activity of essential oils from San Luis-Argentina against *Ascosphera apis*. J Essent Oil Res 15:282–285

Dos Santos FM (2007) Aspectos ecofisiológicos de *Aloysia gratissima* (Gillies et Hook.) Troncoso (Verbenaceae) associados à composição do óleo esencial e sua ação antimicrobiana. Thesis. Universidade Federal de Lavras, Mina Gerais, Brazil, 116 pp

Dos Santos FM, Pinto JEBP, Bertolucci SK, Alvarenga AA, Alvares MN, Duarte MCT, Sartor A (2013) Chemical composition and from the leaves and flowers of *Aloysia gratissima*. Rev Bras Planta Med 15:583–588

Duschatzky CB, Martínez AN, Almeida NV, Bonivardo SL (2004) Nematicidal activity of the essential oils of several Argentina plants against the root-knot nematode. J Essent Oil Res 16:626–628

Elechosa MA (2009) Manual de recolección sustentable de plantas aromáticas nativas de la región central y noreste de la Argentina. Ed. INTA, Ciudad Autónoma de Buenos Aires, Argentina, 64 pp

Franco ALP, Oliveira TB, Ferri PH, Bara MT, Paula JR (2007) Avaliação da composição química e atividade antibacteriana dos óleos essenciais de *Aloysia gratissima* (Gillies & Hook.) Tronc. (alfazema), *Ocimum gratissimum* L. (alfavaca-cravo) e *Curcuma longa* L. (açafrão). Revista Eletrônica de Farmácia 4(2):208–220

Freires IA, Silva BB, LCC G, MCT D, Sartoratto A, Figueira GM, Alencar SM, Rosalen PL (2015) The effect of essential oils and bioactive fractions on *Streptococcus mutans* and *Candida albicans*. Biofilms: a confocal analysis. Evid Based Complement Alternat Med 2015:1–9

Galli MA, Risso OA, Manazza JF (2015) Dinámica de la cadena de valor de especies aromáticas y medicinales en la región de las Sierras Pampeanas. Lilloa 52(Suplemento):30–32

Gálvez CE, Jimenez CM, Gómez AA, Lizarraga EF, Sampietro DA (2018) Chemical composition and antifungal activity of essential oils from *Senecio nutans*, *Senecio viridis*, *Tagetes terniflora* and *Aloysia gratissima* against toxigenic *Aspergillus* and *Fusarium* species. Nat Prod Res 34(10):1442–1445

García C, Talarico L, Almeida N, Colombres S, Duschatzky C, Damonte E (2003) Virucidal activity of essential oils from aromatic plants of San Luis, Argentina. Phytother Res 17(9):1073–1075

García MCF, Soares DC, Santana RC, Saraiva EM, Siani AC, Miranda MFS, MGM D, Souto-Padron TC, Pinto-da-Silva LH (2018) The *in vitro* antileishmanial activity of essential oil from *Aloysia gratissima* and guaiol, its major sesquiterpene against *Leishmania amazonensis*. Parasitology 145(9):1219–1227

Hister CAL, Laughinghouse HD, Da Silva CB, Do Canto Dorow TS, Tedesco SB (2009) Evaluation of the antiproliferative effect of infusions and essential oil of *Aloysia gratissima*. Pak J Biol Sci 12(24):1581–1584

Juliani HR, Koroch A, Simon E, Biurrum FN, Castellano V, Zygadlo J (2004) Essential oils from Argentinean aromatic plants. Acta Hortic (629):491–498

Martínez G (2011) Las plantas en la medicina tradicional de las Sierras de Córdoba. Ed. del Copista, Córdoba, Argentina, 210 pp

Moroni P, O'Leary N, Filloy J (2016) Species delimitation in the *Aloysia gratissima* complex (Verbenaceae) following the phylogenetic species concept. Bot J Linn Soc 180:193–212

Múlgura ME, O'Leary N, Rotman AD (2012) In: Verbenaceae. Flora Argentina. Flora vascular de la República Argentina. Dicotyledoneae, vol 14. IBODA-IMBIV, CONICET, Argentina, 230 pp

Nuñes Gonçalves VR (2016) Espécies utilizadas na medicina popular no município de Dom Pedrito sob os nomes vernáculos de "garupá", "guaçatumba" e "poejo". Universidade Federal do Pampa, Rio Grande do Sul, Brazil, 38 pp

O'Leary N, Calviño C, Martínez S, Olmstead RG, Múlgura ME (2012) Evolution of morphological traits in Verbenaceae. Am J Bot 99:1778–1792

O'Leary N, Lu-Irving P, Moroni P, Siedo S (2016) Taxonomic revision of *Aloysia* (Verbenaceae, Lantaneae) in South America. Ann Mol Bot Gard 101(3):568–609

Pinto J, Cardoso J, Castro E, Bertolucci S, Melo L, Dousseau S (2007) Aspectos morfofisiológicos e conteúdo de óleo essencial de plantas de alfazema-do-Brasil em função de níveis de sombreamento. Hortic Bras 25:210–214

Ricciardi GAL, Ricciardi AIA, Bandoni AL (2000) Fitoquímica de Verbenáceas (*Lippia* y *Aloysia*) del Nordeste argentino. Corrientes, Universidad Nacional del Nordeste, Corrientes, Argentina. Comunicaciones Científicas y Tecnológicas n° 39

Ricciardi GAL, Van Baren CM, Di Leo Lira P, Ricciardi AIA, Lorenzo D, Dellacassa E, Bandoni AL (2006) Volatile constituents from aerial parts of *Aloysia gratissima* (Gillies & Hook.) Tronc. var. *gratissima* growing in Corrientes, Argentina. Flavour Fragr J 21:698–703

Ricco RA, Wagner ML, Portmann E, Reides C, Llesuy S, Gurni AA, Carballo MA (2010) Análisis de polifenoles, actividad antioxidante y genotoxicidad en especies argentinas de *Lippia* y *Aloysia* (Verbenaceae). Boletín Latinoamericano y del Caribe de Plantas Medicinales y Aromáticas 9(5):388–396

Rihl C, Lucas F, Miranda ZS, Leidi L, Cambi VN, Bucciarelli A (2017) Evaluación de la toxicidad aguda de la infusión oral de *Aloysia gratissima* (Verbenaceae). Dominguezia 33(1):63

Risso OA (2018) Caracterización fenotípica de poblaciones de "usillo" *Aloysia gratissima* (Gillies & Hook. ex Hook.) Tronc. del noreste de la provincia de San Luis. Thesis. Universidad Nacional de Cuyo, Mendoza, Argentina, 73 pp

Risso OA, Posadaz A, Lea M, Ardanáz C, Suárez SA, Ojeda MS (2018) Comparativa del aceite esencial de *Aloysia gratissima* var. *gratissima* en poblaciones silvestres del noreste de la provincia de San Luis. 40° Congreso Argentino de Horticultura. Abstracts. In: http://40congresoasaho.com.ar. (Last access, November 2019)

Santos TG, Laemmle J, Rebelo RA, Dalmarco EM, Cruz AB, Scmit AP, Cruz RCB, Zeni AB (2015) Chemical composition and antimicrobial activity of *Aloysia gratissima* (Verbenaceae) leaf essential oil. J Essent Oil Res 27:125–130

Sartoratto A, Augusto E (2003) Application of headspace solid phase microextraction and gas chromatography to the screening of volatile compounds from some Brazilian aromatic plants. Chromatographia 57:351–356

Sérsic A, Cocucci A, Benitez-Vieyra S, Cosacov A, Díaz L, Glinos E, Grosso N, Lazarte C, Medina M, Moré M, Moyano M, Nattero J, Paiaro V, Trujillo C, Wiemer P (2006) Flores del centro de Argentina. Una guía ilustrada para conocer 141 especies típicas. Ed. Academia Nacional de Ciencias de Córdoba, Córdoba, Argentina, 349 pp

Siedo SJ (2006) Systematics of *Aloysia* (Verbenaceae). Thesis. The University of Texas, Austin, USA, 321 pp

Silva DCMN, Meireles MAA, Brat P, Magalhaes PM, Menut C (2007) Chemical composition and biological activity of natural extracts obtained from a Brazilian aromatic plant (*Aloysia gratissima*) by supercritical CO_2 and hydrodistillation. Int J Essent Oil Ther 1:172–178

Singh NP, McCoy MT, Tice RR, Schneider EL (1988) A simple technique for quantitation of low levels of DNA damage in individual cells. Exp Cell Res 175:184–191

Soler E, Dellacassa E, Moyna P (1986a) Composition of *Aloysia gratissima* leaf essential oil. Phytochemistry 25:1343–1345

Soler E, Dellacassa E, Moyna P (1986b) Composition of *Aloysia gratissima* flower essential oil. Planta Med (6):488–490

Souza AA, Wiest JM (2007) Atividade anti-bacteriana de *Aloysia gratissima* (Gill. ex Hook.) Tronc. (garupá, ervasanta), usada na medicina tradicional no Rio Grande do Sul – Brasil. Revista Brasileira de Plantas Medicinais 9(3):23–29

Troncoso NS (1964) Plantas vasculares nuevas o interesantes de la Flora de Entre Ríos I. Darwiniana 13:625–631

Trovati G, Chierice GO, Sanches EA, Galhiane MS (2009) Essential oil composition of *Aloysia gratissima* from Brazil. J Essent Oil Res 21:325–326

Vandresen F, Schmitt E, Kato L, Alves de Oliveira CM, Bersani Amado CA, Conceição da Silva C (2010) Constituintes químicos e avaliação das atividades antibacteriana e antiedematogênica de *Aloysia gratissima* (Gillies & Hook.) Tronc. e *Aloysia virgata* (Ruiz & Pav.) Pers., Verbenaceae. Rev Bras 20(3):317–321

Zeni ALB (2011) Estudo fitoquímico, toxicológico e dos efeitos neuroprotetor e tipo antidepressivo do extrato aquoso de *Aloysia gratissima* (Gill. et Hook.) Tronc. (erva santa). Thesis. Universidade Federal de Santa Catarina, Florianópolis, Brazil. 237 pp

Zeni ALB, Vandresen-Filho S, Dal-Cim T, Martins WC, Bertoldo DB, Maraschin M, Tasca CI (2014) *Aloysia gratissima* prevents cellular damage induced by glutamatergic excitotoxicity. J Pharm Pharmacol 66(9):1294–1302

Zygadlo JA, Lamarque AL, Guzmán CA (1995) Composition of the flower oil of some *Lippia* and *Aloysia* species from Argentina. J Essent Oil Res 7:593–595

Aspidosperma quebracho-blanco Schltdl.

Diego A. Sampietro, M. del Huerto Sánchez Matías, Marta A. Vattuone, and César A. N. Catalán

Aspidosperma quebracho-blanco Schltdl. (Photos: DA Sampietro). Argentina, Tucumán Province, Trancas Department. General view of the tree, January 27, 2010 (above), The same tree showing leaves and fruits (below)

D. A. Sampietro (✉)
Instituto de Estudios Farmacológicos, Facultad de Bioquímica, Química y Farmacia, Universidad Nacional de Tucumán, Tucumán, Argentina
e-mail: dasampietro2006@yahoo.com.ar; diego.sampietro@fbqf.unt.edu.ar

Á. Máthé, A. Bandoni (eds.), *Medicinal and Aromatic Plants of South America Vol. 2*, Medicinal and Aromatic Plants of the World 7,
https://doi.org/10.1007/978-3-030-62818-5_4

Abstract *Aspidosperma quebracho-blanco* Schltdl. is a tree that - at maturity - can reach up to 25 m in height. It has a straight trunk that can reach 0.8–1 m in diameter. It is commonly known as "quebracho blanco" and widely used in the Chaco region as medicine to treat fever, malaria, swellings, stomach upsets, cough, headaches, syphilis, impotence, benign prostatic hypertrophy and asthma related dyspnea. The stem bark of *A. quebracho-blanco* is the most frequently used plant part in medicinal preparations. Terpenoid indole alkaloids, the main biologically active principles of the plant are associated with the medicinal properties of the tree. The bark is rich in aspidospermine, yohimbine, deacetyl-aspidospermine and quebrachamine. Other alkaloids also were isolated. Several of these terpenoid indole alkaloids showed *in vitro* and sometimes *in vivo* pharmacological activities that support the medicinal effects traditionally attributed to *A. quebracho-blanco*. Aspidospermine showed *in vitro* antiplasmodial activity. Aspidospermine, yohimbine and quebrachime had adrenergic blocking activities on a variety of urogenital tissues. Rhazinilam showed mild analgesic activity in mice. The high tannin levels found in the bark, leaves and woods are believed to contribute to the antioxidant and anti-inflammatory activities. Recent research suggests that bark preparations have a potential for the development of antiviral and larvicidal agents. However, further investigations are needed with especial emphasis on the *in vivo* effects, in order to understand the real therapeutic significance of the alkaloids and other compounds of *A. quebracho-blanco*.

Keywords "Quebracho blanco" · Terpenoid indole alkaloids · Aspidospermine · Yohimbine

1 Introduction

Aspidosperma quebracho-blanco is the most abundant native tree in the South american Chaco. It provides a heavy wood which is used to make posts, rods, wheels, rail sleepers, among others. It is also used as source of charcoal and firewood. The stem bark or extracts from the stem bark are sold in some countries due to their traditional uses depicted in folk medicine of South American countries. They are recommended to treat several health disorders such as fever, emphysema and asthma, and illnesses like malaria and syphilis.

The medicinal properties of the bark are attributed to the presence of terpene indole alkaloids. At least 31 of these substances have been identified in this tree.

M. del Huerto Sánchez Matías · M. A. Vattuone
Facultad de Bioquímica, Química y Farmacia, Laboratorio de Biología de Agentes Bioactivos y Fitopatógenos (LABIFITO), Universidad Nacional de Tucumán, Tucumán, Argentina

C. A. N. Catalán
Instituto de Química Orgánica, Facultad de Bioquímica, Química y Farmacia, Universidad Nacional de Tucumán, Tucumán, Argentina
e-mail: ccatalan@fbqf.unt.edu.ar

A. quebracho-blanco is occasionally cultivated as an ornamental tree in some cities of the big Chaco. Presently, it is not yet regarded as an endangered species, the irrational over-exploitation of its wood has, however, already strongly reduced its availability, in the Chaco region where *A. quebracho-blanco* is not as common to find as it used to be.

2 Taxonomic Characteristics

The genus *Aspidosperma* Mart. & Zucc. (Apocynaceae, Rauvolfioideae, Aspidospermateae) occurs from Southern Mexico to Northern Argentina (excluding Chile) and the Antilles (Morales and Zamora 2017). It includes 55 accepted species which are shrubs or big trees (Machate et al. 2016).

The name *Aspidosperma* derives from the particularly winged seeds found in the species of the genus (*aspidos* = a round shield; *perma* = seed). *Aspidosperma quebracho-blanco* Schltdl. is an evergreen tree that can reach at maturity up to 25 m in height with a straight trunk which can reach 0.8–1 m in diameter. It has a uniformly yellow-ochre wood, a thick and corky bark, leathery elliptic to lanceolate leaves, short clusters of fragrant funnel-shaped yellow flowers and capsule fruits with several seeds. The epithet *quebracho* derives from "quiebra hacha" ("breaking axes"), because of the hardness of its wood; and *blanco* (white), in opposition to "quebracho colorado" (red quebracho): *Schinopsis balansae* Engl., of red wood.

Basionyms *Aspidosperma chakensis* Speg.; *A. crotalorum* Speg.; *A. quebracho* Griseb.; *A. quebrachoalba* Schlecht.; *A. quebracho-blanco* var. *pendula* Speg.; *Macaglia quebracho* (Schl.) Kuntze; *Macaglia quebracho-blanco* (Schlecht.) Lyons (Duke et al. 2009).

Common Names "quebracho blanco", "quebracho blanco llorón", "guarirobá" (Barboza et al. 2009); "kacha kacha" (Aymara language), "willka" (Quechua language) (Duke et al. 2009).

3 Crude Drug Used

Infusions prepared of leaves, bark and fruits of *Aspidosperma quebracho-blanco* are used in folk medicine to treat fever and malaria.

The green fruits are crushed and used in poultices against swellings (Venator 1952). The Pilagá Indians (Chaco) prepare bark decoctions to treat stomach upsets, cough, headaches, and syphilis. They also used it as an analgesic and abortive. Tobas Indians (Chaco) attribute to the root decoctions abortive and anticonceptive activities (Martínez Crovetto 1964). Roots and leaves are also used against cardiac diseases (Oblitas Poblete 1969). Preparations of the bark have been traditionally used in many parts of South America to treat impotence, benign prostatic hypertrophy and to obtain relief from cardiac or asthma-related dyspnea (Marzocca 1997).

Extracts of the bark are often used as additives to expectorant phytopreparations. Bark preparations injected directly into the bloodstream increase the frequency and depth of breathing, while the blood pressure drops.

The crude drug is also known to relieve emphysema and asthma. Continuous oral administration reduces the frequency and severity of the attacks (Grieve 1977).

Some known uses of bark include:

= Bark decoctions to relieve asthma spasms are prepared at 1%. Drink two or four cups per day, sweetened with honey (Saggese and Saggese 1950).
= Bark used against fever or malaria: Drink aqueous extract (0.10–0.20 g) or powder tincture prepared in 25% aqueous ethanol (1–3 g). Another option: a daily portion of a combination of 1–2 g of the tincture in water (50 g), peppermint water (40 g) and syrup (15 g) (Dujardin-Beaumetz and Yvon 1894).
= Bark decoctions, for use as wound healer and against venereal diseases: prepare at 20‰ for washings (Saggese and Saggese 1950).
= Tobas Indians use bark decoctions to dye wool and plant fibers (Martínez Crovetto 1964). Bark extract is used in the industry for flavoring alcoholic and non-alcoholic drinks, for milk products and in food technology usually in concentrations of about $3 \times 10^{-3}\%$ (Furia and Bellanca 1975; Leung 1980).

4 Major Chemical Constituents and Bioactive Compounds

Reviews have been published on the chemistry and biological activities of the *Aspidosperma* species, including *A. quebracho-blanco* (Pereira et al. 2007; Oliveira et al. 2009). The medicinal properties of the stem bark of *A. quebracho-blanco* are mainly attributed to the presence of terpenoid indole alkaloids, which are also found in leaves and fruits. The total alkaloid content of barks is of about 6% (w/w). The most abundant alkaloids are usually aspidospermine, yohimbine (also known as aphrodine, quebrachine or corynine), deacetylaspidospermine, and quebrachamine (Biemann et al. 1961; Schnoes et al. 1962; Lyon et al. 1973; Bolzani et al. 1987). Other indol alkaloids reported in lower contents are piryfolidine, 1,2-dehydroaspidospermidine, *N*-methyl-deacetylaspidospermine, aspidospermidine, deacetylpiryfolidine, condilocarpine, *N*-methylaspidospermidine, *N*-Methyl-desacetyldesmethoxyaspidospermine, des-*O*-methylaspidolimidine, fendlerine, acuamicine, condilocarpine, aspidospermatidine, *N*-methylaspidospermatidine, acetylaspidospermatidine, aspidospermatine, 14,19-dihydroaspidospermatine, 14,19-dihydroaspidosperatine, (−)-β-yohimbine, 3α,15α,16α,17β,20α-quebrachine, eburnamenine, eburnamine, β-hydroxyeburnamonine, eburnamonine, rhazidin, aspidoquibine, rhazinilam and 3-oxo-14,15-dehydrorhazinilam (Biemann et al. 1961; Schnoes et al. 1962; Lyon et al. 1973; Bolzani et al. 1987; Aimi et al. 1991; Benoit et al. 1973). The bioactivity reported for these alkaloids, in some cases, support the therapeutic effects traditionally attributed to the bark. Aspidospermine showed *in vitro* antiplasmodial activity against a chloroquine resistant lineage of

Plasmodium falciparum, although it was 100-fold less active than chloroquine (Mitaine-Offer et al. 2002). No tests have been carried out to verify this activity *in vivo*. Yohimbine and *A. quebracho-blanco* bark extract block human penile 2A-adrenergic receptors. In small amounts, yohimbine increases the blood pressure, the cardiac rate, and induces an increase in motor and antidiuretic activities that fully support the use of the bark in the treatment of the erectile dysfunction (Deutsch et al. 1994). In higher amounts it depresses the blood pressure. Aspidospermine and quebrachamine, like yohimbine, have been found to possess adrenergic blocking activities for a variety of urogenital tissues. Rhazinilam showed mild analgesic activity when intravenously administered in mice (Benoit et al. 1973).

A. quebracho-blanco also contains high levels of tannins in the bark, leaves and wood which are considered important contributors to antioxidant and anti-inflammatory activities. The bark of *A. quebracho-blanco* was included until the sixth edition of the National Pharmacopeia of Argentina (Farmacopea Nacional Argentina 1978).

5 Morphological Description

A. quebracho-blanco trees, at maturity, can reach up to a height of 25 m, with a straight trunk of 120 cm in diameter. Tree top cylindrical-globose to obconic in shape with pendular branches. The trunk has a uniformly yellow-ochre wood and a thick and corky bark. Simple perennial leaves lanceolate (20–50 in length and 8–10 mm in width), with leathery consistency, a spinescent apex and an acute base. They arise in numbers of 3 or 2 per node. Fragrant yellowish white flowers in axillar or terminal cymose inflorescences. Flowers are hermaphrodites, actinomorphic, 10 mm in length, shortly pedicelled (2–4 mm), with a campanulate calyx formed by five triangular-ovate sepals. Gamopetalous corolla, hypocrateriform, with a tube about 5 mm long, somewhat widened in the middle, the limbus with 5 linear involute oblong lobules, as long as the tube. Five stamens inserted in the upper part of the corolla tube by means of very short filaments. Two-locular anthers, ovate-lanceolate. Superior ovary, 2-locular, pluriovular, ovoid, about 1 mm high and slightly less in diameter, with cylindrical style 1 mm long and stigma thickened and somewhat sunken. The fruit is a woody capsule, greyish green, dehiscent, with ovate shape, generally flat, smooth, of 7–12 cm long and 4–6 cm wide, and 2–2.5 cm in thickness. It contains several flat-orbicular seeds, surrounded by a yellowish membranous wing, with a total diameter of 3–6 cm (Ezcurra 1981).

The wood is of very good quality, hard and heavy (specific weight of 0.820–0.940). The sapwood is not distinguished from the duramen which is from yellowish brown to pink. The tree flowers from September to January and bears fruits in December onwards (Ezcurra 1981).

6 Geographical Distribution

Aspidosperma quebracho-blanco grows in the dry forests of Southeastern Bolivia, a large part of central and Northern of Argentina, and the Western part of Paraguay and Uruguay, at elevations from around sea level up to 1800 m.a.s.l. (Marzocca 1997).

7 Ecological Requirements

Aspidosperma quebracho-blanco grows in well drained sandy soils that are often saline and rich in humus. It grows the best under sunny locations under temperatures above 15 °C (Bown 1995).

8 Traditional Use (Part(s) Used) and Common Knowledge

Several plant parts of *A. quebracho-blanco* have been described to possess therapeutic properties. The plant parts used, the herbal preparations and the attributed health effects reported include:

Bark, Fruits and Leaves Infusions against malaria and fever (Marzocca 1997).

Bark decoctions used to heal wounds or in sitz baths and steam baths to help in difficult birth labor. Decoctions also can be drunk to "get out" of diseases and as a contraceptive (Martínez Crovetto 1964). The Toba's indigenes get a yellow dye to dye wool and fibers (Martínez Crovetto 1967).

Leaves Prepared as a "tonic" for oral intake provided to weak and rickety patients and as anti-asthmatic (Venator 1952).

Fruits crushed and in cataplasm to cure swellings (Venator 1952).

Roots and Leaves in the form of infusions, decoctions or syrups are used for the treatment of heart disease, asthma and as an aphrodisiac (Oblitas Poblete 1969). Roots: decoction used as abortive (Martínez Crovetto 1964).

9 Modern Medicine Based on Its Traditional Medicine Uses

Bark extracts of *A. quebracho-blanco* showed *in vitro* antimalarial activity on a *Plasmodium falciparum* chloroquine sensitive strain (F32), and inhibition of ferriprotoporphyrin biocrystallization (Bourdy et al. 2004). Ferriprotoporphyrin is generated during *Plasmodium* digestion of hemoglobin and highly toxic for the parasite. Hence, the inhibition of its crystallization is a desirable target in antimalarial chemotherapy (Deharo et al. 2002). Extracts from aerial parts of *A. quebracho-blanco* showed antioxidant activity *in vitro* established by FRAP (Fe^{2+} micromole/g of plant). Infusions from aerial parts orally administered showed also *in vivo* antioxidant activity on mice expressed by an increase in blood phenolics, and redox changes with a decrease of hydroperoxidation in the blood as well as in the levels of nitrites and γ-glutamyltranspeptidase in thymus, blood and spleen (Canalis et al. 2014).

Extracts of the bark powder inhibit smooth muscle contractions in tissues such as human prostate strips, rabbit corpus spongiosum and cavernosum and guinea pig vas deferens (Deutsch et al. 1994). Water infusions of the powdered bark increased cell viability and improve redox markers of murine splenocytes exposed *in vitro* to chlorpyrifos (Scotta et al. 2017). A chloroform extract obtained by Soxhlet extraction of the aerial part of *A. quebracho-blanco* showed a moderate larvicidal activity (DL_{50} = 0.14 mg/ml) on *Culex quinquefasciatus*, mosquito vector of San Luis encephalitis virus (Konigheim et al. 2009). Hexane and chloroform extracts of *A. quebracho-blanco* obtained by Soxhlet extraction exerted a partial inhibition *in vitro* on the replication of Venezuelan equine encephalitis virus (20–30%; 50–70%, respectively), and the herpes simplex type-I virus (50–60%, 50–70%, respectively). A dichloromethane bark extract had a strong antifungal activity on the soybean pathogens *Cercospora kikuchii* and *Septoria glycines* (Sequin et al. 2016). *In vitro* tests of leaf extracts did not affect the growth of phytopathogenic bacteria (Terán Baptista et al. 2020).

10 Conclusions

A review of the manifold traditional uses of the plant parts of the *Aspidosperma quebracho-blanco* confirms several of the traditional medicinal properties attributed to the tree. Recent research into the biological activity of its active ingredients seem to have explored possible new uses. Based on the available information it can be stated that *A. quebracho-blanco* is a valuable, renewable plant resource whose use has been underestimated.

References

Aimi N, Uchida N, Ohya N, Hosokawa H, Takayama H, Sakai S (1991) Novel indole alkaloids from cell suspension cultures of *Aspidosperma quebracho-blanco* Schlecht. Tetrahedron Lett 37:4949

Barboza GE, Cantero JJ, Núñez C, Ariza Espinar L, Pacciaroni A del V (2009) Medicinal plants: A general review and a phytochemical and ethnopharmacological screening of the native Argentine Flora. Kurtziana 34:7–365

Benoit PS, Angry G, Lyon RL, Fong HHS, Farnsworth NR (1973) Biological and phytochemical evaluation of plants XIII: preliminary estimation of analgesic activity of rhazinilam, a novel alkaloid isolated from *Aspidosperma quebracho-blanco* leaves. J Pharm Sci 62:1889
Biemann K, Fierdman-Spiteller M, Spiteller C (1961) An investigation by mass spectrometry of the alkaloids of *Aspidosperma quebracho-blanco*. Tetrahedron Lett 2:485–492
Bolzani VS, Serur LM, Matos FJA, Gottlieb OR (1987) Indole alkaloids evolution in *Aspidosperma*. Biochem Syst Ecol 15:187–200
Bourdy G, Oporto P, Giménez A, Deharo E (2004) A search for natural bioactive compounds in Bolivia through a multidisciplinary approach part VI. Evaluation of the antimalarial activity of plants used by Isoceño-Guarani's Indians. J Ethnopharmacol 93:269–277
Bown D (1995) Encyclopaedia of herbs and their uses. Dorling Kindersley, London, 424 pp
Canalis AM, Cittadini MC, Albrecht C, Soria EA (2014) *In vivo* redox effects of *Aspidosperma quebracho-blanco* Schltdl., *Lantana grisebachii* stuck and *Ilex paraguariensis* A. St.-Hil. On blood, thymus and spleen of mice. Indian J Exp Biol 52:882–889
Deharo E, García R, Oporto P, Sauvain M, Gautret P, Ginsburg H (2002) A non-radiolabeled ferriprotoporphyrin IX biomineralization inhibition test (FBIT) for the high throughput screening of antimalarial compounds. Exp Parasitol 100:252–256
Deutsch HF, Evenson MA, Drescher P, Sparwassers C, Madsen PO (1994) Isolation and biological activity of aspidospermine and quebrachamine from an *Aspidosperma* tree source. J Pharm Biomed Anal 12:1283–1287
Dujardin-Beaumetz G, Yvon P (1894) Formulaire pratique de Thérapeutique et de Pharmacologie. IX Edition. Ed. Octave Doin Éditeur. Paris, France, 704 pp
Duke JA, Bogenschutz-Godwin MJ, Ottesen AR (2009) Duke's handbook of medicinal plants of Latin America. Ed. CRC Press. Boca Raton. 901 + X pp
Ezcurra C (1981) Revisión de las Apocináceas de la Argentina. Darwin 23:367–474
Farmacopea Nacional Argentina (1978) 6th Edition. Editorial Codex SA. Buenos Aires, Argentina
Furia TE, Bellanca N (1975) Fenaroli's handbook of flavor ingredients (ed). CRC Press, Boca Raton, 2135 pp
Grieve M (1977) A modern herbal. Ed. Penguin Books Ltd. WS, 888 pp
Konigheim B, Aguilar J, Batallan G, Almirón W, Contigiani M (2009) Native flora of Cordoba (Argentine) and its importance in public health. Rev Fac Cienc Med 66:35–41
Leung AY (1980) Encyclopedia of common natural ingredients. Ed. Wiley-Interscience Publication. New York, 848 pp
Lyon RL, Fong HHS, Farnsworth NR, Svoboda GH (1973) Biological and phytochemical evaluation of plants XII: isolation of acetylakuammidine from *Aspidosperma quebracho-blanco* leaves. J Pharm Sci 62:833
Machate DJ, Alves FM, Farinaccio MA (2016) *Aspidosperma* (Apocynaceae) in Mato Grosso do Sul state, Brazil. Rodriguésia 67:1011–1024
Martínez Crovetto R (1964) Estudios Etnobotánicos I. Nombres de plantas y su utilidad según los indios Tobas del Este del Chaco. Bonplandia 1:279–333
Martínez Crovetto R (1967) Tintorería Toba. Bonplandia 2(5):103–105
Marzocca A (1997) Vademécum de Malezas Medicinales de la Argentina. Indígenas y Exóticas. Ed. Orientación Gráfica SRL. Buenos Aires, Argentina, 363 pp
Mitaine-Offer AC, Sauvain M, Valentin A, Callapa J, Mallié M, Zèches-Hanrot M (2002) Antiplasmodial activity of *Aspidosperma* indole alkaloids. Phytomedicine 9:142–145
Morales JF, Zamora NA (2017) A synopsis of *Aspidosperma* (Apocynaceae) in Mexico and Central America with a taxonomic clarification of *Aspidosperma cruentum* and a new cryptic species. Phytoneuron 68:1–13
Oblitas Poblete E (1969) Plantas medicinales en Bolivia. Ed. Los Amigos del Libro. Cochabamba, Bolivia, 529 pp
Oliveira VB, Freitas MSM, Mathias L, Braz-Filho R, Vieira IJC (2009) Atividade biológica e alcalóides indólicos do gênero *Aspidosperma* (Apocynaceae): Uma revisão. Revista Brasileira de Plantas Medicinais 11:92–99

Pereira MM, Jácome LRP, Alcántara AFC, Alves RB, Raslan DS (2007) Alcaloides indólicos isolados de espécies do genero *Aspidosperma* (Apocynaceae). Quim Nova 30:970–983

Saggese D, Saggese AA (1950) Yerbas medicinales argentinas. Ed. Antognazzi y Cia. Rosario, Argentina, 189 pp

Schnoes HK, Burlingame AL, Biemann K (1962) Application of mass spectrometry to structure problems: IX. The occurrence of eburnamenine and related alkaloids in *Rhazya stricta* and *Aspidosperma quebracho-blanco*. Tetrahedron Lett 3:993–999

Scotta AV, Bongiovanni GA, Soria EA (2017) *In vitro* modulating activity of aqueous extracts from American plants on chlorpyrifos induced toxicity on murine splenocytes. Rev Fac Cienc Med 74:325–330

Sequin CH, Aceñolaza P, Sampietro D, Gieco A (2016) Especies forestales como fuente de compuestos antifúngicos: Utilidad en el control de *Cercospora kikuchii*. XXIII Congreso Latinoamericano de Microbiología. Rosario, Santa Fe

Terán Baptista ZP, Gómez AA, Kritsanida M, Grougnet R, Mandova T, Aredes Fernandez PA, Sampietro DA (2020) Antibacterial activity of native plants from Northwest Argentina against phytopathogenic bacteria. Nat Prod Res. 34(12):1782-1785.

Venator C (1952) Flora Cordobesa. Plantas medicinales, aromáticas, tintóreas de la provincia de Córdoba. Tomo I. Ed. Flor del Inca. Córdoba, Argentina

Baccharis dracunculifolia DC.

Manuel Minteguiaga, H. Andrés González, Fernando Ferreira, and Eduardo Dellacassa

Baccharis dracunculifolia DC. Photos: M Minteguiaga and D Passarella

M. Minteguiaga (✉)
Espacio de Ciencia y Tecnología Química, Centro Universitario Regional Noreste/Sede Tacuarembó, Universidad de la República, Tacuarembó, Uruguay

Laboratorio de Biotecnología de Aromas, Departamento de Química Orgánica, Facultad de Química, Universidad de la República, Montevideo, Uruguay
e-mail: manuel.minteguiaga@cut.edu.uy

H. A. González
Departamento de Botánica. Museo Nacional de Historia Natural, Ministerio de Educación y Cultura, Montevideo, Uruguay

F. Ferreira
Espacio de Ciencia y Tecnología Química, Centro Universitario Regional Noreste/Sede Tacuarembó, Universidad de la República, Tacuarembó, Uruguay

Laboratorio de Carbohidratos y Glicoconjugados. Departamento de Química Orgánica. Facultad de Química, Universidad de la República, Montevideo, Uruguay

E. Dellacassa
Laboratorio de Biotecnología de Aromas, Departamento de Química Orgánica, Facultad de Química, Universidad de la República, Montevideo, Uruguay

Á. Máthé, A. Bandoni (eds.), *Medicinal and Aromatic Plants of South America Vol. 2*, Medicinal and Aromatic Plants of the World 7,
https://doi.org/10.1007/978-3-030-62818-5_5

Abstract *Baccharis dracunculifolia* DC. (Asteraceae) is an aromatic plant species widely distributed across and along South American countries where it grows in different environments, even being considered, in some cases, as a hard to control weed. Its essential oil is highly appreciated in the fragrance/perfumery industry by its woody, floral, and green notes; mainly attributed to *(E)*-nerolidol and other sesquiterpene alcohols. Recent research has demonstrated a wide range of applications for the essential oil besides its aromatic properties. Although *B. dracunculifolia* cannot be considered as a medicinal plant for most of the indigenous communities in South America (except for Brazil), modern pharmacological research has demonstrated its high potential for the development of new therapeutic drugs, mainly with antiulcer, anti-inflammatory, antiparasitic and anticancer effect. The special literature regarding *B. dracunculifolia* is profuse and thus, this chapter attempts to summarize the principal knowledge about this species.

Keywords "Chilca blanca" · "Vassoura" · Essential oil · Flavonoids · Antiulcerative

1 Introduction

Baccharis L. is one of the most diverse genera of aromatic and medicinal plants from the South American Flora. It has profuse traditional uses in many communities along the continent, as recorded by ethnobotanical surveys. *B. dracunculifolia* is widely recognized as an aromatic plant. Its essential oil is employed by the perfumery industry. Nevertheless, the information related to *B. dracunculifolia* as medicinal plant is scarce and geographically limited, despite its broad distribution range. Recent phytochemical and pharmacological research has demonstrated that there exist several *B. dracunculifolia* bioactivities/properties that are not mentioned by the tradition, so that further research could open up new phytotherapeutical possibilities in its utilization as a renewable resource.

2 Taxonomic Characteristics

Baccharis L. (Asteraceae) is a genus of aromatic and medicinal plants composed of perennial herbs, shrubs, vines, and small trees distributed along the Americas from Southern Canada to Southern Chile and Argentina. Its species grow in a wide variety of ecosystems (Barroso 1976; Müller 2006; Heiden et al. 2009, 2019; Schneider

2009; Giuliano and Plos 2014). The current number of known species is about 440: more than 90% of them are native to South America, mainly to the Andes and Southern Brazilian highlands (Müller 2006; Heiden et al. 2019). The number of *Baccharis* spp. increases every year according to new-new taxonomic reports, demonstrating the huge biodiversity and plasticity related to the genus (Barroso 1976; Müller 2006; Heiden et al. 2019). The infrageneric classification of *Baccharis* continues to be under controversy, being permanently updated (Müller 2006; Heiden et al. 2019). In our region, *Baccharis* spp. biodiversity is represented by at least 150 species described mainly for Brazil, with several endemisms (Barroso 1976; Schneider 2009; Heiden et al. 2009). Ninety nine species are cited for Argentina (Giuliano and Plos 2014), while for Uruguay 39 species are described (Arechavaleta 1906).

A noteworthy characteristic of *Baccharis* spp. is that most of the species are dioecious (that is, male and female occur in separated plant individuals). Dioecious species are scarce in the Asteraceae family (Müller 2006; Budel et al. 2008; Schneider 2009; Campos et al. 2016), conferring to male and female specimens' differences in physiology and metabolism regarding stress tolerance and blooming. Despite recent advances in the understanding of the ecological role of secondary metabolites, the dioecism continues to be scarcely considered in the *Baccharis* L. phytochemical research, to date (Campos et al. 2016; Minteguiaga 2019).

B. dracunculifolia is popularly named in South American Spanish speaking countries as "chilca blanca", "chilca", "chirca", "chilca mata-ojo", "suncho", "thola" or "tola" (Müller 2006; Giuliano and Plos 2014); while across Brazil, it is known as "vassoura", "vassoureira", "vassourinha", "alecrim-de-vassoura", "alecrim-do-campo" and "erva-de-São-João-Maria" (Barroso 1976; Canton and Onofre 2010; Budel et al. 2004; Da Silva Filho et al. 2009). Its common name "vassoura" (broom) derives from Portuguese as, in the past, aerial woody branches of *B. dracunculifolia* were used as rudimental brooms (Queiroga et al. 1990).

Synonyms *Baccharis dracunculifolia* var. *dracunculifolia* DC., *Baccharis dracunculifolia* f. *dracunculifolia* DC., *Baccharis dracunculifolia* var. *integerrima* Kuntze, *Baccharis dracunculifolia* f. *spectabilis* Heering, *Baccharis dracunculifolia* var. *subdentata* Kuntze, *Baccharis dracunculifolia* f. *subviscosa* Kuntze (The Plant List 2019).

Giuliano and Plos (2014) described the subspecies *Baccharis dracunculifolia* subsp. *tandilensis* (Speg.) Giuliano *comb. et. stat. nov.*, which would be an endemism from Tandilia Flora (Buenos Aires Province, Argentina).

3 Crude Drug Used

B. dracunculifolia is broadly known as an aromatic plant, being industrially steam-distilled (aerial parts, mainly leaves; photo 1) in Brazil to obtain its essential oil (commercially known as "vassoura oil"). This oil is used for perfumery purposes

Photo 1 Aerial parts of *B. dracunculifolia* DC. from which it is distilled the vassoura essential oil. The leaves also are employed for medicinal purposes. (Photo: M Minteguiaga and D Passarella)

due to its floral, woody and green aromatic notes attributable to the content of *(E)*-nerolidol and other sesquiterpene alcohols (Queiroga et al. 1990, 2008; Ferracini et al. 1995; Weyerstahl et al. 1996; De Sousa et al. 2009). For medicinal purposes, the aerial parts' infusions or decoctions are consumed in Brazil to treat gastrointestinal and hepatic disorders, as anti-inflammatory, antiulcerative, and antipyretic (Mors et al. 2000; Rodrigues and De Carvalho 2001; Sforcin et al. 2012).

4 Major Chemical Constituents and Bioactive Compounds

Baccharis trimera and *B. dracunculifolia* are the most studied species in the *Baccharis* L. genus regarding their phytochemistry and pharmacological effects (Campos et al. 2016; Sforcin et al. 2012; García et al. 2018). The chemical evaluation of aerial and underground organs demonstrated for *B. dracunculifolia* the occurrence of essential oils, phenolic acids, flavonoids, clerodane diterpenes, triterpenes, and diverse types of glycosides (Abad and Bermejo 2007; Sforcin et al. 2012; Campos et al. 2016). According to Schenkel et al. (2002), chloroform extracts of *B. dracunculifolia* gave positive results in a specific reaction for the presence of peroxides.

B. dracunculifolia essential oil (vassoura oil) composition has been profusely reported in the literature concerning its economic importance. The sesquiterpene acyclic alcohol *(E)*-nerolidol, a compound with antiulcerative activity (Kloppel et al. 2007), is considered a chemomarker of the oil since it is the main compound both in masculine or feminine individuals (Queiroga et al. 1990, 2008; Ferracini et al. 1995; Weyerstahl et al. 1996; Besten et al. 2012: De Sousa et al. 2009; Minteguiaga et al. 2018; Rodrigues et al. 2020). However, in some specific *B. dracunculifolia* populations, *(E)*-nerolidol can be scarce, particularly in Southern Uruguay (Frizzo et al. 2008; Ibáñez and Zoppolo 2011; Minteguiaga 2019). In addition, vassoura essential oil is very rich in quantity and diversity of other volatile sesquiterpene compounds, both hydrocarbons and oxygenated (mainly secondary and tertiary alcohols), as well as monoterpenes (mainly hydrocarbons) (Queiroga et al. 1990, 2008; Weyerstahl et al. 1996; De Sousa et al. 2009; Minteguiaga 2019). In brief, the main compounds described for the species from diverse populations, in their total area of occurrence are α-pinene, β-pinene, limonene, myrcene, β-caryophyllene, germacrene D, bicylogermacrene, δ-cadinene, spathulenol, ledol, globulol and viridiflorol (Queiroga et al. 1990, 2008; Ferracini et al. 1995; Weyerstahl et al. 1996; Frizzo et al. 2008; Minteguiaga et al. 2018; Cazella et al. 2019; Rodrigues et al. 2020; Bonin et al. 2020). Non-terpenic volatile compounds (aliphatics and aromatics) can also be present in the essential oils (Minteguiaga 2019), and eventually, they can also be identified in the hydroalcoholic extracts, such as the case of 2-phenylethanol (Bonin et al. 2020).

The role of volatile compounds in Nature is not yet fully understood, and more chemical ecology studies are needed. However, they seem to play a key role in plant-plant and plant-insect interactions (Ferracini et al. 1995; Rodrigues et al. 2020). The production of such compounds is not stable in the time and, dependence on circadian rhythms, seasonality and in male/female ratio in vegetal populations, could have great influence on the volatile composition (Ferracini et al. 1995; Frizzo et al. 2008; De Sousa et al. 2009; Sforcin et al. 2012; Rodrigues et al. 2020; Minteguiaga 2019). Frizzo et al. (2008) employing multivariate analyses to the composition of the essential oils, proposed two possible *B. dracunculifolia* chemotypes: aromadendrane-type (from South-Southeastern Brazil and Uruguay; with predominance of spathulenol, viridiflorol and globulol) and cadinane-type (from Bolivia; with γ-cadinene, δ-cadinene, ζ-cadinol and α-cadinol as the main compounds).

When medium polar to polar organic and hydroalcoholic extracts of *B. dracunculifolia* are studied, phenolic acids are common components: *p*-coumaric, dihydrocoumaric, ferulic, *(E)*-cinnamic, dihydrocinnamic, hidroxycinnamic, gallic, caffeic, *(E)*-4-(2,3-dihydrocinnamoyloxy) cinnamic, and several caffeoylquinic acids derivatives. Moreover, prenylated derivatives are frequent, such: 3-prenyl-*p*-coumaric (drupanin), 3,5-diprenyl-*p*-coumaric (artepillin C), 3-prenyl-4-dihydrocinnamoiloxy-cinnamic (baccharin), and baccharin-5″-aldehyde (Nagatani et al. 2001, 2002a, b; Midorikawa et al. 2003; Da Silva Filho et al. 2004, 2008, 2009; Park et al. 2004; Lemos et al. 2007; Missima et al. 2007; Resende et al. 2007, 2012; De Funari et al. 2007; Cestari et al. 2011; Bachiega et al. 2013;

Figueiredo-Rinhel et al. 2013; Costa et al. 2019; Rodrigues et al. 2020; Bonin et al. 2020). Other related cinnamic and coumaric acid derivatives, including esters, were identified by Nagatani et al. (2002b) and Midorikawa et al. (2003) from the aerial parts of the species. All these phenolic compounds, able to exert antioxidant activity, are involved in the inhibition of inflammation in colitic rats (Cestari et al. 2011); while *p*-coumaric acid displays strong antiulcer activity (Costa et al. 2019). Ferulic acid and baccharin are moderately active against *Trypanosoma cruzi* (Da Silva Filho et al. 2004), while artepillin C is considered as a strong antimicrobial, especially against methicillin-resistant *Staphylococcus aureus* strains (Veiga et al. 2017).

Flavonoids are one of the most frequently studied secondary metabolite groups for *B. dracunculifolia*. Da Silva Filho et al. (2004, 2009) isolated and purified the flavanone isosakuranetin and the flavonol aromadendrin-4'-*O*-methyl ether (AME) from CH_2Cl_2 extracts of leaves, displaying the former trypanocidal activity. Da Silva Filho et al. (2009) also isolated the methoxylated flavanones acacetin and ermanin, which exhibited antileishmanial activity through *in vitro* assays. Aromadendrin-4'-*O*-methyl ether (AME) and isosakuranetin were also identified in antimicrobial, anti-inflammatory, immunomodulatory, antimutagenic, cytotoxic and antiulcerogenic extracts from the plant species (Lemos et al. 2007; Missima et al. 2007; Resende et al. 2007; De Funari et al. 2007; Da Silva Filho et al. 2008; Cestari et al. 2011; Figueiredo-Rinhel et al. 2013; Bachiega et al. 2013; Costa et al. 2019). Isosakuranetin was also identified in the leaf bud extract, the botanical origin of Brazilian Green Propolis (Park et al. 2004). Depending on the stage of development of the leaves (Midorikawa et al. 2003; Park et al. 2004), other flavonoids were able to be identified in the aerial parts of *B. dracunculifolia*: acacetin, kaempferol, kaempferide, dyhidrokaempferide, pinobanksine, chrysin, naringenin, apigenin, betuletol, galangin, pinocembrin, quercetin, 3-acetoxy-4′,5-dihydroxy-7-methoxyflavanone, 7-*O*-methyl-kaempferol and vicenin 2 (Nagatani et al. 2001, 2002a, b; Midorikawa et al. 2003; Park et al. 2004; Da Silva Filho et al. 2004, 2009; Fukuda et al. 2006; De Funari et al. 2007; Sforcin et al. 2012; Rodrigues et al. 2020; Bonin et al. 2020).

Clerodane-type diterpenes such 15,16-epoxy-19-hydroxy-1,3,13(16),14-clerodatetraen-18-oic acid (Rodrigues et al. 2020), the hautriwaic acid lactone (both active against the parasites *Leishmania donovani* and *Plasmodium falciparum*), and the acetate of this latter were isolated from the leaves of the species (Missima et al. 2007; Da Silva Filho et al. 2008, 2009). Cupressic, isocupressic, agathic, agathalic and other related diterpenic acids also have been identified for *B. dracunculifolia* (Nagatani et al. 2002b; Midorikawa et al. 2003).

Tetracyclic (baccharoxide) and pentacyclic triterpenes (friedelanol) were identified in aerial parts of *Baccharis* spp. by several authors (Abad and Bermejo 2007; Campos et al. 2016). Da Silva Filho et al. (2004) reported the isolation of baccharoxide (presenting high anti-*Trypanosoma cruzi* activity) from *B. dracunculifolia*. Missima et al. (2007) informed that baccharoxide increases the production of H_2O_2 in peritoneal macrophages at a dose of 100μM, acting as an immunomodulator. This also was the case for the triterpene friedelanol (Da Silva Filho et al. 2004; Missima et al. 2007). Other triterpenes isolated from *B. dracunculifolia* aerial parts are ursolic acid, 2-α-hydroxyursolic acid and uvaol; all of them presenting good levels of anti-*Leishmania* activity but the former being the more active (Da Silva Filho et al. 2008, 2009). In addition, both ursolic and 2-α-hidroxyursolic acids also present activity against methicillin-resistant *Staphylococcus aureus* strains (Da Silva Filho et al. 2008).

Nagatani et al. (2001, 2002a, b) performed a systematic study on glycosides and other metabolites from the aerial parts of *B. dracunculifolia*, rendering the identification of 22 new natural products and 34 known compounds. The authors characterized and named 18 dracunculifosides (A to R), which structures include three moieties: a sugar unit, an ester group and an aglycone (Nagatani et al. 2001, 2002a). Dracunculifosides A-E and K were characterized by the same sugar and ester moieties [β-D-glucopyranose and *(E)*-caffeoyl] with the aglycone varying in every one of the named compounds (ionone derivatives, phenolic acids, ethers and ketones, propylene glycol).

Dracunculifoside M, with the same sugar and ester units as the previous compounds, possesses as aglycone a sesquiterpene derivative (eudesmane-type) (Nagatani et al. 2002a). These authors also suggested a mixed biosynthesis for this class of natural products. For dracunculifosides F to I and N to Q, the ester unit was the same as the previous compounds [*(E)*-caffeoyl], but the sugar moiety was a diglycoside [β-D-apiofuranose (1→6) β-D-glucopyranose], and again, a great variation in the aglycone pattern was observed (such as simple aliphatic C_3-C_6 alcohols, benzyl alcohol, phenethyl alcohol, phenol or eugenol derivatives) (Nagatani et al. 2001, 2002a). Dracunculifoside J consisted of a benzofuran-derived aglycone attached to a β-D-glucopyranose (Nagatani et al. 2001), while dracunculifoside L have the same structure with the addition of an *(E)*-caffeoyl group linked to the C-6 position of the sugar moiety (Nagatani et al. 2002a). A different structure was elucidated for dracunculifoside R ($C_{36}H_{44}O_{15}$) consistent with a sesquilignan (pinoresinol derivative) β-D-glucopyranoside, like other new pinoresinol-derived glycoside identified by the authors {pinoresinol *O*-[6-*O*-*(E)*-caffeoyl]-β-D-glucopyranoside} (Nagatani et al. 2002a). Finally, a new terpenic derivative glycoside was also isolated from the aerial parts of *B. dracunculifolia* [(4-*S*)-α-terpineol-*O*-β-D-apiofuranosyl-(1→6)-β-D-glucopyranoside] (Nagatani et al. 2002a). The already-known glycosides obtained in this systematic study included: three flavonoid-type, four caffeoylquinic acid derivatives, four monoterpene-type, five phenolic-type, three C_{13} norisoprenoid-type, a pinoresinol-type and a syringaresinol-type (Nagatani et al. 2001, 2002a).

Miscellaneous compounds identified from *B. dracunculifolia* aerial parts include: viscidone (Midorikawa et al. 2003; Da Silva Filho et al. 2008, 2009), 2,2-dimethyl-6-carboxyethenyl-2H-1-benzopyran acid (DCBEN) (Bachiega et al. 2013) and propolis-benzofuran A and B (Midorikawa et al. 2003). Fukuda et al. (2006), isolated two new low-molecular weight terpenes from an ethanolic extract from the leaves: a monoterpene (*p*-methoxythymol acetate) and an uncommon sesquiterpene structure named baccharisketone. Besides, the authors characterized 17 known natural products, among them: the rare sesquiterpene bisacumol and its related compound

2-methyl-6-(4′-methylphenyl)-3-hepten-2-ol (bisabolane-type), as well as 3,4,3′,4′-tetrahydroxy-5,5′-diisopropyl-2,2′-dimethylbiphenyl (a monoterpene dimer) which inhibits strongly the growth of leukemia cells (Fukuda et al. 2006).

Even less studied, the roots of the plant species yielded compounds related to tremetone (viscidone, 11-hydroxy-10,11-dihydro-euparin and 6-hydroxy-tremetone), as well as the triterpenes baccharoxide and friedelanol (Missima et al. 2007).

5 Morphological Description

B. dracunculifolia was originally described in the Brazilian Flora by the Swiss botanist Augustin Pyramus de Candolle (1778–1841) in his work *Prodromus Systematis Naturalis Regni Vegetabilis* (De Candolle 1836). The species is polymorphic and in general grows as profusely branched woody shrubs or little trees (1–5 m high; globose shaped), with abundant pubescence in the younger shoots and absence of hairs (glabrous) in the older ones (De Candolle 1836; Lombardo 1964; Giuliano and Plos 2014). The leaves are alternated, sessile and narrow or broadly elliptic, obovate or lanceolate shaped, attenuated at the base side and acute at the apex. The typical dimensions are: 2–4 cm long and 0.4–1 cm in diameter; pointed-glandulose. The leaves present 1–4 teeth at both margin sides, or, even entire margin which is dependent on the specific biotype or sub-species considered (De Candolle 1836; Cabrera 1974; Müller 2006; Giuliano and Plos 2014).

The inflorescences (capitula) are numerous, shortly pedunculated and solitary in higher leaves' axils, forming foliated racemes in the younger shoots. Male capitula with broadly campanulate involucra of 3–5 mm long and 2.5–3 mm in diameter; 3-seriated lightly pubescent phyllaries (involucral bracts), ovate-lanceolate shaped with acute apices. Male flowers: 20–50 with short adhered style branches. The female capitula also present campanulate involucra, in general, a little bigger than the males' ones: 4–5 long and 3–5 mm in diameter, 3 or 4-seriated pubescent phyllaries, lanceolate shaped with acute apices. Female flowers: 30–60 with cylindrical and glabrous achenes (1.0–1.5 mm long), and white pappus (De Candolle 1836; Cabrera 1974; Müller 2006; Giuliano and Plos 2014).

A complete morphoanatomic study of *B. dracunculifolia* aerial parts and tissue distribution has been published (Budel et al. 2004; Marchiori and De Oliveira 2007).

6 Geographical Distribution

Baccharis dracunculifolia is a largely distributed South American species from Northern and Center Argentina, Eastern Bolivia, Southern to Northeastern Brazil, Eastern Paraguay, and all Uruguay (Arechavaleta 1906; Barroso 1976; Müller 2006; Giuliano and Plos 2014).

Regarding cultivation, even despite the above described importance of the *Baccharis* L. genus, only two species are commercially important and extracted in Brazil for their essential oils for flavor and fragrance industry: *B. dracunculifolia* and *B. trimera* (Queiroga et al. 1990; Weyerstahl et al. 1996; De Sousa et al. 2009). However, the industrial crops of such species are marginal and incipient (De Sousa et al. 2009; Sforcin et al. 2012; García et al. 2018), and it is frequent the direct harvest from Nature jeopardizing the germplasm source.

7 Ecological Requirements

B. dracunculifolia is a species that grows in a diversity of environments. It is frequently found in scrublands, grasslands, open forests and mountain slopes in moderated-humid to moderated-dry places, from 0–4000 m altitude (typical of the high Bolivian plateau) (Müller 2006). In Uruguay, it is generally found in the low-altitude mountain typical of the Northern and Eastern parts of the country, in grasslands as well as riparian flora near watercourses (Lombardo 1964).

An ecological relevant consideration about *B. dracunculifolia* is its role as host for several gall-inducing insects that birth, feed and develop their larvae into the leaves (localized reaction), changing dramatically the morphology, physiology and metabolism of such plant organs (Fernandes et al. 1996; Coelho De Oliveira et al. 2014). Besten et al. (2015) observed that the level of *(E)*-nerolidol and spathulenol diminishes in galls induced by the psyllid *Baccharopelma dracunculifoliae* (the most frequent gall-inducing insect) compared with healthy leaves, and besides, the total yield of essential oils augmented (0.63% in dried healthy leaves *vs.* 3.5% for fresh galled-leaves). Recently, Rodrigues et al. (2020) demonstrated that male and female plants are differentially preferred by insects: while male plants showed more infestation by galling-inducing insects, female plants are more visited by melliferous bees. These results also are correlated with chemical differences in the volatile and non-volatile profiles from both *B. dracunculifolia* sexes (Rodrigues et al. 2020).

8 Traditional Use (Part(s) Used) and Common Knowledge

Many *Baccharis* spp. are ancestrally employed by indigenous Latin American people as folk medicine in the alleviation of illness and to relieve pain, among other applications; as recorded by several ethnopharmacological surveys (Abad and Bermejo 2007; Budel et al. 2008; Campos et al. 2016). *B. crispa*, *B. articulata,* and *B. trimera* ("carquejas") are the emblematic examples recognized by Brazilian and Argentine Pharmacopeias as medicinal plants by its digestive properties (Bandoni et al. 1978; Abad and Bermejo 2007; García et al. 2018).

The historical references about the folk medicine utilization of *B. dracunculifolia* in South America are scarce, and traditionally it has been hardly recognized as a medicinal species. Reports from Brazil describes its use as infusions or decoctions to treat gastrointestinal/hepatic disorders, physical tiredness, loss of appetite, fever, ulcers and as anti-inflammatory (Mors et al. 2000; Rodrigues and De Carvalho 2001; Da Silva Filho et al. 2004, 2009; Cestari et al. 2011; Sforcin et al. 2012).

Given its woody habit, in the form of small trees (reaching until 4–5 m height), it is frequently employed in South America as fuel wood for heating and cooking (Müller 2006; Giuliano and Plos 2014), especially where other types of such resources are scarcely available. Moreover, *B. dracunculifolia* is considered as a hard to control weed in agricultural fields (diminishing the resources for "more productive" species), and because of that trees (or shrubs) are cut and burned (Negreiros et al. 2014).

9 Modern Medicine Based on Its Traditional Medicine Uses

Regarding the potential employment of *B. dracunculifolia* as a phytotherapeutic, the work that has been performed by the Department of Pharmaceutical Sciences, School of Pharmaceutical Sciences from Ribeirão Preto (FCFRP-University of São Paulo, Brazil) is a key in the future application of *B. dracunculifolia* in medicine. Most of the information presented here has been published by this leading research group, and a summary of its research is available in Sforcin et al. (2012).

9.1 Essential Oil

Antiparasitic Activity In addition to aromatic properties, for the vassoura oil, it has also been reported pharmacological properties as antiparasitic against promastigote forms of *Leishmania donovani* (IC_{50} = 42 μg/ml) and adults' worms of *Schistosoma mansonii* (IC_{50} = 10–100 μg/ml) (Parreira et al. 2010).

Anti-Inflammatory and Immunomodulatory Activity The *in-vitro* work performed by Florão et al. (2012) had a focus on both bioactivities since the immune system and inflammatory response work cooperatively. Through a model with human peripheral blood leukocytes, *B. dracunculifolia* oil caused inhibition on the proliferation of lymphocytes stimulated by phytohemagglutinin (PHA) (not related to cytotoxicity, especially at 0.01 μl/ml dose) reducing the number of lymphoblasts; thus, exhibiting *in vitro* immunomodulatory activity (Florão et al. 2012). At the same time, 0.01 μl/ml dose inhibited the neutrophil migration towards a casein gradient, meaning an anti-inflammatory mechanism (Florão et al. 2012).

Antiulcerogenic Activity The essential oil from this species exhibits antiulcerogenic activity, as is the case for the solvent extracts. Indeed, Kloppel et al. (2007) assessed the *in vivo* activity of the oil [23.6% of abundance of *(E)*-nerolidol] through ethanol-induced gastric lesions in male Wistar rats, and they found that the ulcerative lesion index (ULI) was significantly reduced (42.8–61.6%) after the oral dosage of 50–500 mg/kg (in comparison, omeprazole reduced the ULI by 72.4% at 30 mg/kg). When assessed by this model pure *(E)*-nerolidol (oral doses ranging from 250 to 500 mg/kg), again the results were good and even better than those displayed for omeprazole (ULI reduction of 52.6–87.6% for the terpene against 50.9% for omeprazole at 30 mg/kg), highlighting the gastroprotective and antiulcerogenic properties of *(E)*-nerolidol (Kloppel et al. 2007). These properties have been also confirmed by the work of Massignani et al. (2009), who in addition observed that the oil (dosage: 50–500 mg/kg) reduced the gastric juice volume and its acidity after pylorus-ligature assay (model of gastric secretion).

Antimicrobial Activity Even the information for the vassoura oil is abundant, in some cases is also controversial. Ferronatto et al. (2007) found *in-vitro* activity against *Staphylococcus aureus*, *Escherichia coli* and *Pseudomonas aeruginosa* strains, applying a diffusion model in solid Müeller-Hinton medium. By microdilution assay, Salazar et al. (2018) found activity of the oil against two *S. aureus* strains (an ATCC culture and a resistant clinical isolate) with MIC values of 102 and 512μg/ml respectively, but no activity was observed against *E. coli* neither *P. aeruginosa* strains. However, even when not active by itself, the essential oil can synergistically modulate the activity of antibiotics against several resistant bacterial strains (Salazar et al. 2018). In a systematic antimicrobial screening of the essential oil obtained at flowering plant stage, Cazella et al. (2019) found susceptibility for bacterial strains of *S. aureus*, *P. aeruginosa* and *Bacillus cereus* (MIC ranged from 0.5–1.05 mg/ml) and weak or no activity at all for several species, including *E. coli*; the activity against several fungal strains was negligible compared with the positive controls. Pereira et al. (2011) assessed by a standardized protocol the activity of *B. dracunculifolia* oil against several isolates of the yeasts *Candida* spp. obtained from the oral cavities of infants and their mothers' nipples: all the strains were sensitive to the oil (MIC: 0.2–6.25 mg/ml), included *C. albicans* isolates which were resistant to commercial antifungal drugs. Despite the above-mentioned results, Parreira et al. (2010) did not observe inhibitory activity at all from the *B. dracunculifolia* oil against *C. albicans*, *C. glabrata*, *C. krusei*, *C. neoformans*, methicillin-resistant *S. aureus* and *Mycobacterium intracellulare* ATCC strains, applying a standardized methodology. The described variations in antimicrobial activity of the oil might be a consequence of the different chemical composition of the evaluated oils and methods used. A typical situation described for the secondary metabolism expression in plants originated in different plant collections (geographical origin and phenological stages) and in diverse extraction protocols.

Cytotoxicity *B. dracunculifolia* oil did not demonstrate cytotoxicity (Parreira et al. 2010; Florão et al. 2012) neither antiplasmodial activity (Parreira et al. 2010).

Antioxidant Activity This evaluation for the vassoura oil has demonstrated moderate to weak activity when compared with the *B. dracunculifolia* aqueous extracts (Ferronatto et al. 2006; Paroul et al. 2016; Minteguiaga 2019).

Other Bioactivities Finally, and not related directly to public health applications, the *B. dracunculifolia* essential oil also demonstrated to exert anti-*Ixodidae* spp. (hard ticks) (Lage et al. 2015), anti-insect (Chaaban et al. 2017; Alves et al. 2018), post-harvest antifungal (Pedrotti et al. 2019) and allelopathic (Ibáñez and Zoppolo 2011) properties, highlighting the importance of *B. dracunculifolia* oil as multipurpose product besides aromatic attributes.

9.2 Non Polar and Aqueous Extracts

Antiparasitic Activity This activity was *in vitro* evaluated against the trypomastigote forms of the protozoan *Trypanosoma cruzi*, the etiological agent of Chaga's disease (Da Silva Filho et al. 2004). Crude leaf rinse extract (extracted by CH_2Cl_2), crude branch extract (hydroalcoholic) (both at 3.0 mg/ml concentration) and isolated metabolites (isosakuranetin and baccharoxide) provoked total lysis on *T. cruzi* trypomastigotes (Da Silva Filho et al. 2004). The crude leaf rinse extract and two isolated metabolites from it (ursolic acid, hautriwaic acid lactone) also shown activity against the protozoan parasites *Leishmania donovani* (one of the leishmaniasis etiological agents; IC_{50} = 45.0 μg/ml) and *Plasmodium falciparum* (etiological agent of human malaria; IC_{50} = 25.0 μg/ml) without displaying cytotoxicity against Vero cells (Da Silva Filho et al. 2009).

Antiulcerogenic Activity The potential of *B. dracunculifolia* extracts against ulceration was primarily accomplished by Lemos et al. (2007) following similar *in vivo* protocols as those used for the essential oils (ulcers induced by ethanol, stress and indomethacin) (Kloppel et al. 2007; Massignani et al. 2009). The 70% ethanolic leaves extract [rich in phenolics such caffeic acid, aromadendrin-4′-methyl ether (AME) and baccharin, among others] displayed inhibition on Ulcerative Lesion Index (ULI) (in ethanol-induced ulceration) by 79.9–95.0% when male Wistar rats were administered orally with 50–500 mg/kg of the extract (positive control omeprazole reduced ULI by 98.9% at 30 mg/kg); reducing at the same time the gastric juice volume and its acidity (Lemos et al. 2007). Recently, Costa et al. (2019) have extended the knowledge on the antiulcerogenic activity of *B. dracunculifolia* extracts demonstrating that these can recover (healing) previously ulcerated gastric tissues reducing affected area by 49.4% at 300 mg/kg dose, being *p*-coumaric acid a pivotal component for this bioactivity.

Hepatoprotective Activity Da Silva et al. (2019) reported that *B. dracunculifolia* hydroalcoholic extracts exert *in-vitro* hepatoprotective function (acting in the lipid/retinoid metabolism) assessed by the LX-2 hepatic stellate cell model, without displaying deleterious effects (cytotoxicity, genotoxicity or apoptosis). These results corroborated the previous *in-vivo* findings by Rezende et al. (2014), who observed reduction in the hepatic damage and its serum markers upon pretreatment with a leaves hydroalcoholic extract, in two hepatotoxicity models induced by tetrachloride and acetaminophen.

Anti-Inflammatory and Immunomodulatory Activities Cestari et al. (2011) studied the intestinal anti-inflammatory activity of EtOAc extracts (aerial parts) by means of trinitrobenzene sulfonic (TBNS) methodology, in an *in vivo* model of acute and chronic colitis. The oral administration of such extracts (dose: 5 and 50 mg/kg) to male Wistar rats attenuated the mucosal damage, typical of colonic inflammation after the application of TBNS, also counteracting the depletion of glutathione and the augmentation of myeloperoxidase and alkaline phosphatase activities. In the same set of experiments, the authors demonstrated an inhibitory effect of lipid peroxidation in rat brain membranes by those extracts (rich in phenolic compounds), which in turn may be associated with down-regulation of the bowel inflammation (Cestari et al. 2011). The results obtained by Dos Santos et al. (2010) confirmed the anti-inflammatory and antinociceptive properties of the hydroalcoholic extract by different *in vitro* models. Bachiega et al. (2013) also assessed the *in vitro* anti-inflammatory potential of a hydroalcoholic extract at the same time that assessed the immunomodulatory effects. Missima et al. (2007) found that the CH_2Cl_2 and hydroalcoholic extracts from leaves and roots enhanced the production of H_2O_2 by rat peritoneal macrophages at doses varying from 25–100 μg/ml. Furthermore, the authors demonstrated that pure isolated compounds from those extracts (baccharoxide and friedelanol) also exhibited immunomodulatory action (Missima et al. 2007). Other related works were conducted *in vitro* to determine whether the leaves hydroalcoholic extracts could inhibit the human neutrophil reactive oxygen species (ROS) generation, as well as whether they modulate the neutrophil enzymatic activities (NADPH oxidase and myeloperoxidase) (Figueiredo-Rinhel et al. 2013, 2017).

Antimicrobial Activity Leitão et al. (2004) and Pereira et al. (2016) evaluated the action of a *B. dracunculifolia* leaf CH_2Cl_2 extract or essential oil against *Streptococcus mutans* (which originates dental caries in humans). The botanical extract produced a bacteriostatic effect (0.40 mg/ml) and inhibited the bacterial acid production and the synthesis of glucans (key aspects for producing caries through biofilms), thus, demonstrating anticariogenic effect (Leitão et al. 2004). Da Silva Filho et al. (2008) evaluated the antimicrobial action of CH_2Cl_2 leaf rinsed extract and isolated compounds against a panel of microbes, including several ATCC bacteria and fungi strains. The best results for the extract were obtained in the inhibition of the yeasts *Cryptococcus neoformans* (50% growth inhibitory concentration: IC_{50} = 40 μg/ml) and *Candida krusei* (IC_{50} = 65 μg/ml) growth (positive control, amphotericin B,

IC_{50} = 0.50 μg/ml and IC_{50} = 1.50 μg/ml, respectively); triterpenes ursolic and 2-α-ursolic acid and the flavanone isosakuranetin were active as well (Da Silva Filho et al. 2008). Veiga et al. (2017) worked on hexane and ethanol extract from leaf buds of *B. dracunculifolia*, quantifying the total amount of flavonoids, phenolics and artepillin C, and determining the activity against methicillin-sensitive or resistant *S. aureus* (MSSA or MRSA) isolates from blood cultures obtained from hospitalized patients in Brazil. MIC values presented a wide variation (from 256.7 μg/ml to 770.1 μg/ml for ethanolic extract; 197.0 μg/ml to 394.0 μg/ml for hexane extract) against the different isolates evaluated; but the best results were obtained for isolated artepillin C (Veiga et al. 2017). Casagrande et al. (2018) obtained bactericidal effects against *S. aureus*, *B. cereus* and *E. coli* from ethanolic, acetone and aqueous extracts of the plant material at different concentrations, depending on the solvent employed. The 70% EtOH extract (plant part not specified but probably from aerial parts) was recently evaluated against several Gram (+) and (−) aerobic bacteria by the determination of MIC through the broth microdilution assay and subsequent calculation of minimum bactericidal concentration (MBC) (Bonin et al. 2020). The best results were obtained for Gram (+) ATCC strains of *S. aureus* and *B. subtilis* (MIC = 125 μg/ml; MBC = 250 μg/ml); whilst for *B. cereus* the values were higher, and, the extract resulted inactive against Gram (−) bacteria. The authors also studied quantitatively the diminution in the bacterial growth with increasing concentration of the extract, assessing in parallel the level of damage (morphological alterations) by flow cytometry and electron microscopy (Bonin et al. 2020). The differences in bioactivity obtained from analyzing this information demonstrate not consistent results, which may be assigned to a lack of uniformed protocols to assess the antimicrobial activity properly. In general, the Gram (+) bacteria appear to be more susceptible to the extract action, but results regarding Gram (−) and fungus are controversial.

Antiviral Activity Búfalo et al. (2009) found inhibition of poliovirus type 1 replication from the hydroalcoholic extracts and essential oil from *B. dracunculifolia*, especially at 25μg/100μl concentration (74% of inhibition).

Cytotoxicity and Mutagenic Activity Fukuda et al. (2006) fractionated, isolated and tested separately *in vitro* natural compounds from *B. dracunculifolia* leaves against mouse lymphocytic leukemia cells (L1210); the methanolic fraction produced 100% of cell growth inhibition after 50 μg/ml dosages. The more active compounds were the monoterpene dimer 3,4,3′,4′-tetrahydroxy-5,5′-diisopropyl-2,2′-dimethylbiphenyl and spathulenol, which presented IC_{50} = 9.1 μM and IC_{50} = 13.6 μM, respectively, while other thymol, carvacrol and sesquiterpene derivatives were actives in a lesser extension. Búfalo et al. (2010) demonstrated the *in vitro* cytotoxic action of ethanolic extracts (50–100μg/100 ml) on human laryngeal epidermoid carcinoma (HEp-2) cells, which was at least partially attributed to caffeic and cinnamic acids secondary metabolites. The plant extracts (as well as artepillin C and baccharin) also exhibited a protector effect against colon cancer (Munari et al. 2014) and antiproliferative and selective cytotoxicity against different tumor cell lines (Francielli De Oliveira et al. 2014). The potential mutagenic effect of the aerial parts EtOAc extract was studied by Resende et al. (2007) employing the rat bone marrow and peripheral blood micronucleus test: they did not find any mutagenic activity of the extract and, besides, the DNA damage of animals treated with doxorubicin + extract was diminished, thus, exhibiting antimutagenicity. In the same line, Roberto et al. (2016) did not find genotoxic or mutagenic effects in *in vitro* assessment of hydroalcoholic extracts, and even, it exerts antimutagenic activity protecting against methyl methanesulphonate induced DNA damage over a hepatoma cell line (HCT) model. Similar results were obtained by Resende et al. (2012) who evaluated several mutagens by the Ames test with *Salmonella typhimurium*. Finally, several studies have (i.e., Da Silva Filho et al. 2009; Bonin et al. 2020) demonstrated the absence of cytotoxicity from *B. dracunculifolia* extracts against Vero cells, which highlights the cancer-protective effect of *B. dracunculifolia* extracts and the potential for the development of phytotherapeutic drugs.

Antioxidant Activity Most of the antioxidant assessments using different protocols (DPPH, ABTS, FRAP and superoxide radical scavenger activity, Fe^{2+} chelating activity, among others) corroborate high level of activity for the polyphenolic composition of *B. dracunculifolia* extracts (Guimarães et al. 2012; Figueiredo-Rinhel et al. 2013; Rezende et al. 2014; Veiga et al. 2017; Casagrande et al. 2018; Minteguiaga 2019). In particular, it deserves to be highlighted the work performed by Guimarães et al. (2012) which studied the protection of a glycolic extract of the plant species against oxidative stress in rat liver mitochondria, finding a decrease in the basal local generation of H_2O_2 and other ROS production, preventing lipid and thiol oxidation in mitochondrial membranes.

Other Considerations *B. dracunculifolia* is also the botanical origin of the called "Brazilian Green Propolis" (BGP) (a resinous exudate collected by honeybees from the leaf buds of the plant that includes plant components and compounds derived from insect bowel processing), which is rich in prenylated phenolic acids such artepillin C (Park et al. 2004; Sforcin et al. 2012; Rodrigues et al. 2020). This product exhibits a variety of medicinal properties such: antioxidant, antibacterial and antitumor (Park et al. 2004; Shimizu et al. 2005; Abad and Bermejo 2007; Sforcin et al. 2012), but it can display cytotoxicity against fibroblasts (De Funari et al. 2007). In Brazil, BGP has been incorporated in several foods and beverages to ameliorate the health, as a nutraceutical (Park et al. 2004; Sforcin et al. 2012).

10 Conclusions

Despite the fact that *B. dracunculifolia* is less recognized as a medicinal plant in South America (with the clear exception of Brazil), the research performed in the last years has unraveled a great spectrum of pharmacological activities, suggesting that this species should be considered as a promising raw material for developing phytopharmaceuticals in the next years. In particular, the *in vitro* and *in vivo* results from essential oils, hydroalcoholic extracts and isolated compounds obtained through antiulcer, anti-inflammatory, antiparasitic and anticancer tests are relevant to consider the initiation of standardized pre-clinical assays as initial step in the development of phytotherapeutics. The antimicrobial activity tests of extracts, although controversial, call attention to the importance of starting farther basic research aimed at adopting uniform protocols to assess bioactivity. In the lack of enough information, more work should focus on toxicity issues, especially related to the popular uses of plants in the form of infusions. Regarding its character as an aromatic plant, the fragrance/perfumery industry appreciates the aroma of *B. dracunculifolia* essential oil as an important resource, but more research is needed in order to understand the wide variation in the composition as a result of different geographical origins (with the possible existence of many chemotypes) and phenological stages. Recent publications also suggest that the oil of *B. dracunculifolia* should be considered as a multi-purpose product beyond its aromatic attributes.

Acknowledgments Prof. Diego Passarella (CUT-UdelaR) for his photograph assistance.

References

Abad MJ, Bermejo P (2007) *Baccharis* (Compositae): a review update. ARKIVOC 7:76–96

Alves KF, Caetano FH, García IJP, Santos HL, Silva DB, Siqueira JM, Tanaka AS, Alves SN (2018) *Baccharis dracunculifolia* (Asteraceae) essential oil toxicity to *Culex quinquefasciatus* (Culicidae). Environ Sci Pollut Res 25(31):31718–31726

Arechavaleta J (1906). Flora Uruguaya. Ed. Talleres Barreiro y Ramos, Montevideo, Uruguay. Volume III, pp 219–255. In: facsimile: https://autores.uy/ (Last access, December 2019)

Bachiega TF, de Sousa JPB, Bastos JK, Sforcin JM (2013) Immunomodulatory/anti-inflammatory effects of *Baccharis dracunculifolia* leaves. Nat Prod Res 27(18):1646–1650

Bandoni AL, Medina JE, Rondina RVD, Coussio JD (1978) Genus *Baccharis* L. I: phytochemical analysis of a non polar fraction from *Baccharis crispa* Sprengel. Planta Med 34(7):328–331

Barroso GM (1976) Compositae. Subtribo Baccharidinae Hoffmann. Estudo das espécies ocorrentes no Brasil Rodriguesia 40(28):3–273

Besten MA, Jasinski VCG, Costa AGLC, Nunes DS, Sens SL, Wisniewski A Jr, Simionatto EL, Riva D, Dalmarco JB, Granato D (2012) Chemical composition similarity between the essential oils isolated from male and female specimens of each five *Baccharis* species. J Braz Chem Soc 23(6):1041–1047

Besten MA, Nunes DS, Granato D, Sens SL, Wisniewski A Jr, Edesio LS, Riva-Scharf D (2015) Volatile components from galls induced by *Baccharopelma dracunculifoliae* (Hemiptera: Psyllidae) on leaves of *Baccharis dracunculifolia* (Asteraceae). Quim Nova 38(1):66–70

Bonin E, Carvalho VM, Avila VD, dos Santos NCA, Benassi-Zanqueta E, Lancheros CAC, Previdelli ITS, Ueda-Nakamura T, de Abreu Filho BA, do Prado IN (2020) *Baccharis dracunculifolia*: chemical constituents, cytotoxicity and antimicrobial activity. LWT-Food Sci Technol 120:108920

Budel JM, Duarte MDR, Santos CADM, Farago PV (2004) Morfoanatomia Foliar e Caulinar de *Baccharis dracunculifolia* DC., Asteraceae. Acta Farm Bonaer 23(4):477–483

Budel JM, Matzenbacher N, Duarte MR (2008) Genus *Baccharis* (Asteraceae): a review of chemical and pharmacological studies. In: Singh VK, Govil JN (eds) Recent progress in medicinal plants. Phytopharmacology and therapeutic values III. Ed, vol 21. Studium Press LLC, Houston, USA, pp 1–18

Búfalo MC, Figueiredo AS, De Sousa JPB, Candeias JMG, Bastos JK, Sforcin JM (2009) Antipoliovirus activity of *Baccharis dracunculifolia* and propolis by cell viability determination and real-time PCR. J Appl Microbiol 107(5):1669–1680

Búfalo MC, Candeias JMG, Sousa JPB, Bastos JK, Sforcin JM (2010) *In vitro* cytotoxic activity of *Baccharis dracunculifolia* and propolis against HEp-2 cells. Nat Prod Res 24(18):1710–1718

Cabrera AL (1974) Compositae. Flora Ilustrada de Entre Ríos, Colección Científica del Instituto Nacional de Tecnología Agropecuaria 6(6):106–554

Campos FR, Bressan J, Jasinki VCG, Zuccolotto T, da Silva LE, Cerqueira LB (2016) *Baccharis* (Asteraceae): chemical constituents and biological activities. Chem Biodivers 13(1):1–17

Canton M, Onofre SB (2010) Interferência de extratos da *Baccharis dracunculifolia* DC., Asteraceae, sobre a atividade de antibióticos usados na clínica. Braz J Pharmacogn 20(3):348–354

Casagrande M, Zanela J, Júnior AW, Busso C, Wouk J, Iurckevicz G, Montanher PF, Yamashita F, Malfatti CRM (2018) Influence of time, temperature and solvent on the extraction of bioactive compounds of *Baccharis dracunculifolia*: *In vitro* antioxidant activity, antimicrobial potential, and phenolic compound quantification. Ind Crop Prod 125:207–219

Cazella LN, Glamoclija J, Sokovic M, Gonçalves JE, Linde GA, Colauto NB, Gazim ZC (2019) Antimicrobial activity of essential oil of *Baccharis dracunculifolia* DC (Asteraceae) aerial parts at flowering period. Front Plant Sci 10:27. https://doi.org/10.3389/fpls.2019.00027

Cestari SH, Bastos JK, Di Stasi LC (2011) Intestinal anti-inflammatory activity of *Baccharis dracunculifolia* in the trinitrobenzenesulphonic acid model of rat colitis. Evid Based Complement Alternat Med:524349. https://doi.org/10.1093/ecam/nep081

Chaaban A, Martins CEN, Bretanha LC, Micke GA, Carrer AR, Rosa NF, Ferreira L, Molento MB (2017) Insecticide activity of *Baccharis dracunculifolia* essential oil against *Cochliomyia macellaria* (Diptera:Calliphoridae). Nat Prod Res 32(24):2954–2958

Coelho De Oliveira D, Magalhães TA, Ferreira BG, Teixeira CT, Formiga AT, Fernandes GW, Isaías RMDS (2014) Variation in the degree of pectin Methylesterification during the development of *Baccharis dracunculifolia* kidney-shaped gall. PLoS One 9(4):e94588

Costa P, Boeing T, Somensi LB, Cury BJ, Espíndola VL, França TCS, De Almeida MO, Arruda C, Bastos JK, Da Silva LM, De Andrade SF (2019) Hydroalcoholic extract from *Baccharis dracunculifolia* recovers the gastric ulcerated tissue, and *p*-coumaric acid is a pivotal bioactive compound to this action. Biofactors 45(3):479–489

Da Silva CM, Caetano FH, Pereira FDC, Morales MAM, Sakane KK, Moraes KCM (2019) Cellular and molecular effects of *Baccharis dracunculifolia* DC. And *Plectranthus barbatus* Andrews medicinal plant extracts on retinoid metabolism in the human hepatic stellate cell LX-2. BMC Complement Alter Med 19:222

Da Silva Filho AA, Bueno PCP, Gregório LE, Andrade e Silva ML, Albuquerque S, Bastos JK (2004) *In vitro* trypanocidal activity evaluation of crude extract and isolated compounds from *Baccharis dracunculifolia* D.C. (Asteraceae). J Pharm Pharmacol 56(9):1195–1199

Da Silva Filho AA, Sousa JPB, Soares S, Furtado NJAC, Silva MLA, Cunha WR, Gregório LE, Nanayakkara NPD, Bastos JK (2008) Antimicrobial activity of the extract and isolated compounds from *Baccharis dracunculifolia* DC. (Asteraceae). Z Naturforsch C J Biosci 63(1–2):40–46

Da Silva Filho AA, Resende DO, Fukui MJ, Santos FF, Pauletti PM, Cunha WR, Silva MLA, Gregório LE, Bastos JK, Nanayakkara NPD (2009) *In vitro* antileishmanial, antiplasmodial and cytotoxic activities of phenolics and triterpenoids from *Baccharis dracunculifolia* DC. (Asteraceae). Fitoterapia 80(8):478–482

De Candolle AP (1836). Prodromus Systematis Naturalis Regni Vegetabilis. Ed. Treuttel & Würtz. Paris, France. Volume 5, pp 421

De Funari CS, Ferro VDO, Mathor MB (2007) Analysis of propolis from *Baccharis dracunculifolia* DC. (Compositae) and its effects on mouse fibroblasts. J Ethnopharmacol 111(2):206–212

De Sousa JPB, Jorge RF, Leite MF, Furtado NAJC, Bastos JK, Da Silva Filho AA, Queiroga CL, De Magalhães PM, Soares AEE (2009) Seasonal variation of the *(E)*-nerolidol and other volatile compounds within ten different cultivated populations of *Baccharis dracunculifolia* DC. (Asteraceae). J Essent Oil Res 21(4):308–314

Dos Santos DA, Fukui MDJ, Nanayakkara NPD, Khan SI, Sousa JPB, Bastos JK, Andrade SF, Da Silva Filho AA, Quintão NLM (2010) Anti-inflammatory and antinociceptive effects of *Baccharis dracunculifolia* DC (Asteraceae) in different experimental models. J Ethnopharmacol 127(2):543–550

Fernandes GW, Carneiro MAA, Lara ACF, Allain LR, Andrade GI, Julião GR, Reis TR, Silva IM (1996) Galling insects on neotropical species of *Baccharis* (Asteraceae). Trop Zool 9(2):315–332

Ferracini VL, Paraíba LC, Leitão Filho HL, da Silva AG, Nascimento LR, Marsaioli AJ (1995) Essential oils of seven Brazilian *Baccharis* species. J Essent Oil Res 7(4):355–367

Ferronatto R, Marchesan ED, Bednarski F, Alencar SM, Onofre SB (2006) Atividade antioxidante dos óleos essenciais produzidos por *Baccharis dracunculifolia* DC. e *Baccharis uncinella* DC. (Asteraceae). Arq Ciênc Saúde Unipar 10(2):67–70

Ferronatto R, Marchesan ED, Pezenti E, Bednarski F, Onofre SB (2007) Atividade antimicrobiana de óleos essenciais produzidos por *Baccharis dracunculifolia* DC. e *Baccharis uncinella* DC. (Asteraceae). Braz J Pharmacogn 17(2):224–230

Figueiredo-Rinhel ASG, Kabeya LM, Bueno PCP, Jorge-Tiossi RF, Azzolini AECS, Bastos JK, Lucisano-Valim YM (2013) Inhibition of the human neutrophil oxidative metabolism by *Baccharis dracunculifolia* DC. (Asteraceae) is influenced by seasonality and the ratio of caffeic acid to other phenolic compounds. J Ethnopharmacol 150(2):655–664

Figueiredo-Rinhel ASG, de Melo LL, Bortot LO, Santos EOL, Andrade MF, Azzolini AECS, Kabeya LM, Caliri A, Bastos JK, Lucisano-Valim YM (2017) *Baccharis dracunculifolia* DC (Asteraceae) selectively modulates the effector functions of human neutrophils. J Pharm Pharmacol 69(12):1829–1845

Florão A, Budel JM, Duarte MDR, Marcondes A, Rodrigues RAF, Rodrigues MVN, Santos CADM, Weffort-Santos AM (2012) Essential oils from *Baccharis* species (Asteraceae) have anti-inflammatory effects for human cells. J Essent Oil Res 24(6):561–570

Francielli De Oliveira P, Lima IMDS, Munari CC, Bastos JK, Da Silva Filho AA, Tavares DC (2014) Comparative evaluation of Antiproliferative effects of Brazilian green Propolis, its main source *Baccharis dracunculifolia*, and their major constituents Artepillin C and Baccharin. Planta Med 80(6):490–492

Frizzo CD, Atti-Serafini L, Laguna SE, Cassel E, Lorenzo D, Dellacassa E (2008) Essential oil variability in *Baccharis uncinella* DC. And *Baccharis dracunculifolia* DC. Growing wild in southern Brazil, Bolivia and Uruguay. Flavour Fragr J 23(2):99–106

Fukuda M, Ohkoshi E, Makino M, Fujimoto Y (2006) Studies on the constituents of the leaves of *Baccharis dracunculifolia* (Asteraceae) and their cytotoxic activity. Chem Pharm Bull 54(10):1465–1468

García D, Furlan MR, Ming LC (2018) *Baccharis trimera* (less.) DC. In: Albuquerque UP, Patil U, Máthé Á (eds) Medicinal and aromatic plants of South America. Brazil. Springer, Dordretch, Germany, pp 129–138

Giuliano DA, Plos A (2014) *Baccharis* L. In: Zuloaga FO, Belgrano MJ and Antón AM (eds.): Flora vascular de la República Argentina: Dicotyledoneae-Asteraceae (Anthemideae a Gnaphalieae),

1st Edition, Instituto de Botánica Darwinion. Ed. Estudio Sigma SRL, Buenos Aires, Argentina 7(1):43–123

Guimarães NSS, Melo JC, Paiva JS, Bueno PCP, Berretta AA, Torquato RJ, Nantes IL, Rodrigues T (2012) *Baccharis dracunculifolia*, the main source of green propolis, exhibits potent antioxidant activity and prevents oxidative mitochondrial damage. Food Chem Toxicol 50(3–4):1091–1097

Heiden G, Iganci JRV, Macías L (2009) *Baccharis* sect. Caulopterae no Rio Grande do Sul. Brasil Rodriguésia 60(4):943–983

Heiden G, Antonelli A, Pirani JR (2019) A novel phylogenetic infrageneric classification of *Baccharis* (Asteraceae: Astereae), a highly diversified American genus. Taxon 68(5):1048-1081

Ibáñez F, Zoppolo R (2011) Assessment of allelopathic properties of *Baccharis dracunculifolia* DC. In laboratory and field conditions. Allelopathy J 28(1):77–86

Kloppel FC, Lemos M, Sousa JPB, Comunello E, Maistro EL, Bastos JK, De Andrade SF (2007) Nerolidol, an antiulcer constituent from the essential oil of *Baccharis dracunculifolia* DC. (Asteraceae). Z Naturforsch C J Biosci 62(7–8):537–542

Lage TCDA, Montanari RM, Fernandes SA, Monteiro CMDO, Senra TDOS, Zeringota V, Matos RDS, Daemon E (2015) Chemical composition and acaricidal activity of the essential oil of *Baccharis dracunculifolia* De Candole (1836) and its constituents nerolidol and limonene on larvae and engorged females of *Rhipicephalus microplus* (Acari: Ixodidae). Exp Parasitol 148:24–29

Leitão DPDS, Da Silva Filho AA, Polizello ACM, Bastos JK, Spadaro ACC (2004) Comparative evaluation on *in vitro* effects of Brazilian green propolis and *Baccharis dracunculifolia* extracts on cariogenic factors of *Streptococcus mutans*. Biol Pharm Bull 27(11):1834–1839

Lemos M, de Barros MP, Sousa JPB, da Silva Filho AA, Bastos JK, de Andrade SF (2007) *Baccharis dracunculifolia*, the main botanical source of Brazilian green propolis, displays antiulcer activity. J Pharm Pharmacol 59(4):603–608

Lombardo A (1964) Flora Arbórea y Arborescente del Uruguay, 2nd edn. Concejo Departamental de Montevideo, Montevideo, Uruguay, p 134

Marchiori JNC, De Oliveira AS (2007) Anatomia da madeira de três espécies brasileiras de *Baccharis* L. (Asteraceae-Astereae). Balduinia 10:1–10

Massignani JJ, Lemos M, Maistro EL, Schaphauser HP, Jorge RF, Sousa JPB, Bastos JK, De Andradre SF (2009) Antiulcerogenic activity of the essential oil of *Baccharis dracunculifolia* on different experimental models in rats. Phytother Res 23(10):1355–1360

Midorikawa K, Banskota AH, Tezuka Y, Matsushige K, Message D, Huertas AAG, Kadota S (2003) Buds of *Baccharis dracunculifolia*: potent source of biologically active caffeoylquinic acids and labdane-type diterpenes of Brazilian propolis. J Trad Med 20:187–194

Minteguiaga M, González A, Cassel E, Umpiérrez N, Fariña L, Dellacassa E (2018) Volatile constituents from *Baccharis* spp. L. (Asteraceae): chemical support for the conservation of threatened species in Uruguay. Chem Biodivers 15(5):e1800017

Minteguiaga M (2019) Fitoquímica de *Baccharis* spp. L. (*Asteraceae*): Metabolitos Secundarios, Semi-Síntesis y Bioactividad. Thesis. Universidad de la República, Montevideo, Uruguay. 577 pp. Available at: http://riquim.fq.edu.uy/items/show/5917. (Last access, December 2019)

Missima F, Da Silva Filho AA, Nunes GA, Bueno PCP, De Sousa JPB, Bastos JK, Sforcin JM (2007) Effect of *Baccharis dracunculifolia* DC. (Asteraceae) extracts and its isolated compounds on macrophage activation. J Pharm Pharmacol 59(3):463–468

Mors WB, Rizzini CT, Pereira NA (2000) Medicinal plants of Brazil. In: De Filipps RA (ed) Series: medicinal plants of the world, vol 6. Reference Publications Inc., Algonac, USA. 501 pp

Müller J (2006) Systematics of *Baccharis* (Compositae-Astereae) in Bolivia, including an overview of the genus. Syst Bot Monogr 76:1–341

Munari CC, Furtado RA, Santiago ML, Manhas SS, Bastos JK, Tavares DC (2014) Inhibitory effects of *Baccharis dracunculifolia* on 1,2-dimethylhidrazine-induced genotoxicity and preneoplastic lesions in rat colon. Eur J Cancer Prev 23(4):240–245

Nagatani Y, Warashina T, Noro T (2001) Studies on the constituents from the aerial part of *Baccharis dracunculifolia* DC. Chem Pharm Bull 49(11):1388–1394

Nagatani Y, Warashina T, Noro T (2002a) Studies on the constituents from the aerial parts of *Baccharis dracunculifolia* DC. II. Chem Pharm Bull 50(5):583–589

Nagatani Y, Warashina T, Noro T (2002b) Studies on the constituents from the aerial parts of *Baccharis dracunculifolia* DC. III. Natural Medicines 56(5):195–197

Negreiros D, Esteves D, Fernandes GW, Berbara RLL, Oki Y, Vichiato M, Chalub C (2014) Growth-survival tradeoff in the widespread tropical shrub *Baccharis dracunculifolia* (Asteraceae) in response to a nutrient gradient. Trop Ecol 55(2):167–176

Park JK, Paredes-Guzmán JF, Aguiar CL, Alencar SM, Fujiwara FY (2004) Chemical constituents in *Baccharis dracunculifolia* as the main botanical origin of southeastern Brazilian propolis. J Agric Food Chem 52(5):1100–1103

Paroul N, Rosa RLD, Piazza SP, Bertella T, Pouton BMS, Falcão L, Backes GT, Cansian RL (2016) Composição quimica e atividade antioxidante de *Baccharis trimera* Pers. e *Baccharis dracunculifolia* DC (Asteraceae). Perspectiva Erechim 40(151):55–64

Parreira NA, Magalhães LG, Morais DR, Caixeta SC, De Sousa JPB, Bastos JK, Cunha WR, Silva MLA, Nanayakkara NPD, Rodrigues V, Da Silva Filho AA (2010) Antiprotozoal, schistosomicidal and antimicrobial activities of the essential oil from the leaves of *Baccharis dracunculifolia*. Chem Biodivers 7(4):993–1001

Pedrotti C, Ribeiro RTS, Schwambach J (2019) Control of postharvest fungal rots in grapes through the use of *Baccharis trimera* and *Baccharis dracunculifolia* essential oils. Crop Prot 125:104912

Pereira CA, Da Costa ACBP, Machado AKS, Beltrame M Jr, Zöllner MSAC, Junqueira JC, Jorge AOC (2011) Enzymatic activity, sensitivity to antifungal drugs and *Baccharis dracunculifolia* essential oil by *Candida* strains isolated from the oral cavities of breast feeding infants and in their mothers' mouths and nipples. Mycopathologia 171(2):103–109

Pereira CA, Da Costa ACBP, Liporoni PCS, Rego MA, Jorge AOC (2016) Antibacterial activity of *Baccharis dracunculifolia* in planktonic cultures and biofilms of *Streptococcus mutans*. J Infect Public Heal 9(3):324–330

Queiroga CL, Fukai A, Marsaioli AJ (1990) Composition of the essential oil of Vassoura. J Braz Chem Soc 1(3):105–109

Queiroga CL, Bastos JK, Sousa JPB, Magalhães PM (2008) Comparison of the chemical composition of the essential oil and the water soluble oil of *Baccharis dracunculifolia* DC. (Asteraceae). J Essent Oil Res 20(2):111–114

Resende FA, Alves JM, Munari CC, Senedese JM, Sousa JPB, Bastos JK, Tavares DC (2007) Inhibition of doxorubicin-induced mutagenicity by *Baccharis dracunculifolia*. Mutat Res 634(1–2):112–118

Resende FA, Munari CC, Monteiro Neto MDAB, Tavares DC, Bastos JK, Da Silva Filho AA, Varanda EA (2012) Comparative studies of the (anti) mutagenicity of *Baccharis dracunculifolia* and Artepillin C by the bacterial reverse mutation test. Molecules 17:2335–2350

Rezende TP, Corrêa JODA, Aarestrup BJV, Aarestrup FM, De Sousa OV, Da Silva Filho AA (2014) Protective effects of *Baccharis dracunculifolia* leaves extract against carbon tetrachloride- and acetaminophen-induced hepatotoxicity in experimental animals. Molecules 19:9257–9272

Roberto MM, Matsumoto ST, Jamal CM, Malaspina O, Marin-Morales MA (2016) Evaluation of the genotoxicity/mutagenicity and antigenotoxicity/antimutagenicity induced by propolis and *Baccharis dracunculifolia*, by *in vitro* study with HTC cells. Toxicol In Vitro 33:9–15

Rodrigues VEG, De Carvalho DA (2001) Levantamento etnobotânico de plantas medicinais no domínio do Cerrado na região do Alto Rio Grande-Minas Gerais. Ciênc Agrotec 25(1):102–123

Rodrigues DM, De Souza MC, Arruda C, Pereira RAS, Bastos JK (2020) The role of *Baccharis dracunculifolia* and its chemical profile on green Propolis production by *Apis melífera*. J Chem Ecol 46:150–162

Salazar GJT, Sousa JP, Lima CNF, Lemos ICS, Da Silva ARP, Freitas TS, De Freitas TS, Coutinho HDM, Da Silva LE, Do Amaral W, Deschamps C (2018) Phytochemical characterization of the

Baccharis dracunculifolia DC (Asteraceae) essential oil and antibacterial activity evaluation. Ind Crop Prod 122:591–595

Schenkel EP, Rücker G, Manns D, Falkenberg MB, Matzenbacher NI, Sobral M, Mentz LA, Bordignon SAL, Heinzmann BM (2002) Screening of Brazilian plants for the presence of peroxides. Rev Bras Ciênc Farm 38(2):192–196

Schneider AA (2009) Estudo taxonômico de *Baccharis* L. sect. *Caulopterae* DC. (Asteraceae: Astereae) no Brasil. Thesis. Universidade Federal do Rio Grande do Sul, Porto Alegre. 197 pp

Sforcin JM, De Sousa JPB, Da Silva Filho AA, Bastos JK, Búfalo MC, Tonuci LRDS (2012). *Baccharis dracunculifolia*. Umas das principais fontes vegetais da própolis brasileira. Editora UNESP. São Paulo, Brazil. 100 pp. Available at: http://editoraunesp.com.br/catalogo/9788539303762,baccharis-dracunculifolia. (Last access, December 2019)

Shimizu K, Das SK, Hashimoto T, Sowa Y, Yoshida T, Sakai T, Matsuura Y, Kanazawa K (2005) Artepillin C in Brazilian propolis induces G0/G1 arrest via stimulation of Cip1/p21 expression in human colon cancer cells. Molec Carcinog 44(4):293–299

The Plant List (2019) Version 1.1 Published on the Internet: http://www.theplantlist.org. (Last access, December 2019)

Veiga RS, De Mendonça S, Mendes PB, Paulino N, Mimica MJ, Lagareiro Netto AA, Lira IS, López BGC, Negrão V, Marcucci MC (2017) Artepillin C and phenolic compounds responsible for antimicrobial and antioxidant activity of green propolis and *Baccharis dracunculifolia* DC. J Appl Microbiol 122(4):911–920

Weyerstahl P, Christiansen C, Marschall H (1996) Constituents of Brazilian Vassoura oil. Flavour Fragr J 11(1):15–23

Baccharis tola Phil., *B. boliviensis* (Wedd.) Cabrera

María Inés Isla, Mariana Leal, Silvana L. Carrizo, and Iris C. Zampini

Baccharis tola ex *Baccharis incarum* (a. and b.) and *Baccharis boliviensis* (c. and d.). Photos: I Catiana Zampini, during the sample collection. Antofagasta de la Sierra, Catamarca, Argentina

M. I. Isla (✉) · M. Leal · S. L. Carrizo · I. C. Zampini
Instituto de Bioprospección y Fisiología Vegetal (INBIOFIV, CONICET- Facultad de Ciencias Naturales e Instituto Miguel Lillo-Universidad Nacional de Tucumán), San Miguel de Tucumán, Argentina
e-mail: misla@csnat.unt.edu.ar; zampini@csnat.unt.edu.ar

Á. Máthé, A. Bandoni (eds.), *Medicinal and Aromatic Plants of South America Vol. 2*, Medicinal and Aromatic Plants of the World 7,
https://doi.org/10.1007/978-3-030-62818-5_6

Abstract The genus *Baccharis,* with approximately 500 species, is one of the largest genera of the family Asteraceae. Species of this genus are distributed mainly in the warm temperate and tropical regions of Brazil, Argentina, Colombia, Chile and Mexico. In Argentina, 96 species have been described. *Baccharis tola* and *Baccharis boliviensis* are used in traditional medicine, in Argentina. They are popularly known as "tola", "tolilla", "lejía" or "romerillo". The occurrence and identification of flavonoids and terpenoids have been reported from these species. The scientific validation of its medicinal properties as antioxidants, anti-inflammatory and antibiotics/antifungal agents has been done. Some phytotherapeutic formulations for topical applications have been developed with the ethanolic extract of *B. tola.* Both *B. boliviensis* and *B. tola* seem to be promising species for regional development, in these countries, although farther studies are needed in order to elaborate the technology for its sustainable domestication and subsequent cultivation.

Keywords *Baccharis boliviensis* · *Baccharis tola* · Argentine medicinal plant · Anti-inflammatory · Antibiotic · Antioxidant · Antifungal

1 Introduction

The present chapter shows the advances aimed at exploring the medicinal potential of two Argentinian *Baccharis* species. The ultimate aim is to promote the conservation and sustainable use of this valuable natural resource.

2 Taxonomic Characteristics

The genus *Baccharis,* with approximately 500 species, is one of the largest genera of the Family Asteraceae. Species of this genus are distributed mainly in the warm temperate and tropical regions of Brazil, Argentina, Colombia, Chile and Mexico. *Baccharis* spp. are abundant in Argentina, where 96 species have been described (Abad and Bermejo 2007). In this chapter, *Baccharis tola* (ex *B. incarum*) and *Baccharis boliviensis* are included.

***Baccharis tola* Phil. var. *incarum* (Wedd.) Joch. Müller** has as synonyms: *B. incarum* (Wedd.) Perkins (The Plant List, 2020). This species is known by the Aymara-Quechua names of "tola", "t'ula", "ñakat'ula", "ñakat'ola", "ñaka", "femala ñaka", "pacha thaya", "waka t'ula"; also, with the Spanish names of "baila buena", "tola", "tola macho", "tola hembra", "sacha tola", "tola lejía", "lejía", "le,a lejia" (Beck, 1985; De Lucca 1987; De Lucca and Zalles 1992; Galafassi, 1997; Cazés Camarero 2000; Villagrán et al. 2003; Borgnia et al. 2006). The denomination of "lejía" have their etymology in the Latin *aqua lixiva*: water in which alkalis or their carbonates have dissolved. According to Murillo (1861), "several species of the genus *Baccharis*, contain in their ashes a large amount of potash". For these reasons, *B. tola* ashes serve to peel quinoa and corn.

***Baccharis boliviensis* (Wedd.) Cabrera**. Synonyms: *Psila boliviensis* (Wedd.) Cabrera, *Heterothalamus boliviensis* Wedd. and *Baccharis boliviensis* var. *latifolia* (R.E.Fr.) Cabrera, among others (The Plant List, 2020). *B. boliviensis* is known by the Aymara names of "kulkuma", "kulkut'ula" and the Quechua names of "peskotola" or "tola", also with Spanish names "romero","manzanilla", "tola", "tola limón", "tolilla", "tolita", "tola chica","tola de pájaro", "chijua", "chisqua", "romerillo", "monte alpaca", "monte de paloma". The name "tola de pájaro" or "monte paloma" refers to the birds nesting in this bush, due to its leafy and low foliage. In Quechua, *peskko* means bird, so it is called "peskotola" because when this species blooms it is filled with birds that eat its flowers. The name "tola limón" owes its etymology to the color provided by this dye species, green and yellow. The name "manzanilla" is due to the similarity of its flowers with those of orchard chamomile (Lira 1945; Haber 1992; Cazés Camarero 2000; Villagrán et al. 2003; Borgnia et al. 2006).

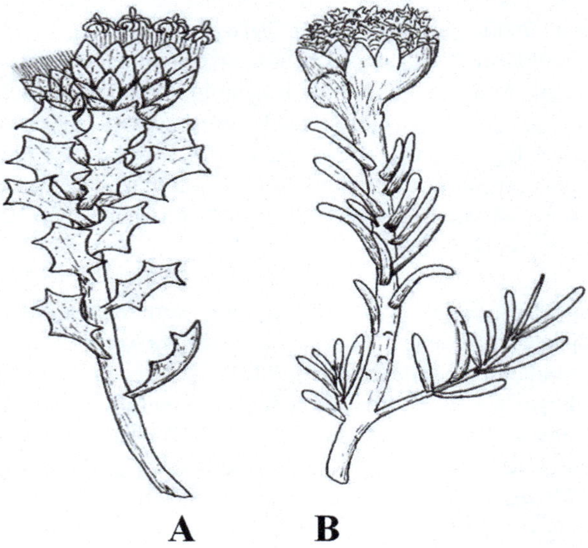

Fig. 1 Parts employed in popular medicine of (**a**) *B. tola*; (**b**) *B. boliviensis*. Pictures: M Leal

3 Crude Drug Used

Plant parts used for crude drug production are: leaves, branches, and flowers (Fig. 1). In general, the crude drug is the dried biomass that is processed into powder. In very few cases the fresh drug is used.

4 Major Chemical Constituents and Bioactive Compounds

Baccharis tola Previous phytochemical studies on the leaves and top parts (aerial parts) of *B. tola* described the isolation and structural elucidation of diterpenes, triterpenes, flavonoids and other phenolic compounds. Flavones such as 5,4′-dihydroxy-6,7,8,3′-tetramethoxyflavone; 5,4′-dihydroxy-3′,3,6,7,8-pentamethoxy flavone; 5,7,4′-trihydroxy-6,8,3′-trimethoxyflavone; 5,7,4′-trihydroxy-3′,3,6,8-tetramethoxyflavone; 3′,4′,5,7-tetrahydroxyflavone; 3′,4′,5,7-tetrahydroxy-3,6-dimethoxyflavone; 3′,4′,5,7-tetrahydroxy-3,6,8-trimethoxyflavone; 3,3′,4′,5,5′,7-hexahydroxyflavone (myricitin) and derivatives compounds have been identified (Faini et al. 1982; Zampini et al. 2006, 2007, 2009a; Herrera et al. 2007; Nuño et al. 2012; Carrizo et al. 2020). Phenolic acids such as chlorogenic, ferulic, quinic, dicaffeoyl quinic, and 1-caffeoyl-5-feruloylquinic were identified (Zampini et al. 2007; Nuño et al. 2012; Simirgiotis et al. 2016; Carrizo et al. 2020). Other simple phenolic compounds such as scopoletin; 4-hydroxy-3-methoxyacetophenone were also identified (San Martín et al. 1980). Terpenoid compounds, such as erythroxylol-A; oleanolic acid; erythroxylol-A oxide; *ent*-beyer-14-en-18-ol; 19-hydroxy-13-epimanoyl oxide, bacchalineol; barticulidiol, bincatriol, hawtriwaic acid, solidagoiol A and derivatives compounds, and *ent*-clerodane derivatives compounds were identified (San Martín et al. 1980, 1983, 1985; Givovich et al. 1986; Simirgiotis et al. 2016; Carrizo et al. 2020).

Baccharis boliviensis Terpenoids and flavonoids were mainly isolated from aerial parts of this species. Labdanes, clerodanes (Gonzaga Verdi et al. 2005) and *ent*-clerodanes were identified (Zdero et al. 1986,1989; Carrizo et al. 2020). Flavones, such as 5,4′-dihydroxy-6,7,8-trimethoxyflavone (xanthomicrol), 5,4′-dihydroxy-6,7,8,3′-tetramethoxyflavone, 5,4′-dihydroxy-3,3′,5′,7-tetramethoxyflavone, 5,3′-dihydroxy-6,7,8,4′-tetramethoxyfllavone, 5,4′-dihydroxy-3,6,7,8,3′-pentamethoxyfllavone, 5,3′-dihydroxy-3,6,7,8,4′-pentamethoxyflavone, 7,4′-dihydroxy-5,6,8-trimethoxyflavone, 4′,5,7-trihydroxy-3′,3,6,8-tetramethoxyflavone, 5,7,3-trihydroxi-6,8,3′,4′-tetramethoxyflavone, 3′,4′,5,7-tetrahydroxyflavone, 3′,4′,5,7-tetrahydroxy-3,6-dimethoxyflavone, 3′,4′,5,7-tetrahydroxy-3,6,8-trimethoxyflavone, 3′,4′,3,5,7-pentahydroxyflavone, 3′,4′,5,7-tetrahydroxy-6-methoxyflavone-3-rhamnoside, 5,6,7,3′,4′-pentahydroxyflavone, 3,3′,4′,5,5′,7-hexahydroxyflavone (myricitin) and derivatives compounds; flavanones, such as 5,7,3′,4′-tetrahydroxyflavanone (eriodictiol), 7-*O*-glucoside eriodictiol and 3′,4′,5,7- tetrahydroxy-6-methoxyflavanone were identified (Morales et al. 1990; Cazón et al. 2002; Gonzaga Verdi et al. 2005; Zampini et al. 2007, 2009a; Calle et al. 2012; Carrizo et al. 2020). Chlorogenic, isochlorogenic, ferulic, *O*-dicaffeoylquinic, 1-caffeoyl-5-feruloylquinic acid and quinic acids were isolated and identified by HPLC-MS and UHPLC-OT-MS from the methanolic and ethanolic extract of *B. boliviensis* aerial parts collected in Antofagasta de la Sierra, Catamarca, Argentina (Cazón et al. 2000; Carrizo et al. 2020).

5 Morphological Description

Baccharis tola Intricate-resinous bush, no more than half a meter high. It has rounded branches, very glandulous and densely leafy to the apex. The leaves are alternate, sessile, oblanceolate-spatulated, obtuse at the apex (rarely acute) and attenuated at the bottom, whole or with one or two triangular teeth on each side, dotted-glandular and, generally, covered with resin droplets, 6–12 mm long by 2–4 mm wide. At the ends of very short and sessile twigs, it has numerous solitary chapters. The feminine chapters have bell-shaped engagement of 5–6 mm high by 4 mm in diameter, filarial in 4–5 series, numerous flowers with filiform corolla. The masculine chapters have the same involvement as the feminine ones, numerous flowers with corolla pentasecta and style of very short branches, not separated, white crested and widened pappus at the apex of the bristles. The fruit is an acyloid, glabrous achene, 2.2 mm long; white and deciduous pappus (Cabrera, 1978).

Baccharis boliviensis Shrubs 0.3–1 m high, erect, bunny bush 30–100 cm tall, with thin twigs, sides, glabrous, glandulous-dotted, densely leafy to the apex. The leaves are alternate, sessile, linear lanceolate or elliptical, acute or semi-blunt, whole, glabrous and dotted-glandular on both sides. It presents numerous chapters, arranged in paucicephal terminal tops. It presents feminine chapters with bell-shaped involvement of about 3 mm high by 4 mm in diameter, filarial in 2–3 series, the oblong obtuse external with laciniated hyaline margin, resinous on the back, the interiors, acute lanceolate. Hemispheric receptacle, covered with lanceolate, hyaline, as long as flowers, deciduous, numerous flowers, female with briefly linked corolla, 1 mm long tube and 0.5 mm ligule, white pappus, longer than the corolla, achenes sericeous-pubescent oblanceolate, 1.5 mm long. It also presents male chapters with bell-shaped engagement of 2–2.5 mm in height by 3 mm in diameter, filarial in 2 series, oblong, obtuse, hyaline in the margin and resinous in the back, flowers with deeply pentalobate corolla, short style branches and little separated (Cabrera, 1978).

6 Geographical Distribution

Baccharis tola is geographically distributed in the high mountains and highlands of Southern Peru and Bolivia, Northern Chile and Northwestern Argentina from Provinces of Jujuy to La Rioja and San Juan (Cabrera, 1978). In Argentine grows between 1900 and 4800 m.a.s.l.

Baccharis boliviensis is distributed geographically in Southern Peru and Bolivia, Northern Chile and Northwestern Argentina (Jujuy, Salta, Tucumán and Catamarca) at 1500–4700 m.a.s.l.

Both species are characteristic of the phytogeographic province of Puna and high streams of Prepuna (Cabrera, 1978; Cuello 2006; Carilla et al. 2018). At present, there is no information about the cultivation and domestication of *B. boliviensis* or *B. tola.*

7 Ecological Requirements

B. tola and *B. boliviensis* grow under conditions of high ultraviolet light, low temperature, high salinity and low atmospheric pressure (Cabrera, 1978).

8 Traditional Use (Part(S) Used) and Common Knowledge

B. tola In Argentina, Chile and Bolivia is widely used as a medicine. Roasted and boiled leaves are used as an infusion for fever, to combat cough, bronchitis, pneumonia, flu and colds, digestive disorders and "empachos" (indigestion), purgative, sweaty, stomatic. It is prepared as a poultice for stomatic pain. The decoction is used against dysentery. The infusion and decoction of the aerial parts of *B. tola* are used in traditional medicine for the treatment of ulcers, burns and wounds, to restore blood circulation and reduce inflammatory processes, as antiseptic and anti-inflammatory and to relieve muscle and bone pain. The leaves and stems are macerated in ethanol for seven days and the solution is rubbed to relieve rheumatism and inflammation. It is appreciated for the resin it contains, in cases of bruises, wounds and to consolidate dislocations and cracks, since it is very sticky and is used in the form of strips or patches (Castro et al. 1982; De Lucca and Zalles, 1992; Wickens, 1993; Abad and Bermejo, 2007). According to Galafassi (1997), it has veterinary use, it is medicinal for cattle, it is used to smoke cattle, donkeys and forage it when they are sick. *B. tola* is also used in human food: the resin is consumed as sweet in the winter (Aldunate et al. 1981); lumps on the branches (gills, sometimes called fruits) serve as a lemon substitute. In its roots live a parasitic plant (*Ombrophytum subterraneum* (Asplund) Hansen, Balanophoraceae) that forms a radical thickening or "sichas" that are edible (Muñoz Pizarro, 1973; Aldunate et al. 1981; Wickens,

1993). It is also consumed as fodder by camelids, goats and sheep (Alzirreca and Cardozo, 1991; Haber, 1992; Wickens, 1993). On the other hand, *B. tola* is used to dye wool and the whole plant is used as fuel, it is a hardwood and is considered the best wood in the mountains (Beck, 1985; Vidaurre et al. 2006).

Baccharis boliviensis The aerial part is used as a medicinal product in Argentina, Chile, Bolivia and Peru. The infusion is used as an antispasmodic, febrifuge and helps liver function. The resin is used in patches, plasters or poultices to heal fractures and, in general, to relieve muscle aches. It is used as an anti-inflammatory, antipyretic, and antiseptic, to heal wounds, burning and ulcers, for rheumatism, stomach pain, gastric protector, liver and prostate illness (Abad and Bermejo, 2007). It is also a good fuel and is preferred as firewood since they heat clay ovens better. It constitutes a forage resource for grazing llamas and sheep (Villagrán et al. 2003; Vidaurre et al. 2006).

9 Modern Medicine Based on its Traditional Medicine Uses

Several studies using different methods have demonstrated the health benefits attributable to these *Baccharis* species. These reports also describe the development and characterization of phytopharmaceutical products.

Antimicrobial Activity The ethanolic extracts (ethanol 60° and 80°) of *Baccharis boliviensis* and *B. tola* inhibited the growth of one or more of the following antibiotic resistant strains: *Staphylococcus aureus, Enterococcus faecalis, Escherichia coli, Klebsiella pneumoniae, Proteus mirabilis, Enterobacter cloacae, Morganella morganii,* and *Pseudomonas aeruginosa*. Ethanol extracts (tinctures) of aerial parts of these species showed the highest levels of antibacterial activity on methicillin, oxacillin and gentamicin resistant *Staphylococcus aureus* with MIC values from 20 to 150μg/ml (Feresin et al. 2000; Eraso et al. 2002; Mahady 2005; Zampini et al. 2006, 2007, 2009a, b). *Baccharis boliviensis* were more active than *B. tola* on *Enterococcus faecalis*. The ethanolic extracts exhibited stronger activity and broader spectrum of action than aqueous extracts. The extracts show a bactericidal effect. Moderate anti-*Trypanosoma cruzi* activity for 5,4′-dihydroxy-3,6,7,8,3′-pentamethoxyflavone isolated from *Baccharis* species was demonstrated (González et al. 1990). *B. tola* and *B. boliviensis* were actives on *Candida* species and on dermatophytes fungi species, principally on *Microsporum canis* and *Trichophyton rubrum* (CIM < 100μg/ml), (Carrizo et al. 2010, 2020; Carrizo 2011). These results are relevant considering that *T.rubrum* is the etiological agent of 80-93% of all clinical infections produced by dermathophytes.

The presence of antibacterial activity and antifungal activity in *B. tola* and *B. boliviensis* extracts against bacteria and fungi resistant to commercial products give support to their traditional use and would seem promising for the treatment of infections in humans and animals.

Antioxidant Capacity The antioxidant capacity of *Baccharis boliviensis* and *Baccharis tola* aerial parts alcoholic extracts collected at 3800 m.a.s.l. in the Argentine Puna were studied (Herrera et al. 2007; Zampini et al. 2006, 2007, 2008, 2009a). Extracts of both *Baccharis* species exerted high hydrogen donating ability in the presence of DPPH radicals. The TEAC values of 15,000 and 7833μmol Trolox equivalent/100 g dry weight to *B. boliviensis* and *B. tola,* respectively, were found (Zampini et al. 2008). The tinctures were active as scavengers of reactive oxygen species such as superoxide anion and hydroxyl radical (Zampini et al. 2008). In the non-enzymatic system of O_2^- generation, extracts were scavengers of superoxide radicals with IC_{50} values between 10 and 20μg of total phenolic compounds/ml. *B. boliviensis* and *B. tola* extracts were also effective hydroxyl radical scavengers with IC_{50} values of 6 and 13.5, respectively. *B. tola* was the best H_2O_2 scavenger with IC_{25} values of 30μg/ml (Zampini et al. 2008).

Numerous studies have linked xanthine oxidase activity to inflammatory process. This enzyme produces superoxide anion radical and uric acid from xanthine and elicits the upregulation of inflammatory markers (Romagnoli et al. 2010). The xanthine oxidase inhibitory effect of ethanolic extracts was described by Zampini et al. (2008).

Aqueous extracts exhibit also antioxidant activity on ABTS cation radical, on the production of reactive oxygen species (ROS) and inhibit the expression of hsp72 in human neutrophils stimulated by different agents (Pérez-García et al. 2001; Zampini et al. 2009a). The oxidation of red blood cells (RBC) by molecular oxygen with the azo-compound AAPH as a free radical initiator was inhibited by *B. tola* infusion (Rojo et al. 2009). The ethanolic and aqueous extracts from aerial parts of *B. tola* and *B. boliviensis* could be exploited as natural medicine to reduce oxidative stress and for the prevention and treatment of hyperuricemia.

Anti-Inflammatory Activity Inflammatory and pathogenic conditions activate secretory phospholipase A2 (sPLA2), cyclooxygenase (COX) and lipoxygenase (LOX), two key enzymes in the synthesis of prostanoids and eicosanoids from polyunsaturated fatty acids, which are involved in various inflammatory process. Nitric oxide (NO) is one of the inflammatory mediators that is synthesized by inducible nitric oxide synthase (iNOS) in macrophages and is induced by different inflammatory stimuli such as bacterial endotoxic and inflammatory cytokines (Achike and Kwan, 2003). Although NO is required in immunological defense mechanisms and to maintain the dilation of blood vessels, high concentrations of NO can cause oxidative damage. Data obtained suggest that the inhibition of NO and prostaglandin (PG) production by *B. tola* extracts in macrophages occurs at the level of enzyme expression and/or enzyme activity. Hydroalcoholic extracts from *B. tola* and *B. boliviensis* were showed also inhibition of COX-2 activity. *B. tola* extract also decreases the COX-2 protein expression (43%). The induction of RAW 264.7 cells into an inflammatory state by treatment with lipo-polysaccharides (LPS) caused a significant increase in NO production. Pretreatment of cells with the *B. tola* and *B. boliviensis* extracts inhibited NO overproduction in a concentration-dependent manner with IC_{50} values of 167 and 195μg/ml, respectively. The iNOS protein expression was significantly decreased by *Baccharis tola* (58%). *B. boliviensis* and *B. tola* were inhibitors of sPLA2 activity (IC_{50} around 200μg/ml) and LOX activity (IC_{50} around 300μg/ml) (Alberto et al. 2009, Torres Carro et al. 2015, 2017, 2019). Therefore, these extracts might be an important therapeutic target to treat various inflammatory diseases or for cancer chemoprevention. It was also demonstrated that the dichloromethane extracts of *B. tola* have an anti-inflammatory effect with anti-edema action (44.54% inhibition) in the carrageenan-foot edema test in rats (Pérez et al. 1995). Intraperitoneal administration of extracts was performed 30 min before subcutaneous injection of 0.1 ml of 1% v/v i-carrageenan in normal saline (Pérez et al. 1995).

Phytotherapeutic Formulations A standardized extract of *Baccharis tola* with antimicrobial, antioxidant and anti-inflammatory activities were included in a topical phytopharmaceutical formulation (Hydrogel/Carbopol W934) (Nuño et al. 2012). The formulation showed antimicrobial and antioxidant activities *in vitro* assays. The hydrogel showed microbiological, chemical, physical and functional stability during storage at room temperature. Studies that measure drug release as a determination of bioavailability were also carried out. The results demonstrated the release of two bioactive compounds (chlorogenic acid and 4′,5-dihydroxy-3′,3,6,7,8-pentamethoxyflavone) from the phytotherapeutic preparation. In consequence, the preparation applied topically could be used to treat skin and soft tissue infection produced by methicillin-resistant *Staphylococcus aureus* (MRSA) or *Enterococcus faecalis* strains and opens new opportunities for the use of active natural ingredients in the cosmeceutical field as anti acne and antioxidant agent.

Recently, *B. tola* and *B. boliviensis* were cited to have medicinal properties that make these species economically attractive in Argentina by its medicinal properties (Cantero et al. 2019).

10 Conclusions

Baccharis tola and *B. boliviensis* that grow in arid regions of Argentina, are species popularly used as medicinal plants. They are reported to show scientifically validated antibacterial, antifungal, antioxidant and anti-inflammatory activity. Both species are promising for the development of pharmaceutical and cosmetic products. The sustainable use of *Baccharis* species could be favorable for the regional economies. Therefore, it is important to promote their domestication (cultivation) and the development of value chains for their production.

Acknowledgments The authors acknowledge the cooperation of the inhabitants of the studied areas and the financial support from the SCAIT-Universidad Nacional de Tucumán (PIUNT2018 G-637), the Agencia Nacional de Promoción Científica y Técnica (ANPCyT, PICT 2014- 3136 and PICT 2017- 4436) and the Consejo Nacional de Investigaciones Científicas y Técnicas (CONICET, PUE 2018- 0011).

References

Abad MJ, Bermejo P (2007) *Baccharis* (Compositae): a review update. ARKIVOC 7:76–96

Achike FI, Kwan C (2003) Nitric oxide, human disease and the herbal products that affect the nitric oxide signaling pathway. Clin Exp Pharmacol Physiol 30:605–615

Alberto MR, Zampini IC, Isla MI (2009) Cyclooxygenase enzyme inhibitory activity of standardized hydroalcoholic extracts of four Asteraceae species from the argentine Puna. Brazilian J Med Biol Res 42(9):776–780

Aldunate C, Armesto J, Castro V, Villagrán C (1981) Estudio Etnobotánico en una Comunidad precordillerana de Antofagasta: Toconce. Bol Mus Nac Hist Nat Chile 38:183–223

Alzirreca H, Cardozo A (1991) Valor de los alimentos para la ganadería andina. Serie técnica IBTA-CRSP. Convenio Universidad de California, Davis -IBTA La Paz, Bolivia. 35 pp

Beck S (1985) Flórula ecológica de Bolivia. Ecol Boliv 6:1–41

Borgnia M, Maggi A, Arriaga M, Aued B, Vilá B, Cassini MH (2006) Caracterización de la vegetación en la Reserva de Biosfera Laguna Blanca (Catamarca, Argentina). Ecol Austral 16:29–45

Cabrera AL (1978) Flora de la Provincia de Jujuy. Colección Científica del INTA. Buenos Aires, Argentina. Parte X – Compositae, 231 pp

Calle A, Yupanqui J, Flores Y, Almanza G (2012) Flavonoides de *Baccharis boliviensis*. Rev Boliv Quím 20:155–160

Cantero JC, Nuñez CO, Bernardello G, Amuchástegui A, Mulko J, Brandolin P, Palchetti MV, Iparraguirre J, Virginil N, Ariza Espinar L (2019) Las plantas de importancia económica en Argentina. First Edition. Ed. UniRío Editora. Universidad Nacional de Río Cuarto, Córdoba, Argentina, 937 pp. https://botanicaargentina.org.ar/wp-content/uploads/2019/06/978-987-688-332-0-baja-resoluci%C3%B3n.pdf. (Last access, January 2020)

Carilla J, Grau A, Cuello S (2018) Vegetación de la Puna argentina. Serie Conservación de la Naturaleza 24: La Puna argentina: naturaleza y cultura 143–156

Carrizo L (2011) Evaluación "*in vitro*" de la actividad antifúngica de extractos provenientes de especies del género *Baccharis*. Thesis. Universidad Nacional de Tucumán. Tucumán, Argentina. 116 pp

Carrizo SL, Zampini C, Isla MI, van Gelderen A (2010) Evaluation "*in vitro*" of antifungal activity of *Baccharis* extracts on yeasts. BIOCELL 34(2)-ISSN 0327–9545 (print). ISSN 1667–5746 (electronic)

Carrizo SL, Zampini IC, Sayago JE, Simirgiotis MJ, Bórquez J, Cuello AS, Isla MA (2020) Antifungal activity of phytotherapeutic preparation of *Baccharis* species from argentine Puna against clinically relevant fungi. J Ethnopharmacol 251(6):112553

Castro M, Villagrán C, Arroyo M (1982) Estudio Etnobotánico en la precordillera y altiplano de Los Andes del norte de Chile (18–19 °S). In: Veloso A, Bustos E (ed) El Hombre y los Ecosistemas de Montañas. Tomo II. Ed. UNESCO, Santiago, Chile. pp 33–199

Cazés Camarero PL (2000) Nombres vulgares de Asteraceae del herbario del Museo "Juan A. Domínguez" (Facultad de Farmacia y Bioquímica. UBA). Dominguezia 16(1):32

Cazón A, de Viana M, Gianello J (2000) Identificación de un compuesto alelopático de *Baccharis boliviensis* (Asteraceae) y su efecto en la germinación de *Trichocereus pasacana* (Cactaceae). Rev Biol Trop 48(1):47–51

Cazón A, de Viana M, Gianello JC (2002). Efecto fitotóxico de aleloquímicos de *Baccharis boliviensis*. Ecol Austral 12:73–78

Cuello S (2006) Guía ilustrada de la Flora de Antofagasta de la Sierra, Catamarca. Ed. Facultad de Ciencias Naturales-Universidad Nacional de Tucumán, Tucumán, Argentina. 165 pp

De Lucca M (1987) Diccionario Práctico Aymara-Español, Español-Aymara. Ed. Los Amigos del Libro, La Paz, Bolivia, 288 pp

De Lucca M, Zalles J (1992) Flora medicinal boliviana. Diccionario enciclopédico. Ed. Los Amigos del Libro, La Paz, Bolivia, 498 pp

Erazo S, Negrete R, Zaldívar M, Backhouse N, Delporte C, Silva I, Belmonte E, López-Pérez JL, San Feliciano A (2002) Methyl psilalate: a new antimicrobial metabolite from *Psila boliviensis*. Planta Med 68:66–67

Faini FA, Castillo M, Torres MR (1982) Flavonoids of *Baccharis incarum*. J Nat Prod 45:501–502

Feresin GE, Tapia A, Bustos DA (2000) Antibacterial activity of some medicinal plants from San Juan, Argentina. Fitoterapia 71:429–432

Galafassi G (1997) Actividad recolectora en pastores de altura (Laguna Blanca, Catamarca). Sup Antropol 32(1–2):397–426

Givovich A, San Martín A, Castillo M (1986) Neo-clerodane diterpenoids from *Baccharis incarum*. Phytochemistry 25:2829–2831

Gonzaga Verdi L, Costa Brighente IM, Pizzolatti MG (2005) Gênero *Baccharis* (Asteraceae): Aspectos químicos, econômicos e biológicos. Quim Nova 28(1):85–94

González J, Sagua H, Araya J, Loyola A, Morales G, Pereira J, Estrada M (1990) *In vitro* activity of natural products against the trypomastigote form of *Trypanosoma cruzi*. Phytother Res 4:1–5

Haber A (1992) Pastores y pasturas. Recursos forrajeros en Antofagasta de la Sierra (Catamarca), en relación a la ocupación formativa. Shincal 2:15–23

Herrera SY, Zampini IC, D'Almeida R, Boguetti H, Isla MI (2007) Supercritical fluid extraction of antioxidants compounds from *Baccharis incarum*: comparison with conventional extraction methods. Bol Latinoam Caribe Plant Med Aromat 6(5):250–252

Lira J (1945) Diccionario Kechuwa-Español. Publicación 369, Instituto de Historia, Lingüística y Folklore, Ed. Universidad Nacional de Tucumán, Tucumán, Argentina. pp 748

Mahady G (2005) Medicinal Plants for the Prevention and Treatment of Bacterial Infections. Curr Pharm Design 11(19):2405–2427

Morales G, Mancilla A, Gallardo O, Trujillo R, Loyola L, Breton JL (1990) Diterpenoids and flavonoids from *Baccharis boliviensis*. Bol Soc Chil Quim 35:257–263

Muñoz Pizarro C (1973) Chile: Plantas en Extinción. Ed. Universitaria. Santiago, Chile. 344 pp

Murillo A (1861) Memoria sobre las plantas medicinales de Chile y uso que de ellas se hace en el país del país. Ed. Imprenta del Ferrocarril, Santiago, Chile. pp 583–645

Nuño G, Zampini I, Ordóñez R, Alberto M, Arias M, Isla MI (2012) Antioxidant/antibacterial activities of a topical phytopharmaceutical formulation containing a standardized extract of *Baccharis incarum*, an extremophile plant species from argentine Puna. Phytother Res 26:1759–1767

Pérez García F, Marín E, Adzet T, Cañigueral S (2001) Activity of plant extracts on the respiratory burst and the stress protein synthesis. Phytomed 8:31–38

Pérez F, Marin E, Adzet T (1995) The antiinflammatory effect of several composite from South America. Extracts in rats. Phytother Res 9:145–146

Rojo L, Benítez J, López J, Rojas M, Díaz P, Ordóñez J, Pastene E (2009) Antioxidant capacity and polyphenolic content of twelve traditionally used herbal medicinal infusions from the south American Andes. Bol Latinoam Caribe Plant Med Aromát 8(6):498–508

Romagnoli M, Gómez-Cabrera MC, Perrelli MG, Biasi F, Pallardo FV, Sastre J, Poli G, Viña J (2010) Xanthine oxidase-induced oxidative stress causes activation of NF-κB and inflammation in the liver of type I diabetic rats. Free Rad Biol Med 49:171–177

San Martín AS, Rovirosa J, Becker R, Castillo M (1980) Diterpenoids from *Baccharis tola*. Phytochemistry 19:1985–1987

San Martín AS, Rovirosa J, Castillo M (1983) Diterpenoids from *Baccharis tola*. Phytochemistry 22:1461–1463

San Martín A, Givovich A, Castillo M (1985) Neo-clerodane diterpenoids from *Baccharis incarum*. Phytochemistry 25:264–266

Simirgiotis MJ, Quispe C, Bórquez J, Mocan A, Sepúlveda B (2016) High resolution metabolite fingerprinting of the resin of *Baccharis tola* Phil. From the Atacama Desert and its antioxidant capacities. Ind Crop Prod 94:368–375

The Plant List (2020) Version 1.1. Published on the Internet: www.theplantlist.org. (Last access, January 2020)

Torres Carro R, Isla MI, Ríos JL, Giner RM, Alberto MR (2015) Anti-inflammatory properties of hydroalcoholic extracts of argentine Puna plants. Food Res Int 67:230–237

Torres Carro R, Isla MI, Thomas Valdes S, Jiménez Aspee F, Schmeda-Hirschmann G, Alberto MR (2017) Inhibition of pro-inflammatory enzymes by medicinal plants from the Argentinean highlands (Puna). J Ethnopharmacol 205:57–68

Torres-Carro R, Yamile A, Rojas-Márquez JD, Stempin CC, Zampini IC, Isla MI, Alberto MR (2019) Argentinean Puna plants with in vitro antioxidant and anti-Inflammatory activities as a potential nutraceutical. J Food Sci 84(12):3352–3363

Vidaurre PJ, Paniagua N, Moraes RM (2006). Etnobotánica en los Andes de Bolivia. In: Moraes M, Øllgaard B, Kvis LP, Borchsenius F, Balslev H: Botánica Económica de los Andes Centrales. Universidad Mayor de San Andrés, La Paz, Bolivia. pp 224–238

Villagrán C, Romo M, Castro V (2003) Etnobotánica del sur de los Andes de la primera región de Chile: un enlace entre las culturas altiplánicas y las de quebradas altas del Loa superior. Chungara, Rev Antrop Chil 35(1):73–124

Wickens G (1993) Vegetation and Ethnobotany of the Atacama Desert and adjacent Andes in northern Chile. Opera Bot 121:291–307

Zampini IC, Vera N, Bardón A, Isla MI (2006) Antioxidant potential of *Baccharis incarum*: isolation of bioactive metabolites. Mol Med Chem 11:56–57

Zampini IC, Vera N, Bardón A, Isla MI (2007) Antioxidant and antimicrobial flavonoid isolated from *Baccharis incarum*. Bol Latinoam Caribe Plant Med Aromát 6(5):292–294

Zampini IC, Meson Gana J, Ordóñez RM, Sayago JE, Nieva Moreno MI, Isla MI (2008) Antioxidant and xanthine oxidase inhibitory activities of plant species from the argentine Puna (Antofagasta, Catamarca). Recent Progress in Med Plants 21:95–110

Zampini IC, Isla MI, Schmeda-Hirschmann G (2009a) Antimicrobial and antioxidant compounds from the infusion and methanolic extract of *Baccharis incarum* (Wedd.) Perkins. J Chil Chem Soc 54(4):289–293

Zampini IC, Cuello S, Alberto MR, Ordóñez RM, D Almeida R, Solórzano E, Isla MI (2009b) Antimicrobial activity of selected plants from "the argentine Puna" against sensitive and multiresistant bacteria. J Ethnopharmacol 124(3):499–505

Zdero CC, Bohlmann H, King RM, Robinson H (1986) Diterpene glycosides and other constituents from argentinian *Baccharis* species. Phytochemistry 25(12):2841–2855

Zdero CC, Bohlmann H, Solomon J, King RM, Robinson H (1989) Ent-clerodanes and other constituents from bolivian *Baccharis* species. Phytochemistry 28(2):531–542

Berberis microphylla G. Forst.

Víctor Fajardo Morales, Loreto Manosalva, and Marisel Araya

Fruits of *Berberis microphylla* G. Forst. (Photo: Victor Fajardo)

Abstract The genus *Berberis* includes about 500 different species, which commonly grow in Europe, the United States, South Asia, some northern areas of Iran and Pakistan and South America. Plant species of the genus *Berberis* (Berberidaceae) are of particular phytochemical interest because they have been found to be important sources of alkaloids. In Chile, the Berberidaceae family is represented only by the *Berberis* genus, which includes 50 species. Only some of them have been chemically studied, as they are an important source of bisbenzyl-isoquinoline alkaloids (BBIAs), which are biogenetically derived from tyrosine. Various species of the

V. Fajardo Morales (✉) · L. Manosalva · M. Araya
Faculty of Science, Universidad de Magallanes, Punta Arenas, Chile
e-mail: victor.fajardo@umag.cl; marisel.araya@umag.cl; loreto.manosalva@umag.cl

Á. Máthé, A. Bandoni (eds.), *Medicinal and Aromatic Plants of South America Vol. 2*, Medicinal and Aromatic Plants of the World 7,
https://doi.org/10.1007/978-3-030-62818-5_7

genus *Berberis* have been used in folk medicine for the treatment of some diseases, highlighting its hypoglycemic value. The roots of *Berberis* (especially *B. microphylla*) have a high content of alkaloids, especially berberine, which appears to be responsible for its beneficial effects. Several studies have determined that the use of this natural product reduces the glycemic levels in healthy subjects and diabetes.

Keywords "Calafate" · Benzylisoquinoline alkaloids (BIAs) · Bisbenzylisoquinoline alkaloids (BBIAs) · Anthocyanins · Folk medicine

1 Introduction

In recent years, some species of *Berberis* from Chile and Argentina have been studied due to interesting biological properties related to their alkaloids and polyphenols content. *Berberis microphylla*, is a native plant commonly known as "calafate" that grows wildly in Southern of Chile and Argentina, and used by aboriginal ethnic groups in traditional medicine. Therefore, research has been focused on identification and isolation of new compounds in the *Berberis* genus with prominent biological activity.

2 Taxonomic Characteristics

Berberis microphylla G. Forst. of common name "calafate" is a thorny evergreen shrub endemic to Argentine and Chilean Patagonia. It is a species belonging to the subfamily Berberidoideae, the family Berberidaceae.

For a long time, the name *Berberis buxifolia* was the most commonly used name for this species, but in accordance with the rules of botanical nomenclature, presently, an older, less frequently used name, *Berberis microphylla* must be used. Other authors have subdivided it into numerous entities, but after field and herbarium studies Landrum in 1999, has published that cannot verify any satisfactory specific divisions. Other binomials assigned to *Berberis microphylla*, with corresponding authors are: *Berberis buxifolia* Lam., *Berberis inermis* Pers., *Berberis heterophylla* Juss. ex Poir., *Berberis cuneata* DC., *Berberis marginata* Gay, *Berberis buxifolia* var. *spinosissima* Reiche, *Berberis heterophylla* var. *pluriflora* Reiche, *Berberis buxifolia* Lam. var. *gracilior* Albov, *Berberis buxifolia* var. *papillosa* C.K. Schneid., *Berberis buxifolia* var. *nuda* C.K. Schneid., *Berberis buxifolia* var. *antarctica* C.K. Schneid., *Berberis antucoana* C.K. Schneid., *Berberis parodii* Job, *Berberis michay* Job, *Berberis barilochensis* Job (Landrum 1999).

Photo 1 The typical root of yellow color of *Berberis microphylla*, used to prepare aqueous decoctions and to be drunk as hypoglycemic agent especially as a popular medicine

3 Crude Drug Used

Normally a root decoction is prepared with boiling water (photo 1).

4 Major Chemical Constituents and Bioactive Compounds

From a chemical point of view, *Berberis microphylla* shows a pattern of accumulation of alkaloids (from stems and roots), anthocyanins and flavonoids (from fruits). Fruits of "calafate" were studied in order to know the phenolic composition and antioxidant activity and also compared with data obtained for other berry fruits from southern Chile, including "maqui" (*Aristotelia chilensis*) and "murtilla" (*Ugni molinae*). Polyphenolic compounds in "calafate" fruits were essentially present in glycosylated form, 3-glucoside conjugates being the most abundant anthocyanins. The anthocyanin content in "calafate" berries and flavonol levels are comparable with those found in "maqui"; however, "maqui" shows lower flavan-3-ol concentration than "calafate". "Maqui" and "calafate" show high antioxidant activity, which correlates highly with total polyphenol content and with anthocyanin concentration (Ruiz et al. 2010). Fig. 1 shows the different anthocyanins found in *Berberis microphylla (*Pomilio 1973; Ruiz et al. 2014).

The root and bark of this species are characterized by producing a very special arrangement of isoquinoline alkaloids, all of which are biogenetically derived from the aminoacid tyrosine (Fajardo et al. 1979a, b). Figure 2 considers only those alkaloids that would be present in aqueous decoctions used by some people living in the Southern region of Chile and Argentina. It includes protoberberine alkaloids, with common names: berberine (main alkaloid of *Berberis* species in the world), jatrorrhizine, columbamine, palmatine, isocorydine, jatrorrhizine, palmatine, reticuline, scoulerine and tetrahydroberberine (Fajardo 1987; Freile et al. 2003; Manosalva et al. 2016).

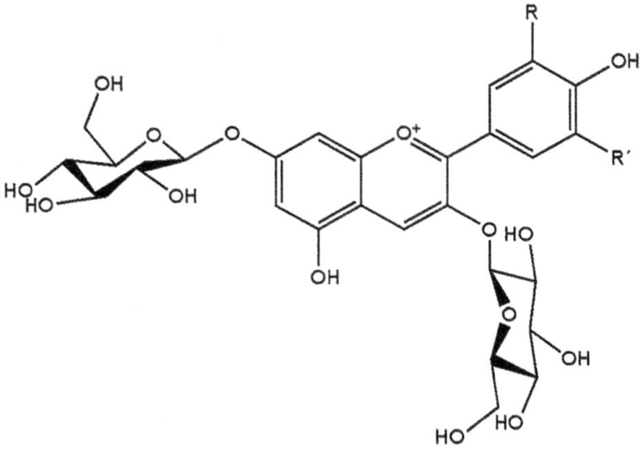

1. R = R` = OH (delphinidin-3,7-β-O-diglucoside)
2. R = H; R` = OH (cyanidin-3,7-β-O-diglucoside)
3. R = OH; R` = OCH_3 (petunidin-3,7-β-O-diglucoside)
4. R = H; R` = OCH_3 (peonidin-3,7-β-O-diglucoside)
5. R = R` = OCH_3 (malvidin-3,7-β-O-diglucoside)

Fig. 1 Anthocyanins in fruits of *Berberis microphylla* G. Forst

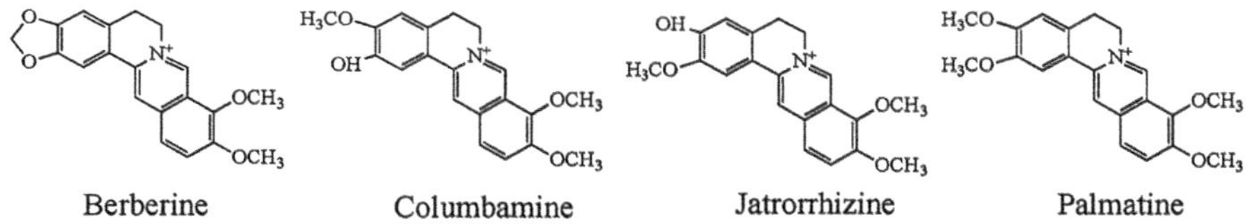

Fig. 2 Main protoberberines alkaloids from stems and roots of *Berberis microphylla* G. Forst

The first report related to bisbenzylisoquinoline alkaloids in *Berberis microphylla* was calafatine (Fajardo et al. 1979a, b). Later, *B. microphylla* has yielded the bisbenzylisoquinolines, chillanamine and (−)-osornine (Fig. 3). Alkaloid chillanamine is the likely precursor of osornine and calafatine (Leet et al. 1983). In the course of an investigation of the alkaloids of *B. microphylla* collected near the town of Punta Arenas, in Chilean Patagonia, were isolated two amorphous bisbenzylisoquinoline N-oxides (Leet et al. 1984).

Damascos et al. (2008) found a significant quantity of Zn and low content of Na and Br in these berries.

5 Morphological Description

Shrub up to *ca.* 2 m high, glabrous or the twigs and young growth pubescent to minutely papillate-puberulent; twigs reddish brown, tan, or gray, with longitudinal ridges, the bark smooth, with age slightly fibrous; spines 3-parted, the arms 3–12 mm

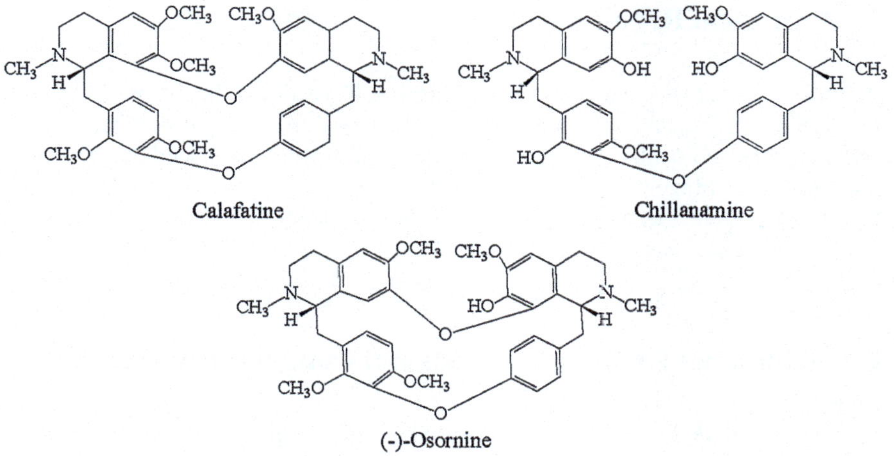

Fig. 3 Main bisbenzylisoquinolines from stems and roots from *Berberis microphylla* G. Forst

long, or often the lateral arms absent or insignificant, especially near the branch tips; bracts ovate to broadly rounded, 1–2 mm long, membranous to submembranous, reddish brown to yellowish tan. Leaves obovate, oblanceolate, or oblong, the blade 5–18 mm long, 1.5–9 mm wide, 2–3.3 times as long as wide, membranous to submembranous, drying gray-green, dull to slightly lustrous above, the margin normally entire; apex acute to rounded, usually without an apiculum; base acute to acuminate; petiole in living leaves usually insignificant (Landrum 2003). Style short; stamens with or without tooth-like appendages; spines 3- parted or simple; inflorescence usually a solitary flower; leaves obovate or oblanceolate to ovate, obtuse to subacute, with sharp rigid apical spine, attenuate at base into short petiole, the margin thickened, entire, coriaceous, glabrous, bright green above, pale beneath, with prominent yellowish veins. Berry subglobose, pruinose, lunate, smooth, dark brown. Flowers October to January (Moore 1983).

6 Geographical Distribution

Berberis microphylla is a native species, endemic to Southwestern South America, from the Andes of Curicó at central Chile at 2500 m.a.s.l. in Tierra del Fuego (Landrum 1999). In Patagonia, *B. microphylla* grows wild in the under-forest, steppe, and forest–steppe ecotones (Correa 1984; Bottini et al. 2000).

7 Ecological Requirements

According to our own experience, working for 40 years in Chilean and Argentine Patagonia, it is possible to mention that *Berberis microphylla* grows in slopes, valleys, canyons and banks of bodies of water. Some authors published that this often growing in the Magellanic subpolar forest ecoregion, in coastal scrub, *Nothofagus* forest margins and clearings, moister areas in grass steppes, and along streams and rivers (Radice and Arena 2015).

8 Traditional Use (Part(s) Used) and Common Knowledge

Berberis microphylla roots have been used by the people in the Austral zone of Chile, for years in folk medicine to treat diabetes because it is thought that they exhibit hypoglycemic activity. Its roots are also used to reduce high levels of cholesterol. Despite the interesting biological activities reported for these roots, there are no scientific studies to support this assumption. It is important to mention that *Berberis microphylla* was used by indigenous people for the treatment of fever, inflammations, stomachaches, diarrhea and urinary tract infection (Zin and Weiss 1998). Additionally, the Aonikenk people used yellow scraping of the bark as tobacco, due to its hallucinogen effect, probably caused by the alkaloid berberine, whereas the fruit of the plant was used by Kawésqar people as a source of food (Domínguez et al. 2012). Aónikenk or Tehuelches, inhabited Patagonia, north of the Strait of Magellan, characterized by being a nomadic land people. The ancient Kawésqar or Alakalufes inhabited the fjords and canals of the extreme south of the South American continent. According to Martínez Crovetto (1968), the Onas in Patagonia used the fruits (named as "kór") as food, and the wood (named as "mich" or thorn) to make arrows.

9 Modern Medicine Based on its Traditional Medicine Uses

In recent years, some studies have corroborated the medicinal use of *B. microphylla*, attributing interesting pharmacological and antimicrobial properties based on the presence of isoquinoline alkaloids. In this context, root extracts of *Berberis microphylla* showed hypoglycemic effects and stimulate glucose uptake in Hep G2 cells with and without resistance by activating AMPK protein (Furrianca et al. 2017). As well, leaves, stems and roots extracts of *Berberis microphylla* possess important antimicrobial activity against human Gram positive pathogenic bacteria. It is interesting that the root extract showed similar activity against *Bacillus cereus* and *Bacillus epidermidis* when compared to ampicillin and cephalothin, commercial antibiotics (Manosalva et al. 2016). Muñoz et al. (2013) found a mild antitrypanosomal activity of extracts of this species.

On the other hand, berberine and its derivatives are used in pharmacy or as a dye substance (Romeo and Sánchez 2005). Related to the used of pure berberine, it has been used in traditional Chinese medicine and Ayurvedic medicine and current research evidence support its use for various therapeutic areas (Singh and Mahajan 2013). It is widely available as a dietary supplement in the different countries and has demonstrated efficacy in the treatment of type 2 diabetes mellitus (Pang et al. 2015). By the way, the genus *Berberis*, also, is used since ancient times for curing eye disease (Srivastava et al. 2015). In addition, there has been a growing interest in the commercial use of *Berberis microphylla* fruits, being consumed fresh or processed as marmalades and jams, in non-alcoholic beverages and in ice creams (Arena et al. 2018). It is known that in Chilean-Argentinean Patagonia the "calafate" syrup is used familiarly and commercially to prepare a "calafate" liquor, which is sold as "calafate sour" (a mixture of syrup with grape brandy, lemon juice and sugar).

10 Conclusions

There is a growing interest to study extracts of plants and/or pure compounds, which historically have shown some biological activity about which many related papers have been published. It should be mentioned that even greater progress is required in this process, especially in relation to trials that may lead to the production of drugs and their use in human beings. In the case of *Berberis microphylla* and other species of the genus *Berberis*, one of the points under discussion is to determine the optimum effective dose for the functions for which its applicability has been indicated and to avoid the collateral effects that might occur. Similarly, highly reliable clinical trials should be conducted.

Acknowledgments This work has been supported by the Vice-Rectory of Research and the Faculty of Sciences of the University of Magallanes, Chile.

References

Ahrendt LWA (1961) *Berberis* and *Mahonia*. A taxonomic revision. J Linnean Soci Botany 57(369):1–410

Arena M, Lencinas M, Radice S (2018) Variability in floral traits and reproductive success among and within populations of *Berberis microphylla* G. Forst., an underutilized fruit species. Sci Hortic 241:65–73

Bottini M, Greizerstein E, Aulicino M, Poggio L (2000) Relationships among genome size, environmental conditions and geographical distribution in natural populations of NW Patagonian species of *Berberis* L. (Berberidaceae). Ann Bot 86:565–573

Correa MN (1984) Berberidaceae. In: Flora Patagónica IV, vol. VIII. Ed. INTA. Buenos Aires, Argentina, pp 325–348

Damascos MA, Arribere M, Svriz M, Bran D (2008) Fruit mineral contents of six wild species of the North Andean Patagonia, Argentina. Biol Trace Elem Res 125(1):72–80

Domínguez E, Aguilera O, Villa-Martínez R, Aravena J, Henríquez JM (2012) Ethnobotanical survey of Kalau Island, ancestral Kawesqar territory, Magallanes region, Chile. Anales del Instituto de la Patagonia 40:19–35
Fajardo V (1987) Estudio químico de las especies chilenas del género *Berberis*. Revista Latinoamericana de Química 18(1):46–50
Fajardo V, Urzúa A, Cassels B (1979a) Calafatine, an unusual bisbenzylisoquinoline alkaloid. Heterocycles 12(12):1559
Fajardo V, Urzúa A, Torres R, Cassels B (1979b) Metabolitos secundarios de *Berberis buxifolia*. Revista Latinoamericana de Química 10:131–134
Freile M, Giannini F, Pucci G, Sturniolo A, Rodelo L, Pucci O, Balzaretti V, Enriz R (2003) Antimicrobial activity of aqueous extracts and of berberine isolated from *Berberis microphylla*. Fitoterapia 74:702–705
Furrianca M, Alvear M, Zambrano T, Fajardo V, Salazar L (2017) Hypoglycemic effect of *Berberis microphylla* G Forst. Root extract. Trop J Pharm Res 16(9):2179–2184
Landrum L (1999) Revision of *Berberis* (Berberidaceae) in Chile and adjacent southern Argentina. Ann Mo Bot Gard 86:793–834
Landrum L (2003) Berberidaceae. In: Marticorena C, Rodríguez R (eds): Flora de Chile. Ed. Universidad de Concepción, Concepción, Chile 2(2):1–23
Leet J, Fajardo V, Shamma M (1983) Some dimeric Benzylisoquinoline alkaloids with an unusual oxygenation pattern. J Nat Prod 46(6):908–912
Leet J, Freyer A, Minard R, Shamma M, Fajardo V (1984) Stereochemistry of Bisbenzylisoquinoline N-oxides. Calafatine 2α-N-oxide and Calafatine 2β-N-oxide. J Chem Soc Perkin Trans 1:651–652
Manosalva L, Mutis A, Urzúa A, Fajardo V, Quiroz A (2016) Antibacterial activity of alkaloids fraction from *Berberis microphylla* G Forst. And study of synergism with ampicillin and cephalothin. Molecules 21(1):76
Martínez Crovetto RN (1968) Estudios Etnobotánicos. IV. Nombres de plantas y su utilidad según los indios Onas de Tierra del Fuego. Etnobiológica (3):1–20
Moore D (1983) Flora of Tierra del Fuego. Ed. Anthony Nelson (England) and Missouri Botanical Garden, USA, pp 66–68
Muñoz OM, Maya JD, Ferreira J, Ph C, San Martin J, López-Muñoz R, Morello A, Kemmerling U (2013) Medicinal plants of Chile: evaluation of their anti-*Trypanosoma cruzi* activity. Zeitschrift fuer Naturforschung, C: J Biosci 68(5/6):198–202
Pang B, Zhao L, Zhou Q, Zhao T, Wang H, Gu C, Tong X (2015) Application of Berberine on treating type 2 diabetes mellitus. Int J Endocrinol 2015:905749
Pomilio A (1973) Anthocyanins in fruits of *Berberis buxifolia*. Phytochemistry 12:218–220
Radice S, Arena M (2015) Environmental effect on the leaf morphology and anatomy of *Berberis microphylla* G. Forst. Int J Plant Biol 6:5677
Romeo R, Sánchez A (2005) Aportes Botánicos de Salta - Ser. Flora Berberidaceae Juss 7:9):1–9):8
Ruiz A, Hermosín-Gutiérrez I, Mardones C, Vergara C, Hertlitz E, Vega M, Dorau C, Winterhalter P, von Baer D (2010) Polyphenols and antioxidant activity of calafate (*Berberis microphylla*) fruits and other native berries from southern Chile. J Agric Food Chem 58(10):6081–6089
Ruiz A, Mardones C, Vergara C, von Baer D, Gómez-Alonso S, Gómez V, Hermosín Gutiérrez I (2014) Isolation and structural elucidation of Anthocyanidin 3,7-β-O-diglucosides and caffeoyl-glucaric acids from calafate berries. J Agric Food Chem 62(29):6918–6925
Singh I, Mahajan S (2013) Berberine and its derivatives: a patent review (2009 – 2012). J Exp Opin Therap Patent 23(2):215–231
Srivastava S, Srivastava M, Misra A, Pandey G, Rawat AKS (2015) A review on biological and chemical diversity in *Berberis* (Berberidaceae). EXCLI J 14:247–267
Zin J, Weiss C (1998) La Salud por Medio de Plantas Medicinales, 8th edn. Don Bosco, Santiago, Chile, pp 70–71

Berberis darwinii Hook.

Victor Fajardo Morales, Marisel Araya, and Loreto Manosalva

Fruits and flowers of *Berberis darwinii* Hook. (Photos: V Fajardo)

Abstract *Berberis darwinii* Hook. has its natural origin in both slopes of the Andes Mountains, in Southern Chile and Argentina, but the species has been naturalized also in other continents. In traditional medicine, it is used by the Mapuche ethnic group for the treatment of inflammatory processes, feverish states and stomach pain. The secondary metabolites in *B. darwinii*, primarily alkaloids and polyphenols, show a great diversity. There are several scientific reports and ethnographic antecedents about the traditional uses. From a medicinal point of view, this seems to be a promising plant species.

V. Fajardo Morales (✉) · M. Araya · L. Manosalva
Faculty of Science, Universidad de Magallanes, Punta Arenas, Chile
e-mail: victor.fajardo@umag.cl

Á. Máthé, A. Bandoni (eds.), *Medicinal and Aromatic Plants of South America Vol. 2*, Medicinal and Aromatic Plants of the World 7,
https://doi.org/10.1007/978-3-030-62818-5_8

Keywords "Michay" · Benzylisoquinoline alkaloids · Anthocyanins · Hydroxycinnamic acids · Flavonols · Folk medicine

1 Introduction

Berberidaceae is a plant family of about 15 genera and 650 species that is widespread in the Northern Hemisphere. With the single genus of *Berberis,* it extends into temperate South America and the Andean region. Fruits of *B. darwinii* are consumed either in a raw state or cooked. It is popular and used in preserves: although acidic, still they have a very pleasant flavor. The fully ripe fruit loses most of its acidity and makes very pleasant eating. Due to its relatively broad ecological range, and content of berberine-alkaloids in its roots and stems, with favorable reported pharmacological activities (antimicrobial, antidiarrhea, anti-inflammatory, antitumor effects, antituberculosis, etc.), it is also used in traditional medicine.

2 Taxonomic Characteristics

Berberis darwinii Hook. is an evergreen thorny shrub, commonly known as "michay" belonging to the subfamily Berberidoideae, family Berberidaceae. The family Berberidaceae, was first established by A.L. Jussieu in 1789 as *Berberides* and was considered as one of the most primitive families of Angiosperms having a high number of disjunction or discontinuous genera.

Synonyms *Berberis costulata* Gand. (Landrum 1999), *Berberis knightii* (Lindl.) K. Koch, *Mahonia knightii* Lindl. (Rodríguez et al. 2018).

Since the times of the native peoples it has received different popular names, among which it is possible to mention "michay" (Hoffmann 2005; Rodríguez et al. 2018) and "quelung" (Rodríguez et al. 2018). The generic name is the Arabic name for the fruit of these plants; the specific name in honor of Charles Darwin (Muñoz 1980).

3 Crude Drug Used

In natural medicine, the most frequent use of *Berberis darwinii* is in decoctions and infusions. Leaves and fruit-infusion are used against inflammation, the roasted and powdered root bark is used as a wound healing agent and through decoction of leaves a soft and febrifugal purgative is obtained (Bisheimer 2012; Cordero et al. 2017).

4 Chemical Constituents and Bioactive Compounds

Berberis darwinii contains different types of phytochemicals in its stems, fruits, leaves and roots. Of primary importance is the presence of different alkaloids and polyphenols (Núñez et al. 2018a, b). The phenolic profiles, antioxidant activity, and inhibitory effects of (enzymes α-glucosidase, α-amylase, and pancreatic lipase) of three Argentinean Patagonia berries were investigated by Chamorro et al. (2019). The most complex polyphenol profile among the studied species was found in the *Berberis darwinii* samples (Maza and Miniati 1993; Chamorro et al. 2019): delphinidin-3-*O*-glucoside, delphinidin rutinoside, cyanidin-3-*O*-glucoside, cyanidin-3-*O*-rutinoside, petunidin hexoside, petunidin rutinoside, peonidin hexoside, peonidin rutinoside, malvidin-3-*O*-glucoside and malvidin rutinoside; the following hydroxycinnamic acids: caffeoylglucaric acid, caffeoylquinic acid, dicaffeoyl glucaric acid, caffeoylquinic acid, feruloylquinic acid, caffeoyl glucaric acid, dicaffeoylquinic acid; and flavonols: quercetin hexoside, quercetin glucuronide, quercetin rutinoside, quercetin acetylhexoside, quercetin rhamnoside, isorhamnetin rutinoside, isorhamnetin acetylhexoside and isorhamnetin acetylhexoside.

On the other hand, many isoquinoline bases have been identified in *Berberis darwinii*, among them the alkaloids berberine, magallanesine, dihydrorugosinone, rugosinone, nuevamine, santiagonamine and chiloenamine (Valencia et al. 1984a, b, c, 1985). A general aspect to highlight in any studied *Berberis*, is the presence of the berberine alkaloid, a situation that is also reflected in *Berberis darwinii*. Berberine bearing special quaternary ammonium salt in the benzylisoquinoline alkaloid core, is a major active component of some Chinese herbal medicines, and multiple therapeutic actions of berberine analogs have been continuously reported including antimicrobial, antidiarrhea, anti-inflammatory, antitumor effects, antituberculosis activities, etc. (Seng et al. 2019).

It is important to mention that some alkaloids of *Berberis darwinii* have been of particular interest among organic chemists. Similarly, various heterocyclic bases, natural products containing the isoindolin-1-one system such as (±)-nuevamine, (±)-lennoxamine and magellanine are very important since their extensive occurrence in nature is known. These architecturally sophisticated chemical structures include five-eight membered rings fused with different aromatic moieties and differently oxygenated substituents. Thus, analogs of these pentacyclic systems incorporating a skeleton of isoindolin-[1,2-*a*]-5-one are of great interest in drug research because of their biological activity (Vázquez-Vera et al. 2017). Therefore, their unique structural features have recently attracted the attention of many organic research groups and several synthetic strategies have been pursued towards these attractive and challenging synthetic targets. (±)-Nuevamine was the first naturally occurring isoindole-[1,2-*a*]-isoquinolinone, whose chemical structure is very interesting from a pharmacological perspective, due to the potential and promising biological activity of many of its analogs, *e.g.* as anti-inflammatory, antimicrobial, antileukemic and antitumor properties (Mertens et al. 1993).

5 Morphological Description

"Michay" is a moderately branched shrub that grows 1–3 meters high. Leaves abundant, variously shaped, margins entire or with fewer than 10 teeth per side, born from 3 to 5 in the armpits of strong thorns, smooth, shiny coriaceous, sessile, with toothed-spiky edge 1–2 cm long, dark green, the underside is somewhat lighter, new shoots have a reddish tone. The inflorescence a raceme; style on fruit 1–3 mm long. Flowers gathered in clusters, with 6 orange sepals, located on two levels; 6 petals, 6 stamens and a pistil in the shape of a small bottle, with discoid stigma. The fruit is a round berry that during the ripening process goes from green to black, passing through red; it contains 3–4 seeds (Muñoz 1980; Landrum 1999).

6 Geographical Distribution

Berberis darwinii is a native shrub to temperate forest of Southern South America (Chalcoff et al. 2006), specifically it is located in southern Chile, from Ñuble to Aysen (Muñoz 1980), and mountains areas of the Argentinean Patagonia (Landrum 1999); however, it has been distributed in Europe and other continents after its Discovery by Charles Darwin (Sykes 1982; Habtemariam 2013). In addition, it is important to note that *B. darwinii* is considered as an invasive species in New Zealand, because the plant invades and persists in different light environments, from grazed pasture to intact forest (Mc Alpine and Jesson 2007), where prolific seed production and seed dispersal in high light environments is considered a key feature contributing to the success of this species' invasion (Mc Alpine and Jesson 2008).

7 Ecological Requirements

Berberis darwinii is highly tolerant to drought, frost and shade (Timmins and Mackenzie 1995), and can also occupy a wide range of soil types in relation to its texture (sandy and clay soil) and pH (acid, neutral and basic), as well as, it can grow in nutritionally poor soils (Allen 1991).

A good bee plant, and birds also love its fruit and will happily eat it before it reaches full maturity. If you want to experience the fully ripe fruit, then it might be necessary to find ways of keep the birds off the plants.

8 Traditional Use (Part(s) Used) and Common Knowledge

Berberis darwinii is recognized as a healing plant, where leaves and fruits have been used by the ethnic group due to effects febrifuge, astringent and anti-inflammatory (Hirschmann et al. 2019). *B. darwinii* also is known for stomach pains, indigestion, and colitis (Montes and Wilkomirsky 1987). In addition, the fruits of this berry have been consumed raw and cooked, *e.g.* in pastry as decorative, juices, sweets and syrups (Cordero et al. 2017).

9 Modern Medicine Based on Its Traditional Medicine Uses

Berberis darwinii has an important pharmacological activity for its potential in the treatment of pathologies associated with inflammation (Núñez et al. 2018a), where the most important active compound is berberine present in high concentrations in root extracts (Núñez et al. 2018b). So, *B. darwinii* shows anti-inflammatory action by inhibiting the production of superoxide anion, the expression of tumor necrosis factor-alpha (TNFα), and interleukin-1-beta (IL-1β) in monocytes activated by lipopolysaccharide (Alarcón et al. 2014). Moreover, it has been shown that the therapeutic potential of the stem-bark extract of *B. darwinii* for Alzheimer's disease, is due to the presence of berberine, which is a potent inhibitor of acetylcholinesterase (ACHE) enzyme thereby increasing the lifespan of the neurotransmitter, acetylcholine (Habtemariam 2011).

10 Conclusions

A search in the SciFinder has resulted more than 20,000 scientific reports on species of the genus *Berberis* or compounds related to isoquinoline alkaloids, i.e. berberine ("berberine" as keyword). It can also be perceived that there is a growing interest in studying plant extracts and/or pure compounds of this genus, which in traditional medicine have demonstrated biological activity. More progress is needed, especially tests that can lead to the development of medicines and their use in humans. In the case of *Berberis microphylla* and *B. darwinii* and other species of the same genus, one of the points under discussion is the optimal effective dose for applicability, as side effects might occur. Similarly, highly reliable clinical trials should be conducted.

Acknowledgments This work has been supported by the Vice-Rectory of Research and the Faculty of Sciences of the University of Magallanes, Chile.

References

Alarcón D, Paredes M, Ramos D (2014) Los extractos acuoso y metanólico de *Berberis darwinii* H. (Berberidaceae) inhiben respuestas celulares innatas en monocitos humanos tratados *in vitro*. BLACPMA 13(1):81–91

Allen RB (1991) A preliminary assessment of the establishment and persistence of *Berberis darwinii* Hook., a naturalized shrub in secondary vegetation near Dunedin, New Zealand. N Z J Bot 29:353–360

Bisheimer M (2012) Flores de la Patagonia Argentina. Flores nativas y exóticas presentes en los ambientes cordilleranos y precordilleranos de la Patagonia argentina. 1° ed. – Neuquén: Serie Patagonia. Ed. Eduardo Marcelo Fernández, Neuquén, Argentina, p 240

Chalcoff V, Aizen M, Galetto L (2006) Nectar concentration and composition of 26 species from the temperate forest of South America. Ann Bot 97:413–421

Chamorro M, Reiner G, Theoduloz C, Ladio A, Schmeda-H G, Gómez S, Jiménez F (2019) Polyphenol composition and (bio) activity of *Berberis* species and wild strawberry from the Argentinean Patagonia. Molecules 24:3331

Cordero S, Abello L, Gálvez F (2017) Plantas silvestres, comestibles y medicinales de Chile y otras partes del mundo. Guía de campo. Ed. Corporación Chilena de la Madera, Concepción, Chile, 292 pp

Habtemariam S (2011) The therapeutic potential of *Berberis darwinii* stem-bark: quantification of Berberine and *in vitro* evidence for Alzheimer's disease therapy. Nat Prod Commun 6(8):1089–1090

Habtemariam S (2013) The hidden treasure in Europe's garden plants: case examples; *Berberis darwinii* and *Bergenia cordifolia*. Med Aromat Plants 2(4):1–5

Hoffmann A (2005) Flora Silvestre de Chile, Zona Araucana Árboles, arbustos y enredaderas leñosas. V ed.. Ed. Fundación Claudio Gay, Santiago, Chile, 257 pp

Landrum L (1999) Revision of *Berberis* (Berberidaceae) in Chile and adjacent southern Argentina. Ann Mo Bot Gard 86:793–834

Maza G, Miniati E (1993) Anthocyanins in Fruits, Vegetables and Grains. Ed. CRC, New York, USA, 362 pp

Mc Alpine K, Jesson I (2007) Biomass allocation, shade tolerance and seedling survival of the invasive species *Berberis darwinii* (Darwin's barberry). N Z J Ecol 31(1):1–12

Mc Alpine K, Jesson I (2008) Linking seed dispersal, germination and seedling recruitment in the invasive species *Berberis darwinii* (Darwin's barberry). Plant Ecol 197:119–129

Mertens A, Zilch H, Koenig B, Schaefer W, Poll T, Kampe W, Seidel H, Leser U, Leinert H (1993) Selective non-nucleoside HIV-1 reverse transcriptase inhibitors. New 2,3-dihydrothiazolo[2,3-a]-isoindol-5(9bH)-ones and related compounds with anti-HIV-1 activity. J Med Chem 36:2526–2535

Montes M, Wilkomirsky T (1987) Medicina Tradicional Chilena. Ed. de la Universidad de Concepción, Concepción, Chile, 205 pp

Muñoz M (1980) Flora del Parque Nacional Puyehue. Ed. Universitaria. Santiago, Chile, 557 pp

Núñez D, Balboa N, Quilaqueo N, Alvear M, Paredes M (2018a) Evaluación de la Actividad Inmunomoduladora de Extractos Metanólicos y de Alcaloides de *Berberis darwinii* H. (Berberidaceae). Int J Morphol 36(2):454–459

Núñez D, Balboa A, Carvajal F, Alvear M, Paredes M (2018b) Efecto del extracto de alcaloides de *Berberis darwinii* Hook. sobre respuestas celulares innatas en fagocitos murinos. BLACPMA 17(3):259–269

Rodríguez R, Marticorena C, Alarcón D, Baeza C, Cavieres L, Finot V, Fuentes N, Kiessling A, Mihoc M, Pauchard A, Ruiz E, Sánchez P, Marticorena A (2018) Catálogo de las plantas vasculares de Chile. Gayan Botanica 75(1):1–430

Schmeda-Hirschmann G, Jiménez-Aspee F, Theoduloz C, Ladio A (2019) Patagonia berries as native food and medicine. J Ethnopharmacol 241:111979

Seng Q, Wang H, Wei W, Guo T, Wang Y, Li Y, Song D (2019) Synthesis and biological evaluation of berberine derivatives as a new class of broad-spectrum antiviral agents against Coxsackievirus B. Bioorg Chem 9(95):103490

Sykes W (1982) Checklist of Dicotyledons naturalized in New Zealand 15. Annonales, Berberidales, Cactales, Fagales, some Geraniales, Juglandales, Laurales, Rutales, Salicales, Sapindales, Tiliales, Nyctaginaceae, and Zygophyllaceae. N Z J Bot 20:333–341

Timmins S, Mackenzie I (1995) Weeds in New Zealand protected natural areas database. Department of Conservation Technical Series No. 8. Wellington, Department of Conservation. New Zealand, 287 pp

Valencia E, Weiss I, Shamma M, Urzua A, Fajardo V (1984a) Dihydrorugosinone, a pseudobenzylisoquinoline alkaloid from *Berberis darwinii* and *Berberis actinacantha*. J Nat Prod 47(6):1050–1051

Valencia E, Patra A, Freyer A, Shamma M, Fajardo V (1984b) Santiagonamine: a new aporphinoid alkaloid incorporating a phenanthridine skeleton. Tetrahedron Lett 25(30):3163–3166

Valencia E, Freyer A, Shamma M, Fajardo V (1984c) Nuevamine, an isoindoloisoquinoline alkaloid, and lennoxamine, an isoindolobenzazepine. Tetrahedron Lett 25(6):599–602

Valencia E, Fajardo V, Freyer A, Shamma M (1985) Magallanesine: an isoindolobenzazocine alkaloid. Tetrahedron Lett 26(8):993–996

Vázquez-Vera O, Sánchez-Badillo J, Islas-Jácome A, Rentería- Gómez M, Pharande S, Cortés-García C, Rincón-Guevara M, Ibarra I, Gámez-Montaño R, González-Zamora E (2017) Efficient Ugi-3CR / aza Diels-Alder / Pommeranz-Fritsch protocol towards novel Aza-analogues of (±)-Nuevamine, (±)-Lennoxamine and Magallanesine: a diversity oriented synthesis approach. Org Biomol Chem 15:2363–2369

Buddleja globosa Hope

Edgar Pastene-Navarrete and Jeniffer Torres-Vega

Buddleja globosa leaves. Chillán, Región de Ñuble, Chile. (Photos: E Pastene, October, 2019)

Abstract *Buddleja globosa* Hope (Buddlejaceae) is a native species cultivated in Chile, Peru and Argentina. In folk medicine, this plant is frequently used for the treatment of different wounds (internal and external), as well as intestinal and liver diseases. There is growing scientific evidence on the growing conditions for *Buddleja globosa* and how its cultivation could be optimized. Methods for its vegetative propagation have also been successfully elaborated. The leaves contain phenylpropanoids,

E. Pastene-Navarrete (✉)
Laboratorio de Síntesis y Biotransformación de Productos Naturales, Universidad del Bío-Bío, Chillán, Chile
e-mail: edgar.pastene@gmail.com

J. Torres-Vega
Laboratorio de Farmacognosia, Departamento de Farmacia, Facultad de Farmacia, Universidad de Concepción, Concepción, Chile

Á. Máthé, A. Bandoni (eds.), *Medicinal and Aromatic Plants of South America Vol. 2*, Medicinal and Aromatic Plants of the World 7,
https://doi.org/10.1007/978-3-030-62818-5_9

iridoids, terpenes and flavonoids. Ethnopharmacological validation studies suggest that phenylpropanoids and flavonoids are the main and most important compounds in defining the therapeutic activity of *Buddleja globosa*. This chemical composition explains its pharmacological activity as antioxidant, anti-inflammatory, antimicrobial and, in particular, its wound healing effects. Due to its undeniable potential as a domesticated, cultivated medicinal plant and its remarkable chemical composition, it is still necessary to develop its properly standardized, clinically validated phytopharmaceutical products.

Keywords "Matico" · Phenylpropanoids · Verbascoside · Wound healing

1 Introduction

Buddleja globosa Hope ("matico") has multiple uses in traditional medicine, such as healing stomach ulcers, colitis, and to improve cicatrization of cutaneous wounds. Plants with wound healing properties are not very abundant. Furthermore, in few cases the pharmacological mechanisms of action are known. Among these species, *Centella asiatica, Curcuma longa, Calendula officinalis*, and *Paeonia suffruticosa* can be mentioned. Remarkably, the chemical composition of these species is different from that of *B. globosa*. Recently, data suggest the possibility that chemical constituents of this species could act jointly on different overlapping events in the wound healing process. So, the effects upon proliferation/migration properties can be added to its antimicrobial, anti-inflammatory and antioxidant properties. It is an important advantage of its utilization that its agronomic aspects are already known and this could contribute to the scaling of its production for pharmaceutical purposes.

2 Taxonomic Characteristics

Buddleja is a generic name given in honor of Adam Buddle, a botanist and rector in Essex, England. "*globose*" is a Latin epithet meaning "spherical, shaped like a balloon". *Buddleja globosa* Hope: This species was described by John Hope (The Plant List 2020). It can be found in Argentina, Bolivia, Chile and Peru.

B. globosa is widely grown as an ornamental in parks and gardens in the United Kingdom. Its decorative globose heads of golden-yellow flowers attract large numbers of honeybees in early summer. In Spanish, *B. globosa* is known as "matico" and as "palñín" or "pañil" in the native tongue of Mapuche (indigenous inhabitants of central Chile). In Quechua's native tongue this plant is known as "palguín" (Malaret 1970; Colmeiro 1871).

Synonyms *Buddleja capitata* Jacq., *Buddleja connata* Ruiz & Pav., *Buddleja globifera* Duhamel (The Plant List 2020).

Photo 1 *Buddleja globosa*. Powdered drug. (Photo: E Pastene, October, 2019)

3 Crude Drug Used

Dried and powdered leaves (Photo 1) are used to prepare infusions, while fresh crushed leaves are used to make poultices that are applied to wounds.

4 Major Chemical Constituents and Bioactive Compounds

The iridoids aucubin, catalpol, and *O*-methyl catalpol (the first two isolated as 7-*p*-methoxycinnamoyl derivatives) are the most important and representative chemical compounds of the Buddlejaceae family. A second group of compounds consists of flavonoids whose main representative is scutellarin. The third group is composed of phenylpropanoids, among them, verbascoside (Pardo et al. 1993), linarine and echinacoside (Muñoz et al. 1999). Backhouse et al. (2008a) extracted "matico" leaves with solvents of increasing polarity to obtain different fractions. In these fractions, verbascoside, luteolin 7-*O*-glucoside and apigenin 7-*O*-glucoside were isolated and identified. In another study of the same group, beta-sitosterol, stigmasterol, stigmastenol, stigmastanol, campesterol and beta-sitosterol-glycoside were isolated (Goity 2007).

The roots of plants in this family produce sesquiterpenoids with caryophyllene skeleton. Several of these compounds were isolated from *Buddleja davidii* Franch and named buddledins A, B, C and E (Yoshida et al. 1976, 1978). Buddledins A, B and C also were isolated from *B. globosa* along with the presence of dehydrobuddledin, zerumbone, buddledone-A and buddledone-B. Moreover, the following constituents have been isolated from *B. globosa:* diterpenes buddlejone, deoxybuddlejone, hydroxybuddlejone and the bisditerpene maytenone (Liao et al. 1999; López et al. 1979). Lupeol and α-amyrin were isolated from the leaves and flowers of *B. globosa* (Marín et al. 1979; López et al. 1979).

5 Morphological Description

Buddleja globosa is an evergreen shrub of 1.5–3 m height, with yellowish sumptuous stems. Opposite leaves, 3–15 cm long by 1–5 cm wide, oval-lanceolate, rough, whitish on the underside, sharp at the tip. Orange, yellow and red flowers arranged in globose heads of 1–2 cm. Fruit in capsule of 3 mm in diameter. Numerous seeds, polyhedral, less than 1 mm long. *Buddleja globosa* is a dioecious shrub attaining 4 m in height which is common in the central and southern regions of Chile. The plant has tomentose branches with subsessile leaves, with conspicuous stipulation, lanceolate or elliptic, subcoriaceous to membranoceous, glabrescent, and decurrent to the base. Flowers are clustered in pairs in pedunculated heads. The tubular calyx is tomentose outside, with acute lobes. The corolla glabrous on the outside and with yellow-orange color. Stamens are sessile, ovary is tomentose in the upper half with a claviform style. The fruit is an ellipsoidal capsule, tomentose and glandular, with septicidal dehiscence and rounded valves (Norman and Ariza Epinar 1995).

6 Geographical Distribution

The plants grow at low elevations in interior valleys. In the Chilean coastal mountain range, this plant grows at 500–2000 m.a.s.l. In Chile, *B. globosa* could be found from Valparaíso, Región Metropolitana de Santiago, O'Higgins, Maule, Biobío, Ñuble, Araucanía, Los Ríos and Los Lagos.

Doll et al. (2003), studied different experimental conditions to achieve vegetative propagation of "matico". These authors reported that *B. globosa* is an easy-rooting species, reaching rooting rates of about 80%. The treatments with 1000–2000 ppm of indole butyric acid (IBA) applied at the base of the cuttings improve rooting process. The authors also found that cuttings from the apical portion of the mother-branch displayed better rooting rates. A mixture (1:1) of perlite and vermiculite resulted to be the most favorable rooting media. Wilckens et al. (2013), determined the growth cycle of *B. globosa* and created a regeneration model in order to establish productive management and sustainable use of this plant. These authors also quantified the growth dynamics of biomass regeneration and identified the effects of harvest methods. In this study, the final average biomass was 2.51 g per branch in the Cordillera, and 18.3 g per branch in the Valley area. The growth rates were significantly lower in the Cordillera than in the Valley. Similar results were obtained also in basal pruning treatments. In the control treatments, the plants in the Cordillera reached growth rates of 4.0% in relation to those of 6.7% in the Valley. In the treatments, the mean pruning reached a regeneration rate of 1.97% in the Cordillera and 10.7% in the Valley, while values for the basal pruning were 0.95% and 8.42%, respectively. The decrement rates of mean and basal pruning showed, significantly higher decrement rates in the Valley than in the Cordillera. According to these authors, December was the month for optimal harvest period for the control, apical and medium pruning. It is in this period when biomass maximization occurs in Cordillera. On the other hand, in the Valleys area, biomass maximization occurred in February only for control treatment, while in the medium and apical treatments the maximum was measured in January. These results suggest an anticipated period of biomass maximization in the Cordillera in relation to the basal pruning.

7 Ecological Requirements

This species grows in areas with permanent rainfall. The plants can withstand short periods of drought, but not beyond one month. The plant grows in water or extends its roots within permanent watercourse. It is common to find this plant in meadows, water courses, lakes edges and swamps. It requires some shade and protection against the sunlight, by low dense vegetation and rocks that help to filter approx. 20–40% of the sunlight. Such conditions can be found also in steep slopes with southern exposure, deep ravines, or under the protection by dense layer of vegetation beneath large trees with a 40–80% sunlight filtration.

In Chile, it can be found from Copiapó to the isle of Chiloé; growing in dry and moist forests, from near sea level to 2000 m, associated with native trees such as *Nothofagus dombeyi, Drimys winteri*, and *Populus pyramidalis*.

8 Traditional Use (Part(s) Used) and Common Knowledge

Buddleja species from Central America, Southern Africa and Eastern Asia have common traditional uses (Houghton and Manby 1985). For instance, "Mi-meng-hua", is a preparation made with the flowers of *B. officinalis* Maxim, which is a traditional herb from Chinese medicine used for the treatment of conjunctival congestion (Chinese Pharmacopoeia Committee 2015). On another hand, the roots of *Buddleja asiatica* Lour. (Dog tail; "Qi-li-xiang") are used as anti-inflammatory (Houghton 1984). Similar use is reported for *B. globosa* Hope from Chile, where leaves and flowers were used to wash wounds and treat gastric ulcers (Houghton and Manby 1985).

B. globosa was official drug included in the first Chilean Pharmacopeia published in 1886. The drug (whole or powdered leaves) is used as an herbal tea for washing wounds. Powdered dried leaves or crushed fresh plant preparation are used to help cicatrization of wounds and skin bruises (Houghton 1984; Houghton and

Manby 1985). Herbal tea is used to treat dysentery, hemorrhoids, hepatitis and catarrh. The juice squeezed from the leaves was also used to treat warts and callous ulcers (Houghton 1984).

9 Modern Medicine Based on Its Traditional Medicine Uses

"Matico" leaves contain phenylpropanoids, iridoids, terpenes and flavonoids (Montes and Wilkomirsky 1985, 1992; Mensah et al. 2000) and show antihepatotoxic, bactericidal, antioxidant, anti-inflammatory, wound healing and analgesic activities (Pardo et al. 1997; Mensah et al. 1998; Liao et al. 1999; Backhouse et al. 2008a, b). Also, diuretic, wound-healing and anti-inflammatory properties have been associated with the presence of tannins (5.7%) and flavonoid glycosides (Varillas and TTito 2018). This medicinal plant presents iridoids with antibacterial activity and phenylpropanoids with analgesic, antibacterial and antihypertensive properties (Recio et al. 1994). In other study, high concentrations from 0.5–1 mM (expressed as gallic acid equivalents) of *B. globosa* leaf extract protect erythrocytes from HClO damage (Suwalsky et al. 2017). Also, using a *Drosophila melanogaster* wing-spot mutagenic model, Carmona et al. (2016) reported that *B. globosa* show desmutagenic effects. This author states that verbascoside and luteolin are bioactive components responsible of this antimutagenic effect. Debenedetti et al. (2002) reported that a methanol extract of *B. globosa* showed antiplasmodial activity with IC_{50} of 6.2μg/ml.

Regarding wound-healing activity, an aqueous extract of *B. globosa* leaves displayed growth-promotive and antioxidant activities upon fibroblasts treated with hydrogen peroxide. Low concentrations of the extract gave an increase in fibroblast growth that was non-significant. Nevertheless, cytotoxicity was observed at concentrations greater than 50 mg/ml (Mensah et al. 2001). In this study, the extract showed a strong antioxidant effect, and further chromatographic fractionation led to the isolation of three flavonoids and two caffeic acid derivatives. All these compounds showed antioxidant effect at concentrations below 10 mg/ml. These activities contribute to wound healing properties (Mensah et al. 2001). Some *in vitro* tests demonstrated antibacterial activity against Gram (+) and (−) germs, which is mainly due to the presence of diterpenes and triterpenes mainly (Alonso 2004). It should be noted that certain compounds, like catalposide and aucubin, described in *B. globosa* and other species an anti-inflammatory effect, which contributes to the wound-healing process (Kim et al. 2004; Park and Chang 2004). In addition, angaroside A is most likely the molecule responsible for antibacterial activity against *Staphylococcus aureus* (Hoffmann et al. 1992; Muñoz et al. 1999). *B. globosa* presents terpenes with antifungal activity against *Trichophyton rubrum*, *Trichophyton interdigitale* and *Epidermophyton floccosum* (Mensah et al. 2000). On the other hand, it is well-known that catalposide -one of the iridoids isolated from *B. globosa*-inhibits NO synthase, an enzyme involved in the pro-inflammatory response (Oh et al. 2002). In line with the effects of catalposide, iridoids such as aucubin, have also been shown to have anti-inflammatory effects that could help the cicatrization process (Recio et al. 1994).

There is a patent (WO2012100365A; US8852654B2): "Use of a standardized dry extract of *Buddleja globosa* leaves, BG-126, for the treatment and prevention of gastrointestinal disorders caused by treatment with nitrofurantoin and other antimicrobials". This invention relates to the use of a composition containing extracts of *Buddleja globosa* ("matico") and pharmacologically accepted additives for the treatment or prevention of various gastrointestinal disorders, in particular those associated with patients treated with nitrofurantoin as a therapy against *Escherichia coli* (Letelier et al. 2014).

In another study, an aqueous extract of *B. globosa* leaves (1 mg/ml) protected hepatocytes treated with carbon tetrachloride, galactosamine and complement-mediated cytotoxic medium (CMC) (Houghton and Hikino 1989). After bioassay-guided fractionation, the main compounds responsible for this effect were iridoids, flavonoids (linarin) and phenylpropanoid glycosides (verbascoside and echinacoside). Linarin (1 mg/ml) was the most active compound giving a 40% reduction of the released glutamic-pyruvic transaminase (GPT) in CCl_4 treated hepatocytes, whereas the reduction reaches 75% in those cells challenged with galactosamine. Isolated echinacoside was also active giving 45% and 66% reduction at the same tested concentrations. This protection in hepatocytes probably could be ascribed to the antioxidant properties common to these molecules. The sesquiterpenes buddledins A–C inhibit both cyclooxygenase and 5-lipoxygenase (5-LOX) at 50 mg/ml, *in vivo*. Among the tested compounds, buddledin A showed the greatest inhibition against COX achieving 89%. In addition, the inhibition against 5-LOX reaches 98% (Mensah et al. 1999).

Antioxidant (DPPH assay) and antiproliferative activity of *B. globosa* aqueous extract (5 mg/ml) have been assessed in T84 (tumoral) and HTR-8/SVneo (non-tumoral) cell lines. The effect was dependent on the cellular context since cytotoxicity was observed mainly in HTR-8/SVneo cells (Gastaldi et al. 2018).

The *in vivo* wound healing properties have been assessed using Sprague-Dawley rats treated with *B. globosa* orally administered for 12 days. Also, the extract was topically administered. The results have shown an improvement in the wound-healing process associated with anti-inflammatory and antioxidant effects. Hence, COX-2 levels and Ki-67 (proliferation biomarker) (Letelier et al. 2012).

Recent work demonstrated that *B. globosa* extracts are able to prevent collagen-induced platelet activation (Fuentes et al. 2017). In this study, *B. globosa* inhibits the phosphorylation of PLC-γ2 and PKC-β2 in human platelets. These kinases are involved in pathways related to collagen induced platelet activation. Considering this mechanism, the author stated that *B. globosa* may be useful in cardiovascular disease (CVD) because platelet hyperreactivity is closely associated with thrombosis related diseases.

Regarding clinical trials, there is only one report of a study that is currently being carried out. However, at the time of the elaboration of this chapter, the results are not yet available (Bustamante et al. 2014, 2015).

10 Conclusions

B. globosa is a medicinal shrub with an important body of scientific evidence including agronomic, phytochemical and experimental and molecular pharmacology studies. Some preclinical studies support the use of this species as a wound healing agent, although the underlying molecular mechanisms are not yet fully elucidated. Many of these studies were carried out in the previous decade, therefore they need to be revisited in order to improve the validation of the ethnomedical uses of this species. On the other hand, it is striking that recent phytochemical studies have not used advanced analytical technology in the identification process of chemical composition of *B. globosa*. Much of the pharmacological properties of this species are attributed to components already isolated from other sources and which, by extrapolation, have been assigned to *B. globosa*. Even to date, there is a lack of properly designed clinical studies. None of the available results appear to categorically support the traditional use of the "matico", therefore the study of this species in the future is guaranteed.

References

Alonso J (2004) Tratado de Fitofármacos y Nutracéuticos, 1st edn. Ed. Corpus Libros, Rosario, Argentina, pp 738–739

Backhouse N, Delporte C, Apablaza C, Farías M, Goïty L, Arrau S, Negrete R, Castro C, Miranda H (2008a) Antinociceptive activity of *Buddleja globosa* (Matico) in several models of pain. J Ethnopharmacol 119(1):1–6

Backhouse N, Rosales L, Apablaza C, Goity L, Erazo S, Negrete R (2008b) Analgesic, anti-inflammatory and antioxidant properties of *Buddleja globosa*, Buddlejaceae. J Ethnopharmacol 116:263–269

Bustamante S, Álvarez N, Mendiburen R, Vergara F, Zárate I, Collado C, Morales M (2015) Fundamentación preclínica del uso etnomédico de matico (*Buddleja globosa* Hope). Rev Fitoter 15(1):37–51

Bustamante S, Collado C, Morales M (2014) Efficacy of Matico (*Buddleja globosa* H.) on healing wounds and ulcers. Randomized, comparative, and open label study. Control Clin Trials. https://doi.org/10.1186/ISRCTN38989659

Chinese Pharmacopoeia Committee (2015) Pharmacopoeia of the People's Republic of China. Ed. Chinese Medical Science and Technology Press, Beijing, China. 1:329

Carmona ER, Escobar B, Obando V (2016) Antimutagenic activity of *Buddleja globosa* extracts in the *Drosophila* wing-spot test. Fresen Environ Bull 25(12A):5758–5764

Colmeiro M (1871) Diccionario de los diversos nombres vulgares de muchas plantas usuales notables del antiguo y nuevo mundo. Ed. Imprenta de G. Alhambra, Madrid, Spain, p 124

Debenedetti S, Muschietti L, van Baren C, Clavín M, Broussalis A, Martino V, Houghton PJ, Warhurst D, Steele J (2002) *In vitro* antiplasmodial activity of extracts of Argentinian plants. J Ethnopharmacol 80(2-3):163–166

Doll H, Vogel H, Jeldres P, Muñoz M (2003) Estudios de Propagación Vegetativa en Matico (*Buddleja globosa*). Ciencia e investigación agraria: revista latinoamericana de ciencias de la agricultura 30(3):211–216

Fuentes M, Sepúlveda C, Alarcón M, Palomo I, Fuentes E (2017) *Buddleja globosa* (matico) prevents collagen-induced platelet activation by decreasing phospholipase C-gamma 2 and protein kinase C phosphorylation signaling. J Tradit Complement Med 8(1):66–71

Gastaldi B, Marino G, Assef Y, Silva Sofrás FM, González SB (2018) Nutraceutical properties of herbal infusions from six native plants of Argentine Patagonia. Plant Foods Hum Nutr 73:180–188

Goity L (2007) Estudio químico y farmacológico de un extracto activo de *Buddleja globosa* Hope, Buddlejaceae, matico y diseño de la metodología analítica. Memoria de Químico Farmacéutico, Universidad de Chile, Facultad de Ciencias Químicas y Farmacéuticas. Santiago, Chile. pp 7–20

Hoffmann A, Farga C, Lastra J, Veghazi E (1992) Plantas Medicinales de Uso Común en Chile. Ed. Fundación Claudio Gay, Santiago, Chile, 134 pp

Houghton PJ, Hikino H (1989) Anti-hepatotoxic activity of extracts and constituents of *Buddleja* species. Planta Medica 55:123–126

Houghton PJ, Manby J (1985) Medicinal plants of the Mapuche. J Ethnopharmacol 13(1):89–103

Houghton PJ (1984) Ethnopharmacology of some *Buddleja* species. J Ethnopharmacol 11(3):293–308

Kim SW, Choi SC, Choi EY, Kim KS, Oh JM, Lee HJ, Oh HM, Kim S, Oh BS, Kimm KC, Lee MH, Seo GS, Kim TH, Oh HC, Woo WH, Kim YS, Pae HO, Park DS, Chung HT, Jun CD (2004) Catalposide, a compound isolated from *Catalpa ovata*, attenuates induction of intestinal epithelial proinflammatory gene expression and reduces the severity of trinitrobenzene sulfonic Acid-induced colitis in mice. Inflamm Bowel Dis 10(5):564–572

Letelier ME, Jones R, López C, Palma K, Aracena P, Razmilic I, Polanco X et al (2012) Safety profile and wound healing properties of a standardized *Buddleja globosa* Hope (matico) extract in Sprage-Dawley rats. Rev Farmacol Chile 5(2):13–19

Letelier ME, Salas ME, Polanco X (2014) Use of a standardized dry extract of *Buddleja globosa* leaves, BG-126, for the treatment and prevention of gastrointestinal disorders caused by treatment with nitrofurantoin and other antimicrobials. Patent USA Appl US8852654B2

Liao Y-H, Houghton PJ, Hoult JRS (1999) Novel and known constituents from *Buddleja* species and their activity against leukocyte eicosanoid generation. J Nat Prod 62:1241–1245

López J, Sierra J, Vegazo ME, Cortés M (1979) Chemical constituents of *Buddleja globosa* Lam. Fitoterapia 5:195–198

Malaret A (1970) Lexicón de Fauna y Flora. Comisión Permanente de la Asociación de Academias de la Lengua Española. Madrid, Spain, vii + 569 pp

Marín G, Giménez M, Cortés F, Pardo F, Núñez J, Naranjo J (1979) Estudio fitoquímico de *Buddleja globosa* Lam. (Buddlejaceae). Revista Latinoamericana de Química 10:19–21

Mensah AY, Houghton PJ, Bloomfield S, Vlietinck A, Vanden Berghe D (2000) Known and novel terpenes from *Buddleja globosa* displaying selective antifungal activity against dermatophytes. J Nat Prod 63(9):1210–1213

Mensah AY, Houghton PJ, Bloomfield S, Vlietinck A, Vanden Berghe D (1999) Novel and Known Constituents from *Buddleja* Species and Their Activity against Leukocyte Eicosanoid Generation. J Nat Prod 62:1241–1245

Mensah AY, Houghton PJ, Hughes M, Cherry G (1998) *In vitro* investigation of the wound-healing properties of *Buddleja globosa*. J Pharm Pharmacol 50:83

Mensah AY, Sampson J, Houghton PJ, Hylands PJ, Westbrook J, Dunn M, Hughes MA, Cherry GW (2001) Effects of *Buddleja globosa* leaf and its constituents relevant to wound healing. J Ethnopharmacol 77(2-3):219–226

Montes M, Wilkomirsky T (1985) Medicina Tradicional Chilena. Ed. de la Universidad de Concepción, Concepción, Chile, 205 pp

Montes M, Wilkomirsky T, Valenzuela L (1992) Plantas medicinales. Ediciones Universidad de Concepción, Concepción, 197 pp

Muñoz O, Montes M, Wilkomirsky T (1999) Plantas medicinales de uso en Chile: Química y Farmacología. Ed. Universitaria S.A, Santiago, Chile, pp 69–72

Norman E, Ariza Epinar L (1995) Buddlejaceae. In: Hunziker AT (ed.): Flora Fanerogámica Argentina, vol 10. Ed. CONICET, Buenos Aires, Argentina, pp 1–14
Oh H, Pae H-O, Oh G-S, Lee SY, Chai K-Y, Song CE, Kwon T-O, Chung H-T, Lee H-S (2002) Inhibition of inducible nitric oxide synthesis by catalposide from *Catalpa ovata*. Planta Medica 68:685–689
Pardo F, Perich F, Torres R (1997) Un nuevo glicósido de *Buddleja globosa* con actividad bactericida. Boletín de la Sociedad Chilena de Química 42(1):101–104
Pardo F, Perich F, Villarroel L, Torres R (1993) Isolation of verbascoside, an antimicrobial constituent of *Buddleja globosa* leaves. J Ethnopharmacol 39:221–222
Park KS, Chang IM (2004) Anti-inflammatory activity of aucubin by inhibition of tumor necrosis factor-alpha production in RAW 264.7 cells. Planta Med 70(8):778–779
Recio MC, Giner RM, Máñez S, Ríos JL (1994) Structural considerations on the iridoids as anti-inflammatory agents. Planta Med 60:232–234
Suwalsky M, Duguet J, Speisky H (2017) An *in vitro* Study of the Antioxidant and Antihemolytic Properties of *Buddleja globosa* (Matico). J Membr Biol 250(3):239–248
The Plant List (2020) Version 1.1 Published on the Internet: http://www.theplantlist.org. (Last access, January 2020)
Varillas AR, TTito DM (2018) Actividad Diurética del extracto etanólico de las hojas de Matico (*Buddleja globosa*) en ratas. Thesis. Universidad Inca Garcilaso de la Vega, Lima, Perú. 167 pp
Wilckens P, Fernández MP, Gomez M, Peña I, Montenegro G (2013) Matico *Buddleja globosa* sustainable management and use of a medicinal plant in Chile. Phyton 82:81–90
Yoshida T, Nobuhara J, Uchida M, Okuda T (1976) Buddledin A B, and C, Piscidal sesquiterpenes from *Buddleja davidii* Franchet. Tetrahedron Lett 41:3717–3720
Yoshida T, Nobuhara FN, Okuda T (1978) Studies on constituents of *Buddleja* species II. Buddledin C, D and E, new sesquiterpenes from *Buddleja davidii* Franch. Chem Pharm Bull 26:2543–2549

Chiliotrichum diffusum (G. Forst.) Kuntze

Sandra Marcela Alcalde Bahamonde, María Luján Flores, Analía Verónica Uhrich, Miriam Delma Escobar Daza, and Osvaldo León Córdoba

Chiliotrichum diffusum (G. Forst.) Kuntze in its habitat (28 de Noviembre city, Santa Cruz, Argentina). (Photo: SM Alcalde Bahamonde and M Errendasoro)

Abstract *Chiliotrichum diffusum* (G. Forst.) Kuntze belonging to the Asteraceae family, is a shrub that is native to South America. *C. diffusum*, like other members of the family Asteraceae are widely distributed in Argentina and comprise a large number of species used in folk medicine. such as anti-inflammatory, digestive, and cholagogues, among others. In Argentina, *C. diffusum* inhabits from the south of the 45° parallel to the province of Tierra del Fuego, Antártida e Islas del Atlántico Sur.

S. M. Alcalde Bahamonde · M. L. Flores · A. V. Uhrich · M. D. Escobar Daza
O. L. Córdoba (✉)
Cátedra de Química Biológica II; Departamento de Bioquímica; Centro Regional de Investigación y Desarrollo Científico Tecnológico (CRIDECIT), Facultad de Ciencias Naturales y Ciencias de la Salud (FCNyCS) - Sede Comodoro Rivadavia, Universidad Nacional de la Patagonia San Juan Bosco (UNPSJB),
Comodoro Rivadavia, Chubut, Argentina
e-mail: okylola@gmail.com

Á. Máthé, A. Bandoni (eds.), *Medicinal and Aromatic Plants of South America Vol. 2*, Medicinal and Aromatic Plants of the World 7,
https://doi.org/10.1007/978-3-030-62818-5_10

Popularly known as "mata negra" or "kòor" in the Ona or Selk´nam language, it is a shrub that grows predominantly in SW Argentina, particularly in areas called wetlands. It has been described as a forage and medicinal plant, highlighting its use by the Onas peoples, who used the flowers to "*clarify the view*" and the branches in cure rituals and tattoos ("lóiste"). In Chilean folk medicine, it is used to improve memory, treat headaches, cramps and varicose veins. *C. diffusum* biosynthesizes bioactive metabolites where the major group corresponds to flavonoids (principal groups), phenolic acids, coumarins, carbohydrates and terpenoids. These metabolites are related to the pharmacological properties evidenced for this species, as antibacterial, cytotoxic, antitumoral, antioxidant, hypotensive, analgesic, anti-inflammatory, gastroprotective and antiulcer. These, in part, are related to the traditional uses of this species.

Keywords "Mata negra" · Onas or Selk´nam · Bioactive metabolites · Patagonia Argentina

1 Introduction

The Patagonian Phytogeographic province of Argentina is inhabited by many native species of the Asteraceae family, which have been used for medicinal purposes by the original peoples, especially the Tehuelches, Tehuelches-Mapuches, Puelches, Onas and Yamanes. Some of these species have survived in the folk medicine of today. In particular, *Chiliotrichum diffusum* (G. Forst.) Kuntze, is a dominant native shrub in the wet Magellan steppe, mainly in the Argentinean provinces of Santa Cruz and Tierra del Fuego, Antártida e Islas del Atlántico Sur, where they were used as traditional medicine by the Ona and Yamanes, as well as the nowadays native peoples of Chile.

Considering the ethnomedical relevance of this species and its abundance in the region, it has been the target of comprehensive research aimed at scientific validation. Both chemical composition and pharmacological properties support the traditional uses and demonstrate the potential of this species in health related applications. At the same time, these studies yield data for the chemotaxonomy of Asteraceae family, and to the interrelationship of species within the particular habitat in which they grow.

2 Taxonomic Characteristics

In Argentina, Asteraceae represent one of the most numerous families with 227 genus and 1400 species. The largest number of native species occurs in the High Andean and Patagonian phytogeographic provinces. In the Santa Cruz province (Patagonia Argentina) they represent 25% of the species (Cabrera 1971).

Chiliotrichum spp. is an informal taxonomic group located in the Astereae tribe that consists of shrubby daisies with densely set, coriaceous leaves with white tomentum abaxially. The species grow mainly in the Andean and Patagonian regions of South America (Bonifacino and Funk 2012). According to Bonifacino (2009) now includes 31 species in 11 genera: *Aylacophora* Cabrera, *Cabreraea* Bonif., *Chiliotrichiopsis* Cabrera, *Chiliophyllum* Phyl., *Chiliotrichum* Cass., *Haroldia* Bonif., *Katinasia* Bonif., *Lepidophyllum* Cass., *Llerasia* Triana, *Nardophyllum* (Hook. & Arn.) Hook. & Arn., and *Ocyroe* Phil.

According to The Plant List (2019) *Chiliotrichum* comprises the following four species: *C. angustifolium* Phil., *C. rosmarinifolium* Less., *C. tenue* Phil. and *C. diffusum* (G. Forst.) Kuntze. However, Bonifacino (2009) reduced *C. rosmarinifolium, C. angustifolium* and *C. tenue* claiming that they are the synonyms of *Chiliotrichum diffusum.* Thus, the genus *Chiliotrichum* is composed of two species: *C. fuegianum* (O. Hoffm.) Bonif., a species confined to the northernmost shores of Tierra del Fuego, and *C. diffusum*, a species occurring from Cape Horn to Mendoza (Argentina) on both sides of the Andes.

Chiliotrichum diffusum (G. Forst.) Kuntze (syn.: *Amellus diffusus* G. Forst.; *Chiliotrichum amelloides* Cass.; *Chiliotrichum amelloides* DC.; *Chiliotrichum amelloideum* Cass.; *Chiliotrichum feliciae* Decne. and *Chiliotrichum virgatum* Phil) (The Plant List 2019), has been classified based on the Cronquist system (Cronquist 1988) and the APG III system (Angiosperm Phylogeny Group) (Cabrera 1971).

Common names of this species are: in Spanish “mata negra”, “mata verde”, “romerillo”, “margarita”, “incienso”; in Ona or Selk´nam language “S q´ò?r” (“arbusto de matanegra” or black bush), “kòor”, “gorhj”; in English, fachine, fascine, fachima, fachina, fachine bush (Martínez Crovetto 1968; Vera 1991; Hince 2000; Abraham 2016). *Chiliotrichum,* from the Greek *chilio*, chilias, chiliados = thousand and *trichum*, thrix, trichos = hairs, alluding to the very bristly pappus (Quattrocchi 1999; Domínguez 2016). The epithet *diffusum*, derived from Latin, refers to the irregular and widely spreading branching pattern that characterizes the species (Bonifacino 2009).

3 Crude Drug Used

The main parts of the plant are used in traditional medicine by the Onas or Selk´nam peoples are the flowers to “clarify the view” and the branches in rituals and tattoos (“lóiste”) (Martínez Crovetto 1968; Vera 1991).

Photo 1 Aerial parts including flowers of *Chilliotrichum diffusum* in 28 de Noviembre city (Santa Cruz, Patagonia Argentina; 51°35′02″ S, 72°12′52″ W). Photo: SM Alcalde Bahamonde and M Errendasoro

The aerial parts are used in the Chilean folk medicine to improve memory, treat headaches, cramps and varicose veins (Asociación Indígena Willi Lafken de Punta Arenas 2010; Domínguez Díaz 2010). Photo 1 shows the aerial parts of this plant.

4 Major Chemical Constituents and Bioactive Compounds

Flavonoids, tannins, steroids, terpenoids, anthraquinones, carbohydrates, proteins and cyanoglycosides were detected in the aerial parts of *C. diffusum* (Alcalde et al. 2008). These chemical groups are related to medicinal properties. The main flavonoids were flavonols, free and glycosylated flavones. Kaempferol-3-*O*-glycoside, apigenin-7-*O*-glycoside, vitexin, gossypin, herbacetin-8-methyl ether, isorhamnetin, quercetin, diosmetin, luteolin, rutin and caffeoylquinic acids stood out. In addition, the presence of glucose, sucrose, fucose, sorbitol, ursolic acid, chlorogenic acid, caffeic acid and betaine were identified (Alcalde Bahamonde 2014; Flores et al. 2015). Coumarins such as umbelliferone, scopoletin, isofraxidin and xanthotoxin were also found (Alcalde Bahamonde et al. 2019a).

In the aqueous extract (decoction) of the flowers of this species were identified hydrolyzable tannins (gallotannins); organic acids (caffeic, chlorogenic and ferulic acid); flavonoids (quercetin 3-*O*-β-D-galactoside, isoquercitrin, quercitrin, formononetin, afzelin, quercetin, rutin, diosmin, rhamnetin, apigenin, kaempferol, myricetin) and carbohydrates (galactose, glucose, xylose, fucose, arabinose, rhamnose and sucrose) (Alcalde Bahamonde 2014; Alcalde Bahamonde et al. 2019b). The proportion of total phenols corresponded to 7.3% expressed as gallic acid and of total flavonoids corresponded to 2.3% expressed as rutin. Also was possible to determine 38.6% of total carbohydrates expressed as rhamnose and 27.0% expressed as galactose (Alcalde Bahamonde 2014).

Other substances have been isolated from the flowers of *C. diffusum*: plumbagin, 3,5-dicaffeoylquinic acid, proline, betaine, asparagine, glycylproline, sorbitol and terpenes such as ursolic acid, spathulenol, bisabolol, azulene and farnesene (Flores et al. 2015; Alcalde Bahamonde et al. 2017). In general, the chemical composition of *C. diffusum* flowers is similar to *Matricaria chamomilla* (Srivastava et al. 2010).

Braun et al. (2015) described substances such as linoleic, palmitic, quinic and propionic acid; benzaldehyde; steroids; phenolic compounds; ferulic alcohol; proline; cyclopentenone and vinyl phenolic compounds.

5 Morphological Description

The genus *Chiliotrichum* includes bush, with alternate, whole leaves, often discolored; radiated, medium, solitary chapters at the apex of the twigs, pedunculated; white marginal flowers and yellow disc. *Chiliotrichum* is the only genus to present white rays in the *Chiliotrichum* group (Bonifacino and Funk 2012).

C. diffusum is perennial, is the tallest life form on the steppe (0.6–2 m), is a slow-growing species, and has leaves with a long life-span (Braun et al. 2015). Tall globose; branches densely disposed, ascending, not ribbed, tomentose and gland-dotted, internodes 1–3 mm long. Leaves densely arranged, persistent, alternate, patent, 10–40 × 1–5 mm, linear to narrowly elliptic or narrowly obovate, base attenuated, petiole broadened towards the stem and constricted above, apex acute to obtuse, margins revolute, densely tomentose on abaxial side, glabrous to arachnoid on adaxial side, gland-dotted in both sides, midvein impressed on adaxial side, discolorous, coriaceous (Bonifacino 2009).

Flowers with the bell-shaped involvement, formed by 3–4 series of imbricated, scarious, oval or lanceolate bracts, the smaller exteriors. Capitula solitary, terminal, notably pedunculate, peduncles 25–40 mm long, heterogamous, radiate. Involucres cylindrical campanulate; phyllaries arranged in 4–5 series; outer phyllaries triangular, apex acute, tomentose abaxially, coriaceous, margins membranaceous, fimbriate; inner phyllaries narrowly elliptic, apex acute, with a tuft of woolly trichomes towards the apex, membranaceous, margins membranaceous, fimbriate. Receptacle is flat or convex, partially paleate, with some linear-lanceolate pallets between the flowers, not embracing the achenes, apex acute, with a tuft of woolly trichomes towards the apex, margins membranaceous, fimbriate (Bonifacino 2009). Dimorphic flowers; the marginal female, with a white corolla linked; those of the disc, hermaphrodites, with tubular yellow corolla pentadentate in the limbus, and anthers with lanceolate connective appendages and rounded teak at their base. Achenes 3.5–6.5 mm long, narrowly obovate, terete, 4- to 6-ribbed, gland-dotted, sometimes with few twin trichomes sparsely distributed towards the apex, carpopodium present, conspicuous. Pappus 3.5–7 mm long, 2 to 3-seriate of rigid, uneven hairs composed of 50–70 linear scales, outer series slightly shorter, scabrid along the margins (Cabrera 1971; Bonifacino 2009). The species shows a wide blooming season, and flowering specimens have been collected from October to April.

6 Geographical Distribution

Chiliotrichum diffusum is native of South America, occurring in Argentina and Chile. It can occur in rather scattered populations or forming denser associations, constituting the dominant species. In Argentina, it is distributed along the Andes in the provinces of Mendoza, Neuquén, Río Negro, Chubut, Santa Cruz and Tierra del Fuego, Antártida e Islas del Atlántico Sur; in Chile, from VI to XII Region. Their distribution includes the immense archipelago West of Chile, Isla de los Estados and Islas Malvinas in Argentina (Bonifacino 2009). It is an important vegetable in the steppe of southern Santa Cruz and northern Tierra del Fuego, from sea level to about 400 m (Chester 2010). Cabrera (1971) also noted that it is a frequent species in the thickets of the Sub-Antarctic Province, from the 45° parallel to Tierra del Fuego and Islas Malvinas. Moore (1968) points out that *Chiliotrichum diffusum* is common in Islas Malvinas, on wet soils beside streams together with *Cortaderia pilosa* (d'Urv.) Hack., growing from sea level to 300 m, but it can also grow on sandy soils near the coast and on occasion forms dense thickets away from streams. In Tierra del Fuego, a considerable part of the Patagonian steppe is covered by associations of *Chiliotrichum diffusum* and *Festuca gracillima* Hook. It can grow scattered or form dense associations intermingled with other shrubs typical of subantarctic forest such as *Berberis buxifolia* Lam., *Ribes magellanicum* Poir., and *Pernettya mucronata* Gaudich. ex G. Don. (Bonifacino 2009). In the Islas Malvinas, at one time it was common but since the arrival of livestock on the islands, it now mainly occurs in valley bottoms, ravines and other inaccessible places (Hince 2000). In Southern Chile and Argentina, it is the dominant plant on the fringes of the *Nothofagus* forest (Chester 2010). This species is cultivated and commercialized in the United Kingdom for ornamental purposes (Bonifacino 2009).

7 Ecological Requirements

Chiliotrichum diffusum inhabits the steppe grasslands in Chile, Argentina and Islas Malvinas, where it is one of the dominant shrubby plants in a plant community that includes *Trevoa* spp, *Schinus polygama*, *Paleaepappus patagonicus*, *Berberis microphylla*, various grasses and sedges (Reynolds and Frame 2005). In Argentina, it grows in the Patagonian phytogeographic provinces, preferably in acid soils, in areas with precipitations from 300 to 450 mm/year (Oyarzábal et al. 2018). It shows a wide adaptation to low temperatures (between 7 and −5 °C) and the drying effect of strong and constant winds (Arroyo et al. 1992). In the Ecotone Fueguino (an ecological unit that represents a transition between the Magellan steppe and the Andean complex), it is associated with the "ñire" (*Nothofagus antarctica*) (Vera 1991). *C. diffusum* represents the greatest floristic wealth in the Magallanes area (Oliva and González 2001). This species does not show effects of decreased growth under the influence of high levels of UV-B radiation (Rousseaux et al. 2001). This could be explained by its physiological acclimatization to withstand harmful radiation. In the case of other species living with *C. diffusum* it has been established that the presence of phenolic compounds in their aerial parts increases in response to UV-B radiation (Vera 1991). This author also highlighted its high values of sociability or ecological integration and abundance-dominance along with *Azorella trifurcata*, *Pernetya mucronata*, some mushrooms, *Nothofagus antarctica* and *Embothrium coccineum*, all considered as indicators of the potential of resources in the environment of Cabo Negro (Magallanes, Chile).

8 Traditional Use (Part(s) Used) and Common Knowledge

Chiliotrichum diffusum (G. Forst) Kuntze is popularly known as "mata negra", "kȍor" (in Selk´nam or Ona language). It has been used, like many other Asteraceae species, in traditional medicine, especially by the Onas peoples, who mentioned its use in cure rituals and tattoos ("lȍiste"), as well as the importance of the flowers that "rubbed in the eyes serve to clarify the view" (Zardini 1984; Vera 1991; Domínguez Díaz 2010). In Chile its medicinal use has been described to improve memory, treat headaches and cramps; it also has been used as a component of creams or oils, sometimes mixed with *Matricaria chamomilla* to treat varicose and cramps (Asociación Indígena Willi Lafken de Punta Arenas 2010). Moreover, it is commonly used as food for sheeps (Ballaré et al. 2001).

9 Modern Medicine Based on Its Traditional Medicine Uses

Different studies of the main metabolites synthesized by *C. diffusum* support its use in traditional medicine. The ethanolic extract of the aerial parts showed activity against *Artemia salina* cytotoxicity bioassay (LD_{50} = 12μg/ml). Their benzene and ethyl acetate fractions were very bioactive (LD_{50} = 3 and 10μg/ml respectively) (Alcalde et al. 2010). These properties are related to the presence of terpenoids and phenolic compounds. The decoction of the flowers showed inhibition of root elongation of wheat of 56% and 69% (dilutions 0.05% and 0.5% respectively) with an IC_{50} of 58.1μg/ml. On the other hand, the aqueous purified fraction from decoction showed inhibition dose dependent activity of 31.4% and 76.5% (dilutions 0.05% and 0.5% respectively); nevertheless, the IC_{50} was > 1000μg/ml. Decoction also showed antiproliferative activity on the cell line SH-SY5Y (IC_{50} 0.62 ± 0.1 mg/ml) and on fibroblast (IC_{50} 1.35 ± 0.84 mg/ml, dose since 1.25 mg/ml). Consequently, the selectivity index of the antiproliferative activity was 2.1 ± 1.6 (Alcalde Bahamonde et al. 2016). The antimicrobial activity was moderate against *Staphylococcus aureus* (Alcalde et al. 2010). Bhakuni et al. (1974) described antibacterial activity against *Escherichia coli*. Ethanol extract from aerial parts and decoction of the flowers showed free radical scavenging capacity on DPPH•

(Alcalde et al. 2008). The results of the quantitative determination of this activity in ethyl ether and the ethyl acetate fraction from the ethanol extract of aerial parts showed SC_{50} of 5.2 and 3.5μg/ml, respectively. Dichloromethane and ethyl acetate fractions of the aqueous extracts from flowers showed SC_{50} values of 35.0 and 9.5μg/ml (Alcalde Bahamonde et al. 2019b). Also, the antioxidant activity was analyzed by bioautography study in relation to the phenolic compounds, fundamentally flavonoids and phenolic acids (Alcalde Bahamonde et al. 2018). The decoction of the flowers had *in vivo* a dose-dependent depressor effect (ΔMAP: 3 $mg.kg^{-1}$ −18 ± 3 mm Hg; 10 $mg.kg^{-1}$ −26 ± 3 mm Hg; 30 $mg.kg^{-1}$ −41±4 mm Hg; n=5) without cardiac effects. This extract showed a dual muscarinic and β-adrenergic depressor effect; this effect of 10 $mm.kg^{-1}$ was blocked by the β-adrenergic antagonist propranolol (ΔMAP: 1 ± 3 mm Hg, n= 5, $p<0,05$) and by atropine muscarinic blockade (ΔMAP: −10 ± 3 mm Hg, n=5, $p<0,05$). A similar dual depressor effect was observed for the ethanol extract of the aerial parts, ED_{50}= 1.71 ± 0.43 mg/kg and a greater hypotensive response of −40.4 ± 4.1 mm Hg (Alcalde Bahamonde 2014).

The ethanolic extract of the aerial parts (30 μg/ml) was negative inotropic in agreement with flavonoids (quercetin and their glycosides) and phenolic acids content, but it was not cardioprotective in I/R (ischemia and reperfusion). It reduced the PICR (postischemic contractile recovery) to 32 ± 11% *vs* 77 ± 3% in the control group after 20 min I-15 minutes R, and finally recovered about 65%, similarly to it. It also increased ΔLVEDP (intraventricular pressure, LVP, in mm Hg). During this intervention, hearts decrease their contractility (P) and increase the diastolic tone (ΔLVEDP), in a characteristic dysfunction (Consolini et al. 2018a).

The ethanolic extract of the aerial parts of the *C. diffussum* also showed antispasmodic effects in *ex vivo* experiments in intestine, bladder and uterus. The extract was about 5 times more potent in intestines (IC_{50} of 9.5 ± 4.2 μg/ml) than in bladder or uterus. The mechanism was a non-competitive inhibition of contraction and of Ca^{2+} influx to smooth muscle, as well as verapamil showed, but with about 15 times more potency than it in intestine (Consolini et al. 2018b). This activity is related to the phytochemical composition, fundamentally phenolic acid and flavonoids.

The decoction of the flowers of this species showed gastroprotective activity dose dependent against gastric damage induced by ethanol in mice, reaching an inhibition of 61, 67 and 75% (100, 500 and 1000 mg/kg of body weight, respectively), while omeprazole reached an inhibition of 75.9%. This extract administered orally did not show toxicity by acute exposure in mice; the histopathological study did not show alterations in the analyzed animal organs (control and treated animal) (Bucciarelli et al. 2018). The presence of flavonoids and phenolic acids with antioxidant activity was observed by bioautography, mainly kaempferol-3-*O*-glycoside, apigenin-7-*O*-glycoside, chlorogenic acid, ferulic acid and caffeic acid. The gastroprotective activity is related to the main phenols present in the extract, which also have antioxidant capacity.

A decoction of the flowers induced a significant anti-inflammatory effect (inhibition of 56.0% at 3 h) and produced significant inhibition on nociception in the acetic acid test (ED_{50} 35 mg/kg i.p.; ED_{50} 709 mg/kg *p.o.*). In the hot plate test, the antinociceptive activity of the extract employed at 500 mg/kg i.p. was significantly suppressed by pretreatment with naloxone (5 mg/kg). This extract showed the presence of chlorogenic acid, caffeic acid, hyperoside, isoquercitrin, quercitrin, afzelin, quercetin, apigenin and kaempferol; antinociceptive and anti-inflammatory effects that may be related to the presence of these metabolites. The opioid system seems to be involved in the mechanism of antinociception of the extract (Alcalde Bahamonde et al. 2013).

10 Conclusions

Chiliotrichum diffusum is widely used in traditional medicine in the Patagonian Region of Argentina and Chile. The studies on its chemical composition (fundamentally flavonoids, phenolic acids, terpenoids and coumarins) and pharmacological properties such as antibacterial, cytotoxic, antioxidant, antitumoral, hypotensive, analgesic, anti-inflammatory, gastroprotective and antiulcer agents validate its principal uses in traditional medicine.

This species presents a metabolic profile similar to *Matricaria chamomilla* ("chamomile"), another plant drug of the Asteraceae widely used medicinally. These results, added to the absence of acute toxicity, demonstrate the pharmacological potential of *C. diffusum*.

Acknowledgments The authors are especially grateful to the ChemBioFight IRSES Project, SPU, UNPSJB and FONCYT Argentina, for financial support.

References

Abraham R (2016) Torres del Paine: Trekking in Chile's Premier National Park. Ed. Cicerone Press Limited, China, 192 pp

Alcalde Bahamonde SM, Flores ML, Córdoba OL, Taira CA, Gorzalczany S (2013) Antinociceptive and anti-inflammatory activities of an aqueous extract of *Chiliotrichum diffusum*. Braz J Pharmacogn 23(4):699–705

Alcalde Bahamonde SM (2014) Actividad farmacológica y estudio fitoquímico de *Chiliotrichum diffusum* (Asteraceae). Thesis. Universidad Nacional de la Patagonia San Juan Bosco. Comodoro Rivadavia, Chubut, Argentina, 254 pp

Alcalde Bahamonde SM, Weisntein-Oppenheimer C, Flores ML, Córdoba OL (2016) Advances in knowledge of *Chiliotrichum diffusum* (Asteraceae), a native species from Patagonia used in traditional medicine of the Onas. Pharmacologyonline 1:156

Alcalde Bahamonde S, Uhrich A, Arancibia L, Flores M, Córdoba O (2017) *Chiliotrichum diffusum* (Asteraceae): new contributions to pharmacognostic knowledge in relation to the traditional use of Onas and Yamanes. In: Abstract of the XXVI Congress SILAE. IX Congreso Colombiano de Cromatografía. 25–29 September. Cartagena de Indias, Colombia

Alcalde Bahamonde SM, Uhrich AV, Gorzalczany SB, Taira CA, Bucciarelli A, Córdoba OL (2018) Perfil fitoquímico y farmacológico de las flores de *Chiliotrichum diffusum* (G. Forst.) Kuntze (Asteraceae). 1° Reunión Conjunta, 5° Reunión Internacional de Ciencias Farmacéuticas

(RICiFa), 50° Reunión Anual de la Sociedad Argentina de Farmacología Experimental (SAFE). 14–16 November. San Luis, Argentina

Alcalde Bahamonde SM, Uhrich AV, Córdoba OL, Flores ML (2019a) Coumarins from extracts of aerial parts of *Chiliotrichum diffusum* (Asteraceae), a medicinal plant used by the Onas. In: Abstract of the XXVIII Congress SILAE. 16-20 September. La Habana, Cuba

Alcalde Bahamonde SM, Taira CA, Flores ML, Córdoba OL (2019b) Antioxidant activity of the aerial parts and the flowers of *Chiliotrichum diffusum* (G. Forst.) Kuntze (Asteraceae), a medicinal plant used by the Onas. Infarma 31:113–120

Alcalde SM, Taira C, Córdoba OL, Flores ML (2008) Contribución al conocimiento farmacognóstico de *Chiliotrichum diffusum* (Asteraceae). Nat pat 4(1):37–45

Alcalde SM, Taira CA, Höcht C, Córdoba OL, Flores ML (2010) Fitoquímica y bioactividad de *Chiliotrichum diffusum* (Asteraceae), una Especie Nativa de la Patagonia Andina. Lat Am J Pharm 29(2):284–288

Arroyo MTK, von Bohlen CP, Cavieres, Marticorena C (1992) Survey of the alpine flora of Torres del Paine National Park, Chile. Gayana Botanica 49:47-70

Asociación Indígena Willi Lafken de Punta Arenas (2010) Catastro de Flora nativa usada con fines medicinales por la Agrupación en la Reserva Nacional Magallanes en la Región de Magallanes y Antártica Chilena. Ed. In: Fondo de Protección Ambiental. Ministerio del Medio Ambiente, Chile. (Disclosure form)

Ballaré C, Rousseaux M, Searles P, Zallerb J, Giordano C, Robson T et al (2001) Impacts of solar ultraviolet-B radiation on terrestrial ecosystems of Tierra del Fuego (Southern Argentina). An overview of recent progress. J Photoch Photobio B 62:67–77

Bhakuni DS, Bittner M, Marticorena C, Silva M, Weldt E, Melo ME et al (1974) Screening of Chilean plants for antimicrobial activity. Lloydia 37:621–632

Bonifacino JM (2009) Taxonomic revision of the *Chiliotrichum* Group *sensu stricto* (Compositae: Asteraceae). Smithsonian Contr Bot (92):1–128

Bonifacino JM, Funk VA (2012) Phylogenetics of the *Chiliotrichum* group (Compositae: Astereae): The story of the fascinating radiation in the paleate Astereae genera from southern South America. Taxon 61(1):180–196

Braun K, Vitale A, Collantes M (2015) Land use and Abiotic Effects on the Biosynthesis of Special Metabolites of the Scrub *Chiliotrichum diffusum* (Asteraceae) in Southern Patagonia. International J Res Pharm Biosc 2:27–37

Bucciarelli A, Alcalde Bahamonde S, Uhrich A, Flores ML, Córdoba O (2018) Gastroprotective activity and toxicological evaluation of the aqueous extract from *Chiliotrichum diffusum* (G. Forst.) Kuntze (Asteraceae). In: Abstract of the XXVII Congress SILAE. 9–13 September. Milazzo, Italia

Cabrera AL (1971) Compositae. In: Correa MN (ed) Flora Patagónica. Part VII. Colección Científica del INTA, Argentina, pp 37–38

Consolini AE, Matera S, Bayley M, Colareda GA, Ragone MI, Bonazzola P, et al (2018a) Ethnopharmacology of South American medicinal plants with cardiovascular effects. In: Abstract of the XXVII Congress SILAE. 9–13 September. Milazzo, Italia

Consolini AE, Matera S, Colareda GA, Bayley M, Alcalde Bahamonde SM, Flores ML et al (2018b) Intestinal, vesical and uterine antispasmodic effects of the patagonic plant *Chiliotrichum diffusum* (Asteraceae). In: Abstract of the XXVII Congress SILAE. 9–13 September. Milazzo, Italia

Chester S (2010) A wildlife guide to Chile: Continental Chile, Chilean Antarctica, Easter Island, Juan Fernandez Archipelago. Ed. Princeton University Press, Princeton/Oxford, p 88

Cronquist A (1988) The evolution and classification of flowering plants, 2nd edn. Ed. New York Botanical Garden, Bronx, USA, 555 pp

Domínguez Díaz E (2010) Flora de interés etnobotánico usada por los pueblos originarios: Selk'nam, Kawésqar, Yagan y Haush en la Patagonia Austral. Dominguezia 26(2):19–29

Domínguez E (2016) Libro flora Torres del Paine. Ed. Ocho libros editores, Santiago, Chile, 175 pp

Flores ML, Alcalde Bahamonde S, Escobar Daza MD, Córdoba OL, Wilson E, Choi YH (2015) Perfil metabólico de *Chiliotrichum diffusum* (Asteraceae) mediante Espectroscopía de Resonancia Magnética Nuclear. In: Abstract of the XLVII Reunión Anual de SAFE, VI Congreso Iberoamericano de Ciencias Farmacéuticas, Reunión de la Comisión Permanente de la COIFFA, III Congreso Sudamericano de Biofarmacia y Farmacocinética (BFFC). 4–6 November. Universidad Nacional de Córdoba, Córdoba, Argentina

Hince B (2000) The Antarctic Dictionary: A Complete Guide to Antarctic English. Ed. Csiro Publishing and Museum of Victoria, Collingwood, Australia, 404 pp

Martínez Crovetto RN (1968) Estudios Etnobotánicos. IV. Nombres de plantas y su utilidad según los indios onas de Tierra del Fuego. Etnobiológica (3):1–20

Moore DM (1968) The vascular flora of the Falkland Islands. Brit Antarct Surv Sci Rep 60:1–202

Oliva G, González L (2001) Áreas ecológicas de Santa Cruz y Tierra del Fuego. Ganadería Ovina Sustentable en la Patagonia Austral 1:41–82

Oyarzábal M, Clavijo J, Oakley L, Biganzoli Z, Tognetti P, Barberis I et al (2018) Unidades de vegetación de la Argentina. Ecol Austral 28:40–63

Quattrocchi U (ed) (1999) World dictionary of plant names: common names, scientific names, eponyms, synonyms, and etymology, vol 1. Ed. CRC Press, Washington DC, USA, 728 pp

Reynolds SG, Frame J (2005) Grasslands: developments, opportunities, perspectives. Ed. Science Publishers Inc., Enfield, UK, 467 pp

Rousseaux MC, Scopel AL, Searles P, Caldwell MM, Sala OE, Ballarea CL (2001) Responses to solar ultraviolet-B radiation in a shrub dominated natural ecosystem of Tierra del Fuego (southern Argentina). Glob Chang Biol 7:467–478

Srivastava JK, Shankar E, Gupta S (2010) Chamomile: a herbal medicine of the past with a bright future. Mol Med Rep 3:895–901

The Plant List (2019) Version 1.1. Published on the Internet; http://www.theplantlist.org/ (Last access, December 2019)

Vera J (1991) Uso alimentario de recursos vegetales entre cazadores recolectores de Bahía Laredo, Cabo Negro (Magallanes). Anales del Instituto de la Patagonia. Universidad de Magallanes, Punta Arenas, Chile. Serie Ciencias Sociales 20:155–168

Zardini EM (1984) Etnobotánica de Compuestas Argentinas con especial referencia a su uso farmacológico. Primera Parte. Acta Farm Bonaerense 3(1):77–79

Cissampelos pareira L.

Bárbara Verónica Ricciardi Verrastro, Ana María Torres, Gabriela Ricciardi, and Eduardo Santiago Dellacassa Beltrame

Cissampelos pareira growing in Paso de la Patria, Corrientes, Argentina. Photos: BV Ricciardi Verrastro

Abstract *Cissampelos pareira* L. (Menispermaceae) is a widely used medicinal plant, also used to treat bites of venomous animals, particularly snakes. Ophidian accidents are a serious public health problem in Argentina where the *Bothrops*

B. V. Ricciardi Verrastro (✉) · A. M. Torres · G. Ricciardi
Laboratorio de Productos Naturales Prof. Armando Ricciardi – LabProdNat – Facultad de Ciencias Exactas y Naturales y Agrimensura, IQUIBA - Universidad Nacional del Nordeste, Corrientes, Argentina
e-mail: barbieriricciardi87@gmail.com

E. S. Dellacassa Beltrame
Laboratorio de Biotecnología de Aromas, Departamento de Química Orgánica, Facultad de Química, Universidad de la República, Montevideo, Uruguay

Á. Máthé, A. Bandoni (eds.), *Medicinal and Aromatic Plants of South America Vol. 2*, Medicinal and Aromatic Plants of the World 7,
https://doi.org/10.1007/978-3-030-62818-5_11

genus of snakes is responsible for 97% of these accidents, peculiarly *B. diporus* ("yarará chica") responsible for 80% of them. In the northeast of the country (Corrientes Province), *C. pareira* is commonly used against the venom of *B. diporus*; its use is described in almost all-ethnobotanical literature of the countries where the plant grows. *In vitro* and *in vivo* antivenom activities of *C. pareira* extracts from plants collected at two different phytogeographic regions of Corrientes (Argentina) were evaluated against *B. diporus* venom. The seasonal influence on the chemical composition of extracts was also studied to determine the associated range of variability and its influence on the antivenom activity. Besides, an evaluation of the chemical composition of the volatile extract was performed to assess its chemical stability. These results support the ethnopharmacological use of this species. Moreover, the presented data demonstrate that certain flavonoids may mitigate some of the venom-induced local tissue damages.

Keywords Antivenom activity · Phytochemical · *In vitro* and *in vivo* activities · "Ka'apevá"

1 Introduction

Production of secondary metabolites by plants represents an adaptive capacity to environmental challenges and the changing of growth conditions. For centuries, medicinal plants' usage evolved as a result of the many years of humanity struggle against illnesses and its learning to pursue drugs in different parts of the plants. In time, many plants have been used as antidotes for snake envenomations (Inoue et al. 2019). Usually, antivenoms are hyper immune sera collected from animals which bind and inactivate venom components. Over the years many attempts have been made for the development of snake venom antagonists especially from plant sources, so many reports can be found in the literature related to the use of medicinal plants against snakebite by the different ethnic communities throughout the world, especially in rural parts from tropical and subtropical countries. Some of the plant extracts were found to have a potency as antivenom *in vitro* but failed to show venom neutralizing ability *in vivo* (Gupta and Peshin 2012).

Generally, local effects of a snakebite occur in the first 10–30 min; and there may be numbness around the bite with bleeding, or a purpuric rash, and/or necrosis or gangrene (De la O Cavazos et al. 2012). These local reactions are not effectively neutralized by conventional antivenom serum therapy. In severe cases, local effects of envenoming may lead to permanent tissue loss, disability, or amputation (Gutiérrez 2002).

The lack of medical infrastructure in the rural areas, human ignorance, the side effects of animal-based antivenoms are issues that explain the necessity to develop alternative therapies, so that snake bite investigators engaged in finding out the scientific basis of certain plants' use as antiophidian ethnomedicine.

The current chapter discusses the pharmacognostic characteristics of *Cissampelos pareira* L. (Menispermaceae) with the aim of providing useful information for its correct identity. However, the information is focused on scientifically explaining the ethnobotanical use of *C. pareira* against venom of *Bothrops diporus* in Northeast Argentina by identifying those compounds responsible for the alexiteric activity.

2 Taxonomic Characteristics

The name of the *Cissampelos* genus derives from the Greek *kissos* which means ivy and *ampelos*, vine, referring to its fruits arranged in form of clusters. In Argentina, the species is commonly known as "ka'apevá", "ka'á-pevá", "ysypó-morotí", "caá-pebá", "zarza", "pareira brava" or "mil hombres" (Ricciardi et al. 1996).

C. pareira is a climbing plant with alternating simple leaves that are usually zygomorphic in this genus, usually small, greenish, or white (Photo 1). The fruit is an aggregate of drupes.

Synonymous: *Cissampelos pareira* L. var. *australis* (A.St.-Hil.) Diels, *Cissampelos pareira* L. var. *caapeba* (L.) Eichler, *Cissampelos pareira* L. var. *gardneri* Diels, *Cissampelos pareira* L. var. *tamoides* (Willd. ex DC.) Diels, *Cissampelos auriculata* Miers, *Cissampelos hederacea* Miers, *Cissampelos monoica* A. St.-Hil., *Cissampelos pareira* L. var. *monoica* (A.St.-Hil.) Eichler, *Cissampelos australis* A. St.-Hil., *Cissampelos litoralis* A. St.-Hil., *Cissampelos pareira* L. *f. reniformis* Chodat & Hassl., *Cissampelos pareira* L. *f. emarginato-mucronata* Chodat & Hassl., *Cissampelos caapeba* L, among others (Zuloaga and Belgrano 2014; The Plant List 2019).

Photo 1 Aerial parts of *Cissampelos pareira*. (Photo: B Ricciardi Verrastro)

3 Crude Drug Used

C. pareira leaves and roots are used in aqueous infusions (Leonti et al. 2001; Jain et al. 2005; Namsa et al. 2011; Heinrich et al. 2014). In certain circumstances leaves are also applied topically or macerated in wine (Abbasi et al. 2010; Haq et al. 2011).

4 Major Chemical Constituents and Bioactive Compounds

Despite the large amount of information on *Cissampelos pareira* phytochemistry, this review focuses on the results obtained with the study of plant extracts on local effects of *Bothrops diporus* envenomation.

Volatile Fraction Semwal et al. (2014) only cite the presence of thymol in roots. By simultaneous distillation extraction (SDE), we found the volatile extracts showed the following chemical composition by GC-MS (Table 1) (full data not published).

The volatile extract from aerial parts was characterized by β-pinene (23.2%), limonene (22.6%), α-pinene (8.8%), myrcene (6.3%) and (*E*)-β-ocimene (6.3%). While in roots, the main volatile compounds were limonene (34.8%), stearic acid (14.2%), myrcene (4.9%), methyl palmitate (4.1%), α-pinene (3.3%) and carvone (3.4%). On the other hand, the presence of oxygenated monoterpenes and fatty acids in the volatile fraction of the roots was remarkable (Table 1).

Nonvolatile Fraction Extracts of *C. pareira* have been widely studied in different countries, and their components are divided into two main chemical groups: alkaloids (Table 2) and non-alkaloids (Table 3).

Some components have biological properties: tetrandrine (analgesic, antipyretic, anti- inflammatory, cardioactive, and hypotensive effects), pareirubines A and B (antileukemic), alkaloids (febrifuges and curarizes), berberine (hypotensive, antimicrobial, and antifungal effects) (Sánchez-Medina et al. 2001) and cissampelin (muscle relaxant) (Semwal et al. 2014).

Table 1 (%) volatile compounds families

Compound family	% aerial parts	% roots
Monoterpenes hydrocarbons	76.4	66.5
Oxygenated monoterpenes	2.8	11.9
Sesquiterpenes hydrocarbons	4.0	0.6
Oxygenated sesquiterpenes	0.1	0
Nor isoprenoids	1.9	0.1
Fatty acids	0	24.7
Fatty acid esters	0	4.0
Others	0.5	0.3
Not identified	17.1	3.7

Table 2 Alkaloid components from *C. pareira*

Alkaloid	Organ	Reference
Pelosine (bisbenzylisoquinoline)	Roots	Wiggers (1840)
Hayatine, (−)-curine	Roots	Scholtz (1896)
Hayatinine	Whole plant	Bhattacharji et al. (1952, 1956, 1962), Bhatnagar et al. (1967), Bhatnagar and Popli (1967), and Haynes et al. (1966)
Isochondodendrine, cissamperine	Roots; fruits; whole plant	Kupchan et al. (1960, 1965, 1966)
Hayatine, berberine, curine, isochondodendrine	Roots	Boissier et al. (1965)
Cissamine hydrochloride, cyclanoline, isochondodendrine, dicetrin	Roots	Anwer et al. (1968)
Dehydrodicentrine, cicleanine, insularine, berberine, hayatidine	Roots	Dwuma-Badu et al. (1975)
Isochondodendrine, cicleanine, sepeerine	Whole plant	Bhakuni et al. (1987)
Warifteine, methyl-warifteine	Leaves	Aguirre-Galvis (1988)
Laudanosine, nuciferine, bulbocarpine, corituberine, magniflorine hydrohloride	Leaves and stems	Ahmad et al. (1992)
Pareirubrines A and B, grandirubrine, isoimerubtine	Whole plant	Morita et al. (1993a, b)
Azafluoranthene, norimeluteine, norruffscine	Whole plant	Morita et al. (1993c)
Pareitropone	Whole plant	Morita et al. (1995)
Berberine, reserpine, cissampeline (pelosine)	Roots	Sharma et al. (2004), Stepp (2004), Bafna and Mishra (2010)
Magnoflorine, magnocurarine	Roots	Bala et al. (2017)

Table 3 Non alkaloid components from *C. pareira*

Non-alkaloid component	Active organ	References
Quercitol	Roots	Srivastava (1955), Chowdury (1972), Dwuma-Badu et al. (1975)
Sterols	Aerial parts	Ramírez et al. (2003)
Cissampeloflavone	Leaves	Singthong et al. (2005)
Chalcone flavone dimer	Whole plants	Amresh et al. (2007a)
Pectins	Whole plants	Ramasubramaniaraja and Badu (2010)
Galacturonic acid (70–75%) and neutral sugars	Leaves	Vardhanabhuti and Ikeda (2006)

For native materials from Paso de la Patria (Corrientes, Argentina) of this species, whose ethanolic extracts showed to be active as alexiteric, the phytochemical analysis showed the stability of phenols and anthraquinones in different vegetative states of the species, while alkaloids were only found in extracts from autumn and saponins in extracts obtained from spring and summer. After bio-guided fractionation of the ethanol extract of aerial parts, following alexiteric activity, the phytochemical analysis resulted in presence of alkaloids, phenols and flavonoids in the most alexiteric active fraction. Possibly the alkaloids and flavonoids have not been identified in the whole extract due to a dilution phenomenon. In addition, steroids would not be related to the biological activity of the extracts, due to their presence in the less active fractions. The presence of polyhydroxylated flavonoids was detected due to its UV spectra by HPLC DAD, in the active fraction analysis. The structural elucidation of flavonoids was carried out by UPLC-MS, identifying quercetin-3-*O*-sophoroside [quercetin-3-*O*-β-D-glucosyl-(1→2)-β-D-glucoside], naringenin-7-*O*-β-D-glucoside, eriodictyol-7-*O*-β-D-glucoside, galangin-7-glucoside, and baicalein-7-*O*-glucoside (oroxin A) (Ricciardi-Verrastro et al. 2018).

5 Morphological Description

C. pareira is a perennial climbing plant, branched, striate, 2–5 m high with a thickened root and pubescent or sub-glabrous. Leaves are peltate and alternate, 3.8 to 14 cm in diameter orbicular or reniform often slightly broader than long, cordate or sometimes truncate at the base. The plant is dioecious with flowers arranged in bunches green, white or yellow; the staminate (male) has 4 free sepals and 4 fused petals; pistillate (female) flowers have 1 sepal, 1 petal, and a solitary carpel. The fruits are small red or yellow drupes (Don 1831; Rhodes 1975; Stevens 2001; Tamaio et al. 2010; Singh et al. 2013a).

6 Geographical Distribution

Cissampelos is one of 70 genera belonging to the Menispermaceae family established by Linne in 1753 in the first edition of the *Species Plantarum* (Granda-Lorenzo and Fuentes-Fiallo 1991). It has a global distribution throughout the five continents, in regions of low altitude rainforests (up to 2100 m.a.s.l.) (Singh et al. 2013a).

In Argentina, there are 3 genera belonging to the Menispermaceae family: *Cissampelos*, *Hyperbaena*, and *Odontocaraya* (Zuloaga and Belgrano 2014). Their geographical distribution comprises the Argentinean provinces of Catamarca, Chaco, Corrientes, Formosa, Jujuy, Misiones, Salta, Santa Fe, Tucumán, Santiago del Estero for *Cissampelos* species, while *Hyperbaena* genus is located in Chaco and Formosa Provinces and *Odontocaraya* species have been reported for Corrientes,

Misiones, Chaco, Formosa, Misiones, Salta and Tucumán (Zuloaga and Belgrano 2014).

7 Ecological Requirements

Cissampelos pareira L. has worldwide distribution, occurring in tropical and subtropical regions of the Americas, Africa and Asia (Ortiz 2001). In Brazil, it is encountered from Caatinga, Atlantic Forest and Amazon forest. In Africa, this species occurs in subtropical forest, savannah, deciduous shrubs, often persisting in cleared land and plantations, also in secondary vegetation and near rock outcrops. It is the most popular species of *Cissampelos* not only for its wide distribution, but mainly because its leaves and roots are widely used as medicinal.

8 Traditional Use (Part(s) Used) and Common Knowledge

The popular uses of *C. pareira* are mentioned in almost all ethnobotanical literature from countries where the plant grows, including South America, Asia, and Africa. As per traditional knowledge, it is used as a carminative, febrifuge, for liver disorders, constipation, menstrual pain, colic, and rheumatism. It also has been used for cough, delirium, madness, epilepsy, seizures, as antiplasmodial, stimulant, sedative, analgesic, antioxidant, tonic and narcotic. (Gessler et al. 1994, 1995; Sudarsanam and Prasad 1995; Taylor 1996; Antoun et al. 2001; Leonti et al. 2001; Galicia et al. 2002; Kakrani and Saluja 2002; Rajan et al. 2002, 2003; Sharma et al. 2004; Chhetri et al. 2005; Jain et al. 2005; Kufer et al. 2005; Kumar et al. 2006; Bora et al. 2007; Mendes and Carlini 2007; Muthaura et al. 2007; Pattanaik et al. 2008; Rukunga et al. 2009; Abbasi et al. 2010; Ramasubramaniaraja and Badu 2010; Giorgetti et al. 2007; Gupta et al. 2011; Haq et al. 2011; Nagarajan et al. 2011; Namsa et al. 2011; Samanta and Bhattacharya 2011; Basha and Sudarsanam 2012; Kaur et al. 2012; Sharma et al. 2012; Singh et al. 2013b; Semwal et al. 2014; Heinrich et al. 2014).

According to ethnopharmacological surveys, decoctions of leaves and roots of *C. pareira*, as well as aqueous or alcoholic infusions, are traditionally used against ophidian venom (alexiteric plants: those which can relieve one or more complex symptoms such as pain, bleeding, inflammation, infection or even the same poisoning). In the Northeast of Argentina (Ricciardi et al. 1996), Paraguay (Jolis 1972; Manfred 1977; Montenegro 1979; González-Torres 2013) and many tropical countries, root decoction in water is used to treat snake bites and other poisonous animals (Morton 1981). In Mexico, the whole plant is used against snake venom (Ramos-Hernández et al. 2007) and its use also extends to Amazon (Ecuador and Peru) and Central America (Barranco-Pérez 2010). In India, root decoction or the whole plant (Chakraborty and Bhattacharjee 2006; Jabeen et al. 2009; Sankaranarayanan et al.

2010; Dey and De 2012); in Pakistan also use leaves or the entire plant in poultices applied on the bite site (Butt et al. 2015).

A 50% hydroalcoholic roots extract (2 g/kg) administered orally did not present acute or subacute toxicity (Amresh et al. 2008). Wipawee and Jintanaporn (2012), studied the acute toxicity of a 50% hydroalcoholic extract obtained from a commercial mixture of *C. pareira* and *Anethum graveolens* (1:5). Up to 5000 mg/kg body weight did not cause death or toxic rat symptoms. According to the guidelines of the Organization for Economic Cooperation and Development (OECD) for acute oral toxicity, an LD_{50} of 2000 mg/kg body weight is classified as unclassified and, therefore, the product is considered safe.

Piero et al. (2015) performed a histopathological analysis of liver, kidney, heart, spleen, brain, lungs, eyes and testicles of albino mice Swiss White after administration of aqueous extract of *C. pareira* leaves (1 g/kg body weight) for 28 days. When the administration was oral, no histopathological changes were evidenced in any of the organs analyzed. However, when the administration was intraperitoneal, changes in the liver were observed, with proliferation of fibrous tissue in the serous layer and presence of mixed inflammatory cells, which would be indicative of peritonitis; and in the spleen, a reduction in cell density of lymphoid follicles.

9 Modern Medicine Based on Its Traditional Medicine Uses

Some of the pharmacological properties of *Cissampelos pareira* have been scientifically investigated by different researchers, as shown in Table 4.

There are many citations on the alexiteric activity scientifically validated for *C. pareira*. Badilla et al. (2008) in Costa Rica, reported antihemorrhagic and antiproteolytic activity from an aqueous infusion of the entire plant against *B. asper* venom. Saravia-Otten et al. (2015) studied the ethanolic extracts of roots collected in Guatemala against *B. asper*; finding antiproteolytic activity but this extract failed to inhibit the hemolytic or coagulant activities of the venom. In Argentina, as above mentioned, the *Bothrops* species is responsible for 97% of the ophidian accidents, being the species *B. diporus* responsible for 80% of them. In order to scientifically evaluate this plant application, the *in vitro* and *in vivo* antivenom activities of *C. pareira* extracts were evaluated against *B. diporus* venom, with a focus on the local effects associated with envenoming.

The seasonal influence on the chemical composition of the active extracts was also studied; in order determine the associated range of variability and its influence on the antivenom activity. The research was conducted using aerial parts (leaves, flowers, tender stems) and roots of *Cissampelos pareira* collected from two different phytogeographic regions of Corrientes (Argentina): Paso de la Patria (PP) and Lomas de Vallejos (LV). In addition, to perform a seasonal analysis and to evaluate the metabolic stability, material was collected at three different growth stages.

In vivo and *in vitro* antisnake venom activities were tested, and a bio-guided chromatographic separation was performed in order to determine the active

Table 4 Biological activities of *C. pareira* scientifically validated

Biological activity	Responsible extracts	References
Antidiarrheal	Hydroalcoholic, roots	Amresh et al. (2004)
Anti-inflammatory	Hydroalcoholic 50%, roots; ethanolic	Amresh et al. (2007a, 2007b)
Analgesic and antipyretic	Ethanolic	Reza et al. (2014)
Antifertility	Methanolic and water, leaves	Ganguly et al. (2007, 2018) and Ampa et al. (2010)
Anthelmintic	Aqueous aerial parts	Padmani et al. (2012)
Antioxidant	Hydroalcoholic 50%, roots; alcoholic and ethyl acetate	Amresh et al. (2007c) and Gul et al. (2016)
Hepatoprotective	Hydroalcoholic 50%, roots	Surendran et al. (2011)
About memory and learning	Hydroalcoholic 50%, roots	Pramodinee et al. (2011)
Anxiolytic	Hydroethanolic 70%, leaves; alkaloids (berberine), terpenoids and phenolic compounds	Priyanka (2013)
Cardioprotective	Ethanolic roots; alkaloids (berberine); flavonoids	Patnaik et al. (1973), De Freitas et al. (1996), Amresh et al. (2007a) and Singh et al. (2013b)
Immunomodulatory	Methanolic, roots; *bis*-benzyl isoquinoline and berberine alkaloids	Bafna and Mishra (2010)
Antitumor	Protein and polysaccharide extract; methanolic extract	Meng et al. (2002) and Thavamani et al. (2014)
Antileukemic	Pareirubrins A and B; grandirubrins and isoimeribrin: Alkaloids	Morita et al. (1993a)
Anticancer	Hexane extract; oleanolic and oleic acids	Bala et al. (2014)
Gastric cancer protector	Hydroalcoholic 50%, roots	Amresh et al. (2007d)
Hypoglycemic	Water, leaves	Kuldeep et al. (2013)and Piero et al. (2015)
Antidengue	Ethanolic	Sood et al. (2015)
Simile curare	Hayatinine methyl hydrochloric	Basu (1970) and Bhatnagar and Popli (1967)
Nephroprotective	Hydroalcoholic of whole plant	Danduga et al. (2015)
Antiurolithic	Ethanolic, leaves	Babu et al. (2014)
Diuretic	Ethanolic, roots	Sayana et al. (2011)

chemicals involved. The fractions obtained were analyzed by sodium dodecyl sulfate polyacrylamide gel electrophoresis (SDS-PAGE) and the chemical profile of the most active constituent was analyzed by ultra-high-performance liquid chromatography coupled to quadrupole/high-resolution mass spectrometry (Q-Orbitrap). (UHPLC-MS). The extracts analyzed showed significant differences in the modification of the venom band pattern of *B. diporus*. These differences depended on their vegetative state (season in which they were collected), on edaphological factors (geographic region of collection) and also on the constituent organ of the plant (roots / aerial parts). In general, aerial parts collected in PP were more active than those collected in LV. However, in the LV, roots were more active.

Respect the vegetative state of the species in summer (flowering stage) the extracts of aerial parts were more active than in spring and autumn. The polar extract (ethanolic) showed greater activity than the aqueous and hexane. When results were analyzed, the alcoholic extract was found to be the most active. The bio-guided fractionation allowed selection one fraction to be analyzed by UHPLC-MS in order to identify the components responsible for the activities found; this study identified five possible flavonoids (Ricciardi-Verrastro et al. 2018).

The entire work on the activity of *C. pareira* against the venom of *B. diporus* allowed to confirm that this species possesses inhibitory effects in both *in vitro* and *in vivo* models. Moreover, these data demonstrate that certain flavonoids may mitigate some of the venom-induced local tissue damage. Considering the results of *C. pareira* from other regions of the world, these results are consistent with those obtained by Saravia-Otten et al. (2015) in Guatemala, who also found no activity in the ethanolic root extracts for *B. asper* venom.

The most alexiteric active fraction found was used to prepare a heat-sensitive gel formulation (1% plant extract, 10% ethanol 96°, 10% DMS, 80% poloxamer) and BALB/c mice were used *in vivo* tests to evaluate the ability to neutralize the local effect of *B. diporus* venom. Histopathological analysis showed differences between groups with and without dermal application. Group treated with heat-sensitive gel showed areas with absence of epidermis with mild inflammatory infiltrate and neutrophilic absence of hemorrhage, while the group without heat-sensitive gel treatment evidenced dermonecrosis and intense bleeding with inflammatory infiltrate. After 3 days of gel application, it was possible to obtain a remarkable enhancement in tissue renewal, demonstrating the activity and usefulness of the preparation. Even when it is essential to improve the permeation of the bioactive compounds, the formulation seems to be a valid option for local treatment (Ricciardi-Verrastro et al. 2016).

10 Conclusions

Investigations into the activity of *C. pareira* against the *B. diporus* venom confirmed that this species possesses inhibitory effects in both *in vitro* and *in vivo* models. The screening of the alexiteric activity by SDS-PAGE showed that plant material collected in different geographical regions can present different activity levels, indicating probable secondary metabolism expression differences. The aerial parts were more active than roots in experiments where a total clearance of the venom bands was observed after treatment with the ethanolic extracts of aerial parts collected in summer by SDS-PAGE. Considering that the snake bites occur in a higher incidence in persons and moderately high incidence in children, the verification of this property is particularly important (Chippaux 1998). Although local effects of snake bites appear in the first 10–30 mins, these local reactions are usually not effectively neutralized by conventional antivenom serum therapy (Lomonte et al. 1994; Ávila-Agüero et al. 2001). In severe cases, the local effects of poisoning may lead to permanent tissue loss, disability, or amputation (Gutiérrez 2002).

The results presented show that flavonoids isolated from this species can mitigate part of the local tissue damages induced by venom. Toxicological studies should be continued with the aim to use this crude drug as an adjuvant in cases of snake accidents.

References

Abbasi A, Khan M, Ahmad M, Zafar M, Jahan S, Sultana S (2010) Ethnopharmacological application of medicinal plants to cures kin diseases and in folk cosmetics among the tribal communities of North-West Frontier Province, Pakistan. J Ethnopharmacol 128(2):322–335

Aguirre-Galvis L (1988) Mass spectra of alkaloids from *Cissampelos pareira* L. Acta Biol Colombiana 1(4):17–23

Ahmad R, Arif Malik M, Zia-Ul-Haq M (1992) Alkaloids of *Cissampelos pareira*. Fitoterapia 63:682

Ampa L, Wanchat S, Jatupon P (2010) Antifertility activity of the aqueous leaf extract of *Cissampelos pareira* in male albino mice. ABAH Bioflux 2(2):59–64

Amresh G, Rao C, Shirwaikar A (2004) Ethnomedical value of *Cissampelos pareira* extract in experimentally induced diarrhoea. Acta Pharma 54(1):27–35

Amresh G, Reddy G, Rao C, Singh P (2007a) Evaluation of anti-inflammatory activity of *Cissampelos pareira* root in rats. J Ethnopharmacol 110(3):526–531

Amresh G, Zeashan H, Ca R, Singh P (2007b) Prostaglandin mediated anti-inflammatory and analgesic activity of *Cissampelos pareira*. Acta Pharm Sci 49(2):153–160

Amresh G, Rao C, Singh P (2007c) Antioxidant activity of *Cissampelos pareira* on benzo (a) pyrene-induced mucosal injury in mice. Nutr Res 27(10):625–632

Amresh G, Rao C, Singh P (2007d) Evaluation of *Cissampelos pareira* against gastric cancer and enzymes associated with carcinogen metabolism. Pharm Biol 45(8):595–603

Amresh G, Rao C, Singh P (2008) Toxicological screening of traditional medicine Laghupatha (*Cissampelos pareira*) in experimental animals. J Ethnopharmacol 116(3):454–460

Antoun MD, Ramos Z, Vazques J, Oquendo I, Proctor GR, Gerena L, Franzblau SG (2001) Evaluation of the flora of Puerto Rico for *in vitro* antiplasmodial and antimycobacterial activities. Phytother Res 15:624–638

Anwer F, Popli S, Srivastava R, Khare M (1968) Studies in medicinal plants. Part III. Protoberberine alkaloids from the roots of *Cissampelos pareira* Linn. Experientia 24(10):999

Ávila-Agüero M, Valverde K, Gutierrez J, Paris M, Faingezicht I (2001) Venomous snake bites in children and adolescents: a 12 years retrospective review. J Venomous Anim Toxins 7(1):69–84

Babu S, Chitra C, Rao N, Vasant R, Ramesh B, Naveen S (2014) Evaluation of antiurolithic activity of alcoholic extract of roots of *Cissampelos pareira* in albino rats. J Clin Diagn Res 8(7):1–4

Badilla B, Chaves F, Jiménez S, Rodríguez G, Poveda L (2008) Effects of an extract of *Cissampelos pareira* on the hemorrhagic and proteolytic activities from *Bothrops asper* venom. Pharmacogn Mag 4(13):27–31

Bafna A, Mishra S (2010) Antioxidant and immunomodulatory activity of the alkaloidal fraction of *Cissampelos pareira*. Linn Sci Pharm 78(1):21–31

Bala M, Pratap K, Verma P, Padwad Y, Singh B (2014) Cytotoxic agents for KB and SiHa cells from n-hexane fraction of *Cissampelos pareira* and its chemical composition. Nat Prod Res 29(7):686–691

Bala M, Kumar S, Pratap K, Verma PK, Padwad Y, Singh B (2017) Bioactive isoquinoline alkaloids from *Cissampelos pareira*. Nat Prod Res 33(5):620–627
Barranco-Pérez W (2010) Especies vegetales de uso antiofídico en las estribaciones de la Sierra Nevada de Santa Marta: inventario, etnobotánica y evaluación biológica, Thesis. Universidad Nacional de Colombia, Sede Medellín. Facultad de Ciencias Agropecuarias, Medellín, Colombia, 45 pp
Basha S, Sudarsanam G (2012) Traditional use of plants against snake bite in Sugali tribes of Yerramalais of Kurnool district, Andhra Pradesh, India. Asian Pac J Trop Med 2:575–579
Basu D (1970) Studies on curariform activity of hayatinin methochloride, an alkaloid of *Cissampelos pareira*. Jpn J Pharmacol 20(2):246–252
Bhakuni D, Jain S, Chaturvedi R (1987) The biosynthesis of the alkaloids of *Cissampelos pareira* Linn. Tetrahedron 43(17):3975–3982
Bhatnagar A, Popli S (1967) Chemical examination of the roots of *Cissampelos pareira* Linn. Structure and stereochemistry of hayatidin. Experientia 23(4):242–243
Bhatnagar A, Bhattacharji S, Roy A, Popli S, Dhar M (1967) Chemical examination of the roots of *Cissampelos pareira*. IV structure and stereochemistry of hayatin. J Org Chem 32(3):819–820
Bhattacharji S, Sharma V, Dhar M (1952) Chemical examination of the roots of *Cissampelos pareira* Linn. J Sci Ind Res India -B 11:81–82
Bhattacharji S, Sharma V, Dhar M (1956) Chemical examination of the roots of *Cissampelos pareira* Linn. J Sci Ind Res India -B 15:363–368
Bhattacharji S, Roy A, Dhar M (1962) Chemical examination of the roots of *Cissampelos pareira*. Part II – isolation and structure of hayatinin. J Sci Ind Res India B 21:428–433
Boissier J, Combes G, Pernet R, Dumont C (1965) Contribution à l'étude des alcaloides de quelques Menispermaceae de Madagascar: *Cissampelos pareira*, *Cyclea madagascariensis*, *Anisocyclea grandidieri* et *Spirospermum penduliflorum*. J Nat Prod-Lloydia 28(3):191–198
Bora U, Sahu A, Saikia A, Ryakala V, Goswami P (2007) Medicinal plants used by the people of North East India for curing malaria. Phytother Res 21(8):800–804
Butt M, Ahmad M, Fatima A, Sultana S, Zafar M, Yaseen G, Ashraf M, Shinwari Z, Kayani S (2015) Ethnomedicinal uses of plants for the treatment of snake and scorpion bite in northern Pakistan. J Ethnopharmacol 168:164–181
Chakraborty M, Bhattacharjee A (2006) Some common ethnomedicinal uses of various diseases in Purulia district, West Bengal. Indian J Tradit Knowl 5(4):554–558
Chhetri D, Parajuli P, Subba G (2005) Antidiabetic plants used by Sikkim and Darjeeling Himalayan tribes, India. J Ethnopharmacol 99(2):199–202
Chippaux J (1998) Snake-bites: appraisal of the global situation. Bull World Health Organ 76(5):515–524
Chowdury A (1972) Chemical investigation on *Cissampelos pareira* Linn. Sci Cult 38:358–359
Danduga R, Siva K, Kumar K, Vrushabendra S, Vijaya K (2015) Nephroprotective activity of *Cissampelos pareira* Linn. Extract against cisplatin induced nephrotoxic rats. Am J Pharm Res 5(3):480–488
De Freitas M, Cortes S, Thomas G, Barbosa Filho J (1996) Modification of Ca^{2+} metabolism in the rabbit aorta as a mechanism of spasmolytic action of warifteine, a bisbenzylisoquinoline alkaloid isolated from the leaves of *Cissampelos sympodialis* Eichl. (Menispermaceae). J Pharm Pharmacol 48(3):332–336
Cavazos ME De la O, Treviño Garza C, Guajardo-Rodríguez G, Hernández-Montelongo BA, Montes-Tapia FF (2012) Snake bites in pediatric patients, a current view. In: Ö Özdemir (ed.): Complementary pediatrics. Ed. InTech, Rijeka, Croatia, pp 123–136
Dey A, De J (2012) Traditional use of plants against snakebite in Indian subcontinent: a review of the recent literature. Afr J Tradit Complement 9(1):153–174
Don G (1831) A General System of Gardening and Botany Vol 1. Ed. Longman and Co, London, UK, pp 109–112
Dwuma-Badu D, Ayim J, Mingle C, Tackie A, Slatkin D, Knapp J, Schiff P Jr (1975) Alkaloids of *Cissampelos pareira*. Phytochemistry 14:2250–2251

Galicia E, Aguilar-Contreras A, Aguilar-Santamaría L, Roman-Ramos R, Chávez-Miranda A, García-Vega L, Flores-Sáenz J, Alarcón-Aguilar F (2002) Studies on hypoglycemic activity of Mexican medicinal plants. Proc West Pharmacol Soc 45:118–124
Ganguly M, Borthakur M, Devi N, Mahanta R (2007) Antifertility activity of the methanolic leaf extract of *Cissampelos pareira* in female albino rats. J Ethnopharmacol 111(3):688–691
Ganguly M, Mahanta R, Hazarika J, Pai K (2018) Study of the effect of *Cissampelos pareira* leaf extract on estrous cycle of Albino mice. Res J Pharm Biol Chem Sci 9(2):64–69
Gessler MC, Nkunya MHH, Mwasumbi LB, Heinrich M, Tanner M (1994) Screening Tanzanian medicinal plants for antimalarial activity. Acta Trop 56(1):65–77
Gessler MC, Msuya DE, Nkunya MHH, Mwasumbi LB, Schar A, Heinrich M, Tanner M (1995) Traditional healers in Tanzania: the treatment of malaria with plant remedies. J Ethnopharmacol 48(3):131–144
Giorgetti M, Rossi L, Rodrigues E (2007) Brazilian plants with possible action on the central nervous system- a study of historical sources from the 16th to 19th century. J Ethnopharmacol 109(2):338–347
González-Torres D (2013) Catálogo de Plantas Medicinales (y alimenticias y útiles) Usadas en Paraguay. Ed. Servi Libro, Asunción, Paraguay, 480 pp
Granda-Lorenzo M, Fuentes-Fiallo V (1991) El género *Cissampelos* L en Cuba. Revista del Jardín Botánico Nacional (Cuba) 12:35–42
Gul M, Chandrasekaran S, Bhat M, Maurya B, Qureshi I, Ghazi I (2016) Antioxidant and enzyme inhibitory activities of *Cissampelos pareira* L leaf extracts. Ann Phytomed 5(1):91–98
Gupta YK, Peshin SS (2012) Do herbal medicines have potential for managing snake bite envenomation? Toxicol Int 19(2):89–99
Gupta A, Pandey S, Shah D, Seth N, Yadav J (2011) Pharmacognostical and phytochemical evaluation of leaves of *Cissampelos pareira*. Pharm J 3(21):25–28
Gutiérrez J (2002) Comprendiendo los venenos de serpientes: 50 años de investigaciones en América Latina. Rev Biol Trop 50(2):377–394
Haq F, Ahmad H, Alam M (2011) Traditional uses of medicinal plants of Nandiar Khuwarrcatchment (District Battagram). J Med Plant Res 5(1):39–48
Haynes L, Hebert E, Plimmer J (1966) 4”-O-methylcurine from *Cissampelos pareira* L. J Chem Soc C 615–617
Heinrich M, Haller B, Leonti M (2014) A perspective on natural products research and ethnopharmacology in Mexico: the eagle and the serpent on the prickly pear Cactus. J Nat Prod 77(3):678–689
Inoue M, Hayashi S, Craker LE (2019) Role of medicinal and aromatic plants: past, present, and future, Pharmacognosy. In: Perveen S, Al-Taweel A (ed). Medicinal plants. Ed IntechOpen:1–13. https://doi.org/10.5772/intechopen.82497
Jabeen A, Khan M, Ahmad M, Zafar M, Ahmad F (2009) Indigenous uses of economically important flora of Margallah Hills National Park, Islamabad, Pakistan. Afr J Biotechnol 8(5):763–784
Jain A, Katewa S, Galav P, Sharma P (2005) Medicinal plant diversity of Sitamata wildlife sanctuary, Rajasthan. India. J Ethnopharmacol 102(2):143–157
Jolis J (1972) Ensayo sobre la historia natural del Gran Chaco. Ed. Universidad Nacional del Nordeste, Facultad de Humanidades, Instituto de Historia UNNE (Resistencia). Resistencia, Chaco, Argentina, 393 pp
Kakrani H, Saluja A (2002) Traditional treatment of gastrointestinal tract disorders in Kutch district, Gujaratstate, India. J Nat Remedies 2(1):71–75
Kaur A, Nain P, Nain J (2012) Herbal plants used in treatment of rheumatoid arthritis: a review. Int J Pharm Pharm Sci 4(4):44–57
Kufer J, Forther H, Poll E, Heinrich M (2005) Historical and modern medicinal plant uses - the example of the Ch’orti’ Maya and Ladinosin Eastern Guatemala. J Pharm Pharmacol 57(9):1127–1152

Kuldeep S, Narayan P, Karuna S, Shiny C, Nidhi M, Sinha P (2013) Assessment of antidiabetic potential of *Cissampelos pareira* leaf extract in streptozotocin-nicotinamide induced diabetic mice. J Pharm Res 6(8):874–878

Kumar V, Chauhan N, Padh H, Rajani M (2006) Search for antibacterial and antifungal agents from selected Indian medicinal plants. J Ethnopharmacol 107(2):182–188

Kupchan S, Yokoyama N, Beal J (1960) Menispermaceae alkaloids I. The alkaloids of *Cissampelos pareira* Linn. and the origin of *radix pareirae brave*. J Am Pharm Assoc 49:727–731

Kupchan S, Patel A, Fujita E (1965) Tumor inhibitors VI. Cissamperine, new cytotoxic alkaloid from *Cissampelos pareira*. Cytotoxicity of bisbenzylisoquinoline alkaloids. J Pharm Sci 54:580–583

Kupchan S, Kubuta S, Fujita E, Kobayashi S, Block J, Telang S (1966) Tumor inhibitors XV. The structure and configuration of cissamperine, a novel bisbenzylisoquinoline alkaloid. J Am Chem Soc 88:4212–4218

Leonti M, Vibrans H, Sticher O, Heinrich M (2001) Ethnopharmacology of the Popoluca, Mexico: an evaluation. J Pharm Pharmacol 53(12):1653–1669

Lomonte B, Moreno E, Tarkowski A, Hanson L, Maccarana M (1994) Neutralizing interaction between heparins and myotoxin II, a lysine 49 phospholipase A2 from *Bothrops asper* snake venom. J Biol Chem 269(47):29867–29873

Manfred L (1977) 7000 Recetas botánicas a base de plantas medicinales. 11a ed. Ed. Kier, Buenos Aires, Argentina, 672 pp

Mendes F, Carlini E (2007) Brazilian plants as possible adaptogens: an ethnopharmacological survey of books edited in Brazil. J Ethnopharmacol 109(3):493–500

Meng X, Riordan N, Casciari J, Zhu Y, Zhong J, González M (2002) Effects of a high molecular mass *Convolvulus arvensis* extract on tumor growth and angiogenesis. P R Health Sci J 21(4):323–328

Montenegro P (1979 - Reedition 2007) Materia Médica Misionera. Ed. Universitaria, Universidad Nacional de Misiones, Misiones, Argentina, 436 pp

Morita H, Matsumoto K, Takeya K, Itokawa H, Iitaka Y (1993a) Structures and solid state tautomeric forms of two novel antileukemic tropoloisoquinoline alkaloids, pareirubrines A and B, from *Cissampelos pareira*. Chem Pharm Bull 41(8):1418–1422

Morita H, Matsumoto K, Takeya K, Itokawa H (1993b) Conformation of tropolone ring in antileukemic tropoisoquinoline alkaloids. Chem Pharm Bull 41(8):1478–1480

Morita H, Matsumoto K, Takeya K, Itokawa H (1993c) Azafluoranthene alkaloids from *Cissampelos pareira*. Chem Pharm Bull 41(7):1307–1308

Morita H, Takeya K, Itokawa H (1995) A novel condensed tropone-isoquinoline alkaloid, pareitropone, from *Cissampelos pareira*. Bioorg Med Chem Lett 5(6):597–598

Morton J (1981) Atlas of medicinal plants of middle America: Bahamas to Yucatan. Ed. Charles C Thomas Pub Ltd., Springfield, USA, 1420 pp

Muthaura CN, Rukunga GM, Chhabra SC, Mungai GM, Njagi ENM (2007) Traditional phytotherapy of some remedies used in treatment of malaria in Meru district of Kenya. S Afr J Bot 73(3):402–411

Nagarajan K, Chauhan N, Mittal A, Singh V, Bodla R, Tiwari R (2011) Phytochemical extraction, optimization and physicochemical characterization of two bioactive isolates from the leaves and stem of *Cissampelos pareira*. Der Pharma Chemica 3(1):327–337

Namsa N, Mandal M, Tangjang S (2011) Anti-malarial herbal remedies of Northeast India, Assam: an ethnobotanical survey. J Ethnopharmacol 133(2):565–572

Ortiz R (2001) Menispermaceae. In: Stevens WD, Ulloa C, Pool A, Montiel OM (eds) Flora de Nicaragua. Monographs in Systematic Botany from the Missouri Botanical Garden, Missouri, USA, pp 1432–1442

Padmani S, Prabodh S, Gopalakrishna B (2012) Investigation of *in vitro* anthelmintic activity of *Cissampelos pareira* Linn. against *Pheretima posthuman*. Int J Pharm Sci Res 3(1):265–267

Patnaik G, Pradhan S, Vohra M (1973) Effects of hayatin methochloride and (+)-tubocurarine chloride on autonomic ganglia of cats. Indian J Exp Biol 11(2):89–94

Pattanaik C, Sudhakar Reddy C, Murthy M (2008) An ethnobotanical survey of medicinal plants used by the Didayi tribe of Malkangiri district of Orissa, India. Fitoterapia 79(1):67–71

Piero N, Eliud N, Susan K, George O, Murugi N, David M, Sakyi A, Peter G, Stanley K, Ngeranwa JNJ (2015) *In vivo* antidiabetic activity and safety in rats of *Cissampelos pareira* traditionally used in the management of *Diabetes mellitus* in Embu county, Kenya. Drug Metab Toxicol 6(3):1–11

Pramodinee D, Mahesh M, Niranjan D, Poournima S (2011) Memory enhancing activity of *Cissampelos pareira* in mice. Int J Pharm Sci 3(2):206–2011

Priyanka T (2013) Effect of *Cissampelos pareira* leaves o anxiety-like behavior in experimental animals. J Tradit Complement Med 3(3):188–193

Rajan S, Sethuraman M, Mukherjee P (2002) Ethnobiology of the Nilgiri Hills, India. Phytother Res 16(2):98–116

Rajan S, Jayendran M, Sethuraman M (2003) Medico-ethnobotany: a study on the Kattunayaka tribe of Nilgiri Hills, Tamil Nadu. J Nat Remedies 3(1):68–72

Ramasubramaniaraja R, Badu N (2010) Anthelmintic studies and medicinal herbs-an overview. Int J Pharm Sci Rev Res 5(3):39–47

Ramírez I, Carabot A, Medez P, Carmona J, Jimenez M, Patel A, Crab T, Blunden G, Cary P, Croft S, Costa M (2003) Cissampeloflavone, a chalcone-flavone dimer from *Cissampelos pareira*. Phytochemistry 64(2):645–647

Ramos-Hernández M, Avila-Bello C, Morales-Mávil J (2007) Etnobotánica y ecología de plantas utilizadas por tres curanderos contra la mordedura de serpientes en la región de Acayucán, Veracruz, México. Bol Soc Bot Méx 81:91–102

Reza H, Shohel M, Aziz S, Pinaz F, Uddin M, Amin M, Khan I, Jain P (2014) Phytochemical and pharmacological investigation of ethanol extract of *Cissampelos pareira*. Indian J Pharm Sci 76(5):455–458

Rhodes D (1975) A revision of the genus *Cissampelos*. Phytologia 30:415–483

Ricciardi A, Caballero E, Chifa C (1996) Identificación botánica de plantas descriptas en "Materia Médica Misionera" usadas en accidentes ofídicos. Rojasiana 3(2):239–245

Ricciardi-Verrastro B, Ricciardi G, Teibler P, Lozina L, Maruñak S, Dellacassa E, Torres A (2016) Inhibición de la actividad del veneno de *Bothrops diporus* mediante una formulación de aplicación tópica con extractos de *Cissampelos pareira*. Rev Medicina 76(1):280

Ricciardi-Verrastro B, Torres A, Ricciardi G, Teibler P, Maruñak S, Barnaba C, Larcher R, Nicolini G, Dellacassa E (2018) The effects of *Cissampelos pareira* extract on envenomation induced by *Bothrops diporus* snake venom. J Ethnopharmacol 212:36–42

Rukunga G, Gathirwa J, Omar S, Muregi F, Muthaura C, Kirira P, Mungai G, Kofi-Tsekpo W (2009) Anti-plasmodial activity of the extracts of some Kenyan medicinal plants. J Ethnopharmacol 121(2):282–285

Samanta J, Bhattacharya S (2011) *Cissampelos pareira*: a promising antifertility agent. Int J Res Ayurveda Pharma 2(2):439–442

Sánchez-Medina A, García-Sosa F, May-Pat L, Peña-Rodríguez M (2001) Evaluation of biological activity of crude extracts from plants used in Yucatecan traditional medicine. Part I. Antioxidant, antimicrobial and β-glucosidase inhibition activities. Phytomedicine 8(2):144–151

Sankaranarayanan S, Bama P, Ramachandran J, Kalaichelvan P, Deccaraman M, Vijayalakshimi M, Dhamotharan R, Dananjeyan B, Sathya Bama S (2010) Ethnobotanical study of medicinal plants used by traditional users in Villupuram district of Tamil Nadu, India. J Med Plant Res 4(12):1089–1101

Saravia-Otten P, Hernández R, Gutiérrez J, Mérida M, Cáceres A (2015) Evaluación de la capacidad neutralizante de extractos de plantas de uso popular en Guatemala como antídotos para el envenenamiento por la mordedura de *Bothrops asper*. Cienc Tecnol Salud 2(1):25–37

Sayana S, Tambi C, Medabala T, Praveen S (2011) Study of diuretic activity of ethanolic extract of leaves of *Cissampelos pareira* in rats. Asian J Pharm Clin Res 7(4):157–159

Scholtz M (1896) Ueber Bebirin. Ber Dtsch Chem Ges 29(2):2054–2058

Semwal D, Semwal R, Vermaak I, Viljoen A (2014) From arrow poison to herbal medicinal- the ethnobotanical, phytochemical and pharmacological significance of *Cissampelos* (Menispermaceae). J Ethnopharmacol 155:1011–1028

Sharma P, Chauhan N, Lal B (2004) Observations on the traditional phytotherapy among the inhabitants of Parvati valley in western Himalaya, India. J Ethnopharmacol 92:167–176

Sharma J, Gairola S, Gaur R, Painuli R (2012) The treatment of jaundice with medicinal plants in indigenous communities of the sub-Himalayan region of Uttarakhand, India. J Ethnopharmacol 143:262–291

Singh S, Singh V, Mishra G (2013a) Morphotaxonomic and pharmacoanatomical studies on *Cissampelos pareira* L. Geophytology 43(1):29–40

Singh B, Pillai K, Kohli K, Haque S (2013b) Effect of *C. pareira* root on isoproterenol-induced cardiac dysfunction. J Nat Med 67(1):51–60

Singthong J, Ningsanond S, Cui S, Douglas Goff H (2005) Extraction and physicochemical characterization of Kureo Ma Noy pectin. Food Hydrocolloid 19:793–801

Sood R, Raut R, Tyagi P, Pareek P, Barman T, Singhal S, Shirumalla R, Kanoje V, Subbarayan R, Rajerethinam R, Sharma N, Kanaujia A, Shukla G, Gupta Y, Katiyar C, Bhatnagar P, Upadhyay D, Swaminathan S, Khanna N (2015) *Cissampelos pareira* Linn.: natural source of potent antiviral activity against all four dengue virus serotypes. PLoS Negl Trop Dis 9(12):e0004255

Srivastava G (1956) Propagation of *Cissampelos pareira* L by root cuttings. Sci Cult 21:601–603

Stepp J (2004) The role of weeds as sources of pharmaceuticals. J Ethnopharmacol 92:163–166

Stevens P (2001 onwards) Angiosperm phylogeny website Version 9, June 2008 http://wwwmobotorg/MOBOT/research/APweb/ (Last access, June 2009)

Sudarsanam G, Prasad GS (1995) Medical ethnobotany of plants used as antidotes by Yanadi tribes in South India. Int J Geogr Inf Syst 3:57–66

Surendran S, Eswaran M, Vijayakumar M, Rao C (2011) *In vitro* and *in vivo* hepatoprotective activity of *Cissampelos pareira* against carbon-tetrachloride induced hepatic damage. Indian J Exp Biol 49(12):939–945

Tamaio N, Joffily A, Braga J, Rajput K (2010) Stem anatomy and pattern of secondary growth in some herbaceous vine species of Menispermaceae. J Torrey Bot Soc 137(2):157–165

Taylor L (1996) Tropical plant database entry for: ABUTA – *Cissampelos pareira*. http://www.rain-tree.com. (Last access, May 2020)

Thavamani S, Mathew M, Dhanabal S (2014) Anticancer activity of *Cissampelos pareira* against Dalton's lymphoma ascites bearing mice. Pharmacogn Mag 10(39):200–206

The Plant List (2019) Version 1.1 Published on the Internet: http://www.theplantlist.org. (Last access, November 2019)

Vardhanabhuti B, Ikeda S (2006) Isolation and characterization of hydrocolloids from monoi (*Cissampelos pareira*) leaves. Food Hydrocoll 20:885–891

Wiggers A (1840) Ueber das Pelosin. Liebigs Ann Chem 33:81–92

Wipawee T, Jintanaporn W (2012) Evaluation of safety and protective effect of combined extract of *Cissampelos pareira* and *Anethum graveolens* (PM52) against age-related cognitive impairment. Evid Based Complement Alternat Med:Article ID 674101

Zuloaga F, Belgrano M (2014) Catálogo de las Plantas Vasculares del Cono Sur. http://www.darwineduar/Proyectos/FloraArgentina/FAhtm. (Last access, May 2020)

Colliguaja integerrima Gillies & Hook.

Graciela Pinto Vitorino, Diana Paula Quezada, María Soledad Namuncurá, Osvaldo León Córdoba, and María Luján Flores

Colliguaja integerrima Gillies & Hook. in its habitat (Comodoro Rivadavia, Chubut, Argentina). Photo: ML Flores and OL Córdoba

Abstract *Colliguaja integerrima* Gillies & Hook. (Euphorbiaceae) is an evergreen shrub native to Argentina and Chile, widely distributed in the Patagonia region. Tehuelches-Mapuches, the native people of Patagonia, have used this abundant plant to treat corns, warts and toothache and used the "colihuai" latex from the leaves and young branches, considered very poisonous, to poison their spearheads and arrowheads. Also, these were used as plasters for their analgesic properties. A decoction of the plant was used by Tehuelches for vaginal infections.

G. Pinto Vitorino · D. P. Quezada · M. S. Namuncurá · O. L. Córdoba · M. L. Flores (✉)
Centro Regional de Investigación y Desarrollo Científico Tecnológico (CRIDECIT), Departamento de Farmacia, Facultad de Ciencias Naturales y Ciencias de la Salud (FCNyCS), Universidad Nacional de la Patagonia San Juan Bosco (UNPSJB), Comodoro Rivadavia, Chubut, Argentina
e-mail: mlujanflo@gmail.com

Á. Máthé, A. Bandoni (eds.), *Medicinal and Aromatic Plants of South America Vol. 2*, Medicinal and Aromatic Plants of the World 7,
https://doi.org/10.1007/978-3-030-62818-5_12

Carbohydrates (polysaccharides with glucose, galactose, sulfate groups and pinitol), proteins, lipids (containing ricinoleic acid), flavonoids (datiscetin, kaempferol, pelargonidin and delfinidin), tannins, steroids and triterpenes (betulin) are the principal compounds. The chemical composition of this plant and the demonstrated antibacterial and antioxidant activities are evidences that could broadly support the traditional use. Methanolic and aqueous extracts evidenced important cytotoxicity against *Artemia salina*. The dichloromethane fraction, rich in betulin, generated 100% inhibition (at 100μg/ml) against promastigotes of *Leishmania donovani* (causal of visceral leishmaniasis). In view of its background in traditional medicine, this chapter reviews scientific studies on the chemical and biological properties of this species.

Keywords "Colihuai" · Euphorbiaceae · Tehuelches · Patagonia Argentina · Bioactive metabolites

1 Introduction

The Golfo San Jorge District (Patagonia Argentina) has a great vegetal diversity, mainly species typical of semi-desert and saline areas, with the ability to adapt to intense solar radiation and soils containing significant amounts of hydrocarbons.

Some species were used medicinally by aboriginal peoples, mainly by Tehuelches and Tehuelches-Mapuches. However, many of them are still less investigated from an integral perspective that allows validating their traditional uses.

Colliguaja integerrima Gillies & Hook. (Euphorbiaceae), is a shrub native to Argentina and Chile, widely distributed in the Golfo San Jorge District. This species was traditionally used to treat diseases such as calluses, warts and toothaches. Its very poisonous latex was used to poison spearheads and arrowheads.

The wide variety of already identified metabolites with demonstrated biological activities seem to be consistent with the traditional uses reported. The possible unexplored valuable properties of the species seem to call for further investigations.

2 Taxonomic Characteristics

Colliguaja integerrima Gillies & Hook. (syn.: *Colliguaja bridgesii* Müll. Arg.; The Plant List 2019), with the native name "colihuai", is a South American species belonging to the family Euphorbiaceae. This family, with around 220 genera and more than 7200 species, is found widely distributed in tropical and subtropical regions with largest centers in America and Africa. It is one of the largest and more diverse families of flowering plants. The species ranked here are trees, shrubs or herbs, annual or perennial, in some cases woody, with or without latex. Various examples are found in the special literature on species of this family that are used in traditional medicine: many of these has been classified as toxic (Ragonese and Milano 1984).

The genus *Colliguaja* is a South American native (Chile and Argentina) and comprises five species: *C. brasiliensis* Klotzsch ex Baill, *C. dombeyana* A. Juss, *C. integerrima* Gillies & Hook., *C. odorifera* Molina and *C. salicifolia* Gillies & Hook. (The Plant List 2019).

Colliguaja integerrima was described by Gillies and Hook. and published in Botanical Miscellany in 1830 (The Plant List 2019). *Colliguaja* is a word of Mapuche origin ("colli" = reddish, "huai" = shrub), relating to the color of the reddish branches of the plant.

Ferreyra and Lorraine (2012) published a correction indicating that *Colliguaja integerrima* would mean "completely whole", in Mapuche language. Other common names are: in Spanish, "duraznillo", "duraznillo patagónico", "coliguay", "colihuai", "colihui", "colihuasi", "colliguay", "colliguaya", "coliguay del cerro", "lechón", "vegetal artillero"; Mapuche: "collihuai", "collihuayu"; Araucanian: "kolliwai", "collihuaye"; Tehuelche: "akanaiúwütr". The similarity of the fruit with small peaches ("duraznos" in Spanish) has earned it the popular name of "duraznillo" (Barboza et al. 2009; Desmarchelier and Alonso 2015).

3 Crude Drug Used

The main parts of the *Colliguaja integerrima* plant used in traditional medicine are the aerial organs (photo 1) and the latex exudate from the leaves and young branches, a white and sticky juice (de Mösbach 1992). These were used as plasters for analgesic properties (Ciampagna 2014).

4 Major Chemical Constituents and Bioactive Compounds

Carbohydrates, phenols and terpenes are the principal constituents of the aerial parts, fundamentally, they are present in the leaves. Latex contains phenols and terpenes. These chemical groups are related to medicinal properties. The following compounds were identified in the methanolic extract of aerial parts: condensed tannins, flavonoids (kaempferol, datiscetin, pelargonidin and delfinidin), hydroxycinnamic acids, predominantly synaptic acid. Delfinidin is a flavonoid with high antiperoxidative power, rare and has not been previously described for the genus *Colliguaja* (Pinto Vitorino et al. 2004a, b). Alcalde et al. (2005) described the presence of wax, palmitic alcohol and mono and di-palmitins in aerial parts. The main fatty acids identified were myristic, palmitic, palmitoleic and ricinoleic acids.

Photo 1 *Colliguaja integerrima* Gillies & Hook. (**a**) Plant detail. (**b**) Inflorescence. (**c**) Fruit. (**d**) Leaves with latex. (Photos: G Pinto Vitorino, ML Flores and OL Córdoba)

Gnecco et al. (1996) described for *Colliguaja integerrima* some hydrocarbons; the main is the *n*-heptaeicosane (C27) with 87% followed by C25, C29, C26, C31. Bittner et al. (2001) described that the whole plant presents lupeol, ursolic acid, oleanolic acid, β-sitosterol and carbohydrates.

The leaves of *C. integerrima* contain carbohydrates, proteins, lipids, flavonoids, tannins, steroids and triterpenes as principal compounds. Ethyl acetate fraction of the methanol extract from the leaves showed flavonoids and tannins as main metabolites, particularly kaempferol, datiscetin and myricetin, among others; in particular, datiscetin was described for the first time in this genus (Pinto Vitorino et al. 2004a, b, 2014). Dichloromethane fraction presented several terpenes, being betulin the main component (Quezada et al. 2015).

The principal polysaccharides obtained with aqueous extraction contained glucose followed by galactose, mannose, arabinose and rhamnose; sulfate groups are very important too. Also, in the leaves was found pinitol (*O*-methyl inositol) (Carrizo et al. 2018).

The inflorescences show the same metabolites groups that the leaves, except triterpenes, and larger amounts of proteins. The monosaccharides identified were galactose, mannose and xylose. The aqueous extract obtained at room temperature contains 7.7% of total sugars and 11.8% sulfate groups. In the aerial parts, the content of total sugars was 5.7% and sulfate groups 8.4% (Pinto Vitorino et al. 2004a, b).

The total phenolic content was 12.9%, both in aerial parts and in inflorescences, of that quantity the percentage of flavonoids corresponds to 6.2% and 18.2% respectively (Pinto Vitorino et al. 2004a, b).

The seeds of *C. integerrima* contain glucose, galactose, proteins, triterpenes and lipids. The total content of lipids obtained was 30.7% (Pinto Vitorino et al. 2004a, b). Malec et al. (1986) reported for seeds a 54.9% of crude oil (not refined). It is possible to assume that these variations could be due, between other factors, to the extraction technique applied, the time of sample collection, as well as to climatic and edaphic differences of the collection area (Pinto Vitorino et al. 2004a, b). *C. integerrima* seeds are a source of drying fixed oils. Riganti et al. (1947) and Malec et al. (1986) determined the chemical composition of fatty acids of the seeds. The major components were oleic, linoleic, linolenic and palmitic acid, and in less quantity were present myristic, stearic, arachidonic, behenic, and palmitoleic acid, with traces of eicosenoic and docosenoic acids. Cholesterol, campesterol, stigmasterol and sitosterol were found too. Ravetta et al. (1991) carried out a study on this species, in order to determine the production and oil content of *C. integerrima*. Proximal analysis of the seeds indicated a composition of 35% oil, with a content of 51% proteins, and a lysine quantity equal to 3.17 g/16 g N_2.

5 Morphological Description

Colliguaja integerrima is a native monoecious evergreen shrub, containing a milky white latex. It reaches 1–2 m height and 1.5 m high diameter. The branches are of reddish-brown color, erect and glabrous. The leaves are sessile, opposite, simple, leathery, linear up to 5 cm in length, obtuse and mucronated, of whole margins, with a prominent central rib of whitish hue (Riedemann et al. 2014). Unisexual flowers gathered in a single inflorescence. Flowers are lacking petals and sepals. Male flowers are terminal spikes of 2–3 cm in length and one or two female flowers are at the base of the male-spikes. The fruit is a capsule with 2 or 3 lobes (Ragonese and Milano 1984). Its flowering season occurs from September to November (Navas 1976). The vegetative growth occurs from September to February, the fruiting season occurs during March–April, and fruits are 3-seeded dry capsules (Navas 1976; Ravetta et al. 1991; Hoffman 1998). The fruits are capsules, strongly lignified, burst abruptly under the sunstroke, projecting the seeds (de Mösbach 1992).

6 Geographical Distribution

The genus *Colliguaja* is endemic in South America. It is represented by five species (Pinto Vitorino et al. 2014). *C. integerrima* is native to Argentina and Chile. In Argentina, it grows wild in the phytogeographical provinces of Patagonia (provinces of Mendoza, Neuquén, Chubut and Santa Cruz). In Central Patagonia, it is widely distributed in the Distrito Golfo San Jorge, zones where the species grows in a wild state (Ramírez 2002).

Ravetta et al. (1991) proposed that the domestication and cultivation of this species could be promising, especially in view of its adaptation to semi-deserts weather and its production of seed oil and by-products. Also, the production of hydrocarbon and petrochemical substitutes would increase interest in the species. However, due to its low economic value, there are no records of domestication experiments related to this species for commercial purposes. In the area of the Golfo San Jorge, Province of Chubut, and in order to recompose the vegetation cover before oil spills, they were carried out experiences with *C. integerrima* for recovery of the hydrocarbon through biodegradation *in situ*, land conditioning and plantations with shrub and tree species (Desmarchelier and Alonso 2015).

7 Ecological Requirements

C. integerrima inhabits the Patagonian steppe, where it grows in communities of different sizes. Usually in windy places with sandy, poor and rocky soil abundant in calcium carbonate; in sunny and "huaycos" (bad lands) zones, with precipitation of 100 mm/year and drought periods from 6 to 10 months. The individuals can survive at low temperatures (until -8 °C) and can tolerate the snow cover for several weeks, during the winter season (Quintana 2014; Gandullo et al. 2016).

There are also reports that indicate that *Colliguaja* species suffer a lower rate of defoliation caused by chewing insects, as compared to other shrub species in central Chile (Montenegro et al. 1980).

8 Traditional Use (Part(s) Used) and Common Knowledge

Tehuelches-Mapuches, the native people of Patagonia used this abundant shrub to treat corns, warts and to soothe the toothache (Ragonese and Milano 1984; Ravetta et al. 1991). The rural population of Patagonia used this plant to cure sheep mange, taking care that the leaves do not touch the eyes, as they could cause serious eye irritation. Pehuelches (Argentina) used "colihuai" latex, considered very poisonous, as a poison of their spearheads and arrowheads (Casamiquela 1999). Aerial parts rich in latex were used for medicinal purposes by Tehuelches. Also, these were used as plasters for their analgesic properties. The decoction of the plant was used for vaginal infections (de Mösbach 1992). Its internal use is dangerous because this plant is cited as toxic to cattle, sheep and horses (Ratera and Ratera 1980; Bittner et al. 2001).

9 Modern Medicine Based on Its Traditional Medicine Uses

Several uses in popular medicine have been supported by different studies *in vitro* about the active principles. However, *in vivo* conditions and clinical trials in humans would be required, mainly because it has been reported as a toxic plant.

Bhakuni et al. (1976), described antibacterial activity against *Bacillus subtilis, Escherichia coli, Pseudomonas aeruginosa, Staphylococcus epidermidis* and *Streptococcus faecalis.* The ethyl acetate extract of the aerial parts of *C. integerrima* demonstrated inhibitory activity *in vitro* on nasopharyngeal cancer cell cultures KB human. Comparatively, the same extract of the related species *C. odorifera* demonstrated cell growth inhibitory activity of mouse lymphocytic leukemia (Bhakuni et al. 1976).

By another way, the latex obtained by the incision of the cortex of *C. integerrima* showed no proteolytic properties (Sequeiros et al. 2003).

Ethyl acetate fraction obtained from the methanolic extract of the leaves of *C. integerrima*, enriched in flavonoids, was active against *E. coli*, *S. aureus* and *P. aeruginosa*, with MIC values of 0.25, 0.50 and 0.25 mg/ml, respectively (Pinto Vitorino et al. 2014). The inhibition growth of the bacteria *Pseudomonas aeruginosa* has been relevant in this study. Despite improvements in antibiotic therapy, *P. aeruginosa* is intrinsically resistant to a great number of antimicrobial agents, frequently it is necessary to employ multiples classes of antimicrobial agents.

The antioxidant activity of the ethyl acetate fraction was very significant, with a concentration required to scavenge 50% of the free radicals (SC_{50}) of 5.7μg/ml. Also, against *Artemia salina,* it presented an $LD_{50} = 77.3$μg/ml. At high doses (1%), this fraction inhibited root length in a higher way, as vinblastine sulphate solution of 0.02% (Pinto Vitorino et al. 2014).

Dichloromethane fraction, rich in betulin, generated 100% inhibition (at 100μg/ml) against promastigotes of *Leishmania donovani* (causal of visceral leishmaniasis) (Quezada et al. 2015).

Álvarez and Borkowski (2007) described that the "collihuai" infusion contains permeation enhancers, substances used to promote the transdermal administration of drugs, which has significant advantages over other forms of administration, although they cause irritation (among other damages) in the skin.

10 Conclusions

The chemical composition of *Colliguaja integerrima*, and its proven antibacterial and antioxidant activity, are evidences that broadly support the traditional uses in conditions of toothache, which is frequently related to infection, inflammatory processes and ROS-induced tissue damage. The capacity of *C. integerrima* extracts to inhibit radicular growth and important antioxidant activity have been related to cytotoxicity and antitumor effects. These capabilities could justify the popular use by the native peoples of Patagonia to treat corns and warts. All these results show the prospective medicinal potential of this species.

Acknowledgments We are especially grateful to the ChemBioFight IRSES Project, UNPSJB and FONCYT Argentina, for financial support. DPQ and MSN are supported by CONICET and CIN, respectively.

References

Alcalde D, Arratia C, Merlo R, Pinto Vitorino G, Córdoba OL, Flores ML (2005) Bioactividad y composición química de un extracto purificado de *Colliguaya integerrima* (Euphorbiaceae). In: Abstracts of the XV Simposio Nacional de Química Orgánica. 6–8 November. Mar del Plata, Argentina

Álvarez ME, Borkowski E (2007) Efecto de infusiones de *Euphorbia schickendantzii* y *Colliguaja integerrima,* en la permeabilidad de piel de sapo. Bol Latinoam Caribe Plant Med Aromát 6(6):321–322

Barboza GE, Cantero JJ, Núñez C, Pacciaroni A, Ariza Espinar L (2009) Medicinal plants: a general review and a phytochemical and ethnopharmacological screening of the native Argentine Flora. Kurtziana 34:7–365

Bhakuni DS, Bittner M, Marticorena C, Silva M, Weldt E, Hoeneisen M (1976) Screening of Chilean plants for anticancer activity I. Lloydia 39(4):225–243

Bittner M, Alarcón J, Aqueveque P, Becerra J, Hernández V, Hoeneisen M et al (2001) Estudio Químico de Especies de la Familia Euphorbiaceae en Chile. Bol Soc Chil Quím 46(4):419–431

Carrizo AC, Pérez NB, Fuentes AV, Quezada DP, Pinto Vitorino G, Uhrich AV et al (2018) Hidratos de carbono en hojas de *Colliguaja integerrima* (Euphorbiaceae): primeros resultados para la especie. Dominguezia 34(Supl):59–60

Casamiquela R (1999) Proyecto etnobotánico de la Patagonia. Primer Informe. In: Aspectos Técnicos, Culturales, Políticos y Legales de la Bioprospección en Argentina. Ed. INTA-CENPAT (CONICET)-UNP-UA. Buenos Aires, Argentina, pp 91–134

Ciampagna ML (2014) Estudio de la interacción entre grupos cazadores recolectores de Patagonia y las plantas silvestres: el caso de la costa norte de Santa Cruz durante el Holoceno medio y tardío. Thesis. Facultad de Ciencias Naturales y Museo, Universidad Nacional de La Plata. La Plata, Argentina, 537 pp

de Mösbach E (1992) Botánica Indígena de Chile. Ed. Andrés Bello. Villarrica, Chile, 140 pp

Desmarchelier CJ, Alonso J (2015) Plantas medicinales autóctonas de la Argentina. Bases científicas para su aplicación en atención primaria de la salud – 1° ed. Ed. Corpus Libros Médicos y Científicos, Buenos Aires, Argentina, 749 pp

Ferreyra M, Lorraine G (2012) Flores de la Estepa Patagónica: Guía para el reconocimiento de las principales especies de plantas vasculares de estepa. First edition. Ed. Vázquez Mazzini. Buenos Aires, Argentina, 288 pp

Gandullo R, Siffredi G, Velasco V (2016) Guía para el reconocimiento de especies del norte neuquino. Ed. INTA. San Carlos de Bariloche, Argentina, 172 pp

Gnecco S, Pérez C, Bittner M, Becerra J, Silva M (1996) Distribución de n-alcanos en especies chilenas pertenecientes a la familia Euphorbiaceae. Bol Soc Chil Quim 41(3):229–233

Hoffman AJ (1998) Flora silvestre de Chile. Cuarta edición. Ed. Fundación Claudio Gay. Santiago, Chile, 142 pp

Malec LS, Vigo MS and Cattaneo P (1986) Semilla de *Colliguaya integerrima*. Composición química del aceite crudo de extracción y de la harina residual. An Asoc Quím Argent 74:229–235

Montenegro G, Jordan N, Alfaro ME (1980) Interactions between Chilean matorral shrubs and phytophagous insects. Oecologia 45:346–349

Navas LE (1976) Flora de la cuenca de Santiago de Chile. Tomo II: Dicotiledoneae – Archichlamydeae. Ed. Universidad de Chile. Santiago, Chile, 559 pp

Pinto Vitorino G, Alcalde D, Arratia C, Merlo R, Córdoba O, Flores ML (2004a) Caracterización química y bioactividad de extractos polares de *Colliguaja integerrima* (H.) G. et H. (Euphorbiaceae). In: Abstracts of the VIII° Simposio Argentino and XI° Simposio Latinoamericano de Farmacobotánica. Buenos Aires, Argentina

Pinto Vitorino G, Toledo I, Flores M, Cabrera J (2004b) Análisis fitoquímico de *Colliguaya integerrima*, una planta de la Patagonia. Acta Farm Bonaerense 23(4):459–465

Pinto Vitorino G, Alcalde DP, Arratia C, Merlo R, Córdoba OL, Flores ML (2014) Biological activity of extracts from leaves of *Colliguaja integerrima* (Hook.) Gill. & Hook. (Euphorbiaceae). J Pharmacogn Phytother 2(1):22–27

Quezada D, Flores M, Pinto Vitorino G, Córdoba O, Smirlis D, Michel S *et al.* (2015). Extracción e identificación de betulina a partir de las hojas de *Colliguaja integerrima* Gillies & Hook. (Euphorbiaceae). In: Abstracts of the VI Congreso Iberoamericano de Ciencias Farmacéuticas, XLVII Reunión Científica Anual de la Sociedad Argentina de Farmacología Experimental y III Congreso Sudamericano de Biofarmacia y Farmacocinética. Córdoba, Argentina

Quintana D (2014). Plantas de la Patagonia Árida. Nativas y exóticas del noroeste del Chubut. Ed. Fondo Editorial; Secretaría de Cultura. Chubut, Argentina, 304 pp

Ragonese A, Milano V (1984) Enciclopedia Argentina de Agricultura y Jardinería. Vegetales y sustancias tóxicas de la Flora argentina. 2nd edn, vol 2. Ed. ACME, Buenos Aires, Argentina, 413 pp

Ramírez C (2002) Usos tradicionales de las plantas en la meseta patagónica. Jardín Botánico de la Patagonia Extra andina. Ed. Dirección de Impresiones Oficiales de la Provincia del Chubut. Chubut, Argentina, 51 pp

Ratera E, Ratera M (1980) Plantas de la Flora argentina empleadas en Medicina popular. Ed. Hemisferio Sur. Buenos Aires, Argentina, 189 pp

Ravetta D, Soriano A, Cattaneo P (1991) *Colliguaya integerrima* as a possible new crop. Econ Bot 45:288–290

Riedemann MMP, Aldunate NG, Teillier AS (2014) Arbustos nativos ornamentales del Centro Sur de Chile. Ed. Corporación Chilena de la Madera. Concepción. Chile, 308 pp

Riganti AR, Cattaneo P, Karman G (1947) Composición química del aceite de *Colliguaya integerrima* (duraznillo). An Asoc Quím Argent 35:21–33

Sequeiros C, López LMI, Caffini NO, Natalucci CL (2003) Proteolytic activity in some Patagonian plants from Argentina. Fitoterapia 74(6):570–577

The Plant List (2019) Version 1.1. Published on the internet; http://www.theplantlist.org/. (Last access, November 2019)

Cyclolepis genistoides D. Don

César A. N. Catalán

Cyclolepis genistoides. (Photos by the author).

Abstract The aerial parts of *Cyclolepis genistoides*, commonly called "palo azul" or "matorro negro", are widely used in folk medicine in Paraguay, Northern and central Argentina, as a diuretic and for the treatment of kidney diseases, urinary tract irritation, kidney pain, bone pain (analgesic). It is also used as an antipyretic, as a blood purifier, against liver diseases and as a hypotensive. The main metabolites identified in the chloroform extract of aerial parts are oleanolic, ursolic and betulinic acids and other closely related triterpenes along with the sesquiterpene lactones deacylcynaropicrin and its 11,13-dihydroderivative. Most of the uses of this plant could be justified by the known biological effects of above metabolites. Very little is known about the water-soluble metabolites of "palo azul", and as such, research on the content of polyphenols, flavonoid glycosides, chlorogenic acid, dicaffeoylquinic acids and phenyletanoids are welcome.

C. A. N. Catalán (✉)
Instituto de Química Orgánica, Facultad de Bioquímica, Química y Farmacia, Universidad Nacional de Tucumán, San Miguel de Tucumán, Argentina
e-mail: ccatalan@fbqf.unt.edu.ar; cancatalan@gmail.com

Á. Máthé, A. Bandoni (eds.), *Medicinal and Aromatic Plants of South America Vol. 2*, Medicinal and Aromatic Plants of the World 7,
https://doi.org/10.1007/978-3-030-62818-5_13

Keywords "Palo azul" · Triterpenes · Sesquiterpene lactones · Anti-inflammatory activity · Diuretic effect

1 Introduction

Cyclolepis genistoides (Asteraceae) is a shrub known as "palo Azul" ("blue stick"). The name refers to the observations according to which when its stems are submerged in water the solution acquires color or blue fluorescence. The decoction of its aerial parts is used in traditional medicine for kidney problems, elimination of uric acid, as a diuretic, antiarthritic, hypoglycemic and antitussive. Oleanolic acid, ursolic acid, betulinic acid and other closely related triterpenes, together with the sesquiterpene lactones deacylcynaropicrin, zaluzanin C, their corresponding 11,13-dihydroderivatives and isolippidiol, are the main metabolites identified in the chloroform extract of aerial parts. The reported pharmacological properties of compounds identified in this species agree with the use in folk medicine. No adverse effects or toxic effects have been reported from the use of this plant.

2 Taxonomic Characteristics

Cyclolepis genistoides D. Don is a medicinal plant belonging to the Asteraceae family, subfamily Mutisioideae, tribe Mutisieae (Katinas et al. 2008). Shrubs that grow up to a height of 2.5 m, are densely branched, with short branches modified in thorns. Its habitats are in the sandy soils of saline margins and salty rivers from the Paraguayan Chaco, northern and central Argentina to Northern Patagonia. In Northern Argentine, *C. genistoides* blooms in early spring.

The species was first described by the Scottish botanist David Don, in 1832 (Don 1832). It is known by the vernacular names "palo azul" or "matorro negro" (Zuloaga and Morrone 1996, 1999). It has a synonym**:** *Gochnatia genistoides* (D. Don) Hook. & Arn.

3 Crude Drug Used

Cyclolepis genistoides is widely used in folk medicine as a diuretic, antirheumatic and antispasmodic agent. A decoction of aerial parts (tender branches and leaves) is used (Filipov 1994; Giberti 1983; Scarpa 2004; Alonso and Desmachelier 2006; Barboza et al. 2009).

A decoction prepared with three 20 cm long branches in 5 liters of water is recommended to treat bone pain; use as a warm bath until symptoms decrease (Scarpa

2004). For other uses (diuretic, urinary tract irritation, kidney pain) the following preparations are reported: (i) a decoction is made with a handful of leaves in 2 liters of water. Let cool and keep in the refrigerator. Drink once a day for three consecutive days (Scarpa 2004); (ii) a decoction is prepared with five apical parts of branches in 2 liters of water. After decanting or filtering, the liquid is taken "like water" (Scarpa 2004); (iii) prepare a decoction (5 min) with 25 g of tender branches in one liter of water. Filter or decant and drink 2–3 cups per day (Alonso and Desmachelier 2006).

4 Major Chemical Constituents and Bioactive Compounds

A chemical investigation reported several triterpenoids with ursolic acid and oleanolic acid in a 3:2 ratio as main components, accompanied by lesser amounts of betulin, betulinic acid, methyl betulinate, β-amyrin, dihydro-β-amyrin, oleanonic acid, taraxasterol, 12α,13α-epoxyoleanolic acid, 30-hydroxylupeol and 3β,28-dihydroxylup-20(29)-en-30-al; also, significant amounts of the sesquiterpene lactones deacylcynaropicrin, 11(13)-dihydrodeacylcynaropicrin and isolippidiol were present (De Heluani et al. 1997). In addition, recently we have also identified zaluzanin C and 11(13)-dihydrozaluzanin C in the sesquiterpene lactones mixture of this plant (Catalán et al. 2020). Oleanolic acid, ursolic acid and their derivatives are pentacyclic tritepenoids with confirmed pharmacological properties (Liu 1995; Wójciak-Kosior 2013; Mlala et al. 2019). The pharmacological effect of decoctions of "palo azul" was attributed to these triterpenic acids (Alonso and Desmachelier 2006; Sosa et al. 2011). Reviews on the pharmacology of the main components of the lipophilic extract of "palo azul", *i.e.*, oleanolic acid (Liu 1995; Wójciak-Kosior 2013) and ursolic acid (Liu 1995; Wójciak-Kosior 2013; Mlala et al. 2019), are available. Oleanolic and ursolic acids displayed hepatoprotective effect (Ma et al. 1982, 1986; Shukla et al. 1992; Liu 1995; Liu et al. 1995; Sosa et al. 2011) and anti-inflammatory activity in carrageenan (Singh et al. 1992) and dextran-induced edema in rats (Sosa et al. 2011). Oleanolic acid elicited marked anti-arthritic action in adjuvant-induced polyarthritis in rats and mice and in formaldehyde-induced arthritis in rats (Liu 1995; Mlala et al. 2019 and references cited therein) and exerts an antidiabetic effect in rats (Wang et al. 2011). On the other hand, it has been reported that deacylcynaropicrin -the main sesquiterpene lactone of "palo azul"- inhibits RANKL-induced osteoclastogenesis by inhibiting NF-κB and MAPK and promoting M2 polarization of macrophages (Li et al. 2019) while 11,13-dihydrozaluzanin C showed anti-inflammatory activity in mice (Piornedo et al. 2011). The reported bioactivities of components isolated from "palo azul" seem to validate the folk uses of this plant.

5 Morphological Description

A gynodioecius shrub up to 2.5 m high, densely branched, rigid branches almost perpendicular to the stems with the short branches modified in thorns. Alternate leaves, briefly petiolated, deciduous, oblong-lanceolate, acute at the apex and contracted at the base, 6–20 mm long by 2.5–6 mm wide; pubescent on both surfaces. Capitula subsessile, homogamous, densely bracteolate; receptacle epaleate; involucre three- to six-seriate. Florets isomorphic in bisexual capitula, corolla actinomorphic, tubular funnelform, deeply five-lobed, lobes coiled; anther apical appendages apiculate, tails slightly papillose; style bifid, branches dorsally smooth; female capitula with florets without staminodes, corolla tubular-filiform, shallowly five-lobed, tube long, up to 2/3 of corolla length, lobes straight. Cypselae villose; pappus of scabrid bristles. Cylindrical achenes, densely sericeous-pubescent, 3 mm long (Cabrera 1978).

6 Geographical Distribution

This species has a wide distribution in saline areas throughout Paraguay, Northern and central Argentina to Northern Patagonia. Possibly it also grows in the bordering areas to the southeast of Bolivia.

7 Ecological Requirements

It grows on the banks of salt flats and salty rivers. This species is threatened by habitat loss. No studies on the domestication of this species have been made.

8 Traditional Use (Part(s) Used) and Common Knowledge

Aerial parts: branches, leaves and (rarely) flowers (photo). *Cyclolepis genistoides* is commonly known as "palo azul" (blue stick) or "matorro negro" (black bush) in Argentina; and "jupoty", "ñuati hu" (Guarani language) and "matorro negro" in Paraguay (Giberti 1983; Alonso and Desmachelier 2006). Infusions and decoctions of aerial parts are widely used in traditional medicine as a diuretic, analgesic, antispasmodic, antiarthritic, antirheumatic and to treat renal ailments and liver diseases (Filipov 1994; Giberti 1983; Scarpa 2004; Alonso and Desmachelier 2006; Barboza et al. 2009). Orally ingested decoction has also been reported to suppress cough and to relieve waist pain (Filipov 1994; Alonso and Desmachelier 2006).

9 Modern Medicine Based on Its Traditional Medicine Uses

It has been shown that two of its major components, namely, oleanolic acid and desacylcynaropicrin, displayed significant anti-inflammatory activity in the carrageenan-induced inflammation test (Sosa et al. 2011). As previously mentioned, reviews on the pharmacology of oleanolic acid (Liu 1995; Liu et al. 1995; Wójciak-Kosior 2013) and ursolic acid (Liu 1995; Wójciak-Kosior 2013; Mlala et al. 2019) are available. However, it should be noted that these triterpenic acids and the other triterpenoid components of *Cyclolepis genistoides* are present only in small amounts in the decoction due to the water insolubility of these compounds (Catalán et al. 2020). The hydro-alcoholic extract of "palo azul" exhibited a weak depressant effect on CNS and low acute toxicity in mice (Montalbetti Moreno et al. 2018). The diuretic effect of 10% infusions of *Cyclolepis genistoides* on Wistar rats was moderate but significant in relation to the control (Sosa et al. 2007) whilst the ethanol extract promotes differentiation of adipocytes and regulates adipokine expression in 3 T3-L1 adipocytes by modulation of peroxisome proliferator–activated receptor (PPAR) (Sato et al. 2013) and induces the formation of myotubes by differentiating C2C12 myoblast cells (Sato et al. 2016). It is worth noting that the amount (concentration) of oleanolic acid, ursolic acid and other triterpenes in the ethanol extract is much higher than in the aqueous decoction due to the greater solubility of these triterpenic acids in alcohol. On the other hand, sesquiterpene lactones are well-known bioactive compounds (Sülsen and Martino 2018) with antitrypanosomal, antileishmanial, antifungal, antibacterial, antiviral, cytotoxic and anti-inflammatory activity.

The anti-inflammatory activity of 11,13-didhydrozaluzanin C in mice has been demonstrated (Piornedo et al. 2011) whilst desacylcynaropicrin, the main sesquiterpene lactone of "palo azul", in addition to its anti-inflammatory properties (Sosa et al. 2011) showed to be effective to suppress RANKL-induced inflammation and osteoclastogenesis and therefore it can be used as a potential treatment for osteoporosis and arthritis (Li et al. 2019). There is a Japanese patent registering the use of *Cyclolepis genistoides* ("palo azul") as an alpha-glucosidase inhibitor (Hasegawa 2005).

10 Conclusions

Little is known about the water-soluble components of "palo azul". Additional research on the presence of other bioactive metabolites such as polyphenols, flavonoids (aglycones and glycosides), chlorogenic acid, caffeoylquinic acids, phenyletanoids, etc., is needed for a better evaluation of this plant.

A recent study to determine the acute toxicity of *Cyclolepis genistoides,* as well as its influence on the general behavior and barbiturate-induced sleeping time in mice, has revealed that the use of hydroalcoholic extract is safe, well tolerated and exhibits significant hypnotic properties (Montalbetti Moreno et al. 2018). Based on the well-known uses of this plant in traditional medicine and the successful isolation / identification of its lipophilic bioactive metabolites (triterpenic acids, sesquiterpene lactones), *C. genistoides* seems to have an excellent potential as a phytotherapeutic.

Acknowledgments The author thanks Consejo de Investigaciones de la UNT (CIUNT) for financial support.

Rerences

Alonso J, Desmachelier C (2006) Plantas medicinales autóctonas de la Argentina. Ed. Fitociencia, Buenos Aires, Argentina, pp 383–385

Barboza GE, Cantero JJ, Núñez C, Pacciaroni A, Ariza Espinar L (2009) Medicinal plants: a general review and a phytochemical and ethnopharmacological screening of the native argentine Flora. Kurtziana 34(1–2):7–365

Cabrera AL (1978) Flora de la provincia de Jujuy. República Argentina. Parte X. Compositae. Ed. Colección Científica del INTA, Buenos Aires, Argentina, p 592

Catalán CAN, Sánchez V, Díaz F (2020) In preparation

De Heluani CS, De Boggiato MV, Catalán CAN, Díaz JG, Gedris TE, Herz W (1997) Triterpenes and sesquiterpenes lactones from *Cyclolepis genistoides*. Phytochemistry 45:801–805

Don D (1832) Mag. Ann. Chem. 11:392. In: The Plant list. http://www.theplantlist.org/tpl1.1/record/gcc-126919. (Last access, March 2020)

Filipov A (1994) Medicinal plants of the Pilagá of Central Chaco. J Ethnopharmacol 44(3):181–193

Giberti GC (1983) Herbal folk medicine in northwestern Argentina: Compositae. J Ethnopharmacol 7(3):321–341

Hasegawa H (2005) Alpha glucosidase inhibitor comprising Palo Azul as active ingredient. Patent of Japan: JP2005263629 (A)-2005-09-29

Katinas L, Pruski J, Sancho G, Tellería MC (2008) The subfamily Mutisioideae (Asteraceae). Bot Rev 74:469–716

Li Z, Xiaodong Z, Xu R, Wang Y, Hu R, Xu W (2019) Deacylcynaropicrin inhibits RANKL-induced Osteoclastogenesis by inhibiting NF-κB and MAPK and promoting M2 polarization of macrophages. Front Pharmacol 10:article 599

Liu J (1995) Pharmacology of oleanolic acid and ursolic acid. J Ethnopharmacol 49:57–68

Liu J, Liu YP, Klaassen CD (1995) The effect of oleanolic acid on chemical-induced liver injury in mice. Acta Pharmacol Sin 15:16–23

Ma XH, Zhao YC, Yin L, Han DW, Ji CX (1982) Studies on the effect of oleanolic acid on experimental liver injury. Acta Pharm Sin 17:93–97

Ma XH, Zhao YC, Yin L, Han DW, Wang MS (1986) Studies on the preventive and therapeutic effects of ursolic acid on acute hepatic injury in rats. Acta Pharm Sin 21:332–335

Mlala S, Oyedeji AO, Gondwe M, Oyedeji OO (2019) Ursolic acid and its derivatives as bioactive agents. Molecules 24:2751

Montalbetti Moreno Y, Ibarrola DA, Heinichen O, Alvarenga N, Dolz Vargas JH, Hellion-Ibarrola MC (2018) Acute toxicity and potentiation of barbiturate-induced sleep in mice orally treated with hydro-alcoholic extract of *Cyclolepis genistoides* D. Don (Asteraceae). J Appl Pharm Sci 8(11):42–47

Piornedo RR, de Souza P, Alves Stefanello ME, Strapasson RLB, Zampronio AR, Leite Kassuya CA (2011) Anti-inflammatory activity of extracts and 11,13-dihydrozaluzanin C from *Gochnatia polymorpha* ssp. *floccosa* trunk bark in mice. J Ethnopharmacol 133:1077–1084

Sato H, Ishikawa M, Funaki A, Kimura Y, Yoshida H, Fukata H, Hasegawa H, Ueno K (2013) *Cyclolepis genistoides* D. Don (palo azul) promotes differentiation of adipocytes and regulates adipokine expression. Nutr Res 33(11):922–931

Sato H, Funaki A, Kimura Y, Sumitomo M, Yoshida H, Fukata H, Ueno K (2016) Ethanol extract of *Cyclolepis genistoides* D. Don (palo azul) induces formation of myotubes, which involves differentiation of C2C12 myoblast cells. Nutr Res 36(7):731–741

Scarpa GF (2004) Medicinal plants used by the Criollos of Argentine Chaco. J Ethnopharmacol 91:115–135

Shukla B, Viser S, Patnaik GK, Tripathi SC, Srimal RC, Day S, Dobhal PC (1992) Hepatoprotective activity in the rat of ursolic acid isolated from *Eucalyptus* hybrid. Phytother Res 6:74–19

Singh GB, Singh S, Bani S, Gupta D, Banerjee SK (1992) Antiinflammatory activity of oleanolic acid in rats and mice. J Pharm Pharmacol 44:456–458

Sosa A, Fusco MR, Petenatti ME, Juárez A, Del Vitto LA, Petenatti E (2007). Estudios farmacognósticos y farmacológicos comparativos sobre tres especies diuréticas de amplio uso popular en el centro-oeste argentino. Bol Latinoam Caribe Plant Med Aromát (BLACPMA) 6(6):386–387

Sosa A, Fusco MR, Rossomando P, Juárez A, Robles S, Petenatti E, Pelzer L (2011) Anti-inflammatory properties from isolated compounds of *Cyclolepis genistoides*. Pharm Biol 49(7):675–678

Sülsen VP, Martino VS (eds) (2018) Sesquiterpene lactones. Advances in their chemistry and biological aspects. Ed. Springer International Publishing AG, Switzerland. Chapters 8, 9 and 10, and references cited therein. 175–240

Wang X, Li YL, Wu H, Liu JZ, Hu JX, Liao N, Peng J, Cao PP, Lian X (2011) Antidiabetic effect of oleanolic acid: a promising use of a traditional pharmacological agent. Phytother Res 25(7):1031–1040

Wójciak-Kosior M (2013) Effect of different extraction techniques on quantification of oleanolic and ursolic acid in *Lamii albi flos*. Ind Crop Prod 44:373–377

Zuloaga FO, Morrone O (eds) (1996/1999) Catálogo de las plantas vasculares de la República Argentina. I. Pteridophyta, Gymnospermae y Angiospermae (Monocotyledonae), II. Dicotyledonae. Monogr Syst Bot Missouri Bot Gard 60:74

Eugenia uniflora L.

Leonardo M. Anconatani, Ignacio J. Agudelo, Rafael A. Ricco, and Marcelo L. Wagner

Eugenia uniflora flower. (Photo: LM Anconatani)

Abstract *Eugenia uniflora* L. is a medicinal plant widely employed in South America. Its evidence of use can be traced back to the Jesuit and Franciscan priests and the first South American botanists. The medicinal and food uses of this species give a proper background for the scientific research of its properties, the validation of its therapeutic utility and the exploration of its taxonomy, morphology, anatomy and phytochemistry. Nowadays, this plant has been extensively studied because of its hypoglycemiant and antihypertensive properties and its edible fruits.

Keywords "Pitanga" · Ethnobotany · Pharmacological activity

L. M. Anconatani (✉) · I. J. Agudelo · R. A. Ricco · M. L. Wagner
Universidad de Buenos Aires, Facultad de Farmacia y Bioquímica, Cátedra de Farmacobotánica, Museo de Farmacobotánica "Juan Aníbal Domínguez", Buenos Aires, Argentina
e-mail: lmanconatani@ffyb.uba.ar

Á. Máthé, A. Bandoni (eds.), *Medicinal and Aromatic Plants of South America Vol. 2*, Medicinal and Aromatic Plants of the World 7,
https://doi.org/10.1007/978-3-030-62818-5_14

1 Introduction

Eugenia uniflora L. is a plant with a long history of uses and meanings by a great diversity of peoples of South America. This fact implies that health sciences have dedicated many studies which not only endorse traditional uses but have proposed other potential pharmaceutical uses. The growing use of natural medicine as an alternative or complement to biomedicine demands integrative and interdisciplinary studies from the scientific community and that is why this unique plant species will be discussed here.

2 Taxonomic Characteristics

Eugenia uniflora L., known as "ñangapirí" or "pitanga", was described by Linneus in 1753 in the first volume of his book Species Plantarum (1753). The genus was named after Prince Eugenio de Saboya, a military nobleman who was also a nature enthusiast (Baeza 1930). It belongs to the Myrtales order and the Myrtaceae family, with more than 1300 species in 140 genera around the world (Johnson and Briggs 1984).

The genus *Eugenia* is native to tropical and subtropical South America and contains more than 500 tree species (Braga 1985). *Eugenia uniflora* L. has only one homonym according to the Kew Royal Botanic Garden database (n.d.): *Stenocalyx uniflorus* (L.) Kausel. and there are about 42 accepted synonyms.

Synonyms *Myrtus brasiliana* L., *M. brasiliana* var. *diversifolia* Kuntze, *M. brasiliana* var. *lanceolate* Kuntze, *M. brasiliana* var. *lucida* (O. Berg) Kuntze, *M. brasiliana* var. *normalis* Kuntze, *M. willdenowii* Spreng., *M. willdenowii* var. *portoriccensis* Spreng. ex DC., *Plinia rubra* L., *P. tetrapetala* L., *P. petiolata* L., *P. pedunculata* L. *f.*, *Eugenia michelii* Lam., *E. myrtifolia* Salisb., *E. zeylanica* Willd., *E. willdenowii* (Spreng.) DC., *E. costata* Cambess., *E. dasyblasta* (O. Berg) Nied., *E. oblongifolia* (O. Berg) Nied., *E. oblongifolia* (O.Berg) Arechav., *E. strigosa* (O. Berg) Arechav., *E. decidua* Merr., *E. arechavaletae* Herter, *E. uniflora* var. *atropurpurea* Mattos, *Stenocalyx costatus* (Cambess.) O. Berg, *S. michelii* (Lam.) O. Berg, *S. affinis* O. Berg, *S. brunneus* O. Berg, *S. dasyblastus* O. Berg, *S. glaber* O. Berg, *S. impunctatus* O. Berg, *S. lucidus* O. Berg, *S. michelii* var. *membranacea* O. Berg, *S. michelii* var. *rigida* O. Berg, *S. oblongifolius* O. Berg, *S. strigosus* O. Berg, *S. rhampiri* Barb. Rodr., *S. ruber* (L.) Kausel, *Syzygium michelii* (Lam.) Duthie, *Luma arechavaletae* (Herter) Herter, *L. costata* (Cambess.) Herter, *L. dasyblasta* (O. Berg) Herter, *L. strigosa* (O. Berg) Herter (Kew science 2020).

Photo 1 *Eugenia uniflora* fruit. (Photo: LM Anconatani)

Photo 2 Commercial sample of fresh leaves and stems of *E. uniflora*. (Photo: LM Anconatani)

3 Crude Drug Used

In folk medicine, the leaves of *Eugenia uniflora* L. are used in infusions, decoctions, and alcoholic extracts for treating diarrhea, stomachache, colic, intestinal infection, verminosis, fever, flu, cough, bronchitis, anxiety, high blood pressure, and diabetes (Photo 1). Occasionally, besides the leaves also fruits mainly, as well as bark and roots are also employed (Photo 2).

4 Major Chemical Constituents and Bioactive Compounds

The leaves of *E. uniflora* contain furanoelemene and elina-1,3,7(11)-trien-8-one, which is responsible for its typical aroma (Weyerstahl et al. 1988). Other authors (Melo et al. 2007) identified furanodiene, furanoelemene, β-elemene and (*E,E*)-germacrone as the most responsible odorant in leaves. The essential oil obtained through distillation of leaves contains mostly atractilone and curzerene (Lagos et al. 2011). According Weyerstahl et al. (1988) it also contains sesquiterpenic compounds such as β-elemene, β-caryophyllene, germacrene D, bicyclogermacrene, β-selinene and germacrene B, in addition to defuranodiene, eugenol, 1,8-cineol, linalool, spathulenol, globulol, viridiflorol and selin-11-en-4a-ol. Schmeda-Hirschmann et al. (1987) have reported the presence of the flavonoids quercitrin and miricitrin and, in lesser amounts, quercetin y myricetin. Einbord et al. (2003) identified cyanidin-3-glucoside and delphinidin-3-glucosie in the fruits. Lee et al. (1997, 2000) have isolated phenolic compounds, tannins and eugeniflorin D1 and eugeniflorin D2, with potential antiviral effects. Some of the volatile compounds found in the essential oil of the leaves by Weyerstahl et al. (1988) are also present, in the fruit extract (Oliveira et al. 2006).

The fruits of *E. uniflora* contain a wide variety of volatile monoterpenes in which *trans*-β-ocimene, *cis*-β-ocimene, and β-pinene predominate (Oliveira et al. 2006). Curzenene and bergaptene have also been reported (Pino et al. 2003).

The intense red color of the mature fruits indicates a higher concentration of anthocyanin pigments compared to the unripe fruit. Anthocyanin pigments have a concentration of 0.03% and flavonoids have a concentration of 0.02%; carotenoids are present in fewer amounts in the fruit epicarp. (Lima et al. 2002). Also, provitamin A (0.012%) (Freyre et al. 2000) and vitamin C (0.017–0.024%) have been reported (Poier Helt et al. 2018).

Ricciardi et al. (1999) highlight the importance of the chemotaxonomic study of this species after the infrequency of the solid (3,6,10-trimethyl-4,7,8,11-tetrahydrocyclodeca-[b]-furan) isolated from volatile oil and highlights the phytochemical stability of this species since this compound was found in other individuals of this species from diverse geographical areas.

5 Morphological Description

E. uniflora is a ramified tree up to 4 m high with glabrous stems. It has opposite subsessile leaves with ovate or ovate-oblong shape with a cuneate base, 3–5 cm height and 1.3–2 cm wide. The leaf lamina has a membranaceus texture and a 2 a 4 mm petiole. The inflorescence is a raceme and flowers have a deciduous ovate bracteole. The sepals are lanceolate and pubescent in its outer face while the petals are white, ovate and pubescent. Stamens have a length of 4–6 mm and style has the same size. The fruit is a globe like a berry orange to red (Dimitri 1978; Bacigalupo 2005).

The leaf presents a unistratified hypostomatic epidermis without trichomes and paracytic stomata. Both epidermises have glandular structures. The mesophyll has a dorsiventral structure with mono stratified palisade parenchyma adjacent to its adaxial face and spongy parenchyma in its abaxial face. Calcium oxalate druses can be seen. The vascular bundle has interxylemic phloem surrounded by pericyclic fibers and annular collenchyma adjacent to the epidermis (Alves et al. 2008; González and González 2011).

6 Geographical Distribution

E. uniflora is native to subtropical America and is distributed in Brazil (Paraná, Rio Grande do Sul and Santa Catarina), Paraguay (Alto Paraguay, Amambay, Caaguazú, Caazapá, Canindeyú, Central, Concepción, Cordillera, Guairá, Ñeembucú, Paraguarí, Presidente Hayes and San Pedro), Uruguay (Durazno, Paysandú, Rivera, Rocha, Salto, Soriano, Tacuarembó y Treinta y Tres), and Argentina (Catamarca, Chaco, Corrientes, Entre Ríos, Formosa, Jujuy, Misiones, Salta, Santa Fe and Tucumán) (Catálogo de PlantasVasculares del Cono Sur 1999). According to Jørgensen et al. (2014), it is also present in Bolivia (Chuquisaca, La Paz, Santa Cruz de la Sierra and Tarija). It is cultivated in almost all the tropical and subtropical word as ornamental or for its fruits, and in many countries is known as "Surinam cherry" (Einbord et al. 2003).

7 Ecological Requirements

E. uniflora lives in different types of environments, both in sunny and shady places. It prefers moist grounds, it is selective hygrophyte (Legrand and Klein 1969). It is a tree-shrub that adapts to different types of soils, except for saline ones. This species grows very well both in tropical and subtropical climates. It is very resistant to droughts and brief floods (Vignale et al. 2018). It blooms from August to November and fructifies until January in our Hemisphere (Peña-Cocarro et al. 2006). Its fruits provide food for mammals, birds and insects.

Furthermore, *E. uniflora* is ecologically important as a colonizing species in disturbed areas and as a food source for local fauna (Margis et al. 2002). Studies have shown that *E. uniflora* is a high-light-demanding species and moderately sensitive to soil flooding (Mielke et al. 2010).

8 Traditional Uses and Common Knowledge

The first reports regarding the use of *E. uniflora* are from the manuscripts of the Jesuit priests distributed among the missions established in Paraguay, Argentina, Bolivia and Brazil. J.F. Pedro de Montenegro wrote in 1710 the first mention of this species for our region, which he named "añangapirí" and describes that its dried and toasted leaves are used to increase blood fluidity and a stomach relaxant when administered with salt. He also mentions that it is much more pleasant to the stomach than "yerba mate" (*Ilex paraguariensis* L., Aquifoliaceae). There is also a report about the use of its bark with a similar purpose (Montenegro [1710] 1945). J.F. José Sánchez Labrador employs the same phytonym "añangapiry" but also indicates that there are other names such as "inapictam", "ibaachoó", "ibaachuá" and "burro nambí". He reports the use of the fruits as a carminative and against "weak stomach" and nausea (Ruiz Moreno 1948). Although at the time the works of Montenegro ([170] 1945) and Sánchez Labrador were published, the binomial taxonomic classification did not exist, it is possible to attribute the taxon *Eugenia uniflora* L. to the phytonyms mentioned by these authors, according to the descriptions proposed in theirs works, coinciding with the referred species (Scarpa and Anconatani 2019). Storni (1944) interprets that the phytonym "añangapyrí" would be formed by the following particles of the Guaraní language: "A": fruit; "ña"; to harvest; "nga": to put, to be and "pyrí": red, incarnate. So, it would mean: "fruit to be harvested when it is red, incarnate", Another interpretation is given by Cadogan (1972) who indicates that "ñangapirí" would be a current variant of the ancient phytonym of this Myrtaceae which could be explained as; "Añanga": name that means demon (Aña), "pi": of the skin and "ry", tree.

In the Brazilian cultures, Martius (1854) indicates that *E. uniflora* is known as "pitangueira" o "pitanga", and the fruit is recommended for consumption because of its color, aroma and sweet and acidic taste. Weyerstahl et al. (1988) indicate the phytonyms "ibitanga", "pitangatuba" and "pitanga" and the use of its fruits as an ingredient in jams, juices and alcoholic beverages. The leaves are used in infusion as a febrifuge, antirheumatic and for stomach aches. González Torres (2018) mentions the phytonyms "pitanga" o "pitangueira" in Brazil and "arrayán" in Argentina. According to this author, the infusion of leaves and flowers are used as a digestive, antispasmodic, astringent and carminative, and can also be used in mouthwashes to treat amygdalitis and sore throat. Bark decoction its use as antidiarrheal, antidysenteric and for the treatment of venereal diseases. Barks and roots can be mixed with "mate" for the treatment of bad digestion, hypertension and diabetes.

In Paraguay, Gatti (1985) indicates the use of leaves and fruits of this species for consumption in refreshing drinks; the fruits are also employed for the elaboration of jams and desserts. Regarding the medicinal uses, fresh or dried leaves can be used in decoctions or infusions as an antihypertensive (Basualdo et al. 2004; Suárez and Mereles 2006; Benítez et al. 2008; González et al. 2013; Soria and Ramos 2015), hypocholesterolemic (Basualdo et al. 2004; Benítez et al. 2008; González et al. 2013) and hypoglycemiant (Ibarrola and Degen de Arrúa 2011). Mashed leaves are added to cold "yerba mate" drinks (known as "tereré") for the treatment of hepatic ailments and as a digestive (Suárez and Mereles 2006; Ibarrola and Degen de Arrúa 2011). There are reports of the use of the leaves's infusion for the treatment of gastritis (Suárez and Mereles 2006) and as an antispasmodic and astringent agent; the use in mouthwashes for the treatment of amygdalitis and sore throats is also mentioned. Mixes of this species with "ñandypá" (*Genipa americana* L., Rubiaceae) have carminative properties (Suárez and Mereles 2006; Ibarrola and Degen de Arrúa 2011).

In Bolivia, the evidence of ethnobotanical use of *E. uniflora* is scarce. Its phytonyms are "ubajay" and "sarassa", the decoction of its fruits is used in asthenic patients and the leaves infusion is used as a febrifuge (De Lucca and Zalles 1992).

The first report in Argentina is present in the book Plantae Lorentzianae (Grisebach 1875).

Hieronymus (1882) employed the phytonym "arrayán" for *Eugenia uniflora* and indicates its use as an odorant in "mate" and in alcoholic beverages, and the use of the fruits as food. Parodi (1886) uses the phytonym "añangá-pirí-mirí" and also mentions the edible fruits. In Central Chaco, Baldrich (1890) reports the use of a decoction of leaves as a treatment for leucorrhea. In Entre Ríos and Corrientes provinces, Rojas Acosta mentions the edible fruits for the elaboration of wines and liquors, and the use of mashed leaves in baths against insolation and headaches and the use of the roots as an antidiarrheal and diuretic agent (Rojas Acosta 1897, 1907, 1914). Martínez Crovetto (1981) reports the use of its leaves in infusions for rheumatism, against coughs, fevers, and as an antihypertensive, hepatic and digestive. He also mentions that *E. uniflora* leaves can be employed as an *I. paraguariensis* substitute. The use of the decoction of its leaves against throat aches is also mentioned by this author.

In fieldwork carried out by one of the authors in eastern Formosa, it was observed that the "criollos" employed this species as an antihypertensive, hypolipemiant and antidiarrheal agent in infusions or mixed in "mate" (Anconatani, personal observation).

There are many studies regarding the use of *E. uniflora* by the original people of Northeastern Argentina. Qom people from the Chaco region employ the phytonyms "taikó" and "taikók" for this species and use the fruit as food and the wood to make arrow tips (Martínez Crovetto 1964); Qom people from the Province of Formosa use it in the same way and also employ the leaves as a refreshing agent (Vuoto, 1981). In the same province, in Misión Laishi, the Qom people and the Franciscans employed this plant as an antispasmodic, cardiotonic and as a digestive agent against "empacho" (Scarpa and Anconatani, 2017). East Chaco Vilela people call this plant "sukcsukelét", "sukesukelí", "taiekóksukesukelét" and "taiekóksukesukelí" and eat the fruit as food (Martínez Crovetto 1968). Moqoit from Chaco name this species "taik'ók" (Scarpa and Rosso 2014) and use it for the same purpose (Martínez Crovetto† 2012).

9 Modern Medicine Based on Its Traditional Medicine Uses

A lyophilized hydroalcoholic extract made of 765 g of dried leaves until exhaustion that yielded 230 g *Eugenia uniflora* dry extract was employed to perform a toxicological analysis in BALB/C mice. The LD_{50} of this extract was determined as 220 mg/kg i.p. This is a high dose since the extract was employed completely dried (Schmeda-Hirschmann et al. 1987).

The aqueous extract of *E. uniflora* was prepared according to ethnobotanical information for the assessment of its hypotensive activity in normotensive rats (Consolini et al. 1999; Consolini and Sarubbio 2002). This extract was made with 0.7 grams of dried leaves per liter, and another two extracts (6.25 and 60 grams per liter respectively) were also made for the evaluation of its diuretic activity. Every extract was dried (yield: 0.17 g/g of dried leaves) and administered intraperitoneally. The aqueous extract prepared according to ethnobotanical information in a dose of 0.12 mg of dry extract/kg decreased the blood pressure of normotensive animals in a dose-dependent relation; the decrease was 47% compared to the control animals. The dose response curves performed by administering phenylephrine to normotensive animals and the extract showed a noncompetitive inhibition of 80% compared to the control. There was also a vasorelaxant effect in hindquarters of animals previously treated with potassium rich Krebs medium.The diuretic activity was seen at higher doses with a decrease in sodium excretion. The authors conclude that the hypotensive effect could be due to an increase in vasodilation with a mild diuretic effect induced by an increase in renal blood flow.

The diuretic effect of these extracts of *E. uniflora* was assessed again with a dose of 0.24 mg of dry extract/kg. This extract induced a peak of diuresis within the hour, which decreased past that time (Amat et al. 1999).

Schapoval et al. (1994), confirmed the anti-inflammatory activity, a significant increase of pentobarbital sleeping time and antidiarrheal properties of the fresh leaf infusion.

Wazlawik et al. (1997) evaluated, from a hydroalcoholic extract, the effects on the metabolism of nitric oxide in the thoracic rat aorta. The extract was made by macerating 1 gram of dried leaves in 80% ethanol and was administered to a bath of Krebs medium in a dose range of 1μg/ml to 300μg/ml, as a result, caused a dose-dependent relaxation in intact endothelial aortic rings. In the aortic rings without endothelium there was no effect. Also, the addition of nitric oxide synthase inhibitors such as N″-nitro-L-arginine and N″-nitro-L-arginine methyl ester inhibited the relaxation; the addition of L-arginine reversed the inhibition caused by the N″-nitro-L-arginine. The addition of the guanylate cyclase inhibitor methylene blue reduced the relaxation. The relaxation induced by the hydroalcoholic extract of *E. uniflora* seems to be endothelium-dependent and mediated by the release of nitric oxide and activation of the cGMP pathway (Wazlawik et al. 1997).

The hypoglycemic and hypotriglyceridemic effect of a hydroalcoholic extract of dried leaves of *E. uniflora* was studied. The extract was made with a Soxhlet apparatus to exhaust the plant material and partitioned with solvents (4 fractions). The remaining aqueous fraction was partitioned by ultracentrifugation. In an oral glucose tolerance test, the fractions of hexane and the final precipitate obtained after the fraction had an inhibitory effect in the increase of glycemia; this effect was not observed when the glucose was administered intraperitoneally. In an oral sucrose tolerance test, all fractions had an inhibitory effect in the plasma glucose level. Regarding the effect on plasmatic lipids, a corn oil tolerance test was performed. The aqueous fraction and the precipitate had an inhibitory effect on the increase of triglyceridemia. Fractions obtained from the ultracentrifugation procedure (compounds with a molecular weight higher than 10,000 and compounds with a molecular weight lower than 10,000), as well as the aqueous fraction and the precipitate, had an inhibitory effect on the enzymes maltase and sucrose tested *in vitro*. Also, all the fractions, except the hexanic one, had an inhibitory effect in lipase activity. These results indicate that *E. uniflora* has compounds with inhibitory activity over the digestive enzymes (Arai et al. 2000). In this direction, an aqueous extract of

E. uniflora was tested to identify inhibitors of alpha glucosidase. This enzyme is a pharmacological target for the treatment of diabetes. The crude extract had an inhibitory effect on the increase of glycaemia in a glucose tolerance test in mice. The active fraction of this extract had two active pentahydroxy indolizidine alkaloids named uniflorine A and B, and tri-hydroxylated methyl piperidine alkaloid (Matsumura et al. 2000).

The contractile effects of extracts of *E. uniflora* in isolated rat duodenum were tested. The extracts were made with a Soxhlet apparatus following an eluotropic series until exhaustion. There was a maximum effect in the contractile activity with the ethyl acetate extract compared to acetylcholine. Fractions obtained from other solvents did not have an effect (Gbolade et al. 1996).

Another of the documented pharmacological effects, although less studied, postulate the antioxidant activity of *E. uniflora*. Victoria et al. (2012) demonstrated through different tests the antioxidant activity that the essential oil of the leaves, having the ability to reduce lipid peroxidation in the kidney of mice. With the same purpose, Velázquez et al. (2003) have shown that a methanolic extract of its leaves can decrease enzymatic and non-enzymatic lipoperoxidation in liposomal membranes. In line with the above, Griffis et al. (2009) highlighted that mature fruits used in the form of juices or jams among other commercial forms, provide essential nutrients for a complete diet, but mainly significant amounts of antioxidants.

10 Conclusions

Eugenia uniflora L. is a species widely used in traditional medicine by various peoples of South America. In addition to the extensive use by the native peoples of the region, the Jesuit order were in charge of documenting and expanding its use through their missionary manuscripts and practices that toured much of South America. The anatomical and morphological studies of its useful organs, as well as the extensive botanical knowledge of this species and even of the Myrtaceae family, in general, provide clear characteristics for its recognition and identification, thus counting on reliable standards for quality controls.

As for its usefulness, the use of its fruits as food transcends borders, as well as the preparation of various drinks such as juices, smoothies or ferments. Its use as an antihypertensive or as a hypoglycemic agent is widely documented by the ethnobotanical and pharmacobotanical bibliography. It is well studied from a chemical and pharmaceutical point of view by scientists from all over the world.

The biological activities mentioned would allow the use of this species in primary health care, as a therapeutic alternative.

References

Alves ES, Tresmondi F, Longui EL (2008) Análise Estrutural de folhas de *Eugenia uniflora* L. (Myrtaceae) coletadas em ambientes rural e urbano, SP, Brasil. Acta Bot Bras 22(1):241–248

Amat AG, De Battista GA, Uliana RF (1999) Diuretic activity of *Eugenia uniflora* L. (Myrtaceae) aqueous extract. Acta Hortic (501):155–158

Arai I, Amagaya S, Komatsu Y, Okada M, Hayashi T, Kasai M, Arisawa M, Momose Y (2000) Improving effects of the extracts from *Eugenia uniflora* on hyperglycemia and hypertriglyceridemia in mice. J Ethnopharmacol 68:307–314

Bacigalupo N (2005) Flora ilustrada de Entre Ríos. Dicotiledóneas Arquiclamídeas. Geraniales a Umbeliflorales. Continuación de la obra de Arturo Burkart Ed INTA Buenos Aires, Argentina, 627 pp

Baeza VM (1930) Los nombres vulgares de las plantas silvestres de Chile y su concordancia con los nombres científicos. 2ª Edición aumentada. Ed. Imp. El globo, Santiago, Chile, 270 pp

Basualdo I, Soria N, Ortiz M, Degen R (2004) Plantas medicinales comercializadas en los mercados de Asunción y Gran Asunción. Parte I. Rojasiana 6:95–114

Benítez B, Bertoni S, Pérez S, González F (2008) Principales especies medicinales comercializadas en los mercados del departamento de Paraguarí. Diversidad y determinación de la importancia relativa. Rojasiana 8(1):45–61

Braga R (1985) Plantas do Nordeste, especialmente do Ceará. 4° edition. Ed. Universitaria da UFRN. Rio Grande do Norte, Brazil, 540 pp

Cadogan L (1972) Ta-ngypuku. Aportes a la etnobotánica de algunas especies arbóreas del Paraguay Oriental. Suplemento antropológico 7(1–2):7–59

Catálogo de Plantas Vasculares del Cono Sur - Instituto de Botánica Darwinion (1999). http://www.darwin.edu.ar/Proyectos/FloraArgentina/fa.htm. (Last access, January 2020)

Consolini AE, Sarubbio MG (2002) Pharmacological effects of *Eugenia uniflora* (Myrtaceae) aqueous crude extract on rat's heart. J Ethnopharmacol 81:57–63

Consolini AE, Baldini OAN, Amat AG (1999) Pharmacological basis for the empirical use of *Eugenia uniflora* L. (Myrtaceae) as antihypertensive. J Ethnopharmacol 66:33–99

De Lucca M, Zalles J (1992) Flora medicinal boliviana. Diccionario enciclopedia. Ed. Los amigos del libro. Cochabamba, Bolivia, 498 pp

Dimitri M (1978) Enciclopedia Argentina de Agricultura y Jardinería. Tomo 1. Descripción de las plantas cultivadas. Ed. Acme. Buenos Aires, Argentina, 1161 pp

Einbord LS, Reynerston KA, Luo X-D, Basile MJ, Kennelly EJ (2003) Anthocyanin antioxidants from edible fruits. Food Chem 84(1):23–28

Freyre M, Baigorria C, Rozycki V, Bernardi C, Charpentier M (2000) Vegetales silvestres sub explotados del Chaco argentino y su potencial como recurso alimenticio. Arch Latinoam Nutr 50(4):394–399

Gatti C (1985) Enciclopedia Guaraní-Castellano de Ciencias Naturales y conocimientos Paraguayos. Ed. Arte nuevo editores. Asunción, Paraguay, 329 pp

Gbolade AA, Ilesanmi OR, Aladesanmi AJ (1996) The contractile effects of the extracts of *Eugenia uniflora* on isolated rat duodenum. Phytother Res 10:613–615

González AB, González AM (2011) Anatomía foliar de especies arbóreas del Chaco Húmedo. Rojasiana 10(2):15–30

González Torres DM (2018) Catálogo de plantas medicinales (y alimenticias y útiles) usadas en Paraguay. Ed. Servi libro. Asunción, Paraguay, 456 pp

González Y, Degen R, González M, Delmás G (2013) Especies medicinales, su estado de conservación y usos, de la compañía Pikysyry, departamento de Cordillera, Paraguay. Rojasiana 12(1–2):105–115

Griffis JL, McDonald TG, Smith VE, Manners MM (2009) *Eugenia uniflora* -a nutritous, easy-to-grow fruit for the tropics. Acta Hortic (806):277–284

Grisebach A (1875) Plantae Lorentzianae. Bearbeitung der ersten und zweiten Sammlung argentinischer Pflanzen. Ed. Dieterich'schen Buchhandlung. Göttingen, Germany, 239 pp

Hieronymus J (1882) Plantae diaphoricae. Florae argentinae. Ed. Kraft. Buenos Aires, Argentina, 404 pp
Ibarrola DA, Degen de Arrúa, RL (2011) Catálogo ilustrado de 80 plantas medicinales del Paraguay. Facultad de Ciencias Químicas, Agencia de Cooperación Internacional del Japón. Asunción, Paraguay, 178 pp
Johnson LAS, Briggs BG (1984) Myrtales and Myrtaceae - a phylogenetic analysis. Ann Mo Bot Gard 71:700–756
Jørgensen PM, Nee MH, Beck SG (2014) Catálogo de las plantas vasculares de Bolivia. Monogr Syst Botany Missouri Botan Garden 127(1–2):1–1744
Kew science (2020) The royal botanic gardens https://wcsp.science.kew.org. (Last access, January 2020)
Lago JHG, Souza ED, Mariane B, Pascon R, Vallim MA, Martins RCC, Baroli AA, Carvalho BA, Soares MG, dos Santos RT, Sartorelli P (2011) Chemical and biological evaluation of essential oils from two species of Myrtaceae - *Eugenia uniflora* L. and *Plinia trunciflora* (O. Berg) Kausel. Molecules 16(12):9827–9837
Lee M-H, Nishimoto S, Yang LL, Yen KY, Hatano T, Yoshida T, Okuda Y (1997) Two macrocyclic hydrolysable tannin dimers from *Eugenia uniflora*. Phytochemistry 44:1343–1349
Lee M-H, Chiou J-F, Yen K-Y, Yang L-L (2000) EPV DNA polymerase inhibition of tannins from *Eugenia uniflora*. Cancer Lett 154:131–136
Legrand D, Klein R (1969) Flora Ilustrada Catarinense. Mirtáceas. Ed. Livreria Blumenauense S.A. Itajeí, Santa Catarina, Brazil, 113 pp
Lima VLAG, De Almeida Mélo E, Lima DE dS (2002) Fenólicos e carotenoides totai sem Pitanga. Sci Agric 59(3):447–450
Linneo C (1753) Species plantarum. Tomus 1. Imp. Laurentii Salvii. Holmiae, 560 pp
Margis R, Felix D, Cldas JF, Salgueiro F, De Araujo D, Breyne P, Van Montagu M, De Oliveira D, Margis-Pinheiro M (2002) Genetic differentiation among three neighboring Brazil-cherry (*Eugenia uniflora* L.) populations within the Brazilian Atlantic rain forest. Biodivers Conserv 11(1):149–163
Martínez Crovetto RN (1964) Estudios etnobotánicos I. Nombres de Plantas y su utilidad, según los indios tobas del este del chaco. Bonplandia 1(4):279–333
Martínez Crovetto RN (1965) Estudios etnobotánicos II. Nombres de Plantas y su utilidad, según los indios vilelas del chaco. Bonplandia 2:1):1–1)28
Martínez Crovetto RN (1968) La alimentación entre los indios guaraníes de Misiones (República Argentina). Etnobiológica (4):1–24
Martínez Crovetto RN (1981) Plantas utilizadas en medicina en el NO de Corrientes. Miscelánea 69:1–139
Martínez Crovetto RN† (2012) Estudios etnobotánicos V. Nombres de plantas y su utilidad según los mbya guaraní de Misiones, Argentina. Bonplandia 21(2):109–133
Martínez Crovetto RN (2014) Algunos datos sobre etnobotánica mocoví. Bonplandia 23(2):119–131
Matsumura T, Kasai M, Hayashi T, Arisawa M, Momose Y, Arai I, Amagaya S, Komatsu Y (2000) α-Glucosidase inhibitors from paraguayan natural medicine, ñangapiry, the leaves of *Eugenia uniflora*. Pharm Biol 38(4):302–307
Melo RM, Correa FS, Amorim CL, Miranda APL, Rezende CM (2007) Identification of impact aroma compounds in *Eugenia uniflora* L. (Brazilian Pitanga) leaf essential oil. J Braz Chem Soc 18(1):179–183
Mielke MS, Schaffer B, Li C (2010) Use of a SPAD meter to estimate chlorophyll content in *Eugenia uniflora* L. leaves as affected by contrasting light environments and soil flooding. Photosynthetica 48:332–338
Montenegro, P ([1710] (1945)) Materia Médica Misionera. Imprenta de la Biblioteca Nacional. Buenos Aires, Argentina, 477 pp
Oliveira AL, Lopes RB, Cabral FA, Eberlin MN (2006) Volatile compounds from Pitanga fruit (*Eugenia uniflora* L.). Food Chem 99:1–5

Parodi D (1886) Notas sobre algunas plantas usuales del Paraguay, de Corrientes y de Misiones. Ed. Coni e hijos. Buenos Aires, Argentina, 123 pp

Peña-Chocarro MC, De Ejea Juvinel J, Vera M, Maturo H, Knapp S (2006) Guía de árboles y arbustos del Chaco Húmedo. The Natural History Museum, Guyra Paraguay, Fundación Moisés Bertoni y Fundación Hábitat y Desarrollo. Asunción, Paraguay, 291 pp

Pino JA, Bello A, Urquiola A, Agüero J, Marbot R (2003) Fruit volátiles of Cayena cherry (*Eugenia uniflora* L.) from Cuba. J Essent Oil Res 15:70–71

Poier Helt KM, Navas R, Gonçalves EM (2018) Características físico-químicas e compostos antioxidantes de frutos de pitanga da região de Capão Bonito, SP. Rev Ciênc Agro Amb 16(1):97–102

Ricciardi AIA, Ruiz Díaz JJJ, Agrelo de Nassiff AE, Ricciardi GAL (1999) Sesquiterpeno aislado de los aceites esenciales de *Eugenia uniflora* L. y *Eugenia uniflora* var. *vermillion* (ñangapirí). Trabajo Especial UNNE Resúmenes ampliados en Actas Reun Comun Cient Argentina 8:97–99

Roja Acosta N (1914) Addenda ad floram Chaco Australis –1909. Bull Geogr Bot 23(295–297):211–219

Rojas Acosta N (1897) Historia Natural de Corrientes: catálogo. Ed. Imp. El hogar y escuela. Argentina, 214 pp

Rojas Acosta N (1907) Catálogo de las plantas medicinales del Chaco Austral. Revista Farmacéutica 47:214–225. 247-258, 281-298, 315-327

Ruiz Moreno A (1948) La medicina en "el Paraguay natural" (1771–1776) del P. José Sánchez Labrador S.J. Exposición comentada del texto original. Ed. Universidad Nacional del Tucumán, Argentina, 348 pp

Scarpa GF, Anconatani LM (2017) Etnobotánica histórica de las misiones franciscanas del este de Formosa II: identificación y análisis de datos inéditos y reelaboración integral de fuentes ya publicadas a partir de hallazgos documentales. Dominguezia 33(2):37–79

Scarpa GF, Anconatani LM (2019) La "Materia Médica Misionera" atribuida al jesuita Pedro de Montenegro en 1710: Identificación, sistematización e interpretación de los usos medicinales de las plantas y sus implicancias para la etnobotánica actual. IHS Antiguos Jesuitas en Iberoamérica 7(1):27–46

Scarpa GF, Rosso CN (2014) La etnobotánica moqoit inédita de Raúl Martínez Crovetto I: Descripción, actualización y análisis de la nomenclatura indígena. Bol Soc Argent Bot 49(4):623–647

Schapoval EES, Silveira SM, Miranda ML, Alice CEW, Henriques AT (1994) Evaluation of some pharmacological activities of *Eugenia uniflora* L. J Ethnopharmacol 44:137–142

Schmeda-Hirschmann G, Theoduloz C, Franco L, Ferro E, de Arias A (1987) Preliminary pharmacological studies on *Eugenia uniflora* leaves: xanthine oxidase inhibitory activity. J Ethnopharmacol 21:183–186

Soria N, Ramos P (2015) Uso de plantas medicinales en la atención primaria de salud en Paraguay: algunas consideraciones para su uso seguro y eficaz. Mem Inst Investig Cienc Salud 13(2):8–17

Storni JS (1944) Hortus guaranensis. Flora. Ed. Universidad Nacional de Tucumán. Tucumán, Argentina, 268 pp

Suárez MO, Mereles MF (2006) Los árboles medicinales utilizados en la comunidad de Paso Jovái, departamento de Guairá, Paraguay. Rojasiana 7(2):91–115

Velázquez E, Tournier H, Mordujovich de Buschiazzo P, Saavedra G, Schinella G (2003) Antioxidant activity in Paraguayan plant extracts. Fitoterapia 74:91–97

Victoria FN, Lenardão EJ, Savegnago L, Perin G, Jacob RG, Alves D, Padilha da Silva W, de Souza da Motta A, da Silva Nascente P (2012) Essential oil of the leaves of *Eugenia uniflora* L.: antioxidant and antimicrobial properties. Food Chem Toxicol 50:2668–2674

Vignale B, Jolochin G, Cabrera D (2018) Pitanga *Eugenia uniflora* L. Ed. Instituto Interamericano de Cooperación para la Agricultura. http://www.procisur.org.uy. (Last access, January 2020)

Martius KFP von (1854) Systema de matéria medica vegetal brasileira. Tradução de Henrique Velloso d'Oliveira. Laemmert. Rio de Janeiro, Brazil, 284 pp

Vuoto P (1981) Plantas útiles entre los toba-Taksék. Entregas del Instituto de Tilcara 10:1–10

Wazlawik E, Da Silva M, Peters R, Correia J, Farías M, Calixto J, Ribeiro do Valle R (1997) Analysis of the role of nitric oxide in the relaxant effect of the crude extract and fractions form *Eugenia uniflora* in the rat thoracic aorta. J Ethnopharmacol 49(4):433–437

Weyerstahl P, Marschall-Weyerstahl H, Christiansen C, Oguntimein BO, Adeoye AO (1988) Volatile constituents of *Eugenia uniflora* L. leaf oil. Planta Med 6:546–549

Eupatorium buniifolium Hook. ex Hook. & Arn.

Daniel Lorenzo, Daniel Paz, Philip Davies, José Villamil, Manuel Minteguiaga, and Eduardo Dellacassa

Eupatorium buniifolium. Plant (Zuloaga 9497, SI). Branches (Zuloaga 14,568, SI). Capitula (Zuloaga 14,568, SI). Photos: Fernando O. Zuloaga, courtesy Instituto de Botánica Darwinion (CONICET - ANCEFN)

D. Lorenzo · D. Paz · E. Dellacassa (✉)
Laboratorio de Biotecnología de Aromas, Departamento de Química Orgánica, Facultad de Química, Universidad de la República, Montevideo, Uruguay
e-mail: edellac@fq.edu.uy

P. Davies · J. Villamil
Ingeniero Agrónomo, Botany Lecturer, Montevideo, Uruguay

M. Minteguiaga
Espacio de Ciencia y Tecnología Química, Centro Universitario de Tacuarembó, Universidad de la República, Tacuarembó, Uruguay

Á. Máthé, A. Bandoni (eds.), *Medicinal and Aromatic Plants of South America Vol. 2*, Medicinal and Aromatic Plants of the World 7,
https://doi.org/10.1007/978-3-030-62818-5_15

Abstract *Eupatorium buniifolium* Hook. ex Hook. & Arn. (Asteraceae) is an aromatic and medicinal plant. Its area is spreading from the Southern parts of Bolivia, Brazil, Paraguay and Uruguay to the northern and central parts of Argentina. It is known as "chilca negra", "romerillo", "chilca", "chirca", "romero colorado". The infusion of leaves and stems is used as a digestive for stomachache, against diarrhea and for liver inflammation. Some studies have confirmed the digestive, hepatoprotective, anti-inflammatory and antioxidant potential of this plant. The main chemical components responsible for its known biological activities have been identified as flavonoids. From *E. buiniifolium* leaves also an essential oil has been obtained that is useful due to its potential insecticidal, antifungal and varroacide activities, both by direct contact and their vapors against the tested organisms. The essential oil has been demonstrated to be bioactive against the Chagas' disease vector *Triatoma infestans*.

Keywords "Chilca" · Asteraceae · Aromatic · Flavonoids

1 Introduction

The genus *Eupatorium* (tribe Eupatorieae, Asteraceae) is represented by almost 1200 species growing in the tropical regions of Americas, Europe, Africa, and Asia. Some species exert a pleasant and have been considered as aromatics. Several species in this genus have also been used in folk medicine as antimalarial, antibacterial, antifungal and anti-inflammatory (Nogueira Sobrinho et al. 2017). *Eupatorium* species is a source of an important number of bioactive metabolites, mainly terpenes, phytosterols and sesquiterpene lactones (being the latest chemotaxonomic markers of the group) which could be a promising bioresource for the development of potential drugs and value-added products (Liu et al. 2015; Zhanga et al. 2008). In the current review, we summarize the progress in the phytochemical isolation of compounds from *Eupatorium buniifolium,* over the last decades. The biological activities of compounds isolated from this species are also included.

2 Taxonomic Characteristics

The genus *Eupatorium* belongs to the Eupatorieae, which is one of the 13 tribes of the Asteraceae (Frohne and Jensen 1979; Woerdenbag 1992). The Index Kewensis listed over 1000 species of *Eupatorium,* but the genus has been taxonomically revised to contain 44 species (Hooker and Jackson 1960; Robinson and King 1985). Species of *Eupatorium* show luxuriant growth, exert allelopathic action and have become major intractable weeds in many parts of the world. The fast growth of members of this genus is disturbing the overall ecology by encroaching upon pastures and replacing forest biomass (Sharma and Dawra 1994; Sharma et al. 1998).

The overt impact is in terms of loss of pastures. In addition, the huge biomass of *Eupatorium* is not used in agriculture and animal husbandry operations, since several species are known to exhibit adverse interactions with livestock and humans.

E. bunififolium has several common names: "chilca negra", "romerillo", "chilca", "chirca", "romero colorado". However, "chilca" seems to be the most common.

Synonyms *Acanthostyles buniifolius* (Hook. & Arn.) R.M. King & H. Rob., *Acanthostyles saucechicoënse* (Hieron.) R.M. King & H. Rob., *Eupatorium buniifolium* Hook. & Arn. var. *buniifolium*, *Eupatorium buniifolium* Hook. & Arn. var. *saucechicoënse*, *Eupatorium buniifolium* Hook. & Arn. var. *bakerii*, *Eupatorium buniifolium* Hook. & Arn. var. *hieronymi*, *Eupatorium crithmifolium* Griseb., *Eupatorium pinnatifidum* DC. *nom. Illeg*., *Eupatorium pinnatifidum* DC. var. *virgata*, *Eupatorium saucechicoënse* Hieron., *Eupatorium virgatum* D. Don ex Hook. & Arn., *Acanthostyles buniifolius* (L.) R.M. King & H. Rob. var. *saucechicoënse*, *Eupatorium pinnatifissum* H. Buek (Zuloaga et al. 2008).

3 Crude Drug Used

Its crude drug consists of the dried leaves harvested independently of their maturation stage. Aqueous and hydroalcoholic extracts of *E. buniifolium* have an aromatic and pleasant odor and a slightly bitter flavor, being the main application as digestive (Lombardo 1959; Alonso Paz et al. 1992).

4 Major Chemical Constituents and Bioactive Compounds

The aerial parts *Eupatorium buniifolium* contains a great number of biologically active compounds, principally flavonoids, in their different hydroxylated and glycosylated options, the most studied components: 5,7,5′-trihydroxy-3,6,2′,4′-tetramethoxyflavone (Muschietti et al. 1994; Zhang et al. 2008), 6-methoxy flavonoids (Caula et al. 1991), hydroxycinnamic acids and their derivatives (Muschietti et al. 1990; Caula et al. 1991), flavonoid glycosides (Muschietti et al. 1990) and two 2′-oxygenated flavonoids (5,7,5′-trihydroxy-3,6,2′,4′- tetramethoxyflavone and 5,2′-dihydroxy-3,6,7,4′,5′-pentamethoxyflavone (Muschietti et al. 1993,1994). *Eupatorium buniifolium* was also investigated for diterpenoid (Caula et al. 1991) and triterpenoid contents. Methyl *ent*-labd-8(17)-en-18-oic acid-15-oate, 15-hydroxy-*ent*-labd-8(17)-en-18-oic acid, 15,16-epoxy-15-methoxy-*ent*--labd-8(17)-en-18-oic acid and 15-methoxy- ent-labd-8(17)-13-dien-18-oic acid methyl ester-16,15-olide were isolated and spectroscopically characterized (Carreras et al. 1998). Visintini et al. (2013) also isolated and characterized from this species the compound euparin, a benzofuranic structure: 1-[6-hydroxy-2-(1-methylethenyl)-5-benzofuranyl] ethenone, possessing antipoliovirus activity. From *E. buniifolium* leaves has also been obtained an essential oil by steam distillation. By GC-MS, 44 components were identified. Monoterpene and sesquiterpene hydrocarbons, especially α-pinene (14.7%), β-elemene (12.2%), germacrene D (11.5%), *trans*-β-guaiene (6.5%) and (*E*)-caryophyllene (4.3%), were the major constituents found in the oils. The metabolomic characterization of *E. buniifolium* was performed by enantioselective GC by evaluation of the enantiomeric ratios of α-pinene, sabinene, β-pinene, limonene, terpinen-4-ol and germacrene D (Lorenzo et al. 2005).

E. buniiflium oil was used as substrate of endophytic microorganisms in order to diversify its chemical composition, mainly by increasing the oxygenation degree of monoterpene hydrocarbons. The essential oil was modified, containing most valuable oxygenated monoterpenes (Cecati et al. 2018).

5 Morphological Description

Native to Bolivia, Brazil, Uruguay and Argentina (Malme 1931; Cabrera 1963, 1974, 1978; King and Robinson 1971; Sayagués et al. 2000; Grossi et al. 2011), this plant is a perennial shrub resinous that can reach 50–100 cm in high. Stems striate, glabrous. Leaves resinous, mainly opposite, rarely alternate in some terminal branches, usually petiolate, blades entire, narrowly lobed to pinnatifid or bipinnatifid, surface glandular, punctate. Inflorescences are pendent or suberect, long, paniculate cyms. Capitula discoid, homogamous. Involucres cylindrical, rarely campanulate; phyllaries subimbricata, 3–5 seriate, the outer surface usually glandular; receptacles convex, epaleate, glabrous, with irregular surface. Florets usually 5 per capitulum; corolla tubular funnelform, purplish or whitish; lobes ovate–triangular, outer surface glandular; anther collar elongate, cylindrical; anther appendages ovate to triangular; anther bases rounded; style bases not enlarged, glabrous, nectary lacking; style branches longlinear, with stigmatic papillae, elongated in two lines, flagelliform apical sterile appendages with large non-septate sweeping hairs scattered along the entire surface. Cypselae prismatic, 4- or 5-costate, with short twin hairs, carpopodium inconspicuous or minutely annuliform; pappus of ~30–40 bristles, persistent, uniseriate, basally fused in a ring, barbellate, not enlarged at the apex, yellowish or whitish. Pollen grains with noticeably short spines (King and Robinson 1971). Stems rich in leaves; *E. buniifolium* differs from other *Eupatorium* spp. by polymorph leaves, sometimes simple and linear, others with pinnate leaves with linear sections.

6 Geographical Distribution

Eupatorium buniifolium is a South American species that is spreading from the southern parts of Bolivia, Brazil, Paraguay and Uruguay to the northern and central parts of Argentina, growing as a weed in overgrazed fields.

The "chircales" in Uruguay, whose predominant species is *E. buniifolium*, occupy areas of several hundred hectares, with little or no pastoral utility, being mechanical and chemical control methods expensive and difficult to apply (Bayce and Del Puerto 1989).

7 Ecological Requirements

Typically, *E. buniifolium* prefers sandy, rocky soils, on stream embankments, hillsides and roadsides. It occurs in shrubby vegetation, forests and grasslands (Alonso Paz and Bassagida 1999; Marchesi 2005; Brussa and Grela 2007; Marchesi et al. 2013).

8 Traditional Use (Part(s) Used) and Common Knowledge

E. buniifolium has been employed in folk medicine as a tincture for its hepatoprotective and disinfectant properties (Saggese 1959; Zardini 1984; Muschietti et al. 1994). The ethanolic extracts of the plant also show good antiherpetic activity against herpes simplex virus (García et al. 1990; Zanon et al. 1999). The decoction of the aerial parts has been used as a disinfectant (Rojas Acosta 1905), for the treatment of liver and nervous diseases (Saggese 1959) and as digestive and antirheumatic (Ríos et al. 1993; Marchesi and Davies 2004). Occasionally, it has been mentioned as "a very interesting plant for honey production" (Marchesi and Davies 2004).

9 Modern Medicine Based on Its Traditional Medicine Uses

The essential oil (EO) was demonstrated to be active against Varroa in laboratory assays. Although activity was less than that for oxalic acid (the positive control), this EO was less toxic to bees than the control, encouraging further studies (Umpiérrez et al. 2013).

Some of the main popular uses of *E. buniifolium* recorded in the scientific literature are as digestive, hepatoprotective (Saggese 1959; Zardini 1984; Muschietti et al. 1994) and antioxidant (Paya et al. 1996; Gorzalczany et al. 2008).

Many laboratory studies on *E. buniifolium* have proved its pharmacological potential as antiviral, using a Herpes simplex virus-VERO cells system (García et al. 1990; Visintini et al. 2013); being also active on murine normal splenocytes proliferation (Fernández et al. 2002). Methanol extract from *Eupatorium buniifolium* was assayed *in vitro* for antifungal activity against yeasts, hialohyphomycetes as well as dermatophytes with the microbroth dilution method showing strong effect

(Muschietti et al. 2001, 2005). Anti-inflammatory activity was detected in the CH_2Cl_2 extract of the aerial parts of *Eupatorium buniifolium* using the TPA-mouse ear model. Three compounds isolated from this extract, by bioassay-guided fractionation, significantly inhibited the inflammatory response (5,7,5′-trihydroxy-3,6,2′,4′-tetramethoxyflavone, scopoletin and centaureidin) (Muschietti et al. 2001).

Preliminary studies indicate that some active principles of *E. buniifolium* act provoking antinociceptive effect was not reversed by pretreatment with naloxone. Inhibition of only the second phase of the formalin test and no significant effects observed in the hot plate test suggest that the antinociceptive activity is unrelated to the activation of the opioid system (Miño et al. 2005). Pharmacological investigations of aqueous extract of this plant evaluated the neuropharmacological profile in mice. The results found suggest that the activity of *E. buniifolium* may be a CNS-depressant (Miño et al. 2007).

In search for new strategies in pest and disease control, essential oils, as botanical pesticides, can provide a potential resource to develop more environmentally friendly and less toxic products. In this sense, the essential oil from *E. buniifolium* was also evaluated in view of its possible insecticidal potential, as well as antifungal and varroacide activities both in direct contact and with their vapors against the tested organisms (Umpiérrez et al. 2012, 2013, 2017).

Guerreiro et al. (2018) also studied bioactivities toward the Chagas' disease vector *Triatoma infestans* (Klug) (Hemiptera: Reduviidae) by the essential oil of *E buniifolium*. The results provided an interesting scope in relation to the potential use of this oil for the control of this insect at the nymph stage as repellent as well as for decreasing the population by the ovicidal effect. Considering the abundance of the wild plant under study and the fact that its essential oil is easy to obtain, it is suggested that it could be an adequate natural resource to control this vector in a sustainable way as a complementary approach to conventional methods.

In another study, Tasso de Souza (2007) evaluated the antioxidant and anticholinesterasic activities on the essential oil finding positive results as antioxidant in a bioautographic assay against DDPH. When the oil was evaluated in a bioautographic assay against acetylcholinesterase enzyme, the activity found was also positive and correlated to the high percentage of monoterpenes present in the oil.

Lancelle et al. (2009) investigated toxic and repellent properties of the essential oils from four *Eupatorium* species toward *Tribolium castaneum* Herbst adults. Contact toxicity assays showed that all the evaluated essential oils were toxic, and mortality was dose dependent. The main repellency was observed for *E. buniifolium* essential oil.

10 Conclusions

E. buniifolium is a species with a wide range of reported ethnomedicinal uses, as well as many biological activities. These activities can be attributed to its both volatile and non-volatile components. The invasion of this species impacts in terms of loss of pastures. Remarkably, the huge biomass of *E. buniifolium* is not yet used in agriculture and animal husbandry. Several *Eupatorium* species are known to exhibit adverse interactions with livestock and humans. The ecological behavior of this species represents a paradigmatic situation, as there are no difficulties in the availability and reproduction of plant material. *E. buniifolium* remains to be underexploited, as aromatic or medicinal species.

References

Alonso Paz E, Bassagoda MJ, Ferreira F (1992) Yuyos. Uso racional de las plantas medicinales. Ed. Fin de Siglo. Montevideo, Uruguay, 158 pp

Alonso Paz E, Bassagoda MJ (1999) Los bosques y los matorrales psamófilos en el litoral platense y atlántico del Uruguay. VI (113). Comunicaciones Botánicas del Museo de Historia Natural de Montevideo. Montevideo, Uruguay, 12 pp

Bayce D, Del Puerto O (1989) Observaciones sobre control de chircales de *Eupatorium buniifolium* mediante pastoreo y quema, vol 6. Notas Técnicas Facultad de Agronomía, Montevideo, Uruguay, pp 1–8

Brussa C, I Grela (2007) Flora arbórea del Uruguay, con énfasis en las especies de Rivera y Tacuarembó. COFUSA. Montevideo, Uruguay, 553 pp

Cabrera AL (1963) Compositae. Flora de la Provincia de Buenos Aires. Colecc Cient INTA 4(6):1–443

Cabrera AL (1974) Compositae. Flora de la Provincia de Entre Ríos. Colecc. Cient. INTA 6(6a):106–554

Cabrera AL (1978) Compositae. Flora de la Provincia de Jujuy. Colecc. Cient. INTA 13(10):1–726

Carreras CR, Rossomando PC, Giordano OS (1998) *ent*-Labdanes in *Eupatorium buniifolium*. Phytochemistry 48(6):1031–1034

Caula S, Villar S, Martino V, Coussio J, Ferraro G (1991) Polyphenols isolated from *Eupatorium buniifolium*. Rev Latinoamer Quim 22:1–3

Cecati FM, Magallanes-Noguera C, Ardanaz CE, Tonn CE, Kurina-Sanz M (2018) Ecofriendly chemical diversification of *Eupatorium buniifolium* essential oil by endophytic fungi. Process Biochem 64:93–102

Fernández T, Cerdá Zolezzi P, Risco E, Martino V, López P, Clavín M, Hnatyszyn O, Cañigueral S, Hajos S, Ferraro G, Álvarez E (2002) Immunomodulating properties of argentine plants with ethnomedicinal use. Phytomedicine 9(6):546–552

Frohne D, Jensen U (1979) Systematik des Pflanzenreichs. Ed. Gustav Fischer, Stuttgart, Germany, pp 218–255

García GH, Campos R, Torres RA, Broussalis A, Ferraro G, Martino V, Coussio J (1990) Antiherpetic activity of some argentine medicinal plants. Fitoterapia 41(6):542–546

Gorzalczany S, Muschietti L, Repetto M, Kartin L, Ferraro G, Martino V, Acevedo C (2008) Choleretic and antioxidant effect of *Eupatorium buniifolium*. Pharmacologyonline 1:139–147

Grossi MA, Gutiérrez DG, Berrueta PC, Martínez JJ (2011) *Acanthostyles* (Asteraceae, Eupatorieae): a revision with a multivariate analysis. Austral Syst Bot 24:87–103

Guerreiro AC, Cecati FM, Ardanáz CE, Donadel OJ, Tonn CE, Sosa ME (2018) Assessment of the insecticidal potential of the *Eupatorium buniifolium* essential oil against *Triatoma infestans* (Hemiptera: *Reduviidae*). A chiral recognition approach. Neotrop Entomol 47(3):418–428

Hooker JD, Jackson BD (1960) Index Kewensis, vol 1. Clarendon Press, Oxford, pp 915–921

King RM, Robinson HE (1971) Studies in the Eupatorieae (Asteraceae). LII. A new genus *Acanthostyles*. Phytologia 22(2):111

Lancelle HG, Giordano OS, Sosa ME, Tonn CE (2009) Chemical composition of four essential oils from *Eupatorium* spp. biological activities toward *Tribolium castaneum* (Coleoptera: Tenebrionidae). Rev Soc Entomol Argent 68(3–4):329–338

Liu PY, Liu D, Li WH, Zhao T, Sauriol F, Gu YC, Shi QW, Zhang ML (2015) Chemical constituents of plants from the genus *Eupatorium* (1904-2014). Chem Biodivers 12:1481–1515

Lombardo A (1959) Contribución al mejor conocimiento de las plantas indígenas. Almanaque del Banco de Seguros del Estado. Montevideo, Uruguay, pp 124–136

Lorenzo D, Paz D, Davies P, Villamil J, Vila R, Cañigueral S, Dellacassa E (2005) Application of multidimensional gas chromatography to the enantioselective characterization of the essential oil of *Eupatorium buniifolium* Hooker et Arnott. Phytochem Analysis 16(1):39–44

Malme GOA (1931) Die Compositen der zweiten Regnellschen Reise. I. Rio Grande do Sul Ark Bot 24(6):1–89

Marchesi E, Davies P (2004) Especies vegetales autóctonas e introducidas presentes en el territorio de Uruguay consideradas como aromáticas y/o medicinales en relación con los datos de uso popular. In: P Davies (Ed.): Estudios en domesticación y cultivo de especies medicinales y aromáticas nativas. INIA Serie FPTA, Montevideo, Uruguay, pp 181–194

Marchesi E (2005) Flora y vegetación del Uruguay. Project Orion. Environmental Impact Assessment. Cap. 5: Características del ambiente receptor. International Finance Corporation, Montevideo, Uruguay

Marchesi E, Alonso E, Brussa C, Delfino L, García M, Haretche F (2013) Plantas vasculares. In: Soutullo A, Clavijo C, Martínez-Lanfranco JA (eds.): Especies prioritarias para la conservación en Uruguay. Vertebrados, moluscos continentales y plantas vasculares. Ed. SNAP/DINAMA/MVOTMA Y DICYT/MEC, Montevideo, Uruguay, 222 pp

Miño J, Muschietti L, Ferraro G, Martino V, Acevedo C (2005) Antinociceptive activity of *Eupatorium buniifolium* aqueous extract. Fitoterapia 76(1):100–103

Miño J, Muschietti L, Ferraro G, Martino V, Acevedo C (2007) Neuropharmacological activity of *Eupatorium buniifolium* aqueous extract in mice. Phytother Res 21(4):362–365

Muschietti L, Martino V, Ferraro G, Coussio J (1994) 5,7,5′-trihydroxy-3,6,2′,4′-tetramethoxyflavone from *Eupatorium buniifolium*. Phytochemistry 36(4):1085–1086

Muschietti L, Gorzalczany S, Ferraro G, Acevedo C, Martino V (2001) Phenolic compounds with anti-inflammatory activity from *Eupatorium buniifolium*. Planta Med 67:743–744

Muschietti L, Derita M, Sülsen V, Muñoz JDD, Ferraro G, Zacchino S, Martino V (2005) *In vitro* antifungal assay of traditional argentine medicinal plants. J Ethnopharmacol 102(2):233–238

Muschietti L, Martino V, Ferraro G, Coussio J (1993) 2'-oxygenated flavonoids from *Eupatorium buniifolium*. Planta Med 59(7 SUPPL):A606

Muschietti L, Martino V, Ferraro G, Coussio J (1990) Polyphenolic compounds isolated from *Eupatorium buniifolium*. An Asoc Quím Argent 78:329–331

Nogueira Sobrinho AC, de Morais SM, Bezerra de Souza E, Oliveira dos Santos Fontenelle R (2017) The genus *Eupatorium* L. (Asteraceae): a review of their antimicrobial activity. J Med Plants Res 11(3):43–57

Payá M, Silla M, Vayá E, Alcaraz MJ, Coussio J, Ferraro G, Martino V, Hnatyszyn O, Debenedetti S, Broussalis A, Muschietti L (1996) Inhibitory effects of various extracts of argentine plant species on free-radical-mediated reactions and human neutrophil functions. Phytother Res 10(3):228–232

Ríos J, Condori L, Muñoz S, Loayza I, Soriano J (1993) Actes du colloque de Chicoutimi 111:22–25

Robinson H, King RM (1985) Comments on the generic concepts of the Eupatorieae. Taxon 34:11–16

Rojas Acosta N (1905) Plantas Medicinales de Corrientes. Ed. Imprenta y Enc. de P. Gadola. Buenos Aires, Argentina, pp 28

Saggese D (1959) Yerbas Medicinales Argentinas, 10th edn. Rosario. Rosario, Argentina, p 111

Sayagués L, Graf E, Delfino L (2000) Análisis de la información publicada sobre composición florística de montes naturales del Uruguay. Agrociencia 4:96–110

Sharma OP, Dawra RK (1994) Toxicity of some problematic weeds in the hills. National Seminar on Weed Management in hills, Himachal Pradesh Agricultural University, Palampur, India, 8 pp

Sharma OP, Dawra RK, Kurade NP, Sharma PD (1998) A review of the toxicosis and biological properties of the genus *Eupatorium*. Nat Toxins 6:1–14

Tasso de Souza TJ (2007) Determinaçao da composiçao química e avaliaçao preliminar das atividades antioxidantes e anticolinesterásica dos óleos voláteis de espécies de *Eupatorium* L. (Asteraceae). Thesis. Universidade Federal do Rio Grande do Sul, Porto Alegre, Brazil, 232 pp

Umpiérrez ML, Lagreca ME, Cabrera R, Grille G, Rossini C (2012) Essential oils from Asteraceae as potential biocontrol tools for tomato pests and diseases. Phytochem Rev 11(4):339–350

Umpiérrez ML, Santos E, Mendoza Y, Altesor P, Rossini C (2013) Essential oil from *Eupatorium buniifolium* leaves as potential varroacide. Parasitol Res 112(10):3389–3400

Umpiérrez ML, Paullier J, Porrini M, Garrido M, Santos E, Rossini C (2017) Potential botanical pesticides from Asteraceae essential oils for tomato production: activity against whiteflies, plants and bees. Ind Crop Prod 109:686–692

Visintini Jaime MF, Campos RH, Martino VS, Cavallaro LV, Muschietti LV (2013) Antipoliovirus activity of the organic extract of *Eupatorium buniifolium*: isolation of euparin as an active compound. Evid-Based Compl Alt Article ID 402364

Woerdenbag HJ (1992) *Eupatorium* species. In: De Smet PAGM, Keller K, Hansel R and Chandler RF (eds): Adverse effects of herbal drugs, vol 2. Springer, Berlin, Germany, pp 171–194

Zanon SM, Ceriatti FS, Rovera M, Sabini LJ, Ramos BA (1999) Search for antiviral activity of certain medicinal plants from Cordoba, Argentina. Rev Latinoam Microbiol 41(2):59–62

Zardini E (1984) Etnobotánica de Compuestas argentinas con especial referencia a su uso farmacológico. Acta Farm Bonaer 3:77–95

Zhanga ML, Wu M, Zhang JJ, Irwin D, Gu YC, Shi QW (2008) Chemical constituents of plants from the genus *Eupatorium*. Chem Biodivers 5(1):40–55

Zuloaga FO, Morrone O, Belgrano MJ, Marticorena C, Marchesi E (2008) Catálogo de las plantas vasculares del Cono Sur (Argentina, Sur de Brasil, Chile, Paraguay y Uruguay). Monographs in Systematic Botany. 3 vol. American Society of Plant Taxonomists. University of Wyoming. Laramie. WY, USA, 3486 pp

Fabiana imbricata Ruiz & Pav.

Silvia Beatriz González

Fabiana imbricata Ruiz & Pav. flowers. (Photo by the author)

Abstract *Fabiana imbricata* Ruiz & Pav. (Solanaceae), is popularly known as "palo piche", "palo pichi", "romero-pichi", etc. The infusion and decoction obtained from the bark and stems of this shrub has gastroprotective, liver stimulant, antiseptic and diuretic effects. This infusion is also used to treat *Fasciola hepatica* infections in goats and sheep, in Chilean folk medicine. The aerial parts of *Fabiana imbricata* have yielded a great variety of non-polar and polar compounds, as n-alkanes, mono, sesqui and triterpenes, alkaloids, anthraquinones, flavonoids and sugars. The plant displays several biological activities including diuretic effect, inhibition of the enzyme β-glucuronidase, antifeedant activity and gastroprotective effect in animal models. It is recommended to domesticate this species by elaborating appropriate

S. B. González (✉)
LIPAM (Laboratorio de Investigación de Plantas Aromáticas y Medicinales), Facultad de Ciencias Naturales y Ciencias de la Salud, Universidad Nacional de la Patagonia San Juan Bosco sede Esquel, Esquel (9200), Chubut, Argentina
e-mail: quim-esq@unpata.edu.ar; silviaesquel@speedy.com.ar

Á. Máthé, A. Bandoni (eds.), *Medicinal and Aromatic Plants of South America Vol. 2*, Medicinal and Aromatic Plants of the World 7,
https://doi.org/10.1007/978-3-030-62818-5_16

propagation, cultivation and agronomic practices that would ultimately ensure the good pharmaceutical quality of this valuable medicinal plant.

Keywords *Fabiana imbricata* · "Palo piche" · Diuretic · Gastroprotective

1 Introduction

Fabiana imbricata has a long history as medicinal plant in the South American traditional medicine and presently, it is still widely used. Hipólito Ruiz and José Pavón (1798) were the first botanists who identified and described it in the XVIII century. The common name "pichi", given by local people, refers to its urination increasing effect after consumption.

2 Taxonomic Characteristics

The Solanaceae Family comprises about 96 genera and 3000 species. They show a cosmopolitan distribution with the main center of taxonomic diversity and endemism in South America. *Fabiana* is a genus of flowering plants native to dry slopes in western South America, it comprises 15 species usually growing in South American Andes Mountains. *Fabiana imbricata* Ruiz & Pav. (Solanaceae), is popularly known as "palo piche", "palo pichi", "romero-pichi and "tola" in the North of Chile (de Mösbach 1992; Hoffmann et al. 1992; Barboza et al. 2009; Guerra et al. 2012).

Synonyms
Fabiana biflora J. Rémy, *Fabiana imbricata* Ruiz & Pav. var. *biflora* (J. Rémy) Reiche, *Fabiana araucana* Phil., *Fabiana lutescens* Phil. (The Plant List 2019).

3 Crude Drug Used

Fabiana imbricata is a species widely known both in the Argentinean Patagonia and throughout Chile. Several chroniclers from the colonial era cited it. Murillo (1889) appointed it for effective treatment of urinary problems and lithiasis.

The bark and stems of this shrub (Photo 1) are used as infusion, decoction or added to the "mate", a traditional drink of the region mainly composed by *Ilex paraguariensis* (Casamiquela et al. 2002). The fruits, in a dehydrated form, are used in desserts as sweeter (Cordero et al. 2017). In Chile, wild plants are harvested and sold to flower shops. In other countries, this shrub is used as an ornamental plant in gardens or as a pot flower (Fischer et al. 2011). Zin (1930) described that the plants

Photo 1 Stems with flowers, twigs, and bark of *F. imbricata*. (Photos by the author)

from the South of Chile have more resin than those from the North, so they are more active.

F. imbricata was an official drug in the Farmacopea Argentina, until the 3th edition (Bandoni 2001) and in the Farmacopea Chilena II ed. (Imbessi 1964). Presently, it is official in the Homeopathic Pharmacopeia from India (Tiwari et al. 2013) and German (German Homeopathic Pharmacopoeia, 2000/2005) and it is registered as skin conditioner and tonic, in cosmetic products in Europe, with the INCI: *Fabiana imbricata* extract. In a French patent it was registered as skin smoothing agent, for irritation and inflammations (Pegeon and Pelletier 2010).

4 Major Chemical Constituents and Bioactive Compounds

In the family Solanaceae some alkaloids are frequently unique to one or only a few species: this is the case also with: fabianine (volatile tetrahydroquinoline alkaloid) in *F. imbricata* (Barboza et al. 2016). Fabianine is - almost certainly - synthesized in nature from two isoprene units, linked head to tail, and a C_4N unit, probably derived from aceto-acetate (Edwars and Elmore 1962).

The aqueous extract of *F. imbricata* tops has yielded D-manno-heptulose, the closely related perseitol (D-glycero-D-galactoheptitol), and D-glycero-D-manno-octulose; D-arabinitol, D-mannitol, galactitol, myo-inositol, D-xylose, D-galactose, and primeverose (6-*O*-β-D-xylopyranosyl-D-glucose) were also obtained (Richtmyer 1970).

Knapp et al. (1972) found in the ethanolic extracts of aerial parts the following compounds: n-alkanes, fatty acids, 6-methoxyanthraquinones: erythroglaucin and physcion; and acetovanillone.

Flavonoids as quercetin, kaempferol and quercetin-3-*O*-rhamnoglucosid (rutin) were reported by Hörhammer et al. (1973). Silva et al. (1962) isolated from the leaves and stalks phytosterol.

Fourteen sesquiterpenes with muurolane and amorphane skeletons were isolated from the petroleum ether extract of the aerial parts of *F. imbricata* by Brown (1994a; Brown and Shill 1994). They seem to fit into two biogenetic classes: a 7-oxygenated muurolane and an 11-oxygenated amorphane. The norcadinane, α-muurolene and γ-amorphene derivatives were identified in samples from Chile (Schmeda-Hirschmann and Papastergiou 1994).

A novel seco-amorphane sesquiterpene incorporating a fully saturated furo[2,3-d]-l,3-dioxole system was isolated from the aerial parts of *F. imbricata* and was named fabianane. Such functionality is unique within the sesquiterpene family (Brown 1994a, b).

F. imbricata yielded coumarin-derivatives, such as scopoletin and its 7-prenyl-derivative in its exudates, differing from the other studied Solanaceae (Wollenweber et al. 2005). Also, fabiatrin after identified as the *O*-β-D-glucoside of scopoletin and a probable saponin (Edwards and Rogerson, 1927; Chaudhury et al. 1947).

Chlorogenic acid is one of the major components of the polar extracts of *F. imbricata* (Quispe et al. 2012) and also has been demonstrated the presence of oleanolic acid and *p*-hydroxyacetophenone (Schmeda-Hirschman and Papastergiou 1994).

Some secondary metabolites of *F. imbricata* seem to show toxicity towards the greenbug *Rhopalosiphum pudi*: fabiaimbricatan-15-oic acid, oleanolic acid, rutin, scopoletin and *p*-hydroxyacetophenone may protect from aphid infestation by acting as toxicants or feeding deterrents depending on the doses (Schmeda-Hirschman et al. 1995).

The essential oils that were analyzed in a population from Argentina (Guerra et al. 2012) contained the following main compounds: tricyclene, α-pinene, camphene, *p*-cymene, limonene and terpinen-4-ol.

5 Morphological Description

F. imbricata plants are chamaephytes or microphyllous shrubs, of homoblastic or heteroblastic growth; stems and leaves with dense resiniferous indumentum (Photo 1). Leaves sessile, imbricate, fasciculate or rosulate. Flowers solitary, 5-merous, actinomorphic; calyx lobes shorter than tube; corolla whitish, rarely lilac or bluish, funnel-shaped or salverform, aestivation contorted-conduplicate (Alaria and Peralta 2013). *Fabiana* is the only genus where the stomata are placed on projections of the epidermis (Barboza et al. 2016). Leaves emitting when crushed an aromatic and resinous odor. *F. imbricata* is a non-resprouting long-lived woody shrub that reaches sexual maturity in approximately 6 years.

The flowering period extends from September to January. Seed production was estimated at 200,000 seeds per adult forming persistent banks. Very longlived individuals were found up to 140 years old (de Torrres Curth et al. 2012). The drug comprises leaves and parts of stems, with or without bark. Luján and Barboza (1999) published the macroscopic and microscopic characters for the identification of this plant.

6 Geographical Distribution

The species of the South American plant genus *Fabiana* grow along arid mountainous area between 16° S and 51° S latitude, between 1000 and 4900 m.a.s.l. There are 15 species, ten are present in Argentina, seven in Chile, four in Bolivia and one in Peru (Cuello et al. 2011). *Fabiana imbricata* occurs from Mendoza to Chubut Provinces in Argentina, including the Patagonia steppe where it forms shrublands. In Chile, it can be found from Atacama to the Región de los Ríos, covering a wide range of dry, Mediterranean landscapes reaching down to the rainier places in Southern Chile (de Torrres Curth et al. 2012). Also occurs in Brazil, Bolivia and Peru (Rätsch 2005).

7 Ecological Requirements

F. imbricata grows in soils with varying chemical and physical properties, such as sandy, loamy sand and loam soils. Most of natural populations of "palo pichi" are adapted to grow in the foothills or in the valleys. *F. imbricata* is a long-lived shrub distributed via seeds. Fire and wind, followed by post-fire high precipitation in the early spring are requirements for successful propagation of the species (Ghermandi et al. 2013). Germination rate of the very tiny seeds is very low, therefore, the effect of gibberellic acid soaking on the germination percentage should be studied (Fischer et al. 2011). The species is a seeder shrub that forms conspicuous monospecific shrublands providing a degree of landscape heterogeneity within a grassland matrix (Ghermandi et al. 2013).

8 Traditional Use (Part(s) Used) and Common Knowledge

It has been used to treat kidney and bladder infections, liver flukes of goats and sheep (Kunz Krauze 1899). Infusion of flowers is diuretic, and infusion of stems is used to treat kidney stones, cystitis, hepatic abscesses and bronchial infections. As plant infusion it is consumed for kidney and urinary duct problems. (San Martín 1983; Houghton and Manby 1985). It is recommended as a diuretic, digestive and to treat kidney complaints (Razmilic et al. 1994; Muñoz et al. 1991). *F. imbricata* is reported for renal and respiratory diseases in Mapuche rural and semi-rural populations in Argentina (Eyssartier et al. 2013). It is used as diuretic, blood depurative, for liver ailments and for hair washing (Schmeda-Hirschmann and Theodoluz 2019). The tips of the branches are dried and sometimes chopped into little pieces. This herbage is then burned as incense or thrown over burning charcoal. This causes the plant to give off resinous smoke that can easily be inhaled. It has a sweetish scent like that of pine (Rätsch 2005).

9 Modern Medicine Based on Its Traditional Medicine Uses

The plant displays several biological activities including diuretic effect, inhibition of the enzyme β-glucuronidase (IC_{50} 6.2–10μg/ml) and antifeedant activity (Schmeda-Hirschmann et al. 1992, 1993, 1994, 1995). The diuretic effect was assessed at 250 mg/kg in rats and resulted in a 47.8% increase in urine output, compared with untreated animals. Hydrochlorothiazide (25 mg/kg) was used as a reference compound (Schmeda-Hirschmann et al. 1994).

In a homoeopathic journal, Mehnert (1989) describes the drug as sovereign for the treatment of migraine and sciatica, as a muscle relaxant and pain reliever in cervical root syndrome, to resolve spasms and thus relieve pain after trauma and in neuropathies.

The main terpenes of the *F. imbricata* exudate were evaluated for gastroprotective effect in animal models reducing the gastric lesions (Reyes et al. 2005). According to the pharmacological evidence from the Chilean collections of the plant, *F. imbricata* is a safe crude drug in the amounts used in traditional medicine. (Schmeda-Hirschmann and Theodoluz 2019).

According to Gastaldi (2012) aerial parts of *F. imbricata* infusion has a high ARP (antiradical power) value and then it would be an important source of antioxidants.

10 Conclusions

Fabiana imbricata has a long history of use in folk medicine, it is abundant in the Argentinean Patagonia; most chemical and bioactivity studies were performed in Chile (Schmeda-Hirschmann and Theodoluz 2019). *F. imbricata* shows activity as a diuretic, while in animal models, its main constituents demonstrated gastroprotective effects. This species possesses a wide diversity of chemical constituents, ranging from non-polar hydrocarbons and terpenes to highly polar compounds such as phenolics and sugars. In view its possible domestication, it seems important to refer micropropagation experiments suggesting that it might be possible to obtain large number of clonal plants in a short time by this procedure (Razmilic et al. 1994; Schmeda-Hirschmann et al. 2004). Further information about the domestication, propagation, cultivation and agronomic practices are needed to ensure good pharmaceutical quality.

Acknowledgements Facultad de Ciencias Naturales y Ciencias de la Salud, Universidad Nacional de la Patagonia San Juan Bosco sede Esquel.

References

Alaria AS, Peralta IE (2013) Las especies de *Fabiana* Ruiz et Pav. (Solanaceae) que crecen en Chile. Chloris chilensis 16(1):1–24

Barboza GE, Cantero JJ, Núñez C, Pacciaroni A, Ariza Espinar L (2009) Medicinal plants: a general review and a phytochemical and ethnopharmacological screening of the native Argentine Flora. Kurtziana 34(1–2):7–365

Barboza GE, Hunziker AT, Bernardello G, Cocucci AA, Moscone AE, Carrizo García E, Fuentes V, Dillon MO, Bittrich V, Cosa MT, Subils R, Romanutti A, Arroyo S, Anton A (2016) Solanaceae. In: Kubitzki K (ed.): The family and genera of vascular plants, vol 14. Ed. Springer, Hamburg, Germany, 412 pp

Brown GD (1994a) Fabianane, an unusual secoamorphane from *Fabiana imbricata*. J Nat Prod 57(2):328–330

Brown GD (1994b) The sesquiterpenes of *Fabiana imbricata*. Phytochemistry 35(2):425–433

Brown GD, Shill J (1994) Isolation of 3,11-amorphadiene from *Fabiana imbricata*. Planta Med 60:495–496

Casamiquela RM, Beeskow AM, Gavirati M, Stanganelli M, Mavrek V (2002) Usos Tradicionales de las Plantas en la Meseta Patagónica CENPAT–CONICET–ICBG, 1–51

Chaudhury DN, Holland RA, Bertson A (1947) The syntheses of glycosides. Part XII. Fabiatrin. J Chem Soc (London):1671–1672

Cordero S, Abello L, Gálvez F (2017). Plantas silvestres comestibles y medicinales de Chile y otras partes del mundo. Guía de Campo Ed Corporación Chilena de la Madera, Concepción, Chile, 292 pp

Cuello S, Alberto MR, Zampini IC, Ordóñez RM, Isla MI (2011) Comparative study of antioxidant and anti-inflammatory activities and genotoxicity of alcoholic and aqueous extracts of four *Fabiana* species that grow in mountainous area of Argentina. J Ethnopharmacol 137(1):512–522

de Mösbach EW (1992). Botánica Indígena de Chile. Ed. Fundación Andes y Ed. A. Bello, Santiago, Chile, 140 pp

de Torres Curth MI, Ghermandi L, Biscayart C (2012) Are *Fabiana imbricata* shrublands advancing over northwestern Patagonian grasslands? A population dynamics study involving fire and precipitation. J Arid Environ 83:78–85
Edwards OE, Elmore NF (1962) Fabianine. Can J Chem 40(2):256–264
Edwards GR, Rogerson H (1927) CXXXIV. The constituents of *Fabiana imbricata*. Biochem J 21:1010–1011
Eyssartier C, Ladio AH, Lozada M (2013) Traditional horticultural and gathering practices in two semi-rural populations of Northwestern Patagonia. J Arid Environ 97:18–25
Fischer S, Berti M, Wilckens R, Baeza M, Pastene E, Inostroza L, Tramón C, González W (2011) Characterization and propagation of some medicinal plants in the central-south region of Chile. Ind Crop Prod 34(2):1313–1321
Gastaldi B (2012). Actividad antioxidante en extractos y aceites esenciales de plantas medicinales nativas de la Patagonia. Thesis. Universidad Nacional de la Patagonia San Juan Bosco, Chubut, Argentina, 55 pp
German Homeopathic Pharmacopoeia (2000/2015) Ed. Medpharm Sci Pub. Stuttgart, Germany, vol 2
Ghermandi L, Franzese J, González SL, de Torres Curth MI, Ruete A (2013) Disentangling *Fabiana imbricata* (Solanaceae) regeneration: the importance of disturbance and rainfall. J Arid Environ 97:9–13
Guerra PE, González SB, Kirner H, Retta D, Di Leo Lira P, Gómez MF (2012) Aspectos anatómicos del leño y composición de los aceites esenciales de especies arbustivo-leñosas del ecotono y la estepa del noroeste de la Provincia del Chubut. Dominguezia 28(1):13–44
Hoffman A, Farga C, Lastra J, Veghazi E (1992) Plantas Medicinales de uso común en Chile. Ed. Fundación Claudio Gay, Santiago de Chile, 274 pp
Hörhammer L, Wagner H, Wilkomirsky MT, Iyengar MA (1973) Flavonoide in einigen chilenischen heilpflanzen. Phytochemistry 12(8):2068–2069
Houghton PJ, Manby J (1985) Medicinal plants of the Mapuche. J Ethnopharmacol 13(1):89–103
Imbessi A (1964) Index Plantarum. Ed. from the author. Messina, Italy, 771 pp
Knapp JE, Farnsworth NR, Theiner M, Schiff PL (1972) Anthraquinones and other constituents of *Fabiana imbricata*. Phytochemistry 11:3091–3092
Kunz-Krause H (1899) Beiträge zur Kenntnis der *Fabiana imbricata* Ruiz und Pavon (Pichi-Pichi) und ihrer chemischen Bestandteile. Arch Pharm 237(1):1–35
Luján MC, Barboza GE (1999) Contribution to the study of some Argentinian medicinal plants and commercial quality control. Acta Hortic (503):141–154
Mehnert H (1989) Fabiana imbricata. Br Homeopath J 78:115–117
Muñoz O, Montes M, Wilkomirsky T (1991) Plantas medicinales de uso en Chile. Química y Farmacología. Ed. Universitaria, Santiago de Chile, 330 pp
Murillo A (1889) Plantes Médicinales du Chili. Ed Exposition Universelle de Paris, Section Chilienne. Paris, France, 234 pp
Pegeon A, Pelletier A (2010) Huile essentielle de *Fabiana imbricata* à titre d'agent apaisant. Patent FR 2.968.988 A1
Quispe C, Viveros-Valdez E, Schmeda-Hirschmann G (2012) Phenolic constituents of the Chilean herbal tea *Fabiana imbricata* R. et P. Plant Foods Hum Nutr 67(3):242–246
Rätsch C (2005) The encyclopedia of psychoactive plants: Ethnopharmacology and its applications. Edition in English. Ed. Park Street Press, Vermont, USA, 2251 pp
Razmilic I, Schmeda-Hirschmann G, Dutra-Behrens M, Reyes S, López I, Theoduloz C (1994) Rutin and scopoletin content and micropropagation of *Fabiana imbricata*. Planta Med 60(02):140–142
Reyes M, Schmeda-Hirschmann G, Razmilic I, Theoduloz C, Yáñez T, Rodríguez JA (2005) Gastroprotective activity of sesquiterpene derivatives from *Fabiana imbricata*. Phytother Res 19(12):1038–1042

Richtmyer NK (1970) The isolation of D-manno-heptulose, perseitol, D-glycero-D-manno-octulose, and other compounds from pichi tops (*Fabiana imbricata* Ruiz and Pav.). Carbohydr Res 12(2):233–239

San Martín JA (1983) Medicinal plants in Central Chile. Econ Bot 37(2):216–227

Schmeda-Hirschmann G, Papastergiou F (1994) Sesquiterpenes from *Fabiana imbricata*. Phytochemistry 36(6):1439–1442

Schmeda-Hirschmann G, Theoduloz C (2019) *Fabiana imbricata* Ruiz et Pav. (Solanaceae), a review of an important Patagonian medicinal plant. J Ethnopharmacol 228:26–39

Schmeda-Hirschmann G, Loyola JI, Sierra J, Retamal R, Rodriguez J (1992) Hypotensive effect and enzyme inhibition activity of mapuche medicinal plant extracts. Phytother Res 6(4):184–188

Schmeda-Hirschmann G, Loyola JL, Razmilic I, Reyes S, Rodríguez J, Pacheco P, Teoduloz C (1993) La Farmacopea Mapuche, una fuente de productos biológicamente activos. Revista Universum de la Universidad de Talca, Chile 8(1):153–179

Schmeda-Hirschmann G, Loyola JI, Reyes S, Hubert E, Rodriguez M, Rodriguez J, Dutra-Behrens M (1994) β-Glucuronidase inhibition and diuretic activity of *Fabiana imbricata* R. & P. (Solanaceae). Phytother Res 8(8):485–487

Schmeda-Hirschmann G, Román P, Theoduloz C, Donoso BC, Corcuera LJ (1995) Effect of *Fabiana imbricata* constituents on *Rhopalosiphum padi* and *Heliothis zea*. Phytother Res 9(3):219–221

Schmeda-Hirschmann G, Jordan M, Gerth A, Hormazabal E, Tapia AA, Wilken D (2004) Secondary metabolite content in *Fabiana imbricata* plants and *in vitro* cultures. Zeitschrift für Naturforschung C 59(1–2):48–54

Silva M, Stuec R, Mancinell P (1962) Chemical study of *Fabiana imbricata*. Boletín de la Sociedad Chilena de Química 12(1):29–30

The Plant List (2019) Published on the Internet: http://www.theplantlist.org/tpl1.1/record/kew-2806430. (Last access, July 2019)

Tiwari L, Rai N, Sharma NK (2013) Regulatory standards on homoeopathic drugs: Indian perspective. Int J Adv Pharm Sci Technol 1:1):1–1)20

Wollenweber E, Dörsam M, Dörr M, Roitman JN, Valant-Vetschera KM (2005) Chemodiversity of surface flavonoids in Solanaceae. Zeitschrift für Naturforschung C 60(9–10):661–670

Zin J (1930) La Salud por medio de las plantas medicinales. Ed. Salesiana, Santiago de Chile, 702 pp

Fabiana punensis S.C. Arroyo, *F. bryoides* Phil., *F. densa* Remy, *F. patagonica* Speg.

María Inés Isla, Mariana Leal, Daniela González, Eugenia Orqueda, Iris Catiana Zampini, and Ana Soledad Cuello

(**a**) *F. punensis* (**b**) *F. bryoides*, (**c**) *F. densa* and (**d**) *F. patagonica,* from Argentine Puna. Antofagasta de la Sierra, Catamarca, Argentina. (Photos: S. Cuello and C. Zampini)

M. I. Isla (✉) · M. Leal · D. González · E. Orqueda · I. C. Zampini · A. S. Cuello
Instituto de Bioprospección y Fisiología Vegetal (INBIOFIV, CONICET- Facultad de Ciencias Naturales e Instituto Miguel Lillo-Universidad Nacional de Tucumán), San Miguel de Tucumán, Argentina
e-mail: misla@csnat.unt.edu.ar; zampini@csnat.unt.edu.ar

Á. Máthé, A. Bandoni (eds.), *Medicinal and Aromatic Plants of South America Vol. 2*, Medicinal and Aromatic Plants of the World 7,
https://doi.org/10.1007/978-3-030-62818-5_17

Abstract *Fabiana* Ruiz & Pav. (Solanaceae) is a South American genus, growing especially in arid mountainous areas. In Argentina, 10 species are present in the High Andean, Pre-Puna, Puna, and Monte biogeographical regions, in addition to Patagonia. In this chapter reference is made to four *Fabiana* species, *F. punensis, F. patagonica, F. bryoides* and *F. densa.* They are described as medicinal plants under the common names of "tola" or "tolilla". Progress is reported regarding the scientific validation of medicinal properties, as well as the study of the chemical composition of these species. The potential as antioxidants, anti-inflammatory, and antibiotics agents from the hydroalcoholic or aqueous extracts from the aerial parts was described. Antioxidant and diuretic properties of infusions and essential oils was demonstrated. The safety of these preparations has also been demonstrated, although further *in vivo* studies would be necessary to fully demonstrate the safety of the extracts Presently, prolonged traditional use by local communities is still recognized to support their safe use. Studies are needed about the agronomic conditions for its cultivation so that these species could be used as raw material in phytotherapic and phytocosmetic preparations.

Keywords "Tola" · Argentine medicinal plant · Anti-inflammatory · Antibiotic · Antioxidant · Diuretic

1 Introduction

Fabiana is a native medicinal plant genus with several underutilized species, in Argentina. Four *Fabiana* species, *F. punensis, F. densa, F. bryoides,* and *F. punensis,* that grown in arid regions, have pharmacological properties and as such can be regarded as possible candidates for domestication. Also, as a raw material source for phytotherapic and phytocosmetic preparations they could be used to improve the quality of life.

2 Taxonomic Characteristics

Fabiana R. et P. is a South American genus of the Family Solanaceae that comprises 15 species, some present in Peru, Chile and Bolivia. In Argentina, inhabits the High Andean, Pre-Puna, Puna and Monte biogeographical regions, including in Patagonia (Cabrera 1957, 1968, 1983; Barboza and Hunziker 1993, 1998; Barboza and Romanutti 1999; Cuello 2006; Carilla et al. 2018). The largest number of species, are distributed in Argentina, 10 of the 15, with five endemic species (*F. friesii, F. nana, F. peckii, F. foliosa* and *F. punensis*), followed by Chile with two

(*F. squamata* and *F. viscosa*) and Bolivia with one (*F. fiebrigii*). Argentina shares the species *F. patagonica, F. densa* with Bolivia and *F. bryoides, F. imbricata* and *F. denudata* with Chile.

This genus was described by Hipólito Ruiz López and José Antonio Pavón and the name *Fabiana* is a tribute to Archbishop of Valencia, Spain, Francisco Fabián y Fuero (1719–1801), patron of botany. In this chapter reference will be made to four *Fabiana* species, *F. punensis, F. patagonica, F. bryoides* and *F. densa.*

While the genus *Fabiana* was described at the end of the 18th century, the species *F. punensis* S.C. Arroyo was known in Arroyo 1976, by Silvia Arroyo, from specimens collected in the department of Cochinoca, in the Puna from Jujuy. It is known under the common name of "tolilla" (Barboza and Hunziker 1993, 1998; Barboza and Romanutti 1999). *F. patagonica* Speg. popularly known as "tolilla", "checal" or "tola". *F. bryoides* Phil. is called "pata de perdíz" or "pata de loro" by the Puna people. And *F. densa* Rémy (The Plant List 2019), is known as "tolilla" or "checal" (Bolivia) (Barboza and Hunziker 1993, 1998) just like *F. patagonica*, The common name "tola o tolilla" is widely used by the inhabitants from Puna region to name the bushes. The word "tola" belongs to the Aymara language, which means strong, resistant, "who does not know diseases" and "tolilla" is the diminutive of the word "tola".

3 Crude Drug Used

In Argentina, traditional medicine uses the leaves, branches and flowers of the four *Fabiana* species described in this chapter. Generally, the plant material is dried and pulverized prior to use. In Fig. 1, the picture of aerial parts of each *Fabiana* species are illustrated. In the case of *F. densa*, the resinous exudate from the leaves, is used.

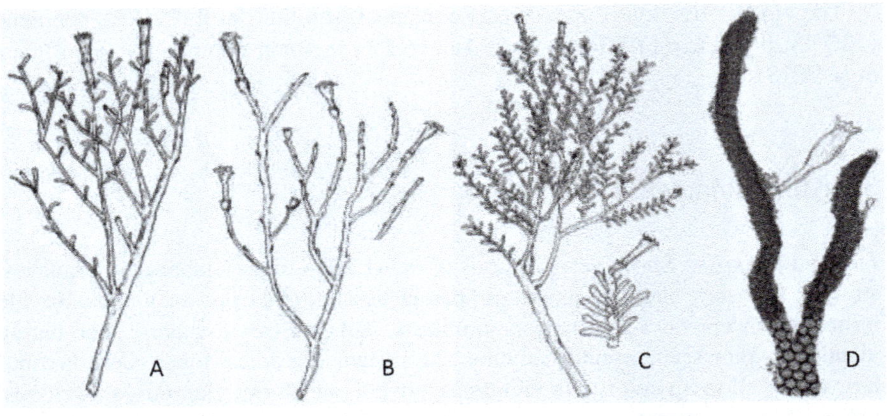

Fig. 1 Used parts for medicines of (**a**) *F. patagonica;* (**b**) *F. punensis;* (**c**) *F. densa*; (**d**) *F. bryoides*. (Pictures: M Leal)

4 Major Chemical Constituents and Bioactive Compounds

Fatty acids composition of seeds from *F. densa* and *F. punensis* was determined. Linoleic acid was dominant with 61.3 and 59.6% of the total fatty acids content to *F. densa* and *F. punensis*, respectively. Palmitic, and oleic acids constituting 13–16%. Other fatty acids (myristic, palmitoleic, linolenic, arachidonic, behenic and lignoceric) were also detected. Stigmasterol, sitosterol and Z~S-avenasterol were identified. Sitosterol was dominant in both species (around 50%). Cholesterol would be present in comparatively large quantities (0.4–14.4%) (Maestri and Guzmán 1995).

From the aerial parts of *F. densa* the following compounds have been isolated: two diterpenes, *ent*-beyer-15-en-18-*O*-succinate and *ent*-beyer-15-en-18-*O*-oxalate (Erazo et al. 2002) with antimicrobial activity. Recently the isolation and structural elucidation of other tetracyclic *ent*-beyerene diterpenes derivatives was reported. In addition to the succinoyl and the oxaloyl esters, the alcohol and the malonoyl ester were also isolated. Both diterpenes demonstrate a selective activity against Gram (+) bacterial strains with negligible cytotoxicity towards human keratinocytes (Quaglio et al. 2020). Triterpenes as ursolic and oleanolic acids were identified in *F. densa* (Quaglio et al. 2020).

Oleanolic acid was also isolated and identified in a *F. patagonica* infusion and could be a responsible factor of the diuretic activity demonstrated for this plant (Álvarez et al. 2002).

The ethanolic extracts of *F. punensis* and *F. patagonica* contain phenolic components, principally rutin (Cuello et al. 2011; Cuello 2012; Orqueda 2014). In aqueous extracts of *F. punensis* and *F. densa* were identified phenolic components. Chlorogenic acid and rutin were identified in *F. bryoides* and quercetin in *F. punensis* (Cuello et al. 2011; Cuello 2012). Chlorogenic acid is important member of hydroxycinnamic acid with natural antioxidant, cardioprotective, anti-Alzheimer, antidiabetic, anti-inflammatory properties and inhibits DNA damage (Agunloye et al. 2019).

The volatile fraction of *F. punensis* contains sesquithuriferol (29.4%), α-pinene (12.7%), β-pinene (11.5%) terpinen-4-ol (6.2%) as main compounds (Rodríguez et al. 2018).

5 Morphological Description

Fabiana punensis Shrub very resinous of 0.5–1.20 m high. Glabrescent branches, straight, with few alternate leaves as spatula, sessile, glabrous, deciduous. Sessile terminal flowers on very short braquiblasts. Tubular calyx, glabrous or barely glandulous-pubescent, cylindrical tube, short triangular apical lobes. Corolla tubular, white-yellowish color with violet base, hypocrateriform, glabrous, short lobes, patents, heterodynamic stamens, adhered to (3–) 4–6 (–7) mm from the base of the corolla, the 2 longer filaments of 6 to 8 mm (almost the same length as the corolla tube) and 3 more 4–6 mm short, 1–1.2 × 0.3–0.5 mm anthers; obovoid ovary, 1.5–2 × 0.8–1 mm, 7–9 mm style, truncated stigma in the shape of a saddle. Obovoid capsule, 5–6 (−7) × 2–3 mm; globose or ellipsoid seeds, brown (Barboza and Hunziker 1993, 1998).

Fabiana bryoides Low shrubs of ± 50 cm high. Macroblasts that are often born from knots very close to each other, coated completely by braquiblasts with squamiform and glabrescent leaves. Lonely lowers in each braquiblast; calyx, urceolate tube, segments triangular; infundibuliform corolla with the tube quite thinned in its lower half; very small anthers (0.5–0.7 mm), almost as long as wide, confluent teak; semilunar stigma, just low cut. Capsule with small seeds (Barboza and Hunziker 1993, 1998.).

Fabiana densa Shrubs up to 1 m high, characteristic for its gemmiferous roots, the density of its foliage and the glandular pubescence. Flowers on macroblasts or braquiblasts; pedicels up to 4.5 mm; calyx 3–5.7 mm, urceolate tube at its base, 2.2–3.7 (4) mm, 1–2 mm segments, narrow-triangular or triangular; infundibuliform corolla, 10–13 (13.5) mm, often with stretch marks reddish or violet on a background yellowish; androecium heterodynamic, 0.7–1 mm anthers; chair-shaped stigma of mount. 6–7 mm capsule. Seeds 1.4–1.5 × 0.5–0.6 × 0.3 mm (Barboza and Hunziker 1993, 1998)

Fabiana patagonica Shrubs of 2 m high. Macroblasts straight or flexible. Leaves solitary or, in the apical branches, fasciculate on brachiblasts. Flowers lonely on brachiblasts, rare terminals on macroblasts; cylindrical tube, hypocraterimorph, with the straight tube that it ends in a subapical widening; androecium homodynamic, ellipsoid anthers, 0.8–1.4 × 0.6–1.3 mm, confluent teak, capitized stigma, something depressed. 6–8 mm capsule. Seeds 1.3–1.8 × 0.6–0.9 × 0.5 mm (Barboza and Hunziker 1993, 1998).

6 Geographical Distribution

The South American genus *Fabiana* (Solanaceae) grows along arid mountainous area between 16° and 51° S, between 1200 and 4900 m.a.s.l. There are 15 species in the region, ten are present in Argentina, seven in Chile, four in Bolivia and one in Peru. There are five species in the Altoandina, Puna, Pre-Puna and Monte phytogeographic regions of the Argentinean Northwest (Cabrera 1957, 1968, 1983; Barboza and Hunziker 1993, 1998; Cuello 2006; Carilla et al. 2018). *F. punensis,* in Argentina, is located in the Provinces of Catamarca, Jujuy, La Rioja, Salta, San Juan and Tucumán, between 2200 and 3700 m.a.s.l. *F. bryoides* grows in Jujuy, Salta and Catamarca between 3200–4900 m.a.s.l.; *F. densa* in Jujuy, Salta, Catamarca and Tucumán between 2800–4300 m.a.s.l.; and *F. patagonica* in Jujuy, Salta, Tucumán, Catamarca, La Rioja, San Juan, Mendoza, San Luis, Río Negro, Neuquén, Chubut and Santa Cruz, between 1200–4000 m.a.s.l. At present, there is no major concern about the cultivation and domestication of species of this *Fabiana* species from Argentina (Barboza and Hunziker 1993, 1998).

7 Ecological Requirements

The native populations of these four *Fabiana* species grow on arid regions, poor quality soils, with limited water supply (Cuello et al. 2011; Carilla et al. 2018). The four *Fabiana* species discussed in this chapter grow in Argentina, Puna, Pre-Puna and Monte. regions characterized by their aridity. The *Fabiana* species are generally found in sandy, rocky soils, with very low fertility values, little organic matter and variable salt content. The flowering of these species occurs in summer and they bear fruit in summer-autumn. In very cold and high climates, flowering occurs early, from October to December in this Hemisphere, while in hot climates, flowering lasts until early Autumn (Alaria and Peralta 2013).

8 Traditional Use (Part(s) Used) and Common Knowledge

These species have been employed as medicines, building material, forage, fuel and elements in spiritual activities (Villagrán and Castro 2000; Villagrán et al. 2003; Rodríguez 2004, 2005; Pérez 2006; Barbarán 2008). Resinous exudate from leaves of *Fabiana* species is used in traditional medicine to immobilize fractured extremities (De Lucca and Zalles 1992; Erazo et al. 2002; Pérez 2006) while the infusion for cough and illness of the lungs (Munizaga 1988), as vulnerary and anti-inflammatory (Pérez 2006). There is archaeological evidence of the use of the woody stems of these species in incenses (Rodríguez 2013a, 2013b; García et al. 2018). Particularly its use as incense stick, where its bioactive compounds can be volatilized and efficiently incorporated into the body through smoke inhalation and cause effects on the health of living beings. In this sense, along with the boiled, crushed, roasted, baths or infusions, scrubs or plasters, are also the smokehouses (Aldunate et al. 1981; Fernández 1995; Villagrán and Castro 2004).

9 Modern Medicine Based on Its Traditional Medicine Uses

Several reports demonstrate the potential health benefits and the scientifically validated of popular medicinal uses of *Fabiana* species.

Antioxidant and Anti-inflammatory Activities Lipoxygenase plays a key role in the biosynthesis of leukotrienes, pro-inflammatory mediators. Hyaluronidase is determinant in chronic inflammations and allergic processes in humans. Thus, lipoxygenase and hyaluronidase inhibitors initially attracted attention as potential agents for the treatment of inflammatory and allergic diseases, but their therapeutic potential has been expanded to certain types of cancer and cardiovascular diseases. Most lipoxygenase and hyaluronidase inhibitors are also antioxidants or free radical scavengers (Torres-Carro et al. 2019). Cuello et al. (2011) have demonstrated inhibitory effects of ethanolic extracts of *F. punensis, F. patagonica, F. densa* and *F. bryoides* on lipoxygenase and hyaluronidase activities. The alcoholic extract of *F. punensis* was an active scavenger of DPPH and ABTS radicals as well as *F. densa, F. patagonica* and *F. bryoides* extracts (Cuello et al. 2011; Isla et al. 2018). *Fabiana patagonica* extracts exhibited the highest peroxyl radical scavenging activity compared with the other three taxa (Cuello et al. 2011). Antioxidant activity of *F. densa* tea and *F. punensis* essential oil was also demonstrated (Rojo et al. 2009; Rodríguez et al. 2018).

Antibacterial Activity The phenolic compounds enriched ethanolic extracts of *F. bryoides, F. densa, F. punensis* and *F. patagonica* inhibited the growth of one or more of the commercial antibiotic resistant human pathogenic strains: *Staphylococcus aureus, Enterococcus faecalis, Escherichia coli, Klebsiella pneumoniae, Proteus mirabilis, Enterobacter cloacae, Morganella morganii, Pseudomonas aeruginosa* (Zampini et al. 2009; Isla et al. 2018). The ethanolic extracts exhibited stronger activity and broader spectrum of action than aqueous extracts. The extracts were bactericidal in most cases and more active on Gram (+) than Gram (−) bacteria (Zampini et al. 2009). Antibacterial activity of two diterpenes isolated (the succinoyl and the oxaloyl esters of the *ent*-beyer-15-en-18-ol) from *F. densa* exudates against Gram (+) and Gram (−) bacteria were also demonstrated (Erazo et al. 2002). Recently, a key role played by the acidic group at C18 of the tetracyclic *ent*-beyerene for the antibacterial effects was described. It was also highlighted how the length and flexibility of the alkyl chain between the two carbonyl groups are crucial factors for the biological activity of the molecule (Quaglio et al. 2020). These finding support the usage of these species for the development of new anti-infective drugs.

Diuretic Activity A moderated diuretic activity of *F. patagonica* extract was reported by Álvarez et al. (2002) and it was suggested to be associated with the presence of oleanolic acid, which was isolated as the major metabolite in *F. patagonica* infusion.

According to above mentioned, these *Fabiana* species have a potential to serve as a raw material for medicinal products. Based in our review, *F. punensis*, *F. bryoides*, *F. densa* and *F. patagonica* are likely candidates as species of economic interest in Argentina (Cantero et al. 2019).

10 Conclusions

The *Fabiana* species herein reviewed (*F. bryoides, F. densa, F. punensis* and *F. patagonica*) that grow in arid regions of Argentina, are popularly used as medicinal plants. The effectiveness of these plants as inhibitors of inflammatory mediator liberation, or as anti-infective and diuretic, described in their traditional use, are supported by recent investigations. As medicinal plants, they could be used to promote the development of new chemical derivatives, as well as products. More chemical studies are needed to standardize their bioactive extracts and to elaborate good practices for their production. Typically, these should include various stages of production, from the harvesting in wild growing populations until the selection of germplasm for cultivation trials.

Acknowledgments The authors acknowledge the cooperation of the inhabitants of the studied areas and the financial support from the SCAIT-Universidad Nacional de Tucumán (PIUNT2018 G-637), the Agencia Nacional de Promoción Científica y Técnica (ANPCyT, PICT 2014, 3136 and PICT 2017 4436) and the Consejo Nacional de Investigaciones Científicas y Técnicas (CONICET, PUE2018 0011).

References

Agunloye O, Oboh G, Oluwaseun A, Ayokunle A, Ademosun O, Afolabi A, Ademola A, Oyagbemi A, Olutayo T, Temitayo O, Ajibade O, Adedapo A (2019) Cardio-protective and antioxidant properties of caffeic acid and chlorogenic acid: Mechanistic role of angiotensin converting enzyme, cholinesterase andarginase activities in cyclosporine induced hypertensive rats. Biomed Pharmacother 109:450–458

Alaria A, Peralta I (2013) The genus *Fabiana* (Solanaceae) in Chile. Chloris Chilensis 16(1):1–24

Aldunate C, Armesto J, Castro V, Villagrán C (1981) Estudio etnobotánico en una comunidad precordillerana de Antofagasta: Toconce. Bol Mus Nac Hist Nat 38:183–223

Álvarez ME, Maria AO, Saad JR (2002) Diuretic activity of *Fabiana patagonica* in rats. Phytother Res 16(1):71–73

Arroyo S (1976) Novedades en el género *Fabiana* (Solanaceae). Hickenia 1:49–54

Barbarán FR (2008) Medicinal plants of the Argentinean Puna: a common property resource and an opportunity for local people. In: Proceedings of the Twelfth Biennial Conference of the International Association for the Study of Commons. Governing Shared Resources: Connecting Local Experience to Global Challenges. July 14–18, 2008. Cheltenham, England. Digital Library of the Commons. Indiana University, USA. http://dlc.dlib.indiana.edu/dlc/handle/10535/2227 (Last access, December 2019)

Barboza GE, Hunziker AT (1993) Estudios en Solanaceae XXXIV. Revisión taxonómica de *Fabiana*. Kurtziana 22:109–153

Barboza GE, Hunziker AT (1998) *Fabiana*. In: Hunziker AT (ed) Flora Fanerogamica Argentina 56, pp 3–10

Barboza GE, Romanutti A (1999) Solanaceae. In: Zuloaga FO, Morrone O (eds) Catálogo de las Plantas Vasculares de la República Argentina. II. Dicotyledoneae. Monogr. Syst. Bot. Missouri Bot. Gard. 74: St. Louis, USA. pp 1059–1114

Cabrera AL (1957) La vegetación de la Puna Argentina. Revista de Investigaciones Agrícolas 11:317–512

Cabrera AL (1968) Ecología vegetal de la Puna. In: Geo-ecology of the mountain regions of the tropical Americas. pp 91–116

Cabrera AL (1983) Solanaceae. In: Cabrera AL (ed) Flora de la Provincia de Jujuy. Colecc. Científica, Ed. INTA, Buenos Aires, Argentina 13(8):292–493

Cantero JC, Núñez CO, Bernardello G, Amuchastegui A, Mulko J, Brandolín P, Palchetti MV, Iparraguirre J, Virginil N, Ariza Espinar L (2019) Las plantas de importancia económica en Argentina. First Edition. Ed. UniRío Editora. Universidad Nacional de Río Cuarto, Córdoba, Argentina, 937 pp. https://botanicaargentina.org.ar/wp-content/uploads/2019/06/978-987-688-332-0-baja-resoluci%C3%B3n.pdf (Last access, December 2019)

Carilla J, Grau A, Cuello S (2018) Vegetación de la Puna argentina. Serie Conservación de la Naturaleza 24: La Puna argentina: naturaleza y cultura. 143–156

Cuello S (2006) Guía ilustrada de la Flora de Antofagasta de la Sierra, Catamarca. Facultad de Ciencias Naturales-Universidad Nacional de Tucumán, Tucumán, Argentina, 165 pp

Cuello S (2012) Potenciales usos biotecnológicos de metabolitos secundarios aislados de especies del género *Fabiana* resistentes a radiación ultravioleta. Thesis. Universidad Nacional de Tucumán, Tucumán, Argentina, 197 pp

Cuello S, Alberto MR, Zampini IC, Ordóñez RM, Isla MI (2011) Comparative study of antioxidant and anti-inflammatory activities and genotoxicity of alcoholic and aqueous extracts of four *Fabiana* species that grow in mountainous area of Argentina. J Ethnopharmacol 137(1):512–522

De Lucca M, Zalles J (1992) Flora medicinal Boliviana. Ed. Amigos del Libro, La Paz, Bolivia. p 498

Erazo S, Zaldívar M, Delporte C, Backhouse N, Tapia P, Belmonte E, Delle Monarche F, Negrete R (2002) Antibacterial diterpenoids from *Fabiana densa* var. *ramulosa*. Planta Medica 68:361–363

Fernández G (1995) Ofrenda, ritual y terapia: las mesas aymaras. Revista Española de Antropología Americana REAA 25:153–180

García G, Gili F, Echeverría J, Belmonte E, Figueroa V (2018) K'oa, andean entity of a plant and other bodies. An interpretative possibility for funerary offerings in the archaeology of Arica. Chungara Rev Antrop Chil 50(4):537–556

Isla M I, Zampini IC, Alberto MR, Cuello S (2018) Plantas de la Puna: fitoquímica y su uso en el cuidado de la salud. Serie Conservación de la Naturaleza 24: La Puna argentina: naturaleza y cultura. pp 383–384

Maestri D, Guzmán C (1995) A Comparative Study of Seed Lipid Components of Nicotianeae (Solanaceae). Biochem Syst Ecol 23(2):201–207

Munizaga C (1988) Algunos enfoques de la importancia de la práctica de la actual medicina popular aborigen en Chile. Atención primaria 3:33–38

Orqueda E (2014) Estudios de composición química y actividad antioxidante de extractos de partes aéreas de *Fabiana patagonica*. Thesis. Facultad de Ciencias Naturales e IML. Universidad Nacional de Tucumán, Tucumán, Argentina, 60 pp

Pérez E (2006) Las plantas utilizadas por la comunidad de Antofagasta de la Sierra, Puna catamarqueña, Argentina. Ed. Universidad Nacional de Tucumán, Facultad de Ciencias Naturales e Instituto Miguel Lillo. San Miguel de Tucumán. Tucumán, 90 pp

Quaglio D, Corradi S, Erazo S, Vergine V, Berardozzi S, Sciubba F, Cappiello F, Crestoni ME, Ascenzioni F, Imperi F, Delle Monache F, Mori M, Loffredo MR, Ghirga F, Casciaro B, Botta B, Mangoni ML (2020) Structural elucidation and antimicrobial characterization of novel diterpenoids from *Fabiana densa* var. *ramulosa*. ACS Med Chem Lett. https://doi.org/10.1021/acsmedchemlett.9b00605

Rodríguez MF (2004) Woody plant resources in the Southern Argentina Puna: Punta de la Peña 9 archaeological site. J Archaeol Sci 31:1361–1372

Rodríguez MF (2005) Human evidence from the mid-Holocene in the salty Argentine Puna: analysis of the archaeobotanical record. Quatern Int 132:15–22

Rodríguez MF (2013a) Acerca de la flora de Antofagasta de la Sierra, Catamarca, Argentina. Ambiente y paleoambiente. Darwiniana 1:295–323

Rodríguez MF (2013b) Los grupos humanos y las plantas en la Puna meridional argentina: arqueobotánica de Antofagasta de la Sierra. Intersecciones en Antropología 14:315–339

Rojo LE, Benites J, López J, Rojas M, Díaz P, Ordóñez J, Pastene E (2009) Antioxidant capacity and polyphenolic content of twelve traditionally used herbal medicinal infusions from the South American Andes. Bol Latinoamer Car Plant Med Aromát 8:498–508

Rodríguez F, Pérez J, Álvarez M, Zampini C, van Baren C, Bandoni A, Isla MI (2018) Caracterización química y funcional de aceites esenciales de *Fabiana punensis*, una planta medicinal de los Valles Calchaquíes. VI Jornadas Nacionales de Plantas Aromáticas Nativas y sus Aceites Esenciales. II Jornadas Nacionales de Plantas Medicinales Nativas. November 26–28, 2018. Buenos Aires. Argentina

The Plant List (2019) Version 1.1. Published on the Internet: www.theplantlist.org (Last access, December 2019)

Torres-Carro R, Yamile A, Rojas-Márquez J, Stempin C, Zampini IC, Isla MI, Alberto MR (2019) Argentinean Puna plants with *in vitro* antioxidant and anti-inflammatory activities as a potential nutraceutical. J Food Sci 84:3352–3363

Villagrán C, Castro V (2000) Ciencia indígena de los Andes de Chile. Ed. Universitaria, Santiago de Chile, Chile, 244 pp

Villagrán C, Romo M, Castro V (2003) Etnobotánica del sur de los andes de la primera región de Chile: un enlace entre las culturas altiplánicas y de las quebradas altas del Loa superior. Chungará Rev Antrop Chil 35:73–124

Villagrán C, Castro V (2004) Ciencia Indígena de los Andes del Norte de Chile. Ed. Universitaria, Santiago de Chile, Chile, 343 pp

Zampini IC, Cuello S, Alberto MR, Ordóñez RM, Almeida RD, Solórzano E, Isla MI (2009) Antimicrobial activity of selected plants from "the Argentine Puna" against sensitive and multiresistant bacteria. J Ethnopharmacol 124(3):499–505

Gentianella multicaulis (Gillies ex Griseb.) Fabris

Jelena L. Nadinic

Gentianella multicaulis. (Photos: Andrea Cocucci)

Abstract *Gentianella multicaulis*, known as “nencia” or “bitter grass”, is an Argentine native plant that grows in the Andean and pre-Andean zone of Bolivia, Chile and Argentina. The infusion prepared from the plant is used in Argentine folk medicine for stomach disorders, as an appetite stimulant, a digestive and a bitter tonic for liver problems; it is also used in bitter drinks as a substitute for the European gentian. Other endemic *Gentianella* species from Argentina, also known with the same vernacular name, are used in Argentine traditional medicine, as well. Phytochemical studies carried out, mainly in *Gentianella multicaulis*, revealed the presence of xanthones, secoiridoids, flavonoids and triterpenes, which account for its traditional use. Some biological activities have been demonstrated. The content of two bisxanthones: 3-*O*-demethylswertipunicoside and swertiabisxanthone-I, as well as mangiferin, bellidifoline, oleanolic acid and ursolic acid have also been established.

J. L. Nadinic (✉)
Facultad de Farmacia y Bioquímica, Universidad de Buenos Aires,
Ciudad Autónoma de Buenos Aires, Argentina
e-mail: jnadinic@gmail.com

Á. Máthé, A. Bandoni (eds.), *Medicinal and Aromatic Plants of South America Vol. 2*, Medicinal and Aromatic Plants of the World 7,
https://doi.org/10.1007/978-3-030-62818-5_18

Keywords "Nencia" · Xanthones · Secoiridoids · C-glucoflavones · Triterpenes

1 Taxonomic Characteristics

The genus *Gentianella* was first described by Moench in 1794 (Glenny 2004). Kusnezov was the first to recognize Moench's name *Gentianella* in 1896, but as a subgenus of *Gentiana*. The recognition of *Gentianella* as a genus by Fabris (1953, 1955) and Gillett (1957) was part of a trend to recognize at genus level what were originally described as sections of *Gentiana* (Gillett 1957; Glenny 2004). Many nomenclatural transferences were performed since then, and today *Gentianella* comprises approximately 200 species with global distribution, mainly in the Southern Hemisphere, South America and New Zealand (Ho and Liu 1993). There are 29 recognized Argentine *Gentianella* species, thirteen of which are endemic (Zuloaga and Morrone 1999; Filippa and Barboza 2006).

Gentianella multicaulis (Gillies ex Griseb.) Fabris (ex *Gentianella achalensis* (Gilg) T.N. Ho & S.W. Liu) is locally known as "nencia" or "pasto amargo" (bitter grass). Other synonyms are: *Gentiana achalensis* Hieron., *Gentiana achalensis* Hieron. ex Gilg, *Gentiana coerulescens* Gillies ex Griseb., *Gentiana coerulescens* Gillies, *Gentiana gilliesii* Gilg, *Gentiana mendocina* Gilg, *Gentiana multicaulis* Gillies ex Griseb., *Gentiana stuckertii* Briq., *Gentianella achalensis* (Gilg) T.N.Ho & S.W. Liu, *Gentianella gilliesii* Martic. & Arroyo, *Gentianella mendocina* (Gilg) J.S. Pringle (The Plant List 2019).

Many *Gentianella* species are also known as "pasto amargo" or "nencia", and some comparative studies were done with all the other species (Nadinic et al. 1996, 1997; Nadinic 2000; Rosella et al. 2007; Nadinic and Debenedetti 2008). For instance, *Gentianella riojae* (Gilg) Fabris ex J.S.Pringle (1986) is found as an endemic herb at 2000–3000 m above sea level in the province of La Rioja (Zuloaga and Morrone 1999). Its medicinal uses are the same as those of other bitter drugs (Palma 1973). *Gentianella florida* (Griseb.) Holub, locally known as "nencia", "pasto amargo" or "genciana", is a small herb distributed throughout northwest and central Argentina, growing at an altitude of 1500–2000 m in the pre-Andean region (Fabris 1953, 1983). The aerial parts of the plant are used as a stomachic, as an appetite stimulant or as a bitter tonic, and it is also used for its febrifugal properties (Saggese 1959; Toursarkissian 1980; Ratera and Ratera 1980; Nadinic et al. 1996, 1997). *Gentianella parviflora* (Griseb.) T.N. Ho is found as an endemic small herb in the province of Córdoba, growing at an altitude of 1000–2500 m. Locally known as "pasto blanco" and "genciana", it has tonic and corroborant properties (Toursarkissian 1980; Filippa et al. 2002; Chiapella and Demaio 2015). *Gentianella cosmantha* (Griseb.) P.S. Pringle is endemic in the biogeographic province named Las Yungas, in Northwest Argentina. There is no report on its ethnopharmacological use, although it is possibly used as a substitute for other medicinal *Gentianella* species (Filippa 2006).

2 Crude Drug Used

The literature describes several argentine native species of the genus *Gentianella*, traditionally known as "nencia" or "bitter grass", which are used indistinctly (Saggese 1959; Toursarkissian 1980; Ratera and Ratera 1980; Del Vitto et al. 1997; Nadinic 2000).

The whole plant, roots, aerial part or flowering tops, varying between authors, are normally dried. The preparation of aromatic wines, known as gentian wines, with the fresh or dry whole plants, has also been described.

It is an excellent digestive that eliminates flatulence; it also helps fighting liver diseases and draining bile. An infusion of 30 g in 1000 ml can be drunk, two cups a day after lunch or dinner (Burgstaller 1984). Also used as a remedy to tone the nerves and invigorate the stomach; to combat muscle weakness, chlorosis, rickets, diarrhea, loss of appetite, fevers, earthworms, dropsy, gout and arthritis (Saggese 1959).

G. multicaulis is a bitter and tonic plant containing a substance called gentiopicrin. Infusion of 10–20 g in 1000 ml of bitter grass stems is used as a stomachic, as a febrifuge, etc. Mammals ingesting this plant produce a bitter milk, considered to be very healthy for those having digestive disorders. Wine is elaborated by macerating 30 g of flowering tops in one litter generous wine for 10 days; then it is filtered, and it can be drunk in a small glass before and after each meal to stimulate appetite, facilitate digestion and, above all, in atonic dyspepsia (Saggese 1959; Ratera and Ratera 1980). Currently found in bitter aperitif drinks, along with other species, in the Argentine market.

3 Major Chemical Constituents and Bioactive Compounds

Species of the Gentianaceae family are well known for the presence of xanthones and secoiridoids (Hostettmann and Hostettmann 1989; Wolfender and Hostettmann 1993; Lacaille-Dubois et al. 1996). Phytochemical investigations on the plants of *Gentianella* spp. have resulted in the isolation of xanthones, *C*-glucoflavones, and terpenoids (iridoids, sesterterpenoids, and triterpenes). The occurrence of iridoids, xanthones, and *C*-glucoflavones is typical of the genus *Gentianella* (Jensen and Schripsema 2002; Janković et al. 2005; Liu et al. 2016), and the medicinal value of the genus is just due to presence of these compounds. Some plants of *Gentianella* spp. are very rich in these compounds and have been studied extensively (Li et al. 2010).

Although there are reports on the identification of some secoiridoids (sweroside, gentiopicroside, swertiamarine) from *Gentianella* species, there are still not enough results to attribute a chemotaxonomic value (Wolfender et al. 2015).

As one of the main constituents in plants belonging to Gentianaceae family, xanthones are widely distributed among plants from two genera: *Gentiana* and *Gentianella*. Comparing the structure diversity of the reported xanthones obtained from the above-mentioned genera, those reported in plants of *Gentianella* Moench were with 1,3,5,8-, 1,3,4,5,8- or 1,3,4,7,8-substitution patterns, the predominant oxidation pattern so far being mainly 1,3,5,8-, and, to a lesser extent, 1,3,7,8-substitution; meanwhile, those of *Gentiana* genus were usually with a 1,2,6,8-substitution pattern. Moreover, the xanthone dimmers were only found in the plants of *Gentianella* genus. Thus, the xanthone pattern should have chemotaxonomic value and facilitate the classification of Gentianaceae using phytochemical data (Li et al. 2010; Liu et al. 2016).

C-glucoflavones appear to have limited distribution in the Gentianaceae family in comparison with iridoids and other xanthones (Jensen and Schripsema 2002; Janković et al. 2005), but are commonly found in plants of *Gentianella* spp. (Li et al. 2010).

Pentacyclic triterpenes are widely distributed in plants, oleanane- and ursane-type triterpenoids are ubiquitously occurring in Gentianaceae family, reported in many species, namely *Swertia japonica, S. chirata, S. petiolata, S. mileensis, S. paniculata,* and many other species of *Swertia*, *Tripterospermum taiwanense, Gentianella acuta* (Liu 1995; Li-Juan and Min-Hui 2009), as well as many species of *Gentiana* (Pan et al. 2016).

From dichloromethane and methanol extracts of the whole plant of *Gentianella multicaulis*, the separation and identification of 18 known compounds was carried out: three secoiridoids (gentiopicroside, sweroside and amarogentin), three C-glucoflavones (isoorientin, swertisin and isovitexin), three xanthone glycosides (mangiferin, demethylbellidifolin 8-*O*-glucoside and bellidifolin 8-*O*-glucoside), five xanthone aglycones (demethylbellidifolin, bellidifolin, isobellidifolin, swerchirin and swertianine), two bisxanthones (3-*O*-demethylswertipunicoside and swertiabisxanthone-I) and two triterpenes (oleanolic acid and ursolic acid) (Fig. 1).

Oleanolic acid, ursolic acid, sweroside, mangiferin and bellidifolin are also described in *G. riojae*, *G. cosmantha*, *G. florida* and *G. parvifolia* (Nadinic and Debenedetti 2008). Demethylbellidifolin and its 8-*O*-glucoside were also found in the above-mentioned species, except for *G. riojae*. The other isolated compounds of *G. multicaulis* have been found in several of the other species, but amarogentin and the bisxanthones 3-*O*-demethylswertipunicoside and swertiabisxanthone-I have only been found in *G. multicaulis*, and they represent markers of the species (Nadinic 2000; Nadinic and Debenedetti 2008). Another secoiridoids, swertiamarine, was found in *G. florida* and *G. cosmantha*, but not in *G. multicaulis* (Rosella et al. 2007). Mangiferin content was determined in authentic and commercial samples of *G. multicaulis* by HPLC-UV, allowed to establish a concentration of about 1,0% of mangiferin in all the analyzed samples, providing a parameter for the normalization of the extracts. Mangiferin is also present in all other mentioned species (Nadinic and Debenedetti 2008).

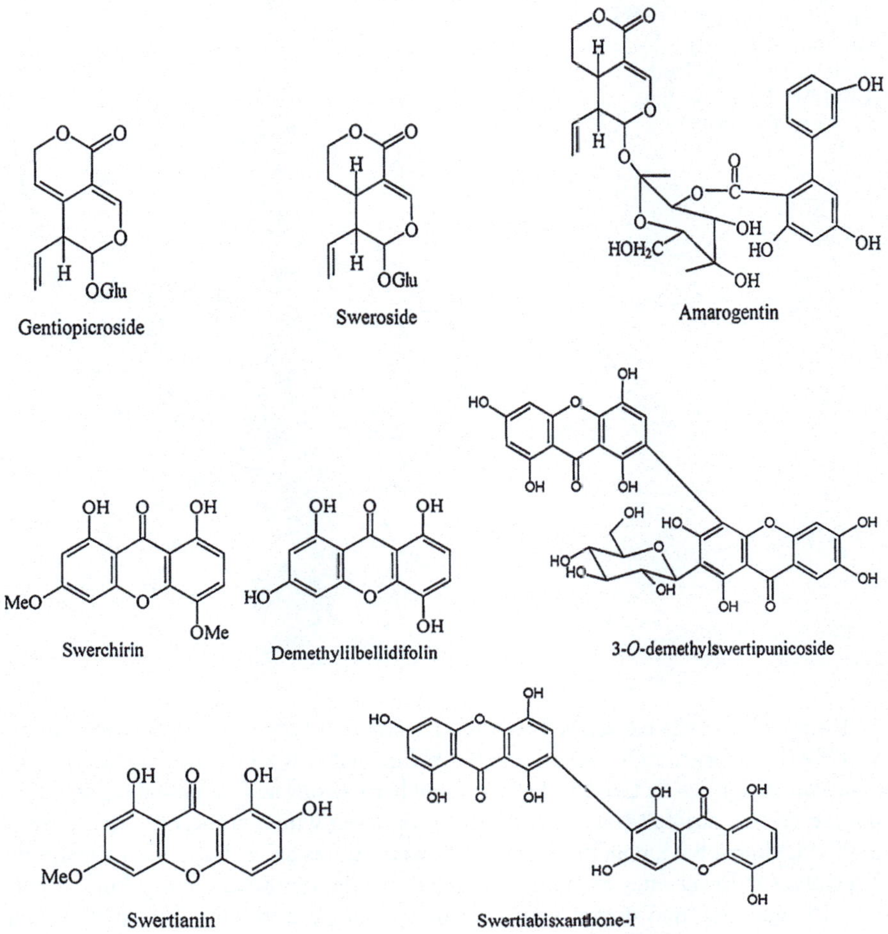

Fig. 1 Main compounds isolated from *G. multicaulis*

4 Morphological Description

The *Gentianella* genus consists of annual, biennial, and more seldom perennial plants differing from the species of genus *Gentiana* for having corolla throat or lobes which are usually ciliate, whereas appendages between corolla lobes are absent (Shiyan 2014).

G. multicaulis (Photo 1) is a very variable species, both in its vegetative and floral characters. The range of variation is observed in the number, thickness and height of the stems, in the presence and quantity of basal leaves, in the number of flowers per inflorescence and in the corolla color (Filippa 2006).

Photo 1 *Gentianella multicaulis* (formerly identified as *G. achalensis* (Gilg) T.N. Ho y S.W. Liu). (Photo: by the author)

Domínguez (1928) described *Gentiana achalensis* Hieron. or *Gentiana achalensis* Gilg (later *Gentianella achalensis* Ho & Liu, and now recognized as *Gentianella multicaulis* (Gillies ex Griseb.) Fabris) as an herbaceous and perennial plant, with a thick tortuous, striated-rough cylindrical root, from which numerous ascending or inclined, almost bare, pauciflorous (1–7 flowers) stems arise. The leaves are simple, opposite; the lower ones are long spatulated, whole, obtuse, attenuated in the petiole, 15 mm long and 4 mm wide; the upper ones are oblong, 15 mm long and 2–3 mm wide.

The flowers have an 8-mm-long campanulate calyx, split into 5 long triangular sharp or obtuse lobes. Corolla sub-rotate, without flange, 16 to 18 mm long, white in the upper part and somewhat violet in the margin, yellowish in lower part. The stamens are inserted into the corolla tube, the anthers are intrinsic. Ovary free, unilocular, crowned by a short style. Capsular fruit (Domínguez 1928). Recently Filippa and Barboza (2006) added some characteristics for its differentiation from other species of Argentine *Gentianella* genus:

Plant monoecious-monoclinous. Stems never scape, simple or branched. Leaves elliptic to linear-elliptic. Flowers terminally solitary, or pauciflorous or pluriflorous cymes; axillary, solitary, geminated or cymes, 2–4-flowers. Corolla tube shorter than the lobes. Stamens very exserted, extending beyond the corolla tube. Corolla white, yellowish white, bluish white, blue or lilacin, with elliptical, ovate or obovate lobes. Leaves of *G. parviflora* are hypostomatic, those of *G. multicaulis* are amphistomatic (Rosella et al. 2007). The presence of large amounts of mangiferin, bisxanthone 3-*O*-demethylswertipunicoside and seco-iridoid amarogentin in the methanolic extract of *G. multicaulis* are the main differences found among these species (Nadinic and Debenedetti 2008). Macro- and micrographic analyses provide a rapid method of identification, for the whole plant and for powdered drugs. These two methods, performed together, allow the recognition of both species.

Comparison with other Argentine *Gentianella* species of their phytochemical content showed that amarogentin and bisxanthones 3-*O*-demethylswertipunicoside and swertiabisxanthone-I were exclusively found in *G. multicaulis* (Nadinic 2000; Rosella et al. 2007; Nadinic and Debenedetti 2008).

5 Geographical Distribution

The Gentianaceae second largest genus is *Gentianella* Moench, with about 300 species, nearly worldwide in distribution, but absent from low-altitude tropical regions and all but northernmost Africa and all but eastern Australia, with its greatest diversity in the Andes of South America (Pringle 2014). *Gentianella multicaulis* inhabits the tall grasslands of Bolivia (Dept. Tarija), Chile (Regions III and VI), and, in Argentina, the provinces of San Juan, La Rioja, Catamarca, Córdoba, Mendoza and San Luis, at 1000–3700 m.a.s.l. (Filippa and Barboza 2006; SIB 2020).

6 Ecological Requirements

Gentianella multicaulis is considered an endemic plant that grows on wet or snowy soils, pastures, ridges, on washed or acid limestone soils, in mountain areas between 1500 and 4000 m. It appears as part of high coverage grasslands that are dominated by hemicryptophytes and chamaephytes. It grows in full light although it can resist shade and prefers cool temperatures. It is also found in low forests and in lands with no or almost no vegetation cover (SIB 2020).

Frequently found at the foot of the slopes along the margins of rivers, streams and springs, in the wide ravines, it grows between microscopic and clasts of various sizes, in unstable and not swamps, and, in stable and swampy terrain in the Andes of Mendoza, with micro-environments characterized by the humidity and high humidity, respectively, wind and solar radiation. It is described in plant communities of *Poaceae*, *Nastanthus* spp., *Carex* spp., *Oxychloe mendocina* and *Eleocharis* spp., adapted to the general environment (Wingenroth 1990).

It blooms in the meridional summer (between January and March), when the plant is harvested. As it is wild collected, this species can be considered endangered, but there are no records of attempted domestication. In fact, there are currently only a few reports in literature regarding the cultivation of *Gentianella* species (Huo and Zheng 2002; Janković et al. 2011; Krstić-Milošević et al. 2015).

7 Traditional Use (Part(s) Used) and Common Knowledge

The infusion of the aerial parts of *G. multicaulis*, locally known as 'nencia', are used in Argentina for stomach disorders, as an appetite stimulant, digestive and bitter tonic for liver problems, and as a substitute for European gentian, in bitter drinks (Saggese 1959; Toursarkissian 1980; Ratera and Ratera 1980; Del Vitto et al. 1997; Nadinic 2000). Other species also described as "nencia" or "pasto amargo" are used for the same purposes, as shown above.

The bitterness of all *Gentianella* Moench is well known and mostly due to the presence of secoiridoids. The bitterness index was determined on 3 botanically identified samples and 6 commercial samples of *G. multicaulis* from different origins, together with a *Gentiana lutea* sample. Results indicated a high homogeneity among all samples of *G. multicaulis*, and its value was found to be 6–10 times higher than that of *G. lutea*. It can be established that the bitterness index of *Gentianella multicaulis* should be higher than 1800 under the conditions of the method proposed by the WHO (1980, 1992; Nadinic et al. 1999b; Nadinic 2000).

Acute toxicity of *G. multicaulis* was determined with the lyophilized infusion administered orally to mice (1.87 g/kg body weight of extract concentration, representing 13.40 g of dry plant per kg of body weight). Neither changes in animal behavior nor toxicity, deaths or macroscopic alterations in vital organs were found, 5 days following administration, which represents 100 times the traditionally usual dose (Gorzalczany et al. 1999).

8 Modern Medicine Based on Its Traditional Medicine Uses

The phytochemical groups found in *Gentianella* spp. have been reported with many pharmacological activities. They are best known for their bitter taste, which can be related to their content on secoiridoids, such as gentiopicroside and amarogentin, the most known and the bitterest compound, respectively. Bitters have been traditionally used as remedies for loss of appetite and fever, and their bitterness value is related to their activity.

Gentiopicroside, which is a ubiquitous secoiridoid constituent in plants of *Gentianella* spp., has shown hepatoprotective activities on two hepatic injury models: carbon tetrachloride-induced and lipopolysaccharide-induced hepatitis (Kondo et al. 1994), and is being used as an anti hepatitis drug. Gentiopicroside showed significant analgesic effects against persistent pain stimuli in mice (Chen et al. 2008). Apart from being important chemotaxonomic markers, xanthones also display interesting pharmacological properties (Hostettmann and Hostettmann 1989; Fotie and Bohle 2006; El-Seedi et al. 2010). Amarogentin, a characteristic bitter compound of *Gentiana lutea*, induces calcium influx in keratinocytes and promotes keratinocyte differentiation by inducing keratin 10, involucrin, and transglutaminase expression; it also activates bitter taste receptors TAS2R1 and TAS2R38 in human keratinocytes (Wölfle et al. 2015). *Gentiana lutea* extract in a cream formulation was found to modulate lipid synthesis *in vivo*. A clinical trial performed on the volar forearms of 33 volunteers having total skin surface lipids measured with a sebumeter showed a significantly increased lipid content as compared to placebo. The enhancement of lipid synthesis in human keratinocytes is essential for building an intact epidermal barrier, and it might be used to improve skin disorders, such as very dry skin and atopic eczema (Wölfle et al. 2017).

Mangiferin and sweroside, two common constituents in plants of *Gentianella* spp., were reported to show hepatoprotective activities (Zhou 1991; Dar et al. 2005; Das et al. 2012).

Mangiferin isolated from different sources (Imran et al. 2017) possesses several health-endorsing properties such as antioxidant, antimicrobial, antidiabetic, antiallergic, anticancer, hypocholesterolemic and immunomodulatory activity. It also has cytotoxic activity against different tumor cells (Pérez G 2001; Gold-Smith et al. 2016). Mangiferin has a neurocytoprotective role related, at least in part, to an antioxidant and anti-inflammatory mechanism, which could be explored for more effective therapies of schizophrenia and other neurodegenerative diseases (Rao et al. 2012). Mangiferin was found to induce apoptosis in the human acute myeloid leukemia cell line HL-60, and to enhance apoptotic effects of hesperidin in the human cervix adenocarcinoma cell line (HeLa) (Bartoszewski et al. 2014). An important pharmacological activity of mangiferin is its strong antioxidant properties (Ghosal and Rao 1996; Muruganandan et al. 2002). Dermal acute studies of mangiferin in rodents did not show toxic effects in mice and rats after exposure to 2 g/kg mangiferin doses (Prado et al. 2015). Many studies have been focused on the preparation and implementation of mangiferin formulation on the skin. Ochocka et al. (2017) analyzed the penetration of mangiferin into the human skin and through the skin and evaluated the influence of mangiferin on two extracellular matrix (ECM) enzymes: collagenase and elastase. Mangiferin could penetrate through the stratum corneum barrier and to the living skin layers, where collagenase and elastase occur. The *in vitro* analysis revealed that mangiferin is capable of dose-dependent inhibition of neutrophil elastase and collagenase activity. In view of the positive effect of mangiferin on the condition and protection of the skin, and its high concentration found in *G. multicaulis* extracts, it could be worth to subject this species to further specific studies.

Antimicrobial activity was reported for the hydro-alcoholic and methanol extracts and for the infusion of *G. achalensis* against *Bacillus subtilis*. The fractions with antibacterial activity showed the presence of sweroside, gentiopicroside and demethylbellidifolin as main constituents (Nadinic et al. 2002).

Gentianella achalensis extracts were tested in two models of inflammation: TPA (12–0-tetradecanoylphorbol-13-acetate)-induced ear edema in mice upon topical administration, and the carrageenan-induced paw edema in rats upon oral administration (Nadinic et al. 1999a). Dichloromethane extract produced a significant effect with 30% inhibition in the ear edema test, and bioguided fractionation showed that the activity was concentrated in a single fraction. Oleanolic and ursolic acids were isolated and identified as the major constituents, thus being the compounds responsible for the topical anti-inflammatory activity. No activity was observed when extracts were administered orally in carrageenan-induced paw edema. Both oleanolic acid (Ayeleso and Matumba 2017) and ursolic acid (Mlala et al. 2019) were described as having strong health benefits, including hepatoprotective, wound healing, antibacterial, anti-inflammatory, antitumoral, and antiviral properties, together with low toxicity and effectiveness in protecting against chemically induced liver injury in laboratory animals (Liu 1995; Safayhi and Sailer 1997; Chen et al. 2015).

Swertisin, a C-glycosyl flavonoid isolated from *Enicostemma hyssopifolium* and from *G. multicaulis*, stimulated insulin secretion from a rat insulinoma cell line and acted via other mechanisms by virtue of which the plant exhibited hypoglycemic activity (Patel and Mishra 2011). Hepatoprotective activity was also reported for swertisin, isovitexin and other *C*-glucosylflavones (Hoffmann-Bohm et al. 1992).

Patents containing flavone-6-C-glucose derivatives from *Swertia japonica* or *Swertia pseudochinensis* extracts as active ingredients were registered, and claim for the preparations a functional effect of effectively treating or preventing cognitive dysfunction disorders such as delirium, dementia or amnesia, stroke, palsy, attention disorders, anxiety disorders or sleep disorders. The pharmaceutical compositions contained the flavone-6-C-β-D-glucose derivatives, selected from a group consisting of swertisin, isovitexin, isoorientin, swertiajaponin, and mixtures (Ryu et al. 2016).

Xanthones are certainly the most interesting compounds biosynthesized by Gentianaceae both from a chemotaxonomic and a pharmacological viewpoint. Xanthone derivatives can inhibit acetylcholinesterase (AChE) and block the acetylcholinesterase-induced β-amyloid aggregation. These compounds are also known for their antiproliferative effect and their ability to inhibit several molecular targets in the tumor cells, including kinases, cyclooxygenases and DNA polymerases (Guedes et al. 2019). Demethylbellidifolin, bellidifolin, and their respective 8-*O*-β-glucopyranosides, all found in *G. multicaulis*, were found to have antiacetylcholinesterase (AChE) effects in this enzyme assay. Bellidifolin showed similar activity to galanthamine, a commercial AChE inhibitor used for Alzheimer's disease (Urbain et al. 2004, 2008). Strong hypoglycemic activity has been reported for bellidifolin and swerchirin (Basnet et al. 1994, 1995; Tian et al. 2010), both present in *Gentianella multicaulis*. Demethylbellidifolin inhibited proliferation and activation of hepatic stellate cells (Li et al. 2011). Some of the xanthones described, isolated from *Gentianella acuta*, were found to have a significant reducing effect on intestine contraction tension, which indicated they may have potential as leading compounds for diarrhea treatment in the development of new drugs for gastroenteritis (Liu et al. 2016).

Xanthones are reported to stimulate the central nervous system (Bhattacharya et al. 1972), and this effect can be explained by their MAO A and B inhibitory activity (Suzuki et al. 1981; Nadinic et al. 1998a, b; Gnerre et al. 2001; Urbain et al. 2008). Bellidifolin, isolated from *Gentianella lactea*, a South American species, was found to be a selective MAO A inhibitor (Suzuki et al. 1981; Schaufelberger and Hostettmann 1988; Urbain et al. 2008).

The antioxidant activity of *G. multicaulis* extracts was assayed in the DPPH radical and in the FRAP assays, and bioguided fractionation showed the activity was concentrated in the xanthone fraction, demethylbellidifolin showing the strongest activity, followed by bellidifolin, swerchirin and isobellidifolin. These xanthones exhibited antifungal activities against *Trichophyton rubrum, Trichophyton mentagrophytes* and *Microsporum gypseum* (Lima et al. 2012) suggesting their clinical use could be beneficial for chronic and recurrent dermatophyte infections in humans.

Dimeric bisxanthones are distinctive of *G. multicaulis* and not described in other Argentine *Gentianella* species, but present in *Swertia* species and New Zealand *Gentianella* species (Glenny 2004; Benn et al. 2009; Kumar and van Staden 2016). Some biological properties were demonstrated; however, further studies are required to establish their value in therapeutics.

9 Conclusions

Gentianella multicaulis is a medicinal plant used in Argentine folk medicine, mainly due to its digestive properties, in the treatment of stomach disorders, as an appetite stimulant, and as a bitter tonic for liver problems, strongly supported by the ethnomedicine and by its phytochemical constituents. Many of the isolated compounds described above were found also in other Gentianaceae species and in other families. Among numerous bioactive compounds, those of *G. multicaulis* emerge as very promising due to their expected antidiabetic, hepatoprotective, antimicrobial, anti-inflammatory, antidepressive and dermatological effects.

Thus, further research is justified, aimed at discovering new approaches of drugs used in phytotherapy. Six different extracts of *Gentianella* species have their INCI names, and cosmetic industry has already taken advantage of them in preparations for skin conditioning (European Commission 2019). Isolated compounds or extracts rich in xanthones, triterpenes, glucosylflavones and secoiridoids showed beneficial bioactivity used in dermatology and phytocosmetics (Lim et al. 2007; Nadinic 2012; Aye et al. 2020). According to its chemical composition, *Gentianella multicaulis* could be attractive as a source of medicinal raw material for extraction of mangiferin and other bioactive compounds. However, since its population is small, efforts should be focused on finding alternative ways for biomass production.

Using only the aerial parts of *Gentianella multicaulis*, instead of the whole plant, in therapy, could have a protective effect on this species endangered by potential overexploitation.

References

Aye A, Song Y-J, Jeon Y-D, Jin J-S (2020) Xanthone suppresses allergic contact dermatitis *in vitro* and *in vivo*. Int Immunopharmacol 78:106061

Ayeleso T, Matumba M (2017) Oleanolic acid and its derivatives: biological activities and therapeutic potential in chronic diseases. Molecules 22(11):1915
Bartoszewski R, Hering A, Marszaøø M, Stefanowicz-Hajduk J, Bartoszewska S, Kapoor N, Ochocka R (2014) Mangiferin has an additive effect on the apoptotic properties of hesperidin in *Cyclopia* sp. tea extracts. PLoS One 9(3):e92128. https://doi.org/10.1371/journal.pone.0092128. PMID: 24633329
Basnet P, Kadota S, Shimizu M, Namba T (1994) Bellidifolin -a potent hypoglycemic agent in streptozotocin (stz)-induced diabetic rats from *Swertia japonica*. Planta Med 60:507–511
Basnet P, Kadota S, Shimizu M, Takata Y, Kobayashi M, Namba T (1995) Bellidifolin stimulates glucose-uptake in rat-1 fibroblasts and ameliorates hyperglycemia in streptozotocin (stz)-induced diabetic rats. Planta Med 61:402–405
Benn M, Joyce NI, Lorimer SD, Perry NB, van Klink JW, Wu Q (2009) Xanthones and bisxanthones in five New Zealand and subantarctic *Gentianella* species. Biochem Syst Ecol 37:531–534
Bhattacharya SK, Sanyal AK, Ghosal S (1972) Monoamine oxidase-inhibiting activity of mangiferin isolated from *Canscora decussata*. Naturwissenschaften 59:651
Burgstaller JA (1984) 700 Hierbas Medicinales, Ed. Color-EFE, Buenos Aires, Argentina, p 105
Chen L, Liu J-C, Zhang X-N, Guo Y-Y, Xu Z-H, Cao W, Sun X-L, Sun W-L, Zhao M-G (2008) Down-regulation of NR2B receptors partially contributes to analgesic effects of gentiopicroside in persistent inflammatory pain. Neuropharmacology 54:1175–1181
Chen P, Li J, Fan X, Zeng H, Deng R, Li D, Huang M, Bi H (2015) Oleanolic acid attenuates obstructive cholestasis in bile duct-ligated mice, possibly via activation of NRF2-MRPs and FXR antagonism. Eur J Pharmacol 765(70177):131–139
Chiapella JO, Demaio PH (2015) Plant endemism in the sierras of Córdoba and San Luis (Argentina): understanding links between phylogeny and regional biogeographical patterns. PhytoKeys 47:59–96. https://doi.org/10.3897/phytokeys.47.8347
Dar A, Faizi S, Naqvi S, Roome T, Zikr-ur-Rehman S, Ali M, Firdous S, Moin ST (2005) Analgesic and antioxidant activity of mangiferin and its derivatives: the structure activity relationship. Biol Pharm Bull 28:596–600
Das J, Ghosh J, Roy A, Sil PC (2012) Mangiferin exerts hepatoprotective activity against D-galactosamine induced acute toxicity and oxidative/nitrosative stress via Nrf2-NFκB pathways. Toxicol Appl Pharmacol 260(1):35–47
Del Vitto LA, Petenatti EM, Petenatti ME (1997) Recursos herbolarios de San Luis (República Argentina); primera parte: plantas nativas. Multequina 6:49–66
Domínguez JA (1928) Contribuciones a la Materia Médica Argentina (Primera Contribución) Trabajos del Instituto de Botánica y Farmacología. Facultad de Ciencias Médicas de Buenos Aires. Buenos Aires, Argentina. pp 44,114
Du X-G, Wang W, Zhang S-P, Pu X-P, Zhang Q-Y, Ye M, Zhao YY, Wang B-R, Khan I-A, Guo D-A (2010) Neuroprotective Xanthone glycosides from *Swertia punicea*. J Nat Prod 73(8):1422–1426
El-Seedi HR, El-Barbary MA, El-Ghorab DMH, Bohlin L, Borg-Karlson A-K, Goransson U, Verpoorte R (2010) Recent insights into the biosynthesis and biological activities of natural xanthones. Curr Med Chem 17:854–901
European Commission (2019) Cosing. https://ec.europa.eu/growth/tools-databases/cosing/. (Last access, January 2020)
Fabris HA (1953) Sinopsis preliminar de las Gencianáceas argentinas. B Soc Argent Bot 4:233–259
Fabris HA (1955) Nuevas especies de *Gentianella* del Perú. B Soc Argent Bot 6(1):45–50
Fabris HA (1983) Gentianaceae. In: Cabrera AL: Flora de la Provincia de Jujuy 13(8). Colección Científica del INTA. Buenos Aires, Argentina, pp 55–83
Feng C-Y, Wu Q, Yin D-D, Li B, Li S-S, Tang Z-Q, Xu Y-J, Wang L-S (2018) Determination of xanthones and flavonoids of methanol extracts obtained from different parts of the plants of three *Gentianaceae* species. J Pharm Biomed Anal 161:455–463

Filippa E (2006) Gentianaceae. In: Barboza GE, Cantero JJ, Nuñez C, Ariza Espinar L (eds), Flora Medicinal de la Provincia de Córdoba (Argentina). Ed. Museo Botánico, Córdoba, Argentina, pp 767–768

Filippa EM, Barboza GE (2006) Novedades en Gentianaceae para América del Sur. SIDA 22(1):129

Filippa E, Rosella M, Rossi, Barboza G, Spegazzini E, López P, Nadinic J, Debenedetti S (2002) I Congreso latinoamericano de Fitoquímica and IV Reunión de la Sociedad Latinoamericana de Fitoquímica, Buenos Aires, Argentina, 2/IV

Fotie J, Bohle DS (2006) Pharmacological and biological activities of xanthones. Anti Infect Agents Med Chem 5:15–31

Ghosal S, Rao G (1996) A plausible chemical mechanism of the bioactivities of mangiferin. Indian J Chem 35B:561–566

Gillett JM (1957) A revision of the North American species of *Gentianella* Moench. Ann Mo Bot Gard 44:195–269

Glenny D (2004) A revision of the genus *Gentianella* in New Zealand. N Z J Bot 42(3):361–530

Gnerre C, Thull U, Gaillard P, Carrupt PA, Testa B, Fernandes E, Silva F, Pinto M, Pinto MMM, Wolfender JL, Hostettmann K, Cruciani G (2001) Natural and synthetic xanthones as monoamine oxidase inhibitors: biological assay and 3D-QSAR. Helv Chim Acta 84:552–570

Gold-Smith F, Fernández A, Bishop K (2016) Mangiferin and cancer. Nutrients 8(7):396

Gorzalczany S, Rojo A, Rondina R, Debenedetti S, Acevedo C (1999) Estudio de toxicidad aguda por vía oral de plantas medicinales argentinas. Acta Farm Bonaer 18(3):221–224

Guedes L, Reis PBPS, Machuqueiro M, Ressaissi A, Pacheco R, Serralheiro ML (2019) Bioactivities of *Centaurium erythraea* (Gentianaceae) decoctions: antioxidant activity, enzyme inhibition and docking studies. Molecules 24(20):3795

Ho TN, Liu SW (1993) New combinations, names and taxonomic notes on *Gentianella* (Gentianaceae) from South America and New Zealand. Bull Brit Mus Nat Hist, Bot 23(2):61–65

Hoffmann-Bohm K, Lotter H, Seligmann O, Wagner H (1992) Antihepatotoxic C-glycosylflavones from the leaves of *Allophyllus edulis* var. *edulis* and *gracilis*. Planta Med 58(06):544–548

Hostettmann K, Hostettmann M (1989) Xanthones. In: Harborne JB (ed) Methods in plant biochemistry. Plant polyphenolics, vol 1. Ed. Academic Press, London, UK, pp 493–508

Huo L, Zheng X (2002) Studies on tissue culture of *Gentianella albiflora*. Chin Pharm J 37:415–418

Imran M, Arshad MS, Butt MS, Kwon J-H, Arshad MU, Sultan MT (2017) Mangiferin: a natural miracle bioactive compound against lifestyle related disorders. Lipids Health Dis 16(1):84

Janković T, Krstić D, Aljančić I, Šavikin-Fodulović K, Menković N, Vajs V, Milosavljević S (2005) Xanthones and C-glucosides from the aerial parts of four species of *Gentianella* from Serbia and Montenegro. Biochem Syst Ecol 33(7):729–735

Janković T, Vinterhalte B, Krstić-Milošević D, Nikolić R, Vinterhalter D, Milosavljević S (2011) Xanthone compounds in shoot cultures of *Gentianella bulgarica*. Acta Physiol Plant 33(4):1515–1520

Jensen SR, Schripsema J (2002) Chemotaxonomy and pharmacology of Gentianaceae. In: Struwe L and Albert VA (ed): Gentianaceae: Systematics and natural history, Ed. Cambridge University Press, Albert Cambridge, UK, pp 573–631

Kondo Y, Takano F, Hojo H (1994) Suppression of chemically and immunologically induced hepatic injuries by gentiopicroside in mice. Planta Med 60:414–416

Krstić-Milošević D, Vinterhalter B, Janković T, Vinterhalter D (2015) Biotechnology and Phytochemistry of *Gentianella* species from the Central Regions of the Balkan Peninsula. In: Rybczyński J, Davey M, Mikuła A (eds). The Gentianaceae – Volume 2: Biotechnology and Applications. Ed. Springer, Berlin, Germany, pp 319–347

Kumar V, van Staden J (2016) A review of *Swertia chirayita* (Gentianaceae) as a traditional medicinal plant. Front Pharmacol, Jan 12 6(308). https://doi.org/10.3389/fphar.2015.00308. 26793105; PMCID: PMC4709473

Lacaille-Dubois MA, Galle K, Wagner H (1996) Secoiridoids and Xanthones from *Gentianella nitida*. Planta Med 62:365–367

Li MH, Li L, Yang YM, Zhang N, Song XL, Xiao PG (2010) Genus *Gentianella* Moench: a phytochemical and Ethnopharmacological review. Chin Herbal Med 2(4):262–271
Li JF, Lu GF, Zou YY (2011) Demethylbellidifolin inhibits proliferation and activation of hepatic stellate cells. J Investig Surg 24:171–177
Li-Juan L, Min-Hui L (2009) Terpenoids, flavonoids and xanthones from *Gentianella acuta* (Gentianaceae). Biochem Syst Ecol 37:497–500
Lim SW, Hong SP, Jeong SW, Kim B, Bak H, Ryoo HC, Lee SH, Ahn SK (2007) Simultaneous effect of ursolic acid and oleanolic acid on epidermal permeability barrier function and epidermal keratinocyte differentiation via peroxisome proliferator-activated receptor. J Dermatol 34(9):625–634
Lima B, Sánchez M, Luna L, Agüero MB, Zacchino S, Filippa E, Palermo JA, Tapia AE, Feresin GE (2012) Antimicrobial and antioxidant activities of *Gentianella multicaulis* collected on the Andean slopes of San Juan Province, Argentina. Z Naturforsch 67c:29–38
Lin C-M, Chen C-S, Chen C-T, Liang Y-C, Lin J-K (2002) Molecular modeling of flavonoids that inhibits xanthine oxidase. Biochem Biophys Res Commun 294:167–172
Liu J (1995) Pharmacology of oleanolic acid and ursolic acid. J Ethnopharmacol 49(2):57–68
Liu Y, Ni Y, Ruan J, Qu L, Yu H, Haiyan Y, Lifeng H, Zhang Y, Wang T (2016) Bioactive gentixanthone and gentichromone from the whole plants of *Gentianella acuta* (Michx.) Hulten. Fitoterapia 113:164–169
Mikuła A, Tomiczak K, Domżalska L, Rybczyński JJ (2015) Cryopreservation of Gentianaceae: trends and applications. In: Rybczyński J, Davey MR, Mikuła A (eds) The Gentianaceae – Volume 2: Biotechnology and applications. Ed. Springer, Berlin, Germany, 267–286
Mirzaee F, Hosseini A, Jouybari HB, Davoodi A, Azadbakht M (2017) Medicinal, biological and phytochemical properties of *Gentiana* species. J Tradit Complement Med 7(4):400–408
Mlala S, Oyedeji AO, Gondwe M, Oyedeji OO (2019) Ursolic acid and its derivatives as bioactive agents. Molecules 24:2751
Muruganandan S, Gupta S, Kataria M, Lal J, Gupta PK (2002) Mangiferin protects the streptozotocin-induced oxidative damage to cardiac and renal tissues in rats. Toxicology 176(3):165–173
Nadinic EL (2000) Pharmacognostic Study of an Argentine medicinal plant known as "Nencia": *Gentianella achalensis* (Gilg) Ho & Liu. Thesis. Facultad de Farmacia y Bioquímica. Universidad de Buenos Aires, Buenos Aires, Argentina, 228 pp
Nadinic J (2012) Los fitoingredientes en el cuidado de la piel. Revista de Fitoterapia 12(2):159–170
Nadinic J, Debenedetti SL (2008) Bioactive compounds from Argentine *Gentianella* species in South American Medicinal Plants as a Potential Source of Bioactive compounds. In: Martino VS, Muschietti LV (eds) South American medicinal plants as a potential source of bioactive compounds. Ed. Research Signpost, Kerala, India,115–131
Nadinic E, Mongelli E, Debenedetti S, Sánchez P, Ciccia G, Giulietti A, Coussio J (1994a) Bioactive molecules from *Gentianella achalensis*. Abstract at Annual Meeting Phytochemical Society of North America, Mexico, p 54
Nadinic E, Wilson E, Debenedetti S, Coussio J (1994b) Abstract at XV Congreso Panamericano de Farmacia y Bioquímica, Buenos Aires, Argentina. November 1994
Nadinic E, Saavedra C, Rojo A, Bandoni A, Debenedetti S (1996) Revalorización de Nuestra Flora Autóctona. *Gentianella florida*: Su potencial uso en la formulación de Medicamentos Herbarios Hepatoprotectores y Bebidas Amargas. Premio Félix de Azara (1995–1996) de la Facultad de Farmacia y Bioqumica, Universidad de Buenos Aires, Buenos Aires, Argentina, 65 pp
Nadinic E, Saavedra C, Di Leo Lira P, Bandoni A, Debenedetti S (1997) Tetraoxygenated xanthones from *Gentianella florida*. Int J Pharm 35:379–381
Nadinic E, Saavedra C, López P, Rubio M, Debenedetti S (1998a) Xanthones with IMAO activity from *Gentianella achalensis*. Abstract at 46th Annual Congress of the Society for Medicinal Plant Research, Vienna, Austria, pp 95
Nadinic E, Saavedra C, van Baren C, Debenedetti S (1998b) Abstract at XII Congreso Nacional de Recursos Naturales Aromáticos y Medicinales, Córdoba, Argentina

Nadinic E, Gorzalczany S, Rojo A, van Baren C, Debenedetti S, Acevedo C (1999a) Topical anti-inflammatory activity of *Gentianella achalensis*. Fitoterapia 70:166–171

Nadinic EL, López PG, Saavedra C, Spegazzini E, Castro MT, Nájera MT, Bandoni AL, Debenedetti SL (1999b) Analytical assays of argentine herbal drugs known as 'Nencia'. Acta Hortic 503:155–161

Nadinic EL, Penna C, Saavedra C, Coussio J, Gutkind G, Debenedetti S (2002) Aislamiento de los compuestos con actividad antimicrobiana de extractos de "nencia": *Gentianella achalensis* (Gilg) Ho & Liu (Gentianaceae). Lat Am J Pharm 21(2):123

Ochocka R, Hering A, Stefanowicz HJ, Cal K, Barańska H (2017) The effect of mangiferin on skin: penetration, permeation and inhibition of ECM enzymes. PLoS One 12(7):1–16

Palma H (1973) Estudio Antropológico de la Medicina Popular de la Puna Argentina. Ed. Cabargón. Buenos Aires, Argentina. p 100

Pan Y, Zhao Y-L, Zhang J, Li W-Y, Wang Y-Z (2016) Phytochemistry and pharmacological activities of the genus *Gentiana* (Gentianaceae). Chem Biodivers 13:107–150

Patel MB, Mishra SH (2011) Hypoglycemic activity of C-glycosyl flavonoid from *Enicostemma hyssopifolium*. Pharm Biol 49(4):383–391

Peres V, Nagem TJ, de Oliveira FF (2000) Tetraoxygenated naturally occurring xanthones. Phytochemistry 55(7):683–710

Pérez G RM (2001) Anti-inflammatory activity of compounds isolated from plants. Sci World J 1:713–784

Prado Y, Merino N, Acosta J, Herrera JA, Luque Y, Hernandez I, Prado E, Garrido G, Delgado R, Rodeiro I (2015) Acute and 28-day subchronic toxicity studies of mangiferin, a glucosyl xanthone isolated from *Mangifera indica* L. stem bark. J Pharm Pharmacogn Res 3(1):13–23

Pringle JS (1986) A new species and taxonomic notes on *Gentianella* (Gentianaceae) in South America. SIDA 11(4):357–369

Pringle JS (2014) Morphological characteristics of the family Gentianaceae. In: Rybczyński J, Davey M, Mikuła A (eds) The Gentianaceae -Volume I – characterization and ecology. Ed. Springer, Berlin, Germany, pp 1–12

Rao VS, Carvalho AC, Trevisan MTS, de Souza AL, Trevisan MT, Andrade GM, Nobre Júnior HV, Moraes MO, Magalhães Iur HI, Morais TC, Santos FA (2012) Mangiferin ameliorates 6-hydroxydopamine-induced cytotoxicity and oxidative stress in ketamine model of schizophrenia. Pharmacol Rep 64:848

Ratera EL, Ratera MO (1980) Plantas de la Flora Argentina empleadas en Medicina Popular. Ed. Hemisferio Sur. Buenos Aires, Argentina, p 134

Rosella MA, Spegazzini ED, Debenedetti SL (2007) Micrographic and phytochemical parameters to identify two *Gentianella* species (Gentianaceae). Bol Latinoam Caribe Plant Med Aromát 6(6):384–385

Ryu JH, Cheong JH, Shin CY, Jang DS, Lee HE, Kim HJ, Ryu B, Jung IH, Jo YW (2016) Pharmaceutical composition for treating or preventing neuropsychiatric disease, containing flavone-6-C-glucose derivatives as active ingredients. Patents EP 3 034 082 A1; US 10,251,902 B2

Safayhi H, Sailer ER (1997) Anti-inflammatory actions of Pentacyclic triterpenes. Planta Med 63:487–493

Saggese D (1959) Yerbas Medicinales Argentinas. Ed. Antognazzi & Co., Rosario, Argentina, pp 71,97

Šavikin K, Aljančić IS, Vajs VE, Milosavljević SM, Jadranin M, Đorđević I, Menković, NR (2015) Bioactive secondary metabolites in several genera of Gentianaceae species from the central regions of the Balkan Peninsula. The Gentianaceae – Volume 2: Biotechnology and Applications, pp 319–347

Schaufelberger D, Hostettmann K (1988) Chemistry and pharmacology of *Gentiana lactea*. Planta Med 54:219–221

Shiyan NM (2014) A review of the taxonomy and distribution of the Gentianaceae in the Ukraine. In: Rybczyński J, Davey M, Mikuła A (eds) The Gentianaceae – Volume 1: characterization and Ecology. Ed. Springer, Berlin, Germany, pp 149–168

SIB (2020) (Sistema de Información de la Biodiversidad). https://sib.gob.ar/especies/gentianella-multicaulis?tab=habitat. (Last access, May 2020)

Suzuki O, Katsumata Y, Oya M, Chari VM, Vermes B, Wagner H, Hostettmann K (1981) Inhibition of type a and type B monoamine oxidases by naturally occurring xanthones. Planta Med 42:17–21

The Plant List (2019). http://www.theplantlist.org/tpl1.1/record/kew-2822186. (Last access, December 2019)

Tian LY, Bai X, Chen XH, Fang JB, Liu SH, Chen JC (2010) Anti-diabetic effect of methylswertianin and bellidifolin from *Swertia punicea* Hemsl. and its potential mechanism. Phytomedicine 17:533–539

Toursarkissian M (1980) Plantas Medicinales de la Argentina. Ed. Hemisferio Sur, Buenos Aires, Argentina. p 55

Urbain A, Marston A, Queiroz EF, Ndjoko K, Hostettmann K (2004) Xanthones from *Gentiana campestris* as new acetylcholinesterase inhibitors. Planta Med 70:1011–1014

Urbain A, Marston A, Grilo LS, Bravo J, Purev O, Purevsuren B, Batsuren D, Reist M, Carrupt PA, Hostettmann K (2008) Xanthones from *Gentianella amarella* ssp. *acuta* with acetylcholinesterase and monoamine oxidase inhibitory activities. J Nat Prod 71:895–897

WHO (1980) Determination of bitterness value In: Quality control methods for medicinal plants. WHO.PHARM/80.502:62–67

WHO (1992) Determination of bitterness value. In: Quality control methods for medicinal plants materials. WHO.PHARM/92.559:31–33

Wingenroth MC (1990) Historia de la vegetación y del clima en la Quebrada Benjamín Matienzo, Cordillera de Los Andes, Mendoza, Argentina. Thesis. Facultad de Ciencias Exactas y Naturales. Universidad de Buenos Aires. Argentina. http://digital.bl.fcen.uba.ar/Download/Tesis/Tesis_2340_Wingenroth.pdf. (Last access, May 2020)

Wolfender J, Hostettmann K (1993) Liquid-chromatographic UV detection and liquid-chromatographic thermospray mass-spectrometric analysis of *Chironia* (Gentianaceae) species- a rapid method for the screening of polyphenols in crude plant-extracts. J Chromatogr A 647:191–202

Wolfender JL, Urbain A, Hostettmann K (2015) Profiling, Isolation, Chemical Characterisation and Distribution of Gentianaceae Constituents. In: Rybczyński J, Davey M, Mikuła A (eds) The Gentianaceae – Volume 2: Biotechnology and applications. Ed. Springer, Berlin, Germany, pp 349–381

Wölfle U, Elsholz FA, Kersten A, Haarhaus B, Müller WE, Schempp CM (2015) Expression and functional activity of the bitter taste receptors TAS2R1 and TAS2R38 in human keratinocytes. Skin Pharmacol Physiol 28:137–146

Wölfle U, Haarhaus B, Seiwerth J, Cawelius A, Schwabe K, Quirin KW, Schempp CM (2017) The herbal bitter drug *Gentiana lutea* modulates lipid synthesis in human keratinocytes *in vitro* and *in vivo*. Int J Mol Sci 18(8):1814

Zhou J (1991) Bioactive glycosides from Chinese medicines. Mem Inst Oswaldo Cruz 86:231–234

Zhou H-M, Liu Y-L, Blaskó G, Cordell GA (1989) Swertiabisxanthone-I from *Swertia macrosperma*. Phytochemistry 28(12):3569–3571

Zuloaga F, Morrone O (1999) Catálogo de las Plantas Vasculares de la República Argentina II. Ed. Missouri Botanical Garden Press, St. Louis, p 749

Haplopappus baylahuen J. Rémy

Edgar Pastene-Navarrete

Haplopappus baylahuen. Chillán, Región de Ñuble, Chile. (Photos by the author (October, 2019)

Abstract *Haplopappus baylahuen* J. Rémy (Asteraceae) is a medicinal plant growing in Chile and Argentina. This is one of seventeen native *Haplopappus* spp. species commercialized in Chile under the name "bailahuén". In folk medicine, these plants are mainly used for the treatment of intestinal and liver diseases. Leaves, flowers and stems contain flavonoids, coumarins, phenolic acids and diterpenes. Although several chemical profiling studies have provided important background that allows deducing some biological effects of specific "bailahuén" species, it is clear that it is necessary to re-study some chemical-biological aspects using advanced bioanalytical techniques. Hitherto, only part of the isolated compounds explains the antioxidant, diuretic, anti-inflammatory and antimicrobial activities. Additionally, just a few *in vivo* studies have been carried out, exclusively in murine models. To date, no clinical studies have been conducted in Chile with any of the species representing this genus.

E. Pastene-Navarrete (✉)
Laboratorio de Síntesis y Biotransformación de Productos Naturales, Universidad del Bío-Bío, Chillán, Chile
e-mail: edgar.pastene@gmail.com

Á. Máthé, A. Bandoni (eds.), *Medicinal and Aromatic Plants of South America Vol. 2*, Medicinal and Aromatic Plants of the World 7,
https://doi.org/10.1007/978-3-030-62818-5_19

Keywords "Bailahuén" · Flavonoids. · Coumarins. · Choleretic.

1 Introduction

Haplopappus baylahuen is one of the most used plant species in Chilean herbal medicine. Due to this, the Ministry of Health of Chile (MINSAL), decided to include its monograph among the 103 medicinal species used in Chile. MINSAL grouped these monographs in the compendium called "Medicamentos Herbarios Tradicionales (MHT)" published in 2009 (de Salud 2010).

In this chapter, we will review its beneficial effects on the gastrointestinal system. Its phytochemical composition includes several families of well-known compounds such as flavonoids and terpenoids plus some compounds reported for the first time in *Haplopappus* spp. Many of these chemical principles have antioxidant, anti-inflammatory, diuretic, antimicrobial and hepatoprotective activities, while several of these compounds are part of the resinous exudate that covers the surface of the leaves of the plant, which has an intense and characteristic aroma (Vogel et al. 2005a).

2 Taxonomic Characteristics

Haplopappus baylahuen J. Rémy belongs to Asteraceae (Compositae) family. Its common names are "bailahuén", "baylahuen", "vai-lahuén", "chejchjra", "chechenaja", "baylahuina", "guaylaven", "bailabuena", "failawen" or "goat horn" ("cacho de cabra"). "Baylahuen" may be the name originally assigned by the Mapuche people (indigenous inhabitants of south-central Chile and Argentina) to a cluster of plants of the same genus (Silva et al. 1995; De la Peña 1997). According to Fabrés (cited by Gunckel 1966), the etymology of the name comes from "vain" = boil and "lahuen" = medicinal plant. In Chile, this family is represented by 157 genera. Among them, *Haplopappus* contributes with 61 species. Besides, *H. baylahuen*, *H. foliosus* DC. *H. multifolius* Reiche and *H. angustifolius* (DC.) Reiche, are also used in Chile with the same therapeutic applications (Marticorena and Quezada 1985).

Synonyms *Aster baylahuen* (J. Rémy) Kuntze, *Haplopappus domeykoi* Phil., *Haplopappus lastarrianus* (J. Rémy) Kuntze, *Haplopappus medicinalis* Phi *Haplopappus domeykoi* Phil., *Haplopappus fluehmannii* Phil. and *Haplopappus glutinosus* Cass (The Plant List 2020).

3 Crude Drug Used

Dried and powdered leaves (Photo 1), stems and flowers are used in individual tea bags (2.0 g) to prepare infusions or decoctions as a stimulant of digestive function. Also whole leaves, flowers and stems are sold to prepare medicinal teas. Due to the presence of resin, this powdered herbal drug has a characteristic and very persistent odor allowing users to identify the plant.

4 Major Chemical Constituents and Bioactive Compounds

Early studies performed on this plant species have reported the presence of anthraquinone heterosides (emodin, chrysophanic acid) and the flavonoids quercetin, quercetin 3-*O*-methyl ether, rhamnetin, isorhamnetin and kaempferol. The presence of different methylated derivatives of quercetin and kaempferol in the resinous exudate is characteristic of the different species of *Haplopappus* that are grouped under the vernacular name "baylahuen". Coumarins (esculetin, prenyletin and haplopinol) have also been isolated from the plant (Chiang et al. 1982; Núñez-Alarcón et al. 1993). *H. multifolius* is the most abundant species and often is used in replacement of *H. baylahuen*. In this species, the presence of known coumarins such as haplopinol; hernianin; *O*-prenyl-umbelliferone; prenyletin; 6-hydroxy-7-[(*E*)-3′,7′-dimethyl-5′-hydroxy-2′,6′-octadienyloxy]-coumarin and 6-hydroxy-7-[(*E,E*)-3′,7′-dimethyl-7′-hydroxy-2′,5′-octadienyloxy]-coumarin have been reported. In the same study, authors elucidate the structure of two new coumarins: 6-hydroxy-7-[(*E,E*)-3′,7′-dimethyl-2′,4′,7′-octatrienyloxy]-coumarin and

Photo 1 – *Haplopappus baylahuen*. Powdered drug. Photo by the author (November 2019)

7-[(*E*)-3′-methyl-4′-hydroxy-2′-butenyloxy]-coumarin (6-deoxyhaplopinol) (Torres et al. 2006). Diterpenes such as friedelan-3-one, epifriedelinol, haplopappic acid have been isolated from *H. angustifolius*, which is also used to replace *H. baylahuen* in the North-Central region of Chile. According to Becerra et al. (2010), the essential oil of *H. baylahuen* has mainly azulene, pinene, α-cadinol, cyclohexadiene, bergamotol and eicosane. This essential oil prepared at 1% concentration shown a 38% of inhibitory activity against *Aspergillus niger*. González et al. (2017) assessed the biomass resin and essential oil production in *Haplopappus rigidus, Haplopappus baylahuen, Haplopappus multifolius* and *Haplopappus taeda*. However, due to the high variability of essential oil contents among plants was not possible to differentiate *Haplopappus* species. In a recent study performed in *H. multifolius, H. taeda, H. baylahuen* and *H. rigidus*, chemical profiling was assessed by HPLC-DAD-ESI-MS method allowing tentative identification of different phenolics present in the herbal teas and methanol extracts (Schmeda-Hirschmann et al. 2015). In this analysis, 27 phenolics were identified. Among these compounds, 10 caffeoylquinic and feruloyl quinic acid derivatives were reported for the first time. Interestingly, this study evidences a central concern regarding these plants. Thus, the phytochemical composition differs significantly between the different species studied, which raises doubts regarding the ethnopharmacological validation of the published results. As will be discussed later in the text (section 8), different bioassays have been carried out with species of the same genus without comprehensive and thorough chemical profiling using high-end analytical technologies.

5 Morphological Description

Haplopappus baylahuen is a rhizomatous shrub, cushion-shaped, resinous, with spatulate leaves. Leaves have coriaceous texture and bright surface, deeply serrate to pinnatisect, 1–3.5 cm long, glabrous or with sparse hairs. Flowers are yellow and grouped in solitary chapters. The chapters are radiated with bracts arranged in 4–5 series. Yellow ligulate flowers, with 6–7 mm long ligule, 3-dentate at the apex with 5 mm pappus. Tubular flowers 6.5 mm long, corolla glabrous, pappus of 5 mm. The fruit is a villous achene with white or straw-color pappus (Muñoz et al. 1999; Berti et al. 2006). As several *Haplopappus* species are used for similar curative purposes it is important to develop taxonomic identification strategies that allow for proper differentiation between species.

Recently, González et al. (2017) proposed a micromorphological analysis of the main characteristics of *H. baylahuen, H. multifolius, H. taeda*, and *H. rigidus* Phil. leaves. He reports that these species have different leaf thicknesses: *H. baylahuen* the highest value, whereas *H. multifolius* presented thinnest leaves. *H. taeda* on the other hand, is the species that has the longest leaves (7.4 cm) but with the thinnest cuticle (14.5μm). The other species have a cuticle that ranging from 23–27μm. Regarding the secretory gland density, *H. baylahuen* shown the highest value compared with *H. multifolius* and *H. taeda. H. baylahuen* also has a higher stomatal index in the upper surface than other species. Secretory glands could be observed on both sides of the leaf with higher density on the upper side. These glands have a multicellular head.

6 Geographical Distribution

H. baylahuen is distributed mainly in Northern Chile, where its area can be found close to 2000 m.a.s.l. It is also possible to find it in the central-south regions at 800–1500 m.a.s.l. Despite most *H. baylahuen* seeds are vain, propagation studies of Vogel et al. (2005b) have demonstrated that viable seeds germinate without much difficulty. These authors also reported that when harvesting a large part of the plant, it ends up drying. Although *ex situ* cultivation was successfully achieved in the VII region of Chile, the wild collection will continue to be the main source of this shrub. The form of propagation and the response of the different species of *Haplopappus* to cultivation conditions were determined in domestication assays (Vogel et al. 2005b). However, the results of this study suggest that plants growing in their natural habitat exude more resin than *ex situ* cultivated plants.

7 Ecological Requirements

H. baylahuen can be found on poorly evolved soils, with light textures, in general on rocky places, with a low concentration of organic matter, nitrogen, phosphorus and potassium, with a pH of approximately 5.7. It grows in the clear ground with high incident UV radiation and high-stress conditions such as drought, low soil fertility and grazing. Such conditions limit their natural regeneration (San Martín et al. 2005; Vogel et al. 2007). The populations are distributed irregularly, but also in pure groups and the plants reach heights between 6 and 20 cm (Berti et al. 2006).

8 Traditional Use (Part(s) Used) and Common Knowledge

In folk medicine, the leaves, stems and flowers are used for the preparation of teas. These medicinal preparations are used for nonspecific gastrointestinal and hepatobiliary (choleretic and cholagogue) conditions. The plant also has an aphrodisiac reputation. Emmenagogue properties have also been reported, although their use during pregnancy is restricted by the risk of inducing bleeding (Muñoz et al. 1981; Montes and Wilkomirsky 1987; Hoffmann et al. 1992). The plant can also be used in other types of disorders that can cause diarrhea, dysentery, flu and urinary infections (Muñoz et al. 1999; Vogel et al. 2005b). Aymara people (Northern Chile) uses this plant to treat renal colic pain. *H. baylahuen* has been used to treat wounds in horses (Hoffmann et al. 1992).

9 Modern Medicine Based on Its Traditional Medicine Uses

Most of the scientific studies carried out with this medicinal plant have been designed to validate its beneficial effects on the gastrointestinal tract. Such studies have contemplated *in vitro* and *in vivo* approaches, but no clinical trials in humans. As an example, the intragastric administration of an infusion prepared with the whole plant to dogs with liver damage decreased the activity of the transaminase enzyme alanine aminotransferase (ALAT). Also, the formation of gallstones and the retention time of sufobromophthalein (BSP) were moderately decreased (Silva et al. 1995). In rats intoxicated with CCl_4, a dose of 4 mg/kg of the aqueous extract of *H. baylahuen* promoted a hepatoprotective effect. This effect was attributable to the presence of flavonoids such as 7-methyl aromadendrine (Núñez-Alarcón et al. 1993), which was superior to that of the silymarin, complex flavolignans from *Silybum marianum*.

On the other hand, the presence of prenyletin, haplopine and haplopinol explain the antimicrobial effects observed for *H. multifolius* (Chiang et al. 1982). Adzet et al. (1991) reported the anti-inflammatory effect (63–68%) of an infusion of *H. baylahuen* (100 mg/kg i.p) using the carrageenan-induced rat paw edema. The antimicrobial effect was also confirmed by Urzúa et al. (1995), who evaluated this activity in the resinous exudates of *H. diplopappus, H. anthylloides, H. schumannii, H. cuneifolius, H. velutinus, H. uncinatus, H. multifolius, H. illinitus* and *H. foliosus*, being more active on Gram (+) bacteria. However, when studying the resinous exudates of other *Haplopappus* species, the latter research group suggests that the presence of lipophilic flavonoids would be more associated with the antimicrobial effect (Urzúa et al. 2012).

The antioxidant effect of baylahuen has also been investigated in various articles. According to Speisky et al. (2006), *H. baylahuen* was the species with the highest antioxidant indexes amongst 13 selected medicinal herbs. Thus, this plant showed superiority in the hypochlorite quenching tests, peroxynitrite quenching and the ability to trap the ABTS free radical as one of the models to determine the Trolox®-equivalent antioxidant capacity (TEAC). Morales et al. (2009), reported that 5,3′,4′-trihydroxy-7-methoxyflavanone isolated from *H. rigidus* was able to scavenge DPPH and ABTS radicals and protect liver microsomes from the oxidative damage induced by iron/ascorbate. Also, they assessed the cytotoxicity of this compound upon A-549 (human lung carcinoma), MCF-7 (human breast carcinoma) and HT-29 (human colon adenocarcinoma) cell lines. In all cases, 5,3′,4′- trihydroxy-7-methoxyflavanone showed an ED_{50} ranged from 3 to 5.66µg/ml. In another study, the neutralizing activity of the DPPH radical was evaluated for aqueous and methanolic extracts of *H. multifolius, H. taeda, H. baylahuen* and *H. rigidus* (Schmeda-Hirschmann et al. 2015). These authors found that the DPPH SC50 (DPPH scavenging concentration of 50%) for infusions of *H. deserticola*, *H. multifolius* and *H. taeda* was 21.2, 19.5 and 20.0μg/ml, respectively. On the other hand, methanol extracts of *H. multifolius* and *H. taeda* display a DPPH SC50 of 21.3μg/ml. This result was coherent with the higher total flavonoids (TF) and total phenolics (TP) content of methanol extracts compared with the infusions. However, the authors point out the fact that high flavonoid and phenolic contents observed in other species (*e.g. H. baylahuen* and *H. rigidus*), would not be correlated with their poor antioxidant activity. Previously, Torres et al. (2006), reported that isolated coumarins from *H. multifolius* also have an antioxidant effect on DPPH with IC_{50} ranging from 109–251μM. These values were much higher compared to that of quercetin (9.8μM), so, likely, the cytoprotective effects of coumarins are better explained by their interaction with other cell targets associated with intrinsic antioxidant mechanisms and that appear at lower concentrations (Kapp et al. 2017; Zhu and Jiang 2018).

Faini et al. (2008) reported that *Haplopappus taeda* and ethanolic extract of the fresh plant has an anti-inflammatory effect in the arachidonic acid-induced acute ear edema assay (20μl, 3.6×10^{-4} M/ear). Also, the radical scavenging activity upon DPPH, xanthine oxidase inhibition and superoxide anion scavenging properties were evaluated. In this study, significant anti-inflammatory activity (72%) was associated with the presence of the monoterpene taedol, followed by clerodane diterpene ent-5α-18-acetoxy-*cis*-cleroda-3,13-*E*-dien-15-oic acid and the flavanone sakuranetin. The same group investigated the anti-inflammatory, antioxidant and cytotoxic activity (CCRF-CEM tumor cell line) of *H. remyanus* (Faini et al. 2011). They report a moderate activity in these assays and associated such activity to the presence of flavonoids. In particular, these authors suggested that the high concentration of pinostrobin (flavanone) could support, at least in part, the tested bioactivities (Roman Junior et al. 2017; Jadaun et al. 2019). In a thesis work, Graff (2008) found that *H. remyanus* 20% infusion scavenge DPPH radical but was unable to protect histological changes promoted by UVB light and benzoyl peroxide oxidative damage on rat skin.

Although the dose used was high (1000 mg/kg), *H. baylahuen* shown a diuretic effect in rats, which was comparable to the effect induced by a dose of 20 mg/kg of hydrochlorothiazide and furosemide (Méttola et al. 2018).

10 Conclusions

The literature considered for this monograph allows us to state that "bailahuén" has been intended as a complex group of plants that gather several species within the same genus *Haplopappus*. Although in Chile important efforts have been made to understand the chemical composition of the genus, it is perceived that it is necessary to incorporate studies with high-end technologies to improve the identification of minor unknown compounds. The different groups of researchers who have worked with these plants recognize the complexity of their ethnopharmacological validation because several species are used as "bailahuén". For such reason, the assignment of a pharmacological activity can only be attributed to specific *Haplopappus* species and not extrapolated to the entire genus. The reason for this statement is based on the fact that the chemical composition can vary significantly among species and it is not correct to assume that all *Haplopappus* species have a conserved phytochemical profile.

This plant may be a useful therapeutic resource in various gastrointestinal conditions. These therapeutic uses are partially supported by the experimental pharmacology data. However, it is expected that the lack of clinical studies will be overcome in the future to establish a minimum level of evidence that supports all the uses recommended by folk medicine.

References

Adzet T, Marín E, Gene RM (1991) Estudio de la actividad antiinflamatoria de especies vegetales de origen centro y sudamericano. Dominguezia 9(1):17–23

Becerra J, Bittner M, Hernández V, Brintrup C, Becerra J, Silva M (2010) Actividad de aceites esenciales de Canelo, Queule, Bailahuén y Culén frente a hongos fitopatógenos. BLACPMA 9(3):212–215

Berti M, Tramón C, Fischer S, Wilckens R, Inostroza L, Baeza M, Tapia M, Pastene E, del Pozo A (2006) Avances en la multiplicación y caracterización de algunas plantas medicinales nativas. Ministerio de Agricultura, Servicio Agrícola y Ganadero (SAG), Chile, p 15

Chiang MT, Bittner M, Silva M, Moncada A, Zemelmann R, Sammes P (1982) Coumarin constituents in *Haplopappus* genus. Phytochemistry 21:2753–2755

De la Peña R (1997) Catálogo de nombres vulgares de la Flora Argentina. (lista preliminar). Ed. Univ. Nacional del Litoral. Centro de Publicaciones. Santa Fe, Argentina, 195 pp

de Salud M (2010) MHT Medicamentos Herbarios Tradicionales, 103 Especies Vegetales. Ed. MINSAL, Santiago, Chile, p 29

Faini F, Labbé C, Delporte C, Backhouse N, Castro C, Torres R (2008) Anti-inflammatory and antioxidant activities of *Haplopappus taeda* extracts. Planta Med 74(09):74–PA78

Faini F, Torres R, Rodilla JML, Labbé C, Delporte C, Jaña F (2011) Chemistry and bioactivity of *Haplopappus remyanus* ("bailahuen"), a Chilean medicinal plant. J Braz Chem Soc 22(12):2344–2349

González B, Vogel H, San Martín J (2017) Histological leaf characterization of four Bailahuen species (*Haplopappus* spp.) for taxonomic identification for quality control of the drug. J Appl Res Med Aromatic Plants 4:41–45

Graff E (2008) Determinación Histopatológica en Piel de Ratas del daño ocasionado por la radiación UVb y el Peróxido de Benzoflo cuando se les administra una infusión de *Haplopappus remyanus*. Tesis. Universidad Austral de Chile, Facultad de Ciencias, Escuela de Química y Farmacia, Valdivia, Chile, 65 pp

Gunckel H (1966) Fitonomía atacameña, especialmente Cunza. Anales de la Academia Chilena de Ciencias Naturales. Rev Universitaria 52:30

Hoffmann A, Farga C, Lastra J, Veghazi E (1992) Plantas Medicinales de Uso Común en Chile. Ed. Fundación Claudio Gay. Santiago, Chile, p 49

Jadaun A, Sharma S, Vermaa R, Dixit A (2019) Pinostrobin inhibits proliferation and induces apoptosis in cancer stem-like cells through a reactive oxygen species-dependent mechanism. RSC Adv 9:12097–12109

Kapp E, Visser H, Sampson SL, Malan SF, Streicher EM, Foka GB, Warner DF, Omoruyi SI, Enogieru AB, Ekpo OE, Zindo FT, Joubert J (2017) Versatility of 7-substituted coumarin molecules as antimycobacterial agents, neuronal enzyme inhibitors and neuroprotective agents. Molecules 22:1644

Marticorena C, Quezada M (1985) Flora vascular de Chile: *Haplopappus*. Gayana Bot 42:33–34

Méttola RM, Brodkewics IY, Vera NR, Reynoso MA (2018) Actividad diurética y antioxidante de los extractos etanólicos de especies vegetales usadas en la medicina popular argentina. Revista Cubana de Plantas Medicinales 23(4) Published on the Internet: http://www.revplantasmedicinales.sld.cu/index.php/pla/article/view/638 (Last access, February 2020)

Montes M, Wilkomirsky T (1987) Medicina Tradicional Chilena. Ed. de la Universidad de Concepción, Concepción, Chile, p 74

Morales G, Paredes A, Sierra P, Loyola L (2009) Cytotoxicity, scavenging and lipid peroxidation - inhibiting activities of 5,3′,4′-trihydroxy-7-methoxyflavanone isolated from *Haplopappus rigidus*. J Chil Chem Soc 54(2):105–107

Muñoz M, Barrera E, Meza I (1981) El uso medicinal y Alimenticio de Plantas Nativas y Naturalizadas de Chile. Museo Nacional de Historia Natural. Publicación ocasional, Santiago, Chile, (33):29

Muñoz O, Montes, M, Wilkomirsky T (1999) Plantas medicinales de uso en Chile: Química y Farmacología. Ed. Universitaria S.A, Santiago, Chile, pp 143–147

Núñez-Alarcón J, Dolz H, Quiñonez MH, Carmona MT (1993) Epicuticular flavonoids from *Haplopappus baylahuen* and the hepatoprotective effect of the isolated 7-methyl aromadendrin. Boletín de la Sociedad Chilena de Química 38:15–22

Roman Junior WA, Gomes DB, Zanchet B, Schönell AP, Diel KAP, Banzato TP, Ruiz ALTG, Carvalho JE, Neppel A, Barison A, Santos CAM (2017) Antiproliferative effects of pinostrobin and 5,6-dehydrokavain isolated from leaves of *Alpinia zerumbet*. Rev Bras 27(5):592–598

San Martín J, Vogel H, González B (2005) Distribución, abundancia y vegetación acompañante de cuatro especies nativas arbustivas promisorias. Biol Res 38:2–3

Schmeda-Hirschmann G, Quispe C, González B (2015) Phenolic profiling of the south American "baylahuen" tea (*Haplopappus* spp., Asteraceae) by HPLC-DAD-ESI-MS. Molecules 20:913–928

Silva M, Alarcón J, Bittner M, Becerra J, Sanhueza L, Marticorena C (1995) In: Gupta M (ed.): 270 Plantas Medicinales Iberoamericanas. Ed. Presencia Ltda. Santafé de Bogotá, Colombia, p 119

Speisky H, Rocco C, Carrasco C, Lissi EA, López-Alarcón C (2006) Antioxidant screening of medicinal herbal teas. Phytother Res 20:462–467

The Plant List (2020) Published on the Internet: http://www.theplantlist.org/tpl1.1/record/kew-2806430. (Last access, March 2020)

Torres R, Faini F, Modak B, Urbina F, Labbé C, Guerrero J (2006) Antioxidant activity of coumarins and flavonols from the resinous exudates of *Haplopappus multifolius*. Phytochemistry 67:984–987

Urzúa A, Torres R, Muñoz M, Palacios Y (1995) Comparative antimicrobial study of the resinous exudates of some Chilean *Haplopappus* (Asteraceae). J Ethnopharmacol 45(1):71–74

Urzúa A, Echeverría J, Espinoza J (2012) Lipophilicity and antibacterial activity of flavonols: antibacterial activity of resinous exudates of *Haplopappus litoralis, H. chrysantemifolius* and *H. scrobiculatus*. BLACPMA 11:369–376

Vogel H, González M, Faini F, Razmilic I, Rodríguez J, San Martín J, Urbina F (2005a) Antioxidant properties and TLC characterization of four Chilean *Haplopappus*-species known as bailahuén. J Ethnopharmacol 97:97–100

Vogel H, Razmilic I, San Martín J, Doll U, González B (2005b) Plantas medicinales chilenas-Experiencias de domesticación y cultivo de boldo, matico, bailahuén, canelo, peumo y maqui. Ed. Universidad de Talca. Talca, Chile, 192 pp

Vogel H, González B, San Martín J, Razmilic I, Villalobos P, Schneider E (2007) Estudio de mercado y sustentabilidad de la recolección silvestre de bailahuén, una planta medicinal chilena. Med Plant Conserv 13:15–21

Zhu JJ, Jiang JG (2018) Pharmacological and nutritional effects of natural Coumarins and their structure–activity relationships. Mol Nutr Food Res 62:e1701073. https://doi.org/10.1002/mnfr.201701073

Hydrocotyle bonariensis Comm. ex Lam.

Adriana Graciela Ouviña and María Luján Flores

Hydrocotyle bonariensis Comm. ex Lam. in its habitat (Miramar, Buenos Aires, Argentina). (Photo: AG Ouviña)

A. G. Ouviña (✉)
Cátedra de Farmacognosia, IQUIMEFA (UBA-CONICET), Facultad de Farmacia y Bioquímica, Universidad de Buenos Aires, Buenos Aires, Argentina
e-mail: adrianaouvi@gmail.com

Á. Máthé, A. Bandoni (eds.), *Medicinal and Aromatic Plants of South America Vol. 2*, Medicinal and Aromatic Plants of the World 7,
https://doi.org/10.1007/978-3-030-62818-5_20

Abstract *Hydrocotyle bonariensis* Comm. ex Lam. (Araliaceae) is an herbaceous medicinal plant commonly known as "redondita de agua" or "paragüita" that is widely distributed in South America. In Argentina it inhabits the Northwest, Northeast and Central regions and it is frequently found in the province of Buenos Aires. This species was formerly used, in South America, to treat skin erythema. The infusion of aerial parts is used for their anti-inflammatory, diuretic, stimulants and antiseptic properties. Chlorogenic acid, caffeic acid, hyperoside, isoquercitrin, quercitrin, quercetin, apigenin and kaempferol are the principal phenols identified in the methanolic and aqueous extracts (infusion) of the aerial parts. These metabolites contribute to the anti-inflammatory and antioxidant properties evidenced. Total carbohydrates of the aerial parts were extracted through infusion. The main polysaccharides identified are galactorhamnans, galacturonans, rhamnogalacturonans and sulfated galactans; they are related to the diuretic, anti-inflammatory and immunomodulatory activity of this species. The identified phenols and carbohydrates are related to the principal ethnopharmacological properties, similarly, to the anti-inflammatory properties to treat skin conditions, such us, psoriasis.

Keywords *Hydrocotyle bonariensis* · Phenols derivatives · Carbohydrates · Psoriasis

1 Introduction

Throughout the history of humanity, plants have been a valuable resource for treating human diseases. Currently they still offer very interesting perspectives for the pharmaceutical industry, as possible sources of novel bioactive molecules. Argentina has a great biodiversity that includes an extensive list of plants used in traditional medicine by different cultures. However, there are still numerous species to be scientifically validated. With such aspects in mind, *Hydrocotyle bonariensis* Comm. ex Lam. (Araliaceae), a native medicinal plant, has been selected as a subject of our review and as a possible source of new agents of pharmacological importance. The phytochemical study and the determination of biological activities are based on traditional uses. The research results, regarding also the interrelationship among the identified metabolites, the determined biological properties and the described uses, seem to demonstrate its therapeutic potential.

M. L. Flores
Cátedra de Farmacognosia, Centro Regional de Investigación y Desarrollo Científico Tecnológico (CRIDECIT), Facultad de Ciencias Naturales y Ciencias de la Salud, Universidad Nacional de la Patagonia San Juan Bosco (UNPSJB), Comodoro Rivadavia, Chubut, Argentina

2 Taxonomic Characteristics

Hydrocotyle bonariensis Comm. ex Lam. (syn.: *Hydrocotyle multiflora* Ruiz & Pav.; *Hydrocotyle yucatanensis* Millsp.; The Plant List 2019), is an herbaceous plant commonly known as "redondita de agua" or "paragüita", belonging to the family Araliaceae, order Apiales. This family includes approximately 43 genera and 1450 species, several of them used in traditional medicine. In Argentina, 7 genera with 31 species have been described (Fiaschi 2008).

Hydrocotyle is a widely distributed genus of more than 130 species with notable morphological variations both within and between species (Nicolas and Plunkett 2009). It is mainly recognized for presenting perennial herbs, with stems rooted in the nodes, generally pelted leaves, stipules and umbel inflorescences; its fruits are compressed laterally, with 5 ribs, without carpophores and have seeds with woody endosperm.

In Argentina, 13 species have been described: *H. bonariensis*, *H. bowlesioides*, *H. callicephala*, *H. chamaemorus*, *H. cryptocarpa*, *H. exigua*, *H. filipes*, *H. indecora*, *H. leucocephala*, *H. modesta*, *H. pusilla*, *H. ranunculoides* and *H. verticillata* (Zuloaga and Morrone 1999).

For *H. bonariensis*, other common names are in Spanish: "cucharita", "oreja de agua", "tembladerilla", "perejil de agua", "sombrilla de sapo", "corona de Santa", "caá ay"; Portuguese: "erva-capitão"; English: largeleaf pennywort, large-leaved pennywort, beach pennywort, coastal plain pennywort, dollar weed. *Hydrocotyle* derives from the Greek, *Hydro* = water and *kotyle* = cup, small container, in relation to the shape of the leaves and the type of habitat. As for *bonariensis*, it corresponds to a geographical epithet, referring to the location in the province of Buenos Aires (Argentina).

3 Crude Drug Used

The crude drug used in traditional medicine is prepared from the aerial parts of *Hydrocotyle bonariensis* (Photo 1). The leaves and flowers are used for their anti-inflammatory, diuretic, stimulant and antiseptic properties (Hieronymus 1882; Marzocca 1997).

4 Major Chemical Constituents and Bioactive Compounds

Phenolic acids, flavonoids and carbohydrates are the principal constituents of the aerial plant parts (Ouviña 2019). These chemical groups are responsible for the medicinal properties attributed to the species. They are also noteworthy for their importance in human health and for their role as chemotaxonomic markers.

Photo 1 *Hydrocotyle bonariensis* Comm. ex Lam. (Photo: AG Ouviña)

The phytochemical analysis of the methanol extract of *H. bonariensis* allowed to reveal a broad phenolic profile. The major phenolic acids are the 3,4,5-trihydroxybenzoic acid (gallic acid), methyl 3,4,5-trihydroxybenzoate (methyl gallate, gallicin), 3-*O*-caffeoylquinic (chlorogenic acid) and 5-*O*-caffeoylquinic acid (isochlorogenic acid). The principal flavonoids are the quercetin, rutin, isoquercitrin, quercitrin and hyperoside (Ouviña 2019). Hyperoside is one of the most important phenolic derivatives, which allows us to propose this metabolite as a chemotaxonomic marker for the species. It represents 0.02% with respect to the dry weight of the aerial parts (Ouviña 2019).

In the methanol extract, the total polyphenols were determined by the Folin-Ciocalteu colorimetric method; they represented 48.5 mg GAE/g expressed as equivalents of gallic acid per gram of dry drug (Ferraro et al. 2008; Ouviña 2019).

In the infusion of the aerial parts were identified mainly carbohydrates, flavonoids, phenolic acids and saponins. This aqueous extract present 25% soluble carbohydrates consisting of rhamnose, glucose, galactose and uronic acids; in addition, 7% sulfate groups and 5% uronic acids were determined (Ouviña et al. 2010). The infusion was purified by dialysis in order to separate the constituents according to their molecular size. The purified fraction (molecular weight > 8000) was enriched with polysaccharides, mainly galactorhamnans, galacturonans, rhamnogalacturonans and sulfated galactans (Ouviña 2019). On the other hand, dialysis waters (fraction of the infusion constituted by metabolites of smaller molecular size) contained flavonoids (hyperoside, quercitrin, rutin), phenolic acids (chlorogenic and gallic acids) and other polar metabolites (Ouviña et al. 2010, 2012).

Saponins were determined in both infusion and methanol extract. Asiaticoside was evident in this extract but not in the methanol extract; in this last extract were present bonarienosides (oleanane-type triterpenoid saponins) (Ouviña 2019). Other authors described the presence of five bonarienosides in the roots of *H. bonariensis* (Tabopda et al. 2012). Maulidiani et al. (2014) reported flavonoids, phenolic acids and triterpenoid saponins in the leaves. Lien Huong et al. (2016) identified two alkaloids (tetrahydropalmatine and xylopinine) and ethyl 2-*O*-α-fructofuranoside in samples collected in Vietnam.

All the metabolites identified are of importance due to their biological activities, their ecological implications and chemotaxonomic markers, among others. Flavonoids have been among the secondary metabolites, the most commonly used in chemotaxonomic determinations, given their structural characteristics and given the particular abundance they present in the Araliaceae family (Harborne 1994).

5 Morphological Description

The species of *Hydrocotyle* are perennial herbs, glabrous or hairy, without fibrous foliar remains in the base, with stems rooted in the nodes. Leaves usually pelted, alternate, petiolate, from orbicular to obovate; stipules and umbels inflorescence simple or proliferate. Fruits from orbicular to ellipsoids, compressed laterally, glabrous; semi-orbicular to triangular mericarps in cross-section, with 5 thin ribs, without carpophores. Seeds with woody endosperm (Medina 2003).

Hydrocotyle bonariensis is a perennial herb, glabrous; with rhizomes, stems creeping and rooting at the nodes. Leaves peltate with orbicular blades, erect petiole, 2–10 cm diameter. Very small yellowish-white flowers, umbels inflorescence, 1–6 cm diam.; rays many, spreading, with the flowers in whorls in the form of interrupted or sometimes branched spikes and clustered at the base of the rays. Solitary peduncle in each knot of the stem, to 40 cm long, but usually shorter, as long as or usually longer than the leaves; pedicels 2–20 mm long; involucral bracts lanceolate, acute. Orbicular, reniform or elliptical fruits, very laterally compressed, 2 mm in diameter, marinated at the base with rudimentary oil channels and without carpophore (Cabrera 1965; Marzocca 1997; National Herbarium of NSW 2000).

6 Geographical Distribution

H. bonariensis is a species of wide distribution in South America, mainly in Brazil, Chile, Uruguay and Argentina. In Argentina, it inhabits the Northeast, Northwest and Central regions: it is frequent in the province of Buenos Aires (USDA 2020).

It has been observed to inhabit Southeastern USA, Mexico, Central America, Cuba, Puerto Rico, sub-Saharan Africa, South Africa, Madagascar and the Mascarenes. *H. bonariensis* is widely naturalized in the coastal districts of Southern and Eastern Australia (USDA 2020).

7 Ecological Requirements

In Argentina, particularly in the province of Buenos Aires, several species of the genus *Hydrocotyle* grow in the riparian forests of the Delta del Río de la Plata and in the grass of the riverbank (Ratera and Ratera 1980). It is frequently found in humid grasslands, grasslands and dunes of the east and south of the province of Buenos Aires. It inhabits park and garden lawns, prefers sandy, damp and still watery or swampy, shady soils, grasslands and dunes. Its creeping stems, which root at the joints (i.e. nodes), allow it to quickly colonize areas and suppress the growth of other plants. *H. bonariensis* invades areas ranging from coastal sand dunes to cliff faces, coastal riparian habitats, brackish wetlands and open woodlands. Its vegetation season lasts throughout the year; blooms from mid-spring to mid-autumn; propagates by seeds and rhizomes (Cabrera 1965; Ratera and Ratera 1980).

8 Traditional Use (Part(s) Used) and Common Knowledge

Hydrocotyle bonariensis was formerly used in South America to treat skin erythema. Its leaves are administered in the form of poultices to heal inflamed wounds and skin inflammations. Fresh water can be distilled from the fresh plant that is used in cosmetics to "erase" the freckles. The "juice" is considered an emetic and liver protector. Infusions prepared with the leaves, flowering tops and stems are used in traditional medicine as a diuretic, stimulant, emmenagogue and antiseptic (Hieronymus 1882; Marzocca 1997).

9 Modern Medicine Based on Its Traditional Medicine Uses

Hydrocotyle bonariensis is less studied from the pharmacological and phytochemical point of view; even so, it is popularly used as an infusion prepared from the plant, as an anti-inflammatory to treat skin disorders (Lucena et al. 2001). Among the plants used to treat skin diseases, *H. bonariensis* has been used in traditional medicine to treat psoriasis (Natarajan and Paily 1973). In addition, it is used for the treatment of tuberculosis, relieving the pain of rheumatism and arthritis, to increase brain capacity and for longevity (Masoumian et al. 2011). Obaseki et al. (2016) described that oral administration of the hexane extract of the *H. bonariensis* leaves (250 mg/kg body weight) having potent anti-inflammatory action.

Dichloromethane extract of this species shows a broad inhibitory activity over the exponential growth stage of *Chlamydia trachomatis* and *Chlamydia pneumoniae,* independently of the chlamydial strain and the cell line (Entrocassi et al. 2013; Catalano et al. 2018).

Flavonoids are a large family of secondary plant metabolites that are present in plant tissues in relatively high concentrations as sugar conjugates and are common in leaves, flowering tissues and pollens (Andersen and Markham 2006). The methanolic extract of the aerial parts of *H. bonariensis* enriched in these metabolites evidenced anti-inflammatory (58% inhibition of mouse ear edema), antioxidant and antiproliferative activities (Ouviña et al. 2009; Ouviña 2019). This extract inhibited the peroxidation of egg phospholipids indicating antioxidant activity with an EC_{50} = 812.83μg/ml, while the antiproliferative activity in the murine lymphoma cell line showed an EC_{50} = 25.7μg/ml (Ouviña 2019).

Tempone et al. (2008) described the activity of methanol extract of this species against *Leishmania amazonensis* and *L. chagasi* promastigotes in which 100% of parasites were killed at the highest concentration (500μg/ml). *H. bonariensis* also present a marked antifungal activity against FCZ-resistant strain *Candida kruse* (Lorenzi and Matos 2002). Vivot et al. (2012) also described activity against *Bacillus subtilis*.

Ajani et al. (2009, 2011) suggested that *H. bonariensis* protects against galactose-induced cataract, and that administration of the extract after cataract onset reduced cataract progression but did not reverse cataractogenesis; and it had not renal or hepatotoxicity in long term administration of an aqueous extract. The phytochemical studies showed the presence of alkaloids, phenolic compounds and saponins as bioactive components.

Aqueous extract of the aerial parts (infusion) inhibited mouse ear edema (local swelling) with a value of 30% (Ouviña et al. 2009). Infusion showed carbohydrates as principal constituents in addition to phenolics derivatives and saponins. The main polysaccharides identified were galactorhamnans, galacturonans, rhamnogalacturonans and sulfated galactans); they are in relation with anti-inflammatory and immunomodulatory activities of this species (Ouviña et al. 2017). The study of the activity on cell proliferation in the murine lymphoma cell line showed an EC_{50} = 5.0μg/ml for infusion, while the analysis of cell viability in normal lymph node lymphocyte cells showed for this extract less toxicity with a CC_{50} > 1000μg/ml (Ouviña 2019).

Antioxidant activity by DPPH inhibition showed that de purified fraction of the infusion (dialysate) is the most active (EC_{50} = 41 ± 3μg/ml) than dialysis waters (EC_{50} = 239 ± 10μg/ml). SOD activity was similar, obtaining for dialysate an EC_{50} = 22 ± 3μg/ml and for dialysis waters, EC_{50} = 239 ± 10μg/ml. Regarding to the ability to reduce Fe^{3+} to Fe^{2+}, although it is low, dialysate with an EC_{50} = 224 ± 20μg/ml is also more active, while for dialysis waters EC_{50} = 575 ± 40μg/ml (Ouviña et al. 2015). These results demonstrate that carbohydrates play an important role in antioxidant activity.

Extracts of *H. bonariensis* are used topically, in folk medicine, against cutaneous erythema. In traditional medicine, many plants have been used to treat various types of diseases, including external and internal wounds and inflammations; many of these plants have been shown to contain polysaccharides that exhibit different biological activities. In addition, many of these polysaccharides can form gels or viscous solutions of great industrial and ecological value (Paulsen and Barsett 2005). All this is related to the diuretic properties, anti-inflammatory and antiseptic properties of the skin attributed to this species (Hieronymus 1882; Toursarkissian 1980; Ouviña et al. 2009; Ouviña 2019). In addition, flavonoids and phenolic acids contribute to the anti-inflammatory and antioxidant properties (Ouviña 2019).

Silva et al. (2010) found an allelopathic activity in the EtAcO extract, inhibiting the germination and growth of lettuce, tomato, onion and wheat.

10 Conclusions

Metabolites biosynthesized by *Hydrocotyle bonariensis* and the biological activities demonstrated are related to the described ethnopharmacological properties and as such, they constitute a valuable contribution to their validation. Carbohydrates are responsible for the immunomodulatory activity they exert their immunosuppressive or anti-inflammatory activity without being cytotoxic to normal cells. This might suggest that they could modulate the immune system. Carbohydrates are closely related to the traditional uses of the species as anti-inflammatory at the skin level and antiseptic. Also, the characterized phenolic derivatives, mainly hyperoside and chlorogenic acid, are related to the reported important antioxidant activity. The lower polarity extract shows activity against *Chlamydia*'s strains. The results surveyed demonstrate the pharmaceutical potential of *Hydrocotyle bonariensis* and thus motivate to continue the chemical and pharmacological studies.

Acknowledgments *In memoriam* Dr. Graciela Ferraro (Farmacognosia, IQUIMEFA-UBA-CONICET). We are especially grateful to the UBA, CONICET, UNPSJB and FONCYT Argentina, for financial support.

References

Ajani EO, Salako AA, Sharlie PD, Akinleye WA, Adeoye AO, Salau BA et al (2009) Chemopreventive and remediation effect of *Hydrocotyle bonariensis* Comm. ex Lam (Apiaceae) leave extract in galactose-induced cataract. J Ethnopharmacol 123(1):134–142

Ajani EO, Salau BA, Olooto EW, Adebayo OL, Adenuga GA (2011) Toxicological implications of continuous administration of aqueous leaves extract of *Hydrocotyle bonariensis* in rats. Arch Appl Sci Res 3(5):471–478

Andersen M, Markham KR (2006) Flavonoids: chemistry, biochemistry, and applications. Ed. Taylor & Francis Group, Boca Raton, New York, USA, 1212 pp

Cabrera AL (1965) Flora de la Provincia de Buenos Aires. Parte IV. Colección Científica del INTA, Buenos Aires, pp 375–381

Catalano V, Entrocassi OA, Wilson E, López P, Rodríguez Fermepín M (2018) Actividad anti clamidial de una planta medicinal argentina: *Hydrocotyle bonariensis* Lam. Dominguezia 34(Suppl):72–73

Entrocassi CA, Ouviña AG, Gallo Vaulet ML, Catalano AV, López PG, Rodríguez Fermepin M (2013) Inhibición *in vitro* de *Chlamydia trachomatis* y *Chlamydia pneumoniae* por extractos de *Hydrocotyle bonariensis* Lam., efecto inhibitorio sobre un sistema persistentemente infectado. In: Abstracts of the XI Simposio Argentino. XIV Simposio Latinoamericano de

Farmacobotánica. I Congreso Latinoamericano de Plantas Medicinales. 20–22 November. Rosario, Argentina

Ferraro G, Anesini C, Ouviña A, Retta D, Filip R, Gattuso M et al (2008) Total phenolic content and antioxidant activity of extracts of *Achyrocline satureioides* flowers from different zones in Argentina. Lat Am J Pharm 27(4):626–628

Fiaschi P (2008) In: Souza VC, Lorenzi H (eds) Botânica Sistemática. Guía ilustrada para identificação das famílias de Fanerógamas nativas e exóticas no Brasil, baseado em APG II, 2ª Edição. Instituto Plantarum de Estudos da Flora, Nova Odessa, Brazil, pp 326–328

Harborne JB (1994) The flavonoids: advances in research since 1986. Ed. Chapman & Hall, London, UK, p 339

Hieronymus J (1882) Plantae Diaphoricae Florae Argentinae. Bol Acad Nac Cienc 4:199–589

Lien Huong TN, Khoanh TTY, Van Vu N (2016) Two new alkaloids and a carbohydrate compound from the species *Hydrocotyle bonariensis* Comm. ex Lam., family Apiaceae. Can Tho Univ J Sci 3:116–119

Lorenzi H, Matos FJA (2002) Plantas medicinais no Brasil: nativas e exóticas cultivadas, 1st edn. Ed. Instituto Plantarum, Nova Odessa, Brazil, 511 pp

Lucena ID, Novara LJ, Cuezzo AR (2001) Apiaceae. Aportes Botánicos de Salta – Serie Flora 6(11):1–75

Marzocca A (1997) Vademécum de Malezas Medicinales de la Argentina. Indígenas y Exóticas. Ed. Orientación Gráfica Editora, Buenos Aires, Argentina, pp 225–226

Masoumian M, Arbakariya A, Syahida A, Maziah M (2011) Effect of precursors on flavonoid production by *Hydrocotyle bonariensis* callus tissues. Afr J Biotechnol 10(32):6021–6029

Maulidiani AF, Khatib A, Shaari K, Lajis NH (2014) Chemical characterization and antioxidant activity of three medicinal Apiaceae species. Ind Crop Prod 55:238–247

Medina L (2003) *Hydrocotyle* L. In: Castroviejo S et al (eds) Flora Ibérica, vol 10. Ed. RJB, CSIC, Madrid, Spain, pp 24–28

Natarajan S, Paily PP (1973) Effect of topical *Hydrocotyle asiatica* in psoriasis. Indian J Dermatol 18:82–85

National Herbarium of NSW, Royal Botanic Gardens and Domain Trust (2000) Mrs Macquaries Road, Sydney, Australia

Nicolas AN, Plunkett GM (2009) The demise of subfamily Hydrocotyloideae (Apiaceae) and the re-alignment of its genera across the entire order Apiales. Mol Phylogenet Evol 53(1):134–151

Obaseki OE, Adesegun OI, Anyasor GN, Abebawo OO (2016) Evaluation of the anti-inflammatory properties the hexane extract of *Hydrocotyle bonariensis* Comm. ex Lam. leaves. Afr J Biotechnol 15(49):2759–2771

Ouviña AG (2019) Estudio fitoquímico y evaluación de las actividades antiinflamatoria y antiproliferativa de extractos de *Hydrocotyle bonariensis* Lam. (Apiaceae). Thesis. Facultad de Farmacia y Bioquímica, Universidad de Buenos Aires. CABA, Argentina, 179 pp

Ouviña AG, Gorzalczany S, Acevedo C, Ferraro G (2009) Actividad antiinflamatoria tópica de extractos de *Hydrocotyle bonariensis* Lam. (Apiaceae). Lat Am J Pharm 28(6):941–944

Ouviña AG, Escobar Daza MD, Alcalde D, Córdoba OL, Ferraro GE, Flores ML (2010) Carbohydrates and phenolics compounds from leaves of *Hydrocotyle bonariensis* Lam. (Apiaceae). In: Abstracts of the Primera Reunión Internacional de Ciencias Farmaceúticas. 24–25 June. Córdoba, Argentina

Ouviña AG, Escobar Daza MD, Córdoba OL, Ferraro GE, Flores ML (2012) Carbohydrates from leaves of *Hydrocotyle bonariensis* Lam. (Apiaceae). In: Abstracts of the Segunda Reunión Internacional de Ciencias Farmacéuticas. 22–23 November. Rosario, Argentina

Ouviña AG, Flores ML, Córdoba OL, Peralta I, Gorzalczany S, Anesini C et al (2015) Actividad antiinflamatoria y antioxidante de *Hydrocotyle bonariensis* Lam. (Araliaceae) en relación a polifenoles e hidratos de carbono constituyentes. In: COIFFA Congress. 3–6 November. Córdoba, Argentina

Ouviña AG, Córdoba OL, Flores ML (2017) *Hydrocotyle bonariensis* Lam. (Araliaceae): chromatographic analysis of carbohydrates from aerial parts. In: Abstracts of the SILAE XXVI. IX Congreso Colombiano de Cromatografía. 25–29 September. Cartagena de Indias, Colombia

Paulsen BS, Barsett H (2005) Bioactive pectic polysaccharides. Adv Polym Sci 186:69–101

Ratera LE, Ratera M (1980) Plantas de la flora Argentina empleadas en Medicina Popular. Ed. Hemisferio Sur, Buenos Aires, Argentina, p 139

Silva CB, Silva Cândido AC, Simionatto E, Faccenda O, de Paula Quintão Scalon S, Lopes Pereira Peres MT (2010) Atividade alelopática, antioxidante e teor de fenóis totais de *Hydrocotyle bonariensis* Lam. (Araliaceae). Acta Sci Technol Maringá 32(4):413–420

Tabopda TK, Mitaine-Offer AC, Miyamoto T, Tanaka C, Mirjolet JF, Duchamp O et al (2012) Triterpenoid saponins from *Hydrocotyle bonariensis* Lam. Phytochemistry 73:142–147

Tempone AG, Sartorelli P, Teixeira D, Prado FO, Calixto IARL, Lorenzi H et al (2008) Brazilian flora extracts as source of novel antileishmanial and antifungal compounds. Mem Inst Oswaldo Cruz, Rio de Janeiro, 103(5):443–449

The Plant List (2019) Version 1.1. Published on the Internet; http://www.theplantlist.org/. (Last access, November 2019)

Toursarkissian M (1980) Plantas Medicinales de la Argentina. Ed. Hemisferio Sur, Buenos Aires, Argentina, p 131

USDA, Agricultural Research Service, National Plant Germplasm System (2020) Germplasm Resources Information Network (GRIN-Taxonomy). National Germplasm Resources Laboratory, Beltsville, Maryland. URL: https://npgsweb.ars-grin.gov/gringlobal/taxonomydetail.aspx?id=417524. (Last access, January 2020)

Vivot EP, Sánchez C, Cacik F, Sequin C (2012) Actividad antibacteriana en plantas medicinales de la flora de Entre Ríos (Argentina). Ciencia, Docencia y Tecnología XXIII(45):165–185

Zuloaga FO, Morrone O (1999) Catálogo de las plantas vasculares de la República Argentina. II. Acanthaceae-Euphorbiaceae (Dicotyledoneae). Monogr Syst Bot Missouri Bot Gard 74:1–621

Ibicella lutea (Lindl.) van Eselt.

Álvaro Vázquez

Ibicella lutea plants growing on the roadside in Paysandú, Uruguay. (Photo by the author)

Abstract *Ibicella lutea* (Lindl.) van Eselt. is the main species of the genus *Ibicella*. It belongs to the small Martyniaceae family and it is used, along with less common species of *Ibicella* such as *I. parodii* and *I. nelsoniana*, in remedies of popular medicine. The phytochemical analysis of the plants showed the presence of several fatty acid glycosides with antibacterial activity against *Staphylococcus aureus,* as well as several strains of uropathogenic *Proteus mirabilis*. These facts help explain the widespread ethnopharmacological use of the plant.

Keywords *Ibicella lutea* · Martyniaceae · Antimicrobial · Devil's horn

Á. Vázquez (✉)
Laboratorio de Farmacognosia y Productos Naturales, Departamento de Química Orgánica, Facultad de Química, Universidad de la República, Montevideo, Uruguay
e-mail: avazquez@fq.edu.uy

Á. Máthé, A. Bandoni (eds.), *Medicinal and Aromatic Plants of South America Vol. 2*, Medicinal and Aromatic Plants of the World 7,
https://doi.org/10.1007/978-3-030-62818-5_21

1 Introduction

Ibicella. lutea is a 'quasi-carnivorous' plant native to South America (Brazil, Argentina, Uruguay and Paraguay) from where it spread to neighboring countries. From here, it was later introduced as an ornamental plant to North America and Europe, where it is known as yellow unicorn plant or devil's claw. It has a characteristic morphology, with its seed capsule covered with short spines, and a generally unpleasant smell. In South America, it is used as medicine, especially in the case of eye and skin infections. It seems strange that locals picked a species with its weird appearance, sticky exudate, spiny fruits and stinky smell as a potential remedy. However, its chemical and pharmacological characteristics seem to explain, at least partially, the properties attributed to it in folk medicine.

2 Taxonomic Characteristics

Ibicella lutea (Lindl.) van Eselt. (Martyniaceae Stampf.) (syn.: *Martynia lutea* Lindl., *Martynia montevidensis* Cham., *Proboscidea lutea* Stampf.) is the main species of the genus *Ibicella*: further species are *I. parodii* and *I. nelsoniana* (The Plant List 2019). It belongs to the small Martyniaceae family, comprising just five genera (*Martynia, Proboscidea, Craneolaria, Holoregmia* and *Ibicella*) with about 13–21 species according to different sources (van Eseltine 1929; Watson and Dallwitz 1992; Zapater and Novara 1995; The Plant List 2019). The family is native to the tropical and subtropical zones of America, where it has been used as food (*Proboscidea parviflora*) and as a medicinal plant (*Martynia. annua* and *M. montevidensis, Ibicella* spp.) (Lombardo 1983; Duke 1999).

The taxonomy of the family has been quite troublesome, and many early authors included these species into the Bignoniaceae family and later with the Pedaliaceae into a single family. Later studies including pollen analysis (Bretting and Nilsson 1988) and more recently, analysis using chloroplastic DNA, support the Martyniaceae family hypothesis, as a different family (Gutiérrez 2008).

3 Crude Drug Used

In Argentina and southern Brazil, but not in Uruguay, young fruits of *Craniolaria* and *Ibicella* species are eaten as sweetmeats (Abbiatti 1939) or as pickles (Bretting and Nilsson 1988). Medicinally, aerial parts are used as an infusion for eyewashes for conjunctivitis treatment, and sometimes the seeds are used in ointments for dermal infections (Lahitte et al. 2004) (Photo 1).

Photo 1 Left: Detail of the dry fruit showing the characteristic horns. Right; Typical aerial parts. (Photos by the author)

4 Major Chemical Constituents and Bioactive Compounds

Ibicella lutea, as well as the entire genus and family, has been relatively poorly studied. The major chemical components isolated and identified from the plants are fatty acids (Simirgiotis et al. 2003), dammarane triterpenes as betulatriterpene C-3-acetate, 24-*epi*-polacandrin 1,3-diacetate, 24-*epi*-polacandrin 3-acetate and 20,24-epoxy-1,3,12,25-tetrahydroxydammarane (Simirgiotis et al. 2003; Asai et al. 2010; Wallace and Vázquez 2011), a mixture of 2-*O*-(acetyloxyfattyacyl)-glycerols and hydroxy fatty acid glycosides (Cerdeiras et al. 2000; Asai et al. 2010). The last group is the most interesting one, as it seems that these are the main antimicrobial compounds.

5 Morphological Description

Ibicella lutea plants are annual, viscous-pubescent herbs with reddish roots and crawling or ascending stems 40–80 cm high, looking like a pumpkin. Opposite leaves, long petiolate, with orbicular or ovate leaf, with cordate base and crenate-dentate margin, 5–14 cm long and 7–16 cm wide. Yellow gamopetalous flowers are in dense clusters; 5-match calyx with two sepaloid basal bracteoles; yellow corolla 3–3.5 cm long, externally glandulous-pubescent. The fruit is a drupaceous capsule of 12–20 cm in length, including the two horns, with the oblong-ovoid body of 7–8 cm in length, initially green and fleshy, then dried and bristled with spines and provided with two long horns curved Seeds oblong, compressed, brown-blackish, rough-tuberculated, approximately 1 cm long (INTA 2019) (See Photo 1).

6 Geographical Distribution

The family Martyniaceae is distributed in the Americas, and it is found from the Southern half of the United States southward through Central America to Uruguay and Argentina.

Even though *Ibicella* spp. are native to the grasslands of Southern Paraguay and Brazil, Northern Argentina and Uruguay (Gutiérrez 2011), *Ibicella* has spread around the tropics and subtropics, posing a serious problem to livestock, who are hobbled or killed by the seed pods that, once trodden on, become embedded in their ankles. Nowadays it can be found as far as Australia, California, Florida, Mississippi and Georgia (USA), or even South Africa (Brooks 1992; Bryson et al. 2004).

7 Ecological Requirements

There are no systematic studies about the requirements of the plant regarding soil composition, or other factors. However, this species, as well as others of the same family, are typically found in disturbed habitats in arid and semi-arid areas such as grasslands, deserts, and thorn scrub, including in and along washes and creeks, along roadsides, in pastures and cultivated fields, and along beaches and dunes, at elevations ranging from sea level to medium altitudes (van Eseltine 1929; Bretting 1984). In our own observation, the plants grow quickly and abundantly in these conditions for a couple of years and then they are quickly substituted by other plants as Gramineae and Solanaceae until complete disappearance.

8 Traditional Use (Part(s) Used) and Common Knowledge

Ibicella lutea is a 'quasi-carnivorous' plant native to Brazil from where it spread to neighboring countries and was introduced as ornamental in North America and Europe, where it is known as yellow unicorn plant or devil's claw. Its uncommon feature is that its seed capsules are covered by short spines that emit its unpleasant smell.

The common names given to this species in Latin America make clear reference to the characteristics of the species` peculiar fruit: "cuerno del Diablo" (devil's horn), "garras del diablo" (dewil'sclaws), "pepino silvestre" (wild cucumber), "guampa del diablo" ("guampa": horns of the cattle). The plant was known by the original inhabitants of South America. For example, the Ranqueles indians called it "kalku müta" (sorcerer's horn) due the damage they cause to cattle (Steibel 1997). This continues to happen today, and the plant is specifically removed from cattle raising fields. In other countries, where the plant was introduced it is known as "bicorne", "cornard", "bockshorn", "gemsenhorn", "hornling", "ongles du diable"

(Bretting 1984). In Uruguay, where the plant grows freely it is called 'cuerno del diablo' and its infusion is used in popular medicine as antimicrobial for the treatment of eye and skin infections (Alonso Paz et al. 1995; Arrillaga de Maffei 1996; Lahitte et al. 2004).

9 Modern Medicine Based on Its Traditional Medicine Uses

Several trials of biological activity, particularly antibacterial, have been reported. The activity against *Staphylococcus aureus*, including several MRSA strains, was reported by a Uruguayan group. Both the chloroform and ethyl acetate extracts were active against the susceptible *S. aureus* strain (6538p) and the isolated compounds showed a remarkable MIC of 9μg/ml. The compound showed activity against the resistant strains 13369, 15027 and 19106. Strain 13369 is a hospital acquired isolate and strains 15027 and 19016 are community acquired infections isolates. The MIC's obtained with these strains were 250, 500 and 62.5 μg/ml, respectively (Alonso Paz et al. 1995; Cerdeiras et al. 2000; Wallace and Vázquez 2011). Also, the group of Martins Volcão et al. (2016) found a very interesting antimicrobial activity against *S. aureus* ATCC 12598 using crude extracts but they also found medium cytotoxic activity using macrophages as model.

Some insights into the structure-activity relation of the hydroxy fatty acid glycosides has also been reported. The methylation of the acid leads to complete loss of activity meanwhile the peracetylation as well as the loss of 6-*O*-Acetyl moiety produces a marked drop in its antibacterial activity, with MIC's of 1000, 3312 and 225 mg/ml, respectively (García Da Rosa et al. 2010).

Another research group studied the effect of *I. lutea* extracts on an *in vitro* uropathogenicity test for *P. mirabilis*. They observed that *I. lutea* extract had a significant effect on various attributes associated with *P. mirabilis* urovirulence, including bacterial growth rate, bacterial morphology, swarming differentiation, hemagglutination, and biofilm formation on glass and polystyrene; however, the extract did not exert a significant antibacterial effect (Sosa and Zunino 2009). Nevertheless, in a later *in vivo* test, the results demonstrated the effectiveness of the *I. lutea* extracts to prevent urinary tract infections in a dose effect manner (Sosa and Zunino 2010).

10 Conclusions

There are only few scientific papers dealing with Martyniaceae They deal mostly with taxonomic aspects. The number of studies on *Ibicella lutea* is even less. This species has been described as "quasi carnivorous" plant and it has all the relevant anatomical characteristics: despite of this, it has never been possible to verify the presence of exoenzymes in it (Rice 2008; ICPS 2019), therefore there is no actual proof of it. The studies on its antimicrobial activity, both *in vitro* and *in vivo*, are quite interesting, and could explain its ethnopharmacological use. Farther studies would, however, be needed to verify its activity against multiresistant *S. aureus*, including the mode of action of the 2-*O*-(acetyloxyfattyacyl)-glycerols and hydroxy fatty acid glycosides.

Another possible use of the species could be as a source of fatty acids for the production of biodiesel. The seeds of the plants have a good oil yield and all the species of the genus *Ibicella* can easily grow in semi-arid areas, thus offering an attractive alternative for industrial use (Houachri et al. 2018). From a taxonomical point of view, further genomic studies should be performed in the Martyniaceae family, in order to elucidate the species it contains, and determine whether all other reported species of the *Ibicella* genus actually exist (Bretting and Nilsson 1988; Gutiérrez 2008, 2011; Gormley et al. 2015).

Acknowledgments The author gratefully acknowledges Lic. J.I. Vázquez for his help in editing the document and its grammatical and spelling correction.

References

Alonso Paz E, Cerdeiras MP, Fernández J, Ferreira F, Moyna P, Soubes M, Vázquez A, Vero S, Zunino L (1995) Screening of Uruguayan medicinal plants for antimicrobial activity. J Ethnopharmacol 45(1):67–70

Arrillaga de Maffei B (1996) Plantas usadas en la medicina natural. Ed. Hemisferio Sur, Montevideo, Uruguay, 156 pp

Asai T, Hara N, Fujimoto Y (2010) Fatty acid derivatives and dammarane triterpenes from the glandular trichome exudates of *Ibicella lutea* and *Proboscidea louisiana*. Phytochemistry 71(8–9):877–894

Bretting PK (1984) Folk names and uses for Martyniaceous plants. Econ Bot 38(4):452–463

Bretting PK, Nilsson S (1988) Pollen morphology of the Martyniaceae and its systematic implications. Syst Bot 13:51–59

Brooks AK (1992) Flora of NSW 3 published in Internet: http://plantnet.rbgsyd.nsw.gov.au/cgi-bin/NSWfl.pl?page=nswfl&lvl=sp&name=Ibicella~lutea. (Last access, March 2020)

Bryson C, Koger CH, Byrd D (2004) Biology and ecology of yellow unicorn-plant (*Ibicella lutea*) Published in Internet: https://www.ars.usda.gov/ARSUserFiles/64022000/Posters/2004/2004_SWSS_CTB.pdf, https://tropicBrooksal.theferns.info/viewtropical.php?id=Ibicella+lutea. (Last access, March 2020)

Cerdeiras MP, Fernández J, Soubes M, Vero S, Ferreira F, Moyna P, Olano I, Vázquez A (2000) A new antibacterial compound from *Ibicella lutea*. J Ethnopharmacol 73(3):521–525

García Da Rosa ME et al (2010) Antibacterial activity of *Ibicella lutea* glycosides. Rev Latinoam Quím 38(2):98–102

Gormley I, Bedigian D, Olmstead R (2015) Phylogeny of Pedaliaceae and Martyniaceae and the placement of Trapella in Plantaginaceae *s.l.* Syst Bot 40(1):259–268

Gutiérrez R (2008) Preliminary chloroplast DNA studies in the flowering plant family Martyniaceae (Order Lamiales). J Ariz Nev Acad Sci 40(1):105–110

Gutiérrez R (2011) A phylogenetic study of the plant family Martyniaceae (Order Lamiales). Thesis. Arizona State University, USA, 256 pp

Houachri T, Bolonio D, Llamas A, Rodríguez-Fernández J, El Gazza M, Mittelbach M, Lapuerta M, Canoira L (2018) Fatty acid methyl esters (FAME) from oleaginous seeds grown in arid lands. Part II: *Ibicella lutea*, *Onopordum nervosum*, *Peganum harmala*, *Smyrnium olusatrum* and *Solanum elaeagnifolium*. Energy Source Part A 40(12):1–8

ICPS (International Carnivorous Plants Society) (2019) Published in Internet https://www.carnivorousplants.org/grow/guides/DevilsClaw. (Last access, November 2019)

INTA (2019) Published in Internet. http://rian.inta.gov.ar/atlasmalezas/atlasmalezasportal/DetalleMaleza.aspx?pagante=CXF&idmal.za=22904. (Last access, November 2019)

Lahitte HB, Hurrell JA, Belgrano M, Jankowski LS, Haloua P, Mehltreter K (2004) Plantas medicinales rioplantenses. Ed. L.O.L.A, Buenos Aires, Argentina, 240 pp

Martins Volcão L, Silveira CT, Vázquez A, Cerdeiras MP, da Silva Almeida P, da Silva FMR Jr, Fernandes Ramos D (2016) Avaliação de *Ibicella lutea* como agente antimicrobiano frente à *Staphylococcus aureus*. R Epidemiol Control Infec Santa Cruz do Sul 6(4):191–196

Rice B (2008) Reassessing commensal-enabled Carnivory in *Proboscidea* and *Ibicella*? ICPS (International Carnivorous Plants Society). Carniv Pl Newslett 37(1):15–19

Simirgiotis MJ, Jiménez C, Rodríguez J, Giordano OS, Tonn CE (2003) New Dammarane triterpenes from the aerial parts of *Ibicella lutea* grown in Argentina. J Nat Prod 66(12):1586–1592

Sosa V, Zunino P (2009) Effect of *Ibicella lutea* on uropathogenic *Proteus mirabilis* growth, virulence, and biofilm formation. J Infect Dev Ctries 3(10):762–770

Sosa V, Zunino P (2010) Assessment of effectiveness and safety of *Ibicella lutea* extract in the control of experimental *Proteus mirabilis* urinary tract infection. J Infect Dev Ctries 4(12):814–821

Steibel PE (1997) Nombres y usos de las plantas aplicados por los Indios Ranqueles de La Pampa (Argentina). Rev Fac Agron – ULPampa, Argentina 9(2):1–40

The Plant List (2019) Version 1.1 published on the Internet: http://www.theplantlist.org. (Last access, November 2019)

van Eseltine GP (1929) A preliminary study of the unicorn plants (Martyniaceae), Technical bulletin 149. NYS Agricultural Exp. Station, Geneva, p 41

Wallace F, Vázquez A (2011) Anti-*Staphylococcus* activity of *Ibicella lutea* compounds. Lat Am J Pharm 30(5):1025–1027

Watson L, Dallwitz MJ (1992 onwards) The families of flowering plants: descriptions, illustrations, identification, and information retrieval. Version: 16th December 2019. Published in Internet: https://www.delta-intkey.com. http://www.theplantlist.org. (Last access, November 2019)

Zapater MA, Novara L (1995) Martyniaceae. Aportes Bot Salta Ser Flora 3(12):1–9

Ilex paraguariensis St. Hil.

Nelson Bracesco, Manuel Minteguiaga, Álvaro Vázquez, Verónica Cesio, Horacio Heinzen, Fernando Ferreira, and Eduardo Dellacassa

Ilex paraguariensis. (Photos: Flora nativa de Uruguay. Andrés González. http://floranativadeuruguay.blogspot.com/search/label/Aquifoliaceae)

N. Bracesco
Laboratorio de Radiobiología, Departamento de Biofísica, Facultad de Medicina, Universidad de la República, Montevideo, Uruguay

M. Minteguiaga
Espacio de Ciencia y Tecnología Química. Centro Universitario Regional Noreste/Sede Tacuarembó, Universidad de la República, Montevideo, Uruguay

Laboratorio de Biotecnología de Aromas, Departamento de Química Orgánica, Facultad de Química, Universidad de la República, Montevideo, Uruguay

Á. Vázquez · V. Cesio · H. Heinzen
Farmacognosia, Departamento de Química Orgánica, Facultad de Química, Universidad de la República, Montevideo, Uruguay

F. Ferreira
Espacio de Ciencia y Tecnología Química. Centro Universitario Regional Noreste, Sede Tacuarembó, Universidad de la República, Montevideo, Uruguay

Laboratorio de Carbohidratos y Glicoconjugados. Departamento de Química Orgánica. Facultad de Química, Universidad de la República, Montevideo, Uruguay

E. Dellacassa (✉)
Laboratorio de Biotecnología de Aromas, Departamento de Química Orgánica, Facultad de Química, Universidad de la República, Montevideo, Uruguay
e-mail: edellac@fq.edu.uy

Á. Máthé, A. Bandoni (eds.), *Medicinal and Aromatic Plants of South America Vol. 2*, Medicinal and Aromatic Plants of the World 7,
https://doi.org/10.1007/978-3-030-62818-5_22

Abstract *Ilex paraguariensis* St. Hil., from the family of holy plants, Aquifoliaceae, is a native South American tree that can reach 18 m in height and is used to produce "yerba mate". It is found primarily in the southern regions of South America, namely, Brazil (Mato Grosso do Sul, Minas Gerais, Parana, Rio Grande do Sul, Rio de Janeiro, Santa Catarina, Sao Paulo), Argentina (Corrientes, Misiones), Paraguay (Alto Paraná, Amambay, Caaguazú, Canendiyú, Central, Guaira, Itapuá, Misiones, San Pedro), and Uruguay. "Yerba mate" tea, an infusion made from the leaves of this tree, is a widely consumed nonalcoholic beverage in South America which is gaining rapid introduction into the world market, either as tea itself or as ingredient in formulated foods or dietary supplements. The indigenous people have used it for centuries as a social and medicinal beverage. The container in which it is generally served is a gourd made from the dry and hollowed fruit of *Lagenaria vulgaris* Ser. (Cucurbitaceae), whose dimensions vary according the region and uses. It is called "mate" in the region. "Mate" has a very important social role. The act of offering and sharing it has connotations like those of the tea ceremony for some Oriental cultures.

Keywords Yerba mate · Aquifoliaceae · Mate · Chimarrão · Tereré

1 Introduction

Ilex paraguariensis, from the family of holy plants, Aquifoliaceae, is a native South American tree that can reach 18 m in height. It is used to produce "yerba mate". It is found primarily in the Southern regions of South America, namely, Brazil (Mato Grosso do Sul, Minas Gerais, Parana, Rio Grande do Sul, Rio de Janeiro, Santa Catarina, Sao Paulo), Argentina (Corrientes, Misiones), Paraguay (Alto Paraná, Amambay, Caaguazú, Canendiyú, Central, Guaira, Itapuá, Misiones, San Pedro), and Uruguay (Giberti 2001; Small and Catling 2001; Cansian et al. 2008; Filip et al. 2009).

Also popularly known as "mate", "yerba mate", "hierba mate", "maté", "erva mate", "chimarrão", and "tereré", *I. paraguariensis*, has been popular for centuries, and was adopted from the inhabitants of the region (Guaraníes) who prepared it as a native stimulant beverage or used as medicine (Assunçao 1967). "Mate" leaves have socioeconomic and therapeutic importance in complementary and alternative medicine as CNS stimulant, diuretic, weight reducing, antioxidant and antihipercholesterolemic among other properties, in large regions of South America. It should be noted that there are still many unanswered questions regarding a serious evaluation of the dose/effect relationship of active principles of *I. paraguariensis*. In this chapter, a revision of recent findings is presented about the chemical composition of "yerba mate" and the related biological and toxicological activities is presented. Our aim is to explore and understand the increasing interest in this plant drug used from ancestral origins.

2 Taxonomic Characteristics

Ilex paraguariensis St. Hil. is a dioecious evergreen tree that grows up to 18 m in height. The leaves are alternate, coriaceous and obovate with a serrate margin and obtuse apex. The inflorescences are in corymboid fascicles, the male ones in a dichasium with three to 11 flowers, the female ones with one or three flowers. The flowers are small, and simple, number four or five and have a whitish corolla. The fruit is in a nucule; there are four or five single seed pyrenes (propagules). Giberti (1994) considered that according to morphological characteristics, two varieties of *I. paraguariensis* can be identified: *I. paraguariensis* St. Hil. var. *paraguariensis* (the cultivated species which is in practice completely glabrous) and *I. paraguariensis* var. *vestita* (Reisseck) Loes (densely pubescent and not used industrially). Both species coexist in some regions of Northeastern Argentina and Brazil.

"Yerba mate" flowers in the spring (October to November) in this Hemisphere, it has entomophilous pollination (diptera, hymenoptera) and fruits from March to June; dissemination is endozoic (birds). There is a rudimentary embryo in many externally ripe seeds which causes a long period of germination from the time of sowing (Zuloaga et al. 2008).

This species has several common names: "yerba mate", "hierba mate", "mate", "maté", "té del Paraguay", "té de San Bartolomé", "té de los Jesuitas", "ka'há", "erva mate", "chimarrão", "tereré". All of them were adopted from the native inhabitants of the region (Guaraníes) for the preparation of stimulant beverages or for its medicinal properties.

Synonyms *Ilex mate* A. St.-Hil., nom. superfl., *Ilex curitibensis* Miers, *Ilex domestica* Reissek var. *glabra*, *Ilex sorbilis* Reissek, *Ilex theaezans* Bonpl. ex Miers, *Ilex paraguariensis* A. St.-Hil. f. *latifolia*, *Ilex paraguariensis* A. St.-Hil. f. *parvifolia*, *Ilex paraguayensis* A. St.-Hil. var. *dentata*, *Ilex paraguayensis* A. St.-Hil. var. *idonea*, *Ilex curitibensis* Miers var. *gardneriana*, *Ilex paraguariensis* A. St.-Hil. f. *domestica*, *Ilex paraguariensis* A. St.-Hil. f. *sorbilis*, *Ilex paraguariensis* A. St.-Hil. f. *confusa*, *Ilex paraguariensis* A. St.-Hil. f. *dasyprionata*, *Ilex paraguariensis* A. St.-Hil. var. *ulei*, *Ilex paraguariensis* A. St.-Hil. f. *glabra*, *Ilex paraguayensis* A. St.-Hil. var. *usitata*, *Ilex paraguariensis* A. St.-Hil. var. *genuina*, (Zuloaga et al. 2008).

3 Crude Drug Used

The "yerba mate" crude drug (Photo 1) consists of the processed, dried and fragmented young leaves and stems of *Ilex paraguariensis* A. St. Hil. var. *paraguariensis* (Aquifoliaceae). It must contain no less than 0.8% caffeine, calculated on the dry substance, and no more than 10.0% stems (Farmacopea Argentina 2013; Código Alimentario Argentino 2020). Moreover, at least four species have been identified as frequent adulterants in the commercial "yerba mate" packages, namely: *I. affinis*

Photo 1 Dried crude drug of *Ilex paraguariensis*

Gardner, *I. amara* (Vell. Conc.) Loes., *I. dumosa* Reisseck, and *I. theezans* C. Martius ex-Reisseck (Giberti 1989). Moreover, at least 22 species have been described as occasional adulterants (Giberti 1989; Spegazzini et al. 2002). The problem of "yerba mate" authenticity could be circumscribed even out of Aquifoliaceae family, because several other plant families and their species has been reported as potential adulterants (Anacardiaceae, Celastraceae, Lauraceae, Myrtaceae, and Rosaceae, among others) (Giberti 1989).

4 Major Chemical Constituents and Bioactive Compounds

The quality and physicochemical characteristics of *Ilex paraguariensis* leaves show variations according to the development and plant age, time of harvest handling of plants (native or reforested), growing region, processing and storage. The industrial processing of *I. paraguariensis* involves several stages, which are not necessarily comparable for different regions, and lead to physicochemical and sensory changes that can be summarized as follows. The harvest of the leaves, rapid heating of the green leaves at 250–300 °C for about 2 min ("zapecado"), this process preserves the green color of the leaves, avoiding the blackening which would occur due to the action of oxidizing enzymes and fermentation (approximately 25% moisture is lost in this stage). Hot air drying at 110 °C until 3% of moisture is reached. According to the time and temperature used this process can be fast "secanza rápida" (between 15 and 60 min), or longer, consuming 12–24 h ("secado a barbacuá"). The leaves are then subjected to a coarse grinding process ("canchado"), packed in sack-cloth bags and stored for approximately 1 year in order to allow the aging. This time is necessary for the product to acquire the flavor, aroma and color required by consumers (Barchuk 1998). Finally, the "yerba" is mixed, blended (according to special flavors of different batches) and packed.

When speaking of the chemical components of "yerba mate", the characteristic of the living plant with dried material, the processed material industrially and infusions or how the preparations are consumed, are often combined. While there is no significant qualitative difference in the composition of components, significant variations can exist according to their proportions in the plant. In this sense, the volatile portion ranges from fresh material to processed material. This affects both the taste and aroma in the final product.

The work of Filip et al. (2009) summarize in detail the advances reported in the research of chemical (volatile and non-volatile) composition of *I. paraguariensis* at the publication date. However, the chemical advances in the last 20 years are impressive, indicating the extreme interest on this particular medicinal plant and the search for chemical explanations to the diversity of activities adjudicated to it.

In 2012, Burris et al. performed a retrospective literature review on the composition and bioactive properties of "yerba mate". Information more accessible since then, and the main chemical research data published are summarized by type of component studied.

The main activities attributed to *I. paraguariensis* are related to the antioxidant activity, justifying the number of reports on polyphenols, specially caffeoyl-derivatives and flavonoids (Dugo et al. 2009; Frizon et al. 2015; Zwyrzykowska et al. 2015; Butiuk et al. 2016; da Silveira et al. 2016a, b, 2017a, b; Lima et al. 2016; Matei et al. 2016; Gómez-Juaristi et al. 2018).

I. paraguariensis also contains saponins that are known to bind bile salts and are reported to have antihypercholesterolemic activity, suggesting the potential usefulness of "mate" as a nutraceutical food with anti obese effects and a source of compounds which could be used as preventive of cardiovascular diseases. Consequently, an important number of reports is available in the literature for saponins in *I. paraguariensis* (Ribeiro et al. 2013; da Silva et al. 2018; Petroselli et al. 2019).

Traditionally, xanthine alkaloids (caffeine, theobromine, and theophylline) have been considered as the primary active chemical constituents of "yerba mate" and, surprisingly, the interest on these components continue growing as demonstrated by the data published in the last years (Przygodda et al. 2010; Mazur et al. 2019; Yin et al. 2015; Oellig et al. 2018; Meinhart et al. 2019; Negrin et al. 2019). In addition, being "yerba mate" leaves a rich source of minerals many studies have been dedicated to study their mineral profiles looking both for their nutritional and toxicological importance (Barbosa et al. 2015; Łukomska et al. 2015; Marcelo et al. 2015; Pozebon et al. 2015; Rossa et al. 2015; Pereira et al. 2016; Rusinek-Prystupa et al. 2016; Kahmann et al. 2017).

Reports on the chemical composition of "yerba mate" do not always distinguish between fresh plant, dried material, industrially processed product and the different beverages prepared with it. Although there are no significant qualitative differences in the composition, there is a very important quantitative variation which produced as a result a large number of generic publications about the phytochemical profile of "yerba mate" (Xu et al. 2010; Blum-Silva et al. 2015; Marcelo et al. 2015; Pinto et al. 2015; Souza et al. 2017; Holowaty et al. 2014).

"Mate" consumption is set in different areas, uses different modalities, reaches different socio-economic groups and is sought for pleasure or for resolving a food need. Obviously, consumers have in common their sensory acceptance of the product and, in general, their loyalty to a commercial brand. Many works have focused on the volatile composition of *I. paraguariensis*, and the dynamic of chemical transformation as consequence of the different steps involved in the "yerba mate" production (Márquez et al. 2013; Holowaty et al. 2014; Polidoro et al. 2016).

5 Morphological Description

Ilex paraguariensis St. Hil. (Aquifoliaceae) is a dioecious evergreen tree, growing to a height up to 8–15 m. The 8 cm long olive-green leaves are perennial, alternate, coriaceous, obovate with slightly crenate dentate margins and obtuse apex, and have a wedge-shaped base. The petioles are up to 15 mm long. The flowering stage occurs during spring season, producing small, unisexual flowers which have 4 white petals. In some tropical or subtropical species, the number of petals may be 5, 6 or 7 (Giberti 2001). These may be clustered in groups of 1–15 flowers that appear in the axil of the leaves. The fruits are red colored berries with 4–5 seeds. This tree grows preferably in low humid regions, as part of the medium stratus of woods. Its propagation is by seeds (Giberti 2001).

6 Geographical Distribution

The genus *Ilex* comprises more than 500 species of dioecious trees and shrubs distributed throughout temperate and tropical regions of the world (Galle 1997). It is one of the main genera of the Aquifoliaceae family together with the monospecific genus *Nemopanthus* of Eastern North America (Noud et al. 2000). Many species of *Ilex* are ornamental (holly plant). However, *I. paraguariensis* has been used since pre-Columbian times and later grown as a domesticated crop in the Northeastern region of Argentina and South-east of Brazil where its leaves are processed to produce "yerba mate". Several wild species of *Ilex* are sympatric with genuine "yerba mate" and have been, or are, used to manufacture the product although, up to the present and according to the legislation in force, they are to be considered adulterations.

I. paraguariensis naturally grows in South America in the area comprised by the Atlantic Ocean to the East and the Paraguay River to the West, between 18 and 30° S. It is found in Northeastern Argentina, Southern Brazil and Eastern Paraguay, where it is currently cultivated. In Argentina, the agro-ecological conditions necessary for its cultivation are found only in two provinces: Misiones and Northeast of Corrientes. Uruguay possesses only wild populations of this species and there is also no mention of commercial cultivation.

7 Ecological Requirements

Prominent among the ecological requirements of this subtropical species are climatic conditions, especially annual precipitation means and an even distribution of rainfall throughout the year. This must not be less than 1200 mm annually and, during the driest quarter (which in the region is winter) the minimum must be 250 mm. *I. paraguariensis'* wild distribution area is always unaffected by water shortages. The mean annual temperature of the area is approximately 21–22 °C. The absolute minimum temperature that this species can tolerate is −6 °C, even though winter snows are frequent on the plateaus and mountain regions to the South of Brazil and East of Misiones (Argentina). In addition, the plants require lateritic, acid (pH between 5.8 and 6.8) soils that are of medium to fine texture (Giberti 1994, 2001).

8 Traditional Use (Part(s) Used) and Common Knowledge

"Yerba mate" is massively produced and consumed as an infusion in South America ("chimarrão" in Brazil, "mate" in Argentina, Uruguay and "tereré" when it is drunk at room temperature, in Paraguay) and spreading all over the world. This product is obtained from dried leaves of *Ilex paraguariensis*, mixed with fragments of dried branches (sticks). For its commercialization, "yerba mate" must have a minimum percentage of leaves because its presence determines "yerba mate" quality and price. However, "yerba mate" can be mixed at different proportions with medicinal and aromatic plants, and so being considered and regulated its consumption as an herbal specialty (Ministerio de Salud Pública 2020; Código Alimentario Argentino 2020).

Ilex paraguariensis extracts have been used as a popular folk beverage from pre-Columbian times and in complementary and alternative medicine as a CNS stimulant (Simões et al. 1986; Alonso et al. 1992; González Torres 1992), diuretic, weight reducing, antioxidant, antihypercholesterolemic among other properties, in large regions of South America. This beverage has also gained popularity in the United States, Europe and Asia during the last years. "Yerba mate" has a very important social role and the act of offering it and sharing has connotations like those of the tea ceremony for some Oriental cultures.

Argentina is the first exporter of "mate" and Uruguay is the country that has the highest per capita consumption: 10 kg/person/year, Argentina follows with 7 kg/person/year. In Brazil, "mate" in consumed only in 3 states (almost 1 kg/person/year) (Montevideo Portal 2019). Its popularity is also increasing outside South America due to its pharmacological properties, proven to be beneficial to health.

9 Modern Medicine Based on Its Traditional Medicine Uses

The primary active chemical constituents of "yerba mate" comprise xanthine alkaloids (caffeine, theobromine, and theophylline), saponins, and chlorogenic acid. Sterols are also present in "yerba mate", and novel saponins have been discovered in the leaves (matesaponins). Saponins are plant chemicals with known pharmacological activities, including, as recent research shows, stimulating the immune system. In addition, "yerba mate" leaves are source of vitamins, minerals, and amino acids (Filip et al. 2009).

In the last decade, research on extracts and isolated compounds from "yerba mate" has provided several pharmacological applications which can be summarized as shown in Fig. 1. *Ilex paraguariensis* has been revealed as an antioxidant, diuretic, hypocholesterolemic, and obesity preventive agent.

For practical purposes, and considering the large amount of information published, we will summarize the biological and pharmacological findings for *I. paraguariensis* published in the scientific literature, in the last 5 years.

Several studies have demonstrated the "yerba mate" lipid-lowering properties due to the presence of polyphenols and saponins (Kim et al. 2015; Gamboa-Gómez et al. 2015; Gambero and Ribeiro 2015; Messina et al. 2015; de Resende et al. 2015; Dunlop 2016; Yimam et al. 2016, 2017; Choi et al. 2017; Souza et al. 2017; Santiago et al. 2017; Conceição et al. 2017; de Freitas Junior and de Almeida Jr 2017; Niraula et al. 2018; dos Santos et al. 2018; León Bianchi et al. 2018; Chaves et al. 2018; Greydanus et al. 2018; de Oliveira et al. 2018; Avena Álvarez et al. 2019; Stuby et al. 2019; Izzo 2019; Mazur et al. 2019; Panza et al. 2019; Balsan et al. 2019; Uecker et al. 2019; Chianese et al. 2019; Barroso et al. 2019; Zapata et al. 2020).

Ilex paraguariensis has been also revealed as a potential antioxidant through the use of both entire extracts or certain components in particular, mainly phenolic in

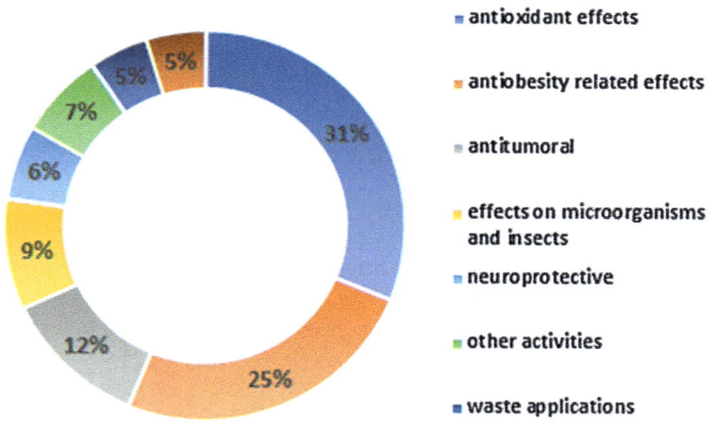

Fig. 1 Pharmacological applications reported in the literature during the 2010–2020 period

nature (de Oliveira et al. 2015; Boaventura et al. 2015; Boado et al. 2015; Luño et al. 2015; Colpo et al. 2016; Portela et al. 2016; Oh et al. 2016; Petrilli et al. 2016; Panza et al. 2016; Piovezan-Borges et al. 2016; Valduga et al. 2016; Santos et al. 2017; Pereira et al. 2017a, b; de Lima et al. 2017; Gerolis et al. 2017; Jang et al. 2018; Santos et al. 2018; Bortoli et al. 2018; Rodríguez-Arzuaga and Piagentini 2018; Baeza et al. 2018; Vieitez et al. 2018; Lopes Machado et al. 2019; Jongberg et al. 2019; Becker et al. 2019; Gremski et al. 2019; Zhou et al. 2019; Correa et al. 2019; Wang et al. 2019).

"Yerba mate" consumption has been associated to high incidence rates of cancers of the upper aerodigestive tract (oral cavity, oropharynx, hypopharynx, larynx, esophagus) which are especially frequent in some parts of Latin America (Lima et al. 2014; De Carvalho et al. 2016; Ronco et al. 2016a, b, 2017; Mello et al. 2018).

Paradoxically, many studies report antitumor effects of "yerba mate" extracts so establishing the need of continuing looking for answers through research activities (Boaventura et al. 2015; Akram and Nawaz 2017; Bracesco et al. 2018; Cuelho et al. 2018; Cittadini et al. 2019a, b; Lodise et al. 2019).

10 Conclusions

"Yerba mate" is primarily consumed as a beverage made by steeping the leaves of the *Ilex paraguariensis* plant in hot water. The growing interest in "mate" products has made it paramount that research on this herbal tea continues, as it has shown extraordinary possibilities not only as a consumer beverage but also in the nutraceutical and pharmaceutical industry. Research should include human-based clinical studies to support the properties verified by the yet – relatively scarce – *in vitro* and *in vivo* models obtained with animals.

References

Akram M, Nawaz A (2017) Effects of medicinal plants on Alzheimer's disease and memory deficits. Neural Regen Res 12(4):660–670

Alonso E, Bassagoda MJ, Ferreira F (1992) Yuyos. Uso Racional de las Plantas Medicinales. Ed. Fin de Siglo, Montevideo, Uruguay, 157 pp

Assunçao FO (1967) El mate. Ed. Arca, Montevideo, Uruguay, 70 pp

Avena Álvarez MV, Messina DN, Corte C, Mussi Stoizik JA, Sáez A, Boarelli P, Pérez Elizalde R (2019) Association between consumption of yerba mate and lipid profile in overweight women. Nutr Hosp 36(6):1300–1306

Baeza G, Sarriá B, Bravo L, Mateos R (2018) Polyphenol content *in vitro* bioaccessibility and antioxidant capacity of widely consumed beverages. J Sci Food Agric 98(4):1397–1406

Balsan G, Pellanda LC, Sausen G, Galarraga T, Zaffari D, Pontin B, Portal VL (2019) Effect of yerba mate and green tea on paraoxonase and leptin levels in patients affected by overweight or obesity and dyslipidemia: a randomized clinical trial. Nutr J 18(5):1–10. https://doi.org/10.1186/s12937-018-0426-y

Barbosa JZ, Zambon LM, Motta ACV, Wendling I (2015) Composition hot-water solubility of elements and nutritional value of fruits and leaves of yerba mate. Cienc Agrotecnol 39(6):593–603

Barchuk RD (1998) Aporte Nutricional de la Yerba Mate. Thesis. Universidad Nacional del Nordeste. Corrientes, Argentina.

Barroso MV, Graça-Reis A, Cattani-Cavalieri I, Gitirana LB, Valenca SS, Lanzetti M (2019) Mate tea reduces high fat diet-induced liver and metabolic disorders in mice. Biomed Pharmacother 109:1547–1555

Becker AM, Cunha HP, Lindenberg AC, de Andrade F, de Carvalho T, Boaventura BCB, da Silva EL (2019) Spray-dried yerba mate extract capsules: clinical evaluation and antioxidant potential in healthy individuals. Plant Food Hum Nutr 74(4):495–500

Blum-Silva CH, Chaves VC, Schenkel EP, Coelho GC, Reginatto FH (2015) The influence of leaf age on methylxanthines total phenolic content and free radical scavenging capacity of *Ilex paraguariensis* aqueous extracts. Braz J Pharmacogn 25(1):1–6

Boado LS, Fretes RM, Brumovsky LA (2015) Bioavailability and antioxidant effect of the *Ilex paraguariensis* polyphenols. Nutr Food Sci 45(2):326–335

Boaventura BCB, Amboni RDDMC, da Silva EL, Prudencio ES, Di Pietro PF, Malta LG, Polinati RM, Liu RH (2015) Effect of *in vitro* digestion of yerba mate (*Ilex paraguariensis* A. St. Hil.) extract on the cellular antioxidant activity antiproliferative activity and cytotoxicity toward HepG2 cells. Food Res Int 77:257–263

Bortoli PM, Alves C, Costa E, Vanin AP, Sofiatti JR, Siqueira DP, Dallago RM, Treichel H, Vargas GDLP, Kaizer RR (2018) *Ilex paraguariensis*: potential antioxidant on aluminum toxicity in an experimental model of Alzheimer's disease. J Inorg Biochem 181:104–110

Bracesco N, Sosa V, Blanc L, Contreras V, Candreva EC, Salvo VA, Hocart S, Mechoso B, Nunes E (2018) Analysis of radioprotection and antimutagenic effects of *Ilex paraguariensis* infusion and its component rutin. Braz J Med Biol Res 51(9):e7404. https://doi.org/10.1590/1414-431X20187404

Burris KP, Harte FM, Michael Davidson P, Stewart CN Jr, Zivanovic S (2012) Composition and bioactive properties of yerba mate (*Ilex paraguariensis* A St-Hil): a review. Chile J Agric Res 72(2):268–274

Butiuk AP, Martos MA, Adachi O, Hours RA (2016) Study of the chlorogenic acid content in yerba mate (*Ilex paraguariensis* St Hil): effect of plant fraction processing step and harvesting season. J Appl Res Med Arom Plants 3(1):27–33

Cansian RL, Mossi AJ, Mosele SH, Toniazzo G, Treichel H, Paroul N, Oliveira JV, Oliveira D, Mazutti M, Echeverrigaray S (2008) Genetic conservation and medicinal properties of mate (*Ilex paraguariensis* St Hil.). Pharmacogn Rev 2:326–338

Chaves G, Britez N, Oviedo G, González G, Italiano C, Blanes M, Sandoval G, Mereles D (2018) Heavy drinkers of *Ilex paraguariensis* beverages show lower lipid profiles but higher body weight. Phytother Res 32(6):1030–1038

Chianese G, Golin-Pacheco SD, Taglialatela-Scafati O, Collado JA, Muñoz E, Appendino G, Pollastro F (2019) Bioactive triterpenoids from the caffeine-rich plants guayusa and mate. Food Res Int 115:504–510

Choi M-S, Park HJ, Kim SR, Kim DY, Jung UJ (2017) Long-term dietary supplementation with yerba mate ameliorates diet-induced obesity and metabolic disorders in mice by regulating energy expenditure and lipid metabolism. J Med Food 20(12):1168–1175

Cittadini MC, Repossi G, Albrecht C, Di Paola Naranjo R, Miranda AR, de Pascual-Teresa S, Soria EA (2019a) Effects of bioavailable phenolic compounds from *Ilex paraguariensis* on the brain of mice with lung adenocarcinoma. Phytother Res 33(4):1142–1149

Cittadini MC, Albrecht C, Miranda AR, Mazzuduli GM, Soria EA, Repossi G (2019b) Neuroprotective effect of *Ilex paraguariensis* intake on brain myelin of lung adenocarcinoma-bearing male BALB/c mice. Nutr Cancer 71(4):629–633

Código Alimentario Argentino. https://www.argentina.gob.ar/anmat/codigoalimentario Art 1193. (Last access, January 2020)

Colpo AC, Rosa H, Lima ME, Pazzini CEF, De Camargo VB, Bassante FEM, Puntel R, Ávila DS, Méndez A, Folmer V (2016) Yerba mate (*Ilex paraguariensis* St Hill)-based beverages: how successive extraction influences the extract composition and its capacity to chelate iron and scavenge free radicals. Food Chem 209:185–195

Conceição EPS, Kaezer AR, Peixoto-Silva N, Felzenszwalb I, De Oliveira E, Moura EG, Lisboa PC (2017) Effects of *Ilex paraguariensis* (yerba mate) on the hypothalamic signalling of insulin and leptin and liver dysfunction in adult rats overfed during lactation. J Dev Orig Health Dis 8(1):123–132

Correa VG, De Sá-Nakanishi AB, Gonçalves GDA, Barros L, Ferreira ICFR, Bracht A, Peralta RM (2019) Yerba mate aqueous extract improves the oxidative and inflammatory states of rats with adjuvant-induced arthritis. Food Funct 10(9):5682–5696

Cuelho CHF, Alves GDAD, Lovatto MO, Bonilha IF, Barbisan F, da Cruz IBM, Oliveira SM, Fachinetto R, do Canto GS, Manfron MP (2018) Topical formulation containing *Ilex paraguariensis* extract increases metalloproteinases and myeloperoxidase activities in mice exposed to UVB radiation. J Photochem Photobiol B 189:95–103

da Silva CGF, Cassel E, Vargas RMF, Marczak LDF (2018) Selective extraction of saponins from *Ilex paraguariensis* St-Hil unripe fruits. Food Bioprod Process 111:72–82

da Silveira TFF, Meinhart AD, De Souza TCL, Teixeira Filho J, Godoy HT (2016a) Phenolic compounds from yerba mate-based beverages - a multivariate optimization. Food Chem 190:1159–1167

da Silveira TFF, Meinhart AD, Coutinho JP, de Souza TCL, Cunha ECE, de Moraes MR, Godoy HT (2016b) Content of lutein in aqueous extracts of yerba mate (*Ilex paraguariensis* St Hil). Food Res Int 82:165–171

da Silveira TFF, Meinhart AD, de Souza TCL, Cunha ECE, de Moraes MR, Godoy HT (2017a) Chlorogenic acids and flavonoid extraction during the preparation of yerba mate based beverages. Food Res Int 102:348–354

da Silveira TFF, Meinhart AD, de Souza TCL, Cunha ECE, de Moraes MR, Filho JT, Godoy HT (2017b) Optimization of the preparation conditions of yerba mate tea beverage to maximize chlorogenic acids extraction. Plant Food Hum Nutr 72(2):219–223

de Carvalho EF, De Oliveira SK, Nardi VK, Gelinski TC, Bortoluzzi MC, Maraschin M, Nardi GM (2016) *Ilex paraguariensis* promotes orofacial pain relief after formalin injection: involvement of noradrenergic pathway. Pharm Res 8:S31–S37

de Freitas Junior LM, de Almeida EB Jr (2017) Medicinal plants for the treatment of obesity: ethnopharmacological approach and chemical and biological studies. Am J Transl Res 9(5):2050–2064

de Lima ME, Colpo AC, Maya-López M, Rosa H, Túnez I, Galván-Arzate S, Santamaría A, Folmer V (2017) Protective effect of yerba mate (*Ilex paraguariensis* St Hill) against oxidative damage *in vitro* in rat brain synaptosomal/mitochondrial P2 fractions. J Funct Foods 34:447–452

de Oliveira CC, De Araújo Calado VM, Ares G, Granato D (2015) Statistical approaches to assess the association between phenolic compounds and the in vitro antioxidant activity of *Camellia sinensis* and *Ilex paraguariensis* teas. Crit Rev Food Sci Nutr 55(10):1456–1473

de Oliveira E, Lima NS, Conceição EPS, Peixoto-Silva N, Moura EG, Lisboa PC (2018) Treatment with *Ilex paraguariensis* (yerba mate) aqueous solution prevents hepatic redox imbalance elevated triglycerides and microsteatosis in overweight adult rats that were precociously weaned. Braz J Med Biol Res 51(6):e7342. https://doi.org/10.1590/1414-431x20187342

de Resende PE, Kaiser S, Pittol V, Hoefel AL, D'Agostini Silva R, Vieira Marques C, Kucharski LC, Ortega GG (2015) Influence of crude extract and bioactive fractions of *Ilex paraguariensis* A St Hil (yerba mate) on the Wistar rat lipid metabolism. J Funct Foods 15:440–451

dos Santos TW, Miranda J, Teixeira L, Aiastui A, Matheu A, Gambero A, Portillo MP, Ribeiro ML (2018) Yerba mate stimulates mitochondrial biogenesis and thermogenesis in high-fat-diet-induced obese mice. Mol Nutr Food Res 62(15):e1800142. https://doi.org/10.1002/mnfr.201800142

Dugo P, Cacciola F, Donato P, Jacques RA, Caramão EB, Mondello L (2009) High efficiency liquid chromatography techniques coupled to mass spectrometry for the characterization of mate extracts. J Chromatogr A 1216(43):7213–7221
Dunlop RA (2016) Claims of yerba mate as a potent anti-obesity agent are not supported by the data. Focus Altern Complement Ther 21(2):124–125
Farmacopea Argentina (2013) VII Ed., vol. 3. Ed. ANMAT, Buenos Aires, Argentina. pp 514–515. http://www.anmat.gov.ar/fna/volumen3asp. (Last access, January 2020)
Filip R, Ferraro GE, Bandoni AL, Bracesco N, Nunes E, Gugliucci A, Dellacassa E (2009) Mate (*Ilex paraguariensis*). In: Imperato F (ed) Recent advances in phytochemistry. Ed. Research Signpost, Kerala, India, pp 113–131
Frizon CNT, Oliveira GA, Perussello CA, Peralta-Zamora PG, Camlofski AMO, Rossa TB, Hoffmann-Ribani R (2015) Determination of total phenolic compounds in yerba mate (*Ilex paraguariensis*) combining near infrared spectroscopy (NIR) and multivariate analysis. LWT Food Sci Technol 60(2):795–801
Galle FC (1997) Hollies: the genus *Ilex*. Ed. Timber Press, Portland, USA, 619 pp
Gambero A, Ribeiro ML (2015) The positive effects of yerba maté (*Ilex paraguariensis*) in obesity. Nutrients 7(2):730–750
Gamboa-Gómez CI, Rocha-Guzmán NE, Moreno-Jiménez MR, Vázquez-Cabral BD, González-Laredo RF (2015) Plants with potential use on obesity and its complications. EXCLI J 14:809–831
Gerolis LGL, Lameiras FS, Krambrock K, Neves MJ (2017) Effect of gamma radiation on antioxidant capacity of green tea yerba mate and chamomile tea as evaluated by different methods. Radiat Phys Chem 130:177–185
Giberti G (1989) Los parientes silvestres de la yerba mate y el problema de su adulteración. Dominguezia 7(1):1–22
Giberti G (1994) Neglected crops: 1492 from a different perspective. In: Hernando Bermejo JE, León J (eds) FAO plant production and protection N° 26. FAO, Rome, pp 245–252
Giberti G (2001) Diferentes aspectos del género *Ilex* (Aquifoliaceae): corología, arquitectura floral, posición sistemática Thesis. Universidad de Buenos Aires. Buenos Aires, Argentina, 154 pp
Gómez-Juaristi M, Martínez-López S, Sarria B, Bravo L, Mateos R (2018) Absorption and metabolism of yerba mate phenolic compounds in humans. Food Chem 240:1028–1038
González Torres D (1992) Catálogo de plantas medicinales usadas en Paraguay. Ed. Servilibro, Asunción, Paraguay, 456 pp
Gremski LA, Coelho ALK, Santos JS, Daguer H, Molognoni L, do Prado-Silva L, Sant'Ana AS, da Silva Rocha R, da Silva MC, Cruz AG, Azevedo L, do Carmo MAV, Wen M, Zhang L, Granato D (2019) Antioxidants-rich ice cream containing herbal extracts and fructooligosaccharides: manufacture functional and sensory properties. Food Chem 298:125098. https://doi.org/10.1016/j.foodchem.2019.125098
Greydanus DE, Agana M, Kamboj MK, Shebrain S, Soares N, Eke R, Patel DR (2018) Pediatric obesity: current concepts. Dis Mon 64(4):98–156
Holowaty SA, Surkan SA, Trela VD, Byczko GD, Schmalko ME (2014) Variation of physicochemical and sensory properties during the aging of yerba mate. Int J Food Stud 3(2):228–238
Izzo AA (2019) Herbal medicinal products and obesity: a PTR virtual issue. Phytother Res 33(7):1751–1753
Jang S-H, Hossain MA, Lee JS, Reza MA, Lee S-P, Kang J, Park S-C (2018) Hepatoprotective effects of *Ilex paraguariensis* St Hilaire (yerba mate) extract in rats. Indian J Tradit Knowl 17(4):707–715
Jongberg S, Racanicci AMC, Skibsted LH (2019) Mate extract is superior to green tea extract in the protection against chicken meat protein thiol oxidation. Food Chem 300:125134. https://doi.org/10.1016/j.foodchem.2019.125134
Kahmann A, Anzanello MJ, Marcelo MCA, Pozebon D (2017) Near infrared spectroscopy and element concentration analysis for assessing yerba mate (*Ilex paraguariensis*) samples according to the country of origin. Comput Electron Agric 140:348–360

Kim S-Y, Oh M-R, Kim M-G, Chae H-J, Chae S-W (2015) Anti-obesity effects of yerba mate (*Ilex paraguariensis*): a randomized double-blind placebo-controlled clinical trial. BMC Complement Altern Med 15(338):1. https://doi.org/10.1186/s12906-015-0859-1

León Bianchi L, Galmarini MV, García-Burgos D, Zamora MC (2018) Time-intensity and reaction-time methodology allied to the dynamic perception and liking of bitterness in relation to body mass index. Food Res Int 109:606–613

Lima NDS, Franco JG, Peixoto-Silva N, Maia LA, Kaezer A, Felzenszwalb I, De Oliveira E, De Moura EG, Lisboa PC (2014) *Ilex paraguariensis* (yerba mate) improves endocrine and metabolic disorders in obese rats primed by early weaning. Eur J Nutr 53(1):73–82

Lima JDP, Farah A, King B, De Paulis T, Martin PR (2016) Distribution of major chlorogenic acids and related compounds in Brazilian green and toasted *Ilex paraguariensis* (Maté). leaves. J Agric Food Chem 64(11):2361–2370

Lodise O, Patil K, Karshenboym I, Prombo S, Chukwueke C, Pai SB (2019) Inhibition of prostate cancer cells by 4,5-dicaffeoylquinic acid through cell cycle arrest. Prostate Cancer 2019:1. Article ID 4520645. https://doi.org/10.1155/2019/4520645

Lopes Machado M, Arantes LP, da Silveira TL, Zamberlan DC, Cordeiro LM, Baptista F, Obetine B, da Silva AF, Mânica da Cruz IB, Antunes Soares FA, Oliveira RP (2019) *Ilex paraguariensis* extract provides increased resistance against oxidative stress and protection against amyloid beta-induced toxicity compared to caffeine in *Caenorhabditis elegans*. Nutr Neurosci 9:1. https://doi.org/10.1080/1028415X.2019.1671694

Łukomska A, Jakubczyk K, Maciejewska D, Baranowska-Bosiacka I, Janda K, Goschorska M, Chlubek D, Bosiacka B, Gutowska I (2015) The fluoride content of yerba mate depending on the country of origin and the conditions of the infusion. Biol Trace Elem Res 167(2):320–325

Luño V, Gil L, Olaciregui M, Jerez RA, de Blas I, Hozbor F (2015) Antioxidant effect of lemon balm (*Melissa officinalis*) and mate tea (*Ilex paraguariensis*) on quality lipid peroxidation and DNA oxidation of cryopreserved boar epididymal spermatozoa. Andrología 47(9):1004–1011

Marcelo MCA, Pozebon D, Ferrão MF (2015) Multivariate analysis of the profile of elements concentrations in the yerba mate. Rev Virtual Quim 7(6):2400–2421

Márquez V, Martínez N, Guerra M, Fariña L, Boido E, Dellacassa E (2013) Characterization of aroma-impact compounds in yerba mate (*Ilex paraguariensis*) using GC-olfactometry and GC-MS. Food Res Int 53(2):808–815

Matei MF, Jaiswal R, Patras MA, Kuhnert N (2016) LC-MSn study of the chemical transformations of hydroxycinnamates during yerba maté (*Ilex paraguariensis*) tea brewing. Food Res Int 90:307–312

Mazur M, Zwyrzykowska-Wodzinska A, Wojtas E, Zachwieja A, Salejda AM (2019) Effect of yerba mate (*Ilex paraguariensis* A St-Hil) supplementation on oxidative stress in ruminants. Chil J Agric Res 79(2):316–322

Meinhart AD, Damin FM, Caldeirão L, Godoy HT (2019) Methylxanthines in 100 Brazilian herbs and infusions: determination and consumption. Emirates J Food Agric 31(2):125–133

Mello FW, Scotti FM, Melo G, Warnakulasuriya S, Guerra ENS, Rivero ERC (2018) Maté consumption association with upper aerodigestive tract cancers: a systematic review and meta-analysis. Oral Oncol 82:37–47

Messina D, Soto C, Méndez A, Corte C, Kemnitz M, Avena V, Del Balzo D, Elizalde RP (2015) Lipid-lowering effect of mate tea intake in dyslipidemic subjects. Nutr Hosp 31(5):2131–2139

Ministerio de Salud Pública, Uruguay. https://www.impo.com.uy/bases/decretos-originales/403-2016. (Last access, January 2020)

Montevideo Portal (2019) https://www.montevideo.com.uy/Negocios-y-Tendencias. (Last access, January 2020)

Negrin A, Long C, Motley TJ, Kennelly EJ (2019) LC-MS metabolomics and chemotaxonomy of caffeine-containing holly *(Ilex)* species and related taxa in the Aquifoliaceae. J Agric Food Chem 67(19):5687–5699

Niraula P, Ghimire S, Lee H, Kim MS (2018) *Ilex paraguariensis* extends lifespan and increases an ability to resist environmental stresses in drosophila. Rejuvenation Res 21(6):497–505

Noud PC, Martínez MADP, Loizeau PA, Spichiger R, Andrews S, Manen JF (2000) Molecular phylogeny and biogeography of the genus *Ilex* L (Aquifoliaceae). Ann Bot London 85:111–122

Oellig C, Schunck J, Schwack W (2018) Determination of caffeine theobromine and theophylline in mate beer and mate soft drinks by high-performance thin-layer chromatography. J Chromatogr A 1533:208–212

Oh K-E, Shin H, Jeon YH, Jo YH, Lee MK, Lee KS, Park B, Lee KY (2016) Optimization of pancreatic lipase inhibitory and antioxidant activities of *Ilex paraguariensis* by using response surface methodology. Arch Pharm Res 39(7):946–952

Panza VP, Diefenthaeler F, Tamborindeguy AC, Camargo CDQ, De Moura BM, Brunetta HS, Sakugawa RL, De Oliveira MV, Puel EDO, Nunes EA, Da Silva EL (2016) Effects of mate tea consumption on muscle strength and oxidative stress markers after eccentric exercise. Br J Nutr 115(8):1370–1378

Panza VP, Brunetta HS, de Oliveira MV, Nunes EA, da Silva EL (2019) Effect of mate tea (*Ilex paraguariensis*) on the expression of the leukocyte NADPH oxidase subunit p47 phox and on circulating inflammatory cytokines in healthy men: a pilot study. Int J Food Sci Nutr 70(2):212–221

Pereira CC, Souza AO, Oreste EQ, Cidade MJA, Cadore S, Ribeiro AS, Vieira MA (2016) Acid decomposition of yerba mate (*Ilex paraguariensis*) using a reflux system for the evaluation of Al Ca Cd Cr Cu Fe K Mg Mn Na Pb and Zn contents by atomic spectrometric techniques. J Braz Chem Soc 27(4):685–693

Pereira AAF, Tirapeli KG, Chaves-Neto AH, da Silva Brasilino M, da Rocha CQ, Belló-Klein A, Llesuy SF, Dornelles RCM, Nakamune ACDMS (2017a) *Ilex paraguariensis* supplementation may be an effective nutritional approach to modulate oxidative stress during perimenopause. Exp Gerontol 90:14–18

Pereira CS, Stringhetta-Garcia CT, da Silva Xavier L, Tirapeli KG, Pereira AAF, Kayahara GM, Tramarim JM, Crivelini MM, Padovani KS, Leopoldino AM, Louzada MJQ, Belló-Klein A, Llesuy SF, Ervolino E, Dornelles RCM, Chaves-Neto AH, Nakamune ACDMS (2017b) *Ilex paraguariensis* decreases oxidative stress in bone and mitigates the damage in rats during perimenopause. Exp Gerontol 98:148–152

Petrilli AA, Souza SJ, Teixeira AM, Pontilho PM, Souza JMP, Luzia LA, Rondó PHC (2016) Effect of chocolate and yerba mate phenolic compounds on inflammatory and oxidative biomarkers in HIV/AIDS individuals. Nutrients 8(5):132. https://doi.org/10.3390/nu8050132

Petroselli G, Parapugna TL, Lagorio MG, Erra-Balsells R (2019) MALDI- and LDI-MS saponin fingerprint of leaves and stick components of commercial yerba mate (*Ilex paraguariensis*). J Mass Spectrom 54(2):195–203

Piovezan-Borges AC, Valério-Júnior C, Gonçalves IL, Mielniczki-Pereira AA, Valduga AT (2016) Antioxidant potential of yerba mate (*Ilex paraguariensis* St Hil) extracts in *Saccharomyces cerevisiae* deficient in oxidant defense genes. Braz J Biol 76(2):539–544

Polidoro AS, Scapin E, Malmann M, do Carmo JU, Machado ME, Caramão EB, Jacques RA (2016) Characterization of volatile fractions in green mate and mate leaves (*Ilex paraguariensis* A St Hil) by comprehensive two-dimensional gas chromatography coupled to time-of-flight mass spectrometry (GC×GC/TOFMS). Microchem J 128:118–127

Portela JL, Soares D, Rosa H, Roos DH, Pinton S, Ávila DS, Puntel RL (2016) *Ilex paraguariensis* crude extract acts on protection and reversion from damage induced by t-butyl hydroperoxide in human erythrocytes: a comparative study with isolated caffeic and/or chlorogenic acids. J Sci Food Agric 97(7):2007–2014

Pozebon D, Dressler VL, Marcelo MCA, de Oliveira TC, Ferrão MF (2015) Toxic and nutrient elements in yerba mate (*Ilex paraguariensis*). Food Addit Contam B 8(3):215–220

Przygodda F, Martins ZN, Castaldelli PA, Minella TV, Vieira LP, Cantelli K, Fronza J, Padoin MJ, Szymańska K, Matos E, Hung RJ, Wünsch-Filho V, Eluf-Neto J, Menezes A, Daudt AW, Brennan P, Boffetta P (2010) Supercritical fluid extraction of alkaloids from *Ilex paraguariensis* St Hil. J Food Eng 100(4):656–661

Ribeiro BD, Coelho MAZ, Rebelo LPN, Marrucho IM (2013) Ionic liquids as additives for extraction of saponins and polyphenols from mate (*Ilex paraguariensis*) and tea (*Camellia sinensis*). Ind Eng Chem Res 52(34):12146–12153

Rodríguez-Arzuaga M, Piagentini AM (2018) New antioxidant treatment with yerba mate (*Ilex paraguariensis*) infusion for fresh-cut ales: modeling optimization and acceptability. Food Sci Technol Int 24(3):223–231

Ronco AL, De Stefani E, Mendoza B, Deneo-Pellegrini H, Vázquez A, Abbona E (2016a) Mate intake and risk of breast cancer in Uruguay: a case-control study. Asian Pac J Cancer Prev 17(3):1453–1461

Ronco AL, Stefani ED, Mendoza B, Vázquez A, Abbona E, Sánchez G, Rosa AD (2016b) Mate and tea intake dietary antioxidants and risk of breast cancer: a case-control study. Asian Pac J Cancer Prev 17(6):2923–2933

Ronco AL, Espinosa E, Calderón JM, Galante EL, De Rosa A, Sánchez G (2017) 'Mate' intake hormone-based risk factors and breast cancer: a case-control study. Asian Pac J Cancer Prev 18(4):941–948

Rossa ÜB, Angelo AC, Nisgoski S, Westphalen DJ, Frizon CNT, Hoffmann-Ribani R (2015) Application of the NIR method to determine nutrients in yerba mate (*Ilex paraguariensis* A St-Hill) leaves. Commun Soil Sci Plant Anal 46(18):2323–2331

Rusinek-Prystupa E, Marzec Z, Sembratowicz I, Samolińska W, Kiczorowska B, Kwiecień M (2016) Content of selected minerals and active ingredients in teas containing yerba mate and rooibos. Biol Trace Elem Res 172(1):266–275

Santiago PG, Gasparotto FM, Gebara KS, Bacha FB, Lívero FADR, Strapazon MA, Junior ELC, Kassuya CAL, de Souza LM, Gasparotto Junior A (2017) Mechanisms underlying antiatherosclerotic properties of an enriched fraction obtained from *Ilex paraguariensis* A St-Hil. Phytomedicine 34:162–170

Santos LPD, Caon T, Battisti MA, Silva CHBD, Simões CMO, Reginatto FH, de Campos AM (2017) Antioxidant polymeric nanoparticles containing standardized extract of *Ilex paraguariensis* A St-Hil for topical use. Ind Crop Prod 108:738–747

Santos JS, Deolindo CTP, Hoffmann JF, Chaves FC, do Prado-Silva L, Sant'Ana AS, Azevedo L, do Carmo MAV, Granato D (2018) Optimized *Camellia sinensis* var *sinensis, Ilex paraguariensis* and *Aspalathus linearis* blend presents high antioxidant and antiproliferative activities in a beverage model. Food Chem 254:348–358

Simões CMO, Auler Mentz L, Schenkel EP, Irgang B, Stehmann J (1986) Plantas da medicina popular no Rio Grande do Sul. Ed. UFRGS, Porto Alegre, Brazil. 173 pp

Small E, Catling PM (2001) Blossoming treasures of biodiversity: 3. Mate (*Ilex paraguariensis*)-better than Viagra, marijuana, and coffee? Biodiversity 2:26–27

Souza SJ, Petrilli AA, Teixeira AM, Pontilho PM, Carioca AA, Luzia LA, Souza JM, Damasceno NR, Segurado AA, Rondó PH (2017) Effect of chocolate and mate tea on the lipid profile of individuals with HIV/AIDS on antiretroviral therapy: a clinical trial. Nutrition 43–44:61–68

Spegazzini ED, Castro MT, Carpano SM, Nájera MT (2002) Taxonomic determination of therapeutic argentine species of *Ilex*. Pharm Biol 40(1):2–15

Stuby J, Gravestock I, Wolfram E, Pichierri G, Steurer J, Burgstaller JM (2019) Appetite-suppressing and satiety-increasing bioactive phytochemicals: a systematic review. Nutrients 11(9):2238. https://doi.org/10.3390/nu11092238

Uecker JN, Schneider JP, Cerqueira JH, Rincón JAA, Campos FT, Schneider A, Barros CC, Andreazza R, Jaskulski IB, Pieniz S (2019) *Ilex paraguariensis* extract prevents body weight gain in rats fed a high-fat diet. Food Sci Technol 39(3):620–626

Valduga AT, Gonçalves IL, Piovezan Borges AC, Mielniczki-Pereira AA, Picolo AP (2016) Cytotoxic/antioxidant activity and sensorial acceptance of yerba-mate development by oxidation process. Acta Sci Technol 38(1):115–121

Vieitez I, Maceiras L, Jachmanián I, Alborés S (2018) Antioxidant and antibacterial activity of different extracts from herbs obtained by maceration or supercritical technology. J Supercrit Fluids 133:58–64

Wang S, Sarriá B, Mateos R, Goya L, Bravo-Clemente L (2019) TNF-α-induced oxidative stress and endothelial dysfunction in EAhy926 cells is prevented by mate and green coffee extracts 5-caffeoylquinic acid and its microbial metabolite dihydrocaffeic acid. Int J Food Sci Nutr 70(3):267–284

Xu GH, Kim YH, Choo SJ, Ryoo I-J Yoo JK, Ahn JS, Yoo ID (2010) Two acetylated megastigmane glycosides from the leaves of *Ilex paraguariensis*. Arch Pharm Res 33(3):369–373

Yimam M, Jiao P, Hong M, Brownell L, Lee Y-C, Hyun E-J, Kim H-J, Kim T-W, Nam J-B, Kim M-R, Jia Q (2016) Appetite suppression and antiobesity effect of a botanical composition composed of *Morus alba*, yerba mate and *Magnolia officinalis*. J Obes 2016:4670818. https://doi.org/10.1155/2016/4670818

Yimam M, Jiao P, Hong M, Brownell L, Lee Y-C, Kim H-J, Nam J-B, Kim M-R, Jia QA (2017) Botanical composition from *Morus alba*, *Ilex paraguariensis* and *Rosmarinus officinalis* for body weight management. J Med Food 20(11):1100–1112

Yin Y, Katahira R, Ashihara H (2015) Metabolism of purine alkaloids and xanthine in leaves of maté (*Ilex paraguariensis*). Nat Prod Commun 10(5):707–712

Zapata FJ, Rebollo-Hernanz M, Novakofski JE, Nakamura MT, González de Mejía E (2020) Caffeine but not other phytochemicals in mate tea (*Ilex paraguariensis* St Hilaire) attenuates high-fat-high-sucrose-diet-driven lipogenesis and body fat accumulation. J Funct Foods 64:103646. https://doi.org/10.1016/j.jff.2019.103646

Zhou F, Jongberg S, Zhao M, Sun W, Skibsted LH (2019) Antioxidant efficiency and mechanisms of green tea rosemary or maté extracts in porcine longissimus dorsi subjected to iron-induced oxidative stress. Food Chem 298:125030. https://doi.org/10.1016/j.foodchem.2019.125030

Zuloaga FO, Morrone O, Belgrano MJ, Marticorena C, Marchesi E (2008) Catálogo de las plantas vasculares del Cono Sur (Argentina, Sur de Brasil, Chile, Paraguay y Uruguay) Monographs in Systematic Botany, 3 vol. Ed. American Society of Plant Taxonomists University of Wyoming, Laramie, USA, 3486 pp

Zwyrzykowska A, Kupczyński R, Jarosz B, Szumny A, Kucharska AZ (2015) Qualitative and quantitative analysis of polyphenolic compounds in *Ilex* sp. Open Chem 13(1):1303–1312

Ligaria cuneifolia (R. et P.) Tiegh.

Marcelo L. Wagner, Cecilia B. Dobrecky, Beatriz G. Varela, Alejandra N. Luquita, Cristina E. Carnovale, María Alejandra Álvarez, and Rafael A. Ricco

Flowers of *Ligaria cuneifolia* (Quilimo, Córdoba, Argentina). (Photo: AA Gurni)

M. L. Wagner (✉) · B. G. Varela · R. A. Ricco
Cátedra de Farmacobotánica y Museo de Farmacobotánica "Juan Aníbal Domínguez", Facultad de Farmacia y Bioquímica, Universidad de Buenos Aires, Buenos Aires, Argentina
e-mail: mlwagner@ffyb.uba.ar; mlwagner550@gmail.com

C. B. Dobrecky
Cátedra de Farmacobotánica y Museo de Farmacobotánica "Juan Aníbal Domínguez", Facultad de Farmacia y Bioquímica, Universidad de Buenos Aires, Buenos Aires, Argentina

Facultad de Farmacia y Bioquímica, Tecnología Farmacéutica I, Universidad de Buenos Aires, Buenos Aires, Argentina

A. N. Luquita
Facultad de Ciencias Médicas, Cátedra de Física Biológica, Departamento Promoción de la Salud, Universidad Nacional de Rosario, Rosario, Argentina

Consejo de Investigaciones de la Universidad Nacional de Rosario (CIURN), Rosario, Argentina

Á. Máthé, A. Bandoni (eds.), *Medicinal and Aromatic Plants of South America Vol. 2*, Medicinal and Aromatic Plants of the World 7,
https://doi.org/10.1007/978-3-030-62818-5_23

Abstract *Ligaria cuneifolia* (R. et P.) Tiegh. (Loranthaceae) is a hemiparasitic species that develops on different hosts. It has a wide geographic distribution ranging from Peru to central Argentina and southern Brazil. In Argentina it is popularly known as: "muérdago criollo", "liga" or "liguilla". This species is traditionally used as a hypotensive agent. Due to its morphological similarity, it constitutes the natural substitute for the European mistletoe (*Viscum album* L. -Santalaceae-). Like other species of the Loranthaceae family, *L. cuneifolia* is rich in flavonoids, especially flavonols that could be determinant of its pharmacological activity. Aqueous extracts proved to be inhibitors of the proliferation of LB tumor cells, alcoholic extracts have a high antioxidant activity, and infusions have the property of lowering plasma cholesterol in patients with a history of hypercholesterolemia. Considering this background information, and the difficulties experienced in its field cultivation, strategies are being developed to produce the metabolites responsible for its different biological activities in *in vitro* cultures.

Keywords "Muérdago criollo" · Polyphenols · *In vitro* cultivation

1 Introduction

"European mistletoe" (*Viscum album* L. -Santalaceae-) has been used in Europe for thousands of years in preparations for the treatment of epilepsy, infertility and weakness, but the most recognized pharmacological uses are its effects on the cardiovascular system (Benigni et al. 1964; Wagner et al. 1986) and blood pressure (Youngken 1951; Font Quer 1962; Paris and Moyse 1981).

In Argentina, the creole mistletoe ("muérdago criollo"), *Ligaria cuneifolia* (R. et P.) Tiegh. (Loranthaceae), is the most used species as substitute for the "European mistletoe". European immigrants and their descendants found that this species had a habitat and similar morphology to that of *V. album*; for this reason, they took it as it's a natural substitute (Nájera 1983; Wagner 1993). Thus, began the use of leaf and

C. E. Carnovale
Facultad de Ciencias Bioquímicas y Farmacéuticas, Departamento de Ciencias Fisiológicas, Universidad Nacional de Rosario, Rosario, Argentina

Consejo Nacional de Investigaciones Científicas y Técnicas, Instituto de Fisiología Experimental (IFISE-CONICET), Buenos Aires, Argentina

M. A. Álvarez
Facultad de Ciencias de la Salud, Carreras de Farmacia y Bioquímica, Cátedra de Farmacobotánica y Farmacognosia, Universidad Maimónides, Buenos Aires, Argentina

Consejo Nacional de Investigaciones Científicas y Técnicas (CONICET), Buenos Aires, Argentina stem infusions -occasionally accompanied by flowers- as therapeutic agents to lower blood pressure (Domínguez 1928; Ratera and Ratera 1980).

Currently, due to advances in phytochemical and pharmacological studies. *L. cuneifolia* is one of the most studied plants in the Argentine medicinal flora.

2 Taxonomic Characteristics

Ligaria cuneifolia (R. et P.) Tiegh. (Loranthaceae), is popularly known as "muérdago criollo", "liga" or "liguilla", and it belongs to the order Santalales. This order is the most important and it has the highest number of species within the hemiparasites. It includes trees, shrubs, woody vines or herbs, mostly hemiparasites of roots or stems from other plants (Takhtajan 1997).

Synonyms *Ligaria emarginata* Tiegh., *Ligaria lanceolata* Tiegh., *Ligaria orbignyana* Tiegh., *Loranthus cuneifolius* Ruiz & Pav., *Loranthus montevidensis* Spreng., *Phrygilanthus cuneifolius* (Ruiz & Pav.) Eichler, *Psittacanthus cuneifolius* (Ruiz & Pav.) G. Don, *Psittacanthus hortonii* Standl. & F.A. Barkley, and *Psittacanthus peruanus* Engl. (The Plant List 2019).

All flowering plants that grow as parasites on the branches of trees or shrubs are generally called "mistletoes" (in Hispanic Latin America, "muérdago"). They represent approximately three quarters of the total parasitic plants (Feuer and Kuijt 1978).

The large family of mistletoes has long been a source of taxonomic difficulty. In 1935 Engel and Krause determined 1300 different species of mistletoes preferably distributed in the tropical and subtropical regions of both hemispheres. These genera were grouped into two subfamilies: Loranthoideae and Viscoideae and the whole group constituted the family of Loranthaceae Don. (Engler and Krause 1935; Wettstein 1944; Abbiatti 1946). Barlow (1964) elevated the two subfamilies to the category of families and divided the Loranthaceae "*sensu latu*" (*s. l.*) into Loranthaceae Don. "*sensu strictu*" (*s. s.*) and Viscaceae Miq. Kuijt's works rescued the Eremolepideae Tiegh tribe from the Viscaceae family and elevated it to the category of family with the name of Eremolepidaceae Kuijt (1988). Taxonomists considered, therefore, that mistletoes were grouped in three families: Loranthaceae (75 genera), Eremolepidaceae (3 genera) and Viscaceae (7 genera) (Barlow et al. 1989). However, many researchers keep the tradition of referring to mistletoes as belonging to Loranthaceae broadly speaking (*s.1.*). According to the current criteria of the Angiosperm Phylogeny Group (APG), the Eremolepidaceae and Viscaceae families do not exist as such but they are included in Santalaceae (Angiosperm Phylogeny Website 2017). The genera *Phoradendron*, *Eubrachion* and *Viscum* (before in Viscaceae) belong to the Santalaceae family.

According to Abbiatti (1946), there are 23 species of mistletoes that grow in the different phytogeographic regions of Argentina, except for the Pampas steppe and the Patagonian and Andean deserts. The most affected phytogeographic region is

Table 1 Mistletoes that grow in Argentina

Family	Genus	Species
Eremolepidaceae Tiegh.	*Eubrachion* Hook. *f.*	*E. ambiguum* (H.et A.) Engler [= *E. andalgalense* Abbiatti]
Loranthaceae D. Don	***Ligaria*** Tiegh.	***L. cuneifolia*** (Ruiz et Pav.) Tiegh.
	Psittacanthus Mart.	*P. cordatus* (Hoffmans) Blume
	Struthanthus Mart.	*S. acuminatus* (Ruiz et Pav.) Blume
		S. uraguensis (Hook. et Arn.) G. Don
	Tripodanthus (Eichl.) Tiegh. [= *Phrygilanthus* Eichl.]	*T. flagellaris* (Cham. et Schlecht) Tiegh.
		T. acutifolius (Ruiz et Pav.) Tiegh.
	Tristerix Mart.	*Trx. corymbosus* (L.) Kuijt
		Trx. verticillatus (Ruiz et Pav.) Barlow et Wiens
Viscaceae Miq.	*Phoradendron* Nutt.	*Ph. affine* (Pohl ex DC) Engl. & Krause
		Ph. argentinum Urb.
		Ph. bathyoryctum Eichler
		Ph. coriaceum Mart.
		Ph. dipterum Eichler
		Ph. falcifrons (Hook. & Arn.) Eichler
		Ph. interruptum (DC.) B.D. Jacks
		Ph. liga (Gill.) Eichl.
		Ph. mucronatum (D.C.) Krug et Urb
		Ph. obtusissimum (Miq.) Eichler
		Ph. paraguari Kuijt
		Ph. piperoides (Kunth) Trel.
		Ph. quadrangulare (Kunth) Griseb
		Ph. reductum Trel.
		Ph. tucumanensis Urb.

Abbiatti (1946), Subils (1984), Wagner (1993), Zuloaga and Morrone (1999), Varela (2003)

the one called 'Formación del Monte' (Mount Formation), where Fabaceae trees abound, which are the plants that mistletoes preferably infect.

From a taxonomic point of view, Argentine mistletoes are distributed in the three families into which Loranthaceae *s.l.* (Table 1) was divided, according to the classification criteria of The South Cone Vascular Plant (Zuloaga and Morrone 1999).

3 Crude Drug Used

The crude drug is formed by the leaves and the young herbaceous stems of *Ligaria cuneifolia*. The ethanol extract contains at least 21.0 ± 0.2% of total phenols (calculated as mg gallic acid/100 mg extract), 14 ± 0.2% total tannins (calculated as mg gallic acid/100 mg extract), 0.47 ± 0.02 condensed tannins (calculated as Absorbance/

mg extract), 1.5 ± 0.1% flavonoids (calculated as mg rutin/100 mg extract), and 1.2 ± 0.1% hydroxycinnamic acids (calculated as mg chlorogenic acid/100 mg extract) (Dobrecky 2016).

4 Major Chemical Constituents and Bioactive Compounds

A galactoside-specific lectin present in the aqueous extract inhibits the proliferation of LB tumor cells. Galactoside-specific lectin is the first isolated compound responsible for a biological action in this species (Varela et al. 2001; Fernández et al. 2003; Cerdá Zolezzi et al. 2005).

Studies performed to specimens that were developed on different hosts and collected in different geographical areas showed the presence of tyramine (Vázquez y Novo et al. 1989).

The concentration detected in most of the specimens did not exceed 10 mg %, although specimens with levels above 100 mg % were detected. In individuals that parasitize *Geoffroea decorticans* (H. et Arn.) Burkart (Fabaceae) (Photo 1), the highest levels were detected, with values ranging from 120 to 360 mg of tyramine per 100 mg dry plant material.

Two major compounds characterized as pentacyclic triterpenes of the lupeol group were isolated in this species: betulin and betulinic acid (Fusco et al. 2004).

Free quercetin and glycoside flavonols were characterized: quercetin-3-*O*-ucoside, quercetin-3-*O*-xyloside, quercetin-3-*O*-arabinopyranoside, quercetin-3-*O*-arabino-furanoside and quercetin-3-*O*-rhamnoside. Moreover, four galloyl derivative compounds: quercetin-3-*O*-α-L-(3″-*O*-gallcyl)-rhamnoside, quercetin-3-*O*-α-L-(2″-*O*-galloyl)-rhamnoside, quercetin-3-*O*-α-D-(2″galloyl)-arabinofuranoside, and quercetin-3-*O*-α-L-(2″-*O*-galloyl)-arabinopyranoside (Graziano et al. 1967; Wagner et al. 1998; Fernández et al. 1998; Arenas Chávez et al. 2011; Soberón et al. 2014; Dobrecky et al. 2017).

Photo 1 *Ligaria cuneifolia* parasitizing *Geoffroea decorticans* (Cuesta de Miranda, La Rioja, Argentina). (Photo: AA Gurni)

In addition, leucoanthocyanidins, catechin-4-ß-ol and proanthocyanidins were detected in different degrees of polymerization: polymers, oligomers and dimers, which, by acid treatment, produced cyanidine (Graziano et al. 1967; Wagner 1993; Wagner et al. 1998; Fernández et al. 1998; Dobrecky et al. 2017).

5 Morphological Description

Ligaria cuneifolia is a hemiparasite species devoid of aerial roots that develops on different hosts. The name *Ligaria* derives from the term league, alluding to the ability of its fruits to "bind" or catch birds. Meanwhile, *cuneifolia* refers to the wedge-shaped appearance of its leaves. It has whole stems, without internodes, glabrous; cylindrical adult branches, sub compressed young ones; both rough striated. Alternate, sessile, cuneiform leaves (Sanzin 1918), linear, lanceolate or linear-spatulated; 1.5–6.5 cm long by 4–15 mm wide; obtuse or acute; fleshy or leathery, rough; basined, slightly visible or 3-veined ribs; apex with a blackish callosity on drying. Flowers in solitary axillary racemes (exceptionally with 2 flowers); floral peduncles 5–15 mm long; calycle with more or less deep teeth. Hexamere flowers, red, orange or sometimes yellow, 3.5–5.5 cm long; pieces of perigonium free or welded at its base; linear-spatulated tepals. In general, the specimens that grow in Argentina are red, although towards the east of the country their flowers acquire a yellow color. Alternately uneven stamens; filaments welded to the tepals in its lower half, the basal region forms a ligule or basal nail (Subils 1984); versatile anthers of longitudinal dehiscence; trilobed pollen with granulated ornamentation exine (Castro and Tellería 1995). Obovoid receptacle, ovary with a lobed disk. Globose or hollow berry, blackish, crowned by the persistent tubular calycle. Embryo with 2 cotyledons; seeds with endosperm (Subils 1984; Ulibarri 1987). Flowering is maximum in spring and lasts until autumn.

6 Anatomical Characteristics of the Plant Drug

Foliar anatomy is characterized by epidermal cells covered by a moderately thick cuticle; paracytic stomata are present in both surfaces; an isobilateral type mesophyll, with radially elongated cells on both epidermis, and shorter cells in the center. In this region, irregular branched sclereids, with calcium oxalate prismatic crystals are observed. The conductive tissue consists of a large central vascular bundle and smaller bundles on each side and are accompanied by arcs of cellulosic cells (Varela and Gurni 1995; Wagner et al. 1998; Varela et al. 2001). The stem anatomy shows a papillose epidermis covered by a thick cuticle, and paracytic stomata. The cortical parenchyma is heterogeneous, with 3–4 layers of radially elongated cells and several layers of tangentially elongated cells. Irregular, branched, crystalliferous sclereids are observed in the first region. In the second region, sclerenchyma fiber groups are present. Central cylinder is formed by open collateral vascular bundles with variable groups of lignified fibers in the xylem. Medullary rays are formed by 2–5 rows of radially elongated parenchymal cells. A distinct perimedullar region appears, formed by cells with simple punctuations. The medulla is formed by parenchyma rounded cells, and some irregular crystalliferous sclereids (Varela and Gurni 1995; Wagner et al. 1998; Varela et al. 2001).

7 Geographical Distribution

Ligaria cuneifolia has a wide geographical distribution from Peru to Central Argentina and Southern Brazil. It inhabits three geographical regions in disjunct form: the Andes region (along the Andes mountain range, from Peru to Argentina and Chile), the Central region (central mountain ranges of Argentina: in the provinces of Córdoba and San Luis) and the East region (the Southeast of South America: Uruguay, Brazil and the Argentine Mesopotamia) (Amico and Vidal-Russell 2019).

In the Andes and Central regions of Argentina, it is found above 700 m.a.s.l., while in the East region, it is at low altitudes. Therefore, *L. cuneifolia* is found in a variety of ecoregions. In the Andes region, it is present in the ecoregions of the Bolivian dry forest, Chilean scrubland, Mount and Puna (Olson et al. 2001; Amico and Vidal-Russell 2019). In the Central region, it is found only in the south of the Chaco Seco ecoregion. In the Eastern region, in the Espinal, Uruguayan Savannah and Pampas, its occurrence is associated with the river corridors (Amico and Vidal-Russell 2019).

8 Ecological Requirements

The range of hosts of *Ligaria cuneifolia* throughout its wide and discontinuous distribution is little known. It is a generalist mistletoe in terms of host use, mainly parasitizing Fabaceae species (Abbiatti 1946; Varela et al. 2001; Amico and Vidal-Russell 2019). According to the observations of Amico and Vidal-Russell (2019), *L. cuneifolia* has been found parasitizing more than 30 native species throughout the whole geographical distribution, preferably in arborescent legumes such as "chañar" (*Geoffroea decorticans* (Gill. ex Hook. & Arn.) Burkart -Fabaceae-) (Photo 1), "algarrobo" (*Prosopis* spp. -Fabaceae-), "churqui" (*Acacia caven* (Molina) Molina -Fabaceae-), *Acacia* spp., *Robinia* spp., "horco-cebil" (*Piptadenia excelsa* (Griseb.) Lillo -Fabaceae -), but also on felling "talas" (*Celtis iguanaea* (Jacq.) Sarg., C. *tala* Gillies ex Planch. -Cannabaceae-), "molles" (*Schinus* spp. -Anacardiaceae-), "quebrachos" (*Schinopsis* spp. -Anacardiaceae-) and cultivated plants such as apple trees, pear trees, peaches, pomegranates, olive trees, among others (Abbiatti 1946; Varela 2003).

L. cuneifolia being a hemiparasitic plant, whose field culture has not been yet elaborated, the establishment of *in vitro* cultures appeared as an interesting alternative for producing its active metabolites without placing the species at risk. In that vein, it was faced the challenge of studying the initiation of *in vitro* cultures of *L. cuneifolia* and analyzing their profile of polyphenols and the influence that factors such as plant growth regulators have both on growth and on polyphenolic content (Ricco et al. 2019).

The presence of flavonols, hydroxycinnamic acids, proanthocyanidins of high degree of polymerization, and catechin was shown both in extracts from plants and from calli; however, in the latter case, at significantly lower levels (Ricco et al. 2018, 2019).

9 Traditional Use and Common Knowledge

In Argentina, "muérdago criollo" ("creole mistletoe"), *L. cuneifolia*, is the most commonly used species as a substitute for the European mistletoe (*Viscum album* L -Viscaceae-), despite belonging to two different families. European immigrants and their descendants found that this species had a habitat and morphology similar to the *V. album*, which is why they took it as a natural substitute for it (Nájera 1983; Wagner 1993) and began to use leaf and stem infusions, occasionally accompanied by flowers, as therapeutic agents to lower blood pressure. This is how it can be found in herbalists, under the name of "muérdago". Hieronymus (1882) mentioned that crushed leaves were used in poultices for bone fractures, a property that was also attributed to the gummy extract of the fruits. The gum was popularly used to hunt birds and insects (Hieronymus 1882; Toursarkissian 1980). The first pharmacological and phytochemical studies on Argentine mistletoes appears in "Contribuciones a la Materia Médica" by Juan A. Domínguez (1928). These studies were carried out with specimens of *Psittacanthus cuneifolius* (Ruiz et Pav.) Engl. (syn. of *L. cuneifolia*) and *Phrygilanthus flagellaris* (Cham. et Schltdl.) Eicher (syn. of *Tripodanthus flagellaris* (Cham. et Schltdl.) Tiegh.), and it was observed that the hypotensive action was verified in specimens that parasitized *Prosopis* species while those collected on *Aspidosperma* acted as hypertensive adrenergic agents. Something similar was observed with species of *Phoradendron flavescens* from the United States (Domínguez 1928). Ratera and Ratera (1980) mentioned the hypotensive action of the "creole mistletoe" (*L. cuneifolia*), which parasites Fabaceae trees, and it was warned that on other hosts (Apocynaceae, for example) it can exert hypertensive adrenergic action due to peripheral vasoconstriction and its action is manifested by a temporary and slight decrease in pressure followed by strong persistent hypertension, accompanied by cardiac acceleration. Burgstaller (2000) mentions the hypotensive action for "league" or for mistletoe besides including its use in arteriosclerosis, saturnism, menopause, glaucoma and for sterility.

The leaves, tender stems and flowers were used in decoction at 20–25%. They are attributed diuretic and tonic-cardiac properties, but especially in hypertension.

The mistletoe juice would have hypotensive and antisclerosis properties (34 drops of juice correspond to 1 g of the plant). The aqueous extract of young leaves and branches is also prepared (100 g in 600 g of water) (González Torres 2018).

The infusion of the leaves was used against vomiting blood, tuberculosis and metrorrhea. The decoction of stems as a vomitive. The infusion of fruits was used as a drink for kidney and bladder conditions (De Lucca and Zalles 1992). The infusion of the leaves and fruits was used as antihemorrhagic and against arteriosclerosis. It has been an excellent food for cattle in times of drought (Pöll 1984; Castellón 1998).

It is exciting for peristaltic contractions and uterine tone in pregnant women. It is also cited as an astringent. The viscin (name given to a mucilaginous, yellowish, tasteless, odorless substance, present in the bark of the stems and fruits of these plants) solution, due to its great binding power, has been recommended for the manufacture of plasters and dermatological medicines (Domínguez 1928; Ratera and Ratera 1980; Ragonese and Milano 1984; Marzocca 1997).

Scarpa and Montani (2011) conducted a study on the medical ethnobotany of mistletoes (Loranthaceae *s.l.*) among the indigenous and creole of Argentina. The objective of this contribution was to compile the medicinal applications of mistletoes by 14 Argentine human groups (7 creole and 7 indigenous groups), including both bibliographic and unpublished data, with an intercultural comparison between creole and indigenous uses. The most commonly used parts of the plant are the leaves and young stems, and among the specific applications of *L. cuneifolia* the following are mentioned: hypotensive, antihemorrhagic, abortive, emmenagogue, oxytocic actions, and against cephalgia, gastralgia and sore throat, in the form of decoction (Martínez-Crovetto 1964, 1981; Pochettino and Martínez 2000; Scarpa 2004, 2009; Menseguez et al. 2007; Mas Serra et al. 2008; Martínez 2008, 2010) . For fractures, ground leaves are placed in the form of a poultice on the condition. For hypothermia, in the form of hot baths and vapors.

10 Modern Medicine Based on Its Traditional Medicine Uses

Regarding the pharmacological action of the extracts, it was established that they have effects on the cardiovascular system and cytostatic and immunomodulatory actions (Taira et al. 1994, 2004; Fernández et al. 1998, 2000; Wagner et al. 1998; Cerdá Zolezzi et al. 2005). Aqueous extracts of *L. cuneifolia* produce inhibition in the growth of activated lymphoid cells. This inhibition would be mediated by the induction of cell death through an apoptotic mechanism. In addition, these extracts can modulate the activity of macrophage cells through induction in the production of nitric oxide. A galactoside-specific lectin present in the aqueous extract inhibits the proliferation of LB tumor cells. Galactoside-specific lectin is the first isolated compound responsible for a biological action in this species (Varela et al. 2001; Fernández et al. 2003; Cerdá Zolezzi et al. 2005).

The cardiovascular effects observed indicate that the infusion of *L. cuneifolia* produces, depending on the host, a pressor effect accompanied or not, of vasodilation and a variable cardiac effect (Taira et al. 1994, 2004; Fernández et al. 2000). This diversity of effects could be due to the qualitative and quantitative differences of the polyphenols present. On the other hand, studies *in vitro* carried out on isolated rat aortic rings demonstrated that polyphenol-enriched fractions have a potent and effective endothelial-dependent vasodilator effect, since the mechanical removal of the endothelium completely nullifies the relaxant activity of the fractions. These results would support the popular use of *Ligaria cuneifolia* as a hypotensive (Gerschcovsky et al. 2018).

It was determined that the phenolic compounds present in *L. cuneifolia* preparations have an important antioxidant activity, especially fractions enriched with flavonoids (Arenas Chávez et al. 2006; Dobrecky et al. 2014). Dobrecky (2016) analyzed the antioxidant capacity of each flavonoid that is part of the methanolic extracts and demonstrated that aglycons and galloyl by-products are more active than glycosides. Antioxidant capacity *in vitro, ex vivo* and *in vivo* by spectroscopic methods, using DPPH and ABTS, were measured, obtaining important values for the fraction enriched with polyphenols (48.1μmol eq. Trolox / g extract for DPPH and 16,279μmol eq. Trolox / g extract for ABTS). As for *ex vivo* assays, the ethyl acetate fraction (polyphenol-enriched fraction) inhibited phospholipid oxidation in rat liver homogenates (IC_{50} = 14.5μg/ml), autoxidation in rat brain homogenates (IC_{50} = 0.27μg/ml), and in the DNA damage induced by H_2O_2. Furthermore, topical application of the enriched fraction decreased skin chemiluminescence by 38% in an *in vivo* model in mice exposed to UVA (control: 49 ± 5 cps/cm^2). Among the identified compounds, quercetin galloyl glycosides mainly account for the observed antioxidant properties and may be one of the central mechanisms by which *Ligaria cuneifolia* enriched fractions have beneficial effects in traditional medicine.

Two major compounds characterized as pentacyclic triterpenes from the lupeol group were also isolated in this species: betulin and betulinic acid (Fusco et al. 2004). Data reported in the literature (Siddiqui et al. 1988) demonstrated the anticancer and anti-HIV activity of betulinic acid and anti-inflammatory activity of betulin. In addition, there are pharmacognostic studies (Deng and Snyder 2002) that support this property, manifested in a strong inhibition of tumor cells *in vitro* (Cui et al. 1994).

Soberón et al. (2014), performed a study of the antibacterial activity of the extracts of leaves of *Ligaria cuneifolia.* A guided fractionation led to the isolation and identification of compounds with activity against phytopathogens and reference strains. Active ingredients were identified as derivatives of flavonoids.

The treatment of rats with crude extract of *L. cuneifolia* intraperitoneally caused increase of blood viscosity, decreased plasma cholesterol and increased biliary excretion of bile salts (Dominighini et al. 2004). On the other hand, it was found that the methanolic fraction, administered in equal parts with polyvinylpyrrolidone, led to a decrease in plasma cholesterol and an increase in the stiffness index that generates a loss of erythrocyte deformability, causing an increase in blood viscosity. Likewise, it was shown that the methanolic extract has an effect on the lipid content of the erythrocyte membrane, which results in a decrease in the deformation of the red blood cell and a change in shape. Both phenomena have an effect on blood viscosity (Ferrero et al. 2007; Dominighini et al. 2015). However, polyphenol-enriched fractions administered to hyperlipidemic rats for 3, 7 and 10 consecutive days, low total cholesterol (Co) and CoLDL, as well as triglycerides. The decrease in Co may be due to the increase in the rate of bile excretion of bile salts, which leads to increased bile flow. In studies with a high cholesterol diet, the histopathological analysis of the liver showed that treatment with polyphenol-enriched fractions leads to a decrease in the fatty vacuoles present in hypercholesterolemic animals (Gonzalvez et al. 2017). It can also be observed that it does not cause alterations in liver enzymes, nor changes in blood viscosity, nor in the erythrocyte shape, that is, it does not alter blood flow (García et al. 2015; Gerschcovsky et al. 2015, 2016, 2017; Fisch et al. 2016, 2017, 2018; Gonzalvez et al. 2017; Giacosa et al. 2018a, b).

According to the results obtained, these studies allow the fraction rich in polyphenols to be considered as a potential tool in the prevention of cardiovascular disease, by reducing plasma CoLDL, one of the main risk factors in the development of atherosclerosis.

In addition, another study was conducted with patients of both sexes (age: 50 ± 15 years) with a history of hypercholesterolemia, ingested infusion of leaves and young stems of *L. cuneifolia* three times a week for 2 months. At the end of the treatment, a decrease in plasma levels of total cholesterol and CoLDL was observed, without showing alterations in blood flow or liver functionalism (Ferrero et al. 2014).

11 Conclusions

Ligaria cuneifolia has started to be used as a substitute for the European mistletoe (*Viscum album*) in the treatment of blood pressure. It has been demonstrated to have differential properties such as inhibiting the proliferation of LB tumoral cells, high antioxidant activity. Its infusions have the property of lowering plasma cholesterol level in patients with a history of hypercholesterolemia. The accumulated knowledge on the biological activity of different plant organs of *L. cuneifolia*, seem to call for the need of establishing *in vitro* cultures that would allow for the production and more frequent use of this species, in therapeutics.

References

Abbiatti D (1946) Las Lorantáceas Argentinas. Rev Mus La Plata (N S) Sec Bot 7:1–110

Amico GC, Vidal-Russell RV (2019) Uso de hospedadores por el muérdago argentino *Ligaria cuneifolia* (Loranthaceae) a lo largo de su distribución geográfica. Bol Soc Argent Bot 54:395–403

Angiosperm Phylogeny Website (2017) Missouri Botanical Garden. http://www.mobot.org/MOBOT/research/APweb/. (Last access, November 2019)

Arenas Chávez C, Pinazo OM, Zegarra Aragón F, Vera Gonzales C, Torreblanca Gonzáles R, Bellido Valdivia A, Chipana M (2006) Evaluación de la capacidad antioxidante e inhibición de la peroxidación lipídica por el extracto metanólico de *Ligaria cuneifolia* (liga-liga). Véritas 10(1):104–112

Arenas Chávez C, Vera Gonzáles C, Valeriano Zapana JA, Yáñez JA (2011) Identificación de polifenoles individuales y evaluación de la actividad anticancerígena de la *Ligaria cuneifolia* (Liga-liga). Rev ECIPERÚ 8(2):158–162

Barlow B (1964) Classification of the Loranthaceae and Viscaceae. Proc Linnean Soc NSW 89(Part 2):268–272

Barlow B, Hawsksworth F, Kuijt J, Polhill R, Wiens D (1989) Genera of mistletoes. Golden Bough 11:1–4

Benigni R, Capra C, Cattorini P (1964) Piante Medicinali, Chimica, Farmacologia e Terapia, vol II. Ed. Inverni della Beffa, Milán, Italy, pp 1760–1782

Burgstaller JA (2000) 700 hierbas medicinales. Ed. Lancelot, Buenos Aires, Argentina, 224 pp

Castellón A (1998) Loranthaceae. In: Flora del Valle de Lerma, Aportes Botánicos de Salta, Serie Flora 5(2), pp 1–15

Castro MT, Tellería MC (1995) Análisis palinológico de las Loranthaceae (*s.s.*) argentinas. Darwiniana 33(1–4):301–307

Cerdá Zolezzi P, Fernández T, Aulicino P, Cavaliere V, Greczanik S, Lopes C, Wagner M, Ricco R, Gurni A, Hajos S, Álvarez E (2005) *Ligaria cuneifolia* flavonoid fractions modulate cell growth of normal lymphocytes and tumor cells as well as multidrug resistant cells. Immunobiology 209:737–749

Cui S, Reichner J, Matero R, Albina J (1994) Activated murine macrophages induce apoptosis in tumor cells through nitric oxide-dependent or independent mechanism. Cancer Res 54:2462–2467

De Lucca D, Zalles J (1992) Flora Medicinal Boliviana. Ed. Los Amigos del Libro, Cochabamba, Bolivia, pp 306–307

de Pöll EP (1984) Contribución al estudio de las Loranthaceae de Guatemala. Rev. Científica, Facultad de Ciencias Químicas y Farmacia, Universidad de San Carlos de Guatemala 2(1):22–32

Deng Y, Snyder J (2002) Preparation of a 24-nor-1,4-dien-3-one triterpene derivative from Betulin: a new route to 24-Nortriterpene. J Org Chem 67:2864–2873

Dobrecky CB (2016) Variación de los perfiles de polifenoles en *Ligaria cuneifolia*, su estudio mediante electroforesis capilar y su relación con la actividad farmacológica. Thesis. Universidad de Buenos Aires, Argentina. 268 pp. https://core.ac.uk/download/pdf/148073051.pdf

Dobrecky CB, Moreno E, Garcés M, Lucangioli S, Ricco R, Evelson P, Wagner ML (2014) Contenido de polifenoles en *Ligaria cuneifolia* y su relación con la capacidad antioxidante. Dominguezia 30(2):35–39

Dobrecky CB, Flor SA, López PG, Wagner ML, Lucangioli SE (2017) Development of a novel dual CD-MEKC system for the systematic flavonoid fingerprinting of *Ligaria cuneifolia* (R. et P.) Tiegh. -Loranthaceae- extracts. Electrophoresis 38:1292–1300

Domínguez JA (1928) Loranthaceae. In: Contribuciones a la Materia Médica Argentina, Trabajos del Instituto de Botánica y Farmacología 44, pp 260–264

Dominighini A, Ferrero M, Mengarelli G, Ronco M, de Luján Alvarez M, Wagner ML, Gurni A, Carnovale C, Luquita A (2004) Hemorheologic changes induced by aqueous extract of *Ligaria cuneifolia*: treatment by two different administration routes. Clin Hemorheol Microcirc 31(2):113–121

Dominighini A, Ferrero M, Crosetti D, Ronco MT, Gonzálvez J, Urli L, Wagner ML, Gurni A, Carnovale CE, Luquita A (2015) Effects of proanthocyanidin enriched extract from *Ligaria cuneifolia* on plasma cholesterol and hemorheological parameters. In vivo and in vitro studies. Clin Hemorheol Microcirc 60(3):317–325

Engler A, Krause K (1935) Loranthaceae. In: Enlger A, Prantl K (eds) Die natürlichen Pflanzenfamilien, 2° Edition, vol 16B. Ed. W. Engelmann, Leipzig, Germany, pp 98–203

Fernández TB, Wagner ML, Varela BG, Ricco RA, Hajos SE, Gurni AA, Álvarez E (1998) Study of an Argentine Mistletoe, the hemiparasite *Ligaria cuneifolia* (R. et P.) Tiegh. (Loranthaceae). J Ethnopharmacol 62:25–34

Fernández TB, Varela BG, Taira CA, Ricco RA, Gurni AA, Hajos SE, Álvarez EMC, Wagner ML (2000) Mistletoes from Argentina. *Ligaria cuneifolia* var. *cuneifolia* as a substitute for the European Mistletoe (*Viscum album* L.). In: Bussing A (ed) Mistletoe. The genus *Viscum*. Ed. CRC Press, London, UK, pp 61–74

Fernández T, Cerdá Zolezzi P, Aulicino P, Caldas Lopes E, Wagner M, Ricco R, Hajos S, Gurni A, Álvarez E (2003) Immunobiological features of the galactoside lectin L-Lc isolated from the Argentine mistletoe *Ligaria cuneifolia*. J Ethnopharmacol 85(1):81–92

Ferrero M, Crosetti D, Dominighini A, de Luján Alvarez M, Ronco MT, Wagner ML, Gurni A, Carnovale C, Luquita A (2007) Effect of the vehicle polyvinylpyrolidone and the methanolic fraction of *Ligaria cuneifolia* (Argentine Mistletoe) extract on hemorheological properties and biliary secretion in rats. Clin Hemorheol Microcirc 36(2):95–104

Ferrero M, Crosetti D, Svetaz MJ, Wagner ML, Scaglione, Petrucci J, Belloscar J, Carnovale CE, Luquita A (2014) Efecto de la infusión de *Ligaria cuneifolia* o "muérdago criollo" sobre el colesterol plasmático, propiedades reológicas y forma eritrocitaria en pacientes con hipercolesterolemia. Publicación Periódica Anual de la Sociedad de Biología de Rosario. Resúmenes del XVI Congreso y XXXIV Reunión Anual: 39. https://www.sbr.org.ar/Libros/libro_resumenes_2014.pdf

Feuer S, Kuijt J (1978) Fine structure of mistletoe pollen. Eremolepidaceae, *Lepidoceras* and *Tupeia*. Can J Bot 56:2853–2864

Fisch J, García G, Dominighini A, Galliano S, Urli L, Crosetti D, Gerschcovsky N, Monti J, Lambertucci F, Ronco MT, Wagner ML, Carnovale CE, Luquita A (2016) Treatment with *Ligaria cuneifolia* proanthocyanidin enriched fraction produces plasma lipids concentrations decrease and increase of bile salts biliary excretion in Wistar rats fed with a high fat diet (hfd). Biocell 40(s4):8

Fisch J, Gerschcovsky N, Dominighini A, Galliano S, Gonzalvez J, Urli L, Crosetti D, Giacosa C, Monti J, Lambertucci F, Ronco MT, Wagner ML, Carnovale C, Luquita A (2017) Effect of *Ligaria cuneifolia* (Lc) proanthocyanidin enriched fraction on plasma lipids levels and bile salts biliary output in Wistar rats fed with a high fat diet (HFD). Medicina 77(sI):339

Fisch J, Giacosa C, Galliano S, Urli L, Dominighini A, Crosetti D, Gonzálvez J, Monti J, Ronco MT, Wagner ML, Carnovale C, Luquita A (2018) Analysis of concentration of plasma cholesterol and evaluation of hemorheological parameters in hyperlipemic Wistar rats treated for 7 and 10 days with *Ligaria cuneifolia* (Lc)-proanthocyanidine enriched fraction. Medicina 78(sIII):87

Font Quer P (1962) Plantas Medicinales: El Dioscórides Renovado. Ed. Labor, Barcelona, Spain, pp 136–139

Fusco M, Petenatti E, Inostroza N, San-Martín A, Rivera P (2004) Principios activos de la *Ligaria cuneifolia*. Una planta Medicinal. Cienc Trabajo 6(12):79–82

García G, Dominighini A, Gonzálvez J, Urli L, Crosetti D, Gerschcovsky N, Monti J, Ronco MT, Wagner ML, Carnovale CE, Luquita A (2015) Efecto del tratamiento de proantocianidinas extraídas de *Ligaria cuneifolia* (Lc) en ratas Wistar con dieta hiperlipémica sobre la concentración de colesterol plasmático y la viscosidad sanguínea. Medicina 75(sII):78

Gerschcovsky N, García G, Dominighini A, Gonzálvez J, Galliano S, Urli L, Crosetti D, Monti J, Ronco MT, Wagner ML, Carnovale C, Luquita A (2015) Effect of *Ligaria cuneifolia* (Lc) proanthocyanidine enriched fraction on plasma cholesterol levels and blood fluidity in Wistar rats fed with hyperlipidemic diet. Biocell 39(s4):14–15

Gerschcovsky N, Galliano S, García G, Dominighini A, Gonzálvez J, Urli L, Crosetti D, Monti J, Lambertucci F, Ronco MT, Wagner ML, Carnovale C, Luquita A (2016) Decrease of plasma lipids without impairment of blood fluidity by *Ligaria cuneifolia* (Lc) protoanthocyanidine enriched fraction in Wistar rats fed with high fat diet (HFD). Medicina 76(sI):104

Gerschcovsky N, Fisch J, Dominighini A, Giacosa C, Gonzálvez J, Urli L, Galliano S, Crosetti D, Monti J, Ronco MT, Lambertucci F, Wagner ML, Carnovale C, Luquita A (2017) *Ligaria cuneifolia* (Lc) protoanthocyanidine fraction produces plasma diminution of cholesterol and triglycerides improvement morphological index in rats fed with high fat diet (HFD). Medicina 77(sI):604

Gerschcovsky N, Knox M, Carnovale C, Wagner ML, Vinet R, Luquita A (2018) Efecto biológico de proantocianidina extraída de *Ligaria cuneifolia* sobre la contractibilidad de aorta aislada de rata. http://solaplamed.org/wp-content/uploads/2019/01/MPC-1-1-4-2018-23-39.pdf. (Last access, November 2019)

Giacosa C, Gerschcovsky N, Galliano S, Fisch J, Dominighini A, Gonzálvez J, Crosetti D, Urli L, Ronco MT, Monti J, Wagner ML, Carnovale C, Luquita A (2018a) Evaluation of the concentration of plasma cholesterol, blood: fluidity, hepatic and renal functions in hyperlipemic Wistar rats treated with proanthocyanidin extracted from *Ligaria cuneifolia* (Lc). Biocell 42(s4):25 https://www.mendoza.conicet.gov.ar/portal//biocell/ROSARIO_BIOLOGY_SOCIETY_XIX_CONGRESS_XXXV_ANNUAL_MEETING.pdf. (Last access, November 2019)

Giacosa C, Fisch J, Galliano S, Dominighini A, Urli L, Roma SM, Monti J, Ronco MT, Wagner ML, Pisani G, Carnovale C, Luquita A (2018b) Analysis of concentration of plasma lipids and histological liver study in hyperlipemic Wistar rats fed with high fat diet (HFD) treated for 7 and 10 days with *Ligaria cuneifolia* (Lc)-proanthocyanidin enriched fraction. Medicina 78(sIII):162–163

González Torres DM (2018) Catálogo de Plantas medicinales (y alimenticias útiles) usadas en Paraguay. Editorial Sevilibro, Asunción, Paraguay, 456 pp

Gonzálvez J, García G, Galliano S, Dominighini A, Urli L, Monti J, Ronco MT, Francés D, Wagner M, Carnovale C, Luquita A (2017) The enriched proanthocyanidin extract of *Ligaria cuneifolia* shows a marked hypocholesterolemic effect in rats fed with cholesterol- enriched diet. Recent Pat Endocr Metab Immune Drug Discovery 11(1):47–53

Graziano NN, Widmer GA, Juliani R, Coussio JD (1967) Flavonoids from the argentine mistletoe *Psittacanthus cuneifolius*. Phytochemistry 6(12):1709–1711

Hieronymus J (1882) Plantae diaphoricae florae Argentinae. Ed. Atlántida, Buenos Aires, Argentina, pp 259–262

Kuijt J (1988) Monograph of the Eremolepidaceae. Syst Bot Monogr Am Soc Plant Tax 18:1–60

Martínez GJ (2008) La farmacopea natural en la etnomedicina de los Tobas del Río Bermejito (Chaco, Argentina). Thesis. Universidad Nacional de Córdoba. Argentina. 288 pp

Martínez GJ (2010) Las plantas en la medicina tradicional de las Sierras de Córdoba. Un recorrido por la cultura campesina de Paravachasca y Calamuchita. Ed. del Copista, Córdoba, Argentina, 212 pp

Martínez-Crovetto R (1964) Estudios etnobotánicos I. Nombres de plantas y su utilidad, según los indios tobas del este del Chaco. Bonplandia 1:279–333

Martínez-Crovetto R (1981) Plantas utilizadas en medicina popular en el Noroeste de Corrientes (República Argentina). Miscelánea 69:1–139

Marzocca A (1997) Vademécum de malezas medicinales de la Argentina, indígenas y exóticas. Ed. Orientación Gráfica Editora, Buenos Aires, Argentina, pp 165–166

Mas Serra MT, Verdú González AM, Trillo C, Bertero HD (2008) Plantas silvestres de Santa Victoria Oeste y sus usos. Norte de la provincia de Salta, Argentina. Universitat Politècnica de Catalunya, Barcelona, Spain, 175 pp

Menseguez P, Galetto L, Anton AM (2007) El uso de plantas medicinales en la población campesina de El Puesto (Córdoba, Argentina). Kurtziana 33(1):89–102

Nájera MT (1983) La Herboristería en la República Argentina. Acta Farm Bonaer 2(1):55–59

Olson DM, Dinerstein E, Wikramanayake ED, Burgess ND, Powell GVN, Underwood EC, D'amico JA, Itoua I, Strand HE, Morrison JC, Loucks CJ, Allnutt TF, Ricketts TH, Kura Y, Lamoreux JF, Wettengel WW, Hedao P, Kassem KR (2001) Terrestrial Ecorregions of the world: a new map of life on earth: a new global map of terrestrial ecoregions provides an innovative tool for conserving biodiversity. Bioscience 51:933–938

Paris RR, Moyse H (1981) Précis de Matière Médicale. 3 tomes. Ed. Masson, Paris, France
Pochettino ML, Martínez MR (2000) Aporte al conocimiento actual de las plantas medicinales en Argentina: Estudio etnobotánico en el Departamento de Molinos, Provincia de Salta, Argentina. In: Amat AG (comp.), Farmacobotánica y farmacognosia en Argentina 1980–1998. Ed. Culturales Argentinas, La Plata, Argentina, p 298
Ragonese A, Milano V (1984) Vegetales y sustancias tóxicas de la flora Argentina. Enciclopedia Argentina de Agricultura y Jardinería, II Fasc. 8–2, 2nd edn. Ed. ACME, Buenos Aires, Argentina, pp 89–92
Ratera E, Ratera M (1980) Plantas de la flora argentina empleadas en medicina popular. Ed. Hemisferio Sur, Buenos Aires, Argentina, pp 82–84
Ricco MV, Bari ML, Ricco RA, Wagner ML, Álvarez MA (2018) Análisis de polifenoles e iniciación de cultivos in vitro de *Ligaria cuneifolia* (R. et P.) Tiegh (Loranthaceae) de ejemplares provenientes de la localidad La Población, provincia de Córdoba, Argentina. Dominguezia 34(1):53–64
Ricco MV, Bari ML, Bagnato F, Cornacchioli C, Laguia- Becher M, Spairani LU, Posadaz A, Dobrecky C, Ricco RA, Wagner ML, Álvarez MA (2019) Establishment of callus cultures of the Argentinean mistletoe, *Ligaria cuneifolia* (R. et P.) Tiegh. (Loranthaceae) and screening of their polyphenolic content. Plant Cell Tissue Organ Cult 138:167–180
Sanzin R (1918) Sobre el *Loranthus cuneifolius*. Rev Chil Hist Nat 22:126–129
Scarpa GF (2004) Medicinal plants used by the Criollos of Northwestern Argentine Chaco. J Ethnopharmacol 91(1):115–135
Scarpa GF (2009) Etnobotánica médica de los indígenas chorote y su comparación con la de los criollos del Chaco semiárido (Argentina). Darwiniana 47(1):92–107
Scarpa GF, Montani M (2011) Etnobotánica médica de las "ligas" (Loranthaceae *sensu lato*) entre indígenas y criollos de Argentina. Dominguezia 27(2):5–19
Siddiqui S, Hafeez F, Begun S, Siddiqui B (1988) Oleanderol, a new pentacyclic triterpene from the leaves of *Nerium oleander*. J Nat Prod 51:229–233
Soberón J, Sgariglia M, Dip Maderuelo M, Andina M, Sampietro D, Vattuone M (2014) Antibacterial activities of *Ligaria cuneifolia* and *Jodina rhombifolia* leaf extracts against phytopathogenic and clinical bacteria. J Biosci Bioeng 118(5):599–605
Subils R (1984) Loranthaceae. Bol Soc Argent Bot 23(176):121–264
Taira CA, Wagner ML, Adrados HM, Pino R, Gurni AA (1994) Estudio farmacológico de un agente vasoactivo presente en *Ligaria cuneifolia* var. *cuneifolia*. Acta Farm Bonaer 13(2):91–95
Taira CA, Wagner ML, Gurni AA (2004) Pharmacological and cardiovascular aspects of *Ligaria cuneifolia* infusions. Pharm Biol 42(3):234–239
Takhtajan A (1997) Diversity and classification of flowering plants. Ed. Columbia University Press, New York, USA, pp 11–19/353–359
The Plant List (2019) Version 1.1 published on the Internet: http://www.theplantlist.org. (Last access, November 2019)
Toursarkissian M (1980) Plantas medicinales de la Argentina. Ed. Hemisferio Sur, Buenos Aires, Argentina, p 79
Ulibarri E (1987) Loranthaceae. In: Flora ilustrada de Entre Ríos (Argentina). Col. Cient. INTA, Buenos Aires, Argentina 6(III):106–121
Varela BG (2003) Anatomía foliar y caulinar comparativa de los muérdagos argentinos. Importancia farmacobotánica. Relación con el muérdago europeo. Thesis. Universidad de Buenos Aires, Argentina, 260 pp
Varela BG, Gurni AA (1995) Anatomía foliar y caulinar comparativa del muérdago criollo y del muérdago europeo. Acta Farm Bonaer 14(1):21–29
Varela BG, Fernández T, Taira C, Cerdá Zolezzi P, Ricco R, Caldas López E, Álvarez E, Gurni A, Hajos S, Wagner ML (2001) El muérdago criollo, *Ligaria cuneifolia* (R. et P.) Tiegh. -Loranthaceae-. Desde el uso popular hacia el estudio de los efectos farmacológicos. Dominguezia 17(1):31–50

Vázquez y Novo SP, Wagner ML, Gurni AA, Rondina RVD (1989) Importancia Toxicológica de la presencia de sustancias aminadas en ejemplares de *L. cuneifolia* var. *cuneifolia* colectados en distintas áreas de la República Argentina. Acta Farm Bonaer 8(1):23–29

Wagner ML (1993) Estudios fitoquímicos comparativos de los flavonoides de Loranthaceae de la flora argentina. Relación con el muérdago europeo. Thesis. Universidad de Buenos Aires, Argentina. 255 pp

Wagner H, Feil B, Seligmann O, Petricic J, Kavogjera Z (1986) Phenylpropanes and lignans of *Viscum album* L. Cardioactive drugs V. Planta Med 2:77–162

Wagner ML, Fernández T, Varela B, Álvarez E, Hajos S, Gurni A (1998) Anatomical, phytochemical and immunochemical studies on *Ligaria cuneifolia* (R. et P.) Tiegh. (Loranthaceae). Pharm Biol 36(2):131–139

Wettstein R (1944) Tratado de Botánica Sistemática. Ed. Labor, Barcelona, Spain, pp 601–605

Youngken HW (1951) Tratado de Farmacognosia. Ed. Atlante, México, pp 365–367

Zuloaga FO, Morrone O (1999) Catálogo de Plantas Vasculares del Cono Sur – Instituto de Botánica Darwinion http://www.darwin.edu.ar/Proyectos/FloraArgentina/fa.htm. (Last access, November 2019)

Lippia integrifolia (Griseb.) Hieron.

César A. N. Catalán, Julieta V. Catalán, and Diego A. Sampietro

Lippia integrifolia (Griseb.) Hieron. (Photos: C Catalán)

Abstract *Lippia integrifolia* (Griseb.) Hieron. is a native aromatic shrub, commonly known as "incayuyo" or "inca tea". It is widely used in traditional medicine in northern and central Argentina, where infusions of leaves and flowers are used as digestive, emmenagogue, antitussive, carminative, and against dyspepsia, stomachache, flu and indigestions. Its essential oil is a rich source of sesquiterpenoids based on unique or very rare skeletons such as integrifoliane, lippifoliane, asteriscane,

C. A. N. Catalán (✉) · J. V. Catalán
Instituto de Química Orgánica, Facultad de Bioquímica, Química y Farmacia, Universidad Nacional de Tucumán, San Miguel de Tucumán, Tucumán, Argentina
e-mail: ccatalan@fbqf.unt.edu.ar; cancatalan@gmail.com

D. A. Sampietro
Facultad de Bioquímica, Química y Farmacia, Laboratorio de Biología de Agentes Bioactivos y Fitopatógenos (LABIFITO), Universidad Nacional de Tucumán, San Miguel de Tucumán, Tucumán, Argentina

Á. Máthé, A. Bandoni (eds.), *Medicinal and Aromatic Plants of South America Vol. 2*, Medicinal and Aromatic Plants of the World 7,
https://doi.org/10.1007/978-3-030-62818-5_24

africanane and 4,5-*seco*-africanane. The essential oil shows a remarkable chemical diversity, with at least five clearly defined chemotypes. Two of the most frequently found chemotypes stand out for their unique components, one called "*trans*-davanone chemotype" that has this ketone as the dominant component (>75%); and the other, called "lippifolienone chemotype", rich in oxygenated sesquiterpenes based on the rare lippifoliane and integrifoliane skeletons. Aqueous and ethanol extracts of "incayuyo" showed antiadhesive properties against *Helicobacter pylori.* In addition, the secretion of IL-8 induced by *Helicobacter pylori* was significantly reduced by co-incubation of AGS cells with "incayuyo" extract. Infusions and decoctions of "incayuyo" displayed strong antioxidant, choleretic and antispasmodic effects, stimulated phagocytosis rate of macrophages and inhibited the LPS-induced NO-secretion. In addition, a significant reduction of contractions induced by acetylcholine, $CaCl_2$ and KCl was observed in isolated rat jejunum. The bioactive components of "incayuyo" extracts are flavonoid glycosides, (poly)-phenolic compounds, phenylethanoid glycosides (acteoside, isoacteoside) and sesquiterpenoids. All experimental data support the use of "incayuyo" in traditional medicine. Comparative studies on the pharmacological efficacy of the different essential oil chemotypes are necessary. The qualitative and quantitative content of non-volatile bioactive metabolites should be used to define quality parameters.

Keywords "Incayuyo" · Essential oil · Chemotypes · Anti-inflammatory · Antiadhesive · Phenolic compounds

1 Introduction

Lippia integrifolia (Gris.) Hieronymus is an aromatic medicinal plant commonly known as "incayuyo" that grows wild in Northwestern and Central Argentina. Infusions of its aerial parts have been used since colonial times mainly for the treatment of dyspepsia, gastric inflammation, as a digestive and as a nervous tonic. Its essential oil shows a high variability in its chemical composition and is characterized by the presence of sesquiterpenoids with rare skeletons. Five chemotypes have been characterized, namely, (i) *trans*-davanone; (ii) lippifolienone; (iii) β-davanone-2-ol; (iv) spathulenol/bicyclogermacrene; and v) *trans*-nerolidol. The aqueous extracts (infusions and decoctions) showed to be rich in flavonoid hexosides, phenylethanoides (acteoside, isoacteoside) and sesquiterpenes derived from lippidulcine. The extracts exhibited antiadhesive properties against *Helicobacter pylori* and anti-inflammatory effect on stomach cells. Furthermore, both infusions and decoctions showed strong antioxidant activity (β-carotene/linoleic acid assay) and free radical scavenging activity (DPPH assay) that were comparable to the synthetic antioxidant BHT. Experimental results validate the use of incayuyo in traditional medicine and indicate that the bioactive components are concentrated in the phenolic fraction.

2 Taxonomic Characteristics

The genus *Lippia* belonging to the Verbenaceae family, Lantaneae tribe, embraces some 160 species growing in warm and temperate regions of South America, Central America and Africa (Jansen-Jacobs 1988; Múlgura de Romero 2003; Denham et al. 2006). The members of this genus are shrubs, sub-shrubs or perennial herbs, many of them are aromatic and widely used in traditional medicine (Terblanché and Kornelius 1996; Pascual et al. 2001; Catalán and de Lampasona 2002; Barboza et al. 2009). The name of this genus is a tribute to Dr. Augusto Lippi (1678–1703), a French doctor and naturalist.

Lippia integrifolia (Griseb.) Hieron. is a native aromatic shrub, perennial, 30–100 cm tall, that grows in the Northwest and Center of Argentina (provinces of Catamarca, Jujuy, Salta, Tucumán, La Rioja, San Juan and Córdoba), Southwest of Bolivia (departments of Cochabamba, Potosí and Tarija) and probably in Northern Chile (Denham et al. 2006). It has been also cited for Peru (Pool 1993).

Synonyms *Lippia boliviana* Rusby, *Lippia boliviana* var. *angusta* Moldenke; *Lippia boliviana* var. *integrifolia* Moldenke, *Lippia integrifolia* var. *beckii* Moldenke, *Lippia turbinata* var. *integrifolia* Griseb. (Denham et al. 2006; The Plant List 2020).

Common names: "incayuyo", "té del inca", "poleo", "puleo", "pulco" "manzanilla" (Múlgura de Romero 2003; Alonso and Desmarchelier 2006; Barboza et al. 2009).

Incayuyo is taxonomically and phytogeographically related to two other aromatic Verbenaceae species, commonly known as "poleo": *Lippia turbinata* Griseb. and *L. fissicalyx* Tronc., that sometimes may be present as contaminants. However, there are anatomo-morphological differences that allow distinguishing the three species (Bassols and Gurni 2000).

3 Crude Drug Used

Infusions of dried leaves or dried leaves mixed with flowers and fruits are used as digestive, emmenagogue, antitussive, carminative, against dyspepsia, stomachache, flu, indigestions, as a sedative and as a soft diuretic (Ratera and Ratera 1980; Rondina et al. 2003; Alonso and Desmarchelier 2006; Gorzalczany et al. 2008; Barboza et al. 2009). "Incayuyo" aqueous extracts and especially its phenolic fraction displayed antiadhesive properties against *Helicobacter pylori*, anti-inflammatory effect on stomach cells and powerful antioxidant activity (Ricco et al. 2010; Marcial et al. 2014).

L. integrifolia has been incorporated into the Argentine Food Code as a seasoning. It is used as an ingredient of some well-known trademarks of soft drinks, appetizers and teas (Marcial et al. 2016).

4 Major Chemical Constituents and Bioactive Compounds

Several authors have investigated the chemical composition of the essential oil obtained from aerial parts of *Lippia integrifolia,* that proved to be rich in oxygenated sesquiterpenoids based on novel and rare sesquiterpene skeletons (Catalán et al. 1983, 1991, 1992, 1993, 1994, 1995; Dartayet et al. 1984; Velasco-Negueruela et al. 1993; Zygadlo et al. 1995; de Lampasona et al. 1999; Fricke et al. 1999; Juliani et al. 2004; Cerda-García-Rojas et al. 2005, 2008; Coronel et al. 2006; Ricco et al. 2010; Gleiser et al. 2011; Lima et al. 2011; Barbieri et al. 2016; Marcial et al. 2016; Leal et al. 2018) such as integrifoliane-1,5-dione (**1**) (structures are shown in Fig. 1), lippifoli-1,6-en-5-one (**2**), lippifoli-1,6-en-4β-ol-5-one (**7**) and 4,5-*seco*-africane-4,5-dione (**10**) based on the novel sesquiterpene skeletons integrifoliane, lippifoliane and 4,5-*seco*-africanane, respectively (Catalán et al. 1991, 1992); the oxygenated derivatives **3** (Fricke et al. 1999), **4** (Coronel et al. 2006) and **6** (Catalán et al. 1995) based on the uncommon africanane skeleton; asteriscane derivative **5** and keto-alcohol **8** (Catalán et al. 1995).

Recently, a detailed study of the essential oil composition of thirty-one wild populations of "incayuyo" covering most of its natural range (Marcial et al. 2016) allowed to establish the presence of five clearly differentiated chemotypes, namely: (i) "lippifolienone chemotype" with ketone **2** as relevant and distinctive constituent accompanied by several africanane derivatives such as **3** and **6**, integrifoliane-1,5-dione (**1**), 4,5-*seco*-africane-4,5-dione (**10**), and caryophyllene oxide; (ii) "*trans*-davanone chemotype" with ketone **11** as dominant component of the essential oil (>75%); (iii) "β-davanone-2-ol chemotype" (= hydroxydavanone chemotype) with keto-alcohol **12** as the main component accompanied by significant amounts of *trans*-davanone (**11**); (iv) "spathulenol/bicyclogermacrene chemotype" characterized by the presence of significant amounts of spathulenol, bicyclogermacrene and β-caryophyllene; and v) "*trans*-nerolidol chemotype" with sesquiterpenic alcohol **13** as the main component. The "*trans*-davanone" chemotype is found mainly in the north of its natural distribution area (Southern Bolivia and Jujuy, Salta and Tucumán provinces in Argentina), while the chemotypes "lippifolienone" and "spathulenol/bicyclogermacrene" are frequent in the center and South of its distribution area (Catamarca, La Rioja, San Juan, San Luis and Córdoba provinces of Argentina). Chemotypes "*trans*-nerolidol" and "β-davanone-2-ol" are rare (Marcial et al. 2016).

Variations in the essential oil composition may be due to the phenological stage, environmental conditions or the existence of chemotypes. In general, variations in the environmental conditions (soil type, water availability, season and so on) produce mostly quantitative variations in the essential oil composition (Coronel et al. 2006) with little or no change in the qualitative chemical profile. In turn, chemotypes frequently display significant qualitative differences that are largely genetically controlled and remain stable over the years (Catalán and de Lampasona 2002; Coronel et al. 2005; González et al. 2012; Mercado et al. 2015; Marcial et al. 2016). In relation to non-volatile metabolites, the content of (poly)-phenolic compounds, flavonoids

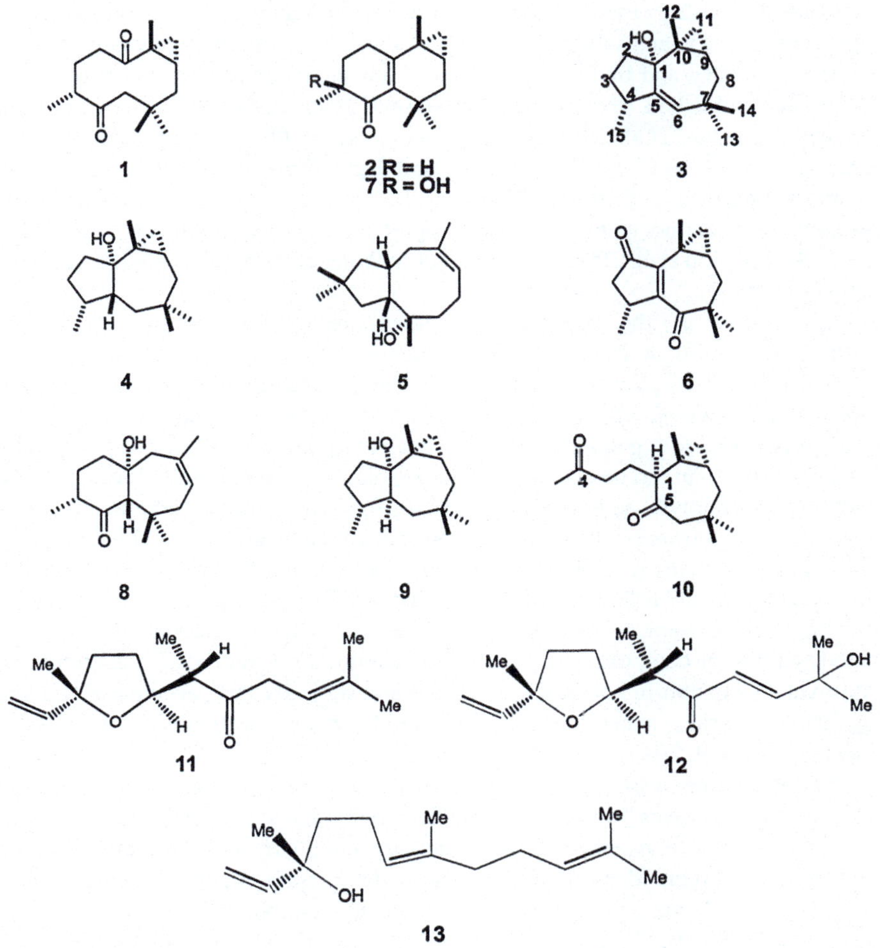

Fig. 1 Structure of some distinctive components of *Lippia integrifolia*

and hydroxycinnamic acids as well as the presence of phenylethanoid glycosides (acteoside and isoacteoside), flavonoid glycosides and sesquiterpenoids in the infusions of *L. integrifolia* have been determined (Ricco et al. 2010; Marcial et al. 2014).

"Incayuyo" infusions and decoctions showed high phenolic content and strong antioxidant activity, while no genotoxicity was observed in the comet-assay (Ricco et al. 2010). In addition, liquid chromatography coupled to mass spectrometry (LC-MS) identified several flavonoid hexosides, phenylethanoids and the sesquiterpenes (epi)-lippidulcin A and peroxylippidulcin in a decoction of aerial parts (Marcial et al. 2014). The following phenolic compounds were identified: two 6-hydroxyluteolin-hexosides, a 6-methoxyluteolin-*O*-hexoside, two 6-methoxyluteolin-hexosides, 6-methylscutellarein-hexoside, phloretin-3′,5′-di-C-hexoside,acteoside, isoacteoside, dimethyl-*seco*-loganoside, a methoxylated apigenin-hexoside and salvigenin (Marcial et al. 2014). Much research and reviews on the chemistry and biological activity of flavonoids and their impact on health have been published (Robak and Gryglewski 1996; Agrawal 2011; Kumar and Pandey 2013; Jucá et al. 2018; and references therein). On the other hand, acteoside and isoacteoside are isomeric phenylethanoid glycosides with cytoprotective and memory improvement effects (Shiao et al. 2017), and potent antioxidant, anti-inflammatory, antinociceptive, and hepatoprotective activity (Ohno et al. 2002; Lee et al. 2004; Chiou et al. 2004; Koo et al. 2006; Wang et al. 2009; Yamada et al. 2010; Jing et al. 2015).

In relation to the essential oil components, *trans*-davanone (**11**) and β-davanone-2-ol both showed potent antifungal activity against *Aspergillus niger*, *A. ochraceus*, *A. versicolor*, *A. flavus*, *A. terreus*, *Penicillium ochrochloron*, *P. funiculosum*, *Trichoderma viride*, *Cladosporium cladosporioides* and *Alternaria alternate*. It is worth noting that the activity shown by β-davanone-2-ol was comparable to that of the commercial antifungal fluconazole (Vajs et al. 2004). Davanone (**11**) exhibited spasmolytic activity (Perfumi et al. 1995) and moderate repellency against *Tribolium castaneum* (Coleoptera: Tenebrionidae) (Hu et al. 2019), whilst hydroxy-davanone (β-davanone-2-ol) and specially hydroperoxide of davanone strongly inhibit cell growth of MCF-7, SK-N-MC, and A2780 cancer cells through induction of apoptosis (Hosseinzadeh et al. 2019). The sesquiterpenic alcohol spathulenol is another relevant component of "incayuyo" essential oil that possesses significant antioxidant and anti-inflammatory activities and showed to be particularly effective against an ovarian cancer cell line. It has moderate antimycobacterial activity (do Nascimento et al. 2018).

Almost nothing is known about the bioactivity of lippifoliane, integrifoliane and africanane derivatives which are distinctive components of lippifolienone chemotype; and, therefore, research on the bioactivity and pharmacological effects of this type of sesquiterpenoids is highly desirable and necessary. A paper reports hypotensive and CNS depressant effects for an africanane hydrocarbon, african-4(15)-ene (Reddy et al. 1999). "Incayuyo" has many bioactive metabolites and therefore several effects should be expected, especially if it is consumed regularly and for prolonged times. Due to the existence of at least five well differentiated chemotypes, one might expect differential pharmacological effects for each of them. However, no comparative studies have been conducted on the pharmacological efficacy of different chemotypes, consequently this is still a promising area of research.

5 Morphological Description

Perennial aromatic bush 30–100 cm tall; branches with reddish-brown bark, exfoliable. Leaves with entire margins, 1–5(7) × 0.5–3 cm, opposite, occasionally trifoliate, linear or narrowly oblong, sub-obtuse at the apex, base with often rumpled

Photo 1 Branches with flowers of *L. integrifolia*. (Photo: C Catalán)

margin, rough-scabrous adaxial surface, glandulous abaxial surface, with remarkable central nerve. Small white flowers in pedunculated spikes clustered on dense heads and small dried fruit included into persistent calyx which becomes two separate nutlets. Capituliform spikes, 1 or 2 per armpit, subglobous, 4–5 mm diam., peduncles of 5–6 mm; imbricated bracts, 2–2.5 mm, rhombic or ovate, concave, accumulated, pubescent and dotted-resinous under pubescence; calyx tubular, 2.5 mm long, cleft in the adaxial and abaxial faces, hispidule and ciliated, persistent and covering the fruits; white or white-pink corolla, *ca.* 5 mm. Piriform fruit, 2 × 2 mm (Photo 1) (Múlgura de Romero 2003; Denham et al. 2006).

A comparative anatomical study of nine Argentine species of the genus *Lippia* used in folk medicine, which includes *L. integrifolia* (Bassols and Gurni 2000) is available.

Also, a cytogenetic study of *Lippia turbinata*, *L. fissicalyx* and *L. integrifolia* showed that the number of chromosomes was 2n = 30 for *L. integrifolia*; 2n = 30 and 2n = 4 × 60 for *L. turbinata*; and 2n = 4 × 60 for *L. fissicalyx*; the meiotic chromosome behavior was regular in diploids and highly irregular in tetraploids. Pollen analysis evidenced a high production of normal pollen in diploid individuals and a low one in tetraploids (Pastoriza and Andrada 2006).

6 Geographical Distribution

Southwestern Bolivia (Cochabamba, Potosí and Tarija Departments), Northwestern and central Argentina (Catamarca, Jujuy, Salta, Tucumán, La Rioja, San Juan and Córdoba provinces) and possibly Northern Chile (Múlgura de Romero 2003; Denham et al. 2006). It has been also cited for Peru (Pool 1993).

7 Ecological Requirements

"Incayuyo" grows mainly in the Biogeographical Provinces of "Monte" and "Chaqueña" in stony hills and dry and arid fields. The biodiversity of this species is threatened by habitat loss and the indiscriminate and uncontrolled collection of wild populations to meet market demands.

The interaction between cool temperature and darkness during alternating temperatures treatment on the germination of photoblastic seeds of *Lippia integrifolia*, as well as other *Lippia* and *Aloysia* species have been investigated (Galíndez et al. 2017).

Reproduction by seeds is the usual form of propagation in nature: after a few years, disturbed habitats like landslides and clearings are usually invaded by "incayuyo" and other species such as pimples (*Acacia caven*) and "molle" (*Schinus molle*, *Schinus polygamus*).

In cultivation, propagation by cuttings is the most frequent method. Cuttings are prepared from woody stems of adult plants (Herbotecnia 2005). Breeding can be done by seeds, and by stakes. Seeds are rarely used for seedling production given the slow growth and the easiness with which it can be reproduced by other methods.

Studies have been carried out on breeding and *in vitro* propagation of *L. integrifolia* (Passera and Ambrosetti 1999) and the effect of polyploidization on the production of essential oil (Iannicelli et al. 2016).

8 Traditional Use (Part(s) Used) and Common Knowledge

Infusions prepared from dried leaves with flowers and fruits are used as digestive, emmenagogue, antitussive, carminative, against dyspepsia, stomachache, against flu, indigestions, as a sedative and as a soft diuretic (Ratera and Ratera 1980; Rondina et al. 2003; Alonso and Desmarchelier 2006; Gorzalczany et al. 2008; Barboza et al. 2009). Leaves and flowers infusions are also used as digestive, to relieve stomachache, as emmenagogue, against flu and as antitussive (Barboza et al. 2009). Dried aerial parts provide infusions that are used as nervine, stomachic, emmenagogue, tonic, diuretic (Barboza et al. 2009). Infusions are prepared at 2%, drinking two to three cups per day (Alonso and Desmarchelier 2006).

9 Modern Medicine Based on Its Traditional Medicine Uses

Several *Lippia* species are widely used in folk medicine and the chemical composition, pharmacologic activity and traditional use of the members of this genus have been reviewed (Bassols and Gurni 1996; Terblanché and Kornelius 1996; Pascual et al. 2001; Catalán and de Lampasona 2002).

Two well-known members of this genus are *L. dulcis* Trev., a herb producing (+)-hernandulcin, a sesquiterpenoid 1000 times sweeter than sucrose (Compadre et al. 1986; Souto-Bachiller et al. 1997) and *L. graveolans* Kunth, an aromatic shrub commonly called "Mexican oregano" that grows in Texas and Central America where its leaves and flowers are sold as a culinary seasoning under the name "whole oregano"; this spice should not be confounded with "oregano" or "European oregano", the leaf of the herb *Origanum vulgare* L. (Lamiaceae) (Leung and Foster 1996).

In the case of *L. integrifolia*, the aqueous extracts showed strong *in vitro* antioxidant capacity and inhibited adhesion to stomach cells up to 40%, while an ethanol-soluble fraction showed inhibition rates of up to 60%. Decoctions of aerial parts increased the cellular viability of AGS cells significantly while the proliferation rate was not influenced. In addition, the secretion of IL-8 induced by *Helicobacter pylori* was significantly reduced by co-incubation of AGS cells with "incayuyo" extracts (Marcial et al. 2014). The content of total phenols, flavonoids and hydroxycinnamic acids in infusions and decoctions of "incayuyo" and other members of the Verbenaceae family (Ricco et al. 2010) were determined. *L. integrifolia* showed the highest antioxidant activity and no genotoxic activity was detected by means of the comet-assay for any of the species tested (Ricco et al. 2010).

Aqueous extracts stimulated phagocytosis rate of macrophages and inhibited the LPS-induced NO-secretion (Marcial et al. 2014). The essential oil exhibited fungicidal activity against the ascomycete fungus *Ascosphaera apis* (Dellacasa et al. 2003), strong repellent activity against *Triatoma infestans* (triatomine bugs), a Chagas disease vector (Lima et al. 2011) and fungitoxic activity against *Aspergillus* spp. (Passone et al. 2012), while the methanol extract showed low antifungal activity against yeasts, hialohyphomycetes and dermatophytes (Muschietti et al. 2005).

The application of the essential oil on maize grains had a significant impact on aflatoxin B1 accumulation produced by the fungus *Aspergillus* section *Flavi* (Bluma and Etcheverry 2008; Bluma et al. 2009). Aqueous and organic extracts of "incayuyo" displayed trypanocidal activity with percentages of inhibition higher than 70% at 100μg/ml (Sülsen et al. 2006). The aqueous extract also showed strong choleretic and antispasmodic effects in rats. Doses of 250, 500 and 750 mg/kg administered orally significantly increased the bile flow and the bile acid output. In addition, a significant reduction of contractions induced by acetylcholine, $CaCl_2$ and KCl was observed in isolated rat jejunum (Gorzalczany et al. 2008). Overall, the experimental data support the use of "incayuyo" in traditional medicine.

10 Conclusions

Experimental data support the uses of "incayuyo" in traditional medicine. Particularly noteworthy are its antiadhesive effects against *Helicobacter pylori* and its strong anti-inflammatory, antispasmodic and choleretic properties. It should be noted here that *Helicobacter pylori* is a persistent pathogenic organism that colonizes approximately half of the world's human population. It is a relevant risk factor for chronic gastritis, peptic ulcer and gastric tumors. Stimulation of the immune system, antioxidant and radical scavenging activity may additionally contribute to the biological activity of "incayuyo" infusions. Since several essential oil chemotypes are known, comparative studies of bioactivity and chromatographic profiles of non-volatile metabolites (flavonoid glycosides, phenylethanoids, chlorogenic acid, caffeoylquinic acids, etc.) are needed to accurately define the chemotype(s) with better bioactivity and pharmacological profile.

References

Agrawal AD (2011) Pharmacological activities of flavonoids: a review. Int J Pharm Sci Nanotechnol 4(2):1394–1398

Alonso J, Desmarchelier C (2006) Plantas Medicinales Autóctonas de la Argentina. Ed. Fitociencia, Buenos Aires, Argentina, pp 267–270

Barbieri N, Costamagna M, Gilabert M, Perotti M, Schuff C, Isla MI, Benavente A (2016) Antioxidant activity and chemical composition of essential oils of three aromatic plants from La Rioja province. Pharm Biol 54:168–173

Barboza GE, Cantero JJ, Núñez C, Pacciaroni A, Ariza Espinar L (2009) Medicinal plants: a general review and a phytochemical and ethnopharmacological screening of the native Argentine Flora. Kurtziana 34(1–2):7–365

Bassols GB, Gurni AA (1996) Especies del género *Lippia* utilizadas en medicina popular latinoamericana. Dominguezia 13:7–25

Bassols GB, Gurni AA (2000) Comparative anatomical study on Argentine species of *Lippia* known as "Poleo". Pharm Biol 38(2):120–128

Bluma RV, Etcheverry MG (2008) Application of essential oils in maize grain: impact on *Aspergillus* section *Flavi* growth parameters and aflatoxin accumulation. Food Microbiol 25(2):324–334

Bluma R, Landa MF, Etcheverry M (2009) Impact of volatile compounds generated by essential oils on *Aspergillus* section *Flavi* growth parameters and aflatoxin accumulation. J Sci Food Agric 89(9):1473–1480

Catalán CAN, de Lampasona MEP (2002) The chemistry of the genus *Lippia* (Verbenaceae). In: Kintzios SE (ed) Oregano. The genera *Origanum* and *Lippia*. Ed. Taylor & Francis Inc., New York, USA, pp 127–149

Catalán CA, Iglesias DI, Retamar JA, Iturraspe JB, Dartayet GH, Gros GE (1983) A sesquiterpene diketone from *Lippia integrifolia*. Phytochemistry 22(6):1507–1508

Catalán CAN, De Fenik IJS, Dartayet GH, Gros EG (1991) Integrifolian-1,5-dione and a revised structure for 'africanone', biogenetically related sesquiterpene ketones from *Lippia integrifolia*. Phytochemistry 30(4):1323–1326

Catalán CAN, De Fenik IJS, De Arriazu PJ, Kokke WCMC (1992) 4,5-*seco*-african-4,5-dione from *Lippia integrifolia*. Phytochemistry 31(11):4025–4026

Catalán CAN, de Lampasona MEP, De Fenik IJS, Cerda-García-Rojas CM, Joseph-Nathan P (1993) Structure and conformation of a humulenedione from *Lippia integrifolia*. J Nat Prod 56(3):381–385

Catalán CAN, de Lampasona MEP, de Fenik IJS, Cerda-García-Rojas CM, Mora-Pérez Y, Joseph-Nathan P (1994) Minor constituents of *Lippia integrifolia*. J Nat Prod 57(2):206–210

Catalán CAN, de Lampasona MEP, Cerda-García-Rojas CM, Joseph-Nathan P (1995) Trace constituents of *Lippia integrifolia*. J Nat Prod 58(11):1713–1717

Cerda-García-Rojas CM, Coronel ADC, de Lampasona MEP, Catalán CAN, Joseph-Nathan P (2005) Absolute configuration of lippifoliane and africanane derivatives. J Nat Prod 68(5):659–665

Cerda-García-Rojas CM, Catalán CAN, Muro AC, Joseph-Nathan P (2008) Vibrational circular dichroism of africanane and lippifoliane sesquiterpenes from *Lippia integrifolia*. J Nat Prod 71(6):967–971

Chiou WF, Lin LC, Chen CF (2004) Acteoside protects endothelial cells against free radical-induced oxidative stress. J Pharm Pharmacol 56:743–748

Compadre CM, Robbins EF, Kinghorn AD (1986) The intensely sweet herb, *Lippia dulcis* Trev.: historical uses, field inquiries, and constituents. J Ethnopharmacol 15:89–106

Coronel AC, de Lampasona ME, Catalán CAN (2005) Variability of essential oils of *Lippia fissicalyx* and *Lippia turbinata* from Argentina. Rev Latinoam Quím 33(1):40–48

Coronel AC, Cerda-García-Rojas CM, Joseph-Nathan P, Catalán CAN (2006) Chemical composition, seasonal variation and a new sesquiterpene alcohol from the essential oil of *Lippia integrifolia*. Flavour Fragr J 21:839–847

Dartayet GH, Catalán CA, Retamar JA, Gros EG (1984) Sesquiterpenoids from *Lippia integrifolia* - africanone, a tricyclic sesquiterpene ketone. Phytochemistry 23(3):688–689

de Lampasona MEP, De Fenik IS, Catalán CAN, Dartayet GH, Gros GE, Cerda-García-Rojas C, Mora-Pérez Y, Joseph-Nathan P (1999) Constituents of the essential oil of *Lippia integrifolia* (Griseb.) Hieron. (Verbenaceae). Acta Hortic 500:81–88

Dellacasa AD, Bailac PN, Ponzi MI, Ruffinengo SR, Eguaras MJ (2003) *In vitro* activity of essential oils from San Luis-Argentina against *Ascosphaera apis*. J Essent Oil Res 15(4):282–285

Denham S, Múlgura de Romero ME, Slanis A, Bulacio E (2006) *Lippia integrifolia* versus *L. boliviana* (Verbenaceae). Darwiniana 44:363–364

do Nascimento KF, Moreira FMF, Alencar Santos J, Kassuya CAL, Croda JHR, Cardoso CAL, Vieirae MC, Góis Ruiz ALT, Foglio MA, de Carvalho JE, Formagio ASN (2018) Antioxidant, anti-inflammatory, antiproliferative and antimycobacterial activities of the essential oil of *Psidium guineense* Sw. and spathulenol. J Ethnopharmacol 210:351–358

Fricke C, Hardt IH, König WA, Joulain D, Zygadlo JA, Guzmán CA (1999) Sesquiterpenes from *Lippia integrifolia* essential oil. J Nat Prod 62(5):694–696

Galíndez G, Seal CE, Daws MI, Lindow L, Ortega-Baes P, Pritchard HW (2017) Alternating temperature combined with darkness resets base temperature for germination (T_b) in photoblastic seeds of *Lippia* and *Aloysia* (Verbenaceae). Plant Biol 19(1):41–45

Gleiser RM, Bonino MA, Zygadlo JA (2011) Repellence of essential oils of aromatic plants growing in Argentina against *Aedes aegypti* (Diptera: Culicidae). Parasitol Res 108(1):69–78

González AM, Tracanna MI, Amani SM, Schuff C, Poch MJ, Bach H, Catalán CAN (2012) Chemical composition, antimicrobial and antioxidant properties of the volatile oil and methanol extract of *Xenophyllum poposum*. Nat Prod Commun 7(12):1663–1666

Gorzalczany S, Sülsen V, Redko F, Vescina C, Muschietti L, Martino V, Acevedo C (2008) Choleretic and antispasmodic effects of *Lippia integrifolia* aqueous extract. Braz J Pharmacogn 18(1):16–20

Herbotecnia (2005) Information on the cultivation of *Lippia integrifolia* (Gris.) Hieron. (incayuyo, té del Inca, poleo). www.herbotecnia.com.ar. (Last access, February 2020)

Hosseinzadeh L, Shokoohinia Y, Arab M, Allahyari E, Mojarrab M (2019) Cytotoxic and apoptogenic sesquiterpenoids from the petroleum ether extract of *Artemisia aucheri* aerial parts. Iran J Pharm Res 18(1):391–399

Hu J, Wang W, Dai J, Zhu L (2019) Chemical composition and biological activity against *Tribolium castaneum* (Coleoptera: Tenebrionidae) of *Artemisia brachyloba* essential oil. Ind Crop Prod 128:29–37

Iannicelli J, Elechosa MA, Juárez MA, Martínez A, Bugallo V, Bandoni AL, Escandón AS, van Baren CM (2016) Effect of polyploidization in the production of essential oils in *Lippia integrifolia*. Ind Crop Prod 8:20–29

Jansen-Jacobs MJ (1988) Verbenaceae. In: Rijn ARA (ed) Flora of the Guianas, series A: Phanerogams, Fascicle 4 (148). Ed. Görts-Van Hoetz Scientific Books, Koenigstein, Germany, p 116

Jing W, Chunhua M, Shumin W (2015) Effects of acteoside on lipopolysaccharide-induced inflammation in acute lung injury via regulation of NF-κB pathway *in vivo* and *in vitro*. Toxicol Appl Pharmacol 285:128–135

Jucá MM, Cysne Filho FMS, de Almeida JC, Mesquita DDS, Barriga JRM, Dias KCF, Barbosa TM, Vasconcelos LC, Leal LKAM, Ribeiro JE, Vasconcelos SMM (2018) Flavonoids: biological activities and therapeutic potential. Nat Prod Res 16:1–14

Juliani HR, Koroch A, Simon JE, Biurrun FN, Castellano V, Zygadlo JA (2004) Essential oils from Argentinean aromatic plants. Acta Hortic 629:491–498

Koo KA, Kim SH, Oh TH, Kim YC (2006) Acteoside and its aglycones protect primary cultures of rat cortical cells from glutamate-induced excitotoxicity. Life Sci 79:709–716

Kumar S, Pandey AK (2013) Chemistry and biological activities of flavonoids: an overview. Sci World J 2013:1–16. Article ID 162750

Leal LE, Alarcón AA, Ortega-Baes P, Cayo F, Alarcón R (2018) Effects of essential oils from two *Lippia* species on growth of phytopathogenic fungi. Bol Latinoam Caribe Plant Med Aromát 17(1):30–35

Lee KJ, Woo ER, Choi CY, Shin DW, Lee DG, You HJ, Jeong HG (2004) Protective effect of acteoside on carbon tetrachloride-induced hepatotoxicity. Life Sci 74:1051–1064

Leung AY, Foster S (1996) Oregano. In: Leung AY, Foster S (eds) Encyclopedia of common natural ingredients used in food, drugs and cosmetics, 2nd edn. Ed. Wiley, New York, USA, pp 398–400

Lima B, López S, Luna L, Agüero MB, Aragón L, Tapia A, Zacchino S, López ML, Zygadlo J, Feresin GE (2011) Essential oils of medicinal plants from the Central Andes of Argentina: chemical composition, and antifungal, antibacterial, and insect-repellent activities. Chem Biodivers 8(5):924–936

Marcial G, Sendker J, Brandt S, de Lampasona MP, Catalán CAN, de Valdez GF, Hensel A (2014) Gastroprotection as an example: Antiadhesion against *Helicobacter pylori*, anti-inflammatory and antioxidant activities of aqueous extracts from the aerial parts of *Lippia integrifolia* Hieron. J Ethnopharmacol 155(2):1125–1133

Marcial G, de Lampasona MP, Vega MI, Lizarraga E, Viturro CI, Slanis A, Juárez MA, Elechosa MA, Catalán CAN (2016) Intraspecific variation in essential oil composition of the medicinal plant *Lippia integrifolia* (Verbenaceae). Evidence for five chemotypes. Phytochemistry 122:203–212

Mercado MI, Lizarraga E, Catalán CAN, Ponessa GI (2015) Composición del aceite esencial de *Xenophyllum poposum* (Asteraceae) de las provincias de Catamarca y Tucumán. Estructuras foliares y caulinares asociadas a su síntesis. Lilloa 52(Suppl):56–61

Múlgura de Romero ME (2003) *Lippia*. In: Anton M, Zuloaga FO (eds) Flora Fanerogámica Argentina, vol 84. Proflora, Córdoba, Argentina, pp 21–40

Muschietti L, Derita M, Sülsen V, Muñoz JDD, Ferraro G, Zacchino S, Martino V (2005) *In vitro* antifungal assay of traditional Argentine medicinal plants. J Ethnopharmacol 102(2):233–238

Ohno T, Inoue M, Ogihara Y, Saracoglu I (2002) Antimetastatic activity of Acteoside, a Phenylethanoid Glycoside. Biol Pharm Bull 25(5):666–668

Pascual ME, Slowing K, Carretero E, Sánchez Mata D, Villar A (2001) *Lippia*: traditional uses, chemistry and pharmacology: a review. J Ethnopharmacol 76:201–214

Passera CB, Ambrosetti JA (1999) *In vitro* propagation of "incayuyo", *Lippia integrifolia* (Gris.) Hier. (Verbenaceae), a medicinal and aromatic plant of Monte Phytogeographical Province, Argentina. Acta Hortic 502:319–324

Passone MA, Girardi NS, Ferrand CA, Etcheverry M (2012) *In vitro* evaluation of five essential oils as botanical fungitoxicants for the protection of stored peanuts from *Aspergillus flavus* and *A. parasiticus* contamination. Int Biodeterior Biodegradation 70:82–88

Pastoriza AV, Andrada AB (2006) Cytogenetics in *Lippia* species (Verbenaceae) from Northwest Argentina. J Basic Appl Genet 17(2):25–32

Perfumi M, Paparelli F, Cingolani ML (1995) Spasmolytic activity of essential oil of *Artemisia thuscula* Cav. from the Canary Islands. J Essent Oil Res 7(4):387–392

Pool A (1993) Verbenaceae. In: Brako L, Zarucchi JL (eds) Catálogo de las Angiospermas y Gimnospermas del Perú, Monographs in systematic botany from the Missouri Botanical Garden, vol 45. Missouri Botanical Garden, Missouri, pp 1164–1180

Ratera E, Ratera M (1980) Plantas de la Flora Argentina Empleadas en Medicina Popular. Ed. Hemisferio Sur, Buenos Aires, Argentina, 189 pp

Reddy BSG, Rao VD, Rao BC, Dhananjaya N, Kuttan R, Babu TD (1999) Isolation and structural determination of new sphingolipids and pharmacological activity of africanene and other metabolites from *Sinularia leptoclados*. Chem Pharm Bull 47:1214–1220

Ricco RA, Wagner ML, Portmann E, Reides C, Llesuy S, Gurni AA, Carballo MA (2010) Análisis de polifenoles, actividad antioxidante y genotoxicidad en especies argentinas de *Lippia* y *Aloysia* (Verbenaceae). Bol Latinoam Caribe Plant Med Aromát 9(5):388–396

Robak J, Gryglewski RJ (1996) Bioactivity of flavonoids. Pol J Pharmacol 48(6):555–564

Rondina R, Bandoni A, Coussio J (2003) Plantas Silvestres Argentinas con Reconocidas Propiedades Medicinales o Tóxicas. Ed. OEA-CYTED, Buenos Aires, Argentina. (in CD)

Shiao YJ, Su MH, Lin HC, Wu CR (2017) Acteoside and Isoacteoside protect amyloid β peptide induced cytotoxicity, cognitive deficit and neurochemical disturbances *in vitro* and *in vivo*. Int J Mol Sci 18(4):895

Souto-Bachiller FA, De Jesús-Echevarria M, Cárdenas-González OE, Acuña-Rodríguez MF, Meléndez PA, Romero-Ramsey L (1997) Terpenoid composition of *Lippia dulcis*. Phytochemistry 44:1077–1086

Sülsen V, Güida C, Coussio J, Paveto C, Muschietti L, Martino V (2006) *In vitro* evaluation of trypanocidal activity in plants used in Argentine traditional medicine. Parasitol Res 98(4):370–374

Terblanché FC, Kornelius G (1996) Essential oil constituents of the genus *Lippia* (Verbenaceae)- A literature review. J Essent Oil Res 8:471–485

The Plant List (2020) Version 1.1 published on the Internet: http://www.theplantlist.org. (Last access, February 2020)

Vajs V, Trifunovic S, Janackovic P, Sokovic M, Milosavljevic S, Tesevic V (2004) Antifungal activity of davanone-type sesquiterpenes from *Artemisia lobelia* var. *conescens*. J Serb Chem Soc 69(11):969–972

Velasco-Negueruela A, Pérez-Alonso MJ, Guzmán CA, Zygadlo JA, Ariza Espinar L, Sanz J, García-Vallejo MC (1993) Volatile constituents of four *Lippia* species from Córdoba (Argentina). J Essent Oil Res 5(5):513–524

Wang H, Xu Y, Yan J, Zhao X, Sun X, Zhang Y, Guo J, Zhu C (2009) Acteoside protects human neuroblastoma SH-SY5Y cells against β-amyloid-induced cell injury. Brain Res 1283:139–147

Yamada P, Iijima R, Han J, Shigemori H, Yokota S, Isoda H (2010) Inhibitory effect of acteoside isolated from *Cistanche tubulosa* on chemical mediator release and inflammatory cytokine production by RBL-2H3 and KU812 cells. Planta Med 76:1512–1518

Zygadlo JA, Lamarque AL, Guzmán CA, Grosso NR (1995) Composition of the flower oils of some *Lippia* and *Aloysia* species from Argentina. J Essent Oil Res 7(6):593–595

Lippia turbinata Griseb.

César A. N. Catalán, Diego A. Sampietro, and Julieta V. Catalán

Lippia turbinata Griseb. (Photos: C Catalán)

Abstract *Lippia turbinata* Griseb. is a native aromatic shrub commonly known as "poleo", "té del país", "té criollo" or "poleo fino". It is widely used in traditional medicine of Northern and central Argentina. Infusions of aerial parts are used as diuretic, tonic, emmenagogue, for dysmenorrhea, as a regulator of the menstrual

C. A. N. Catalán (✉) · J. V. Catalán
Instituto de Química Orgánica, Facultad de Bioquímica, Química y Farmacia, Universidad Nacional de Tucumán, San Miguel de Tucumán, Argentina
e-mail: ccatalan@fbqf.unt.edu.ar; cancatalan@gmail.com

D. A. Sampietro
Facultad de Bioquímica, Química y Farmacia, Laboratorio de Biología de Agentes Bioactivos y Fitopatógenos (LABIFITO), Universidad Nacional de Tucumán, S. M. de Tucumán, Argentina

Á. Máthé, A. Bandoni (eds.), *Medicinal and Aromatic Plants of South America Vol. 2*, Medicinal and Aromatic Plants of the World 7,
https://doi.org/10.1007/978-3-030-62818-5_25

cycle and as abortifacient. Its essential oil shows remarkable variability in chemical composition. Very little research has been conducted on its non-volatile components. Eight oleanane-type triterpenes were identified in a methanol-methylene chloride extract of "poleo". There are also relatively few studies on the biological activity of this herb. Antifungal, antimicrobial, nematocidal and antiviral activities against the Junin-virus have been reported for the essential oil. Piperitenone oxide, one of the main components in several collections, has been evaluated in relation to the following biological activities: cardiovascular, hypotensive, bradycardic, insecticidal, trypanocidal, schistosomicidal, antimicrobial, antinociceptive, analgesic and larvicidal. Infusions are used to stimulate uterine motility and therefore should not be consumed by pregnant women. In order to better assess its pharmacological utility, much additional research is needed on the non-volatile metabolites and biological properties of this species.

Keywords "Poleo" · Essential oil · Digestive · Diuretic · Emmenagogue · Abortifacient

1 Introduction

Lippia turbinata Griseb. is a highly polymorphic species native to Central and Northern Argentina, where it is widely used in traditional medicine as a digestive, diuretic, tonic, emmenagogue and abortifacient. Its essential oil is constituted mainly by monoterpenoids, being limonene the dominant component in most samples. Also α-thujone, carvone, piperitenone oxide, *trans*-sabinol or caryophyllene oxide have been reported occasionally as major components. The flavonoids luteolin, apigenin isorhamnetin, chrysoeriol and diosmetin, free and as glycosides, have been characterized in *L. turbinata* infusions, while other phenolics such as chlorogenic acid, caffeoylquinic acids, acteoside and isoacteoside have not been determined. The oleanane-type triterpenoids lantanilic acid, camaric acid, lantanolic acid, and rehmannic acid along with several closely related novel triterpenes were identified in a methanol-methylene chloride extract of *L. turbinata*. Much additional research on non volatile metabolites, as well as biological assays are needed to better assess the pharmacological properties of this species, particularly those related to the phenolic fraction and with effects on uterine motility attributed to piperitenone oxide.

2 Taxonomic Characteristics

The genus *Lippia* is member of the Verbenaceae family, Lantaneae tribe, and embraces some 160 species distributed mainly in warm and temperate America, from Mexico to central Argentina, with only a few species in Africa (Jansen-Jacobs 1988; Múlgura de Romero 2003). Many members of this genus are aromatic and widely used in folk medicine (Terblanché and Kornelius 1996; Pascual et al. 2001; Catalán and De Lampasona 2002; Barboza et al. 2009). The name of this genus is a tribute to Dr. Augusto Lippi (1678–1703), a French doctor and naturalist.

Lippia turbinata Griseb. is a native aromatic shrub, perennial, 1–2 meters tall that grows in the west and center of Argentina (provinces of Jujuy, Salta, Tucumán, Catamarca, Santiago del Estero, Chaco, La Rioja, San Juan, Córdoba, Mendoza, San Luis, La Pampa y Buenos Aires), Chile and Peru (Múlgura de Romero 2003).

This species has a high polymorphism. Currently, two forms are recognized, namely, *L. turbinata* f. *turbinata* and *L. turbinata* f. *magnifolia*, but it is common to find intermediate individuals that are difficult to classify (Múlgura de Romero 2003). In the older literature, the latter form is treated as a different taxon, namely, *L. fissicalyx* Tronc. (see section 4 below). In addition, in some papers appears the name *L. turbinata* var. *integrifolia* Griseb. which is synonymous of *L. integrifolia* (Griseb.) Hieron. (Múlgura de Romero 2003; Passone et al. 2012a,b, Passone et al. 2013). *Lippia turbinata* Griseb. is taxonomically and phytogeographically related with *L. integrifolia* (Griseb.) Hieron. that sometimes may be present as a contaminant. However, there are anatomical and morphological differences that allow both species to be distinguished (Bassols and Gurni 2000).

Common Names "poleo", "té del país", "té criollo" "poleo fino" (Múlgura de Romero 2003; Alonso and Desmarchelier 2006; Barboza et al. 2009).

Synonyms For *Lippia turbinata* Griseb. f. *turbinata*: *Lippia disepala* Phil.; *Lippia aprica* Phil.; *Lippia globiflora* var. *microphylla* f. *glabriuscula* Kuntze; *Lippia turbinata* f. *angustifolia* Osten ex Moldenke. For *Lippia turbinata* Griseb. f. *magnifolia* Moldenke: *Lippia fissicalyx* Tronc. (Múlgura de Romero 2003).

3 Crude Drug Used

Dried leaves with flowers (+ fruits) and tender branches. Infusions are used as digestive, diuretic, tonic, emmenagogue, for dysmenorrhea, as a regulator of the menstrual cycle and as an abortifacient (Ratera and Ratera 1980; Rondina et al. 2003; Coronel et al. 2005; Alonso and Desmarchelier 2006; Barboza et al. 2009). Infusions are prepared at 2%. Recommended to drink two to three cups per day (Marzocca 1998; Alonso and Desmarchelier 2006). In the mountains of Córdoba province, a decoction of 100 g/liter of aerial parts is used to facilitate the delivery (Marzocca 1998).

4 Major Chemical Constituents and Bioactive Compounds

An outstanding feature of the *Lippia* genus is the disconcerting variability in the essential oil composition of a same species harvested at different locations (Terblanché and Kornelius 1996; Catalán and De Lampasona 2002; Coronel et al. 2005). In *L. turbinata*, the chemical composition of the essential oil obtained from the same plant stock remains essentially constant for several years, which suggests that the chemical profile is mainly under genetic control (Souto-Bachiller et al. 1996; Marcial et al. 2016). As frequently happens in this genus, *L. turbinata* shows a high degree of polymorphism accompanied by significant variations in the composition of its essential oil. As stated above, two forms of this species are currently recognized, namely *L. turbinata* f. *turbinata* and *L. turbinata* f. *magnifolia*, the latter was previously called *L. fissicalyx* and treated as a different species by some authors (Múlgura de Romero 2003; Coronel et al. 2005). Unfortunately, all but one of the papers published so far, (Coronel et al. 2005), only mention that the material analyzed is *L. turbinata* Griseb. without indication of what form it is. The yields of essential oil obtained by steam distillation of aerial parts is 0.2–1.3% based on dry plant material (Juliani et al. 2004; Coronel et al. 2005; Passone et al. 2013; Barbieri et al. 2016). The higher the ratio of leaves and flowers in relation to stems, the higher the yield in essential oil. Table 1 shows the main essential oil components of plants collected in different provinces and localities of Argentina. The chemical composition of the different collections shows remarkable qualitative variations. In six collections (RC, AC, LL, SL, MSL and SA) the dominant component was limonene; α-thujone was the main component in three collections (SC, CD and QO); carvone in two (R38 and RSP); piperitenone oxide was the main component in one (ALC) and the second component in two collections (SL and SML); and *trans*-sabinol (UL) and caryophyllene oxide (LM) in one collection each.

So far, only one document has been published on the essential oil composition of certified collections of *Lippia turbinata* Griseb. f. *magnifolia* Moldenke (= *Lippia fissicalyx* Tronc.) and *L. turbinata* Griseb. f. *turbinata*. (Coronel et al. 2005). Four tetraploids ($2n = 60$) collections of *L. turbinata* Griseb. f. *magnifolia* Moldenke (= *Lippia fissicalyx* Tronc.) were analyzed and their essential oils typically showed two main components accounting for more than 75% of the oil. One of the main components was always limonene and the other was either piperitenone oxide (= lippione), carvone or 1,8-cineole. In addition, two diploids ($2n = 30$) collections and two tetraploids ($2n = 60$) collections of *L. turbinata* Griseb. f. *turbinata* were analyzed comparatively. Though the chromatographic profile of a tetraploid *L. turbinata* Griseb. f. *turbinata* collected in Catamarca province, showed similarities with some collections of *L. turbinata* Griseb. f. *magnifolia* Moldenke, generally the oils from the former are more complex and diverse chemically paralleling its greater cytogenetic variability and enzyme polymorphism (Coronel et al. 2005).

Very few works on non-volatile metabolites of "poleo" have been published. The following flavonoids were identified by HPLC-ESI-Q-TOF-MS: isorhamnetin, luteolin, apigenin, chrysoeriol/diosmetin, chrysoeriol diglycoside, diosmetin

Table 1 Main compounds in the essential oils (expressed as percentage in the essential oil) of *Lippia turbinata* obtained from plants collected in different provinces and localities of Argentina

Compound		Cordoba [a]					La Rioja [b]				San Luis[c]		Chaco	Salta
	UL[d]	RC	AC	SC	CD	LM	ALC	QO	LL	R38	SL	MSL	RSP[e]	SA[f]
Sabinene	8.0	–	–	–	–	–	–	–	–	–	–	–	–	–
Limonene	–[g]	48.8	76.8	6.1	–	11.9	7.2	–	48.1	31.3	43.3	60.6	8.7	84.3
1,8-cineole	–	–	5.0	–	–	–	–	–	–	–	14.7	–	1.7	–
trans-Sabinol	58.3	–	–	–	–	–	–	–	–	–	–	–	–	–
α-Thujone	14.2	–	–	30.2	28.3	–	–	48.3	–	–	–	–	–	–
Piperitenone	–	16.8	–	–	–	–	–	–	–	–	–	–	–	–
Piperitenone oxide	–	–	–	–	–	–	63.0	–	30.1	–	24.8	17.8	–	–
Carvone	–	–	–	10.1	7.4	–	–	17.4	–	55.4	–	–	80.7	–
Camphor	–	–	–	4.0	–	–	–	–	–	–	–	–	–	1.5
Bornyl acetate	–	–	–	–	–	8.1	–	–	–	–	–	–	–	–
β-Caryophyllene	–	18.1	1.3	–	–	–	–	10.0	–	5.1	–	6.4	2.1	6.1
Caryophyllene oxide	–	–	–	–	7.0	17.8	5.6	–	–	–	–	–	–	–
Spathulenol	–	–	–	–	6.1	11.4	–	–	–	–	–	–	–	–
Methyleugenol	–	–	–	–	–	–	–	–	1.8	–	–	–	–	–

[a]*RC* aerial parts purchased at a local market in Rio Cuarto city (Passone and Etcheverry 2014); *AC* Altas Cumbres (Quiroga et al. 2013); *SC* Oil from flowers of plants collected at Saldán, Colon Department (Zygadlo et al. 1995); *CD* Colon Department (Velasco Negueruela et al. 1993); *LM* Dique Los Molinos (Coronel et al. 2005)

[b]*ALC* Valle de Atinaco – Los Colorados (Barbieri et al. 2016); *QO* Quebrada de Olta (Gleiser and Zygadlo 2007); *LL* Los Llanos region (Juliani et al. 2004); *R38* on the side of the road, national highway 38 (Coronel et al. 2005);

[c]*SL* undisclosed location of San Luis province (Dellacassa et al. 2003); *MSL* Merlo (Duschatzky et al. 2004);

[d]*UL* undisclosed location of Argentina (Pellegrini et al. 2017)

[e]*RSP* Roque Sáenz Peña (Perez-Zamora et al. 2016)

[f]*SA* undisclosed location (Leal et al. 2018)

[g]Compound not detected

diglycoside, luteolin-*O*-glycoside, luteolin-*O*-glucuronide and isorhamnetin-*O*-glycoside (Aguado et al. 2016). An extract obtained from infusions of "poleo" (yield 16.8%) had 0.41 ± 0.04μmol of total polyphenols (expressed as μmol of caffeic acid equivalents/mg dry matter) and 0.12 ± 0.01μmol of total flavonoids (expressed as μmol of rutin equivalents/mg dry matter) (Dadé et al. 2009). Other authors report 10.62 ± 0.62 mg tannic acid/g dry material of total phenols for infusions, and 13.36 ± 0.26 mg tannic acid/g dry material for decoctions; and 5.95 ± 0.60 mg rutin/g dry plant of total flavonoids for the infusion; and 8.68 ± 0.90 mg rutin/g dry material for the decoction (Wernert et al. 2009). Many research and reviews are available on the chemistry and biological activity of flavonoids and polyphenols which are well known for their antioxidants properties and favorable impact on health (Robakand Gryglewski 1996; Agrawal 2011; Kumar and Pandey 2013; Jucá et al. 2018; and references therein).

Finally, eight oleanane-type triterpenes were identified in a methanol-methylene chloride extract of "poleo", namely, the novel triterpenoids, 3β,25-epoxy-3α,21α-dihydroxy-22β-(3-methylbut-2-en-1-oyloxy)olean-12-en-28-oic acid; 3β,25-epoxy-3α,21α-dihydroxy-22b-angeloyloxyolean-12-en-28-oic acid; 3β,25-epoxy-3α,21α-dihydroxy-22β-tigloyloxyolean-12-en-28-oic acid; and 3β,25-epoxy-3α-hydroxy-22β-(2-methylbutan-1-oyloxy)-olean-12-en-28-oic acid, together with the known triterpenoids lantanilic acid, camaric acid, lantanolic acid, and rehmannic acid which showed activity against *Mycobacterium tuberculosis* (Wachter et al. 2001).

Limonene is often one of the main components of the essential oil from *L. turbinata*. In addition to its characteristic aroma, it has a wide range of pharmacological properties and low toxicity that make it attractive to be incorporated into medical and pharmaceutical formulations. A review on the pharmacology of limonene reports the following properties: antimicrobial, antioxidant, anti-inflammatory, antinociceptive, anticancer and insecticide (Erasto and Viljoen 2008). Another relevant component of several collections of "poleo" is α-thujone (Velasco-Negueruela et al. 1993; Zygadlo et al. 1995; Gleiser and Zygadlo 2007; Pellegrini et al. 2017).This ketone is the toxic principle in absinthe (Höld et al. 2000; Lachenmeier and Uebelacker 2010) while the β-diastereomer is generally considered less toxic. For regulatory purposes, the sum of both isomers is generally assessed, and the term thujone without specifying any isomer usually refers to the total content of both diastereomers. Thujone was found to correct the lipid profile in diabetic rats (Al-Haj Baddara et al. 2011) and showed antinociceptive activity in mice (Rice and Wilson 1976), the α-isomer being more potent and toxic than the β-isomer. Spathulenol appears as a relevant component in a couple of collections (Velasco-Negueruela et al. 1993; Coronel et al. 2005); it showed to be effective against an ovarian cancer cell line and possesses significant antioxidant and anti-inflammatory activities (do Nascimento et al. 2018). Piperitenone oxide (= lippione), one of the main components of several collections (Dellacassa et al. 2003; Juliani et al. 2004; Duschatzky et al. 2004; Coronel et al. 2005; Barbieri et al. 2016), has been evaluated in relation to the following biological activities: cardiovascular, hypotensive, bradycardic, insecticidal, trypanocidal, schistosomicidal, antimicrobial, antinociceptive, analgesic and larvicidal. A review of the biological activities and chemistry of this ketone is available (Božović et al. 2015). Carvone (unspecified enantiomer) is another significant component of "poleo" that has antioxidant, antimicrobial, insect repellent properties and acts as a sprouting inhibitor (De Carvalho and Da Fonseca 2006). β-Caryophyllene is known to possess antioxidant (Lourens et al. 2004; Singh et al. 2006), antibiotic (Alma et al. 2003; Lourens et al. 2004; Pichette et al. 2006), anti-inflammatory (Tambe et al. 1996; Cho et al. 2006), anticarcinogenic (Kubo et al. 1996) and anesthetic (Ghelardini et al. 2001) activities. As can be seen, "poleo" has many bioactive metabolites, both volatile and non-volatile, and consequently multiple pharmacological effects as well as potential synergies can be expected.

5 Morphological Description

Perennial aromatic bush of 1–2 m tall, very branched. Opposite or ternate leaves, acuminates, attenuated at the base in a short petiole, sawn in the upper 2/3, rarely whole, rough-scabrous to glabrescent on both surfaces. Axillary inflorescences of 4–6 mm diameter; in fruiting up to 1 cm, with peduncles of 3–6 mm; imbricated bracts, the lower 3–4.5 x 1–1.8 mm, narrowly elliptical, the remaining 1.8–2 x 1.7–2 mm oval or rhombic-oval, acuminated, strigose in the middle part and subglabrous towards the apex; calyx with two leaves, sepals of 1.5–2 mm, linear or narrowly elliptical; corolla white to white-lilac, tubular-bilabiate, of reduced limbo and tube of 2.5–3 mm, slightly pubescent towards the apex. Obovoid fruit, nutlets of 2 x 1.8 mm, with dorsal face subglobous and somewhat rough; commissural face flat and smooth. n = 15 (Múlgura de Romero 2003). Currently, two forms of this species are recognized: i) *L. turbinata* f. *turbinata*: distinguished from f. *magnifolia* for having short internodes, 1–2 (3) cm, generally ternate leaves, narrowly elliptical or elliptical, of 1.5–3 x 0.3–0.5 cm and braquiblasts generally developed; and ii) *L. turbinata* f. *magnifolia*, which has more spaced internodes (up to 5 cm), generally opposite, ovate leaves, 1–4.5 x 0.8–2 cm, and braquiblasts in general undeveloped. This form was considered a different species named *L. fissicalyx* Tronc. by some authors. However, the presence of intermediate specimens makes separation difficult and it seems more convenient to consider it as a different form (Múlgura de Romero 2003).

A comparative epidermal analysis between "creole pennyroyal" ("poleo criollo" in Spanish = *L. turbinata* Griseb.) and "European pennyroyal" ("poleo europeo" in Spanish = *Mentha pulegium* L., Lamiaceae) (Bassols and Gurni 2001), as well as an anatomical study of nine *Lippia* species of Argentine used in folk medicine, which includes *L. turbinata* (Bassols and Gurni 2000), are available. A cytogenetic study of *Lippia turbinata* showed both diploid (2n = 30) and tetraploid (2n = 60) specimens; the meiotic chromosome behavior was regular in diploids and highly irregular in tetraploids. Pollen analysis evidenced a high production of normal pollen in diploid individuals and a low one in tetraploids (Pastoriza and Andrada 2006), while the variability in the morphology and anatomy of the leaf and stem have been described by Andersen et al. (2006) and Coll Aráoz and Ponessa (2007).

6 Geographical Distribution

According to Múlgura de Romero (2003), *L. turbinata* f. *turbinata* grows in Central and Western Argentina in the provinces of Jujuy, Salta, Catamarca, Santiago del Estero, San Juan, La Rioja, Córdoba, Mendoza, San Luis, La Pampa and Buenos Aires. It mainly inhabits the Monte and Espinal phytogeographic provinces. It has also been cited for Chile and Peru (Múlgura de Romero 2003). It is particularly abundant in Catamarca, La Rioja and Córdoba provinces of Argentina, where it occupies the alluvial terraces of the mountain streams (Marzocca 1998). On the other hand, *L. turbinata* f. *magnifolia* has been described from material collected in the provinces of Salta, Jujuy, Chaco and Tucumán, where it grows mainly in the phytogeographic province of Las Yungas, on the border with the Chaqueña phytogeographic province.

7 Ecological Requirements

L. turbinata requires deep and fertile soils in full sun. The reproduction can be done by seeds, and by cuttings. Reproduction by seeds is seldom used because it grows very slowly and can be easily reproduced by other methods. Reproduction by cuttings is done using woody stems of adult plants (Herbotecnia 2004). The interaction between cool temperature and darkness during alternating temperatures treatment on the germination of photoblastic seeds of *Lippia turbinata* and other *Lippia* and *Aloysia* species has been investigated (Galíndez et al. 2017).

8 Traditional Use (Part(s) Used) and Common Knowledge

Infusions are prepared using leaves, flowers and twigs. Photo 1 shows a flowery twig of *L. turbinata* Griseb. f. *turbinata* with ternate leaves (left) and another twig with opposite leaves (right). Both types are frequently found on the same plant.

Traditional uses reported for *Lippia turbinata* f. *magnifolia* Moldenke (syn. *Lippia fissicalyx* Tronc.) are the following: infusions of aerial parts used for back pain, digestive, diuretic, tonic, emmenagogue and abortifacient. The essential oil has nematicidal activity (Barboza et al. 2009). Two to three cups per day of 2% infusions are recommended (Marzocca 1998; Alonso and Desmarchelier 2006).

Traditional use reported for *Lippia turbinata* Griseb. f. *turbinata*: infusion of aerial parts is used as stimulant, diuretic, digestive, emmenagogue, bradycardic and abortifacient. Dried entire plant has antimicrobial activity (Barboza et al. 2009). In the province of Córdoba, a decoction of 100 g/liter of aerial parts is used to facilitate delivery (Marzocca 1998). As can be seen, both forms of this species have essentially the same use in traditional medicine.

9 Modern Medicine Based on its Traditional Medicine Uses

Reviews on the chemical composition, pharmacological activity and use in traditional medicine of *Lippia* genus have been published (Bassols and Gurni 1996; Terblanché and Kornelius 1996; Pascual et al. 2001; Catalán and De Lampasona 2002).

Photo 1 – *Lippia turbinata* Griseb. f. *turbinata*, flowery twig with ternate leaves (left) and with opposite leaves (right). (Photos: C Catalán)

The essential oil of *L. turbinata* Griseb. in the vapor phase was highly effective against the phytopathogenic fungi *Sclerotium rolfsii*, *S. sclerotium* and *Rhizoctonia solani* (Leal et al. 2018) and against *Aspergillus* spp. (Bluma and Etcheverry 2008; Bluma et al. 2009; Passone and Etcheverry 2014).

Other biological activities reported for *Lippia turbinata* are presented in Table 2. According to Fester et al. (1960), piperitenone oxide (lippione) or its derivative diosphenolene (lippiaphenol) stimulates uterine motility (Table 2). For this reason, it should not be consumed by pregnant women because of the possibility of causing abortion (Montes 1961; Alonso and Desmarchelier 2006). The activity against *Mycobacterium tuberculosis* of eight triterpenoids isolated by assay-guided fractionation of a methanol-methylene chloride extract of *L. turbinata* was determined. The most active compound was rehmannic acid (MIC 32μM) (Wächter et al. 2001).

10 Conclusions

L. turbinata Griseb. shows variability both in its morphological traits and in the qualitative composition of its essential oil. The essential oil composition reported for different collections of *L. turbinata* clearly indicates the existence of various chemotypes that still await to be characterized. Also, comparative studies of the chromatographic profile of non-volatile metabolites are necessary to precisely define the chemotypes. The abortifacient properties and the effects on uterine motility attributed to piperitenone oxide (lippione) should be experimentally tested to better evaluate the pharmacological utility of this species.

Table 2 Biological activities reported for *Lippia turbinata*

Part or component tested	Activity	References
Essential oil	Nematicidal against root knot nematode (*Meloidogyne* sp.)	Duschatzky et al. (2004)
	Virucidal against Junin virus	García et al. (2003)
	Fungicidal against the ascomycete fungus *Ascosphaera apis*	Dellacassa et al. (2003)
	Repellent effect against *Varroa destructor*	Ruffinengo et al. (2005)
	Antioxidant	Dadé et al. (2009), Quiroga et al. (2013), Aguado et al. (2016), Barbieri et al. (2016)
	Antimicrobial against yeast and bacteria	Demo et al. (2005), Aguado et al. (2016), Pérez-Zamora et al. (2016)
	Antibacterial against *Paenibacillus larvae*	Fuselli et al. (2006), González and Marioli (2010), Pellegrini et al. (2017)
	Insecticidal against *Culex quinquefasciatus*	Gleiser and Zygadlo 2007 , Kembro et al. (2009)
	Insecticidal against *Musca domestica*	Palacios et al. (2009)
	Low activity against herpes simplex virus type 1 (HSV-1) and dengue virus type 2 (DEN-2)	García et al. (2003)
	Antifungal agent on peanut seed	Girardi et al. (2017a,b)
	Antimethanogenic	García et al. (2018a,b)
Piperitenone oxide (lippione)	Cardiovascular, hypotensive, bradycardic, insecticidal, trypanocidal, schistosomicidal, antimicrobial, antinociceptive, analgesic and larvicidal	Božović et al. (2015)
	Uterine motility	Montes (1961) , Alonso and Desmarchelier (2006)
Flavonoid extract	Antibacterial against Gram (+) and Gram (−) bacteria	Hernández et al. (2000)
Rehmannic acid (a triterpenoid)	Activity against *Mycobacterium tuberculosis*	Wächter et al. (2001)

References

Agrawal AD (2011) Pharmacological activities of flavonoids: a review. Int J Pharm Sci Nanotechnol 4(2):1394–1398

Aguado MI, Dudik NH, Zamora CMP, Torres CA, Núñez MB (2016) Antioxidant and antibacterial activities of hydroalcoholic extracts from *Aloysia polystachya* Griseb Moldenke and *Lippia turbinata* Griseb. (Verbenaceae). Int J Pharm Pharm Sci 8(3):393–395

Al-Haj Baddara NW, Aburjaib TA, Tahab MO, Disia AM (2011) Thujone corrects cholesterol and triglyceride profiles in diabetic rat model. Nat Prod Res 25:1180–1184

Alma MH, Mavi A, Yildirim A, Digrak M, Hirata T (2003) Screening chemical composition and *in vitro* antioxidant and antibacterial activities of the essential oils from *Origanum syriacum* L. growing in Turkey. Biol Pharm Bull 26:1725–1729

Alonso J, Desmarchelier C (2006) Plantas Medicinales Autóctonas de la Argentina. Ed. Fitociencia. Buenos Aires, Argentina, pp 267–270

Andersen A, Lucchini F, Moriconi J, Fernández EA (2006) Variability in leaf morphology and anotomy in *Lippia turbinata* (Verbenaceae) in the province of San Luis (Argentina). Phyton 75:137–143

Barbieri N, Costamagna M, Gilabert M, Perotti M, Schuff C, Isla MI, Benavente A (2016) Antioxidant activity and chemical composition of essential oils of three aromatic plants from La Rioja province. Pharm Biol 54:168–173

Barboza GE, Cantero JJ, Núñez C, Pacciaroni A, Ariza Espinar L (2009) Medicinal plants: a general review and a phytochemical and ethnopharmacological screening of the native Argentine Flora. Kurtziana 34(1–2):7–365

Bassols GB, Gurni AA (1996) Especies del género *Lippia* utilizadas en medicina popular latinoamericana. Dominguezia 13:7–25

Bassols GB, Gurni AA (2000) Comparative anatomical study on argentine species of *Lippia* known as "Poleo". Pharm Biol 38(2):120–128

Bassols GB, Gurni AA (2001) Estudio epidérmico comparativo entre el "poleo criollo" (*Lippia turbinata* Griseb. -Verbenaceae-) y el "poleo europeo" (*Mentha pulegium* L. -Lamiaceae-). Act Farm Bonaer 20(2):105–108

Bluma RV, Etcheverry MG (2008) Application of essential oils in maize grain: Impact on *Aspergillus* section *Flavi* growth parameters and aflatoxin accumulation. Food Microbiol 25(2):324–334

Bluma R, Landa MF, Etcheverry M (2009) Impact of volatile compounds generated by essential oils on *Aspergillus* section *Flavi* growth parameters and aflatoxin accumulation. J Sci Food Agric 89(9):1473–1480

Božović M, Pirolli A, Ragno R (2015) *Mentha suaveolens* Ehrh. (Lamiaceae) essential oil and its Main constituent Piperitenone oxide: biological activities and chemistry. Molecules 20:8605–8633

Catalán CAN, De Lampasona MEP (2002) The chemistry of the genus *Lippia* (Verbenaceae). In: Kintzios SE (ed) Oregano. The genera *Origanum* and *Lippia*. Ed. Taylor & Francis Inc., New York, USA, pp 127–149

Cho JY, Chang HJ, Lee SK, Kim HJ, Hwang JK, Chu HS (2006) Amelioration of dextran sulfate sodium-induced colitis in mice by oral administration of beta-caryophyllene, a sesquiterpene. Life Sci 80(10):932–939

Coll Aráoz MV, Ponessa G (2007) Anatomía foliar y caulinar de *Lippia turbinata* f. *magnifolia* Moldenke –Verbenaceae– (Loc. Vipos, Dpto. Trancas, Tucumán, República Argentina). Dominguezia 23(1):5–12

Coronel AC, De Lampasona MP, Catalán CAN (2005) Variability of essential oils of *Lippia fissicalyx* and *Lippia turbinata* from Argentina. Revista Latinoamericana de Química 33(1):40–48

Dadé MM, Fioravanti DE, Schinella GR, Tournier HA (2009) Total antioxidant capacity and polyphenol content of 21 aqueous extracts obtained from native plants of Traslasierra valley (Argentina). Boletín Latinoamericano y del Caribe de Plantas Medicinales y Aromáticas (BLACPMA) 8(6):529–539

De Carvalho CCCR, Da Fonseca MMR (2006) Carvone: why and how should one bother to produce this terpene. Food Chem 95:413–422

Dellacassa ED, Bailac PN, Ponzi MI, Ruffinengo SR, Eguaras MJ (2003) *In vitro* activity of essential oils from San Luis-Argentina against *Ascosphaera apis*. J Essent Oil Res 15(4):282–285

Demo M, Oliva MDLM, López ML, Zunino MP, Zygadlo JA (2005) Antimicrobial activity of essential oils obtained from aromatic plants of Argentina. Pharm Biol 43(2):129–134

do Nascimento KF, Moreira FMF, Alencar Santos J, Kassuya CAL, Croda JHR, Cardoso CAL, Vieirae MC, Góis Ruiz ALT, Foglio MA, de Carvalho JE, Formagio ASN (2018) Antioxidant, anti-inflammatory, antiproliferative and antimycobacterial activities of the essential oil of *Psidium guineense* Sw. and spathulenol. J Ethnopharmacol 210:351–358

Duschatzky CB, Martinez AN, Almeida NV, Bonivardo SL (2004) Nematicidal activity of the essential oils of several Argentine plants against the root-knot nematode. J Essent Oil Res 16(6):626–628

Erasto P, Viljorn AM (2008) Limonene – a review: biosynthetic, ecological and pharmacological relevance. Nat Prod Commun 3(7):1193–1202

Fester G, Martinuzzi E, Retamar J, Ricciardi A (1960) Estudio sobre esencias volátiles argentinas. Revista de la Facultad de Ingeniería Química, Santa Fe (Argentina) 29:21–43

Fuselli SR, de la Rosa SBG, Gende LB, Eguaras MJ, Fritz R (2006) Antimicrobial activity of some argentinean wild plant essential oils against *Paenibacillus larvae*, causal agent of American foulbrood (AFB). J Apic Res 45(1):2–7

Galíndez G, Seal CE, Daws MI, Lindow L, Ortega-Baes P, Pritchard HW (2017) Alternating temperature combined with darkness resets base temperature for germination (T_b) in photoblastic seeds of *Lippia* and *Aloysia* (Verbenaceae). Plant Biol 19(1):41–45

García CC, Talarico L, Almeida N, Colombres S, Duschatzky C, Damonte EB (2003) Virucidal Activity of Essential Oils from Aromatic Plants of San Luis, Argentina. Phytother Res 17(9):1073–1075

García F, Brunetti MA, Lucini E, Scorcione Turcato MC, Moreno MV, Frossasco GP, Colombatto D, Martínez MJ, Martínez Ferrer J (2018a) Essential oils from Argentinean native species reduce *in vitro* methane production. Revista de Investigaciones Agropecuarias 44(1):76–83

García F, Vercoe PE, Martínez MJ, Durmic Z, Brunetti MA, Moreno MV, Colombatto D, Lucini E, Martínez Ferrer J (2018b) Essential oils from *Lippia turbinata* and *Tagetes minuta* persistently reduce *in vitro* ruminal methane production in a continuous-culture system. Anim Prod Sci 59(4):709–720

Ghelardini C, Galeotti N, Di Cesare M-L, Mazzanti G, Bartolini A (2001) Local anesthetic activity of β-caryophyllene. Farmaco 56:387–389

Girardi NS, García D, Passone MA, Nesci A, Etcheverry M (2017a) Microencapsulation of *Lippia turbinata* essential oil and its impact on peanut seed quality preservation. Int Biodeterior Biodegrad 116:227–233

Girardi N, Passone MA, García D, Nesci A, Etcheverry M (2017b). Essential oil microencapsulation technology to provide peanut seeds protection against fungal pathogens. Chapter 6. In: Gorawala P, Mandhatri S (eds) Agricultural Research Updates, vol 21. Ed. Nova Science Publishers, New York USA, pp 213–242

Gleiser RM, Zygadlo JA (2007) Insecticidal properties of essential oils from *Lippia turbinata* and *Lippia polystachya* (Verbenaceae) against *Culex quinquefasciatus* (Diptera: Culicidae). Parasitol Res 101(5):1349–1354

González MJ, Marioli JM (2010) Antibacterial activity of water extracts and essential oils of various aromatic plants against *Paenibacillus larvae*, the causative agent of American Foulbrood. J Invertebr Pathol 104(3):209–213

Herbotecnia (2004) Autóctonas. *Lippia turbinata* Gris. – Flia. Verbenáceas. Poleo, té criollo. www.herbotecnia.com.ar (Last access, March 2020)

Hernández NE, Tereschuk ML, Abdala LR (2000) Antimicrobial activity of flavonoids in medicinal plants from Tafí del Valle (Tucumán, Argentina). J Ethnopharmacol 73(1–2):317–322

Höld KM, Sirisoma NS, Ikeda T, Narahashi T, Casida JE (2000) α-Thujone (the active component of absinthe): α-aminobutyric acid type a receptor modulation and metabolic detoxification. Proc Natl Acad Sci 97(8):3826–3831

Jansen-Jacobs MJ (1988). Verbenaceae. In: Rijn ARA (ed) Flora of the Guianas, Series A: Phanerogams, Fascicle 4 (148). Ed. Görts-Van Hoetz Scientific Books, Koenigstein, Germany, p 116

Jucá MM, Cysne Filho FMS, de Almeida JC, Mesquita DDS, Barriga JRM, Dias KCF, Barbosa TM, Vasconcelos LC, Leal LKAM, Ribeiro JE, Vasconcelos SMM (2018) Flavonoids: biological activities and therapeutic potential. Nat Prod Res 16:1–14

Juliani HR, Koroch A, Simon JE, Biurrum FN, Castellano V, Zygadlo JA (2004) Essential oils from Argentinean aromatic plants. Acta Hortic 629:491–498

Kembro JM, Marín RH, Zygadlo JA, Gleiser RM (2009) Effects of the essential oils of *Lippia turbinata* and *Lippia polystachya* (Verbenaceae) on the temporal pattern of locomotion of the mosquito *Culex quinquefasciatus* (Diptera: Culicidae) larvae. Parasitol Res 104(5):1119–1127

Kubo I, Chaudhuri SK, Kubo Y, Sánchez Y, Ogura T, Saito T, Ishikawa H (1996) Cytotoxic and antioxidative sesquiterpenoids from *Heterotheca inuloides*. Planta Med 62:427–430

Kumar S, Pandey AK (2013). Chemistry and biological activities of flavonoids: an overview. Sci World J, article ID 162750, 16 pp

Lachenmeier DW, Uebelacker M (2010) Risk assessment of thujone in foods and medicines containing sage and wormwood – evidence for a need of regulatory changes? Regul Toxicol Pharmacol 58:437–444

Leal LE, Alarcón AA, Ortega-Baes P, Cayo F, Alarcón R (2018) Effects of essential oils from two *Lippia* species on growth of phytopathogenic fungi. Boletín Latinoamericano y del Caribe de Plantas Medicinales y Aromáticas 17(1):30–35

Lourens AC, Reddy D, Baser KH, Viljoen AM, Van Vuuren SF (2004) *In vitro* biological activity and essential oil composition of four indigenous south African *Helichrysum species*. J Ethnopharmacol 95:253–258

Marcial G, De Lampasona MP, Vega MI, Lizarraga E, Viturro CI, Slanis A, Juárez MA, Elechosa MA, Catalán CAN (2016) Intraspecific variation in essential oil composition of the medicinal plant *Lippia integrifolia* (Verbenaceae). Evidence for five chemotypes. Phytochemistry 122:203–212

Marzocca A (1998) Vademécum de malezas medicinales de la Argentina. Indígenas y exóticas. Ed. Orientación Gráfica Editora. Buenos Aires, Argentina, pp 243–244

Montes A (1961) Analítica de los Productos Aromáticos. Colección Científica del INTA. Buenos Aires, Argentina, 555 pp

Múlgura de Romero ME (2003) Lippia. In: Múlgura de Romero ME, Rotman AD, Atkins S: Verbenaceae, tribu Lantaneae. In: Anton M, Zuloaga FO (eds) Flora Fanerogámica Argentina 84: 21–40. http://www.floraargentina.edu.ar/proflora/. (Last access, March 2020)

Palacios SM, Bertoni A, Rossi Y, Santander R, Urzúa A (2009) Insecticidal activity of essential oils from native medicinal plants of Central Argentina against the house fly, *Musca domestica* (L.). Parasitol Res 106(1):207–212

Pascual ME, Slowing K, Carretero E, Sánchez Mata D, Villar A (2001) *Lippia*: traditional uses, chemistry and pharmacology: a review. J Ethnopharmacol 76:201–214

Passone MA, Etcheverry M (2014) Antifungal impact of volatile fractions of *Peumus boldus* and *Lippia turbinata* on *Aspergillus* section Flavi and residual levels of these oils in irradiated peanut. Int J Food Microbiol 168-169:17–23

Passone MA, Girardi NS, Etcheverry M (2012a) Evaluation of the control ability of five essential oils against *Aspergillus* section Nigri growth and ochratoxin A accumulation in peanut meal extract agar conditioned at different water activities levels. Int J Food Microbiol 159(3):198–206

Passone MA, Girardi NS, Ferrand CA, Etcheverry M (2012b) *In vitro* evaluation of five essential oils as botanical fungitoxicants for the protection of stored peanuts from *Aspergillus flavus* and *A. parasiticus* contamination. Int Biodeterior Biodegradation 70:82–88

Passone MA, Girardi NS, Etcheverry M (2013) Antifungal and antiaflatoxigenic activity by vapor contact of three essential oils, and effects of environmental factors on their efficacy. LWT- Food Sci Technol 53(2):434–444

Pastoriza AV, Andrada AB (2006) Cytogenetics in *Lippia* species (Verbenaceae) from northwest Argentina. J Basic Appl Genet 17(2):25–32

Pellegrini MC, Alonso-Salces RM, Umpiérrez ML, Rossini C, Fuselli SR (2017) Chemical Composition, Antimicrobial Activity, and Mode of Action of Essential Oils against *Paenibacillus larvae*, Etiological Agent of American Foulbrood on *Apis mellifera*. Chem Biodivers 14(4):e1600382

Pérez-Zamora CM, Torres CA, Aguado MI, Bela AJ, Núñez MB, Bregni C (2016) Antibacterial activity of essential oils of *Aloysia polystachya* and *Lippia turbinata* (Verbenaceae). Boletín Latinoamericano y del Caribe de Plantas Medicinales y Aromáticas (BLACPMA) 15(4):199–205

Pichette A, Larouche PL, Lebrun M, Legault J (2006) Composition and antibacterial activity of *Abies balsamea* essential oil. Phytother Res 20:371–373

Quiroga PR, Grosso NR, Lante A, Lomolino G, Zygadlo JA, Nepote V (2013) Chemical composition, antioxidant activity and anti-lipase activity of *Origanum vulgare* and *Lippia turbinata* essential oils. Int J Food Sci Technol 48(3):642–649

Ratera E, Ratera M (1980). Plantas de la Flora Argentina Empleadas en Medicina Popular. Ed. Hemisferio Sur, Buenos Aires, Argentina. pp 88–89

Rice KC, Wilson RS (1976) (-)-3-Isothujone. A small nonnitrogenous molecule with antinociceptive activity in mice. J Med Chem 19:1054–1057

Robak J, Gryglewski RJ (1996) Bioactivity of flavonoids. Pol J Pharmacol 48(6):555–564

Rondina R, Bandoni A, Coussio J (2003). Plantas Silvestres Argentinas con Reconocidas Propiedades Medicinales o Tóxicas. Ed. OEA-CYTED, Buenos Aires, Argentina. (In CD)

Ruffinengo S, Eguaras M, Floris I, Faverin C, Bailac P, Ponzi M (2005) LD_{50} and repellent effects of essential oils from argentinian wild plant species on *Varroa destructor*. J Econ Entomol 98(3):651–655

Singh G, Marimuthu P, de Heluani CS, Catalán CA (2006) Antioxidant and biocidal activities of *Carum nigrum* (seed) essential oil, oleoresin, and their selected components. J Agric Food Chem 54:174–181

Souto-Bachiller FA, de Jesús-Echevarría M, Cárdenas-González O (1996) Hernandulcin is the major constituent of *Lippia dulcis* Trev. (Verbenaceae). Nat Prod Lett 8:151–158

Tambe Y, Tsujiuchi H, Honda G, Ikeshiro Y, Tanaka S (1996) Gastric cytoprotection of the non-steroidal anti-inflammatory sesquiterpene, β-caryophyllene. Planta Med 62:469–470

Terblanché FC, Kornelius G (1996) Essential oil constituents of the genus *Lippia* (Verbenaceae)- a literature review. J Essent Oil Res 8:471–485

Velasco-Negueruela A, Pérez-Alonso MJ, Guzmán CA, Zygadlo JA, Ariza-Espinar L, Sanz J, García-Vallejo MC (1993) Volatile constituents of four *Lippia* species from Córdoba (Argentina). J Essent Oil Res 5(5):513–524

Wächter GA, Valcic S, Franzblau SG, Suárez E, Timmermann BN (2001) Antitubercular activity of triterpenoids from *Lippia turbinata*. J Nat Prod 64(1):37–41

Wernert MF, Wagner ML, Gurni AA, Carballo MA, Ricco RA (2009) Study of polyphenols of infusions and decoctions of "cedrón" (*Aloysia citrodora* Palau) and "Poleo" (*Lippia turbinata* Griseb.) – Verbenaceae –leaves. Boletín Latinoamericano y del Caribe de Plantas Medicinales y Aromáticas (BLACPMA) 8(4):308–311

Zygadlo JA, Lamarque AL, Guzmán CA, Grosso NR (1995) Composition of the flower oils of some *Lippia* and *Aloysia* species from Argentina. J. Essent Oil Res 7(6):593–595

Lithraea molleoides (Vell.) Engler

Paula Gladys López, Patricia Basile, Federico Wallace, Cristina Olivaro, Manuel Minteguiaga, and Fernando Ferreira

Lithraea molleoides (Vell.) Engler. Photo: MÁ López

P. G. López (✉)
Farmacognosia, Facultad de Farmacia y Bioquímica, Universidad de Buenos Aires, C. A. de Buenos Aires, Argentina
e-mail: plopez@ffyb.uba.ar

P. Basile
Espacio de Biología Vegetal del Noreste, Centro Universitario Regional Noreste/Sede Tacuarembó, Universidad de la República, Tacuarembó, Uruguay

F. Wallace · C. Olivaro · M. Minteguiaga
Espacio de Ciencia y Tecnología Química. Centro Universitario Regional Noreste., Sede Tacuarembó. Universidad de la República., Tacuarembó, Uruguay

F. Ferreira
Espacio de Ciencia y Tecnología Química. Centro Universitario Regional Noreste/Sede Tacuarembó, Universidad de la República, Tacuarembó, Uruguay

Laboratorio de Carbohidratos y Glicoconjugados. Departamento de Química Orgánica. Facultad de Química, Universidad de la República, Montevideo, Uruguay

Á. Máthé, A. Bandoni (eds.), *Medicinal and Aromatic Plants of South America Vol. 2*, Medicinal and Aromatic Plants of the World 7,
https://doi.org/10.1007/978-3-030-62818-5_26

Abstract *Lithraea molleoides* (Anacardiaceae) is a tree with a broad area of distribution in the southern region of South America, including Southern Brazil, Southern, and Eastern Bolivia, Southern Paraguay, Northern, and Central Argentina, and reaches its southernmost distribution limit in Northern and Northeastern Uruguay. It presents considerable morphological variability. It is used as medicinal plant and as a sweetener in traditional infusions. There have been reports on its antibacterial, antiviral, antioxidant, anti-inflammatory and antinociceptive activities. It is traditionally known that it may induce allergic contact dermatitis in sensitive persons by direct and/or airborne contact, i.e. when used as firewood. Many of its biological activities have been associated with the reported presence of 5-alkyl/alkenyl resorcinols. Its allergenic activity has been ascribed to 3-alkyl/alkenyl catechols, as it is well known for other species of the Anacardiaceae family. The wide distribution of the species, its morphological plasticity, the traditional medicinal uses, and toxic properties indicate the necessity of a more extensive and systematic study of the species to assess its medicinal and biological potential.

Keywords *Lithraea molleoides* · Alkyl and alkenyl catechols · Alkyl and alkenyl resorcinols · Allergic contact dermatitis · Sweetener

1 Introduction

Lithraea molleoides (Vell.) Engler (Anacardiaceae) is a medium-sized tree from South America used in ethnomedicine. The aerial parts are described as curative for respiratory and digestive diseases. Furthermore, the fruits of *L. molleoides* are used to prepare beverages ("arrope", "aloja de molle"). On the other hand, it is well known to induce direct/airborne contact dermatitis in sensitive people, causing rash and swelling in exposed parts of the body. In general, skin hypersensitivity appears after several contacts with this species and it has been compared with the one that is caused by *Toxicodendron* spp. There are current pharmacological and chemical investigations on both medicinal and poisonous activities of *L. molleoides*, which highlight the necessity to investigate the underlying mechanisms behind.

2 Taxonomic Characteristics

Lithraea (Anacardiaceae) genus was described for the first time in 1826 by John Miers, being taxonomically known as *Lithraea* Miers ex. WJ Hooker et Arnott. (Tropicos 2020; Index Nominum Genericorum database 2020). It includes only three species native from the South American region: *L. caustica* (Molina) Hook. & Arn., *L. brasiliensis* Marchand, and *L. molleoides* (Vell.) Engl. The latter two species grow in Argentina, Bolivia, Brazil, Paraguay, and Uruguay, while *L. caustica* is native from Chile (Hurtado 1986). The type of leaves differentiates *L. molleoides* from *L. brasiliensis*: the former has imparipinnate compound leaves with three to five (eventually, seven) leaflets, while the second has simple leaves (Muñoz 2000; Juárez and Novara 2007). These species were commonly confused due to the morphological variability observed for *L. molleoides* (Pienaar and von Teichman 1998).

Etymologically, *Lithraea* name was originated from "*litre*", noun given in Chile by Mapuche communities to *L. caustica* and dermatitis caused by it (Hurtado 1986). Its specific name is due to its morphological similarity with the "molle" tree (*Schinus molle* L.), which also belongs to the Anacardiaceae family (Muñoz 2000; Juárez and Novara 2007). It is reported that *Lithraea* spp., in particular, the resins from the plant sap, have vesicant properties and provoke severe contact dermatitis and allergies (Arechavaleta 1906; Hurtado 1986; Alé et al. 1997). The capacity to produce allergic responses in sensitive individuals is shared with at least 25% of Anacardiaceae species (tribes Anacardieae, Rhoeae, and Semecarpeae) (Correia et al. 2006). Such detrimental property has converted these trees in a matter of folklore legends and beliefs recorded even as poems and essays (mainly in Chile and Argentina), which tell about "it dangerous inviting shade" as well as its "magical healing power" (Hurtado 1986; Kott et al. 1999).

In Argentina, *L. molleoides* is popularly known as "molle", "chirimolle", "molle de beber", "molle blanco", "molle dulce", and "falso molle" (Northwestern region); "chichita", "chichita colorada", and "árbol malo" (bad tree) (Northeastern region); while in the Central region of the country it is named "molle" or "molle de Córdoba" (Montes 1969; Muñoz 2000; Ruffa et al. 2002; Menseguez et al. 2007; Juárez and Novara 2007; Toledo 2009; Arambarri et al. 2011; Mercado et al. 2014). In Brazil, the traditional names are "aroeira", "aroerinha", "aruera", "aroeira-branca", "aroeira-brava", "aroeira-do-brejo", "aroeira-da-capoeira" and "bugreiro" (Shimizu et al. 2006; Berger 2007; Simoni et al. 2007; Arambarri et al. 2011; Da Silva 2016). Other common names conferred in Bolivia to *L. molleoides* (under the synonym *Lithraea ternifolia*) are "lloque", "lloque negro", "lloke", "lloohke", "lloquei", "montecillo", and "wallpa chaqui" (Rodríguez 2001; López 2010). In Uruguay, this tree species is dominant in the southeastern, northeastern, and northern forests being traditionally known by the names "aruera", "aruera dura", and "laurel del monte"; names sometimes shared with the similar species *L. brasiliensis* which co-occurs in those forests (Arechavaleta 1906; Sayagues Laso et al. 2000; Piaggio and Delfino 2009; Profumo et al. 2017).

Synonyms *Lithraea gilliesii* Griseb.; *Lithraea aroeirinha* March. ex Warm.; *Lithraea ternifolia* (Gillies ex Hook. & Arn.) FA Barkeley; *Lithraea ternifolia* Brkl. & Rom.; *Schinus ternifolia* Gillies ex Hook; *Schinus brasiliensis* March. ex Cabrera; *Schinus molleoides* Vell.; *Schinus leucocarpus* Mart. ex Engl. in Mart. & Eichl.; *Schinus tenuifolius* Steud.; *Schinus terebinthifolia* var. *ternifolia* March.; *Rhus claussеniana* Turcz. (Zuloaga and Morrone 1999; Muñoz 2000; Tropicos 2020; The Plant List 2020). *Lithraea molleoides* var. *lorentziana* Hieron ex Lillo is an accepted sub-species (Tropicos 2020; The Plant List 2020).

3 Crude Drug Used

Numerous authors mention the therapeutic potential of *L. molleoides*, still, its medicinal use is not endorsed in the current legislation from any South American countries. In contrast, its medicinal purposes are preserved in several local cultures.

The leaves and young stems are used in ethnomedicinal practices, and occasionally the resin is also mentioned. The leaves and fruits are employed to prepare beverages and as a sweetener to the "mate", a traditional infusion made with *Ilex paraguariensis* leaves, which is popular in Argentina, Southern Brazil, Paraguay, and Uruguay), conferring at the same time diuretic properties (Sorarú and Bandoni 1978; Rondina et al. 2003; Menseguez et al. 2007; Juárez and Novara 2007).

Lithraea molleoides (Vell.) Engler. (Photos: MÁ López)

4 Major Chemical Constituents and Bioactive Compounds

Phenols, catechols, resorcinol, and hydroquinone derivatives, generally attached to an odd-numbered alkyl or alkenyl side chain, have been described in the Anacardiaceae family. These phenolic lipids have been associated with the vesicant properties and most of the biological activities informed for these plant species (Valcic et al. 2002; López et al. 2005; Correia et al. 2006; Alvarenga et al. 2016; Xie et al. 2016; Dang et al. 2019). The activity of these molecules is frequently ascribed to their amphipathic character, which confers affinity to biological membranes and the capacity to destabilize the native structures. In the case of *Toxicodendron* spp. (formerly *Rhus* spp.), these classes of compounds are stored in the sap as resins named "urushiols" (from the Japanese word *kiurushi* which means sap), being mixtures of several catechol derivatives [1-(alkyl)- or 1-(alkenyl)-2,3-dihydroxybenzenes] where the C_{15}/C_{17} alkyl/alkenyl side chains and one to three double bonds (Correia et al. 2006). After penetration in the skin, the phenolic lipids (haptens) react with proteins giving protein-orthoquinone adducts (real antigens of allergic contact dermatitis) (Roberts and Lepoittevin 1998; Goetz et al. 1999; Rustemeyer et al. 2020).

From the *L. molleoides* aerial parts there have been described 3-alkyl C_{15} and C_{17} catechols (Alé et al. 1997); and 5-alkylene resorcinols containing unbranched side chains of 13 carbon atoms (Valcic et al. 2002; López et al. 2005; Chiari et al. 2010; Carpinella et al. 2011). Alé et al. (1997) identified in the leaves and stem bark extract from *L. molleoides* and *L. brasiliensis* a mixture of four saturated and unsaturated catechols derivatives (3-pentadecyl, 3-pentadecenyl, 3-heptadecenyl, and 3-heptadecadienyl) without assigning positions for the double bonds; the C_{15} monoene catechol was the most abundant compound in both species. This catechol derivatives mixture was tested in humans for allergenic reactions, and the authors found a high percentage of occurrences of skin irritations in sensitive persons after the topical application of the extract (Alé et al. 1997). From the aerial parts of *L. molleoides,* it was also isolated catechol (Garro et al. 2015). The natural products isolated by Valcic et al. (2002) from *L. molleoides* were resorcinol derivatives [(*Z*)-5-(trideca-4′-enyl); (*Z,Z*)-5-(trideca-4′,7′-dienyl); (*Z,Z,Z*)-5-(trideca-4′,7′,10′-trienyl); and (*Z,Z,E*)-5-(trideca-4′,7′,10′-trienyl)], exhibiting nematicidal activity against two experimental models. Later, López et al. (2005), following a bio-guided fractionation, isolated from *L. molleoides* leaves a 1,3-dihydroxy-5-(tridec-4′,7′-dienyl)-benzene [or (*Z,Z*)-5-(trideca-4′,7′-dienyl)-resorcinol]. In additional studies, these authors confirmed the cytotoxic potential, anti-inflammatory, immunomodulatory, and antioxidant properties of such compounds (López et al. 2005; Barbini et al. 2006; Gorzalczany et al. 2011; López et al. 2011). Chiari et al. (2010) and Carpinella et al. (2011) isolated the same compound and demonstrated it inhibited the enzyme tyrosinase and presented antibacterial activity. Similar resorcinol derivatives, with anti-*Schistosoma mansoni* activity, have been found in the cashew (*Anacardium occidentale*) and *Swintonia floribunda* (Anacardiaceae) (Alvarenga et al. 2016; Dang et al. 2019).

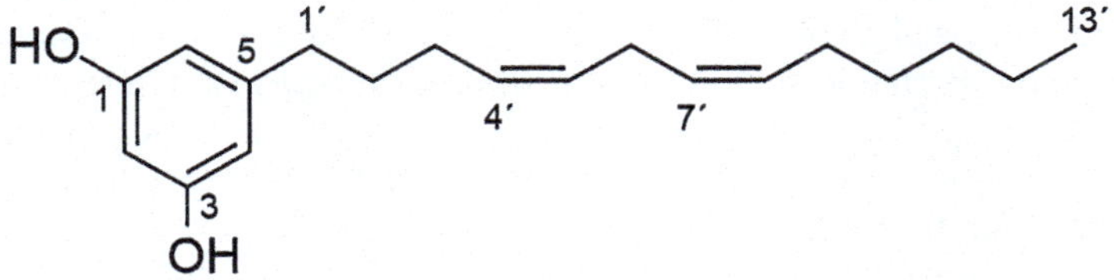

(*Z,Z*)-5-(trideca-4′,7′-dienyl)-resorcinol, the most cited bioactive compound from *L. molleoides*.

Several phenolic acids such: shikimic, gallic, vanillic, syringic, caffeic, and chlorogenic, and also its esters as the case of methyl gallate, have been isolated from the aerial parts (López 2010; Gorzalczany et al. 2011; Garro et al. 2015). Flavonoids also are common components from *L. molleoides*: rutin, quercetin, isoquercetin, luteolin, apigenin, and kaempferol (López 2010). The sugar alcohol mannitol is also a common component of the aqueous infusions obtained from *L. molleoides* aerial parts (López 2010; Garro et al. 2014).

The *L. molleoides* essential oil from the fruits was studied for the first time in Argentina by Montes (1969). He preliminary identified (without specifying the sample collection site): α-pinene, 1,8-cineole, linalool, camphor, α-terpineol, geraniol, and carvone. Later analysis of the essential oil from samples collected in Córdoba Province in Argentina (under the synonym *L. ternifolia*) confirmed the previous results rendering carvone (34.7%), limonene (21.6%), linalool (12.1%), α-terpineol (9.3%), elemol (5.8%) and spathulenol (5.1%) (Maestri et al. 2002). The concordance in these results suggests that the samples belonged to a unique chemotype. In contrast, the essential oil from mature fruits of *L. molleoides* collected in Brazil was rich in monoterpene hydrocarbons: limonene (89.9%), α-pinene (3.5%), β-pinene (2.6%), myrcene (0.6%), camphene (0.2%), and δ-3-carene (0.1%), with minor amounts of oxygenated terpenes: α-terpineol (1.3%), sabinene hydrate (0.5%), and terpinen-4-ol (0.3%) (Shimizu et al. 2006).

5 Morphological Description

L. molleoides is a polygamous-dioic tree having a grey cortex with lenticels, resinous, with dense foliage and globose canopy, semi-deciduous. In general, from two to eight meters high, but exceptionally reaching up to 12 m high (Martijena 1987; Muñoz 2000; Carballo et al. 2005; Juárez and Novara 2007). It is equipped with imparipinnate-compound alternate leaves (sometimes simple at the base of the tree), petioles from 1.5–3 cm long. The three/five leaflets are sessile, glabrous, lanceolate, or oblong/lanceolate (dimensions are highly variable), with entire margin, cuneate base, and acute/mucronate apex; in general with a bigger terminal leaflet (Muñoz 2000; Juárez and Novara 2007).

The yellowish/greenish flowers are arranged in terminal or axillary panicles (5–7 cm; shorter than the leaves) provided with a bract and bracteoles; pentasepal calyx, and pentapetal corolla. The androceous flowers have ten uneven stamens, five of them shorter in staminate flowers, and all of them vestigial in pistillate flowers. Gynoecium composed of three carpels, three styles, and three stigmas, forming a unilocular subglobose ovary; vestigial in the case of staminate flowers (Muñoz 2000; Juárez and Novara 2007).

Fruits are whitish to yellowish/greenish externally lignified drupes (0.5–0.8 cm long) with oleaginous mesocarp, and one seed (having membranaceous external coat) per fruit, fleshy cotyledons (Muñoz 2000; Marco and Páez 2000; Carmello-Guerreiro and Paoli 2005; Juárez and Novara 2007). *L. molleoides* blooms during September–October, and fruiting takes place between November and March (Medri et al. 2007).

There is wide variability in the morphology of *L. molleoides* leaflets and fruits (especially in Bolivia, South Brazil, and Northeastern Argentina), and thus, phenotypic plasticity has been observed (Pienaar and von Teichman 1998; Muñoz 2000; Mercado et al. 2014). The micro-anatomical descriptive analysis of different organs of *L. molleoides* has been performed: leaves (Arambarri et al. 2011; Mercado et al. 2014), wood (as *L. ternifolia* botanical synonym; Martijena 1987), and fruits and seeds (Carmello-Guerreiro and Paoli 2005).

6 Geographical Distribution

L. molleoides is a native species to South America, where it grows and develops in Southern Brazil, Southern and Eastern Bolivia, Southern Paraguay, Northern, and Central Argentina, and reaches its southernmost distribution limit in Northern and Northeastern Uruguay (Muñoz 2000; Brussa and Grela 2007; Juárez and Novara 2007; Mercado et al. 2014; Tropicos 2020). In this last country, it is found both in xerophytic bushland as well as in defiles near watercourses, forming the upper stratum of the canopy (Brussa and Grela 2007). In Argentina, it is a dominant orophilic species (a member of climax community) in the mountain ranges forests and forest edges from the biogeographical Semi-arid Chacoan region with a continental climate and annual precipitations between 500–800 mm (Muñoz 2000; Arambarri et al. 2011). In the Paranaense and Humid Chacoan rainforests, *L. molleoides* is a member of the secondary succession community (Muñoz 2000). *L. molleoides* grows in general from 0 to 500 m altitudes, but in some case, it was found in mountain regions at 2000–3000 m above the sea level (Muñoz 2000; Catálogo de la Plantas Vasculares de la Flora del Cono Sur 2020). Cultivation of *L. molleoides* is restricted to experimental regeneration of woodlands degraded or invaded by exotic species (Verzino et al. 2004; Ferreras et al. 2019).

7 Ecological Requirements

L. molleoides is a plastic species that in general grows in a variety of soils, from those dried to the humid (even in rocky soils), being wide tolerant to those diverse environments partially because of its wood anatomic features (Martijena 1987; Marco and Páez 2000; Medri et al. 2007). The reproduction of this species is by seeds, which optimal germination range is between 22–24 °C, submitted to full sunlight radiation, and moderate humidity (Marco and Páez 2000; Brussa and Grela 2007; Lorenzi 2014; Ross et al. 2018). According to Alves-Costa and Eterovick (2007), coatis (*Nasua nasua*: *Procyonidae*), native South American mammals play a vital role in the seed dispersion of *L. molleoides* and other species among the Atlantic forests in Southeastern Brazil. This plant species presents relative seed dormancy (Berger 2007; Berger et al. 2014). Brussa and Grela (2007) and Lorenzi

(2014) suggest collecting seeds from ripe fruits and removing the epicarp before sowing them to avoid seed dormancy. Piveta et al. (2014) observed that 20 minutes of seed scarification with sulfuric acid, soaking in hot water at 70 °C, and the application of gibberellic acid/potassium nitrate pretreatment remove the seed dormancy successfully. Nevertheless, seeds can be sown directly in tubes with a clayey substrate, in a half-shaded environment, and daily watered (Brussa and Grela 2007; Lorenzi, 2014, Ferreras et al. 2019). According to Lorenzi (2014), the germination rate is up to 80%, the emergency occurs after 8–12 days, and plantlets must be transplanted when they achieved 4–5 cm. The medium to low growth rate of *L. molleoides* has endangered its preservation, and in some cases have been replaced by alien species as *Gleditsia triacanthos* (Fabaceae) and *Ligustrum lucidum* (Oleaceae) among others, in particular in the Montane Forests in Central Argentina (Marco and Páez 2000; Zeballos et al. 2014). Despite its growth rate, *L. molleoides* is a species that sprout very well after a fire allowed by a high accumulation of reserves in below-ground organs (Gurvich et al. 2005), which suggests an ecological strategy to survive in xerophytic forests with long dry seasons and frequent fires. The leaves of *L. molleoides* exposed to shade have physiological and morpho anatomical adaptations that allow them to achieve the same photosynthetic efficiency as leaves exposed directly to sunlight (Dias et al. 2007). These findings show the ecological importance of *L. molleoides* in the initial successional forest stages.

8 Traditional Use (Part(s) Used) and Common Knowledge

Infusions, decoctions or tinctures from the aerial parts (leaves, buds, and young stems) of *L. molleoides* are employed in ethnomedicine against respiratory (colds, coughs, bronchitis, and phlegm), and digestive inflammations and illnesses (Toursarkissian 1980; Ratera and Ratera 1980; Rondina et al. 2003; Ruffa et al. 2002; Arambarri et al. 2011). In Central Argentina, *L. molleoides* is popularly considered as a "hot plant" because its consumption "warms" the body and allows the cure of respiratory diseases (Menseguez et al. 2007). Besides, there are other indications of use for the aerial parts decoctions/infusions such as sedative, hemostatic, diuretic, as a sweetener to improve the flavor (for example, together with the "mate"), and even to treat toothache (Sorarú and Bandoni 1978; Ruffa et al. 2002; Juárez and Novara 2007; Menseguez et al. 2007; Toledo 2009; Arambarri et al. 2011). Moreover, *L. molleoides* resins are cited as antiarthritic (Toursarkissian 1980).

This tree gives appreciable lasting wood rich in tannins (semi-hardwood, difficult to cut, with a specific gravity between 0.55 and 0.87 kg/dm^3 depending on the information source), serving for civil construction, for fences, and eventually as firewood (Hurtado 1986; Martijena 1987; Muñoz 2000; Juárez and Novara 2007). A yellow dye useful for staining is obtained from the leaves (Sorarú and Bandoni 1978; Hurtado 1986).

L. molleoides fruits are employed in Argentina for making traditional beverages named "arrope" (a species of syrup), and "aloja de molle", this last produced by fermentation in cold water (Hieronymus 1882; Muñoz 2000; Menseguez et al. 2007; Juárez and Novara 2007). To this traditional use refers the vernacular names "molle de beber" and "molle de tomar" (literally, "molle for drinking") (Hieronymus 1882; Sola 1942; Rondina et al. 2003; Carballo et al. 2005).

The direct or airborne contact with *L. molleoides* might induce allergenic contact dermatitis (ACD) in sensitive persons, characterized by erythema, edema, vesiculation and bullae reactions in the skin (which also could be accompanied with fever), popularly known as "flechadura" (Hurtado 1986; Alé et al. 1997; Muñoz 2000; Carballo et al. 2005). This sensitization is the origin of many of the common names such as "árbol malo" (evil tree) or "aroeira-brava" (harmful "aroeira") (Hurtado 1986). Similar sensitizing effects are provoked by other Anacardiaceae members, such as *L. caustica* and *Toxicodendron* spp. (Hurtado 1986; Alé et al. 1997; Correia et al. 2006). The species "poison ivy" [*T. radicans* (L.) Kuntze], "poison sumac" [*T. vernix* (L.) Kuntze], and "poison oak" [*T. diversilobum* (Torr. & A Gray) Greene] are famous worldwide as vesicants and causative of severe rash elicited by their "urushiols" resins (Alé et al. 1997; Valcic et al. 2002; Correia et al. 2006; Hodgson 2012). "Urushiols" can cause hypersensitivity by dermal contact or even, through inhalation of the atmosphere surrounding the vegetation or after its burning when used as firewood; and because this they represent severe environmental hazards in outdoors both in rural areas as well as in urban locations (Alé et al. 1997; Correia et al. 2006; Hodgson 2012). As mentioned before, C_{15} and C_{17} alkyl and alkenyl catechols are the sensitizing compounds from *L. molleoides* resins. However, their potential as allergenic is lesser than those of poison oak even for individuals previously sensitized to *Lithraea* spp. (Alé et al. 1997). The mechanisms behind ACD and the corresponding interaction between haptens and proteins have been reviewed (Roberts and Lepoittevin 1998; Rustemeyer et al. 2020).

9 Modern Medicine Based on Its Traditional Medicine Uses

Different extracts and the fruit essential oil from *L. molleoides* have been reported as having several *in vitro* and/or *in vivo* bioactivities, such: antibacterial (Penna et al. 2001; Shimizu et al. 2006; Carpinella et al. 2011; Garro et al. 2015; Ibáñez et al. 2017); antiviral (Kott et al. 1999; Simoni et al. 2007); antifungal (Penna et al. 2001; Muschietti et al. 2005; Shimizu et al. 2006); nematicidal (Valcic et al. 2002); antiulcerogenic (Araujo et al. 2006; Garro et al. 2015); immunomodulatory (Fernández et al. 2002; López et al. 2011); anti-inflammatory (Gorzalczany et al. 2011); antinociceptive (Morucci et al. 2012); antioxidant (López et al. 2007, 2011); enzymatic inhibitory (Chiari et al. 2010); diuretic (Garro et al. 2012); cytotoxic (Ruffa et al. 2002, Lopes et al. 2003; López et al. 2005; Barbini et al. 2006); and genotoxic (Carballo et al. 2005).

Kott et al. (1999) performed an ethnopharmacological screening with Argentine medicinal plants evaluating the antiviral activity by *in vitro* models. The authors found that the aqueous infusions from the *L. molleoides* leaves were active against two viruses (without exerting cytotoxic effects): HSV-1 (herpes virus; $ED_{50} = 51.5$ µg/ml) and RSV (respiratory syncytial virus; $ED_{50} = 78.0$ µg/ml), but were innocuous against ADV-7 (adenovirus) (Kott et al. 1999). Simoni et al. (2007) described the potential of *L. molleoides* aqueous extract to treat the avian reovirus (S133 strain) with an inhibition superior to 99% (accompanied by a variable level of cytotoxicity), while the activity against the BoHV-1 (bovine herpesvirus) was negligible (74% inhibition), and no susceptibility at all was observed for IBDV (infectious bursal disease virus of chickens).

Penna et al. (2001), working on the same plant species studied by Kott et al. (1999) evaluated the antimicrobial properties of the organic and aqueous *L. molleoides* extracts against a set of bacteria and fungi using disc diffusion assays and bioautography. Commercial and isolated strains of *Bacillus subtilis*, *Micrococcus luteus*, *Staphylococcus aureus,* and *Mucor* sp. were susceptible to the *L. molleoides* extracts. At the same time, the fungus *Aspergillus niger* was resistant to all of them (Penna et al. 2001). Extracts in methanol from the leaves also demonstrated antifungal activity against a series of species including all the dermatophytes strains tested by the authors (*Microsporum canis*, *M. gypseum*, *Epidermophyton flocossum*, *Trychophyton rubrum*, and *T. mentagrophytes;* MIC = 250 µg/ml for all of them) (Muschietti et al. 2005). However, no activity was observed against yeasts and filamentous fungi, including *Aspergillus* spp. and *Candida* spp. strains (Muschietti et al. 2005). Interestingly, Shimizu et al. (2006) found activity of the *L. molleoides* fruit essential oil against some strains of *S. aureus*, *S. epidermidis*, *C. tropicalis*, and *C. dubliniensis*; while all the Gram (−) bacteria tested by the authors were resistant. Another evaluation demonstrated *in vitro* resistance of the phytopathogenic fungus *Sphaceloma ampelinum* (agent causative of canker in grapevines) to alcoholic extracts and fractions from leaves of this plant species (Da Silva 2016).

In an attempt to improve the effectiveness against microbes and to avoid resistance, Carpinella et al. (2011) isolated (Z,Z)-5-(trideca-4′,7′-dienyl)-resorcinol from *L. molleoides*, and tested it against the bacteria *Proteus mirabilis* (one of the main causatives of urinary tract infections in humans). The compound inhibited the growth of *P. mirabilis* and provoked its complete death in 24 hours of experiments (assessed by the time-kill assay), effects that are associated with the amphipathic properties which induce membrane disruption. Moreover, (Z,Z)-5-(trideca-4′,7′-dienyl)-resorcinol inhibited significantly *P. mirabilis* swarming motility, a key virulence factor of this pathogen (Carpinella et al. 2011). Other compounds isolated from the *L. molleoides* aerial parts (catechol, mannitol, rutin, ferulic, and caffeic acid) showed antimicrobial activity against *Helicobacter pylori* (NCTC11638 and several isolated strains). Catechol exhibited the lowest MIC value (0.5 µg/ml) for most of the *H. pylori* strains assayed (Garro et al. 2015). Moreover, also the aqueous infusion from the aerial parts of this plant species displayed *in vitro* activity against *H. pylori* (including five clinical isolates) with a MIC range between 16 and 125 µg/ml, depending on the bacterial strain susceptibility (Ibáñez et al. 2017). These authors also confirmed that the infusion inhibited the *H. pylori* film formation and induced morphological changes in the bacteria both in planktonic and biofilm states. Moreover, *L. molleoides* infusion inhibited urease activity (essential colonization and a virulence factor for *H. pylori*), as well as decrease the *ure*A gene expression (Ibáñez et al. 2017).

The organic extract of aerial parts *L. molleoides* and the isolated 5-alkylene resorcinols showed *in vitro* nematicidal activity against *Caenorhabditis elegans* and *Trichostrongylus colubriformis* (Valcic et al. 2002). In *C. elegans*, the crude extract produced complete paralysis after four h (2.4 mg/ml), while the isolated compounds induced paralysis after just 10–25 minutes (50 µg/ml). Despite the activity potency declined after two h, the paralytic effects were sustained until 24 h of treatment) (Valcic et al. 2002). For *T. colubriformis* larvae (important parasites in sheep) *in vitro* assays, the 5-alkylene resorcinols also provoked paralysis but with a later onset and to a higher concentration compared to *C. elegans*; while no effect was observed through *in vivo* assays (rodent test model) (Valcic et al. 2002).

A moderate diuretic effect was observed for the 10% infusion of *L. molleoides* leaves after 60–105 min of treatment (Garro et al. 2012, 2014), which validates partially one of the ethnopharmacological uses. This medicinal property is attributed, at least in part, to the content of the sugar alcohol mannitol, which is found in the aqueous infusions at a concentration around 0.46 g/100 g (Garro et al. 2012, 2014).

The antiulcerogenic activity was informed for the hydroalcoholic *L. molleoides* leaves extract employing male Wistar rats as an *in vivo* model (Araujo et al. 2006). In preliminary phytochemical screening, such extract gives positive results for flavonoids, saponins, and tannins. The extract at an oral dose of 1000 mg/kg inhibited the gastric mucosa damage by 48.1% (indomethacin-induced ulceration model), and 55.8% (absolute ethanol-induced ulceration model) (Araujo et al. 2006). A complementary study performed by Garro et al. (2015) demonstrated that the aqueous infusion and methanol extract of *L. molleoides* aerial parts significantly inhibited *in vivo* the gastric ulceration caused by several necrotizing agents (NaOH, HCl, acetylsalicylic acid, and absolute ethanol). The best results were obtained for the methanol extract at 500 mg/kg oral dose, which inhibited all types of ulceration in the range 75–88%, being even more effective than the positive control omeprazole (60 mg/kg dose) (Garro et al. 2015). Catechol, mannitol, rutin, ferulic, and caffeic acids isolated from the methanol extract demonstrated being even more active inhibitors (100 mg/kg dose) than the extracts in the treatment of absolute ethanol-induced ulceration, having displayed catechol the best results; conversely, gallic acid was inactive (Garro et al. 2015).

Dichloromethane, methanol, and aqueous extracts of *L. molleoides* leaves showed a moderate immune-modulatory activity on tumor and normal cells, splenocytes, and lymphocytes being the aqueous extract slightly more active in the cell proliferation assays. The extracts showed a moderate inhibitory activity on the classical and alternate complement pathways (Fernández et al. 2002).

The effect of *L. molleoides* dichloromethane leaves extract, and (*Z,Z*)-5-(trideca-4′,7′-dienyl)-resorcinol isolated from it on normal and tumoral lymphocytes was assessed *in vitro* by López et al. (2011). Regarding the former cells, the extract and the compound enhanced their proliferation at low concentrations acting as immunostimulatory (up to 1.0 µg/ml; with a maximum stimulation effect of

300% at 0.1 µg/ml extract dose), with the extract being more active than the isolated compound. However, for tumoral murine lymphocytes (lymphoma cells), both the extracts and the compound exerted antiproliferative action. However, in this case, the resorcinol derivative displayed the most activity at all the concentration evaluated by the authors (EC_{50} = 428 ± 40 µg/ml, and 17,4 ± 1 µg/ml, for the extract and isolated compound, respectively) (López et al. 2011).

The anti-inflammatory activity of aqueous, dichloromethane, and methanol extracts and two main compounds isolated from the leaves of *L. molleoides* were investigated by Gorzalczany et al. (2011). Carrageenan-induced rat paw edema and 12-*O*-tetradecanoylphorbol-13-acetate induced mouse ear edema were employed as experimental animal models. The results obtained showed that methanol extract presents a significant anti-inflammatory effect in both tests. Methyl gallate, the main compound in the bioactive fraction from this extract, could explain at least in part the effect found. Also, (Z,Z)-5-(trideca-4′,7′-dienyl)-resorcinol, the main component obtained from the dichloromethane extract, showed a significant topical anti-inflammatory activity.

Different extracts of *L. molleoides* present antioxidant activity, as has been assessed using the DPPH capture and ferric thiocyanate methods (FTC; inhibition of lipid peroxidation) (López et al. 2007, 2011). Furthermore, the most common resorcinol derivative from this species, namely (Z,Z)-5-(trideca-4′,7′-dienyl)-resorcinol, also presents antioxidant activity but in a lesser extension than the extracts (EC_{50} = 8.87 and 3.63 µg/ml, with DPPH and FTC models, respectively) (López et al. 2011). The antioxidant activity, in this case, is attributed to its di-phenolic structure, which facilitates the capture of electrons and free radicals.

In vivo, the antinociceptive effect was evaluated by Morucci et al. (2012) for the aqueous extract of *L. molleoides* leaves through acetic acid-induced abdominal writhing, formalin, and hot-plate tests in mice. The oral and intraperitoneal administration of the extract showed a significant inhibition of nociception induced by acetic acid and formalin. Nevertheless, the results obtained by the hot-plate test showed no significant activity at any doses or any administration route. The phytochemical analysis of the extract performed by HPLC-UV showed the presence of phenolic acids and flavonoids (Morucci et al. 2012). To identify the compounds responsible for the activity, two main compounds, vanillic and shikimic acids, were tested by the same chemical nociceptive model. Both compounds were active but with less activity than the extract, suggesting that other compounds present in the extract could contribute to the activity. Even though the mechanism of nociception was not completely elucidated, the authors obtained evidence supporting the involvement of adrenergic and dopaminergic systems in the activity (Morucci et al. 2012).

The extract of *L. molleoides* in ethanol was reported as an effective tyrosinase inhibitor employing L-tyrosine or L-DOPA as substrate models (Chiari et al. 2010). From this extract, it was isolated the compound responsible for the activity and identified as (Z,Z)-5-(trideca-4′,7′-dienyl)-resorcinol. The inhibition of tyrosinase (also known as polyphenol oxidase) is essential for preventing the browning in stored food products to avoid deterioration in the appearance, and the rejection by consumers (Chiari et al. 2010).

Methanol extracts from *L. molleoides* leaves and fruits exhibited cytotoxic activity against human hepatocellular carcinoma cells (Hep G2) according to an *in vitro* screening of Argentine medicinal plants extracts performed by Ruffa et al. (2002). The results indicated that such extracts inhibited the growth of Hep G2 (IC_{50} = 244 ± 25) significantly. However, the activity level was in a lesser extension than the corresponding extracts of *Schinus molle* (IC_{50} = 50 ± 7), a species closely related to *L. molleoides* (Ruffa et al. 2002). With the same objective but employing other cell line models (colon carcinoma HT-29, and lung cancer NCI-H460), Lopes et al. (2003) confirmed the cytotoxic potential of *L. molleoides* and *L. brasiliensis* 10% aqueous and organic extracts. The best results were for those of both *Lithraea* spp. aerial parts in ethanol, which inhibited the carcinogenic cellular growth by 90–100% (100 μg/ml dose) (Lopes et al. 2003). In an attempt to assess the cytotoxic action of *L. molleoides* secondary metabolites, López et al. (2005) isolated from the leaves dichloromethane extract a resorcinol derivative [1,3-dihydroxy-5-(trideca-4′,7′-dienyl)-benzene]. The cytotoxic activity of this compound was assayed on three human tumoral cell lines (Hep G2, mucoepidermoid pulmonary carcinoma H292, wand mammary gland adenocarcinoma MCF7). Similar cytotoxic activity against Hep G2 and H292 (IC_{50} = 68 ± 2 mM, and IC_{50} = 63 ± 5 mM; respectively) and a weaker activity against MCF7 (IC_{50} = 147 ± 5 mM). were observed. However, the activity of the pure phenolic compound was higher than those of the crude extracts (Ruffa et al. 2002; López et al. 2005), which stimulated further studies. Indeed, Barbini et al. (2006) tested this resorcinol derivative against two human hepatocarcinoma cell lines (HepG2 and Hep3B) again, but besides, some mechanistic aspects were investigated. In this approach, the authors observed that the named phenolic lipid displayed a significant cytotoxic activity in both lines, which morphological effects on cells were compatible with programmed death (apoptosis) in a dose-dependent manner, inducing DNA fragmentation and nuclear condensation. Moreover, the evidence of this study suggests that for Hep3B cells (with mutated p53 and Fas death regulating proteins), the apoptosis would proceed by p53- or Fas-independent pathways, an important fact for the potential development of antitumoral agents (Barbini et al. 2006).

Regarding toxicity, in a screening of Argentine medicinal plants performed by Carballo et al. (2005), it was found that the infusion of *L. molleoides* (presumably aerial parts) displays genotoxicity breaking down one or both lymphocytes DNA helix, evaluation which was made by the classical comet test. In the other hand, an *in vivo* acute toxicity study on the infusion of the aerial parts (5% w/w) from the species did not provoke any evidence of detrimental effects on animals (mortality nor macroscopic alteration in the internal organs) at a 3000 mg/kg oral dose up to fifteen days of evaluation (López 2010). Similarly, Garro et al. (2015) also did not find toxicity symptoms after the application of 2000 mg/kg oral dose of the *L. molleoides* aerial parts infusion.

10 Conclusions

Lithraea molleoides is a South American species used in traditional medicine against respiratory and digestive inflammations and as a sweetener. Several studies, *in vitro* and *in vivo*, have already verified some of its properties and medicinal activities: antibacterial, antiviral, antifungal, nematicidal, antiulcerogenic, immunomodulatory, anti-inflammatory, antinociceptive, antioxidant, enzymatic inhibitory, diuretic and cytotoxic among others. 5-Alkylene resorcinols containing unbranched side chains of 13 carbon atoms were isolated from the aerial parts of *L. molleoides*, being the (*Z*,*Z*)-5-(trideca-4′,7′-dienyl)-resorcinol the most cited bioactive compound. *L. molleoides* may induce allergic contact dermatitis in sensitive persons caused by the resin. 3-Alkyl C_{15} and C_{17} catechols have been identified as the main compounds responsible for this undesirable effect. The alcohol mannitol obtained from *L. molleoides* aerial parts might explain its use as a sweetener. In conclusion, the chemistry and biological activities so far discovered in this species open up new possibilities for the understanding of its known traditional therapeutic activities and the acute contact dermatitis that it causes.

Acknowledgments The authors wish to thank Ing. Miguel Ángel López for the photos included in this chapter.

References

Alé S, Ferreira F, González G, Epstein W (1997) Allergic contact dermatitis caused by *Lithraea molleoides* and *Lithraea brasiliensis*: identification and characterization of the responsible allergens. Am J Contact Dermat 8(3):144–149

Alvarenga TA, de Oliveira PF, de Souza JM, Tavares DC, Andrade ML, Cunha SRL, Groppo M, Januário AH, Magalhães LG, Pauletti PM (2016) Schistosomicidal activity of alkyl-phenols from the cashew *Anacardium occidentale* against *Schistosoma mansoni* adult Worms. J Agric Food Chem 64(46):8821–8827

Alves-Costa CP, Eterovick PC (2007) Seed dispersal services by coatis (*Nasua nasua*, Procyonidae) and their redundancy with other frugivores in southeastern Brazil. Acta Oecol 32(1):77–92

Arambarri AM, Novoa MC, Bayón ND, Hernández MP, Colares MN, Monti C (2011) Anatomía foliar de arbustos y árboles medicinales de la región chaqueña semiárida de la Argentina. Dominguezia 27(1):5–24

Araujo CEP, Bela RT, Bueno LJF, Rodrigues RFO, Shimizu MT (2006) Anti-ulcerogenic activity of the aerial parts of *Lithraea molleoides*. Fitoterapia 77(5):406–407

Arechavaleta J (1906) Flora Uruguaya. Ed. Talleres Barreiro y Ramos, Montevideo, Uruguay. Volume I, pp 292–294. In: facsimile: https://autores.uy/ (Last access, February 2020)

Barbini L, López P, Ruffa J, Martino V, Ferraro G, Campos R, Cavallaro L (2006) Induction of apoptosis on human hepatocarcinoma cell lines by an alkyl resorcinol isolated from *Lithraea molleoides*. World J Gastroenterol 12(37):5959–5963

Berger APDA (2007) Variabilidade intra-específica em *Lithraea molleoides* (Vell.) Eng. (aroeira-branca) a partir dos processos de germinação e emergência. Thesis. Universidade Federal de Uberlândia, Uberlândia, Brazil. 57 pp. Available at: https://repositorio.ufu.br/handle/123456789/12260 (Last access, February 2020)

Berger APDA, Ranal MA, de Santana DG (2014) Variabilidade na dormência relativa dos diásporos de *Lithraea molleoides* (Vell.) Eng. Ciência Florestal Santa Maria 24(2):1–13
Brussa C, Grela A (2007) Flora arbórea del Uruguay, con énfasis en las especies de Rivera y Tacuarembó. Ed. COFUSA. Montevideo, Uruguay, 544 pp
Carballo MA, Cortada CM, Gadano AB (2005) Riesgo y beneficios en el consumo de plantas medicinales. Theoria 14(2):95–108
Carmello-Guerreiro SM, Paoli AAS (2005) Anatomy of the pericarp and seed-coat of *Lithraea molleoides* (Vell.) Engl. (Anacardiaceae) with taxonomic notes. Braz Arch Biol Technol 48(4):599–610
Carpinella MC, De Bellis L, Joray MB, Sosa V, Zunino PM, Palacios SM (2011) Inhibition of development, swarming differentiation and virulence factors in *Proteus mirabilis* by an extract of *Lithraea molleoides* and its active principle (Z,Z)-5-(trideca-4′,7′-dienyl)-resorcinol. Phytomedicine 18(11):994–997
Catálogo de las Plantas Vasculares de la Flora del Cono Sur (2020) Instituto de Botánica Darwinion, Buenos Aires, Argentina. Published on the Internet; http://www.darwin.edu.ar. (Last access, February 2020)
Chiari ME, Joray MB, Ruiz G, Palacios SM, Carpinella MC (2010) Tyrosinase inhibitory activity of native plants from Central Argentina: isolation of an active principle from *Lithraea molleoides*. Food Chem 120(1):10–14
Correia SDJ, David JP, David JM (2006) Metabólitos secundários de espécies de Anacardiaceae. Quim Nova 29(6):1287–1300
Da Silva MM (2016) Obtenção de extratos e frações de *Lithraea molleoides* e *Lithraea brasiliensis* e avaliação da atividad antifúngica contra *Sphaceloma ampelinum*. Thesis. Universidade Federal de Santa Catarina, Curitibanos, Brazil. 36 pp. Available at: https://repositorio.ufsc.br/handle/123456789/165891. (Last access, February 2020)
Dang PH, Nguyen LTT, Nguyen HTT, Le TH, Do TNV, Nguyen HX, Le ND, Nguyen MTT, Nguyen NT (2019) A new dimeric alkylresorcinol from the stem barks of *Swintonia floribunda* (Anacardiaceae). Nat Prod Res 33(20):2883–2889
Dias J, Pimenta JA, Medri ME, Boeger MRT, de Freitas CT (2007) Physiological aspects of sun and shade leaves of *Lithraea molleoides* (Vell.) Engl. (Anacardiaceae). Braz Arch Biol Technol 50(1):91–99
Fernández T, Zolezzi PC, Risco E, Martino V, López P, Clavín M, Hnatyszyn O, Cañigueral S, Hajos S, Ferraro G, Álvarez E (2002) Immunomodulating properties of Argentine plants with ethnomedicinal use. Phytomedicine 9(6):546–552
Ferreras AE, Whitworth-Hulse JI, Tecco PA, Marcora PI, Funes G (2019) Environmental constraints to native woody species recruitment in invaded mountain woodlands of Central Argentina. For Ecol Manag 440:189–201
Garro MF, Petenatti ME, Petenatti EM, Del Vitto LA, Pelzer L, Saad JR, María AO (2012) Pharmacobotanical characterization and diuretic activity of 'molles' of the Cuyo region, Argentina. Biocell 36(3):A-144
Garro MF, Gette MA, Petenatti MA, Popovich MC, Camí G, Saidman E, Del Vitto LA, Petenatti EM (2014) Caracterización por HPLC y FTIR de los extractos acuosos de tres especies diuréticas del centro-oeste argentino. Dominguezia 30(2):41–45
Garro MF, Ibáñez AGS, Vega AE, Sosa ACA, Pelzer L, Saad JR, Maria AO (2015) Gastroprotective effects and antimicrobial activity of *Lithraea molleoides* and isolated compounds against *Helicobacter pylori*. J Ethnopharmacol 176:469–474
Goetz G, Meschkat E, Lepoittevin JP (1999) Synthesis of a water-soluble analog of 6-methyl-3-N-alkyl catechol labeled with carbon 13: NMR approach to the reactivity of poison ivy/oak sensitizers toward proteins. Bioorg Med Chem Lett 9(8):1141–1146
Gorzalczany S, López P, Acevedo C, Ferraro G (2011) Anti-inflammatory effect of *Lithraea molleoides* extracts and isolated active compounds. J Ethnopharmacol 133(3):994–998
Gurvich DE, Enrico L, Cingolani AM (2005) Linking plant functional traits with post-fire sprouting vigour in woody species in Central Argentina. Austral Ecol 30(8):868–875

Hieronymus G (1882) Plantae Diaphoricae florae Argentinae. Ed. Kraft. Buenos Aires, Argentina, p 69
Hodgson E (2012) Toxins and venoms. In: Hodgson E (ed) Progress in molecular biology and translational science, vol 112. Ed. Academic Press, Oxford, pp 373–415
Hurtado I (1986) Poisonous Anacardiaceae of South America. Clin Dermatol 4(2):183–190
Ibáñez AGS, Sosa ACA, Ferramola FF, Paredes J, Wendel G, Maria AO, Vega AE (2017) Inhibition of *Helicobacter pylori* and its associated urease by two regional plants of San Luis Argentina. Int J Curr Microbiol App Sci 6(9):2097–2106
Index Nominum Genericorum database Published on the Internet: https://naturalhistory2.si.edu (Last access, February 2020)
Juárez FC, Novara LJ (2007) Anacardiaceae Lindl. In: Aportes botánicos de Salta-Ser. Flora Flora del Valle de Lerma 8(6):1–35
Kott V, Barbini L, Cruañes M, Muñoz JDD, Vivot E, Cruañes J, Martino V, Ferraro G, Cavallaro L, Campos R (1999) Antiviral activity in argentine medicinal plants. J Ethnopharmacol 64(1):79–84
Lopes RM, Fossá G, Fossá A, Fedrigo CA, Da Matta SG, Machado KRB, Benetti MN, Bordignon SAL, Faria DH, Kurek AG, Tonding LS, Apel MA, Schwartsmann G, Grivicich I, Da Rocha AB (2003) Avaliação da atividade citotóxica de *Lithraea brasiliensis* e *Lithraea molleoides* (Anacardiaceae). Revista de Iniciação Científica da ULBRA 2:28–33
López P (2010) Estudio Químico y Farmacológico de una planta medicinal Argentina: *Lithraea molleoides* (Vell.) Engl. Thesis. Universidad de Buenos Aires, Buenos Aires, Argentina, 135 pp
López P, Ruffa MJ, Cavallaro L, Campos R, Martino V, Ferraro G (2005) 1,3-dihydroxy-5-(tridec-4′,7′-dienyl)-benzene: a new cytotoxic compound from *Lithraea molleoides*. Phytomedicine 12(1–2):108–111
López P, Anesini C, Cogoi L, Ferraro F (2007) Estudio de la actividad antioxidante de *Lithraea molleoides* (Anacardiaceae). Bol Latinoam Caribe Plant Med Aromát 6(6):352–353
López P, Ferraro F, Anesini C (2011) Comparative antioxidant activity of an extract of *Lithraea molleoides* and an isolated 5-alkyl resorcinol derivative. Effects on the proliferation of normal and tumoral lymphocytes. Phytother Res 25(2):271–276
Lorenzi H (2014) Árvores Brasileiras: manual de identificação e cultivo de plantas arbóreas nativas do Brasil, Vol. 1. 6ª Ed. Nova Odessa, Instituto Plantarum. São Paulo, Brazil, 384 pp
Maestri DM, Labuckas DO, Zygadlo JA (2002) Volatile constituents of *Lithraea ternifolia* (Gillies) Barkley fruit. J Essent Oil Res 14(4):293–294
Marco DE, Páez SA (2000) Invasion of *Gleditsia triacanthos* in *Lithraea ternifolia* montane forests of Central Argentina. Environ Manag 26:409–419
Martijena N (1987) Wood anatomy of *Lithraea ternifolia* (Gill.) Barkley & Rom. (Anacardiaceae). IAWA J 8(1):47–52
Medri ME, Ferreira AC, Kolb RM, Bianchini E, Pimenta JA, Davanso-Fabro VM, Medri C (2007) Alterações morfoanatômicas em plantas de *Lithraea molleoides* (Vell.) Engl. submetidas ao alagamento. Acta Sci Biol Sci 29(1):15–22
Menseguez P, Galetto L, Anton AM (2007) El uso de plantas medicinales en la población campesina de El Puesto (Córdoba, Argentina). Kurtziana 33(1):89–102
Mercado MI, Slanis A, Ruiz AI, Ponessa GI (2014) Morfoanatomía y arquitectura foliar de *Lithraea molleoides* (Anacardiaceae). Lilloa 51(2):209–217
Montes AL (1969) La cromatografía gaseosa aplicada en el estudio de algunos aceites esenciales de plantas nativas del centro y norte argentinos. An Soc Cient Argent 187:21–48
Morucci F, López P, Miño J, Ferraro G, Gorzalczany G (2012) Antinociceptive activity of aqueous extract and isolated compounds of *Lithraea molleoides*. J Ethnopharmacol 142(2):401–406
Muñoz JDD (2000) Anacardiaceae Lindley. In: Huzinker AT (ed.): Flora Fanerogámica Argentina, vol 65. Ed. IMBIV-CONICET, Córdoba, Argentina, pp 1–29
Muschietti L, Derita M, Sülsen V, Muñoz JDD, Ferraro G, Zacchino S, Martino V (2005) *In vitro* antifungal assay of traditional argentine medicinal plants. J Ethnopharmacol 102(2):233–238

Penna C, Marino S, Vivot E, Cruañes MC, Muñoz JDD, Cruañes J, Ferraro G, Gutkind G, Martino V (2001) Antimicrobial activity of Argentine plants used in the treatment of infectious diseases. Isolation of active compounds from *Sebastiania brasiliensis*. J Ethnopharmacol 77(1):37–40
Piaggio M, Delfino L (2009) Florística y fitosociología de un bosque fluvial en Minas de Corrales, Rivera, Uruguay. Iheringia 64(1):45–51
Pienaar ME, von Teichman I (1998) The generic position of *Lithraea brasiliensis* Marchand (Anacardiaceae): evidence from fruit and seed structure. Bot J Linn Soc 126(4):327–337
Piveta G, Muniz MDFB, Reiniger LRS, Dutra CB, Pacheco C (2014) Qualidade sanitária e fisiológica de sementes de aroeira-preta (*Lithraea molleoides*) submetidas a métodos de superação de dormência. Ciência Florestal Santa Maria 24(2):289–297
Profumo L, Portillo M, Mieres I (2017) Crecimiento de *Lithraea molleoides* en dos eco-regiones de Uruguay. In: Brazeiro A (ed) Recientes avances en investigación para la gestión y conservación del bosque nativo de Uruguay. Facultad de Ciencias-UdelaR/MGAP/BMEL, Montevideo, Uruguay, pp 21–23
Ratera EL, Ratera MO (1980) Plantas de la flora Argentina empleadas en medicina popular. Ed. Hemisferio Sur. Buenos Aires, Argentina, p 127
Roberts DW, Lepoittevin JP (1998) Hapten-protein interactions. In: Lepoittevin JP, Basketter DA, Goosens A, Karlberg AT (eds) Allergic Contact Dermatitis. The Molecular Basis. Ed. Springer-Verlag, Berlin/Heidelberg, Germany, pp 81–111
Rodríguez TM (2001) Árboles fuera del bosque en Bolivia. Proyecto: Información y Análisis para el manejo forestal sostenible: integrando esfuerzos nacionales e internacionales en 13 países tropicales en América Latina (GCP/RLA/133/EC). FAO/UE, Santiago, Chile. 39 pp. Published on the internet: http://www.fao.org/tempref/GI/Reserved/FTP_FaoRlc/old/proyecto/rla133ec/AFB-pdf/AFB%20-%20Bolivia.PDF (Last access, February 2020)
Rondina RV, Bandoni AL, Coussio JD (2003) Plantas Silvestres Argentinas con Reconocidas Propiedades Medicinales o Tóxicas. Ed. OEA-CYTED, Buenos Aires, Argentina. CD-ROM
Ross P, Muñoz J, Cracco P (2018) Flora Indígena del Uruguay. Árboles y arbustos ornamentales. 3ª Ed. Ed. Hemisferio Sur . Montevideo, Uruguay, 320 pp
Ruffa MJ, Ferraro G, Wagner ML, Calcagno ML, Campos RH, Cavallaro L (2002) Cytotoxic effect of argentine medicinal plant extracts on human hepatocellular carcinoma cell line. J Ethnopharmacol 79(3):335–339
Rustemeyer T, van Hoogstraten IMW, von Blomberg BME, Scheper RJ (2020) Mechanisms in allergic contact dermatitis. In: John SM, Johansen JD, Rustemeyer T, Elsner P, Maibach HI (eds) Kanerva's occupational dermatology. Ed. Springer Nature, Cham, Swtizerland, pp 151–190
Sayagues Laso L, Graf E, Delfino L (2000) Análisis de la información publicada sobre composición florística de montes naturales del Uruguay. Agrociencia Uruguay 4(1):96–110
Shimizu MT, Bueno LJF, Rodrigues RFO, Sallowicz FA, Azuaya ACHF, Marques MOM (2006) Essential oil of *Lithraea molleoides* (Vell.): chemical composition and antimicrobial activity. Braz J Microbiol 37(4):556–560
Simoni IC, Manha APS, Sciessere L, Hoe VMH, Takinami VH, Fernandes MJB (2007) Evaluation of the antiviral activity of Brazilian Cerrado plants against animal viruses. Virus Rev Res 12(1–2). https://doi.org/10.17525/vrr.v12i1-2.13
Sola W (1942) Árboles y arbustos de Córdoba. Ed. Estrada. Córdoba. Argentina, pp 35, 37, 97
Sorarú SB, Bandoni AL (1978) Plantas de la medicina popular Argentina. Ed. Albatros. Buenos Aires, Argentina, pp 21–25
The Plant List (2020) Version 1.1 Published on the Internet: http://www.theplantlist.org (Last access, February 2020)
Toledo BA (2009) Diversidad de usos, prácticas de recolección y diferencias según género y edad en el uso de plantas medicinales en Córdoba, Argentina. Bol Latinoam Caribe Plant Med Aromát 8(5):389–401
Toursarkissian M (1980) Plantas medicinales de la Argentina. Sus nombres vulgares. Usos y distribución geográfica. Ed. Hemisferio Sur. Buenos Aires, Argentina, pp 4, 5, 178

Tropicos (2020) Missouri Botanical Garden. Published on the Internet: http://www.tropicos.org/Name/40007859 (Last access, February 2020)

Valcic S, Wächter GA, Eppler CM, Timmermann BN (2002) Nematicidal alkylene resorcinols from *Lithraea molleoides*. J Nat Prod 65(9):1270–1273

Verzino G, Joseau J, Díaz MDP, Dorado M (2004) Comportamiento inicial de especies nativas del Chaco Occidental en plantaciones en zonas de pastizales de altura de las Sierras de Córdoba, Argentina. Bosque (Valdivia) 25:53–67

Xie Y, Zhang J, Liu W, Xie N, Feng F, Qu W (2016) New urushiols with platelet aggregation inhibitory activities from resin of *Toxicodendron vernicifluum*. Fitoterapia 112:38–44

Zeballos SR, Tecco PA, Cabido M, Gurvich DE (2014) Composición de especies leñosas en comunidades invadidas en montañas del centro de Argentina: su relación con factores ambientales locales. Rev Biol Trop 62(4):1549–1563

Zuloaga F, Morrone O (1999) Catálogo de las plantas vasculares de la República Argentina II. Dicotyledoneae (Acanthaceae-Euphorbiaceae) y (Fabaceae-Zygophyllaceae). Monographs in Systematic Botany from the Missouri Botanical Garden, vol 74. Missouri Botanical Garden Press. St. Louis, Missouri, USA, pp 41–46

Madia sativa Mol.

Fresia Melina Silva Sofrás

Madia sativa Mol. (Photo: Silvia González Román)

Abstract *Madia sativa,* commonly called "melosa" or "madi", belongs to the Asteraceae family, and is native to America, with a wide range of distribution stretching to both North America and South America. In Argentina, it is present in the Patagonian regions, but it grows also in further eight provinces. It can grow in dry, degraded and eroded soils as a pioneer species. Medicinal properties, as a purgative, are attributed to the herbaceous part of the plant, whereas the nutritional and energizing properties, to the seeds. It is an annual species that can grow from 30 cm to 2 m in height. It is recognized by its pubescence, aroma and stickiness.

Keywords "Madi" · "Melosa" · Lipids · Fixed oil · Energizer

F. M. S. Sofrás (✉)
Facultad de Farmacia y Bioquímica, IQUIMEFA (CONICET), Universidad de Buenos Aires, C.A. de Buenos Aires, Argentina
e-mail: f.silvasofras@conicet.gov.ar

Á. Máthé, A. Bandoni (eds.), *Medicinal and Aromatic Plants of South America Vol. 2*, Medicinal and Aromatic Plants of the World 7,
https://doi.org/10.1007/978-3-030-62818-5_27

1 Introduction

More than 50% of the native medicinal plants of Argentina have not been chemically analyzed nor have their biological activities been evaluated (Barboza et al. 2009). Patagonia is recognized as an environment with little-known medicinal plants of pharmaceutical or industrial interest, which must be urgently studied (Zuloaga 2013).

Madia sativa is a species with extensive records on its ancestral medicinal and nutritional uses, but with little research on it. It is interesting to note that while other species in its genus grow only in North America, *M. sativa* is the only species that has a disruptive distribution, being endemic to both North America and Patagonia. It is a pioneer plant, which grows in any type of environment, matures seeds with nutritive characteristics comparable to those of sunflower.It is therefore regarded a good crop species to produce excellent quality oil (Schmeda-Hirschmann 1995; Matthäus and Angelini 2005) and as a source of bioenergy (San Martín et al. 2011). Considering all the medicinal and nutritional records that have been reported, it is important to summarize the major relevant investigations into this species of Family Asteraceae.

2 Taxonomic Characteristics

The tribe *Madieae*, within the Asteraceae family, comprises five subtribes that include Madiinae with several genera on it. The genus *Madia* includes 12 species, and it has a disjunct distribution between North America and Patagonia, where *M. sativa* is the only one from the genus that grows in South America. They are annual, hirsute-glandular or pubescent plants. *Madia* spp. has medium or small yellow flower heads (capitulum), solitary or located in spiciform, racemiform or corimbiform tops (Zardini 1992). Synonyms: *Madia viscosa* Cav. (Correa 1971), *Madia sativa* subsp. *capitata* (Nutt.) Piper, *Madia sativa* var. *congesta* Torr. & A. Gray (The Plant List 2019).

3 Crude Drug Used

Traditionally the vegetative or herbaceous parts of *Madia sativa* have been used as a tonic against rheumatism, sciatica, gout. It was also cited as a purgative, and to relieve symptoms of arthritis with external application. Nutritional properties are attributed to the seeds, that are postulated as an interesting food source because it contains a high percentage of crude protein, and the composition of its oil is nutritionally valuable. Other sources propose their oil for bioenergetic use (Zardini 1992; Schmeda-Hirschmann 1995).

4 Major Chemical Constituents and Bioactive Compounds

Bohlmann et al. (1982) described for *M. sativa* the presence of diterpenes of the labdane type: melcantane, hexanorlabdane and grindelane. Lupeyl acetate, stigmasterol, chroman derivatives, *p*-hydroxyacetophenone derivatives, thymol epoxide, and a tremetone-hydroxyketone derivative were also extracted from the roots.

Bohm et al. (1992) cited for *M. sativa* three flavonols (rhamnocitrin, axillarin, chrysophanol-D), four flavanones (naringerin, eriodictyol 7-methyl ether, 7,3′-dimethyl ether and 7,3′,4′-trimethyl ether) and one dihydroflavonol (7-methyl ether taxifolin). Wollenweber et al. (1997) indicated the presence of six flavones (bucegin, chryseriol, velutin, luteolin, nepetin and cirsiliol), four other flavonols (quercetin, quercetin-3-methyl ether, tomentin, and gossypetin-3,8-dimethyl ether), and one more flavanone (eriodyctiol).

The extract of *M. sativa* did not present sesquiterpene lactones in significant amounts. Chlorogenic acid, caffeic acid and quercetin were identified by thin layer chromatography of the plant methanolic extract. Gallic acid, rutin, quercitrin were some of the phenolic compounds also identified by HPLC-MS in *M. sativa* infusion (Gastaldi 2019). α-pinene (32,3%) and α-terpineol (31,6%) were the main components of the essential oil (Gastaldi 2019).

M. sativa seeds were also studied for their nutritional properties. According to Angelini et al. (1997), the average amount of glyceride oil in *Madia sativa* seeds was 31.0% with the highest value of 42.0%. Linoleic acid (52.0–72.4%) predominated in fatty acid fraction followed by palmitic and oleic acid (Schmeda-Hirschmann 1995; Angelini et al. 1997; Acevedo et al. 2012). Antova et al. (2017), also studied the content and composition of the biologically active compounds (phospholipids, free and esterified sterols, tocopherols, and fatty acids of the triacylglycerols and sterol esters). The seed meal showed a high crude protein content (28–31%), this is higher than a sunflower seed (19%). The oil composition of *M. sativa* and its adaptability to poor soils suggest considerable potential as a future oil crop (Schmeda-Hirschmann 1995; Antova et al. 2017).

5 Morphological Description

Madia sativa is an annual plant with simple and erect stems. All vegetative parts of the plant are pubescent and glandular. Leaves are sessile. Lower and upper leaves are differently disposed, in a rosette and alternate respectively. Yellow flowers disposed in inflorescences (capitulum) are dimorph. Ligulated and feminine those of the margin. Tubular and hermaphrodites those of the center. Achene is small, curved and black, smooth, without pappus. *M. sativa* is very variable in the size of the flowers and vegetative organs (Correa 1971; Zardini 1992).

6 Geographical Distribution

Madia sativa is native to the American continent, with a wide distribution, occupying both North America and South America. It is considered as an amphitropical species (Raven 1963) and its distribution in the northern hemisphere has been postulated because of the migration routes of birds, and possibly native from North America (Zardini 1992). In Argentina, it is present in the Patagonian region, being found in other eight provinces. According to Zuloaga and Morrone (1999), it is located in the Provinces of San Juan, Mendoza, Neuquén, Río Negro, Chubut and Tierra del Fuego. Though there is also an apparent record of *Madia* in the province of Santa Cruz (Zardini 1992). In Chile, it grows in Regions III to XII, Archipelago Juan Fernández and Santiago Metropolitan Region (Zuloaga and Morrone 1999).

The cultivation of *M. sativa* was very widespread. Native people from South and North America cultivated *M. sativa* for nutritional purpose and, after colonization for the middle of the XIX century, *M. sativa* was remarkably distributed also in Europe (Feuillée 1724). There is not a study about *in vitro* cultivation or domestication, but Yáñez Pinna (2014) proposed a zonification of *M. sativa* production in Chile.

7 Ecological Requirements

M. sativa is a pioneer plant, so its ecological requirements are minimal for its growth. According to the special literature, it grows in dry, degraded and eroded soils, also next to roads and trails, and usually in meadows at elevations of up to 2000 m.a.s.l. (Murillo 1896; Schmeda-Hirschmann 1995).

8 Traditional Use (Part(s) Used) and Common Knowledge

Native people in South America knew *M. sativa* as 'madi' or 'melosa' (like honey) (Picture 1). The common name seems to have inspired the genus name, this probably occurred because the naturalist who described the genus, first discovered the species from South America (de Moesbach 1992). The name "madi" applies only for the seeds, not for the whole plant, which was called "melosa". *M sativa* was also called "pasto aceite" (oil grass) or "yuyo aceite". These names refer to the sticky condition of the plant, that easily stick to the clothes (Zardini 1992). The other common names of *M. sativa* are "hushl" (by Onas from argentine Patagonia, Martínez-Crovetto 1968), "kuradew" ("*Kura*" means "stone", and "*dew*", "action" or "work"). This is related to the oil extraction from the seeds, since native people usually used a stone to crush them on. Also "Madi-fillkun" "madivilcum" and are other common names that could appear associated with *M. sativa (*Febrés 1765). In North America

Picture 1 *Madia sativa* (center). by L. Feuille (1724)

M. sativa is also known as "tarweed" (Carlquist et al. 2003). "Tarweed" denotes not only *M. sativa* but includes a large group of plants with similar characteristics. Native people from North America named this plant as "mushchakili", "muhca" means grain, and "kili" means black, referring to the *M. sativa* seeds (Gifford 1967). Both native cultures from North and South America inspired the name of the plant in the usefulness of the seeds to produce oil. This is a very interesting ethnobotanical character that is shared between the native cultures (Zardini 1992).

The first person that cited *M. sativa* was in 1574 the Spanish cosmograph and historian Juan López de Velasco (1894), who deeply studied the native cultures of South America and their natural resources. He wrote the following sentence: "… You can expect a lot of oil to come from a seed called "madi", which looks like a lettuce seed, also good for eating and enlightening…".

M. sativa was cultivated, even before colonial times with that purpose (Guevara 1929; Parodi 1936). de Ovalle (1646) already quoted the "madi", reportedly offering very good oil. But Molina (1810), who writes extensive comments on their properties, clarifies that it was stopped using it, replacing it with the olive tree, that he did not need to renew the crop periodically "para evitar la fatiga de sembrarlo cada año" (to avoid the fatigue of planting it every year). The act of extracting the oil from the seed is called "maditan", that was a habit widely distributed in Patagonia.

Seeds were crushed in a mortar, then squeezed and the oil produced was collected in a clay pot (Latcham 1909). Another way to extract the oil was to make a decoction of the seeds (de Vidaurre 1889). Lighting with the oil from "madi" was exercised even after colonization (López de Velasco 1894; Guevara 1929).

Native cultures also made "maditun" (South America, according to Febrés 1765), "pinol" or "coast tarweed" (North America), a sweet comestible preparation of seed meal (De Oña 1917; Small 2014). They designed some strategies to make a better "maditun" or "pinole", one of them was to burn the areas where the plant grew, since in that way the achenes did not need a lot of processing to be used al flour.

Louis Feuillée (1724) was interested in the medicinal properties of the plant. He described its use for general pain. In addition to its medicinal uses, he noticed that the oil was more pleasant and sweeter in taste than the olive oil. He took the seeds of "madi" to expand its cultivation in Europe in the XIII century. *Madia sativa* was the name given by the Italian naturalist Giovanni Ignazio Molina (1810), who latter expressed that to cultivate "madi" was appropriate in "hostile" places where olive did not grow.

The idea of cultivating "madi" was widespread at the middle of the XIX century. "Madi" was remarkably distributed, not only in Europe but in South America, with the endorsement of Manuel Belgrano in Buenos Aires, Argentina (Murillo 1896). The first records registers about medicinal applications of "madi" were written by de Rosales (1877), a Spanish chronicler. Other records mention that vegetative parts of the plant produced honey by decoction (de Rosales 1877). This was useful to treat gout, and sciatica pain. It was also cited as purgative and emmenagogue, and to remove or mature tumors. de Rosales described that the plant has the same application as soap and that the oil was used by painters.

9 Modern Medicine Based on Its Traditional Medicine Uses

M. sativa is considered as a purgative plant, ideal to treat indigestions. It is reported that its extracts possess anti-inflammatory capacity (Ekenäs et al. 2008). Gastaldi (2019) reported antioxidant activity by DPPH and beta carotene methods. *M. sativa* did not report general toxicity while its genotoxicity and cytotoxicity against cell lines was not significant (Gastaldi 2019). *M. sativa* aerial parts toxicity was studied by Karchesy et al. (2016), by the brine shrimp (*Artemia salina*) bioassay. The LC_{50} (μg/ml) was >1000.

Two labdanes: 13,14,15-trihydroxylabd-7-ene and 3,14,15-trihydroxylabd-8-ene, were isolated from the resinous exudate of *M. sativa* and provided significant protection against the soil-borne fungal pathogens such as *P. cinnamomi* under *in vitro* conditions (Díaz et al. 2018). Other applications of *M. sativa* are related to the nutritional composition of its seed oil, it appears to be a good crop species to produce excellent quality oil (Schmeda-Hirschmann 1995; Matthäus and Angelini 2005). The oil from the seeds is also proposed for bioenergetic use, to produce biodiesel (San Martín et al. 2011).

10 Conclusions

M. sativa is a species that has been little studied, despite of the rich traditional knowledge that different cultures have attributed to it. It is an annual plant that grows in almost any environment and is characterized mainly as being very sticky and pubescent. The most valued organs of the plant are the seeds. From them, an oil with excellent nutritional characteristics, can be obtained. *M. sativa* has been currently cited to treat indigestion and as a purgative (aerial parts). In addition, it is a possible source of bioenergy. With the exception of the antioxidant activity of its aqueous extracts, its use as a medicinal plant and/or component of cosmetic preparations (due to its lipid and volatile fractions) has not been sufficiently studied.

References

Acevedo F, Rubilar M, Shene C, Navarrete P, Romero F, Rabert C et al (2012) Seed oil bodies from *Gevuina avellana* and *Madia sativa*. J Agric Food Chem 60(28):6994–7004

Angelini LG, Moscheni E, Colonna G, Belloni P, Bonari E (1997) Variation in agronomic characteristics and seed oil composition of new oilseed crops in Central Italy. Ind Crop Prod 6(3–4):313–323

Antova GA, Angelova-Romova MY, Petkova ZY, Teneva OT, Marcheva MP, Zlatanov MD (2017) Biologically active components in *Madia sativa* seed oil. J Food Sci Technol. https://doi.org/10.1007/s13197-017-2739-9

Barboza GE, Cantero JJ, Núñez C, Ariza Espinar L, Pacciaroni ADV (2009) Medicinal plants: a general review and a phytochemical and ethnopharmacological screening of the native argentine Flora. Kurtziana 34(1–2):7–365

Bohlmann F, Jakupovic J, King RM, Robinson H (1982) New labdane derivatives from *Madia sativa*. Phytochemistry 21(5):1103–1107

Bohm BA, Fong C, Hiebert M, Jamal A, Crins WJ (1992) Non-polar flavonoids of *Calycadenia*, *Lagophylla* and *Madia*. Phytochemistry 31(4):1261–1263

Carlquist SJ, Baldwin BG, Carr GD (eds) (2003) Tarweeds & silverswords: evolution of the Madiinae (Asteraceae), vol 1. Ed. Missouri Botanical Garden Press, USA, pp 2–3

Correa M (1971) Flora Patagónica. Parte VII. Compositae. Colección Científica Instituto Nacional de Tecnología Agropecuaria (INTA), Buenos Aires, Argentina. 8(7):457 pp

de Moesbach EW (1992) Botánica indígena de Chile. Ed. Fundación Andes y Ed. A. Bello, Santiago, Chile, 140 pp

De Oña P (1917) Arauco Domado. Edición Crítica de la Academia Chilena Correspondiente de la Real Academia Española. Ed. Imprenta Universitaria, Santiago, Chile, 719 pp

Ovalle A de (1646) Histórica Relación del Reyno de Chile. Roma, Italy, 456 pp + 19 pictures

de Rosales D (1877) Historia general de el Reyno de Chile, Flandes Indiano. Reedition by Mackenna BV. Ed. Imprenta del Mercurio, Valparaíso, Chile, 506 pp

de Vidaurre FG (1889) Historia geográfica, natural y civil del Reino de Chile. (Vol1). Ed. Imprenta Ercilla, Santiago de Chile, Chile, 357 pp

Díaz K, Espinoza J, Espinoza L, González C, Madrid A (2018) Labdanes: antifungal compounds isolated from the resinous exudate of *Madia sativa* against *Phytophthora cinnamomi* Rands. Nat Prod Res 34:1–5

Ekenäs C, Zebrowska A, Schuler B, Vrede T, Andreasen K, Backlund A et al (2008) Screening for anti-inflammatory activity of 12 *Arnica* (Asteraceae) species assessed by inhibition of NF-κB and release of human neutrophil elastase. Planta Med 74(15):1789–1794

Febrés A (1765) Diccionario araucano-español; ó sea calepino chileno-hispano. Reedition from 1882. Ed. JA Alsina. Buenos Aires, Argentina, 104 pp

Feuillée L (1724) Journal des observations physiques, mathématiques et botaniques faites par ordre du Roy sur les côtes orientales de l'Amérique méridionale & dans les Indes Occidentales. Vol. 3. Ed. J. Mariette. Paris, France, p 39 and picture XXVI

Gastaldi B (2019) Análisis de los compuestos fenólicos y volátiles de plantas medicinales y aromáticas del noroeste de la Patagonia Argentina, estudio de sus actividades antioxidante y citotóxica. Thesis. Universidad Nacional de La Patagonia San Juan Bosco, Esquel, Chubut, Argentina. 152 pp

Gifford EW (1967) Ethnographic notes on the southwestern Pomo. Ed. University of California Press. Berkeley, California. USA. 48 pp

Guevara T (1929) Historia de Chile. Chile Preshispano. 2 vol. Ed. Balcells & Co. Santiago. Chile. 389 pp

Karchesy YM, Kelsey RG, Constantine G, Karchesy JJ (2016) Biological screening of selected Pacific Northwest forest plants using the brine shrimp (*Artemia salina*) toxicity bioassay. Springerplus 5(1):510

Latcham RE (1909) Ethnology of the Araucanos. J R Anthropol Inst G B Irel 39:334–370

López de Velasco J (1894) Geografía y descripción universal de las Indias; recopilada por el cosmógrafo-cronista... desde el año de 1571 al de 1574. Ed. Establecimiento Tipográfico de Fontanet, Madrid, Spain, p 516

Martínez-Crovetto R (1968) Estudios Etnobotánicos IV. Nombres de plantas y su utilidad según los Indios. Onas de Tierra del Fuego. Etnobiológica 3:1–20

Matthäus B, Angelini LG (2005) Anti-nutritive constituents in oilseed crops from Italy. Ind Crop Prod 21:89–99

Molina GI (1810) Ensayo sobre la Historia Natural de Chile. (1810). Reedition (1986) by Academia de las Ciencias del Instituto de Bolonia. Ed. Alfabeta Impresores, Santiago, Chile, 385 pp

Murillo A (1896) *Madia sativa* Mol., vulg. Madi, melosa. Actes Soc Sci Chili 6:89–92

Raven PH (1963) Amphitropical relationships in the floras of north and South America. Q Rev Biol 38(2):151–177

San Martín C, Pérez Y, Ramírez C, Veste M, O'Higgins B (2011) El Madi: Una Planta Oleaginosa Endémica de Chile con Potencial Bioenergético. 7° Seminario Internacional "Hacia dónde va la matriz energética en Chile y el mundo", At Fundación Copec – Universidad Católica, Santiago de Chile, Chile

Schmeda-Hirschmann G (1995) *Madia sativa*, a potential oil crop of central Chile. Econ Bot 49(3):257–259

Small E (2014) North American Cornucopia. Top 100 Indigenous Food Plants. Ed. CRC, Boca Raton, USA, 715 pp

The Plant List (2019) Version 1.1. Published on Internet. http://www.theplantlist.org. (Last access, December 2019)

Wollenweber E, Dörr M, Dörsam M, Hassan AEH, Ahmed AA, Hegazy MF, Zeller KP (2003) Flavonoids and terpenoids from the resinous exudates of *Madia* species (Asteraceae, Helenieae). Zeitschrift für Naturforschung C 58(3–4):153–160

Yáñez Pinna IJ (2014) Zonificación agroecológica y productiva para el establecimiento de áreas potenciales de cultivos bioenergéticos de chañar (*Geoffroea decorticans* [Gill. ex Hook. et Arn.] Burk.) y madi (*Madia sativa* Mol.) en Chile. Thesis. Universidad de Chile, Facultad de Ciencias Agronómicas. Santiago de Chile, Chile, pp 69–70

Zardini E (1992) *Madia sativa* Mol. (Asteraceae-Heliantheae-Madiinae): an ethnobotanical and geographical disjunct. Econ Bot 46(1):34–44

Zuloaga F (2013) Ciencia para conocer y proteger la Flora Argentina. Published on Internet. https://www.conicet.gov.ar/ciencia-para-conocer-y-proteger-la-flora-argentina. (Last access, April 2020)

Zuloaga FO, Morrone O (1999) Catálogo de las Plantas Vasculares de la República Argentina, vol II. Ed. Missouri Botanical Garden Press. St-Louis, USA, 1269 pp

Minthostachys verticillata (Griseb.) Epling

H. Rodolfo Juliani, Paula Brunetti, and Adolfina R. Koroch

Abstract The objective of this review is to provide an overall description of the taxonomy, morphology, uses, active principles and future prospects of the endemic Argentinean aromatic plant, locally known as "peperina" (*Minthostachys verticillata*). This species is an important plant that has been frequently enjoyed and used by the people of central Argentina. It was perhaps the initial interest of the indigenous population and settlers paid to this remarkable plant that led to uses and applications in many commercial products. Due to over-collection from the wild by a population increasingly interested in natural plant products, to date "peperina" has become a threatened species. This work seeks to generate and reignite interest in the use and conservation of this important medicinal and aromatic shrub from Argentina.

Keywords "Peperina" · Diversity · Essential oils · Traditional and commercial uses

1 Taxonomic Characteristics

The Lamiaceae family (formerly Labiatae), is traditionally known for the number of aromatic plant species that have been used since ancient times. This statement is especially valid for the representatives of the Nepetoideae subfamily (Scandaliaris 2017). These species are characterized by the presence of glandular trichomes in the

---

H. R. Juliani (✉)
Rutgers University, New Brunswick, NJ, USA
e-mail: hjuliani@rutgers.edu

P. Brunetti
Universidad Nacional de Córdoba. Facultad de Ciencias Agropecuarias, Córdoba, Argentina

A. R. Koroch
The City University of New York. Borough of Manhattan Community College, New York, NY, USA

Á. Máthé, A. Bandoni (eds.), *Medicinal and Aromatic Plants of South America Vol. 2*, Medicinal and Aromatic Plants of the World 7,
https://doi.org/10.1007/978-3-030-62818-5_28

leaves, stems, and flowers, that synthesize and accumulate essential oils that are responsible for the characteristic, sometimes pleasant aroma of this group of plants. Within this group is 'peperina' (*M. verticillata)* one of the most used and intensively harvested species of Argentina (Ojeda and Karlin 2015).

The genus *Minthostachys*, belongs to the subfamily Nepetoidea and the tribe Mentheae, the same tribe as the common mint (*Mentha* spp.). This genus is an important group of aromatic plants, growing wild in mountainous regions of South America, from Venezuela, along the coastal Cordillera, and south through Colombia, Ecuador, Bolivia, and Peru, to mountain areas of central Argentina, particularly the province of Córdoba. All species have a decumbent habit, not able to fully support themselves, they are commonly found growing under other shrubs (Schmidt-Lebuhn 2008a).

Initially, *Minthostachys* sp. was considered as monotypic with the single species *M. mollis* (HBK) Griseb., with this species being recognized as an extremely variable. Through the years, this name had priority over other names. In Central Argentina, a popular species of the genus has been traditionally named as *M. mollis* by recognized botanists, as Barboza and collaborators (2001), following the traditional botanical priority. However, as research has progressed during the last two decades, variations were found in the morphology and ecology of the different *Minthostachys* forms grown in the equally diverse mountainous regions of South America. Consequently, there have been recognized 17–19 species within the genus (Schmidt-Lebuhn 2008a). *Minthostachys verticillata* (Griseb.) Epling is endemic to Argentina, particularly in the center of the country (Barboza et al. 2009; Ojeda and Karlin 2015; Martínez 2015). This species is commonly known as "peperina", "peperita" or "piperita", among others (Barboza et al. 2001). The specific epithet was given because of the structure of the 'verticillated' inflorescences, and the vernacular name probably is derived from the common mint *Mentha* x *piperita*, as both species shared a similar aroma, that is characterized by the presence of menthone, thus given these aromatic plants the pleasant mint-like aroma.

It is interesting to note, that because of the taxonomic confusion many reports of *M. mollis* from Argentina are actually *M. verticillata*. "Peperina" is a strongly aromatic shrub, widely used in Central Argentina, particularly in the province of Córdoba, where it has become so popular as it represents traditional medicinal uses of herbs, in these picturesque and tourist areas of central Argentina. *M. verticillata* has the following synonyms (The Plant List 2020): *Bystropogon kuntzeanum* Briq., *Minthostachys verticillata* var. *eupatorioides* Epling, *Xenopoma verticillatum* Griseb.

2 Crude Drug Used

"Peperina" is perhaps one of the most popular herbs in Argentina, both in popular medicine and culture but also in the herbal industry. In the province of Córdoba, "peperina" is one of the most used medicinal plants in central Argentina (Juliani et al. 2007; Martínez 2015), it is also the most studied species of the genus (Schmidt-Lebuhn 2008b).

The dried aerial parts are used, including leaves, stems, and inflorescences. As stems contain lower amounts of essential oils, thus being much less aromatic, dried materials containing mostly leaves and inflorescences are considered of the highest quality. The fresh leaves are green, with mature leaves turning yellow, perhaps reflecting the poor nutritional status of some soils when the plant is growing on the wild. When properly dried the leaves reach a bright green color, that can turn dark when leaves are not properly dried. The dried leaves have a pleasant, mint-like aroma, with a strong herb/green notes, possible due to the presence of sesquiterpenes and other monoterpenes that differentiate "peperina" from mint.

Other exotic species are also marketed under the same vernacular name as "peperina", including *Calamintha nepeta* (L.) Savi subsp. *nepeta* (Lamiaceae) which creates additional confusion (Scandaliaris et al. 2007) and it was used as adulterant. Although these species have a great polymorphism, it is a taxon very different from *M. verticillata*, easily distinguishable. Both species, of similar aroma, share chemical compounds such as menthol, pulegone, piperitone, but differ in their morphological characters (Xifreda et al. 2005).

3 Major Chemical Constituents and Bioactive Compounds

As the popularity of "peperina" is due to the minty and pleasant aroma of the aerial parts, the aroma is due to the presence of essential oils. Essential oils are relatively easy to extract and analyze, consequently, they have been the most studied secondary metabolites in this species.

Zygadlo and co-workers (1996) conducted an analysis of the essential oil chemistry of different *M. verticillata* varieties from the provinces of Catamarca (3 varieties), Córdoba (3), San Luis (2) and Tucumán (2), they found a remarkable variability in the essential oils. All samples had varying levels of limonene (1–25%), menthone (4–30%) and pulegone (1–47%). One of the varieties from Catamarca showed a remarkable different profile, being dominated by thymol (32%), and carvacrol (34%). The wild varieties from Córdoba showed higher levels of menthone (16–30%) as well as samples from Tucumán (28%) and San Luis (27%), with higher levels of pulegone (31–47%, 21% and 45%, respectively).

More recently, van Baren and collaborators (2014) conducted a comprehensive study of the essential oils yields and composition collected in 40 natural populations in Central and Northwestern Argentina. With the analysis of nine populations, the authors confirmed the presence of the Central Argentina (Córdoba-San Luis) chemotype dominated by menthone, pulegone and piperitenone. While for the Northwestern provinces, they identified five chemotypes: Chemotype 1 dominated by dihydrocarvone (38.4–57.2%) and carvone (28.0–46.5%); Chemotype 2 dominated by pulegone (57.1–76.3%) with menthone absent; Chemotype 3 with carvacryl acetate (24.1–50.9%) and carvacrol (19.2–32.8%) as the main components;

Chemotype 4 dominated by limonene (35.7–43.3%); Chemotype 5 with high content of linalool (56.1–84.2%).

The yield of essential oils is an important indicator for possible utilization. In "peperina", yields depend on areas of collection and growing conditions, as cultivated plants were found to have lower yields (0.7–0.9%) than wild-collected (1.3 to 5%). It is interesting to note, that wild populations growing at higher altitudes (1270–1542 m) in 'Sierras Grandes' had higher yields (3.8–5.3%) than those growing in the 'Sierras Chicas' (1.3–1.9%) at lower altitudes (723–773 m) (Ocaño 2017). The cultivated varieties showed high levels of pulegone (32–43%) and lower levels of menthone (17–27%), similar to the samples of 'Sierras Chicas' for pulegone (35–64%) and menthone (28–32%), while the 'Sierras Grandes' showed an opposite trend with lower levels of pulegone (5–9%) and higher levels of menthone (13–20%) (Ocaño 2017). This author also conducted sensory analysis to determine the level of acceptance of the aroma of the leaves in relation to the levels of pulegone and menthone, panelists seemed to prefer leaves that are high in menthone (78%) and low in pulegone (10%). In the study of van Baren et al. (2014), yields of oils were found to varied, with samples from San Luis (2.6–4.9%) and Córdoba (2.6–4.2%) being the highest, and samples from northwestern provinces (Tucumán, Catamarca, and Salta) showing the lowest yields (0.3–1.6%).

It is known the relationship between the phenotypic state and quality of the essential oil as a result of the different biosynthetic stages of its main constituents (Croteau et al. 2005; Bandoni et al. 2002). The best harvest time can be defined by the simple observation of the phenotypic state of the plant, coinciding with the best proportion between pulegone, isomenthone and menthone in the essential oil to achieve the most typical sensory profile of "peperina": as occurs with *Minthostachys* and other Lamiaceae, this occurs normally in full bloom.

Few studies have focused on the non-volatile secondary metabolites of "peperina" showing that the content of polyphenols, phenolic acids, and flavonoids is less than half as in peppermint (*Mentha* x *piperita*) (Rodríguez-Vaquero et al. 2014) or other native plants from Córdoba (Dade et al. 2009). In addition, variability in the yield and composition of essential oils, has also been reported due to environmental conditions and developmental stages (Zygadlo et al. 1996; Mora et al. 2009; Ojeda 2011).

4 Morphological Description

"Peperina" is a perennial, aromatic, scandent shrub, up to 2 m in height, typical squared stems, twigs, and young leaves are densely haired (Figs. 1 and 2) with glandular and non-glandular trichomes. The leaves range from 1 to 5 cm in length, ovate, margin serrate, apices attenuate with base rounded, petioles 0.5 to 1 cm long. Adaxial surface less hairy than the abaxial. Inflorescences are verticillate, flowers pedunculated, grouped in axillary cymes and subcymes, with numerous flowers, calyces are densely pilose, 2-3 mm long, corolla is white, tubular and typically bilabiate and gynodioecious (Fig. 1, left) (Schmidt-Lebuhn 2008a; Ocaño 2017; Glinos

Fig. 1 Pictures of the aerial part (left) of "peperina" (*Minthostachys verticillata*) with close-ups insert of inflorescences (right) (School of Agronomy, Córdoba, Argentina). Photos: P. Brunetti

Fig. 2 Diseases affecting "peperina" (*Minthostachys verticillata*). Left, aphids producing curling of the leaves. Right, discoloration likely produced by Cucumber mosaic virus (CMV). Photos: P. Brunetti

et al. 2019). Fruits are formed by 4 ellipsoid clusters, brown and finely crosslinked (Scandaliaris et al. 2007).

Stomata are mostly found in the abaxial epidermis surface, they are diacitic, with epidermal cells surrounding the companion cells, with wavy walls. The glandular trichomes are typical of the genus, peltate type with an 8-cell head, located in depressions of the epidermis. Non-glandular trichomes are unicellular with a pluricellular base, a diagnostic character is the presence of diosmin (flavones) crystals (Barboza et al. 2001).

In the wild, plants are easily identified because of their decumbent habit, growing between other shrubs, with green leaves, and with mature leaves turning yellow. Moreover, when leaves are damaged, the characteristic minty and green aroma is released.

5 Geographical Distribution

The genus *Minthostachys* occurs along the Andes from Venezuela to Argentina. *Minthostachys verticillata* is distributed in central, northwestern Argentina, in the provinces of Catamarca, Córdoba, La Rioja, San Luis, Tucumán (Schmidt-Lebuhn 2008b; INTA 2009), Salta, and Jujuy (Ojeda et al. 2001). "Peperina" grows wild in the "Sierras Pampeanas" or Pampa's Mountains, a mountain chain off to the east of the Andes (Ojeda and Karlin 2015).

Wild populations of "peperina" are threatened because of overcollection from the wild and habitat loss. It has been reported that 87% of the commercialized volume of *M. verticillata* (reported as *M. mollis*) is harvested directly from the wild (Ojeda et al. 2001). The current exploitation of this plant exceeds the natural regeneration and the need for developing a conservation strategy has been indicated (Ojeda et al. 2001). *M. verticillata* (reported as *M. mollis*) and *Hedeoma multiflorum* (Lamiaceae) have been reported as native species prioritized for conservation according to an Index of Conservation Priority (ICP) developed by Martínez et al. (2006). The index employed is based on abundance, propagation methodology, origin, and commercial demand for plant species.

To initiate the multiplication of a high number of plants Chebel et al. (1998) described an alternative rapid *in vitro* (micropropagation) method that seemed suitable also for introducing "peperina" into cultivation. Seeds were collected from wild plants (Los Cóndores, Córdoba), with a chemical composition dominated by menthone (46%), pulegone (39.6%), piperitenone (5%), and a total amount of sesquiterpenes of 7.3%. Seeds were introduced to aseptic conditions and multiplied via micropropagation, acclimatized and transferred to field conditions, where the plants successfully developed during one season, when leaves were collected and subjected to steam distillation. The study found some differences in the chemical composition of cultivated plants, with lower levels of menthone (33.7%) and piperitenone (2%), higher levels of pulegone (45.4%) and sesquiterpenes (8.8%). This is in accordance with Ocaño (2017), where the cultivated varieties showed high levels of pulegone (32–43%) and lower levels of menthone (17–27%).

Years later, the School of Agronomy of the National University of Córdoba initiated a research and development program for the domestication of "peperina" (Ojeda et al. 2001; Ojeda 2011). This program, to develop "peperina" into a valuable crop, involved an active collaboration with the industry (Ojeda and Karlin 2015). The team developed the cultivar 'Champaqui FCA' and recently the team developed a comprehensive study to obtain cultivars with improved sensory profiles, leave morphology, disease resistance and essential oil yields (Ocaño 2017). Development of disease resistance still needs to be improved, as the cultivated plants are having diseases such as discoloration of the leaves possibly produced by Cucumber mosaic virus (CMV) and curling of the leaves by aphids (Fig. 2).

6 Ecological Requirements

"Peperina" prefers mountain areas with altitudes between 500 to around 2300–2500 m. It grows quite well in mountainous soils, usually in shallow, sandy and stony soils, and it seems the species grow better in darker, high organic matter content soils. It is found in the hillsides of the mountains protected by the partial shade of other shrubs and native trees such as *Acacia* spp., *Lithraea* spp., and many others. The species is adapted to a broad ecological range, from moist forested valleys (500 m) to cloud forest and dry highland pastures and slopes. It grows in areas where the precipitations are around 600 to 800 mm, up to 1100 mm, since this species is mostly adapted to dry conditions, does not tolerate waterlogged soils. As a perennial shrub, it restarts sprouting in August–September, and flowers mostly during the summer from December to March, up to May when the first frosts appear (Schmidt-Lebuhn 2008a; Ojeda and Karlin 2015; Ocaño 2017).

Interestingly, *M. verticillata* is a gynodioecious species, the coexistence of female and hermaphrodite individuals in the same population. It behaves as dioecious, with hermaphrodite plants functioning mainly as pollen donors and female plants as seed producers. More importantly, "peperina" has an essential dependence on animal pollinators. Although its flowers are visited by bees, wasps, butterflies, and flies, it appears that pollination depends mainly on one fly pollinator species (*Ptilodexia* cf. *cingulipes*), which in turn during its immature stage depends on a parasitic relationship with other insects (Glinos et al. 2019).

7 Traditional Use (Part(s) Used) and Common Knowledge

In traditional medicine, it is used in a myriad of commercial products alone or mix with other medicinal herbs. The leaves and aerial parts are used in infusions mostly for digestive purposes, as antispasmodic, carminative, antidiarrheal and antiemetic.

The infusion is also used in rheumatic conditions, it is sedative, and the leaves applied to wounds (Ojeda and Karlin 2015), for low blood pressure and as abortive (Arias Toledo 2009), aphrodisiac (Martínez and Planchuelo 2003) and digestive tract diseases, spasmolytic and anti-inflammatory (Dade et al. 2009).

"Peperina" has been used in the preparation of alcoholic and non-alcoholic beverages, including bitters and liqueurs (Ojeda et al. 2001). A survey conducted by Juliani and collaborators (2007) found that "peperina" was used mostly as a herbal tea/infusion and blended with the popular infusion 'yerba mate' the typical infusion made from the leaves of *Ilex paraguariensis* that is prepared in the characteristic dried calabash/gourd container and drink with an ornate metal straw.

These trends in product development have continued up to the present. For example, some companies in Argentina have filed documents to commercialize "peperina" blended with other herbs such as peppermint (*Mentha* spp.), chamomile and another indigenous herb, 'poleo' (*Lippia turbinata*) as a dietary supplement used as an infusion to treat digestive problems (ANMAT 2015). These uses were confirmed later by Martínez (2015), with "peperina" being used in infusions, alone or in combination with yerba mate, chamomile and other as an aid for digestion. This author considered this herb as 'warm', as it is used to prepare hot infusions to increase sweating to treat colds and the flu.

"Peperina" used to be listed in the Argentinean National Pharmacopeia, in 1978. (Barboza et al. 2001), however it has been removed from the VII edition of this Pharmacopeia.

Due its vast popularity, and the resulting tremendous collection pressure, "peperina" has disappeared from many areas. The decrease of natural populations and the poor quality of plant materials collected from the wild have led to the domestication of this species which also generates new economic opportunities for producers (Ocaño 2017). "Peperina" is also listed in the Argentinean Food Codex as a food additive of botanical native origin (ANMAT 2019).

8 Modern Medicine Based on Its Traditional Medicine Uses

The essential oil components of *Minthostachys verticillata* have shown antiallergic and lymphoproliferative *in vitro* bioactivities. The combination of limonene, menthone, and pulegone increased the multiplication of clonal lymphocytes, in similar ways of phytohemaglutinine (PHA) and concanavalin A (ConA) (Caridi et al. 2007a). Although there is still a lot of research to be conducted, these results support the use of "peperina" to treat colds and flu (Martínez 2015).

There have been concerns about the carcinogenic activities of pulegone in rodent models, for example, there has been an increased incidence of hepatocellular neoplasms in mice, renal and liver lesions in rats after prolonged 2-year exposure to pulegone (18.75–75 mg/kg) (NTP 2001). Although it is not always appropriate to extrapolate these results to humans, the ingestion of pulegone as a tea or food, it is well below the levels of ingestion recorded in these experiments.

A study conducted on leaf essential oil, containing high levels of pulegone (60.5%) and menthone (18.2%) showed neither cyto-genotoxic *in vitro* (in mice, 25–500 mg of EO/kg body weight) nor *in vivo* (Escobar et al. 2012, 2015). As an example, a tea is prepared with few grams (~2–3 grams) of dried "peperina", with theoretical levels of pulegone of around 75 mg of pulegone per tea (essential oils with 50% of pulegone) this would give a value of 1 mg of pulegone per kg for humans. However, the commercialization of the pure essential oils should not be allowed as the public can be exposed to high levels of pulegone while handling essential oils. It is recommended for the public seeking to use botanical products to treat diseases, to always consult a medical doctor.

Essential oils of "peperina" have been reported as antimicrobial against *Salmonella typhi* and *Bacillus subtilis* (Mora et al. 2009; Lang and Buchbauer 2012). *M. verticillata* essential oil and one of its main components, limonene, have immunomodulatory effects *in vitro* on cells from allergic patients (Cariddi et al. 2007b, 2009). Further research indicates that limonene has the most potent immunomodulatory activity, suggesting it as a promising natural alternative for a novel treatment of allergic diseases (Cariddi et al. 2011).

In the last two decades, there have been growing evidence supporting the use of products containing menthone and other essential oil components to help indigestion, as it is well reported the choleretic and cholagogue properties of menthone that promotes the release of bile by the gallbladder to help in the digestive process. Menthone and other monoterpenes also produced smooth muscle relaxation, which in the gut translates in increasing the time to digest food (Koroch et al. 2007).

Recently, it has been reported that essential oils of "peperina" have insecticidal activity against *Musca domestica* with minimal environmental effects (Palacios et al. 2009; Rossi et al. 2012) and ethanolic extracts of "peperina" showed high levels of toxicity against mites while being harmless to bees (*Apis mellifera*), highlighting the uses of essential oils as promising natural compounds as pest control agents (Damiani et al. 2011).

9 Conclusions

"Peperina" is a remarkable aromatic shrub from Argentina that will continue to generate interest and economic opportunities for farmers and collectors. There should be concerted efforts to conserve the genetic variability of the species throughout Argentina, by either conserving the natural habitats (*in situ*) and by collecting varieties all over the country and conserve them in arboreta and botanical gardens (*ex situ*). Central Argentina has seen the destruction of most of the natural forest, thus the importance to restore those habitats and protect the remaining flora. There is still a lot of research to be done to better explore the characters that define the safety and effectiveness of "peperina" products. This concern, particularly, the high levels of pulegone, that can be of concern in international markets, although studies suggest that the levels of pulegone in the dried leaves would not pose safety risks when used as herbal teas or additives in foods. However, additional research is needed to ensure the safe use of the species. As *Minthostachys verticillata* exhibits a great variability in the essential oil composition, breeding efforts should be focused on developing varieties with lower levels of pulegone. To date, most of the relevant research has focused on the chemistry of essential oils, research should be expanded to include also the non-volatile components, such as flavonoids, phenolic acids, and others, that could provide new opportunities for the further uses and applications for "peperina".

Acknowledgments PCB is thankful to CONICET (Argentinean Council of Scientific and Technical Research) and the Office of Science and Technology of the National University of Córdoba (Argentina) (SECYT/UNC). HRJ and PCB are also thankful to the Fulbright Program (funded by the U.S. Department of State).

References

ANMAT (Administración Nacional de Medicamentos, Alimentos y Tecnología Médica) (2015) Ministerio de Salud, Argentina. Dispos. 8357/2015

ANMAT (Administración Nacional de Medicamentos, Alimentos y Tecnología Médica) (2019) Código Alimentario Argentino. Ministerio de Salud. Argentina

Arias Toledo B (2009) Diversidad de usos, prácticas de recolección y diferencias según género y edad en el uso de plantas medicinales en Córdoba, Argentina. Boletín Latinoamericano y del Caribe de Plantas Medicinales y Aromáticas 8(5):389–401

Bandoni AL, López MA, Juárez MA, Elechosa MA, van Baren C, Di Leo Lira P (2002). Seasonal variations in the composition of the essential oil of peperina (*Minthostachys mollis* (Kunth) Griseb.) from a local population of the province of Córdoba, Argentina. Essenze Derivati Agrumari 72(1):11–14

Barboza GE, Bonzani N, Filippa EM, Luján MC, Morero R, Bugatti M, Decolatti N, Ariza Espinar L (2001) Atlas histo-morfológico de plantas de interés medicinal de uso corriente en Argentina. Editorial de la Universidad Nacional de Córdoba, Córdoba, Argentina, 212 pp

Barboza GE, Cantero JJ, Nunez C, Pacciaroni A, Ariza Espinar L (2009) Medicinal plants: a general review and a phytochemical and ethnopharmacological screening of the native Argentine Flora. Kurtziana 34(1–2):7–365

Cariddi LN, Moser M, Zygadlo JA, Sabini IL, Maldonado A (2007a) Pure compounds isolated of essential oil of *Minthostachys verticillata* (Griseb.) Epling like polyclonal activators *in vitro*. Bol Latinoam Caribe Plant Med Aromáticas 6(5):236–237

Cariddi LN, Panero A, Demo MS, Grosso M, Zygadlo J, Sabini LI, Maldonado AM (2007b) Inhibition of immediate-type allergic reaction by *Minthostachys verticillata* (Griseb.) Epling essential oil. J Essent Oil Res 19:190–196

Cariddi L, Moser M, Andrada M, Demo M, Zygadlo J, Sabini L, Maldonado A (2009) The effect of *Minthostachys verticillata* essential oil on the immune response of patients allergic to dust mites. Boletín Latinoamericano y del Caribe de Plantas Medicinales y Aromáticas 8(3):224–233

Cariddi L, Escobar F, Moser M, Panero A, Alaniz F, Zygadlo J, Sabini L, Maldonado A (2011) Monoterpenes isolated from *Minthostachys verticillata* (Griseb.) Epling essential oil modulates immediate-type hypersensitivity responses *in vitro* and *in vivo*. Planta Med 77(15):1687–1694

Chebel AD, Koroch AR, Juliani HR, Juliani HR Jr, Trippi VS (1998) Micropropagation of *Minthostachys mollis* (H.B.K) Griseb. and essential oil composition of clonally propagated plants. In Vitro Plant Cell Dev Biol 34:249–251

Croteau RB, Davis EM, Ringer K, Wildung MR (2005) Menthol biosynthesis and molecular genetics. Naturwissenschaften 92:562–577

Dade MM, Fioravanti DE, Schinella CR, Tournier HA (2009) Total antioxidant capacity and polyphenol content of 21 aqueous extracts obtained from native plants of Traslasierra valley (Argentina). Boletín Latinoamericano y del Caribe de Plantas Medicinales y Aromáticas 8(6):529–539

Damiani N, Gende LB, Maggi MD, Palacios S, Marcangeli JA, Eguaras MJ (2011) Repellent and acaricidal effects of botanical extracts on *Varroa destructor*. Parasitol Res 108:79–86

Escobar FM, Cariddi LN, Sabini MC, Reinoso E, Sutil SB, Torres CV, Zanon SM, Sabini LI (2012) Lack of cytotoxic and genotoxic effects of *Minthostachys verticillata* essential oil: studies *in vitro* and *in vivo*. Food Chem Toxicol 50(9):3062–3067

Escobar FM, Sabini MC, Cariddi LN, Sabini LI, Manas F, Cristofolini A, Bagnis G, Gallucci MN, Cavaglieri LR (2015) Safety assessment of essential oil from *Minthostachys verticillata* (Griseb.) Epling (peperina): 90-days oral subchronic toxicity study in rats. Regul Toxicol Pharmacol 71:1–7

Glinos E, Condat E, Mulieri P, Ashworth L (2019) Essential dependence on wild pollination service: a medicinal plant under threat *Minthostachys verticillata* (Lamiaceae). Arthropod Plant Interact 13(6):865–874

INTA (Instituto Nacional de Tecnología Agropecuaria) (2009) Manual de recolección sustentable de plantas aromáticas nativas de la región central y noroeste de la Argentina. Ed. INTA, Buenos Aires, Argentina, 48 pp

Juliani HR, Biurrun FN, Koroch AR, De Carli AG, Zygadlo JA (2007) The production of native and exotic herbs, medicinal and aromatic plants in Argentina. In: Janick J, Whipkey A (eds) Issues in new crops and new uses. Ed. Am Soc Hortic Sci (ASHS) Press, Alexandria, USA, pp 316–321

Koroch AR, Zygadlo JA, Juliani HR (2007) Bioactivity of essential oils and their components. In: Berger RG (ed) Flavours and Fragrances. Chemistry, bioprocessing and sustainability. Springer, Berlin-Heidelberg, Germany, pp 87–115

Lang G, Buchbauer G (2012) A review on recent research results (2008–2010) on essential oils as antimicrobials and antifungals. Flavour Fragr J 27:13–39

Martínez G (2015) Las Plantas en la Medicina Tradicional de las Sierras de Córdoba. Editorial Detodoslosmares. Capilla del Monte. Córdoba. Argentina, 210 pp

Martínez GJ, Planchuelo AM, Fuentes E, Ojeda M (2006) A numeric index to establish conservation priorities for medicinal plants in the Paravachasca Valley, Córdoba, Argentina. Biodivers Conserv 15:2457–2475

Mora FD, Araque M, Rojas LB, Ramírez R, Silva B, Usubillaga A (2009) Chemical composition and *in vitro* antibacterial activity of the essential oil of *Minthostachys mollis* (Kunth) Griseb. Vaught from the Venezuelan Andes. Nat Prod Commun 4(7):997–1000

NTP (National Toxicology Program) (2001) Technical report on the toxicology and carcinogenesis studies of Pulegone in: F344/N rats and B6c3f1 mice (gavage studies). National Institutes of Health. NTP TR 563/NIH Publication No. 11-5905. Research Triangle Park, North Carolina, USA, 202 pp

Ocaño SF (2017) Generación de una Población de "Peperina" *Minthostachys verticillata* (Griseb.) Epling, Mejorada por Sanidad, Rendimiento y Calidad de Aceites Esenciales. Thesis. Universidad Nacional de Córdoba. Escuela para Graduados. Facultad de Ciencias Agropecuarias. Córdoba, Argentina, 132 pp

Ojeda MS (2011) La Peperina. Caracterización de poblaciones y avances en la domesticación. Editorial Académica Española. Córdoba, Argentina, 152 pp. ISBN 978-3-8454-8773-1

Ojeda MS, Karlin UOT (2015) Plantas Aromáticas y Medicinales. Modelos para su Domesticación, Producción y Usos Sustentables. Editorial Universidad Nacional de Córdoba. Córdoba. Argentina, 183 pp

Ojeda M, Coirini R, Cosiansi J, Zapata R, Zygadlo J (2001) Evaluation of variability in natural populations of peperina (*Minthostachys mollis* (Kunth.) Griseb.), an aromatic species from Argentina. Plant Genetic Resour Newsletter 126:27–30

Palacios SM, Bertoni A, Rossi Y, Santander R, Urzúa A (2009) Insecticidal activity of essential oils from native medicinal plants of Central Argentina against the house fly, *Musca domestica* (L.). Parasitol Res 106:207–212

Rodríguez-Vaquero MJ, Vallejo CV, Aredes-Fernaández PA (2014) Biological activity of phenolic compounds from Argentinean herbs infusions. Open Confer Proc J 2014(5):1–7

Rossi YA, Canavoso L, Palacios SM (2012) Molecular response of *Musca domestica* L. to *Minthostachys verticillata* essential oil, (4R)(+)-pulegone and menthone. Fitoterapia 83:336–342

Scandaliaris M (2017). Caracterización morfo-histológica de las núculas y su valor taxonómico en Lamiaceae nativas de Argentina. Thesis. Universidad Nacional de Córdoba. Facultad de Ciencias Agropecuarias. Córdoba, Argentina, 218 pp

Scandaliaris M, Fuentes E, Lovey RJ (2007) Dos especies de Lamiáceas comercializadas en Córdoba (Argentina) bajo el nombre de "peperina". Multequina 16:73–81

Schmidt-Lebuhn AN (2008a) Revision of the genus *Minthostachys* (Labiatae). Mem N Y Bot Gard 98:1–78

Schmidt-Lebuhn AN (2008b) Ethnobotany, biochemistry and pharmacology of *Minthostachys* (Lamiaceae). J Ethnopharmacol 118:343–353

The Plant List (2020) In Internet: http://www.theplantlist.org/tpl1.1/record/kew-128642. (Last access, May 2020)

van Baren CM, Di Leo Lira P, Elechosa MA, Molina AM, Juárez MA, Martínez A, Perelman S, Bandoni AL (2014) New insights into the chemical biodiversity of *Minthostachys mollis* in Argentina. Biochem Syst Ecol 57:374–383

Xifreda CC, Mallo AC, Fatta NA (2005) La calaminta (*Melissa calamintha*, Lamiaceae) cultivada como tapizante ornamental. Resúmenes de las VII Jornadas Nacionales de Floricultura. In electronic support. Trevelin, Chubut, Argentina

Zygadlo JA, Maestri DM, Lamarque AL, Guzmán CA, Velasco-Negueruela A, Pérez-Alonso MJ, García-Vallejos MC, Grosso NR (1996) Essential oil variability of *Minthostachys verticillata*. Biochem Syst Ecol 24(4):319–323

Nectandra angustifolia (Schrad.) Nees & Mart.

**Ana María Torres, Francisco José Camargo,
Bárbara Verónica Ricciardi Verrastro,
Eduardo Santiago Dellacassa Beltrame, and Gabriela Ana Leticia Ricciardi**

Nectandra angustifolia. San Isidro, Corrientes, Argentina. (Photo: AM Torres)

In memoriam: Prof. Armando IA Ricciardi.

A. M. Torres (✉) · F. J. Camargo · B. V. Ricciardi Verrastro · G. A. L. Ricciardi
Laboratorio de Productos Naturales Prof. Armando Ricciardi –LabProdNat– Facultad de Ciencias Exactas y Naturales y Agrimensura, IQUIBA-Universidad Nacional del Nordeste, Corrientes, Argentina
e-mail: amtorres39@yahoo.com.ar

E. S. Dellacassa Beltrame
Laboratorio de Biotecnología de Aromas, Departamento de Química Orgánica, Facultad de Química, Universidad de la República, Montevideo, Uruguay

Á. Máthé, A. Bandoni (eds.), *Medicinal and Aromatic Plants of South America Vol. 2*, Medicinal and Aromatic Plants of the World 7,
https://doi.org/10.1007/978-3-030-62818-5_29

Abstract Medicinal plants encompass a rich source of active compounds that can neutralize snake venoms or toxins. *Nectandra angustifolia* (Schrad.) Nees & Mart. (Lauraceae), commonly known as "laurel amarillo", "aju'yhû", "laurel-hu", "laurel del río", is used in traditional medicine as a digestive, for the treatment of rheumatism, arthritis, pain, and as antivenom. Snakebite envenomation was included in category A of neglected tropical diseases (NTD), in 2017, as they affect millions of people in the world (Chippaux, J Venom Anim Toxins 23:38–39: 2017). In Argentina, the snake genus *Bothrops* is responsible for the greatest number of accidents. Presently, the application of serum antivenom is the only specific treatment. This treatment does not protect over local effects, which are the main cause of morbidity. In the search for effective antivenoms, extracts of different polarity obtained from *Nectandra angustifolia* were evaluated both, *in vitro* and *in vivo*, against the venom of *B. diporus* "yarará chica". The effect of seasonal variability in this activity was also studied, to optimize both, the harvest period and the most appropriate type of extract. By successive bio-guided fractionations of the ethanol extract of *N. angustifolia* a fraction enriched in flavonoids responsible for the alexiteric activity of the species has been isolated. The phytochemical study of the volatile fraction of *N. angustifolia* and *N. megapotamica* allowed for the phytochemical characterization of farther interesting compounds.

Keywords "Laurel amarillo" · Alexiteric activity · Phytochemistry

1 Introduction

The Lauraceae family, which includes the genus *Nectandra*, is widely distributed in tropical and subtropical regions of the world, predominantly in Southeast Asia and Brazil (Cronquist, 1992). Ethnobotanical research has determined different uses for many species belonging to genus. These can be justified by the presence of certain types of secondary metabolites determined in different *Nectandra* species, including sesquiterpenes, phytosterols, polyalcohols, arylpropionic acid derivatives, flavonols, arylpropanoids, furofuran lignans, (Macías-Villamizar et al. 2015) and components of essential oils (Torres et al. 2014).

Nectandra angustifolia (Schrad.) Nees & Mart. ex Nees (Lauraceae), a plant found in the fluvial zones of Argentina, and popularly named as "laurel amarillo" or "aju'y hû", is reported to possess anti-inflammatory, antimicrobial and molluscicidal activities (Oliveira de Melo et al. 2006; Truiti et al. 2006). People from Corrientes and Chaco Provinces of Argentina (González-Torres 2018) also use its extracts for treating their domestic animals when bitten by snakes. The present review summarizes the scientific information available on this plant species, as an input to initiate further investigations into its pharmacological properties.

2 Taxonomic Characteristics

The family of Lauraceae encompasses native shrubs to tall trees. The genus *Nectandra* counts on numerous species in Central and South America, distributed almost all in tropical and subtropical regions (Castiglioni 1951). *Nectandra angustifolia* (Schrad.) Nees & Mart. has as synonyms*: Nectandra angustifolia* var. *falcifolia* Nees, *Nectandra falcifolia* Nees J. A. Castigl. ex Mart. Crov. & Piccinini, *Nectandra membranacea* var. *falcifolia* Nees Hassl. *Ocotea angustifolia* Schrad (The Plant List 2019).

Bernardi (2018) compared the descriptions of *N. megapotamica* (Spreng.) Mez (Rohwer 1993) with that of *N. angustifolia* (Schrad.) Nees & Mart. (Rohwer 1993) and found that the descriptions showed no significant differences, within a limited degree of variation. He concluded that they represent a single entity and therefore they might be considered as synonymous. However, the study of the metabolic expression showed significant differences between the species, so in this chapter we will discuss *N. angustifolia* as a single, independent species (Torres et al. 2011a, b, 2014).

Common Names: "laurel Amarillo", "aju'yhû", "laurel-hu", "laurel del río", "laurel mini" (Bertucci et al. 2008), "louro-branco" (Marqués 2001). The seeds of *Nectandra* sp. named as "ishpingo" in the colonial epoch of the XVI century and at present, it is known as "hamalas" (Carod-Artal and Vázquez-Cabrera 2007).

3 Crude Drug Used

N. angustifolia leaves and bark are used as tea or infusion to treat digestive disorders (Martínez Crovetto 1981), rheumatism, arthritis, and pain (Bertucci et al. 2008). Besides, leaves are used as antivenom against viper bites (alexiteric) (González-Torres 2018).

4 Major Chemical Constituents and Bioactive Compounds

N. angustifolia is an aromatic species with an essential oil that was first studied by Fester et al. (1958,1961), from plant materials collected in Santa Fe Province (Alto Verde and Puerto Ocampo, Argentina). The presence of terpenes, mainly α-pinene, 1,8-cineole and safrole were determined. More recently, Torres (2011) and Torres et al. (2011b, 2014) performed a comparative study on the chemical composition of the essential oils from *N. angustifolia* and *N. megapotamica,* in order to chemically evaluate their taxonomic similarities detected in their secondary metabolism expression. The essential oil compositions were observed to maintain almost the same composition during the different sampling seasons, but with differences in the number of single components. *N. angustifolia* oil was characterized by an elevated number of hydrocarbons terpenes (~70%) among which *p*-1(7),8-menthadiene (~25%); α-terpinolene (~20%); α-pinene (~12%) and β-pinene (~3%), δ-elemene (~6%) and germacrene D (~5%). The oxygenated fraction was constituted by a small number of oxygenated monoterpenes (~1%) and sesquiterpenes (~5%). (Torres et al. 2011a, b). By contrast, *N. megapotamica* oil shows a predominance of sesquiterpene hydrocarbons (~65%) mainly bicyclogermacrene (~27%) and germacrene D (~17%) (Torres et al. 2014). The differences found between *N. angustifolia* and *N. megapotamica* oil compositions chemically demonstrates that these two species are not synonymous, even if there is a botanical discussion on that.

The genus *Nectandra* has been characterized by the presence of phytosterols, polyalcohols, arylpropionic acid derivatives, flavonols, arylpropanoids, furofuran lignans, dihydro-benzofuran neo-lignans, sesquiterpenes, nor-lignans, indole alkaloids, tannins, and diterpenes (Macías-Villamizar et al. 2015). Bertucci et al. (2008) found in an ethanolic leaf extract of *N. angustifolia* triterpenoids/steroids, flavonoids and glycosides; acetonic extract, triterpenoids/steroids, low molecular weight terpenes and flavonoids; while chloroformic extract contained triterpenes/steroids, low molecular weight terpenes and alkaloids.

In a bio-guided fractionation of ethanolic leaf extracts, nine fractions were isolated: their analysis showed the presence of flavonoids, saponins, triterpenes, phenols and alkaloids (Torres et al. 2011b). A new LC fractionation yielded a polar subfraction responsible for the alexiteric activity, where HPLC and UPLC-MS analysis allowed to identify four quercetin glycosylated derivatives and the polymethoxylated flavonoid natsudaidain (3-hydroxy-3',4',5,6,7,8-hexamethoxyflavone) (Torres 2011).

5 Morphological Description

N. angustifolia is a tree of 3-12 m tall and 0.15-0.60 cm in diameter. The bark is gray-brown, smooth, with some superficial fissures. It is perennial, with alternate leaves, linear-lanceolated to oblong sheets, with long acuminate apex glabrescent, trilobed cross-section. Leaves show cuneate base, symmetrical to slightly asymmetric; glabrous adaxial face; slightly pubescent abaxial face; did not present domatia (Photo 1). The inflorescences are axillary panicles, generally located at the extremes of the branches. Flowers yellowish-white (Photo 1) and the fruit is an ellipsoid berry light brown colored with dark green spots (Zuloaga et al. 2008; Poszkus et al. 2016).

Photo 1 Leaves and flowers of *Nectandra angustifolia*. (Photo: AM Torres)

6 Geographical Distribution

The Lauraceae family comprising from bushes to tall trees, is widely distributed in tropical and subtropical regions of the world, predominantly in Southeast Asia and Brazil (Macías-Villamizar et al. 2015). It is an excellent and sensible indicator of the rainforest formations and wet forests. The genus *Nectandra* Rol. ex Rottb. (Lauraceae), consists of four spices, of which *Nectandra angustifolia* (Schrad.) Nees & Mart. and *Nectandra megapotamica* (Spreng.) Mez are found in the Northeast of Argentina. Particularly, *N. angustifolia* grows in the Argentinean Mesopotamia (Múlgura 1987) and adjacent regions of Paraguay (Stutz 1983; López et al. 1987; Soria Rey et al. 1994; Basualdo and Soria Rey 1999; Jiménez et al. 2000), Uruguay (Sayagués et al. 2000) and Brazil (Pedralli 1987).

7 Ecological Requirements

Nectandra is considered endemic in the neotropics (Chanderbali et al. 2001) and has the second-highest number of specific entities of the Lauraceae family, in South America (Macías-Villamizar et al. 2015). *N. angustifolia* is a tree found in wet zones, over the sea level to 500 m high; growing on the riverbanks in the Province of Misiones (Argentina), Uruguay and Paraná (Brazil) (Anton and Zuloaga 2018).

Nectandra angustifolia can be recognized by the narrow, linear leaves, rarely reaching 1.5 cm in width, in addition to the secondary ribs on both sides. It has been proposed the width of the leaves could be an adaptation to periods of flooding that the species, riverside, faces (Rohwer 1993; Zanon et al. 2009).

8 Traditional Use (Part(s) Used) and Common Knowledge

This genus has been reported in folk medicine as antifungal, anti-inflammatory, antimalarial, analgesic, treatment of ulcers, and febrifuge, also have been reported as antitumor and antimalarial activity and activity against cardiovascular disease (Macías-Villamizar et al. 2015). The leaves infusion of *N. angustifolia* is used as digestive (Martínez Crovetto, 1981), to treat rheumatism, arthritis, and pain (Bertucci et al. 2008), and as antivenom against viper bites (alexiteric) (González-Torres 2018).

9 Modern Medicine Based on Its Traditional Medicine Uses

The raw leaf-extract of *N. angustifolia* has been reported to possess an anti-inflammatory effect. The topic application on the rat ear inhibits the edema induced by the *Croton* oil by reduction of the myeloperoxidase enzyme activity (Oliveira de Melo et al. 2006).

The oral administration of the ethanolic extract reduces the volume of the exudate on rats induced by the intrapleural injection of carrageenan, possess antimicrobial activity against Gram (+) bacteria and *Bacillus subtilis* (Truiti et al. 2006). Antileishmaniasis activity *versus* promastigotes forms of *Leishmania braziliensis* has also been reported (Truiti et al. 2005).

The recognition of the snakebite accident, as an unattended tropical disease, is very important because it will allow attending a scourge that continues to inflict suffering and chronic disability, as it can be experienced in most of the marginalized communities. The conventional treatment, consisting of the application of antivenom serum, has several disadvantages, such as: immediate or later hypersensibility reactions, limited effectiveness to protect the local immediately damage, low stability, requires use by qualified personnel to avoid adverse reactions, besides the need of specific antivenoms (due to significant geographical variation on the composition of the snake venom). Small molecule enzyme inhibitors of plants could provide several potential advantages as adjuvants, reducing the effects of intoxication while the patient treated in a health care center, which would increase the recovery time, thereby improving the treatment window (Gómez-Betancur et al. 2019).

Looking for snake antivenom activity, a screening with extracts of different polarities (aqueous, ethanolic and hexane) from *N. angustifolia* leaves was performed with material collected in different vegetative stages. The ethanolic leaves extract of *N. angustifolia* was the most active *in vitro*, showing a higher inhibition of biological activities against *Bothrops diporus* ("yarará chica") venom (Torres et al. 2011a, 2014). In addition, this activity presented a seasonal variation: the inhibition of the coagulant activity was more pronounced in autumn and spring while the indirect hemolytic activity was only in the spring. There were no differences in proteolytic activity according to the vegetative state. These results indicate that the collection of the *N. angustifolia* species in spring can be advised and, also, the use of ethanolic leaves extract as an antivenom (Camargo et al. 2005; Torres et al. 2011b). The SDS-PAGE bio-guided fractionation (Camargo et al. 2011) in combination with flash chromatography of the ethanolic leaf extract, allowed to obtain an active polar fraction that not presented toxicity by *in vivo* tests. This fraction protected mice (DE_{50}) of 4 lethal doses (DL_{50}) venom injected intraperitoneally (Torres et al. 2012). Those results occur due to the presence of flavonoids in the active fraction that can work through the formation of a complex of hydrogen bonds between phenolic and amide groups of venom proteins (Liang 1987; Torres 2011).

The *N. angustifolia* essential oil showed selective alexiteric action (on coagulation and proteolysis and not on the phospholipases) (Torres et al. 2014). In parallel, when the *N. angustifolia* volatiles removed from the distillation water (hydrosol) were assayed for coagulation inhibition produced by "yarará chica" venom, samples from autumn and spring showed higher activity for both summer samples and essential oil (Torres et al. 2014). However, the low oil yields and the great amount required to test the interaction with venom, indicates, in practice, that volatile compounds are not suitable to be used as alexiteric.

10 Conclusions

The chemical study of *N. angustifolia* essential oil, which possess a high content in monoterpene hydrocarbons as a characteristic, and the comparison with N. *megapotamica* oil composition, where sesquiterpenes hydrocarbons are the main fraction, supply additional information to the discussion open about the synonymy between both *Nectandra* species.

The ethanolic extracts of *N. angustifolia* leaves exerted inhibitory activity on the main activities of the *Bothrops diporus* venom. This finding is important, since the snake accidents are a serious public health problem. The *Bothrops* genus venom affects the muscular tissues and the blood coagulation, with local tissular damage effects, hemorrhage, myonecrosis, and edema, as main causes of morbidity in this kind of accidents (Gómez-Betancur et al. 2019). The conventional treatment does not eliminate these local effects so the search for complementary alternatives to this treatment is both important and necessary. The studies of Torres (2011) and Torres et al. (2014) provide evidence on the alexiteric activity of *N. angustifolia* against the effects of *Bothrops diporus*, including the inhibition of the lethal activity in experimental animals.

References

Anton A, Zuloaga F (2018) Flora Argentina. In: http://www.floraargentina.edu.ar. (Last access, December 2020)

Basualdo IZ, Soria Rey N (1999) La vegetación de la reserva de recursos manejados Cerro Sarambi. Dpto. Amambay. Rojasiana 5:51–65

Bernardi L (2018) Novedades taxonómicas en *Nectandra* Rottb. y *Ocotea* Aubl. (Lauraceae) de la Flora del Paraguay. Candollea 73(1):105–108

Bertucci A, Haretche F, Olivaro C, Vázquez A (2008) Prospección química del bosque de galería del río Uruguay. Rev Bras Farmacogn 18(1):21–25

Camargo F, Torres A, Tressens S, Dellacassa E, Ricciardi A. (2005). Inhibición de la actividad hemolítica del veneno de *Bothrops neuwiedi diporus* Cope (yarará chica) por extractos de plantas del nordeste argentino. Comunicaciones Científicas y Tecnológicas, UNNE: E009

Camargo F, Torres AM, Ricciardi G, Ricciardi A, Dellacassa E (2011) SDS PAGE: A useful tool for preliminary screening of antisnake activity of plant extracts. BLACPMA 10(5):429–434

Carod-Artal F, Vázquez-Cabrera C (2007) Semillas psicoactivas sagradas y sacrificios rituales en la cultura Moche. Rev Neurol 44:43–50

Castiglioni J (1951) Lauráceas Argentinas. Bol Soc Argentina Bot 4:66–94

Chanderbali A, Van der Werff H, Renner S (2001) Phylogeny and historical biogeography of Lauraceae: evidence from the chloroplast and nuclear genomes. Ann Mo Bot Gard 88(1):104–134

Chippaux J (2017) Snakebite envenomation turns again into a neglected tropical disease. J Venom Anim Toxins 23:38–39

Cronquist A (1992) An integrated system of classification of flowering plants. Ed. The New York Botanical Garden. New York, USA, 641 pp

Fester G, Retamar J, Ricciardi A (1958) Esencias volátiles argentinas. Rev Fac Ing Quim 18:31–32

Fester G, Martinuzzi E, Retamar J, Ricciardi A (1961) Aceites esenciales de la República Argentina. Ed. Academia Nacional de Ciencias. Buenos Aires, Argentina, 113 pp

Gómez-Betancur I, Gogineni V, Salazar-Ospina A, León F (2019) Perspective on the therapeutics of anti-snake venom. Molecules 24(18):3276

González-Torres D (2018) Catálogo de plantas medicinales (y alimenticias y útiles) usadas en Paraguay. Ed. Servilibro, Asunción, Paraguay. 456 pp

Jiménez B, Knapp S, Marín G, Peña-Chocarro MC (2000) Listado preliminar de plantas vasculares de la Reserva Natural del Bosque Mbaracayú, Paraguay. Rojasiana 5:141–277

Liang W (1987) Antisnake bite action of *Picrasma quassioides*. Bull Chin Mat Med 12(4):54

López JA, Little EL, Ritz GF, Rombold JS, Hahn WJ (1987) Árboles comunes del Paraguay. I-IV. Ed. Imprenta Cuerpo de Paz, Asunción, Paraguay, 425 pp

Macías-Villamizar V, Cuca-Suarez L, Coy-Barrera E (2015) Genus *Nectandra*: phytochemistry and biological activity. BLACPMA 14(4):317–342

Marqués C (2001) Importância econômica da família Lauraceae Lindl. Floresta e Ambiente 8:195–206

Martínez Crovetto R (1981). Las plantas utilizadas en medicina popular en el noroeste de Corrientes (República Argentina). Ministerio de Cultura y Educación. Ed. Fundación Miguel Lillo. Tucumán, Argentina, 139 pp

Múlgura ME (1987) Lauraceae. In: Troncoso NS, Bacigalupo NM: Flora Ilustrada de Entre Ríos, Colecc Cient INTA 6(3a):330-337

Oliveira de Melo J, Truiti M, Muscará M, Bolonheis S, Dantas J, Caparroz-Assef S, Cuman R, Bersani-Amado C (2006) Anti-inflammatory activity of crude extract and fractions of *Nectandra falcifolia* leaves. Bio Pharm Bull 29(11):2241–2245

Pedralli G (1987) Lauráceas 6. *Nectandra*. In: Reitz R (ed.): Flora Ilustrada Catarinense. Herbário Barbosa Rodrigues. Itajaí, Brazil, 93 pp

Poszkus Borrero PA, Bohren AV, Keller HA, Grance LA, Dummel CJ (2016) Leaf architecture of native Lauraceae species from Misiones, Argentina. Bol Soc Argentina Bot 51(1):37–57

Rohwer JG (1993) Lauraceae: *Nectandra*. Fl Neotrop Monogr 60:1–332

Sayagués L, Graf E, Delfino L (2000) Análisis de la información publicada sobre composición florística de montes naturales del Uruguay. Agrociencia 4:96–110

Soria Rey N, Basualdo IZ, Zardini EM (1994) Catálogo de la flora vascular del Cerro Mbatovi, Dpto. de Paraguari, Paraguay. Rojasiana 2:20–35

Stutz LC (1983) Etudes floristiques de divers stades secondaires des formations forestières du Haut Parana (Paraguay oriental). Inventaire floristique d'une réserve forestière. Candollea 38:541–573

The Plant List (2019) Version 1.1 Published on the Internet: http://www.theplantlist.org. (Last access, November 2019)

Torres AM (2011) Caracterización quimiotaxonómica de especies pertenecientes al género *Nectandra* en el Nordeste Argentino. Evaluación de su potencial actividad antiveneno en accidentes provocados por *Bothrops neuwiedi diporus* (COPE) "yarará chica". Thesis. Universidad Nacional del Nordeste. Corrientes, Argentina, 159 pp

Torres A, Camargo F, Ricciardi G, Dellacassa E, Ricciardi A (2011a) Neutralizing effects of *Nectandra angustifolia* extracts against *Bothorps neuwiedi* snake venom. Nat Prod Comm 6(9):1393–1396

Torres A, Camargo F, Ricciardi G, Ricciardi A, Guerra M, Dellacassa E (2011b) Fitoquímica de una fracción del extracto alcohólico *de Nectandra angustifolia*, activa contra el veneno de *Bothrops diporus*. Comunicaciones Científicas y tecnológicas UNNE. Resistencia, Chaco, Argentina. CE001. https://www.unne.edu.ar

Torres A, Camargo F, Ricciardi G, Ricciardi A, Dellacassa E, Lozina L, Maruñak S, Acosta de Pérez O (2012) Inhibición de la actividad letal del veneno de *Bothrops diporus* por extractos de *Nectandra angustifolia*. Revista de Fitoterapia 12(S1):124

Torres A, Camargo F, Ricciardi G, Ricciardi A, Dellacassa E (2014) *Nectandra megapotamica* (Spreng.) Mez.: phytochemical characterization and neutralizing effect on *Bothrops diporus* venom. J Essent Oil Res 26(3):197–203

Truiti M, Ferreira I, Zamuner M, Nakamura C, Sarragiotto M, Souza M (2005) Antiprotozoal and molluscicidal activities of five Brazilian plants. Braz J Med Biol Res 38:1873–1878

Truiti M, Versan-Amado C, Dias-Filho B, Sarragiotto M, De Souza M (2006) Screening of five Brazilian plants for anti-inflammatory and antimicrobial activities. Pharm Biol 44:516–521

Zanon MMF, Goldenberg R, Moraes PLR (2009) The genus *Nectandra* Rol. ex Rottb. (Lauraceae) in Paraná State, Brazil. Acta Bot Bras 23(1):22–35

Zuloaga FO, Morrone O, Belgrano MJ, Marticorena C, Marchesi E (2008) Catálogo de las plantas vasculares del Cono Sur (Argentina, Sur de Brasil, Chile, Paraguay y Uruguay). Monographs in Systematic Botany. 3 vol. American Society of Plant Taxonomists. University of Wyoming. Laramie. WY, USA, 3486 pp

Nothofagus antarctica (G. Forster) Oersted

Silvia Beatriz González

Nothofagus antarctica (G. Forster) Oersted, in Autumn. (Photo by the author)

Abstract *Nothofagus antarctica* (G. Forst.) Oerst. (subgenus *Nothofagus*) commonly called "ñire", is a member of the Family Nothofagaceae, native to Argentina and Chile. Flavonoid aglycones and flavonol glycosides were found in the lipophilic leaf exudate of this species. The presence of essential oils has been reported in *N. antarctica* populations from Chubut to Tierra del Fuego, in Argentina and near Talca, in Chile. Among the species of the *Nothofagus* genus, *N. antartica* is the only one which contains a high percentage of the sesquiterpene α-agarofuran. *N. antarctica* is a promising source of antioxidants as well as volatile oils, with potential in pharmaceutical and cosmetic industries.

Keywords Nothofagaceae · "Ñire" · Flavonoids · Essential oils · Patagonia

S. B. González (✉)
LIPAM (Laboratorio de Investigación de Plantas Aromáticas y Medicinales), Facultad de Ciencias Naturales y Ciencias de la Salud, Universidad Nacional de la Patagonia San Juan Bosco sede Esquel, Esquel (9200), Chubut, Argentina
e-mail: quim-esq@unpata.edu.ar; silviaesquel@speedy.com.ar

Á. Máthé, A. Bandoni (eds.), *Medicinal and Aromatic Plants of South America Vol. 2*, Medicinal and Aromatic Plants of the World 7,
https://doi.org/10.1007/978-3-030-62818-5_30

1 Introduction

Nothofagus species are usually either the dominant or one of the important species in their associations, in the Patagonia. *N. antarctica* occurs in the widest range of habitats and it is common at sites which are too harsh for most other tree species (Veblen et al. 1996). It has a great potential because of its pleasant aroma that can be perceived when walking through a "ñire" forest, called "ñirantal". The surface of young leaflets often contains several flavonoids. Although this species is native to South America it has been cultivated in USA and Europe.

2 Taxonomic Characteristics

Nothofagus antarctica is popularly known as "ñire", which means fox in Mapuche language (so called because these animals tend to burrow under them). The same common name is share by *Nothofagus pumilio* (de Mösbach 1992). The Ona named it as "kechárn" or "winchi" (Martínez Crovetto 1968).

Synoniyms : according to The Plant List: *Calucechinus antarctica* (G. Forst.) Hombr. & Jacquinot ex Decne., *Calucechinus montagnei* (Hombr. & Jacquinot) Hombr. & Jacquinot ex Decne., *Fagus antarctica* G. Forst., *Fagus antarctica* f. *latifolia* Kurtz ex Albov, *Fagus antarctica* var. *microphylla* Phil., *Fagus antarctica* var. *palustris* Albov, *Fagus antarctica* var. *subalpina* Albov, *Fagus antarctica* var. *sublobata* A.DC., *Fagus antarctica* var. *uliginosa* A. DC., *Fagus montagnei* Hombr. & Jacquinot, *Nothofagus* × *alpina* (Poepp. ex A. DC.) Oerst., *Nothofagus montagnei* (Hombr. & Jacquinot) Krasser, *Nothofagus uliginosa* (A. DC.) Krasser.

The genus *Nothofagus* (southern beeches) belonging to Nothofagaceae family, occurs in Australia, Tasmania, New Zealand, South America, New Guinea and New Caledonia. It is essentially a Southern Hemisphere genus with both temperate and tropical species. This ancient genus has been an important focus for discussions of biogeographical history because of its distribution in Gondwana, the fact that its nuts are apparently limited in dispersal movement, and the existence of a phylogenetically relevant fossil record of distinctive pollen types (Hill and Read 1991). The phylogenetic relationships within the monophyletic subgenus *Nothofagus* have not yet been resolved. Acosta and Prémoli (2010) found that *N. antarctica* (subgenus *Nothofagus*) was a sister to a clade of evergreen species (*N. betuloides*, *N. dombeyi* and *N. nitida*), while *N. pumilio* likely diverged earlier. Although the subgenus *Nothofagus* is geographically restricted to temperate areas of South America at present, it was widespread among Gondwanan continents in the Tertiary (Dettmann et al. 1990).

3 Crude Drug Used

The crude drug of *N. antarctica* consists of fresh or dry leaves with young stems and flowers (Photo 1). Conspicuous exudates in *Nothofagus* crude drugs result from the colleters (multicellular secretory hairs) at the base of the stipules. These, as illustrated by van Steenis (1953) may also be derived from glands located on the leaves (Hill and Read 1991), or from glands on subtending bracts.

4 Major Chemical Constituents and Bioactive Compounds

The very thin resinous coating observed on young leaflets of *N. antarctica* contain fatty and terpenoid materials and besides, they were identified five flavonoid aglycones and five flavonol glycosides from the phenolic fraction (Wollenweber et al. 1997). The unusual occurrence of flavonoid glycosides in lipophilic plant excretions has been reported only a few times before (Wollenweber et al. 1997). The aglycones were unambiguously identified as galangin, galangin-7-methyl ether, 8-methoxygalangin, myricetin and pinocembrin. The polar portion of the phenolic fraction was found to contain five flavonoid glycosides: 3-*O*-α-L-arabinopyranosides and the 3-*O*-β-D-galactopyranosides each of quercetin and of myricetin. The fifth product has been elucidated to be myricetin 3'-β-D-glucopyranoside (Wollenweber et al. 1997). *N. antarctica* has some unique flavonoids (myricetin and myricetin glycosides) which were not detected in the other species of *Nothofagus* studied (Wollenweber et al. 2003).

González et al. (2016) reported the presence of essential oils in *N. antarctica* populations from Chubut to Tierra del Fuego in Argentina. It is the only species with a high percentage of the sesquiterpene α-agarofuran, up to 84%. However, there

Photo 1 Used parts of "ñire": leaves and green stems. (Photos by the author)

would be chemotypes with low levels of this component (González et al. 2017, 2018a, b). This compound, α-agarofuran, present in *Nothofagus* species from Talca (Chile) was found as attractive to alates aphid oligophagous *Neuquenaphis edwardsi,* which use them as hosts. These results suggest that chemicals play a significant role in the host-plant associations between *Neuquenaphis* spp. and *Nothofagus* spp. (Quiroz et al. 1999; Russell et al. 2004; Webster 2012). Incidentally, it is interesting to know that there are several antifeedant products derived from dihydro-β-agarofuran (González et al. 1997).

Ethanolic extracts were analyzed about phenolic compounds and antioxidants activities, finding high levels both. Moreover, antiproliferative activities were detected against HT-29 and Caco-2 (ATCC) cellular lines of colon cancer (Gastaldi et al. 2018).

5 Morphological Description

N. antarctica is a woody deciduous species that grows in pure stands, reaching up to 17 m total height and more than 1 m in diameter at the better sites (Lencinas et al. 2002). It is a monoecious species, with solitary male flowers appearing first at the base of the shoots, while female flowers with tricarpelar ovaries grow later at the distal extremes. Budburst occurs in late spring and fruits develop in early summer. The fruit has three seeds per cupule, reaching maturity in late March.

Four ecotypes were reported (McQueen 1976) and it was found a strong variation of the leaves between at least two of them (postrate shrub and subalpine shrub) (Gandolfo and Romero 1992) (Photo 2). The cupules of *Nothofagus* species is a massive, robust and often compound envelope of the fruits, and it varies in several features, including: (1) size, shape and number of valves; (2) number and type of

Photo 2 - Upper and back sides of leaves of *N. antarctica.* (Photo by the author)

lamellae on the external surface of the valves; (3) presence and type of trichomes; (4) peduncle morphology, and (5) number and arrangement of fruits within the cupules (Hill and Read 1991). It has membranous, persistent, usually non-glandular lamellae, the outer surface of the cupule valves is covered with simple trichomes, having four cupule valves. The character of deciduousness, classically, is associated with climatic extremes, of cold or of drought.

6 Geographical Distribution

N. antarctica presents a wide ecological distribution along with all Patagonian forests, from Neuquén at 36° 8´ S to Cabo de Hornos at 56° 8´ S (Donoso et al. 2006; Ramírez et al. 1985; Vidal and Prémoli 2004). Most species within *Nothofagus*, although ecologically and genetically distinct, are similarly widespread and encompass the entire distributional range of temperate forests in southern South America. *N. antarctica* displays the greatest morphological variation of the five members of the subgenus, inhabiting low to high elevation environments, valley bottoms with cold air drainage, and wetlands (Veblen et al. 1996; Acosta and Prémoli 2010).

In South Patagonia, these forests grow in the ecotone between rangelands and *Nothofagus pumilio* woods (Roig 1998). It has been introduced in the Pacific west coast of USA, in Scotland and Feroe Islands and is cultivated in many European botanic gardens (Wollenweber et al. 1997).

7 Ecological Requirements

N. antarctica grows preferably in cold places with a snow cover in winter and had colonized also bog and poorly drainages soils (Ramírez et al. 1997). The species shows a great ecological and morphological plasticity (Ramírez et al. 1985), living under extreme conditions such as high-mountain Andean 'Krummholz' (stunted forests occurring at the forest-tundra ecotone), peatbogs and swamps. It is found in the so called 'nadis', in south-central Chile (Ramírez et al. 1985). "Nadis" are places with a thin layer of soil separated from the subsoil by an impervious aluminum-iron rich hard pan which renders surface soil saturated in winter (eight months) and very dry in summer (four months).

N. antarctica had low seed viability (11–17% of the potential seed production), when compared with other *Nothofagus* species, such as *N. pumilio* (up to 60%) (Cuevas 2000; Martínez Pastur et al. 2008). Flowers, immature fruits and seeds are exposed to predation by insects and birds, and when they finally fall to the forest floor, they may be consumed by rodents (*Akodon, Euneomys* and *Oligoryzomys* species) or birds. Surviving seeds germinate in November/December in this Hemisphere. Seedlings can persist up to 10 years in some stands (Soler et al. 2010). Although little is known about the flowering process of *N. antarctica*, some authors reported that extreme environmental conditions (e.g. low temperature, winds, or nutrient shortages) affect the reproductive capabilities of the species (Ramírez et al. 1985; Prémoli 1991).

It rarely grows to more than 10 m high and is multi-branched, deciduous as *N. obliqua*. As described by Weinberger, it behaves in a similarly plastic way in middle latitude Chile as it does in Magellanic Chile, occupying at lower altitudes poorly drained "nadi" soils (Wright 1965).

Seedling establishment was improved in the stands under silvopastoral use compared to primary forests, but the mortality rate was high (or total) in all situations (Bahamonde et al. 2013).

8 Traditional Use (Part(s) Used) and Common Knowledge

The species was mainly used as fuel-wood and rural constructions and in ancient times as bows (Martínez Crovetto 1982). In popular medicine leaves and young buds has been used as a febrifuge (Barboza et al. 2009). Hikers and climbers use the leaves as teas in mountain refuges (personal communication, Gastaldi 2019).

9 Modern Medicine Based on Its Traditional Medicine Uses

N. antarctica stands out because of its antioxidant activity which is higher than that of the green tea and mate herb (*Ilex paraguariensis*) (González et al. 2018a, b). Phenolic compounds and flavonoids, commonly known antioxidants, would be responsible for such activity. Among them, galangin showed various pharmacological activities such as antimutagenic, anticlastogenic, antioxidative, radical scavenging, metabolic enzyme modulating and anticancer activity. *N. antarctica* has some unique flavonoids (myricetin and myricetin glycosides) which were not detected in the other species of analyzed *Nothofagus*, these compounds exhibit a wide range of activities that include strong antioxidant, anticancer, antidiabetic and anti-inflammatory activities. It displays several activities that are related to the central nervous system and numerous studies have suggested that myricetin may be beneficial to protect against diseases such as Parkinson's and Alzheimer's. (Semwal et al. 2016). Although few epidemiological and clinical studies have reported the health benefits of myricetin on diabetes mellitus, increasing evidence from *in vitro* and animal studies have confirmed its hypoglycemic effect (Li and Ding 2012). Moreover, antiproliferative activity *in vitro* of infusions of *N. antarctica* showed promising results (Gastaldi et al. 2018).

10 Conclusions

N. antarctica is a promising source of antioxidants, with potential bioactive compounds as quercetin and its glycosylated derivatives, actives against numerous health problems and diseases, including colon cancer. The essential oils obtained from the crude drug have potential interest in the fragrance industries due to their particular notes. The sustainable use of natural populations is a challenge ahead.

Acknowledgments To Secretaría de Ciencia y Técnica, Universidad Nacional de la Patagonia San Juan Bosco, Project C y T Nro. 1298/15.

References

Acosta MC, Prémoli AC (2010) Evidence of chloroplast capture in South American *Nothofagus* (subgenus *Nothofagus*, Nothofagaceae). Molecular Phylogenetics and Evolution 54(1):235–242

Bahamonde HA, Peri PL, Monelos LH, Martínez Pastur G (2013). Regeneración por semillas en bosques nativos de *Nothofagus antarctica* bajo uso silvopastoril en Patagonia Sur, Argentina. Bosque (Valdivia), 34(1):89–101. https://doi.org/10.4067/S0717-92002013000100011. (Last access, July 2019)

Barboza G, Cantero J, Núñez C, Pacciaroni A, Ariza Espinar L (2009) Medicinal plants: A general review and a phytochemical and ethnopharmacological screening of the native Argentine Flora. Kurtziana 34(1-2):7–365

Cuevas J (2000) Tree recruitment at the *Nothofagus pumilio* alpine timberline in Tierra del Fuego, Chile. Ecology 88:840–855

De Mösbach EW (1992) Botánica Indígena de Chile. Ed. Andrés Bello, Santiago, Chile. 140 pp

Dettmann ME, Pocknall DT, Romero EJ, Zamaola M (1990) *Nothofagidites* Erdtman ex Potonie 1960: a catalogue of species with notes on the paleogeographic distribution of *Nothofagus* Bl. (Southern Beech). N Z Geol Surv Paleontol Bull 60:1–79

Donoso C, Steinke L, Prémoli A (2006). *Nothofagus antarctica*. In: Donoso C (ed.): Las especie arbóreas de los bosques templados de Chile y Argentina. Ed. Marisa Cúneo Ediciones, Valdivia, Chile, 678 pp

Gandolfo MA, Romero EJ (1992) Leaf morphology and a key to species of *Nothofagus* Bl. Bulletin of the Torrey Botanical Club:152–166

Gastaldi B, Marino G, Matenett F, Peri P, González SB (2018) Infusión de *Nothofagus antarctica*: compuestos fenólicos y actividad antiproliferativa en células derivadas de cáncer de colon (HT-29 y Caco-2). Dominguezia 34(S):70–71

González GA, Jiménez IA, Ravelo AG, Coll J, González TJA, Lloria J (1997) Antifeedant Activity of Sesquiterpenes from Celastraceae. Biochemical Systematics and Ecology 25(6):513–519

González SB, Gastaldi B, Mattenet F, Peri P, van Baren C, di Leo Lira P, Retta D, Bandoni AL (2016). Aceites esenciales en partes aéreas de *Nothofagus antarctica* (G. Forst.) Oerst. de diferentes sitios de la Patagonia. Dominguezia 32(2):90-91

González SB, Gastaldi B, Silva Sofrás F, Guajardo J, Argel C, Mattenet F, Peri P, Di Leo Lira P, Retta D, van Baren C, Bandoni A (2017) ¿Quimiotipos en *Notohofagus antarctica* (ñire)? Dominguezia 33(1):50

González SB, Gastaldi B, Mattenet F, Peri P, van Baren C, Di Leo Lira P, Retta D, Bandoni AL (2018a). Análisis fitoquímicos comparativos de *Nothofagus antarctica* proveniente de dos sitios con ambientes diferentes en la provincia de Santa Cruz. Dominguezia 34(S):41-42
González S, Gastaldi B, Mattenet F, Peri P, Di Leo Lira P, Retta D, van Baren C, Bandoni A (2018b) Actas del IV Congreso Nacional de Sistemas Silvopastoriles, Villa la Angostura, Neuquén, Argentina, October 31 to November 2, 2018
Gut B (2017) Árboles en la Patagonia. Ed. Vázquez Mazzini, Buenos Aires, Argentina. 416 pp
Hill RS, Read J (1991) A revised infrageneric classification of *Nothofagus* (Fagaceae). Bot J Linn Soc 105:37–72
Lencinas MV, Martínez Pastur G, Cellini JM, Vukasovic R, Peri P, Fernández C (2002) Incorporación de la altura dominante y la calidad de sitio a ecuaciones estándar de volumen para *Nothofagus antarctica* (Forster *f.*) Oersted. Bosque 23(2):5–17
Li Y, Ding Y (2012) Minireview: Therapeutic potential of myricetin in diabetes mellitus. Food Science and Human Wellness 1(1):19–25
Martínez Crovetto RN (1968) Estudios Etnobotánicos. IV. Nombres de plantas y su utilidad según los indios onas de Tierra del Fuego. Etnobiológica (3):1–20
Martínez Crovetto R (1982) Breve panorama de las plantas utilizadas por los indios de Patagonia y Tierra del fuego. Suplemento Antropológico, Universidad Católica, Asunción, Paraguay 17(1):61–97
Martínez Pastur G, Lencinas MV, Peri PL, Cellini JM (2008) Flowering and seeding patterns in unmanaged and managed *Nothofagus pumilio* forests with a silvicultural variable retention system. Forstarchiv 79:60–65
McQueen DR (1976) The ecology of *Nothofagus* and associated vegetation in South America. Tuatara 22(1):38–68
Patel DK, Patel K, Gadewar M, Tahilyani V (2012) Pharmacological and bioanalytical aspects of galangin -a concise report. Asian Pacific Journal of Tropical Biomedicine 2(1):S449–S455
Prémoli A (1991) Morfología y capacidad germinativa en poblaciones de *Nothofagus antarctica* (Forster) Oerst. del noroeste andino patagónico. Bosque 12:53–59
Quiroz A, Fuentes-Contreras E, Ramírez CC, Russell GB, Niemeyer HM (1999) Host-plant chemicals and distribution of *Neuquenaphis* on *Nothofagus*. Journal of Chemical Ecology 25(5):1043–1054
Ramírez C, Correa M, Figueroa H, San Martín J (1985) Variación del hábito y hábitat de *Nothofagus antarctica* en el sur de Chile. Bosque 6(2):55–73
Ramírez C, San Martín C, Oyarzún A (1997) Morpho-ecological study on the South American species of the genus *Nothofagus*. Plant Ecology 130:101
Roig F (1998) La vegetación de la Patagonia. In: Correa M (ed) Flora Patagónica. Ed. Colección Científica del INTA, Buenos Aires, Argentina
Russell GB, Faundez EH, Niemeyer HM (2004) Selection of *Nothofagus* host trees by the aphids *Neuquenaphis staryi* and *Neuquenaphis edwardsi*. Journal of Chemical Ecology 30(11):2231–2241
Semwal D, Semwal R, Combrinck S, Viljoen A (2016) Myricetin: A dietary molecule with diverse biological activities. Nutrients 8(2):90
Soler Esteban R, Martínez Pastur G, Lencinas MV, Peri PL (2010) Flowering and seeding patterns in primary, secondary and silvopastoral managed *Nothofagus antarctica* forests in South Patagonia. New Zealand Journal of Botany 48(2):63–73
van Steenis CGGJ (1953) Results of the Archbold expeditions Papuan Nothofagus. J Arnold Arbor 34:301–374
Veblen TT, Donoso C, Kitzberger T, Rebertus AJ (1996) Ecology of southern Chilean and Argentinean *Nothofagus* forests. In: Veblen TT, Hill RS, Read J (eds) The Ecology and Biogeography of Nothofagus Forests. Ed. Yale University Press, New Haven, USA, 403 pp
Vidal R, Prémoli AC (2004) Variación en *Nothofagus antarctica*. In: Donoso C, Prémoli A, Gallo L and Ipinza R (eds.). Variación intraespecífica en las especies arbóreas de los bosques templados de Chile y Argentina. Ed. Universitaria, Santiago, Chile, 434 pp

Webster B (2012) The role of olfaction in aphid host location. Physiological Entomology 37:10–18
Wollenweber E, Stuber A, Kraut L (1997) Flavonoid aglycones and flavonol glycosides in lipophilic leaf exudate of *Nothofagus antarctica*. Phytochemistry 44(7):1399–1400
Wollenweber E, Stevens JF, Dörr M, Rozefelds AC (2003) Taxonomic significance of flavonoid variation in temperate species of *Nothofagus*. Phytochemistry 62(7):1125–1131
Wright ACS (1965). The Volcanic Ash Soils of Chile. Ed. FAO. Rome, Italy. Report no. 2017, 201 pp

Parastrephia lucida (Meyen) Cabrera

María Inés Isla, Iris Catiana Zampini, and Ana Soledad Cuello

P. lucida from Argentine Puna. Antofagasta de la Sierra, Catamarca, Argentina. (Photo: Soledad Cuello)

M. I. Isla (✉) · I. C. Zampini · A. S. Cuello
Instituto de Bioprospección y Fisiología Vegetal (INBIOFIV, CONICET- Facultad de Ciencias Naturales e Instituto Miguel Lillo-Universidad Nacional de Tucumán), San Miguel de Tucumán, Argentina
e-mail: misla@csnat.unt.edu.ar

Á. Máthé, A. Bandoni (eds.), *Medicinal and Aromatic Plants of South America Vol. 2*, Medicinal and Aromatic Plants of the World 7,
https://doi.org/10.1007/978-3-030-62818-5_31

Abstract *Parastrephia lucida,* known as “romero”, “tola” o “t’hola”, “chachakoa”, “tola de río”, “tola de agua”, is a medicinal plant found in large areas in the semi-arid Puna of Argentina, Chile, Bolivia and Peru. It is part of the landscape called “tolar”, with other species of *Parastrephia*, *Baccharis,* and *Fabiana*. The “tolares” have been a very important fuel sources and sometimes the only one available. Currently its use as firewood is limited for greater access. Although its resources are protected by both legal and police protection measures, still it is occasionally used as firewood. *P. lucida* is resistant to cold, high ultraviolet radiation, drought and salinity. Since ancient times, *P. lucida* is widely used as medicine by the local communities of the Puna region, mainly for dislocations, traumatism, pain, as anti-inflammatory, febrifuge, antibiotic, acaricide, among others. Progress has been made in the scientific validation of its medicinal properties and in the knowledge of the chemistry of this plant, demonstrating that hydroalcoholic extracts could be used as antioxidants, anti-inflammatory, potent antibiotics, and acaricides. Furthermore, the infusion showed potent antioxidant capacity. No toxic, mutagenicor pro-mutagenic effects have been described. Although *in vivo* studies would be necessary to fully demonstrate the safety of the extracts, prolonged traditional use by local communities could support their safe to use locally. The use of native plant species from South America as an alternative to commercial plant species could have a positive impact on the economies of local communities.

Keywords “Tola” · Argentina · Medicinal plants · Anti-inflammatory · Antibiotic · Antioxidant

1 Introduction

Parastrephia is undoubtedly an underutilized native medicinal species in Argentina, especially if we compare it with major traditional crops. This species has important pharmacological properties and an advantage over cultivated species to tolerate hydric stress, saline stress and intense ultraviolet radiation. In the marginal lands where it lives, the plants have undergone natural selection. According to Cantero et al. (2019) it is a plant species of economic interest in Argentina.

2 Taxonomic Characteristics

The genus *Parastrephia* (Asteraceae) comprises four species, namely *P. lucida, P. lephidophylla, P. quadrangularis* and *P. teretiuscula* (Cabrera 1978). *P. lucida* (Meyen) Cabrera has been described also by others names (synonyms). According to The Plant List (2019), these names include the followings: *Parastrephia phyliciformis* (Meyen) Cabrera, *Baccharis lucida* (Meyen), *Baccharis phyliciformis*

(Meyen), *Dolichogyne glabra* (Phil.), *Dolichogyne rigida* (Wedd.), *Dolichogyne rupestris* (Wedd.), *Parastrephia ericoides* (Nutt.), *Polycladus abietinus* (Phil.), *Lepidophyllum abietinum* (Phil.) Reiche, *Lepidophyllum lucidum* (Meyen) Cabrera, *Lepidophyllum phyliciforme* (Meyen) Hieron. ex R.E. Fr., *Lepidophyllum phyliciforme* (Meyen) Hieron. ex R.E. Fr. var. *resinosum, Lepidophyllum rigidum* (Wedd.) Benth. & Hook., *Vernonia phyliciformis* (Meyen) Walp., *Vernonia phyliciformis* (Meyen) Walp. var. *resinosa, Diplostephium tovari* Cuatrec., *Vernonia phylicaeformis* Walp.

Popularly, this plant species is known under the common name of "romero", "tola" o "t'hola", "unut'hola", "chachakoa", "tola de río", "tola de agua". The names "unut'hola" and "tola del agua" allude to their habitat near water (Castro et al. 1982). "T'hola" is a generic term used to a variety of resinous shrubs (*Parastrephia, Baccharis,* and others), which characterizes the Altiplano landscape. In some regions, the "tolares" are clusters of appreciable magnitude which a height greater more than one meter.

3 Crude Drug Used

Parts of the plant used are leaves, young branches and flowers (Photos 1a, b). In general, the drug is dried and processed into powder. In very few cases the fresh drug is used. The drug is collected in summer.

A B

Photo 1 *P. lucida*. (**a**) Aerial parts and fruits, (**b**) Flowers. (Photos: Soledad Cuello, in Antofagasta de la Sierra, Catamarca, Argentina)

4 Major Chemical Constituents and Bioactive Compounds

Several chemical components were described in *P. lucida, among them* new tremetone derivates (D'Almeida et al. 2012, 2013; D'Almeida 2013; Echiburru-Chau et al. 2017), caffeoylquinic acid, dicaffeoylquinic acid, dicaffeoylquinic acid pentoside, 4-methoxycinnamic acid hexoside pentoside; hydroxycoumaric acid hexoside pentoside, ferulic acid prenyl ester, methyl ferulic acid prenyl ester, ferulic acid 2-phenethyl alcohol, dimethylcaffeic acid 2-phenethyl ester, (quinic acid rhamnoside 3-(3,4-dimethoxyphenyl) propanoate, quinic acid glucuronate, rutin (quercetin rutinoside), kaempferol pentoside hexoside, kaempferol hexoside, quercetin, 5,3′,4′-trihydroxy-7-methoxy flavanone (D'Almeida et al. 2012, 2013; D'Almeida 2013; Nieto et al. 1993; Torres Carro et al. 2015, 2017), aesculetin, 8-hydroxy-7-methoxyscopoletin, 7-methoxy-8-hydroxyesculetin, luteolin-*O*-hexoside and its derivatives, 19-hydroxybacchalineol A acetate and derivatives, and dehydro-19-acetinyl-hawtriwaic acid methyl ester (Echiburru-Chau et al. 2017). Tremetones were also found in *P. lepidophylla* (Bohlmann et al. 1979). The major bioactive compounds identified as antioxidant and anti-inflammatory in *P. lucida* were ferulic acid, cinnamic acid and quinic acid derivates, luteolin-7-methyl ether and tremetone (D'Almeida et al. 2013; Torres Carro et al. 2017, 2019).

5 Morphological Description

Parastrephia lucida is a shrub of 0.5–1.7 m, resinous, branched from the base. The young branches are glabrous with leaves to the apex. The leaves (3–7 mm length by 1 mm in diameter) are sessile with linear and fleshy sheets *arched* out, separating from the stem. The chapters are solitary in the terminal twigs. The flowers are scarce marginal, with the yellow corolla, shortly linked; central flowers with corolla 6 mm long, with 5 lobed. The name *Parastrephia* means "strange" and refers to a particular botanical characteristic of its flowers (Muñoz-Schick et al. 2012). This plant has the peculiarity that the ligules of the border are masculine and those of the middle feminine. Dry fruits are achene sericeous-hairy with whitish pappus and a seed inside (Photo 1a).

6 Geographical Distribution

P. lucida (Meyen) Cabrera is abundant in the eco-region of the Puna, in southern Peru, Bolivia, northern Chile and northwestern Argentina, from Jujuy to Tucumán and Catamarca provinces. This species grows in the Argentinean highlands, characterized by high altitude (3200–4200 m.a.s.l.), high exposure to ultraviolet radiation, low oxygen concentration, high daily temperature variations and mean annual rainfall of 100–400 mm or even less. This plant species is cold resistant (equivalent to USDA climate zone 8, between −7 and −12 °C), it can tolerate snowfall for a couple of weeks a year. After flowering, at the beginning of the seed season, the plants of the genus *Parastrephia* spp. loses the branches and leaves and produce a mulch of organic matter for the soil (IBTA 1995). The "tolar", located on slopes and in totally erosive soils, constitute a very important factor for soil conservation (Montero Terry 2006). There are no references on the phenology of *P. lucida*, but Pérez Mercado (1994) described the changes in the phenological phases of *P. lepidophylla* indicating that they begin in July with the formation of buds. Flowering begins in September and mature seeds are presented from the last week of October and germinate in February. Studies conducted by IBTA (1995) in Bolivia, determined that *Parastrephia lepidophylla* not bloom in the first two years of succession. Seed propagation tests were conducted for other species of *Parastrephia* such as *P. lepidophylla* and *P. quadrangularis* and were tested in silvo-agropecuary systems for the community (Montero Terry 2006).

7 Ecological Requirements

Parastrephia spp. grows in large areas in the semi-arid Puna. Its altitudinal range is from 3200 to 4200 m.a.s.l. (Borel 1990). Tola's prefer a habitat with a shallow water table and very fine texture (Borel 1990; Carilla et al. 2018). *P. lucida* is widely dominant on the river terraces, funds from broken, or depressions with water table located a few meters deep, it develops dense shrub communities. In general, the soils seem to be dry and poor in most cases (Weberbauer 1945), generally acidic soils, with a pH that varies between 4.0 to 5.5; cation exchange capacity, humus content and H^+ changes increase with altitude, while decreasing with the depth of each profile (IIPSC 2001a, b, 2002). The plant tolerates low temperatures (−8 °C) and can tolerate occasional snow cover for up to a couple of weeks per year (ChileFlora database 2020).

8 Traditional Use (Part(s) Used) and Common Knowledge

Parastrephia lucida, is used by the villagers as firewood because it is lit even when it is green (Carilla et al. 2018). Before of use of kerosene, "tola" was used as a first-class fuel (Cárdenas 1989). It is also used to dye the colors green and yellow (Romo et al. 1999). The animals such as alpacas and llamas feed on flowers (Villagran et al. 2003). The plant has reported veterinary use as acaricide on scabies (Ayma Morales

1993; Ayma Morales et al. 1995). In Chile, *P. lucida* had also good ornamental values (ChileFlora database 2020). Species of "tolar" constitute the support of the feeding of the cattle, allow to conserve the soil and to maintain a balance between the components of the ecosystem (Martínez 2002).

Parastrephia lucida is used in popular medicine of the Andes, as an anti-inflammatory agent, to treat toothache by applications of their leaves (poultice). The plant resin is used for rapid healing of wounds, bruises and to consolidate bone dislocation and fractures (vegetable plaster) (Bowman 1983; Villagrán et al. 2003). It serves to prepare a plaster against cracks, dislocations and bruises, consisting of ground buds and beaten with an adult's urine, to form a paste that is used to wrap the affected limb, which, when hardened, is a true "vegetable plaster". Powder from the leaves and buds of *Parastrephia quadrangularis* and *P. lucida*, mixed with salt and egg white weld broken bones (Araya et al. 2003). The plant is also used as a patch, poultice or plaster, mixed with "yareta" (*Azorella compacta*), "kupala" (mineral mixed with vegetable resin), incense and urine (Villagrán et al. 2003). The aerial parts are used for the treatment of stomach pain, as febrifuge and antiseptic agents (Giberti 1983; Villagrán et al. 2003). It is also used in baths and infused drinks, with lemon and burnt sugar. "Unut'ola" is bitter and is also taken for lung and toothache issues, with three garlic cloves (Villagrán and Castro 2004; Villagrán et al. 2003). It was included into the list of Argentinian traditional medicines, largely due to the lack of any reported incidents, and it is considered generally safe.

Parastrephia lucida is also used to prepare "incense/smoke offering" to "cure" some diseases caused by the "aire", such as headache (Giberti 1983; Fernández 1995; Villagrán and Castro 2004; García et al. 2018).

9 Modern Medicine Based on Its Traditional Medicine Uses

Several reports have demonstrated potential health benefits and the popular medicinal uses were scientifically validated.

Antibiotic Activity *P. lucida* is used as antiseptic, for rapid healing of wounds, bruises and to consolidate bone dislocation and fractures (vegetable plaster) (Bowman 1983; Giberti 1983; Villagrán et al. 2003). The antibiotic activity of tinctures and aqueous extracts of *Parastrephia lucida, Parastrephia lepidophylla, Parastrephia phyliciformis* collected in Argentinean Puna, was determined on antibiotic resistant clinical strains isolated from nosocomial infection in human lesions of skin and soft parts. The tinctures inhibited the growth of one or more of the following strains: *Staphylococcus aureus, Enterococcus faecalis, Escherichia coli, Klebsiella pneumoniae, Proteus mirabilis, Enterobacter cloacae, Morganella morganii* and *Pseudomonas aeruginosa* with MIC values from 80 to 300µg/ml on Gram (+) bacteria and 150 to 600µg/ml on Gram (-) bacteria. The tinctures exhibited a stronger activity and a broader spectrum of action than aqueous extracts (Zampini et al. 2009). The antimicrobial effect found in *P. lucida* was associated with mixtures of phenylpropanoids, including prenyl and phenethyl esters of caffeic and cinnamic acids (D'Almeida et al. 2012; D'Almeida 2013). The results support at least in part the traditional use of the plant as local antiseptic and vulnerary. Antimicrobial activity of aqueous and ethanolic extracts of *P. lucida* on strains isolated from infections in animals was also demonstrated (Heredia 2018).

Antioxidant Activity Numerous pathological events such as inflammation processes, aging phenomena and degenerative dysfunction are associated with the generation of reactive oxygen species. Thus, the effectiveness of plant extracts used in folk medicine to suppress inflammatory responses may be due to their capacity to reduce oxidative stress. Antioxidant activity was demonstrated to extracts obtained in different alcoholic grade and infusions of *P. lucida* collected in Argentina and Chile (Zampini et al. 2008; Rojo et al. 2009; D' Almeida 2013; Echiburu Chau et al. 2017; Torres Carro et al. 2017, 2019). The standardized extracts from *P. lucida* showed iron reductive and chelating capacity, ABTS free radical scavenging ability and reactive oxygen scavenging activity, inhibition of production of reactive oxygen species, protection against oxidative hemolysis and lipid oxidation (Zampini et al. 2008; Rojo et al. 2009; D'Almeida 2013; Echiburu Chau et al. 2017; Torres Carro et al. 2017, 2019). Since these plant species contain xanthine oxidase inhibitors, they reduce both uric acid and $O_2^{\bullet-}$ production. Recently was demonstrated that a herbal combination of equal parts of *P. lucida* and *Tessaria absinthioides* exhibit the highest level of inhibition of the enzyme xanthine oxidase, along with a good overall free radical scavenging capacity (Torres Carro et al. 2019). It is an attractive mixture to be used as part of antioxidant nutraceutical formulations. Recently, a Porter-style craft beer enriched with *P. lucida*, for a higher antioxidant capacity was produced (Bustos et al. 2019).

Anti-inflammatory Activity The leaves and stems of these plants maintained in ethanol are used as "rubbing" or topical preparation in order to treat rheumatism, fever and inflammation. Plant resin poultices are used to consolidate luxations and fractures. Inflammatory conditions activate phospholipase (sPLA2), cyclooxygenase (COX) and lipo-oxygenase (LOX), the key enzymes in the synthesis of prostanoids and eicosanoids from polyunsaturated fatty acids. *P. lucida* hydroalcoholic extracts were effective inhibiting COX, sPLA2 and LOX. The main compounds identified in the most active fractions as proinflammatory enzymes inhibitors were 5,4′-dihydroxy-7-methoxyflavanone, apigenin, apigenin methyl ether and apigenin trimethyl ether, methyl and dimethyl ethers from quercetin, kaempferol and luteolin methyl ether, ferulic acid esters, cinnamic acid and vanillin (Alberto et al. 2009; D'Almeida et al. 2013; D'Almeida 2013; Torres Carro et al. 2015, 2017, 2019). NO is one of the inflammatory mediators that is synthesized by iNOS and is induced by different inflammatory stimuli. *P. lucida* has a significant effect on iNOS expression as well as NO scavenging capacity (Torres Carro et al. 2015). The potential use of *P. lucida* in herbal mixtures made with other medicinal plant species from the Argentinean Puna (*Tessaria absinthioides* and *Ephedra multiflora*) as nutraceutical or dietary supplements with anti-inflammatory activities were demonstrated (Torres

Carro et al. 2019). A mix of equal parts of the three selected plant species showed an important inhibitory capacity on proinflammatory enzymes. Its potency on COX-2 was also higher than that of ibuprofen (Torres Carro et al. 2019).

Acaricide Activity The activity of aqueous extracts of *P. lucida* against mites (*Sarcoptes*) was demonstrated by *in vitro* and *in vivo* tests. In concentrations of 0.125 g/mL and 0.25 g/mL are more effective on the mortality of mites in times of 4 and 5 min, respectively (Ayma Morales 1993; Ayma et al. 1995).

Toxicity and Genotoxicity Studies Basma Rajeh et al. (2012) showed that a $LD_{50} > 25\mu g/mL$ for the *Artemia salina* test is correlated to an *in vivo* LD_{50} between 2500 and 8000 mg/kg, which is above the values established as toxic according to the Organization for Economic Cooperation and Development (OECD 2001). *P. lucida* extracts did not exhibit toxicity up to 1000µg/mL on *A. salina* nor mutagenic or promutagenic effect in assays with *Salmonella typhymurium* TA 98 and TA 100 (D'Almeida et al. 2013; Torres Carro et al. 2019).

10 Conclusions

P. lucida has a social use in the Puna region of Argentina, Chile, Peru and Bolivia. It is frequently used as fuel and also in traditional human and animal medicine. Biological properties, chemical composition and toxicity of extracts obtained from the aerial parts of *P. lucida* have been studied and determined by several authors in the last ten years. This review allows to envision *P. lucida* as the source for promising new plant drugs, in medicinal and industrial applications, alone or mixed with other plant species. While *in vivo* studies would be necessary to fully demonstrate the safety of extracts, traditional use by local communities seems to support the safe use of *P. lucida* preparations. Studies on the biological characteristics (incl. phenophases) and the possible - *in vivo* and *in vitro* - propagation methods of *P. lucida* (e.g. *in vitro* micropropagation) are still necessary to promote the sustainable use of this promising species.

Acknowledgments The authors acknowledge the cooperation of the inhabitants of the studied areas and the financial support from the SCAIT-UNT (PIUNT 2018 G-637), the Agencia Nacional de Promoción Científica y Técnica (ANPCyT, PICT 2014- 3136 and PICT 2017- 4436), the Consejo Nacional de Investigaciones Científicas y Técnicas (CONICET, PUE 2018- 0011), Argentina.

References

Alberto MR, Zampini IC, Isla MI (2009) Cyclooxygenase enzyme inhibitory activity of standardized hydroalcoholic extracts of four Asteraceae species from the Argentine Puna. Braz J Med Biol Res 42:776–869

Araya-Presa J, Squeo F, Barrientos L, Belmonte E, Mamani M, Arancio G (2003) Manual de las plantas y canciones aymara. Proyecto Explora-CONICYT. Universidad de La Serena. Chile, 39 pp

Ayma Morales L (1993) Propiedades acaricidas de chilca (*Parastrephia lucida*) en llamas. Thesis. Universidad Técnica de Oruro, Bolivia, 98 pp

Ayma L, Herve D, Sauvain M (1995) Efecto del extracto acuoso de la ch'illka (*Parastrephia lucida*) en el control de la sarna en llamas. In: Genin D, Picht H-J, Lizarazu R, Rodríguez T (eds) Waira Pampa. Un Sistema pastoril camélidos - ovinos del altiplano árido. ORSTOM. CONPAC-Oruro. IBTA, Bolivia, pp 185–194

Basma Rajeh M, Kwan Y, Zakaria Z, Yoga Latha L, Jothy S, Sasidharan S (2012) Acute toxicity impacts of *Euphorbia hirta* L. extract on behavior, organs body weight index and histopathology of organs of the mice and *Artemia salina*. Pharm Res 4(3):170–177

Bohlmann F, Fritz U, King R (1979) Neue tremeton-derivate aus *Parastrephia lepidophylla*. Phytochemistry 18:1403–1405

Borel R (1990) Especies agrosilvopastoriles para la zona altoandina. Pomata, PE, Proyecto Arbolandino. 205 pp

Browman D (1983) Aspectos de nutrición prehistórica en la cuenca del Lago Titicaca. Diálogo Andino 2:27–41

Bustos L, Soto E, Parra F, Echiburu-Chau C, Parra C (2019) Brewing of a Porter Craft beer enriched with the plant *Parastrephia lucida*: a promising source of antioxidant compounds. J Am Soc Brew Chem 77:261–266

Cantero JC, Núñez CO, Bernardello G, Amuchastegui A, Mulko J, Brandolín P, Palchetti MV, Iparraguirre J, Virginil N, Ariza Espinar L (2019) Las plantas de importancia económica en Argentina. First Edition. Ed. UniRío Editora. Universidad Nacional de Río Cuarto, Río Cuarto, Córdoba, Argentina, 937 pp. https://botanicaargentina.org.ar/wp-content/uploads/2019/06/978-987-688-332-0-baja-resoluci%C3%B3n.pdf. (Last access, December 2019)

Cárdenas M (1989) Manual de plantas económicas de Bolivia, 2ed edn. Los Amigos del Libro, Cochabamba, Bolivia, p 234

Carilla J, Grau A, Cuello S (2018) Vegetación de la Puna argentina. Serie Conservación de la Naturaleza La Puna argentina: naturaleza y cultura 24:143–156

Castro M, Villagrán C, Kalin MT (1982) Estudio Etnobotánico en la precordillera y altiplano de los Andes del Norte de Chile (18–19° S) Vol. de síntesis Proyecto M AB-6-UNEP UNESCO Santiago de Chile 2:133–205

Chile Flora Database (2020). http://www.chileflora.com/Florachilena/FloraEnglish/HighResPages/EH2088.htm (Last access, April 2020)

D'Almeida R (2013) Potencialidades funcionales de especies vegetales que crecen en la Puna Argentina. Thesis. Universidad Nacional de Tucumán, Tucumán, Argentina, 267 pp

D'Almeida R, Alberto M, Quispe C, Schmeda-Hirschmann G, Isla MI (2012) Antimicrobial phenylpropanoids from the Argentinean highland plant *Parastrephia lucida* (Meyen) Cabrera. J Ethnopharmacol 142:407–414

D'Almeida R, Isla MI, Vildoza E, Quispe C, Schmeda-Hirschmann G, Alberto M (2013) Inhibition of arachidonic acid metabolism by the Andean crude drug *Parastrephia lucida* (Meyen) Cabrera. J Ethnopharmacol 150:1080–1086

Echiburu-Chau C, Pastén L, Parra C, Bórquez J, Mocan A, Simirgiotis M (2017) High resolution UHPLC-MS characterization and isolation of main compounds from the antioxidant medicinal plant *Parastrephia lucida* (Meyen). Saudi Pharm J 25:1032–1039

Fernández G (1995) Ofrenda, ritual y terapia: las mesas aymaras. Rev Esp Antropol Am 25:153–180

García G, Gili F, Echeverría J, Belmonte E, Figueroa V (2018) K'oa, andean entity of a plant and other bodies. An interpretative possibility for funerary offerings in the archaeology of Arica. Chungará Rev Antropol Chil 50:537–556
Giberti G (1983) Herbal folk medicine in Northwestern Argentina: Compositae. J Ethnopharmacol 7:321–341
Heredia MJ (2018) Potenciales usos de extractos de plantas que crecen en ambientes extremos aplicados en medicina veterinaria. Thesis. Universidad Nacional de Tucumán. Tucumán. Argentina, 200 pp
IBTA (Instituto Boliviano de Tecnología Agropecuaria) (1995) Efectos del pastoreo en el establecimiento y desarrollo de diez especies vegetales en campos agrícolas de descanso en el altiplano central. (Boletín Técnico no. 35), 14 pp
IIPSC QOLLASUYO (Instituto de investigación, producción, servicios y capacitación "Qollasuyo") (2001a) Manejo de suelos con Tholares. (Boletín no. 1), 12 pp
IIPSC QOLLASUYO (Instituto de investigación, producción, servicios y capacitación "Qollasuyo") (2001b) Producción e instalación de plantones de thola. (Boletín no. 2), 8 pp
IIPSC QOLLASUYO (Instituto de investigación, producción, servicios y capacitación "Qollasuyo") (2002) Informe final subcontrato 21.07: Estudio de la tola y su capacidad de soporte para ovinos y camélidos en el ámbito peruano del sistema TDPS. Puno, Peru, 87 pp
Martínez E (2002) Caracterización morfológica y fitosociológica de las especies de thola en el ámbito de la cuenca Titicaca. Puno, Perú, IIPSC QOLLASUYO, 66 pp
Montero Terri S (2006) Propagación sexual de: *Parastrephia quadrangularis*, y *Baccharis tricuneata*, en San José de Aymará – Huancavelica. Thesis. Universidad Nacional Agraria La Molina, Lima, Peru, 108 pp
Muñoz-Schick M, Moreira-Muñoz A, Moreira Espinoza S (2012) Origen del nombre de los géneros de plantas vasculares nativas de Chile y su representatividad en Chile y el mundo. Gayana Bot 69:309–359
Nieto S, Garrido A, Sanhueza J, Loyola L, Morales G, Leighton F, Valenzuela A (1993) Flavonoids as stabilizers of fish oil: an alternative to synthetic antioxidants. J Am Oil Chem Soc 70:773–778
OECD (2001) Guideline for testing of chemicals: acute oral toxicity-fixed dose procedure, No. 420. Retrieved from https://ntp.niehs.nih.gov/iccvam/suppdocs/feddocs/oecd/oecd_gl420.pdf (2020)
Pérez Mercado R (1994) Comportamiento hídrico-fisiológico y fases fenológicas de la Thola. Thesis. Universidad Técnica de Oruro, Bolivia, 134 pp
Rojo LE, Benites J, Lopez J, Rojas M, Díaz P, Ordóñez J, Pastene E (2009) Antioxidant capacity and polyphenolic content of twelve traditionally used herbal medicinal infusions from the South American Andes. Bol Latinoam Caribe Plant Med Aromát 8:498–508
Romo M, Castro V, Villagrán C, Latorre C (1999) La transición entre las tradiciones de los oasis del desierto y de las quebradas altas del Loa superior: Etnobotánica del Valle del Río Grande, Segunda Región, Chile. Chungará Rev Antrop Chil 31(2):319–360
The Plant List (2019) Version 1.1. Published on the Internet: www.theplantlist.org. (Last access, December 2019)
Torres Carro R, Isla MI, Ríos J, Giner R, Alberto M (2015) Antiinflammatory properties of hydroalcoholic extracts of Argentina Puna plants. Food Res Int 67:230–237
Torres-Carro R, Isla MI, Thomas-Valdes S, Jiménez-Aspee F, Schmeda-Hirschman G, Alberto MR (2017) Inhibition of pro-inflammatory enzymes by medicinal plants from the Argentinean highlands (Puna). J Ethnopharmacol 205:57–68
Torres-Carro R, Yamile A, Rojas-Márquez JD, Stempin CC, Zampini IC, Isla MI, Alberto MR (2019) Argentinean Puna plants with *in vitro* antioxidant and anti-inflammatory activities as a potential nutraceutical. J Food Sci 84:3352–3363
Villagrán C, Castro V (2004) Ciencia Indígena de los Andes del Norte de Chile. Ed. Universitaria, Santiago, Chile, 343 pp

Villagrán C, Romo M, Castro V (2003) Etnobotánica del sur de los Andes de la Primera Región de Chile: un enlace entre las culturas altiplánicas y de las quebradas altas del Loa superior. Chungará Rev Antrop Chil 35:73–124

Weberbauer A (1945) El mundo vegetal de los Andes peruanos. 2nd edn. Universidad Nacional Agraria de La Molina, Lima, Peru, 776 pp

Zampini IC, Mesón Gana M, Ordóñez R, Sayago J, Nieva Moreno M, Isla M, Singh V, Govil J (2008) Antioxidant and xanthine oxidase inhibitory activities of plant species from the Argentine Puna (Antofagasta, Catamarca). Recent Progress in Medicinal Plants. Ed. Studium Press, LLC, USA, 21:89–103

Zampini IC, Cuello S, Alberto MR, Ordóñez RM, D'Almeida R, Solorzano E, Isla MI (2009) Antimicrobial activity of selected plant species from the Argentine Puna against sensitive and multi-resistant bacteria. J Ethnopharmacol 124:499–505

Passiflora caerulea L.

Manuel Minteguiaga, Federico Wallace, Patricia Basile, Fernando Ferreira, Cristina Olivaro, and Eduardo Dellacassa

Passiflora caerulea L. Photo: Darío Niz. With permission by the author

M. Minteguiaga (✉)
Espacio de Ciencia y Tecnología Química, Centro Universitario Regional Noreste/Sede Tacuarembó, Universidad de la República, Tacuarembó, Uruguay

Laboratorio de Biotecnología de Aromas, Departamento de Química Orgánica, Facultad de Química, Universidad de la República, Montevideo, Uruguay
e-mail: manuel.minteguiaga@cut.edu.uy

F. Wallace · C. Olivaro
Espacio de Ciencia y Tecnología Química, Centro Universitario Regional Noreste/Sede Tacuarembó, Universidad de la República, Tacuarembó, Uruguay

P. Basile
Espacio de Biología Vegetal del Noreste, Centro Universitario Regional Noreste/Sede Tacuarembó, Universidad de la República, Tacuarembó, Uruguay

F. Ferreira
Espacio de Ciencia y Tecnología Química, Centro Universitario Regional Noreste/Sede Tacuarembó, Universidad de la República, Tacuarembó, Uruguay

Laboratorio de Carbohidratos y Glicoconjugados, Departamento de Química Orgánica, Facultad de Química, Universidad de la República, Montevideo, Uruguay

E. Dellacassa
Laboratorio de Biotecnología de Aromas, Departamento de Química Orgánica, Facultad de Química, Universidad de la República, Montevideo, Uruguay

Á. Máthé, A. Bandoni (eds.), *Medicinal and Aromatic Plants of South America Vol. 2*, Medicinal and Aromatic Plants of the World 7,
https://doi.org/10.1007/978-3-030-62818-5_32

Abstract *Passiflora caerulea* L. (Passifloraceae) ("mburucuyá", "maracujá-laranja"), a widely distributed species in South America, has been traditionally considered as a medicinal, ornamental, and edible plant. Its fruits, besides being well appreciated for their flavor and sweetness, have valuable nutritional properties. The ethnopharmacological use as anxiolytic, antistress and sedative of the non-reproductive aerial parts (leaves and tendrils) from *P. caerulea* is the main reason for its commercialization as an infusion herb. In general, its preparations are rich in flavonoids (chrysin and *C*-glycosides such as vitexin, isovitexin, orientin, isoorientin, and vicenin-2, among others), which were confirmed as responsible for the anxiolytic effect. The Argentine Pharmacopeia includes a monograph on *P. caerulea*, while other Pharmacopeias (Brazilian, European, and British) consider it as a potential adulterant of *P. incarnata* and *P. edulis*. Harmala alkaloids from *P. caerulea* might also contribute to the anxiolytic effect, but no preclinical assays have been conducted to confirm it. Pharmacological research confirmed other bioactivities associated with the aerial parts of *P. caerulea*, such as antimicrobial, gastroprotective, analgesic, and anticonvulsant. Even though, no systematic toxicological assessments have been conducted in order to evaluate the safety in the consumption, especially regarding the presence of cyanogenic glycosides both in the vegetal drug and the unripe fruits.

Keywords Mburucuyá · Maracujá-laranja · Flavonoids · Anxiolytic · Anticonvulsant

1 Introduction

Passiflora L. is a genus which comprises several worldwide-known tropical medicinal species such as *P. alata*, *P. incarnata*, and *P. edulis*. *P. caerulea* is native to the temperate-Southern American region (Argentina, Southern Bolivia, Southern Brazil, Northern Chile, Paraguay, and Uruguay), being the most distributed *Passiflora* species there. It is currently cultivated in many regions of the world, as an ornamental, but also as an edible plant. *P. caerulea* has also been recognized as a medicinal plant by the Argentine Pharmacopeia. Its activity is associated to its traditional role as anxiolytic and sedative which has been validated, to some extent, through the identification of flavonoids as responsible of such activities. In this chapter we summarize the scientific information available for this plant species, as a basis for both already on-going and prospective investigations into its pharmacological properties.

2 Taxonomic Characteristics

Passiflora L. (Passifloraceae, Dicotyledoneae) is a genus widely distributed in the pantropical regions of the world with approximately 400 species described (Deginani 2001; Vieira and Carneiro 2005; Faleiro et al. 2017). The name *Passiflora* ("flower of the passion") was given by Linnaeus after the common name "pasionaria", which in turn is attributed to the appearance of the flower resembling the passion of Jesus Christ including the thorns' crown (corolla), the nails (stamens), and the Trinity (pistils) (Alonso Paz et al. 2008; Faleiro et al. 2017); but the origin of such association is not completely clear. The distribution of *Passiflora* includes North, Central and South America, South Asia, Africa, and Oceania. However, South America is considered the center of diversity because it bears most of 95% of the species (Vieira and Carneiro 2005; Faleiro et al. 2017).

The main diversity of American *Passiflora* spp. occurs in Central and South America tropics. In particular, in Argentina have been described 19 species (Deginani 2001), while in Brazil this number is approximately 150 species (Faleiro et al. 2017). In Uruguay, where the temperate climate prevails, only six species of this genus have been reported (with special occurrence in the Uruguay river islands, in the border with Argentina): *P. misera* Kunth, *P. chrysophylla* Chodat, *P. elegans* Mast., *P. foetida* L., *P. suberosa* L. and *P. caerulea* L. (Deginani 2001; Alonso Paz et al. 2008). This last species (locally known as "mburucuyá") is the most abundant along with the country (Arechavaleta 1906; Alonso Paz et al. 2008), and it is frequently found over the fences in urban and non-urban environments. Owing to the almost absence of tropical and subtropical climates in Chile, *Passiflora* biodiversity there is scarce, and includes only three native species: *P. caerulea*, *P. foetida*, and *P. pinnatistipula* Cav. (Gay 1846; Muñoz Pizarro 1959; Deginani 2001), this latter strongly threatened and included as a priority species for its conservation (Hechenleitner et al. 2005). Other *Passiflora* spp. have been naturalized and currently cultivated in Chile, among them *P. edulis* Sims and *P. tripartita* (Juss.) Poir. (Díaz et al. 2006; Simirgiotis et al. 2013).

P. caerulea has been extensively employed as a model for physiological and plant protection studies, especially with relation to the flower development, the resistance of its characteristic tendrils, and tolerance to pest attacks. Some examples are the works performed by Brush (1912), Gangstad (1938), Larson (1966), Junker (1976), Drew (1991), Staveley and Wolstenholme (1990), and Grech and Rijkenberg (1991), but the related bibliography is extensive.

According to The Plant List (2020), the following scientific names are recognized as synonyms of *Passiflora caerulea*: *P. caerulea* var. *angustifolia* G. Don, *P. caerulea* var. *glauca* Mast., *P. caerulea* var. *glaucophylla* Loudon, *P. caerulea* var. *imbricata* Mast., *P. caerulea* var. *regnellii* Mast. By contrast, *P. caerulea* var. x *colvillii* (Sweet) Loudon is recognized as a subspecies (The Plant List 2020). In the literature, it can be also found as *P. coerulea*, but this is a grammar error intending to refer to *P. caerulea*.

3 Crude Drug Used

In traditional medicine, the aerial parts, the vegetative plant organs (leaves, stems, and tendrils), but not the flowers of *P. caerulea* are employed (Alonso Paz et al. 2008). Instead, as an edible plant, the fruits are consumed (Photo 1).

The use of *Passiflora* spp. herbs as phytomedicines (tinctures, infusions, and capsules) for stress alleviation and sleep aid is largely known around the world. In general, *P. incarnata* L. is the medicinal species reported in Europe, being its monograph available in the British and European Pharmacopeias. In these, thin layer chromatography (TLC) is used in order to differentiate *P. incarnata* from *P. edulis* Sims and *P. caerulea* L. (European Pharmacopeia 2008; British Pharmacopeia 2009). Even if "passion flower" is not recognized by the United States Pharmacopeia (2020), it is present in the American Herbal Pharmacopeia (Upton et al. 2011). In Latin America, Brazil recognizes, as official the species *P. edulis* and *P. alata* Curtis (Farmacopeia Brasileira 2019); while in Argentina *P. caerulea* L. is the accepted medicinal species (Farmacopea Argentina 2018). According to this latter document, the official herb is defined as the desiccated aerial parts collected at full-flowering period, with a total flavonoids content higher than 1.5% expressed as isovitexin. The chemical identification of the vegetal drug is performed by TLC using isoorientin (homoorientin) and isovitexin as standards to differentiate this material from that of other related species, as *P. incarnata*; while the quantification is performed by a spectrophotometric method using boric and oxalic acids (Farmacopea Argentina 2018). However, *Passiflora* spp. officinal drugs can easily be found altered in the market by the addition of strange vegetal material, determining the need of adopting as mandatory, additional and more specific chemical quality control procedures (Pereira et al. 2004; Hajimehdipoor et al. 2007).

Photo 1 Aerial parts of *P. caerulea* L. (leaves, stems, and tendrils) (left). The edible fruit of the species (right). (Photos: Darío Niz, with permission)

Passiflora genus contains almost 50 edible plant species, with *P. edulis* Sims ("maracujá-amarelo", passionfruit) as the most important. In the international market it is popular due to its valuable fruits (Vieira and Carneiro 2005; Faleiro et al. 2017). "Passion-fruit" is mainly produced in Brazil (with more than 44,000 ha dedicated to its production), being consumed fresh or as highly appreciated sweet preparations such juices, candies, yogurts, cakes, ice-creams, and jams; or incorporated in salty cooking recipes (Vieira and Carneiro 2005; Faleiro et al. 2017). Special caution must be taken when collecting the *Passiflora* spp. fruits *in-natura* for their consumption because immature fruits are quite toxic by their cyanogenic glycosides content (Siegler et al. 1982).

4 Major Chemical Constituents and Bioactive Compounds

Bandoni et al. (1972) performed a qualitative phytochemical screening on *P. caerulea* aerial parts and wood from Argentina through typical reactions for different natural product groups. They found positive results for the presence of leucoanthocyanidins and traces of flavonoids, triterpenoids, saponins and alkaloids in the aerial parts. The wood contained alkaloids and cyanogenic glycosides, as well as traces of leucoanthocyanidins and triterpenoids (water-soluble and insoluble). Şesan et al. (2016) obtained positive results for glycosides and triterpenoids from aerial parts of *P. caerulea* growing at greenhouse conditions in Romania, but the clearest results were the presence of saponins (foam test) in the ethanol extract and the absence of tannins and alkaloids. Conversely, El-Askary et al. (2017) performing a phytochemical screening on the leaves (from a *P. caerulea* cultivated in Egypt) found traces of saponins and coumarins, and positive results for flavonoids, carbohydrates/glycosides, sterols/triterpenes, and tannins. Birk et al. (2005) using a TLC fingerprinting on several Brazilian *Passiflora* spp. (including *P. caerulea*) found saponins only in *P. alata* ethanol extract. Moreover, the authors differentiate related species by differential TLC flavonoid patterns (Birk et al. 2005). These preliminary data suggest that this species could display different metabolic expressions in different environmental situations. A common characteristic on plant secondary metabolism expression; although incidences and biases derived from the different screening/analytical methodologies applied cannot be discarded. Meaning that controversies and debate about *P. incarnata* attributed sedative and anxiolytic active principles (benzoflavone-related) could be explained for secondary metabolism variations (Dhawan et al. 2001, 2002; Holbik et al. 2010).

Flavonoids represent one of the groups of secondary metabolites responsible for the anxiolytic activity of the *P. caerulea* aqueous infusions. In particular, chrysin (5,7-dihydroxyflavone) displayed good activity as a ligand compound, acting on the central and peripheral benzodiazepine receptors in the nervous system, as well as anticonvulsant (Medina et al. 1990; Wolfman et al. 1994). The occurrence of *C*-glycosidated flavonoids is frequent in *P. caerulea* aerial parts: apigenin-8-*C*-glucoside (vitexin), apigenin-6-*C*-glucoside (isovitexin), vitexin-2″-*O*-glucoside, vitexin-6″-*O*-glucoside, isovitexin-8-*C*-arabinoside-7-*O*-glucoside, apigenin-6-*C*-glucoside-8-*C*-arabinoside (schaftoside), apigenin-6,8-di-*C*-glucoside (vicenin-2), apigenin-6-*C*-rhamnoside-8-*C*-arabinoside, luteolin-8-*C*-glucoside (orientin), luteolin-6-*C*-glucoside (isoorientin), luteolin-6,8-di-*C*-glucoside (lucenin-2), and chrysin-6-*C*-β-D-glucoside, among others (Pastene et al. 2000; Pereira et al. 2004, 2005; Hajimehdipoor et al. 2007; Anzoise et al. 2016; Farag et al. 2016; El-Askary et al. 2017; Ozarowski et al. 2018). *C*-glycosylated flavonoids can be considered as chemo-markers from *P. caerulea*, and even from the entire *Passiflora* genus. The presence in the aerial parts of other glycosylated flavonoids (no *C*-glycosides) derivatives of luteolin, quercetin, apigenin, myricetin, hesperidin, and chrysin have also been reported (Anzoise et al. 2016; Farag et al. 2016; Ozarowski et al. 2018; Aseervatham et al. 2020). Several analytical methods have been suggested for the screening, identification and quantification of these *C*-glycosylated flavonoids as an alternative to the pharmacopeial spectrophotometric methods, including HPTLC, HPLC-PDA (Pereira et al. 2004; El-Askary et al. 2017), HPLC-MS and CID-MS/MS (Pereira et al. 2005), and capillary electrophoresis (CE) (Pastene et al. 2000). Pereira et al. (2004) demonstrated, by HPLC-PDA, that the orientin, isoorientin and vitexin contents are higher in *P. caerulea* and *P. incarnata* leaves than in the related species *P. edulis* and *P. alata*. Bussilacchi et al. (2008) did not find qualitative differences in a TLC screening of flavonoid profiles between the *in-vitro* micropropagated *P. caerulea* plants and their parental.

The presence of cyclopentenoid cyanogenic glycosides in the leaves of several *Passiflora* spp. (including *P. caerulea*) was reported by Tantisewie et al. (1969), being gynocardin and deidaclin (epimer of tetraphyllin A) the main compounds found. However, later Siegler et al. (1982), working with a leaves methanolic extract, did not find gynocardin nor deidaclin, instead, they identified two novel products: tetraphylline B and *epi*-tetraphyllin B sulfates in an epimeric mixture. These molecules are composed by a cyclopentenoid moiety, where C_1 is linked both to a cyano group and to a glucose unit by a glycosidic linkage; while the sulfate is attached to C_4 (Siegler et al. 1982). In a screening of aerial part extracts from several *Passiflora* spp. (including *P. caerulea*) conducted by TLC and HPLC, Echeverri et al. (2001) identified the cyanogenic glucoside prunasin (mandelonitrile-β-D-glucoside) in all of them; but conversely, passifloricins (α-pyrone polyketides) were exclusives from *P. foetida*.

Harmala alkaloids (β-carboline/indole derivatives) are typical trace-level secondary metabolites identified in many *Passiflora* spp. (Dhawan et al. 2004). Harmine, harmane, harmol and several not identified trace-level harmala alkaloids were isolated and structurally identified in *P. caerulea* by Ambühl (1966). Even though, in a more recent phytochemical research, only one of them (harmine) was found and quantified by HPLC in a sample of aerial parts of the species (Frye and Haustein 2007). These alkaloids are generally associated with anxiolytic and antidepressant activities (Aricioglu and Altunbas 2003), suggesting, to some extent, an additional contribution to the flavonoid's activity.

Patterson et al. (1978) studied the polar lipid (phospholipids) composition of *Passiflora* spp. (including *P. caerulea* as a model of cold tolerance) and their relationship with chill-resistance as a mechanism of adaptation, exploiting the existing temperature differences ranges from the habitats where such species develop and grow. *P. caerulea* displayed a slight, but not significant, trend to contain higher unsaturated phospholipid levels (in particular, the phosphatidyl glycerol fraction) than *P. flavicarpa* (the chill-sensitive model species), which presumably conferred to the former more tolerance to the frosts compared to other related species. Even though, the differences in *Passiflora* spp. chill-resistance appears to be more related to physical differences in the membranes than in the degree of lipid unsaturation (Patterson et al. 1978).

Studies by high-resolution LC-MS allowed to describe other miscellaneous secondary metabolites for *P. caerulea* aerial part extracts. The list includes: amides (palmitamide, oleamide, stearamide, docosenamide) and acidic derivatives (4-hydroxycinnamic, *p*-coumaric, caffeic, dihyroxybenzoic, feruloylquinic) (Farag et al. 2016; El-Askary et al. 2017; Ozarowski et al. 2018).

The fresh and pasteurized juice of *P. caerulea* fruits was studied from the chemical and nutritional point of view by dos Reis et al. (2018), who did not find differences in parameters such total soluble solids and acidity between both juices upon the same storage time, while low variations in color were evidenced. The quantified flavonoids (*epi*-gallocatechin gallate and quercetin) decreased their concentration in both juices as a consequences of the storage (with a simultaneous diminishing in the antioxidant capacity), while for carotenoid concentration (provitamin A, lycopene, lutein, zeaxanthin, β-cryptoxanthin, α- and β-carotene) variations related to such conditions were also evidenced. Other secondary metabolites have been identified by LC-MS in the fruit pulp as naringenin, hesperidin, *epi*-catechin, amentoflavone (a bi-flavonoid), chrysoeriol-8-*C*-glucoside, luteolin-6-*C*-glucoside, 7-*O*-methyl-*iso*-rhamnetin-3-*O*-glucoside, apigenin-6,8-di-*C*-β-D-glucopyranoside, protocatechuic acid, caffeic acid, chlorogenic acid, and a steroidal glycoside (ginsenoside), among others (Aseervatham et al. 2020).

The oil of the *P. caerulea* seeds obtained by Soxhlet extraction using *n*-hexane (29.9% yield) is composed mainly by the following three fatty acids: palmitic, 10.1%, oleic, 17.6% and linoleic, 63.1% (Quiroga et al. 2000). Other specific indexes of fats and oils were also determined to characterize such no-traditional fatty products; the authors also evaluated some nutritional parameters in the residual from seed extraction (meal) (Quiroga et al. 2000). The oil of some *Passiflora* spp. are incipiently employed in Brazil for food and cosmetic purposes (Faleiro et al. 2017).

5 Morphological Description

P. caerulea was described originally by Carl Linnaeus (1753) in the Brazilian Flora, focusing exclusively on its floral and leaves characters. The name *P. caerulea* (literally, the blue *Passiflora* in Latin) was given for the color of the flowers (Linnaeus 1753; Deginani 2001).

P. caerulea is a woody vine, profusely ramified with persistent foliage. The stems are glabrous or subglabrous with simple tendrils (Deginani 2001; Lombardo 1982; Ross et al. 2018). Its leaves are simple, alternate, glabrous, palmate with five lanceolate lobes, 2–13 cm length; intensely green in the adaxial face and glaucous in the abaxial side, exhibiting an entire margin (Deginani 2001; Ross et al. 2018; Álvarez 2019). Nevertheless, *P. caerulea* presents a high variability in the morphology and number of leaves' lobes, from elliptical to linear, with thickened edges and the central lobe being generally longer than the lateral ones (Deginani 2001; Şesan et al. 2016). Regarding petiole (2–4 cm length), it has nectaries, some glands, and reniform stipules (Ross et al. 2018; Álvarez 2019). In some ecotypes, the stipules can be absent. Usually, younger specimens' stems, tendrils, and petioles have a characteristic purple-greenish color (Deginani 2001). A detailed morphoanatomic study of the leaves, stems and tendrils was performed by Şesan et al. (2016).

The emblematic *P. caerulea* flowers are large (5–10 cm in diameter), solitary and hermaphrodite; with a greenish penta-sepal calyx and a greenish penta-petal corolla having five free or fused stamens. Furthermore, the crown has many filaments: purple at the base, blue at the apex, and white in the middle; nectaries are also present (Lombardo 1982; Deginani 2001; Ross et al. 2018; Álvarez 2019). The flowers produce visual and olfactory attractant stimuli (clues) to their insect pollinators, which must be displayed together to increase the number of visitors and ensure the pollination success (Varela et al. 2016).

The species blooms and fruits at the maximum expression by the Spring and the Autumn (September–May in the Southern Hemisphere), but different phenology patterns can be found in the entire area of occurrence (Deginani 2001; Ross et al. 2018; Álvarez 2019). The flowers remain open for approximately two days (both during the day and night) and have a strong pyrethrum scent (Deginani 2001). The fruits are oblong berries (Photo 1), 5–7 cm length, outside orange-colored, with an edible red endocarp and many perfumed seeds. In Uruguay, the fruit ripens during the Summer (Lombardo 1982; Ross et al. 2018).

6 Geographical Distribution

P. caerulea is largely distributed across the South American Continent, covering different tropical to subtropical phytogeographical regions. It is native from Brazil to central Argentina including Paraguay, Uruguay, Bolivia and Northern regions of

Chile (Lombardo 1982; Deginiani 2001; Mondin et al. 2011; Aquino and Garcia 2019).

It is also marginally cultivated in some regions as an edible and ornamental plant. Besides, it serves as cold, salinity, soil humidity and pest-tolerant rootstock for other *Passiflora* fruit-species (especially *P. edulis*) (Staveley and Wolstenholme 1990; Grech and Rijkenberg 1991). For all these reasons *P. caerulea* is incipiently cultivated in Argentina (Bussilacchi et al. 2008), Brazil (Vieira and Carneiro 2005), Egypt (El-Askary et al. 2017), India (Santhoshkumar et al. 2019), Iran (Hajimehdipoor et al. 2007), Mexico, Bermudas, and Guiana (dos Reis et al. 2018), Romania (Şesan et al. 2016), South-Africa (Staveley and Wolstenholme 1990; Grech and Rijkenberg 1991), and the United States (Seigler et al. 1982). In Uruguay, *P. caerulea* is only cultivated at a domestic level.

P. caerulea is propagated by seeds and cuttings, and several experiences on germinative or vegetative propagation are recorded in the literature (Mendiondo and García 2006, 2009; Bussilacchi et al. 2008). Aguirre et al. (2018) suggest collecting seeds from healthy and large ripe fruits, wash and dry them under the shade, and sow after 24–48 h; the germination takes place between 20–25 days. According to Mendiondo and García (2009), this species is orthodox and have physical dormancy as it presents low levels of seed emergence, but the best results in germination percentage are achieved when employing desiccated stored seeds (without arils).

7 Ecological Requirements

Passiflora caerulea develops successfully in xerophytic and wet forests, grasslands as well as in the edge of forests and jungles; and it is tolerant to chill (Patterson et al. 1978; Deginiani 2001; Vieira and Carneiro 2005; Ross et al. 2018). *P. caerulea* vines grow well on clayey, rocky and modified sandy soils, up to 1400 m. It is one of the most typical ruderal species in the region. As a climbing species, it spreads over the vegetation (frequently associated with *Acacia* spp. and *Prosopis* spp.) and wire fences, both in rural and urban areas (Deginiani 2001; Aguirre et al. 2018; Ross et al. 2018).

Regarding biotic interactions, as for other *Passiflora* spp., *Heliconius* spp. butterflies (Lepidoptera) lay their eggs on the *P. caerulea* vines. After hatching, the larvae feeding on the vegetal tissues, but the plant counteracts using different strategies: (1) synthesizing toxic cyanogenic glycosides, (2) altering the leaves morphology, and (3) "sequestering" nectar-rewarded ants that remove the larvae (Siegler et al. 1982; Vieira and Carneiro 2005). When the plant is consumed and crashed by the larvae, cyanide is liberated as a defense mechanism against herbivory (Siegler et al. 1982).

8 Traditional Use (Part(s) Used) and Common Knowledge

P. caerulea receives different vernacular names according to the geographical places where it develops, such: "mburucuyá", "maracuyá", "burucuyá", "pasiflora", "flor de la pasión", "granadilla" or "pasionaria" in the Spanish speakers countries (Bandoni et al. 1972; Ajmat et al. 2003; Alonso Paz et al. 2008; Anzoise et al. 2016). In Brazil, this species is known as "maracujá-laranja" and "maracujá-do-mato" (dos Reis et al. 2018). The infusion of the aerial parts (leaves and stems, rarely flowers) at 1% concentration is employed mainly in the folk medicine as anxiolytic and sedative (Medina et al. 1990; Alonso Paz et al. 2008), but there are also ethnobotanical screening reports on digestive, anti-inflammatory, anthelmintic, diuretic, emmenagogue, antiscorbutic, and anti-icteric usages (Berro 1899; Bandoni et al. 1972; Sorarú and Bandoni 1978; Toursarkissian 1980; Ratera and Ratera 1980; Alonso and Desmarchelier 2005). Even though, the root infusion is also cited for the pneumonia treatment and as anthelmintic (Berro 1899; Bandoni et al. 1972; El-Askary et al. 2017). Besides being edible, the *P. caerulea* fruits can be also consumed as sweet pulp jams (Berro 1899). In Uruguay, this species is employed also as ornamental plant (Vieira and Carneiro 2005; Ross et al. 2018).

9 Modern Medicine Based on Its Traditional Medicine Uses

The pharmacological and phytochemical studies on *P. caerulea* allowed the description of many bioactivities for the extracts and its isolated metabolites. One of the first modern bioactivity research has been conducted by Nicolls et al. (1970, 1973) and Birner and Nicolls et al. (1973), who described the antimicrobial activity of several *Passiflora* spp., being *P. caerulea* the most active of them. In a first approach, Nicolls (1970) assayed the *Passiflora* spp. phytopathogen fungal resistance through the incubation of the fungi on potato dextrose agar in the presence of round pieces of green vegetal tissues (simple diffusion method), measuring afterward the inhibition halos' sizes. Then, the authors extracted the plant material by aqueous/organic solvents, repeating the growth assessment with many fungal and bacterial strains from clinical interest, assayed their toxicity with mice and rabbits, and finally performed a series of physicochemical analysis in an attempt to determine the chemical nature of the inhibiting compound/s (Nicolls 1970; Nicolls et al. 1973; Birner and Nicolls 1973). Despite their efforts, none structure was determined, but they identified a biological active fraction as composed by labile polyacetylene metabolites, and named it passicol (obtained from *P. mollisima* but with an UV-spectra similar to the corresponding fraction from *P. caerulea*) (Nicolls et al. 1973; Birner and Nicolls 1973). The authors hypothesized that this labile compound/s might be associated with the wounding of *Passiflora* spp. organs, and it could play a key role in the inhibition of infection and colonization by phytopathogenic fungi (Nicolls 1970; Nicolls et al. 1973; Birner and Nicolls 1973). No modern phytochemical studies have been performed so far for identifying the chemical structure of passicol. Despite that, no antimicrobial activity was observed for Argentinean *P. caerulea* aerial part infusions against *Aspergillus niger*, *Escherichia coli*, and *Staphylococcus aureus* (Anesini and Pérez 1993). In a recently published work, Santhoshkumar et al. (2019) synthesized silver nanoparticles (Ag-NP) enclosing *P. caerulea* leaves infusions as a technological solution for delivering the active secondary metabolites in a programmed way. These nanoparticles were physicochemically characterized and their activity was confirmed against dermatophytes fungi: *Trychophyton rubrum*, *T. metagrophytes*, *Epidermophyton floccosum*, *Microsporum audouinii*, and *M. canis*; but dose-dependent toxicity was evidenced against zebrafish model (*Danio rerio* Hamilton-Buchanan) (Santhoshkumar et al. 2019).

Medina et al. (1990) confirmed the anxiolytic effects of the flavone chrysin isolated from the aerial parts of *P. caerulea* through *in vitro* assays (radioligand binding benzodiazepine receptors models); and its anticonvulsant potential by an *in vivo* experiment after intracerebral-ventricular administration to adult mice, inhibiting the pentylenetetrazol seizure action. Moreover, Wolfman et al. (1994) demonstrated with mice that chrysin increases the number of entries and the time spent in the elevated plus-maze test of anxiety, consistent with anxiolytic action (without exerting sedation or muscle relaxation) like the commercial drug diazepam. These results are important because chrysin induced reduction of anxiety without depressing the central nervous system as is the case for commercial benzodiazepines (Medina et al. 1990; Wolfman et al. 1994). Chrysin has also been reported as virility enhancer in sexual mature male rats, since after its oral administration (1.0 mg/kg) an increase in the coupling behavior, sperm count, and fertilization rate was recorded (Dhawan et al. 2002); indicating the aphrodisiac potential of *P. caerulea* extracts. The mechanism under this effect could be related to the aromatase inhibition which is responsible for the conversion of androgens to estrogens (Dhawan et al. 2002).

Infusions obtained from commercial standardized capsules of MELIPASS® [a flavonoid-rich phytotherapeutic based on *Melissa officinalis* L. and *P. caerulea*] were evaluated as physiological stress decreasers (Feliú-Hemmelman et al. 2013). The aim of this study was to determine the effect of such infusions on the severity of physiological chronic stress induced by movement restriction, employing CF-1 mice. The infusion treated animals had lower plasma corticosterone levels compared to the control, being this metabolite the most important biomarker associated with physiological stress. In the end, these results demonstrated *in vivo* that the consumption of MELIPASS® infusion diminished the chronic stress (Feliú-Hemmelman et al. 2013). Moreover, Keck et al. (2020) reported a diminishing in the number of benzodiazepine prescriptions for hospitalized psychiatric patients after the usage of commercial phytotherapeutic tablets (Ze 185, Relaxane®) based on of *Petasites hybridus* (L.) Gaertn., B. Mey. & Scherb ("butterbur"), *Valeriana officinalis* L. ("valerian"), *M. officinalis* ("lemon balm"), and *P. incarnata*; all containing the flavone isovitexin (common also in *P. caerulea*) as one of the main components.

El-Askary et al. (2017) performed an extensive pharmacological battery of preclinical assays using male albino Sprague Dawley rats, extracting DNA authenticated *P. caerulea* leaves by water and ethanol. Both crude extracts were active in all the performed assays in some extension, with emphasis on the ethanol extract as an anticonvulsant, anti-inflammatory, and analgesic. Bio-guided fractionation of the ethanol extract allowed the isolation of the flavonoids lucenin-2 and chrysin-6-*C*-β-D-glucoside, besides 4-hydroxycinnamic acid as the putative bioactive compounds (El-Askary et al. 2017). Zarei et al. (2014) confirmed through *in vivo* studies the potential of *P. caerulea* methanol extracts as antinociceptives and analgesics.

The role of *P. caerulea* as gastroprotective has been highlighted by Anzoise et al. (2016) studying the effect of the ethanol extract on colitic rodents by applying an experimental model related to inflammatory bowel disease. The phytochemical analysis of the extract demonstrated the presence of the *C*-glycosylated flavonoids isoorientin, vitexin, isovitexin, and vicenin-2. Moreover, this study showed that *P. caerulea* extracts possessed anti-inflammatory, antidiarrheal, spasmolytic, and antioxidant effects, but the mechanisms of action involved remain to be elucidated (Anzoise et al. 2016). The biological activities found by these authors are in accordance with the popular medicinal use of the species infusions against different pathologies associated with the intestinal tract (Toursarkissian 1980; Ratera and Ratera 1980; Alonso and Desmarchelier 2005) and its use as an anti-inflammatory agent (Soraru and Bandoni 1978). In addition, commercial capsules containing desiccated leaves' fragments of *M. officinalis* and *P. caerulea* (MELIPASS®) are indicated for the treatment of irritable bowel syndrome (KNOP 2020).

The *in vitro* cytotoxic potential on leukemia cell lines was assessed by Ozarowski et al. (2018). These authors compared the activity profiles of three *Passiflora* species (*P. alata*, *P. incarnata*, and *P. caerulea*; all of them obtained under controlled greenhouse conditions) leaves extracts and correlated them with their secondary metabolite composition (mainly *C*-glysosylated flavonoids and C_{13} norisoprenoids such blumenols). The data obtained showed inhibitory activity against human acute lymphoblastic leukemia CCRF-CEM cells for *P. alata* and *P. incarnata* extracts. However, extracts from *P. caerulea* did not show any activity (Ozarowski et al. 2018). The latter occurs, despite the qualitative phytochemical similarities between the profiles obtained from *P. incarnata* and *P. caerulea*, a fact also corroborated by Farag et al. (2016). Thus, differences in the concentration of secondary metabolites between these two extracts can determine a different bioactivity profile, specifically regarding the antileukemic activity. Şesan et al. (2016) found that *P. caerulea* leaves extract does not inhibit *in vitro* the fibroblast growth (NCTC L929 cell line) up to 150µg/ml after 48 h of evaluation, as well as no morphological alterations nor density changes were recorded by microscopy. Thus, inferring the absence of cytotoxicity against normal cells up to such concentration (the opposite is true at >250µg/ml where the damage on the cell membranes was observed) (Şesan et al. 2016).

Antioxidant activity from aerial parts and fruits extracts of *P. caerulea* was also confirmed by several authors applying different evaluation protocols (DPPH, ABTS, glutathione level in blood), which is attributed to their high polyphenolic and flavonoid contents (Şesan et al. 2016; El-Askary et al. 2017; dos Reis et al. 2018).

Recently, Aseervatham et al. (2020) studied the role of *P. caerulea* fruits aqueous extracts as anticonvulsant treating pilocarpine-induced epileptic mice. Likewise, cognitive functions and oxidative stress were studied. The results showed anticonvulsant activity, improving the cognitive function as well, and provoking an oxidative damage reduction; which suggests the therapeutic potential of the extracts to treat epilepsy and neurodegeneration. Additionally, *in vitro* antioxidant activity was assessed, demonstrating strong response in agreement with the presence of phenolic and flavonoid compounds in the extract as its main components (Aseervatham et al. 2020).

Regarding *P. caerulea* toxicity, there were tested infusions of aerial parts in mice, orally and parenterally, at doses up to 40 times higher than those commonly used in an infusion, and *P. caerulea* showed no toxic effects nor killed animals (Álvarez et al. 1990). Other work has been published (El-Askary et al. 2017) in which the authors estimated the $LD_{50} > 5.0$ g/kg body weight from leaves aqueous/ethanolic extracts.

10 Conclusions

P. caerulea is a both edible and medicinal plant species widely distributed along South America. Incipiently it is cultivated in other regions of the world. Modern research is discovering interesting nutraceutical ingredients for the consumption of *P. caerulea* fresh fruits, its juices, and seeds oils. From a medicinal point of view, the vegetative aerial parts are traditionally employed for infusion preparations due to its sedative, anxiolytic, anti-inflammatory, diuretic, and digestive properties; these have been recognized in the Argentine Pharmacopeia as a substitute of the well-known *Passiflorae herb* (*P. incarnata*). Some commercial formulations (capsules) based on *P. caerulea* are available in the market to treat stress and depression, most of the time combining several vegetal drugs. Phytochemical studies performed on *P. caerulea* were able to identify *C*-glycosylated flavonoids (vitexin, orientin and its isomers, among others) together with cyanogenic glycosides, harmala alkaloids and diverse phospholipids. Pharmacological research also confirmed *in vivo* anxiolytic and antistress properties, in addition to antimicrobial, anticonvulsant, analgesic, and gastroprotective effects on animals. However, no systematic toxicological assessments have been performed in order to ensure the consumer safety. Similarly, no data are available to support how *P. caerulea* metabolites interact with other drugs when commercialized together in various formulas.

Acknowledgments Darío Niz, for the photographs included in this chapter.

References

Aguirre C, Perondi H, Buono S, Abdo G, Ansonnaud G (2018) Mburucuyá: Granadilla/ pasionaria, *Passiflora caerulea* L. PROCISUR-IICA. San José, Costa Rica, 13 pp. Published on the Internet: http://www.procisur.org.uy/adjuntos/e90c938ee72e_urucuya-PROCISUR.pdf. (Last access, February 2020).

Ajmat M, Bado S, Coviella M, Pannunzio M (2003) Aspectos morfológicos, biológicos y daño de *Gargaphia lunulata* (Mayr) 1865 (*Heteroptera:Tingidae*) sobre *Passiflora caerulea* L. (Passifloraceae). Bol Sanid Veg Plagas 29(3):339–346

Alonso J, Desmarchelier C (2005) Plantas medicinales autóctonas de la Argentina. Bases científicas para su aplicación en atención primaria de la salud. Ed. LOLA. Buenos Aires, Argentina, pp 409–416

Alonso-Paz E, Bassagoda MJ, Ferreira F (2008) Yuyos. Uso racional de las plantas medicinales. 2ª edition. Ed. Fin de Siglo. Montevideo, Uruguay, 222 pp

Álvarez MJ (2019) Pharmacological properties of native plants from Argentina. Ed. Springer Nature Switzerland AG, Cham, Switzerland, 255 pp

Álvarez R, Del Alamo R, Saldivia M, Dell'Orso M, Rondina RVD (1990) Ensayo de Viabilidad de Altas Dosis de Extractos de Plantas Medicinales Argentinas en Animales de Laboratorio. Acta Farm Bonaer 9(1):29–39

Ambühl H (1966) Anatomische und chemische Untersuchungen an *Passiflora coerulea* L. und *Passiflora incarnata* L. ein Beitrag zur Analytik einiger Harmanalkaloide und ihrer Verbreitung im Pflanzenreich. Thesis. ETH Zürich, Switzerland. Juris Druck, Verlag Z Zürich. 3 pp (Abstract). https://doi.org/10.3929/ethz-a-000115993

Anesini C, Pérez C (1993) Screening of plants used in Argentine folk medicine for antimicrobial activity. J Ethnopharmacol 39(2):119–128

Anzoise ML, Marrassini C, Bach H, Gorzalczany S (2016) Beneficial properties of *Passiflora caerulea* on experimental colitis. J Ethnoparmacol 194:137–145

Aquino DS, García MTA (2019) Pollen dispersal in a population of *Passiflora caerulea*: spatial components and ecological implications. Plant Ecol 220(9):845–860

Arechavaleta J (1906) Flora Uruguaya. Ed. Talleres Barreiro y Ramos, Montevideo, Uruguay. Volume II, pp. 122–125. In: facsimile: https://autores.uy/ (Last access, February 2020)

Aricioglu F, Altunbas H (2003) Harmane induces anxiolysis and antidepressant-like effects in rats. Ann N Y Acad Sci 1009(1):196–201

Aseervatham GSB, Abbirami E, Sivasudha T, Ruckmani K (2020) *Passiflora caerulea* L. fruit extract and its metabolites ameliorate epileptic seizure, cognitive deficit and oxidative stress in pilocarpine-induced epileptic mice. Metab Brain Dis 35(1):159–173

Bandoni AL, Mendiondo ME, Rondina RV, Coussio JD (1972) Survey of Argentine medicinal plants. I. Folklore and phytochemical screening. Lloydia 35(1):69–80

Berro MB (1899) La vegetación Uruguaya. Plantas que se hacen distinguir por alguna propiedad útil ó perjudicial. Anales del Museo Nacional de Montevideo, Vol 11(2):59. Ed. Establecimiento Tipo-Litográfico Oriental. Montevideo, Uruguay. In: facsimile: https://autores.uy/. (Last access, February 2020)

Birk CD, Provensi G, Gosmann G, Reginatto FH, Schenkel EP (2005) TLC fingerprint of flavonoids and saponins from *Passiflora* species. J Liq Chromatogr Relat Technol 28(14):2285–2291

Birner J, Nicolls JM (1973) Passicol, an antibacterial and antifungal agent produced by *Passiflora* plant species: preparation and physicochemical characteristics. Antimicrob Agents Chemother 3(1):105–109

British Pharmacopeia (2009) Passion Flower. Volume III, Herbal Drugs and Herbal Drugs Preparations. (Ph Eur monograph 1459), p 3

Brush WD (1912) The formation of mechanical tissue in the tendrils of *Passiflora caerulea* as influenced by tension and contact. Bot Gaz 53(6):453–477

Busilacchi H, Severin C, Gattuso M, Aguirre A, Di Sapio O, Gattuso S (2008) Field culture of micropropagated *Passiflora caerulea* L.: histological and chemical studies. Bol Latinoam Caribe Plant Med Aromáticas 7(5):257–263

Deginani NB (2001) Las especies argentinas del género *Passiflora* (Passifloraceae). Darwiniana 39(1–2):43–129

Dhawan K, Kumar S, Sharma A (2001) Anti-anxiety studies on extracts of *Passiflora incarnata* Linneaus. J Ethnopharmacol 78(2–3):165–170

Dhawan K, Kumar S, Sharma A (2002) Beneficial effects of chrysin and benzoflavone on virility in 2-year-old male rats. J Med Food 5(1):43–48

Dhawan K, Dhawan S, Sharma A (2004) *Passiflora*: a review update. J Ethnoparmacol 94(1):1–23

Díaz LS, Padilla C, Sepúlveda C (2006) Identificación del Principal Pigmento Presente en la Cáscara del Maracuyá Púrpura (*Passiflora edulis*). Inf Tecnol 17(6):75–84

dos Reis LCR, Facco EMP, Flôres SM, Rios ADO (2018) Stability of functional compounds and antioxidant activity of fresh and pasteurized orange passion fruit (*Passiflora caerulea*) during cold storage. Food Res Int 106:481–486

Drew RA (1991) *In vitro* culture of adult and juvenile bud explants of *Passiflora* species. Plant Cell Tiss Org Cult 26(1):23–27

Echeverri F, Arango V, Quiñones W, Torres F, Escobar G, Rosero Y, Archbold R (2001) Passifloricins, polyketides α-pyrones from *Passiflora foetida* resin. Phytochemistry 56(8):881–885

El-Askary HI, Haggag MY, Abou-Hussein DR, Hussein SM, Sleem AA (2017) Bioactivity-guided study of *Passiflora caerulea* L. leaf extracts. Iran J Pharm Res 16:46–57

European Pharmacopeia (2008) Passion flower. *Passiflorae Herba*. Version 6.0., Volume 1. EDQM, Council of Europe. 01/2005:1459, pp 2192–2193

Faleiro FG, Junqueira NTV, Costa AM, de Jesus ON, Machado CDF (2017) *Maracujá. Passiflora* spp. PROCISUR-IICA. San José, Costa Rica. 31 pp. Published on the Internet: http://www.procisur.org.uy/adjuntos/procisur_maracuja_506.pdf (Last access, February 2020)

Farag MA, Otify A, Porzel A, Michel CG, Elsayed A, Wessjohann LA (2016) Comparative metabolite profiling and fingerprinting of genus *Passiflora* leaves using a multiplex approach of UPLC-MS and NMR analyzed by chemometric tools. Anal Bioanal Chem 408(12):3125–3143

Farmacopea Argentina (2018) *Pasionaria, hierba*. 7ª Ed., 1st Supp. ANMAT. Buenos Aires, Argentina, pp 214-217. Published on the Internet: https://www.boletinoficial.gob.ar/detalleAviso/primera/214347/20190827 (Last access, February 2020)

Farmacopeia Brasileira (2019) *Maracujá-azedo, folha. Maracujá-doce, folha*. 6ª Ed., 2nd Volume, *Plantas Medicinais*. ANVISA. Brasilia, Brazil. PM061–00 & PM062–00. Published on the Internet: http://portal.anvisa.gov.br/farmacopeia-brasileira (Last access, February 2020)

Feliú-Hemmelmann K, Monsalve F, Rivera C (2013) *Melissa officinalis* and *Passiflora caerulea* infusion as physiological stress decreaser. Int J Clin Exp Med 6(6):444–451

Frye A, Haustein C (2007) Extraction, identification, and quantification of harmala alkaloids in three species of *Passiflora*. Am J Undergrad Res 6(3):19–26

Gangstad VB (1938) A morphological study of the leaf and tendril of *Passiflora caerulea*. Am Midl Nat 20(3):704–708

Gay C (1846) Historia Física y Política de Chile. Vol. II Botánica-Flora Chilena. Ed. Imprenta Fain & Thunot. Paris, France, pp 414–417

Grech NM, Rijkenberg FHJ (1991) Laboratory and field evaluation of the performance of *Passiflora caerulea* as a rootstock tolerant to certain fungal root pathogens. J Hortic Sci 66(6):725–729

Hajimehdipoor H, Rezazadeh S, Hadjiakhoondi A, Hamedani MP, Rezaee A, Zahedi H, Khandani NK (2007) Comparative and differential study of *Passiflora incarnata* L. and *Passiflora caerulea* L. in raw plants and pharmaceutical dosage forms. J Med Plants (ACECR) 3(23):65–71. Published on the Internet: http://jmp.ir/article-1-592-en.html. (Last access, February 2020)

Hechenleitner VP, Gardner MF, Thomas PI, Echeverría C, Escobar B, Brownless P, Martínez CA (2005) Plantas amenazadas del Centro-Sur de Chile. Distribución, Conservación y Propagación. Ed. Universidad Austral de Chile & Real Jardín Botánico de Edimburgo. Santiago, Chile,

188 pp. In facsimile: http://dspace.utalca.cl/bitstream/1950/10294/1/Hechenleitner%20%20V.pdf. (Last access, February 2020)
Holbik M, Krasteva S, Mayer N, Kählig H, Krenn L (2010) Apparently no sedative Benzoflavone moiety in *Passiflorae* Herba. Planta Med 76(7):662–664
Junker S (1976) Auxin transport in tendril segments of *Passiflora caerulea*. Physiol Plant 37(4):258–262
Keck ME, Nicolussi S, Spura K, Blohm C, Zahner C, Drewe J (2020) Effect of the fixed combination of valerian, lemon balm, passionflower, and butterbur extracts (Ze 185) on the prescription pattern of benzodiazepines in hospitalized psychiatric patients – a retrospective case-control investigation. Phytother Res 34(6):1436–1445
KNOP Laboratorios (2020) MELIPASS®. Published on the Internet: http://www.knoplabs.com/melipass./ (Last access, February 2020)
Larson DA (1966) On the significance of the detailed structure of *Passiflora caerulea* exines. Bot Gaz 127(1):40–48
Linnaeus C (1753) 22. *Passiflora foliis palmatis intergerrimis, caerulea*. Sp. Pl. 2: 959–960. In: facsimile: https://www.biodiversitylibrary.org/item/13830. (Last access, February 2020)
Lombardo A (1982) Flora Montevidensis. Tomo I. Ed. Intendencia Municipal de Montevideo. Montevideo, Uruguay. 316 pp
Medina JH, Paladini AC, Wolfman C, Levi De Stein M, Calvo D, Díaz LE, Peña C (1990) Chrysin (5,7-di-OH-flavone), a naturally-occurring ligand for benzodiazepine receptors, with anticonvulsant properties. Biochem Pharmacol 40(10):2227–2231
Mendiondo GM, García MTA (2006) Emergence of *Passiflora caerulea* seeds simulating possible natural destinies. Fruits 61(4):251–258
Mendiondo GM, García MTA (2009) Germination of stored and scarified seeds of *Passiflora caerulea* L. (Passifloraceae). Plant Byosist 143(2):369–376
Mondin CA, Cervi AC, Moreira GRP (2011) Sinopse das espécies de *Passiflora* L. (Passifloraceae) do Rio Grande do Sul, Brasil. Rev Bras Biociências 9(S1):3–27
Muñoz Pizarro C (1959) Sinopsis de la Flora Chilena. Ed. Universidad de Chile. Santiago, Chile, 500 pp. Facsimile: http://www.chlorischile.cl/. (Last access, February 2020)
Nicolls JM (1970) Antifungal activity in *Passiflora* species. Ann Bot 34(1):229–237
Nicolls JM, Birner J, Forsell P (1973) Passicol, an antibacterial and antifungal agent produced by *Passiflora* plant species: qualitative and quantitative range of activity. Antimicrob Agents Chemother 3(1):110–117
Ozarowski M, Piasecka A, Paszel-Jaworska A, Chaves DSDA, Romaniuk A, Rybczynska M, Gryszczynska A, Sawikowska A, Kachlicki P, Mikolajczak PL, Seremak-Mrozikiewicz A, Klejewski A, Thiem B (2018) Comparison of bioactive compounds content in leaf extracts of *Passiflora incarnata*, *P. caerulea* and *P. alata* and *in vitro* cytotoxic potential on leukemia cell lines. Braz J Pharmacogn 28(2):179–191
Pastene ER, Bocaz G, Peric I, Montes M, Silva V, Riffo E (2000) Separation by capillary electrophoresis of *C*-glycoylflavonoids in *Passiflora* sp extracts. Bol Soc Chil Quím 45(3):6. https://doi.org/10.4067/S0366-16442000000300017
Patterson BD, Kenrick JR, Raison JK (1978) Lipids of chill-sensitive and -resistant *Passiflora* species: fatty acid composition and temperature dependence of spin label motion. Phytochemistry 17(7):1089–1092
Pereira CAM, Yariwake JH, Lanças FM, Wauters JN, Tits M, Angenot L (2004) A HPTLC Densitometric determination of flavonoids from *Passiflora alata*, *P. edulis*, *P. incarnata* and *P. caerulea* and comparison with HPLC method. Phytochem Anal 15(4):241–248
Pereira CAM, Yariwake JH, McCullagh M (2005) Distinction of the *C*-glycosylflavone isomer pairs orientin/isoorientin and vitexin/isovitexin using HPLC-MS exact mass measurement and in-source CID. Phytochem Anal 16(5):295–301
Quiroga OE, Bou S, Vigo MS, Nolasco SM (2000) Chemical characteristics of *Passiflora caerulea* seed oil and residual seed meal. Molecules 5(3):376–378

Ratera EL, Ratera MO (1980) Plantas de la flora argentina empleadas en medicina popular. Ed. Hemisferio Sur. Buenos Aires, Argentina, p 126

Ross P, Muñoz J, Cracco P (2018) Flora Indígena del Uruguay. Árboles y arbustos ornamentales. 3ª ed. Ed. Hemisferio Sur. Montevideo, Uruguay, p 320

Santhoshkumar J, Sowmya B, Kumar SV, Rajeshkumar S (2019) Toxicology evaluation and anti-dermatophytic activity of silver nanoparticles synthesized using leaf extract of *Passiflora caerulea*. South African J Chem Eng 29:17–23

Seigler DS, Spencer KC, Statler WS, Conn EE, Dunn JE (1982) Tetraphyllin B and epitetraphyillin B sulphates: novel cyanogenic glucosides from *Passiflora caerulea* and *P. alato-caerulea*. Phytochemistry 21(9):2277–2285

Şesan TE, Sârbu A, Smarandache D, Oancea F, Oancea A, Savin S, Toma A, Ştefan L, Negru G, Bira AF, Vlăsceanu G, Ghiurea M, Jecu L, Vasilescu G, Pomohaci CM (2016) Botanical and phytochemical approach on *Passiflora* spp.-new nutraceutical crop in Romania. J Plant Develop 23:97–126

Simirgiotis MJ, Schmeda-Hirschmann G, Bórquez J, Kennelly EJ (2013) The *Passiflora tripartita* (Banana passion) fruit: a source of bioactive flavonoid C-glycosides isolated by HSCCC and characterized by HPLC-DAD-ESI/MS/MS. Molecules 18(2):1672–1692

Sorarú SB, Bandoni AL (1978) Plantas de la medicina popular. Ed. Albatros. Buenos Aires, Argentina, p 85

Staveley GW, Wolstenholme BN (1990) Effects of water stress on growth and flowering of *Passiflora edulis* (Sims) grafted to *P. caerulea* L. Acta Hortic (275):551–558

Tantisewie B, Rujigrok HWL, Hegnauer R (1969) Die Verbreitung der Blausäure bei den Cormophyten. 5. Mitteilung: Über Cyanogene Verbindungen bei den Parietales und bei einigen weiteren Sippen. Pharm Weekbl 104:1341–1355. Cited by Seigler *et al.* (1982)

The Plant List (2020) Version 1.1 Published on the Internet: http://www.theplantlist.org. (Last access, February 2020)

Toursarkissian M (1980) Plantas medicinales de la Argentina: sus nombres botánicos, vulgares, usos y distribución geográfica. Ed. Hemisferio Sur. Buenos Aires, Argentina, pp 96–97

United States Pharmacopeia (USP-NF) (2020) Published on the Internet: https://www.usp.org/ (last access, February 2020)

Upton R, Graff A, Jolliffe G, Laenger R, Williamson E, Swisher D (2011) American herbal pharmacopoeia: botanical Pharmacognosy. Microscopic characterization of botanical medicines. Ed. American Herbal Pharmacopoeia/CRC Press, Boca Raton, USA, pp 513–517

Varela G, Cocucci A, Sérsic AN (2016) Función de la corona en *Passiflora caerulea* (Passifloraceae) como atrayente de sus vectores de polinización. Bol Soc Argentina Bot 51(1):99–110

Vieira MLC, Carneiro MS (2005) Passifloraceae, *Passiflora* spp. Passionfruit. In: Litz RE (ed) Biotechnology of fruits and nuts crops. Biotechnology in agriculture series. vol 29. Ed. CABI Publishing, Cambridge, MA, USA, pp 435–453

Wolfman C, Viola H, Paladini A, Dajas F, Medina JH (1994) Possible anxiolytic effects of chrysin, a central benzodiazepine receptor ligand isolated from *Passiflora coerulea*. Pharmacol Biochem Behav 47(1):1–4

Zarei M, Mohammadi SM, Nematian MA (2014) Evaluation of the antinociceptive effect of methanolic extract of *Passiflora caerulea* L. in adult male rat. Armaghane Danesh 19(1):56–66. Published on the Internet: http://armaghanj.yums.ac.ir/article-1-226-en.html. (Last access, February 2020)

Peumus boldus Mol.

Bruce K. Cassels

Peumus boldus. (Photos by the author)

B. K. Cassels (✉)
Department of Chemistry, Faculty of Sciences, University of Chile, Santiago, Chile
e-mail: bcassels@uchile.cl

Á. Máthé, A. Bandoni (eds.), *Medicinal and Aromatic Plants of South America Vol. 2*, Medicinal and Aromatic Plants of the World 7,
https://doi.org/10.1007/978-3-030-62818-5_33

Abstract *Peumus boldus* is a tree that grows in relatively humid habitats, from the semiarid to the temperate forest regions in Chile. Its leaves are exported in large amounts and are widely used to prepare a digestive tea. Historically they were employed for headaches, rheumatism, and severe but not clearly identified conditions characterized by fever, edema, catarrh, and also sexually transmitted diseases. The most characteristic constituents are a mainly terpenoidal essential oil and a number of isoquinoline and aporphine alkaloids. Remarkably, it is generally not the major alkaloidal component of the leaves, boldine that continues to be the subject of pharmacological investigations: some of these seem to support the traditional uses of boldo leaves.

Keywords Monimiaceae · Boldo · Medicinal · Digestive · Ornamental

1 Taxonomic Characteristics

Peumus boldus Molina (Monimiaceae), is the only species of its genus. It is known worldwide as boldo and is endemic to Chile (although in Brazil it is often called "boldo do Chile" to distinguish it from the unrelated "boldo brasileiro" or "boldo paraguayo": *Plectranthus barbatus* = *Coleus forskohlii* (Lamiaceae). It forms a clade (Monimioideae) with the Australia-New Guinea *Palmeria* and the Mascarenian *Monimia*, but is a relatively distant relative of the other South- and Central American Monimiaceae (*Mollinedia* and *Hennecartia*) which might have arrived by trans-Pacific dispersal from Australasia before the major rise of the Andes mountains (Renner et al. 2010).

2 Crude Drug Used

The air-dried leaves (Photo 1) are the only part currently used for medicinal purposes, to aid digestion, for liver ailments, to prepare herbal teas, and to season "yerba mate" or "erva mate" (*Ilex paraguariensis*) infusions. They are widely consumed in the Southern Cone of South America, also figure in the European Pharmacopoeia as "boldo folium". Both ground leaves and their extracts are sold worldwide, as herbal supplements.

Photo 1 Leaves of boldo. (Photo by the author)

3 Major Chemical Constituents and Bioactive Compounds

The chemical profile of boldo extracts has been rather well studied since the original extraction of boldine from the leaves in 1872 and its more likely, higher yield isolation from twigs a couple of years later (Bourgoin and Verne 1872; Verne 1874). In the leaves, which are the commonly used part of the plant, related alkaloids such as its common precursor coclaurine, laurotetanine, *N*-methyl-laurotetanine, or isocorydine usually seem to be more abundant than boldine (Fuentes-Barros et al. 2018).

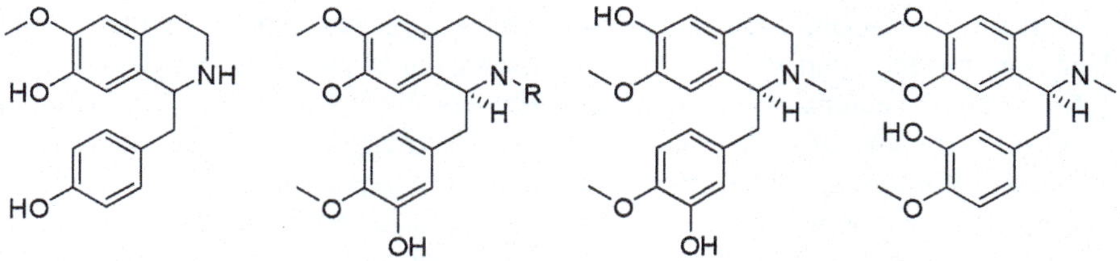

coclaurine laurotetanine (R = H) and *N*-methyllaurotetanine (R = methyl) boldine isocorydine

The belief that boldine is the major or characteristic alkaloid of boldo leaves has led to many publications on its pharmacology, almost exclusively in animals and *in vitro*. Boldine is an excellent free radical scavenger (Speisky et al. 1991; Cassels et al. 1995), and reduces the expression of oxidative stress markers (e.g. Qiu et al. 2017). Of possible relevance to the traditional uses of the plant are its anti-inflammatory and antipyretic activity (Backhouse et al. 1994), which are also manifest in the leaf extract together with the commonly accepted choleretic and cholagogue activities (Lanhers et al. 1991), but may be due mainly to other components. Boldine and related aporphine alkaloids are dopaminergic, and also adrenergic and serotonin receptor antagonists (Madrero et al. 1996; Asencio et al. 1999; Walstab et al. 2014). The inhibition of connexin hemichannel function by boldine has been implicated in a variety of protective effects shown by this alkaloid (e.g. Yi et al. 2017).

Although boldo is not generally recommended for this purpose, a dry extract of boldo leaves has been shown to prolong gastrointestinal transit in humans (Gotteland et al. 1995).

Its characteristic aroma is due to an essential oil with a somewhat variable composition. Ascaridole has usually been considered to be a major component, but it is not always present in detectable amounts, and *p*-cymene, α-terpinene and 1,8-cineole may predominate (Cassels et al. 2019).

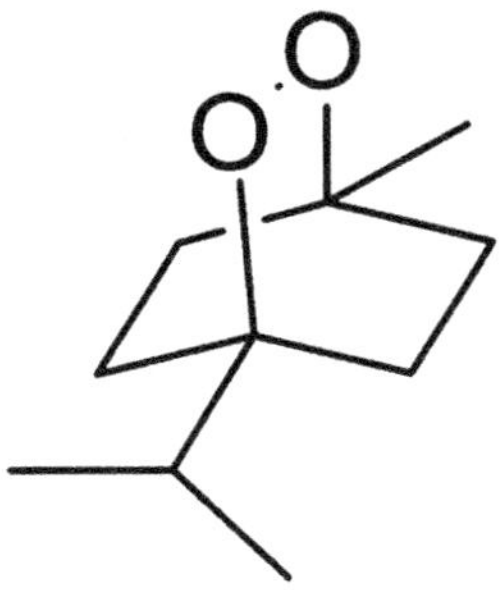

ascaridole

The high content of catechin in boldo leaves (Schmeda-Hirschmann et al. 2003), is probably responsible, rather than boldine, for some of their antioxidative and health-promoting properties (Schmeda-Hirschmann et al. 2003; Quezada et al. 2004). Much smaller amounts of other phenolics, including a series of flavonoid glycosides (mainly of quercetin, kaempferol and their *O*-methylated derivatives) (Krug and Borkowski 1965) are presumably of little importance.

Surprisingly, a recent study of the aqueous infusion found mainly kaempferol glycosides and no catechin, but instead small amounts of its dimer procyanidin B2 (Falé et al. 2012). An extensive review on boldo, its secondary metabolites and their derivatives has been published recently (Cassels et al. 2019).

4 Morphological Description

Peumus boldus is a medium-sized evergreen tree with a rounded crown, attaining a height of about 15 m. However, due to its extensive harvesting for its leaves, and its ability to produce many suckers after the main trunk has been felled, it usually grows as a large, dense bush with many stems rising in a circle around the remnants of the original trunk and it is now managed commercially by partial coppicing (Donoso et al. 2015).

5 Geographical Distribution

This species is a typical component of the Mediterranean-type sclerophyllous forest of Central Chile. However, it grows by wetlands near the mouth of rivers in the semiarid Coquimbo Region as far north as 30° 20' S, is present in relict coastal fog forests around 30°30' S, and extends southwards through progressively more humid regions into the temperate rainforest until about 41° S, from the vicinity of the ocean shore to about 1000 m.a.s.l. in the Andean lake region (Donoso et al. 2015).

6 Ecological Requirements

Peumus boldus grows in full sun, but few of its abundant seedlings survive unless they are shaded by full-grown trees and can send long taproots into relatively humid soil. In the drier, warmer northern parts of its range it is typically restricted to southern-facing slopes where solar irradiation is less intense. It seems to prefer slightly acid soils, as in natural habitats young trees are only found growing in rich forest litter (Pizarro 1989). Internet sites related to the herbal drug industry commonly make fantastic claims as to its provenance, such as "native to Chile, Ecuador, Argentina, Bolivia, and Peru", or the nonexistent "hot, arid heights of Chile", as well as to its vegetative characteristics and health-promoting virtues.

7 Traditional Use and Common Knowledge

The earliest evidence of a presumably medicinal use of boldo is the finding of masticated boldo leaves at least 100 km south of its range in an archeological site in south-central Chile dated to at least 13,000 years before present (Dillehay 1984). Frézier (1716) mentioned the plant in 1716, but only stating that "the leaves smell of incense, and the bark has a pungent flavor somewhat like that of cinnamon". These features suggest that Frézier's "boldu" is in fact *Beilschmiedia miersii*

(Lauraceae), also mentioned as "boldu" by Molina, the author of the scientific binomial *Peumus boldus*. This author only refers to its hard seed that can be used to string rosaries, and to the bark "to perfume barrels before filling them with wine" (Molina 1782). Hipólito Ruiz and José Pavón's account of their botanical expedition to the Viceroyalty of Peru, which included Chile (1777–1788), contains a passage that is possibly the earliest printed reference to the medicinal uses of boldo (called *Ruizia fragrans* by them, a name that had to be changed subsequently because the generic name *Ruizia* had already been given to a different plant): "A light infusion of the leaves, sweetened with sugar, is often taken after meals instead of tea and coffee to avoid indigestions. A decoction in wine or water, applied to the temples or the stomach, calms migraines and headaches, settles the stomach, dissipates flatus and fortifies the nerves. Women believe that a remedy made from one part of powdered leaves and three of Chilean pine (presumably *Araucaria araucana?*) resin, placed on the navel, dissipates uterine suffering. Powders are often sniffed into the nose in order to clear the head" (Ruiz and Pavón 1798). Half a century later Claudio Gay's description of the uses of the leaves (as *Boldoa fragrans*) are very succinct, only mentioning rheumatism, head colds, earache, dropsy and syphilis (Gay 1849). Murillo (1889) may have seized on the latter claim, plus his perception that the aroma of the leaves resembled turpentine, used at the time for inflammations of the urinary tract, to successfully test the administration of an alcoholic extract on soldiers with gonorrhea. More significantly from a historic point of view, Murillo gives a critical account of events leading to the belief that boldo is generally good for the liver (although admitting that it might exhibit anthelmintic properties, which are now attributable to the ascaridole content of the leaves (Murillo 1889).

At the turn of the twentieth century, W. F. Neger, of Munich, spent some years in Chile, and developed a keen interest in boldo. An interesting historical reference is Neger's mention that "Other than for medicinal purposes (already imported to France in 1868!) the leaves could be used for the preparation of an oil ideally suited as a room aromatizer. More than 20 years ago such a product was put on the market as "boldo oil" by the German firm established in Santiago (Chile), H. Gockel i Cia. (*Perfumería, Jabonería y Velería del Progreso*)." Neger's accounts (1901, 1902) centre to some extent on the essential oil. "The leaves emit a much more agreeable odor than those of peumo … (where) there seems to be another substance with an odor reminiscent of something rancid". He also goes to some length describe the anatomical features distinguishing *P. boldus* leaves from those of *Cryptocarya alba* (Lauraceae), which seem to be easily confused by some, and may have been used as an adulterant (Wichtl 2004). An amusing story by Looser (1935) recounts that a "certain foreign professor under contract in one of our universities felt that he was suffering from a liver complaint. He asked a friendly doctor who recommended boldo. Our good professor consulted his botany book and began to take 'boldo' tea regularly, prepared from the leaves of the 'boldos' growing in the yard of his country house. It so happened that his doctor went on a visit and the professor, overjoyed with his improved health, showed his visitor the beneficial trees that he said looked beautiful covered with their scarlet fruit. The doctor was suspicious and looked closer. Our professor had been taking peumo [*Cryptocarya alba*] instead of boldo tea for six months!" The widespread citation "Looser G. Revista Universitaria. 1903; 15:89–104" must be in error, as the author was four years old at the time.

Boldo leaves are included in pharmacopoeias worldwide, including the first edition of the Chilean Pharmacopoeia (Puga Borne and Miranda 1905) and its most recent (fourth) edition (Farmacopea Chilena 2017), the seventh of the Argentine Pharmacopoeia (Farmacopea Argentina 2013), and the sixth of the Brazilian Pharmacopoeia (Farmacopeia Brasileira 2019), representing the countries where they are consumed most. The European, the Spanish and the British Pharmacopoeias also contain articles on boldo leaves, a monograph has been proposed for development in the Herbal Medicines Compendium of the US Pharmacopoeia (USP 2013), and the European Committee on Herbal Medicinal Products has an assessment report on the subject (EMA 2019). The essential oil is analyzed by standard methods, and total alkaloids are assayed and reported as "boldine" due to its wide availability as a standard.

Boldo tea is still widely consumed for its flavor and as a presumed aid for digestion, and it is still considered "good for the liver", although not generally prescribed. The leaves are also a major ingredient of herbal preparations that are sold without any clear evidence of clinical efficacy, mainly for mild abdominal spasms and to stimulate bile flow, but also as a mild sedative, uses which receive support through their publication in prestigious pharmacognosy manuals (e.g. Wichtl 2004; Bruneton et al. 2008; Evans 2009; Heinrich et al. 2018). Aside from the social and medicinal uses of boldo leaves, their rather variable essential oil (Cassels et al. 2019) is insecticidal (Betancur et al. 2010; Urzúa et al. 2010).

Exports of boldo leaves increased from 1,553,027 kg in 2002 to 2,166,158 kg in 2018, with an average export price in the last few years of around 2 U$S/kg, about 12 times the price paid the primary producer. "Other boldo products" amounting to 351,795 kg were shipped in 2018 (Chile Aduana 2019), of which variable quantities of bark (a maximum in 2017 of 16,360 kg, at about 8 U$S/kg) went almost exclusively to Italy (Benedetti and Barros 2018). Considering that all this material is harvested in the wild, together with the relatively slow rate of regeneration of suckers from boldo roots, such figures illustrate the current extractivist threat to this species. For this reason, studies are underway on the cultivation of boldo under various conditions (Vogel et al. 2011).

8 Modern Medicine Based on Its Traditional Medicinal Uses

Boldo essential oil has been mentioned as a treatment of intestinal worms, but its use is strongly discouraged because of its unusually high toxicity, in part due to its ascaridole content (Grassmann and Elstner 2003). Modern studies have focused almost exclusively on boldine, which was shown to stimulate bile production and secretion in experimental animals (Kreitmair 1952) and is also hepatoprotective and anti-inflammatory (Lanhers et al. 1991; Backhouse et al. 1994; Bannach et al. 1996). Despite these and many other interesting, potentially useful properties of boldine, unrelated to its traditional uses and extensively reviewed in 2019 (Cassels et al. 2019), it must be kept in mind that this alkaloid is present in rather low concentrations in boldo leaves (Fuentes-Barros et al. 2018), and that the medicinal properties of boldo herbal preparations, if actually demonstrated (Cordero-Pérez et al. 2013), are probably due to the additive or synergistic actions of a number of different compounds, some of them present in considerably larger amounts than boldine.

9 Conclusions

Boldo leaves have a long tradition of use for a variety of different complaints. However, modern evidence of their efficacy in the forms and doses currently employed (mainly aqueous infusions) or found in commerce (powder, alcoholic extracts) is lacking. As the cost of extensive clinical trials is a limiting factor, it seems reasonable to focus on studying the constituents responsible for its pharmacological effects that can be isolated from the plant in sufficient amounts. Aside from the widespread flavonoids, the major alkaloids, e.g. boldine, laurotetanine and its *N*-methyl derivative, isocorydine, and coclaurine, are probably the best candidates for more advanced studies.

References

Asencio M, Delaquerrière B, Cassels BK, Speisky H, Comoy E, Protais P (1999) Biochemical and behavioral effects of boldine and glaucine on dopamine systems. Pharmacol Biochem Behav 62:7–13. https://doi.org/10.1016/S0091-3057(98)00096-3

Backhouse N, Delporte C, Cassels BK, Valenzuela A, Speisky H (1994) Anti-inflammatory and antipyretic effects of boldine. Agents Actions 42:114–117

Bannach R, Valenzuela A, Cassels BK, Núñez-Vergara LJ, Speisky H (1996) Cytoprotective and antioxidant effects of boldine on tert-butyl-hydroperoxide-induced damage to isolated hepatocytes. Cell Biol Toxicol 12:89–100

Benedetti S, Barros S (eds) (2018) Boldo (*Peumus boldus* Mol.). Avances en la investigación para el desarrollo de modelos productivos sustentables. Ed. INFOR, Santiago, Chile, 144 pp

Betancur RJ, Gonzalo Silva AG, Rodríguez JM, Fischer GS, Zapata SM (2010) Insecticidal activity of *Peumus boldus* Molina essential oil against *Sitophilus zeamais* Motschulsky. Chil J Agric Res 70(3):399–407

Bourgoin E, Verne C (1872) Sur l'existence d'un alcali organique dans le boldo. J Pharm Chim 16:191–193

Bruneton J, Hatton C, Bruneton A (2008) Pharmacognosy, phytochemistry, medicinal plants. Ed. Intercept/Lavoisier, Andover, USA, p 1019

Cassels BK, Asencio M, Conget P, Speisky H, Videla LA, Lissi EA (1995) Structure-antioxidative activity relationships in benzylisoquinoline alkaloids. Pharmacol Res 31:103–107

Cassels BK, Fuentes-Barros G, Castro-Saavedra S (2019) Boldo, its secondary metabolites and their derivatives. Curr Trad Med 5:31–65

Chile Aduana (2019) https://www.aduana.cl/aduana/site/edic/base/port/estadisticas.html. (Last access, August 2019)

Cordero-Pérez P, Torres-González L, Aguirre-Garza M, Camara-Lemarroy C, Guzmán-de la Garza F, Alarcón-Galván G, Zapata-Chavira H, Sotelo-Gallegos M, Torres-Esquivel CN, Sánchez-Fresno E, Cantú-Sepúlveda D, González-Saldívar G, Bernal-Ramírez J, Muñoz-Espinosa LE (2013) Hepatoprotective effect of commercial herbal extracts on carbon tetrachloride-induced liver damage in Wistar rats. Pharm Res 5(3):150–156

Dillehay TD (1984) A late ice-age settlement in Southern Chile. Sci American 251(4):100–109

Donoso S, Peña-Rojas K, Durán S, Pacheco C, Galdames E, Espinoza C (2015) Influence of thinning over growth, water potential and photosynthetic response of *Peumus boldus:* guidelines for their silvicultural management proposal. Bosque (Valdivia) 36(3):457–466

EMA, Committee on Herbal Medicinal Products (2019) Boldi folium (PDF). https://www.ema.europa.eu/en/search/search?search_api_views_fulltext=boldo (last access, December 2019)

Evans WC (2009) Trease & Evans's pharmacognosy, 16th edn. Ed. WB Saunders, London, UK, p 616

Falé PL, Amaral F, Madeira PJA, Silva MS, Florêncio MH, Frazão FN, Serralheiro MLM (2012) Acetylcholinesterase inhibition, antioxidant activity and toxicity of *Peumus boldus* water extracts on HeLa and Caco-2 cell lines. Food Chem Toxicol 50:2656–2662

Farmacopea Argentina, VII Edición (2013) ANMAT, Ministerio de Salud, Buenos Aires, Argentina, p 445

Farmacopea Chilena, IV Edición (2017) Universidad de Valparaíso, Valparaíso, Chile

Farmacopeia Brasileira, VI Edição (2019) ANVISA, Vol II, PM021-00

Frézier AF (1716 or 2nd ed. 1732?) cited by Molina (1782, p 168) Relation du voyage de la mer du Sud, aux côtes du Chili, du Pérou et de Brésil, fait pendant les années 1712, 1713, et 1714. vol I, pp 137–139

Fuentes-Barros G, Castro-Saavedra S, Liberona L, Acevedo-Fuentes W, Tirapegui C, Cassels BK (2018) Variation of the alkaloid content of *Peumus boldus* (boldo). Fitoterapia 127:179–185

Gay C (1849) Historia Física y Política de Chile, vol 5. Museo de Historia Natural, Santiago, Chile, pp 351–353

Gotteland M, Espinoza J, Cassels BK, Speisky H (1995) Efecto de un extracto seco de boldo sobre el tránsito intestinal oro-cecal en voluntarios sanos. Rev Méd Chile 123:955–960

Grassmann J, Elstner EF (2003) In: Caballero B, Finglas P, Toldra F (eds) Encyclopedia of food sciences and nutrition, 2nd edn. Ed. Academic, San Diego, USA, pp 2177–2184

Heinrich M, Barnes J, Prieto-García JM, Gibbons S, Williamson EM (2018) Fundamentals of pharmacognosy and phytotherapy. Ed. Churchill Livingstone, London, UK, p 360

Kreitmair H (1952) Pharmakologische Wirkung des Alkaloids aus *Peumus boldus* Molina. Pharmazie 7:507–511

Krug H, Borkowski B (1965) Flavonoidverbindungen in den Blättern von *Peumus boldus* Molina. Naturwiss 52(7):161

Lanhers MC, Joyeux M, Soulimani R, Fleurentin J, Sayag M, Mortier F, Younos C, Pelt J (1991) Hepatoprotective and anti-inflammatory effects of a traditional medicinal plant of Chile, *Peumus boldus*. Planta Med 57:110–115

Looser G (1935) ¿Cuál es el verdadero nombre botánico del peumo y del boldo? Rev Chil Hist Nat 39(1):203–211

Madrero Y, Elorriaga M, Martínez S, Noguera MA, Cassels BK, D'Ocon P, Ivorra MD (1996) A possible structural determinant of selectivity of boldine and derivatives for the α_{1A}-adrenoceptor subtype. Brit J Pharmacol 119:1563–1568

Molina JI (1782) Saggio sulla storia naturale del Chili. Ed. Stamperia de S. Tomasso d'Aquino. Bologna, Italy, 568 pp

Murillo A (1889) Plantes médicinales du Chili. Exposition Universelle de Paris 1889, Section Chilienne, Paris. France, pp 142–149

Neger WF (1901) Ueber Folia Boldi. Pharmazeut Zentralh 42(31):461–464

Neger WF (1902) Sobre la hoja del boldo. Rev Chil Hist Nat 6(1):56–57

Pizarro S (1989) Estudio fitosociológico estadístico de los bosques de boldo de la cuenca de Río Bueno (Chile). Thesis, Universidad Austral, Valdivia, Chile, 80 pp

Puga Borne F, Miranda JB (1905) Farmacopea Chilena. Ed. Imprenta Barcelona, Santiago, Chile

Qiu X, Shi L, Zhuang H et al (2017) Cerebrovascular protective effect of boldine against neural apoptosis *via* inhibition of mitochondrial Bax translocation and cytochrome C release. Med Sci Monit 23:4106–4116

Quezada N, Asencio M, del Valle JM, Aguilera JM, Gómez B (2004) Antioxidant activity of crude extract, alkaloid fraction, and flavonoid fraction from boldo (*Peumus boldus* Molina) leaves. J Food Sci 69(5):C371–C376

Renner SS, Strijk JS, Strasberg D, Thébaud C (2010) Biogeography of the Monimiaceae (Laurales): a role for East Gondwana and long-distance dispersal, but not West Gondwana. J Biogeogr 37(7):1227–1238

Ruiz H, Pavón J (1798) Systema Vegetabilium Florae Peruvianae et Chilensis. Ed. De Sancha G, Madrid, Spain, pp 267–269

Schmeda-Hirschmann G, Rodríguez JA, Theoduloz C, Astudillo SL, Feresin GE, Tapia A (2003) Free-radical scavengers and antioxidants from *Peumus boldus* Mol. ("boldo"). Free Rad Res 37(4):447–452

Speisky H, Cassels BK, Lissi EA, Videla LA (1991) Antioxidant properties of the alkaloid boldine in systems undergoing lipid peroxidation and enzyme inactivation. Biochem Pharmacol 41:1575–1582

Urzúa A, Santander R, Echeverría J, Villalobos C, Palacios SM, Rossi Y (2010) Insecticidal properties of *Peumus boldus* Mol. essential oil on the house fly, *Musca domestica* L. BLACPMA 9(6):465–469

USP (2013) *Peumus boldus* leaf. Herbal Medicines Compendium https://hmc.usp.org/monographs/peumus-boldus-leaf-0-1. (Last access, December 2019)

Verne C (1874) Étude sur le boldo. Thesis. École Supérieure de Pharmacie, Paris, France, 52 pp

Vogel H, González B, Razmilic I (2011) Boldo (*Peumus boldus*) cultivated under different light conditions, soil humidity and plantation density. Ind Crop Prod 34:1310–1312

Walstab J, Wohlfarth C, Hovius R, Schmitteckert S, Röth R, Lasitschka F, Wink M, Bönisch H, Niesler B (2014) Natural compounds boldine and menthol are antagonists of human 5-HT_3 receptors: implications for treating gastrointestinal disorders. Neurogastroenterol Motil 26(6):810–820

Wichtl M (2004) *Boldi folium*. In: Herbal drugs and phytopharmaceuticals, 3rd edn. (English). Ed. Medpharm Scientific Publishers, Stuttgart, and CRC Press, Boca Raton, USA, pp 91–93

Yi C, Ezan P, Fernández P et al (2017) Inhibition of glial hemichannels by boldine treatment reduces neural suffering in a murine model of Alzheimer's disease. Glia 65:1607–1625

Picrasma crenata Engl. in Engl. & Prantl

Ignacio J. Agudelo, Leonardo M. Anconatani, Rafael A. Ricco, Cecilia B. Dobrecky, and Marcelo L. Wagner

Picrasma crenata. Commercial crude drug. (Photos: R Ricco and M Wagner)

Abstract *Picrasma crenata* Engl. in Engl. & Prantl, Simaroubaceae, is a dioecius tree, native to the South of Brazil and North Argentina. Its wood is used as a drug for the treatment of diabetes and pediculosis; these biological activities are partially validated by *in vivo* assays. It contains a variety of triterpenic bitter compounds, known as quassinoids, which are – at least – partially responsible for its pharmacological activities. Further studies are required for the validation of the underly-

I. J. Agudelo · L. M. Anconatani · R. A. Ricco · M. L. Wagner (✉)
Facultad de Farmacia y Bioquímica, Universidad de Buenos Aires, Buenos Aires, Argentina
e-mail: mlwagner@ffyb.uba.ar; mlwagner550@gmail.com

C. B. Dobrecky
Facultad de Farmacia y Bioquímica, Universidad de Buenos Aires, Buenos Aires, Argentina

Facultad de Farmacia y Bioquímica, Departamento de Tecnología Farmacéutica, Cátedra de Tecnología Farmacéutica I, Universidad de Buenos Aires, Buenos Aires, Argentina

Á. Máthé, A. Bandoni (eds.), *Medicinal and Aromatic Plants of South America Vol. 2*, Medicinal and Aromatic Plants of the World 7,
https://doi.org/10.1007/978-3-030-62818-5_34

ing pharmacological mechanisms that account for its hypoglycemic activity and its phytochemical normalization.

Keywords "Palo amargo" · *Aeschrion crenata* · Simaroubaceae · Quassinoids · Hypoglycemic activity

1 Introduction

Picrasma crenata is the source of a highly employed plant drug for the treatment of pediculosis, in Argentina and Paraguay. It also has activity against diabetes, a chronic disease with a high prevalence worldwide. Based on its traditional uses in South America *P. crenata* could contribute to its treatment.

The mosquito repellent activity, mentioned by the Swiss naturalist Moisés Bertoni (1900) in his works about Guarani people and its culture, may also be of importance against the vector borne diseases of our days (e.g. dengue, zika and chikungunya). Dengue is an epidemic disease in Paraguay and this crude drug could be used for the prevention of mosquito-bites, as it was used in malaria epidemics the previous century.

2 Taxonomic Characteristics

Picrasma crenata Engl. in Engl. & Prantl belongs to the Simaroubaceae DC. family. This family has up to 121 accepted species according to The Plant List (2019), distributed in 19 accepted genera. This family is known for the content of bitter active substances responsible for many pharmacological properties. It is distributed in South America, West Africa, Asia and Australia. The genera present in South America are *Picrasma* Blume, *Quassia* L., *Picrolemma* Hook. f., *Castela* Turpin, *Simaba* Aubl. and *Simarouba* Aubl. (Alves et al. 2014).

P. crenata has been previously known with several names: *Aeschrion crenata* Vell., *Picraena palo-amargo* Speg., *Picraena vellozii* (Planch.) Engl., *Picramnia crenata* (Vell.) Hassl., *Picrasma palo-amargo* Speg. and *Picrasma vellozii* Planch. It is worth to mention that there are many close species within the Simaroubaceae family such as *Quassia amara* L. and *Picrasma excelsa* (Sw.) Planch.

3 Crude Drug Used

Picrasma crenata is known in the folk medicine as "palo amargo" and "cuasia" in Argentina and "quássia do Brasil", "quássia amarga", and "pau-quassia" in Brazil. The crude drug employed in folk medicine is the wood (Photo 1) (Domínguez 1928; Toursarkissian 1980). Up to this moment, *P. crenata* does not have a monograph in any of the pharmacopoeias employed in Latin America.

4 Major Chemical Constituents and Bioactive Compounds

The main constituents of *P. crenata* are a family of triterpenic bitter compounds known as quassinoids, widely distributed within the Simaroubaceae family (Alves et al. 2014). The bark contains the quassinoids paraine, isoparaine, and 12-*nor* quassine, and the β-carboline alkaloids 1-carbomethoxy-β-carboline, 1-ethyl-4-methoxy-β-carboline (crenatine) and 1-ethyl-4,8-dimethoxy-β-carboline (crenatinine) (Sánchez and Comín 1971; Vitagliano and Comín 1972). The constituents isolated from the wood are quassin, neoquassin, paraine, ll-dihydro-12-*nor* neoquassin, 16-β-*O*-methylneoquassin and 16-β-*O*-ethylneoquassin (Novello et al. 2003). There are also reports of other compounds such as coniferyl aldehyde, coniferin, cantin-6-one, 4,5-dimethoxycantin-6-one and (+)-neo-olivil (Krebs et al. 2001).

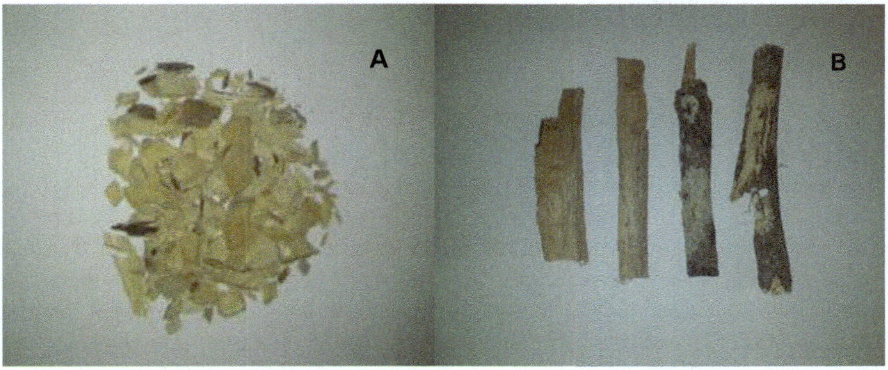

Photo 1 *Picrasma crenata*. (**a**) chopped wood. (**b**) bark. (Photos: I Agudelo)

5 Morphological and Micrographical Description

Picrasma crenata is a small tree up to 6–15 m high. It has a brown lenticellate bark, and compound leaves with oval-elliptical leaflets, 4–10 cm long. Its flowers are arranged in corymbose inflorescences. It is a dioic species with green feminine flowers and yellow male flowers, which have 4–5 sepals and 4–5 petals. Its fruit is a drupe that becomes black when mature (Campagna et al. 2013).

Regarding its micrographic characteristics (Campagna et al. 2013, 2016), the outer cortex has a layer of 5–7 squamous imbricated peridermis, each one 15–40 cork cells width and a pure stratified felodermis. These cork cells have homogeneous walls and have a quadrangular-rectangular form with a polygonal shape when observed in a macerate. The inner cortex has a great number of phloematic radiuses of 1–7 cells width and few fibers and parenchyma with druses and polyhedral crystals of calcium oxalate. The wood has a semi-annular to annular growth and has vases with a circular margin with simple perforation plaques and intervascular areolated punctuations. It also has polyhedral crystals of calcium oxalate and xylematic radiuses of 1–5 cells in width (Márquez et al. 1997).

6 Geographical Distribution

Picrasma crenata is native to the Paranaense forest, which can be found in the South of Brazil and the Northwestern region of Argentina, particularly in the Province of Misiones (Argentina) (Campagna et al. 2013).

7 Traditional Uses and Common Knowledge

The Swiss naturalist Moisés Bertoni reports that, in Paraguay, several species of the Simaroubaceae with a similar morphology were used as a mosquito repellent by the Guarani culture. Among these species, *Picrasma crenata* (*P. palo-amargo* in that time) was the most effective due to its bitter taste. Therefore he recommends its culture. He considers *P. crenata* was more potent in its therapeutic properties than *Quassia amara* L. and reports its use as a tonic, aperitive, gastric stimulant, antimalarial and antisyphilitic in higher doses. It is worthy to mention that this plant drug was used during a malaria epidemic in 1887 in Paraguay (Bertoni 1900). Domínguez, in his book "Contribuciones a la Materia Médica Argentina", reports that these species contain an alkaloid and bitter substances and proposes it as a substituent of *Q. amara* (Domínguez 1928). Toursarkissian reports the use of the wood as an antisyphilitic, tonic and insecticidal plant drug. In Brazil, it is used as a febrifuge, for hepatic and gastric ailments and antiophidic (Toussarkissian 1980). Kujawska and

Hilgert (2014) have reported the use of *P. crenata* in the folk medicine of the Polish migrants in Misiones Province in Argentina, as a digestive and for liver pains and stomach aches. It is also mentioned its use as a pediculicidal agent in lice infestations. Nowadays, its main use is as a pediculicidal agent in hydroalcoholic extracts. Given its similarity with *Q. amara*, these species share many uses between them.

8 Modern Medicine Based on its Traditional Uses

There are several studies about the insecticidal and hypoglycemic activity of *Picrasma crenata*. Regarding the first one, the insecticidal activity of organic extracts against the rice weevil (*Sitophilus oryzae* L. (Coleoptera: Curculionidae)) and *Myzus persicae* Sulzer (Hemiptera: Aphididae) was tested with good results. The ethyl acetate extract was the most effective extract against *M. persicae* and *S. oryzae* (Rodríguez et al. 2008, 2011; Rodríguez 2015). The dichloromethane extract of *P. crenata* had deterrent effects against *Epilachna paenulata* Germar (Coleoptera, Coccinellidae, Epilachninae) and, to a lesser extent, *Spodoptera frugiperda* J.E. Smith (Lepidoptera: Noctuidae). However, when tested against the "cattle tick" *Rhipicephalus microplus* Canestrini (Acari: Ixodidae) the ixocidal effect observed was very weak (Martínez et al. 2013).

Regarding the hypoglycemic activities, Novello et al. (2008) failed to observe an effect in the glycemia in healthy mice after an administration of amylose. Also, antihipertriglyceridemic activity was assessed but not observed. However, an antiulcerogenic effect was present in animals treated with indomethacin and acidified ethanol when a dose of 1000 mg of wood extract/kg animal weight was administered. The extract employed in these studies was hydroalcoholic extracted to exhaustion (Novello et al. 2008).

The antihyperglycemic effect was confirmed in normoglycemic mice as well as mice treated with streptozotocin by Ikegame and Pereira (2003). The extract employed was a decoction (5, 10 or 20 g of wood in 100 ml of water). In a concentration of 10% w/v, a hypoglycemic effect was observed in both experimental models, with an inhibition in the intestinal absorption of glucose and, also, an inhibition in the reabsorption of this sugar by the kidneys (Ikegame and Pereira 2003).

Methanolic and aqueous extracts of *P. crenata* have been tested to regain information about its antimalarial efficacy. These extracts have shown to have a weak antiplasmodial activity when administered *in vitro* to *Plasmodium falciparum* Wellch (Plasmodiidae) pyrimethamine/chloroquine-resistant strain K1 (Debenedetti et al. 2002).

There are no studies about the safety of the extracts made from this plant, except for a report about its genotoxic activity in *Allium cepa* roots. The aqueous extract of *P. crenata* induced an increase of the mitotic index in the roots of the onions that grew within it, though its genotoxic activity in cell cultures and animals must be confirmed through the proper assays (Roldán et al. 2007).

There is a report regarding the obtention of granulates of a hydroalcoholic extract with povidone K30 and lactose. It was mentioned that an extract granulated with a mix of these two excipients could maintain the amount of quassine, paraine and isoparaine stable for 35 days at 37 °C and after direct irradiation of 254 nm UV light. However, after a year of storage at room temperature, only quassine could be detected in traces (Márquez et al. 1997).

9 Conclusions

Picrasma crenata is a species with ethnobotanical antecedents. It is still used today, mainly as a hypoglycemic and insecticidal agent. There are no *in vivo* evidences about these activities; consequently, there is a need for more precise validation of the traditional uses. Randomized, placebo-controlled studies will be necessary for this purpose. It is also worth noting that quassinoids, the main chemotaxonomic character of the Simaroubaceae family, have been widely studied. There is information about the therapeutic efficacy and safety of these compounds. However, the results obtained with isolated molecules and Simaroubaceae species, different from *P. crenata,* cannot be extrapolated to this species. Finally, since this species is a substituent of *Quassia amara*, toxicological studies should also be performed to assess the safety of this change.

References

Alves IA, Miranda HM, Soares LA, Randau KP (2014) Simaroubaceae family: botany, chemical composition and biological activities. Rev Bras Farm 24(4):481–501

Bertoni MS (1900) La civilización guaraní. Parte III Etnografía: Conocimientos. Ed. Imprenta y Edición Ex Sylvis, Asunción, Paraguay, 529 pp

Campagna MN, Di Sapio O, Gattuso M (2013) Parámetros micrográficos para la identificación de hojas, corteza y leño de *Picrasma crenata* (Vell.) Engl. (Simaroubaceae). Dominguezia 29(1):17–28

Campagna MN, Gattuso M, Martinez ML, Rodríguez MV, Sapio OD (2016) Novel micromorphological features of wood and bark of Argentinean Simaroubaceae. NZ J Bot 55(2):134–150

Debenedetti S, Muschietti L, Van Baren C, Clavín M, Broussalis A, Martino V, Houghton PJ, Warhurst D, Steele J (2002) *In vitro* antiplasmodial activity of extracts of Argentinian plants. J Ethnopharmacol 80(2–3):163–166

Domínguez JA (1928) Contribuciones a la materia médica argentina. (Primera contribución). Ed. Peuser, Buenos Aires, Argentina, 433 pp

Ikegame ALM, Pereira NA (2003) Atividade hipoglicemiante do cerne da quassia-do-Brasil (*Picrasma crenata* [Vell.] Engl. -Simaroubaceae). Rev Bras Farm 84:51–53

Krebs HC, Schilling PJ, Wartchow R, Bolte M (2001) Quassinoids and Other Constituents from *Picrasma crenata*. Zeitschrift Für Naturforschung B 56(3):315–318

Kujawska M, Hilgert NI (2014) Phytotherapy of Polish migrants in Misiones, Argentina: legacy and acquired plant species. J Ethnopharmacol 153(3):810–830. Corrigendum: 155:1629–1640

Márquez A, Borri K, Dobrecky J, Gurni AA, Wagner ML (1997) New aspects in quality control of "palo amargo" (*Aeschrion crenata* Velloso -Simaroubaceae). Acta Hortic 503:111–116

Martínez ML, von Poser G, Henriques A, Gattuso M, Rossini C (2013) Simaroubaceae and Picramniaceae as potential sources of botanical pesticides. Ind Crops Prod 44:600–602

Novello CR, Ferreira AG, Marques LC, Cortez DA (2003) Quassinoids from *Picrasma crenata*. Nat Prod Res 17(3):145–148

Novello CR, Bazotte RB, Bersani-Amado CA, Marques LC, Cortez DA (2008) Toxicological and pharmacological studies of *Picrasma crenata* (Vell.) Engler (Simaroubaceae) in mice and rats. Lat Am J Pharm 27(3):345–348

Rodríguez SM (2015) Evaluación del efecto insecticida de *Picrasma crenata* (Vell.) Engl.-Simaroubaceae- sobre coleópteros plaga de granos almacenados. Thesis. Facultad de Ciencias Naturales y Museo, Universidad Nacional de La Plata, La Plata, Argentina, 176 pp. http://sedici.unlp.edu.ar/bitstream/handle/10915/48533/Documento_completo.pdf-PDFA.pdf?sequence=4&isAllowed=y. (Last access November 2019)

Rodríguez SM, Moreira MI, Giménez RA, Russo S, Márquez AM, Ricco RA, Gurni AA, Wagner ML (2008) Acción insecticida de extractos de *Picrasma crenata* (Vell.) Engl. (Simaroubaceae) en el gorgojo del arroz, *Sitophilus oryzae* L. (Coleoptera, Curculionidae). Dominguezia 24(2):95–102

Rodríguez S, Regonat M, Carrizo P, Meilán J, Wagner M, Gurni A (2011) Activity of nonpolar extracts from *Picrasma crenata* (Simaroubaceae) against *Myzus persicae* (Hemiptera: Aphidididae). Rev Latinoam Quím 39(3):113–120

Roldán RM, Noriega MF, Wagner ML, Gurni AA, Bassols GB (2007) Estudio de genotoxicidad de *Picrasma crenata* (Vell.) Engl. -Simaroubaceae. Acta Toxicol Argent 15(2):39–42

Sánchez E, Comín J (1971) Two new β-carboline alkaloids from *Aeschrion crenata*. Phytochemistry 10(9):2155–2159

The Plant List (2019) Version 1.1 published on the Internet: http://www.theplantlist.org. (Last access November 2019)

Toursarkissian M (1980) Plantas medicinales de la Argentina: Sus nombres botánicos, vulgares, usos y distribución geográfica. Ed. Hemisferio Sur, Buenos Aires, Argentina,79 pp

Vitagliano J, Comín J (1972) Quassinoids from *Aeschrion crenata*. Phytochemistry 11(2):807–810

Zank S, Hanazaki N, de Mello AS (2015) Participatory approaches and conservation of medicinal plants: identifying priority species in the community of Areais da Ribanceira (Brazil). Ethnobot Res Appl 14:357–366

Quillaja brasiliensis (A. St.-Hil. & Tul.) Mart.

Fernando Ferreira Chiesa, Federico Wallace, Zohra Bennadji, Manuel Minteguiaga, Patricia Basile, Juliane Deise Fleck, Simone Gasparín Verza, and Cristina Olivaro

Quillaja brasiliensis. Valle Edén, Uruguay. (Photo: Z Bennadji)

F. F. Chiesa (✉)
Espacio de Ciencia y Tecnología Química, Centro Universitario Regional Noreste, Sede Tacuarembó, Universidad de la República (UdelaR), Tacuarembó, Uruguay

Laboratorio de Carbohidratos y Glicoconjugados, Departamento de Química Orgánica, Facultad de Química, Universidad de la República (UdelaR), Montevideo, Uruguay
e-mail: ff@fq.edu.uy

Á. Máthé, A. Bandoni (eds.), *Medicinal and Aromatic Plants of South America Vol. 2*, Medicinal and Aromatic Plants of the World 7,
https://doi.org/10.1007/978-3-030-62818-5_35

Abstract *Quillaja brasiliensis* (A. St.-Hil. & Tul.) Mart. is currently one of the only two species of the only genus in the Quillajaceae D. Don (Dicotyledoneae) family. Recently, remarkable developments have been achieved regarding its domestication, propagation, and biochemical study. The first results on the structural study of the complex mixture of saponins produced by different organs of the tree have been published. Analogously to the related species *Q. saponaria* Molina, its saponins, either alone or in colloidal formulations, have been proved to be very effective as adjuvants in experimental vaccines, being able to elicit an early humoral and cellular response against the co-administered antigens. In consideration of the growing knowledge on *Q. brasiliensis*, this species constitutes a very promising botanical source for the development of valuable products for the biotechnological industry and research.

Keywords "Quillaja" · "Palo de jabón" · Saponins · Quillaic acid · Vaccines adjuvants · Tensoactives

1 Introduction

The traditional use of *Quillaja brasiliensis* (A. St.-Hil. & Tul.) Mart. was originally related to the foam-forming capacity of the aqueous extracts of its aerial organs (leaves and branches). More recently, progress has been made regarding the immunoadjuvant activity of its saponins and aqueous extracts that show promising properties as potential vaccine adjuvants. The phytochemical analysis of the aqueous extracts and fractions from its aerial parts have shown the presence of a complex mixture of triterpenic saponins, to which the adjuvant activity can be ascribed according to the current knowledge on this and the related species *Q. saponaria*. Studies on the propagation and cultivation of this tree species are promising. These and the increasing amount of available biological and phytochemical information

F. Wallace · M. Minteguiaga · C. Olivaro
Espacio de Ciencia y Tecnología Química, Centro Universitario Regional Noreste, Sede Tacuarembó, Universidad de la República (UdelaR), Tacuarembó, Uruguay

Z. Bennadji
Programa Nacional de Investigación en Producción Forestal, Instituto Nacional de Investigación Agropecuaria, Tacuarembó, Uruguay

P. Basile
Espacio de Biología Vegetal del Noreste, Centro Universitario Regional Noreste, Sede Tacuarembó, Universidad de la República (UdelaR), Tacuarembó, Uruguay

J. D. Fleck · S. G. Verza
Molecular Microbiology Laboratory, Institute of Health Sciences, Feevale University, Novo Hamburgo, RS, Brazil

point to the feasibility of *Q. brasiliensis,* as a potential economic source for the production and development of new biologicals of natural origin.

2 Taxonomic Characteristics

Quillajaceae D. Don (Dicotyledoneae) is a numerically very restricted botanical family, which comprises a unique genus (*Quillaja* Molina) with only two, currently accepted, species: *Q. saponaria* Molina and *Q. brasiliensis* (A. St.-Hil. & Tul.) Mart. (Luebert 2014; Tropicos 2020; The Plant List 2020). Earlier, Don (1831) segregated Quillajaceae from Rosaceae, but the scientific support for such separation is only recent. It was based mainly on DNA phylogenetic analysis from the plastid region sequence (*rbcL*) (Morgan et al. 1994; Bello et al. 2009). Moreover, chromosome numbers (Goldblat 1976), crystal morphology in the bark and leaves (Lersten and Horner 2005) and floral traits (Bello et al. 2007) largely have corroborated this separation, building the foundations for a new taxonomy proposal for the genus *Quillaja* (Soltis et al. 2017), as well as for *Q. brasiliensis* (Luebert 2013, 2014). The presence of saponins in the bark of *Q. saponaria* and *Q. brasiliensis* (and even, of some polyphenols such leucodelphinidin) differently from Rosaceae members gives chemotaxonomic additional evidence to confirm such segregation (Bate-Smith 1965; Kensil et al. 1991; Wallace et al. 2019).

Q. brasiliensis (Photo 1) is a South American monoecious and evergreen shrub or tree species that was originally described in the Brazilian Flora by Saint-Hilaire and Tulasne (1842) as *Fontenellea brasiliensis*. Later, it was classified as *Quillaja brasiliensis* by von Martius (1843), but these authors only informed the bark characteristics, and no additional data regarding other organs were provided (Fuks 1982). *Q. brasiliensis* natural populations are distributed in Argentina, Brazil, Paraguay and Uruguay (Fuks 1982; Luebert 2014; Ribeiro et al. 2018; Ross et al. 2018; Tropicos 2020). The vernacular names of the species for the Spanish speakers countries are: "quillaja" "quillay, "árbol del jabón" (soap tree), "palo de jabón" (soap stick), and "jabonera"; in a clear reference to the saponins obtained from the bark and leaves of the tree which are employed traditionally as a soap substitute (Arechavaleta 1901; De Mello and Cantos 2017; Ribeiro et al. 2018; Ross et al. 2018). In Brazil, also it is known with the similar meaning terms "quilaia", "árvore-de-sabão", "saboeiro" "pau-sabão", "saboneteira", "lava-cabelo", and "sabão-de-soldado" (soldier's soap) (Cantos 2013; Ribeiro et al. 2018). However, also in Brazil, this species is popularly named as: "bugreiro-da-várzea", "pau-de-bugre", "aroeira-vermelha", "assa-toucinho", "pirubaúva", "tibura", "timbaúva", "timbau-vão" and "timbuva" (Fuks 1982; Cantos 2013; Ribeiro et al. 2018).

The more common synonymous name is *Fontenellea brasiliensis* A. St.-Hil. & Tul. (Luebert 2013; Tropicos 2020; The Plant List 2020). Luebert (2014) also consider as synonyms the names *Q. lanceolata* D. Dietr., *Q. lancifolia* D. Don, and *Q. sellowiana* Walp. However, there are some taxonomic controversies regarding

Photo 1 Details of the flowers, bark, and leaves of Q. *brasiliensis*. (Photos: Z Bennadji and F Ferreira)

the authenticity of such names (and others related), being currently their status unresolved/untypified (Luebert 2014; Tropicos 2020; The Plant List 2020).

3 Crude Drug Used

The bark and the chopped trunk were commercialized during the nineteenth and early twentieth centuries for "saponification of fat materials" and as a substitute to soap due to the detergent properties of the aqueous extract of these materials. Even though it was more commonly used as a substitute for *Q. saponaria*, natural from Chile and Peru (Arechavaleta 1901; Lombardo 1969). Contemporary, there is a reference regarding the use of the leaves as medicinal for external washes in the Tacuarembó Department, Uruguay (Tabakian 2019).

4 Major Chemical Constituents and Bioactive Compounds

The genus *Quillaja* is rich in saponins with several biological activities described, especially as an immunoadjuvant and immunostimulant, being important compounds for vaccine development (Sjölander et al. 2001; Rajput et al. 2007; Sun et al. 2009; Barbosa 2014; Marty-Roix et al. 2016). This genus has recently been reviewed, covering chemical characteristics and biological activities of their saponins, as well as its main applications (Fleck et al. 2019; Magedans et al. 2019).

In contrast to *Q. saponaria*, the phytochemical studies on *Q. brasiliensis* started later. Just in 2004, a pioneering work reported a prosapogenin composed of quillaic acid glycosylated at C-3 with a β-D-GlcA*p* residue from the products of partial hydrolysis of the leaves aqueous extract (Kauffmann et al. 2004). However, the first complete structural report of *Q. brasiliensis* saponins, from leaves and bark, were presented recently (Wallace et al. 2017a, 2019) using direct infusion and liquid chromatography/electrospray ionization ion trap multiple-stage mass spectrometry in negative ion mode (DI-ESI-IT-MS^n and LC-ESI-IT-MS^2). All those saponins are bidesmosidic and consists of the triterpene aglycone substituted with oligosaccharides at C-3 and C-28. The C-3 oligosaccharide is almost always a branched trisaccharide composed of an acetal-linked β-D-GlcA*p* residue bearing β-D-Gal*p* at its O-2 position and either a β-D-Xyl*p* or an α-L-Rha*p* residue at its O-3 position (R_0). The oligosaccharide substituted at C-28 also has a constant disaccharide consisting of α-L-Rha*p*-(1→2)-β-D-Fuc*p* esterified to the aglycone. This disaccharide is further extended by monosaccharide, oligosaccharide, or acyl substituents at particular positions (R_1 – R_4). The monosaccharide substituents found are D-Glc, D-Xyl, D-Api, and L-Rha. The acyl group (in R_4) may be a simple acetyl substituent or a complex moiety, such as a fatty acid chain linked to an L-Ara residue (Fa-Ara). These authors identified 48 saponins from leaves and 54 saponins from bark, bearing five types of triterpenic aglycones, namely quillaic acid, 22-β-hydroxy-quillaic acid, gypsogenin, phytolaccinic acid, and its O-23 acetate (Fig. 1).

Moreover, the saponins from both species are bidesmosidic, present similar patterns of glycosylation at positions C-3 and C-28 of the aglycone with the same conserved motifs. Despite those mentioned above, some saponins from the bark of *Q. brasiliensis* show new structural motifs not previously reported in the genus *Quillaja*. One of them corresponds to a gypsogenin aglycone attached to a trisaccharide in C-3, with a deoxyhexose at R_0 instead of a xylose residue. The other motifs correspond to the phytolaccinic acid and 23-O-acetyl-phytolaccinic acid attached to a trisaccharide in C-3 instead of a disaccharide, both with a deoxyhexose at R_0 (Wallace et al. 2019).

As is the case for most of the secondary metabolites, the saponin content in the species is highly dependent on the environmental conditions. Indeed, de Costa et al. (2013) evaluated through *in vitro* studies on *Q. brasiliensis* (leaf-disks) the influence of different abiotic and simulate-biotic stress-causing factors such: osmotic concentration, ultrasound, ultraviolet radiation, light intensity and the application of the hormones jasmonic (JA) and salicylic acids (SA). The application of an osmotic

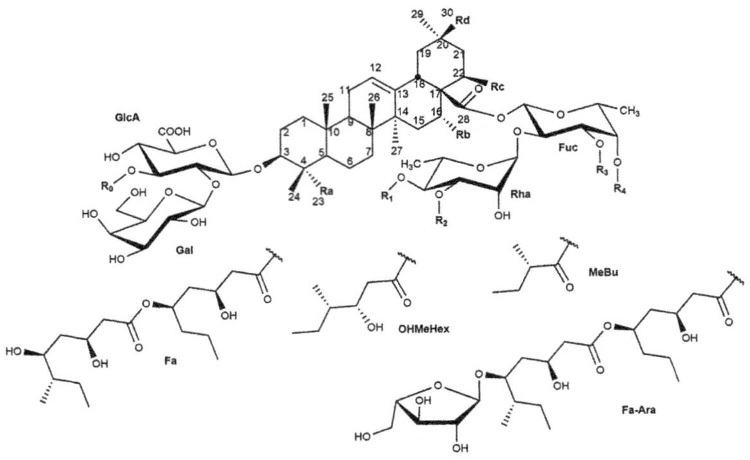

Aglycones	Ra	Rb	Rc	Rd
Quillaic acid (Q)	CHO	OH	H	CH_3
22β-hydroxyquillaic acid (Q-OH)	CHO	OH	OH	CH_3
Gypsogenin (G)	CHO	H	H	CH_3
Phytolaccinic acid (P)	CH_2OH	H	H	$COOCH_3$
23-O-acetylphytolaccinic acid (P-Ac)	CH_2OCOCH_3	H	H	$COOCH_3$

Fig. 1 The general structure of *Q. brasiliensis* saponins. R_0-R_3: different mono- or oligosaccharide substituents. R_4: different acyl chains (Fa, Fa-Ara, MeBu, OHMEHex, etc.). (Adapted from Fleck et al. 2019; Wallace et al. 2019)

stress agent (sodium chloride, sorbitol, or polyethylene glycol) evaluated by the authors triggers a significant enhancement (more than threefold the control level) of the saponin production, mainly after 4 days from the application. These results could be interpreted as an adaptation mechanism of the plant to tolerate drought conditions or salinity excess, which was preliminarily confirmed at field conditions: *Q. brasiliensis* adult trees produced more saponins in the driest months in Southern Brazil (de Costa et al. 2013). Regarding ultrasound exposure (that in general influence the cell redox balance), the saponin concentration level in *Q. brasiliensis* leaf-disks also responded increasing, which also could be associated with a stress defense strategy. Higher visible light irradiance level and UV-C irradiation also augmented the saponin content in a time-dependent manner; in contrast, UV-B did not induce any effect (de Costa et al. 2013). Finally, the authors observed that the application of jasmonic and salicylic acids hormones induced the biosynthesis of saponins in the same line as exerting mechanical damage on the disks, thus, influencing the physiology of the plant as a putative mechanism to lead with an herbivore attack (de Costa et al. 2013). Regarding hormones, despite Fleck et al. (2006) found alteration in the survival, number of roots and its length in *Q. brasiliensis* micro-propagated rooted-cuttings by the action of auxins (3-indoleacetic, 3-indolebutyric, and α-naphtaleneacetic acids); they did not find any influence on the saponin contents which suggests that the hormonal control on the saponins biosynthesis is only partial.

Regarding the saponin organ-specific distribution patterns, Fleck et al. (2013), applying an HPLC method, did not find any significant differences in the content between the bark, branches, and leaves from adult *Q. brasiliensis* trees. However, they found more saponins in the leaves of laboratory seedlings compared to the roots and stems, and this value was fourfold higher compared with adult trees (Fleck et al. 2013). In agreement, preliminary results obtained the afrosimetric method confirmed that seedling leaves accumulate higher levels of saponins compared with adult trees, but a seasonal concentration dynamic was observed (Wallace et al. 2017b).

In addition to the saponins in *Q. brasiliensis,* quercetin, rutin, and a new diterpene, 19-*O*-β-D-glucopyranoside of 16-hydroxylambertic acid have been isolated from the leaves (Kauffmann et al. 2004). Furthermore, 1D and 2D NMR experiments allowed the identification of piscidic acid as the major impurity of Fraction B obtained from *Q. brasiliensis* leaves (Wallace et al. 2016, 2017a, 2019).

5 Morphological Description

Q. brasiliensis is commonly known as soap tree because its leaves and bark produce foam when agitated in water due to the presence of saponins. It is a tree with an average height of 10–20 m, usually with a single straight stem (pristine population) or multiple ones (in secondary formations). The bark is dark grey to black with strong longitudinal grooves, globose canopy, and dark green persistent foliage (Reitz 1996; Marchiori 2000; Brussa and Grela 2007; Ribeiro et al. 2018).

The leaves are simple (1.5–3 cm of width and 4–8 cm of length), alternate, leathery, elliptical to elliptical/lanceolate in shape; these characteristics vary according to the environmental conditions where the plant species grows and its regrowth stage. Furthermore, the adult leaves are glabrous on both sides, with paucidentate or entire margins, acute apex, three nerved starting from the base with a well-defined yellowish central rib (Brussa and Grela 2007; Ross et al. 2018).

The flowers develop in small axillary greenish corymbs, with male and female flowers on the same individual. The calyx is yellowish-green, penta-sepal, while the corolla is dialypetal and penta-petal. The corolla is actinomorphous with ten biseriate stamens and a penta-carpel ovary (Brussa and Grela 2007; Ross et al. 2018). The flowering occurs in summer (January and February), while the fructification is from the final summer to autumn (March–May). The fruits are very characteristic, star-shaped with five follicles joined only by its base, dry, grey when ripe, with longitudinal dehiscence. The fruits have multiple seeds that disperse by the action of the wind after ripening (Brussa and Grela 2007; Ribeiro et al. 2018).

6 Geographical Distribution

According to the Köppen climate classification, *Q. brasiliensis* grows naturally from tropical to subtropical zones (Chen and Chen 2013). This species occurs in North-eastern Argentina, Eastern Paraguay, Northern Uruguay, and Southern Brazil. In Southern Brazil, the area of occurrence of *Q. brasiliensis* is typically the Ombrophilous Mixed Forest (a highly threatened ecosystem, 900–1000 m altitude) where it is a pioneer species, associated frequently with *Araucaria angustifolia* (Bertol.) Kuntze ("Brazilian pine") (Luebert 2014; Ribeiro et al. 2018). In Argentina, this species is associated with grassland and deciduous and semi-deciduous forest, while in Uruguay and Paraguay, it occurs in the riverine forest (Brussa and Grela 2007; Luebert 2013; Ribeiro et al. 2018; Ross et al. 2018; Velazco et al. 2018).

In Uruguay, its phytogeographical area lies in the humid subtropical zone with hot summers (Cfa). In general terms, this area has an annual precipitation average of 1200 mm and an annual media temperature of 25 °C in summer and 15 °C in winter (GRAS-INIA 2019). *Q. brasiliensis* flourishes in riparian vegetal formations, exhibiting a marked pioneer, hygrophilous and heliophilous behavior, *Q. brasiliensis* has also been reported in more rocky and xerophytic zones (Brussa and Grela 2007; Ross et al. 2018). In Brazil and Paraguay, *Q. brasiliensis* has been included in the category "endangered" regarding conservational status (Sakuragui et al. 2013; Ribeiro et al. 2018), which means that currently it is seriously threatened; fortunately, this is not the case for Uruguay where this species is not in the "red list" (Marchesi et al. 2015) and presents good representation in northern Uruguayan native forests (MGAP 2019).

Employing a mathematical model (*Maxent* algorithm), De Mello and Cantos (2017) found a high correlation between the current natural distribution of *Q. brasiliensis* and the predicted by the model, highlighting that the occurrence of rains (mainly in the driest periods) might be the reason for such distribution.

7 Ecological Requirements

Q. brasiliensis is easily propagated by seeds, it does not require pre-germinative treatments and can be sown directly in tubes with an adequate substrate (Brussa and Grela 2007; Velazco et al. 2018). Nevertheless, in nursery, seedlings are highly susceptible to the incidence of pathogens (Velazco et al. 2018).

Seeds have a high germination rate, up to 80% and can be stored for 1.5 years. After this period, the seed viability decreases to 30%. Environmental factors, such as temperature and photoperiod, influence the germination rate: seeds submitted to temperatures around 25 °C and a short-day photoperiod increase germination and synchrony of seedlings. Transplantation does not have any negative effects on seedling survival and growth compared to seedlings from direct sowing (Velazco et al. 2018). The seedlings and juvenile plants present satisfactory growth when cultivated with a substrate consisting of peat, organic compost, sand, and vermiculite (Ribeiro et al. 2018). *Q. brasiliensis* can be propagated vegetatively by micropropagation techniques, using aseptically germinated seeds or shoots cultures of selected germplasm. The use of this technique allows the production of plants that can accumulate the bioactive saponin fraction (Fleck et al. 2006). Fleck et al. (2006, 2013) showed that leaves from young plants accumulated higher amounts of saponins than leaves from adult trees, suggesting a new biomass source of saponins.

8 Traditional Use (Part(s) Used) and Common Knowledge

Besides its traditional use as soap or as a soap component, the use of the *Q. brasiliensis* wood as combustion fuel, or as a structural material in civil constructions and carpentry is also reported (Sakuragui et al. 2013; Ribeiro et al. 2018). *Q. brasiliensis* leaves, branches, and bark have a high saponin concentration with multiple uses. Most recently, its medicinal use in external washes was reported by Tabakian (2019).

It is also used as an insecticide and his wood is exploited by the construction, carpentry industries and as fuelwood (Carvalho 2003).

9 Modern Medicine Based on Its Traditional Medicine Uses

The investigation on the biological activities of *Q. brasiliensis* materials and secondary products were initially encouraged by the known biological activities of the related species *Q. saponaria*, of widespread use in different areas, including industrial use as a surfactant, use as alimentary additives, and in animal and human health.

A comprehensive review of *Q. brasiliensis* has been published, including an extensive revision of the presently known biological activities of its metabolites and different preparations (Fleck et al. 2019).

Aqueous extract and some of the purified fractions obtained from leaves of *Q. brasiliensis* have immunoadjuvant activity comparable with Quil-A®, the main commercial adjuvant product based on *Q. saponaria* saponins. The effectiveness of *Q. brasiliensis* saponins has been confirmed in experimental vaccines against bovine *herpesvirus* type 1 and 5 (BoHV), human poliovirus, bovine viral diarrhea virus (BVDV) and rabies in mice (Fleck et al. 2006; Silveira et al. 2011; de Costa et al. 2014; Cibulski et al. 2016a; Yendo et al. 2016). These saponins are also able to form micellar nanometric structures rod-shaped (Ferreira et al. 2012, 2016, 2018) and ISCOM- type (Quirici et al. 2013), which are even more effective vaccine adjuvants generating both humoral and cellular immune response (Ferreira et al. 2012, 2016, 2018; Cibulski et al. 2016b, 2018, 2019). Purification protocols to obtain an immunoadjuvant saponin-rich fraction from *Q. brasiliensis* have been published (Fleck et al. 2006; Wallace et al. 2017a; Yendo et al. 2017).

QS-21, a mixture of two isomeric saponins from *Q. saponaria*, is also found in leaves and bark extracts from *Q. brasiliensis* (Wallace et al. 2017a, 2019). QS-21 has been profusely tested in clinical studies as a vaccine adjuvant for human use for the prevention or treatment of infectious diseases, cancer, HIV, and Alzheimer's (Fleck et al. 2019). In 2015, a malaria vaccine containing the adjuvant AS01, composed by QS-21, monophosphoryl lipid A and liposomes, was approved for use in children were malaria is endemic (European Medicines Agency 2015). A vaccine also containing the AS01 adjuvant has been recently proved effective in the prevention of tuberculosis in a large human trial (Tait et al. 2019). Thus, *Q. brasiliensis* is also an alternative source of high-added value saponins.

Many other biological activities have only been widely studied for the saponins of the bark of *Q. saponaria*, but since structurally, these compounds from both *Quillaja* spp. are analogous, it would be expected that those biological activities be shared. The main biological properties reported include antiviral, antifungal, antibacterial, antiparasitic, antitumor, and hepatoprotective (Fleck et al. 2019). The *Quillaja* spp. saponins have only be assessed simultaneously in the case of the anti-*Trichomonas vaginalis* activity, demonstrating that both inhibited the viability of the protozoan (Rocha et al. 2012).

The saponins and different aqueous extracts of leaves present are hemolytic and cytotoxic due to their tensioactive properties when administered parenterally in animal models (Silveira et al. 2011; Ferreira et al. 2012; Fleck et al. 2019).

10 Conclusions

Quillaja brasiliensis, traditionally known for its foam-forming properties, has been the subject for ongoing research efforts regarding its botany, ecology, domestication, biochemistry, and the structure of its secondary metabolites, as well as its adjuvant properties as in experimental vaccines. In this regard, thanks to the growing knowledge on this species, there seems to be a clear prospect for its future sustainable production and the development of different biotechnological applications.

References

Arechavaleta J (1901) Flora Uruguaya. Talleres Barreiro y Ramos, Montevideo, Uruguay. Volume I in Anales del Museo de Historia Natural (Tomo III), pp 450–452. In: facsimile: https://autores.uy/. (Last access Jan 2020)

Barbosa ADP (2014) Saponins as immunoadjuvant agent: a review. Afr J Pharm Pharmacol 8:1049–1057

Bate-Smith (1965) Investigation of the chemistry and taxonomy of sub-tribe Quillajeae of the Rosaceae using comparisons of fresh and herbarium material. Phytochemistry 4(4):535–539

Bello MA, Hawkins JÁ, Rudall PJ (2007) Floral morphology and development in Quillajaceae and Surianaceae (Fabales), the species-poor relatives of Leguminosae and Polygalaceae. Ann Bot 100(7):1491–1505

Bello MA, Bruneau A, Forest F, Hawkins JA (2009) Elusive relationships within order Fabales: phylogenetic analyses using *matK* and *rbcL* sequence data. Syst Bot 34(1):102–114

Brussa C, Grela A (2007) Flora arbórea del Uruguay, con énfasis en las especies de Rivera y Tacuarembó. Ed. COFUSA. Montevideo, Uruguay, 544 pp

Cantos AA (2013) Metodologias para condução do teste de germinação e produção de mudas de *Quillaja brasiliensis*. Thesis. Universidade Federal de Pelotas, Pelotas, Brazil, 52 pp

Carvalho PER (2003) Espécies arbóreas brasileiras. Embrapa Informação Tecnológica, Brasilia, Brazil, 1039 pp

Chen D, Chen HW (2013) Using the Köppen classification to quantify climate variation and change: an example for 1901–2010. Environ Dev 6:69–79

Cibulski SP, Silveira F, Mourglia-Ettlin G, Teixeira TF, dos Santos HF, Yendo AC, de Costa F, Fett-Neto AG, Gosmann G, Roehe PM (2016a) *Quillaja brasiliensis* saponins induce robust humoral and cellular responses in a bovine viral diarrhea virus vaccine in mice. Comp Immunol Microbiol Infect Dis 45:1–8

Cibulski SP, Mourglia-Ettlin G, Teixeira TF, Quirici L, Roehe PM, Ferreira F, Silveira F (2016b) Novel ISCOMs from *Quillaja brasiliensis* saponins induce mucosal and systemic antibody production, T-cell responses and improved antigen uptake. Vaccine 34:1162–1171

Cibulski SP, Rivera-Patron M, Mourglia-Ettlin G, Casaravilla C, Yendo ACA, Fett-Neto AG, Chabalgoity JA, Moreno M, Roehe PM, Silveira F (2018) *Quillaja brasiliensis* saponin-based nanoparticulate adjuvants are capable of triggering early immune responses. Sci Rep 8:1–15

de Costa F, Yendo ACA, Fleck JD, Gosmann G, Arthur Germano Fett-Neto AG (2013) Accumulation of a bioactive triterpene saponin fraction of *Quillaja brasiliensis* leaves is associated with abiotic and biotic stresses. Plant Physiol Biochem 66:56–62

de Costa F, Yendo ACA, Cibulski SP, Fleck JD, Roehe PM, Spilki FR, Gosmann G, Fett-Neto AG (2014) Alternative inactivated poliovirus vaccines adjuvanted with *Quillaja brasiliensis* or Quil-A saponins are equally effective in inducing specific immune responses. PLoS One 9:e105374

De Mello LM, Cantos AA (2017) Modelo de la distribución espacial de *Quillaja brasiliensis* (A. St.-Hil. & Tul.) Mart. en el sur de Brasil y la historia de la dispersión del género *Quillaja* (Quillajaceae) en América del Sur. Bol Mus Nac Hist Nat Chile 66(2):297–308

Don D (1831) On the characters and affinities of certain Genera, chiefly belonging to the Flora Peruviana: *Quillaja* and *Kageneckia*. Edin New Philos J 10:229–232. In: facsimile: https://books.google.com/. (Last access Jan 2020)

European Medicines Agency (2015). http://www.ema.europa.eu/ema/index.jsp?curl=pages/news_and_events/news/2015/07/news_detail_002376.jsp&mid=WC0b01ac058004d5c1. (Last access Feb 2020)

Ferreira F, Bessio MI, Quirici L, Silveira F (2012) Adyuvantes de vacunación, preparación y vacunas que lo contienen. Uruguay DNPI, Patent Application 34,506, filed December 18, 2012

Ferreira F, Bessio MI, Quirici L, Silveira F (2016) Vaccination adjuvant, and preparation and vaccines containing the same. Australia Patent AU 2013360504 B2, filed August 12, 2013, issued January 12, 2016

Ferreira F, Bessio MI, Quirici L, Silveira F (2018) Vaccination adjuvant, and preparation and vaccines containing the same. European Patent EP 2929894 B1, filed August 12, 2013, issued February 21, 2018

Fleck JD, Kauffmann C, Spilki F, Lencina CL, Roehe PM, Gosmann G (2006) Adjuvant activity of *Quillaja brasiliensis* saponins on the immune responses to bovine herpesvirus type 1 in mice. Vaccine 24:7129–7134

Fleck JD, de Costa F, Yendo ACA, Segalin J, Dalla Costa TCT, Fett-Neto AG, Gosmann G (2013) Determination of new immunoadjuvant saponin named QB-90, and analysis of its organ-specific distribution in *Quillaja brasiliensis* by HPLC. Nat Prod Res 27(10):907–910

Fleck JD, Betti AH, Pereira da Silva F, Troian EA, Olivaro C, Ferreira F, Verza SG (2019) Saponins from *Quillaja saponaria* and *Quillaja brasiliensis*: Particular chemical characteristics and biological activities. Molecules 24:171–199

Fuks R (1982) O gênero *Quillaja* no Brasil. Arq Jard Bot Rio de Janeiro 26:61–67

Goldblatt P (1976) Cytotaxonomic studies in the tribe Quillajeae (Rosaceae). Ann Mo Bot Gard 63(1):200–206

GRAS-INIA. https://inia.uy/gras. (Last access, October 2019)

Kauffmann C, Machado AM, Fleck JD, Provensi G, Pires VS, Guillaume D, Sonnet P, Reginatto FH, Schenkel EP, Gosmann G (2004) Constituents from leaves of *Quillaja brasiliensis*. Nat Prod Res 18:153–157

Kensil CR, Patel U, Lennick M, Marciani D (1991) Separation and characterization of saponins with adjuvant activity from *Quillaja saponaria* Molina cortex. J Immunol 146(2):431–437

Lersten NR, Horner HT (2005) Macropattern of styloid and druse crystals in *Quillaja* (Quillajaceae) bark and leaves. Int J Plant Sci 166(5):705–711

Lombardo A (1969) Árboles y Arbustos. Colección Nuestra Tierra, vol 27. Ed. Nuestra Tierra, Montevideo, Uruguay, pp 63–64

Luebert F (2013) Proposal to conserve the name *Fontenellea brasiliensis* (*Quillaja brasiliensis*) against *Q. lancifolia* (Quillajaceae). Taxon 62(5):1068–1069

Luebert F (2014) Taxonomy and distribution of the genus *Quillaja* Molina (Quillajaceae). Feddes Repert 124(4):157–162

Magedans YVS, Yendo ACA, de Costa F, Gosmann G, Fett-Neto AG (2019) Foamy matters: an update on *Quillaja* saponins and their use as immunoadjuvants. Future Med Chem 11:1485–1499

Marchiori JNC (2000) Dendrologia das Angiospermas: das Bixáceas às Rosáceas. Universidade Federal de Santa Maria, Santa Maria, Santa Catarina, Brazil, 240 pp

Marty-Roix R, Vladimer GI, Pouliot K, Weng D, Buglione-Corbett R, West K, MacMicking JD, Chee JD, Wang S, Lu S (2016) Identification of QS-21 as an inflammasome-activating molecular component of Saponin adjuvants. J Biol Chem 291:1123–1136

Morgan DR, Soltis DE, Robertson KR (1994) Systematic and evolutionary implications of *rbcL* sequence variation in Rosaceae. Am J Bot 81(7):890–903

Müller LE, Schiedeck G (2018) Physical properties of botanical surfactants. Sci Total Environ 610–611:1133–1137

Quirici L, Verza SG, Mastrogiovanni M, Miraballes I, Casanova G, Soule S, et al (2013) Candidate particulate antigen delivery system based on *Quillaja brasiliensis* saponins. Elsevier, editor. seventh Vaccine & ISV Congress. Elsevier, Sitges,. p 11. https://web.archive.org/web/20170203090417/http://isvcongress.org/images/docs/VC7_Poster.pdf. (Last access, February 2020)

Rajput ZI, Hu S, Xiao C, Arijo AG (2007) Adjuvant effects of saponins on animal immune responses. J Zhejiang Univ Sci B 8:153–161

Reitz R (1996) Flora Ilustrada Catarinense: Rosáceas. Herbário Barbosa Rodrigues, Itajaí, Santa Catarina, Brazil, 135 pp

Ribeiro CDL, Blum CT, Grabias J, Borgo M, Hoffmann PM, Velazco SJE (2018) *Quillaja brasiliensis* (A. St.-Hil. & Tul.) Mart. (Quillajaceae). Manual de Cultivo, first Ed. Boletín Chauá 001. Published on the Internet: http://www.conexaoambiental.pr.gov.br/sites/conexao-ambiental/arquivos_restritos/files/documento/2019-07/quillaja_brasiliensis_-boletim_chaua_001.pdf. (Last access, January 2020)

Rocha TD, De Brum Vieira P, Gnoatto SCB, Tasca T, Gosmann G (2012) Anti-*Trichomonas vaginalis* activity of saponins from *Quillaja*, *Passiflora*, and *Ilex* species. Parasitol Res 110:2551–2556

Ross P, Muñoz J, Cracco P (2018) Flora Indígena del Uruguay. Árboles y arbustos ornamentales. 3ª ed. Ed. Hemisferio Sur. Montevideo, Uruguay, 320 pp

Saint-Hilaire A, Tulasne LR (1842) *Fontenellea brasiliensis*. In: Flora Brésil. Ann Sci Nat Bot 2(17):141. Cited by Fuks (1982) and Luebert (2014)

Sjölander A, Drane D, Maraskovsky E, Scheerlinck JP, Suhrbier A, Tennent J, Pearse M (2001) Immune responses to ISCOM formulations in animal and primate models. Vaccine 19:2661–2665

Soltis D, Soltis P, Endress P, Chase M, Manchester S, Judd W, Majure L, Mavrodiev E (2017) Phylogeny and evolution of the Angiosperms: revised and updated edition. Ed. University of Chicago Press, Chicago, USA, 590 pp

Sun H-X, Xie Y, Ye Y-P (2009) Advances in saponin-based adjuvants. Vaccine 27:1787–1796

Tabakian G (2019) Estudio comparativo de plantas medicinales vinculadas a la tradiciones indígenas y europeas en Uruguay. Bonplandia 28(2):135–158

Tait DR, Hatherill O, Der Meeren V, Ginsberg AM, Van Brakel E, Salaun B, Scriba TJ et al (2019) Final analysis of a trial of $M72/AS01_E$ vaccine to prevent tuberculosis. New Engl J Med 381:2429–2439

The Plant List (2020) Version 1.1 Published on the Internet: http://www.theplantlist.org. (Last access, January 2020)

Tropicos (2020) *Quillaja*. Published on the Internet: http://www.tropicos.org/. (Last access January 2020)

Velazco SJE, Blum CT, Hoffmann PM (2018) Germination and seedlings development of the threatened species *Quillaja brasiliensis*. Cerne 24:90–97

von Martius FCP (1843) *Quillaja brasiliensis*. Syst Mat Med Veg Bras. p 127. Cited by Fuks (1982) and Luebert (2014)

Wallace F, Bennadji Z, Fontana C, Bessio MI, Ferreira F, Olivaro C (2016) QS-21: a highly valuable biotechnological product can be obtained in a sustainable form from *Q. brasiliensis*, a native tree from Uruguay. Memoria digital. Workshop on insights and strategies towards a bio-based economy (Uruguay). pp 114–115

Wallace F, Bennadji Z, Ferreira F, Olivaro C (2017a) Analysis of an immunoadjuvant saponin fraction from *Quillaja brasiliensis* leaves by electrospray ionization ion trap multiple-stage mass spectrometry. Phytochem Lett 20:228–233

Wallace F, Bennadji Z, Mello S, de Souza G, Ferreira F, Olivaro C (2017b) Determinación de saponinas totales en *Quillaja brasiliensis* por el método de la espuma. Memoria digital Congreso ENAQUI 5 (Uruguay 2017). Published on the Internet: http://www.ainfo.inia.uy/digital/bitstream/item/7398/1/Determinacion.pdf. (Last access, January 2020)

Wallace F, Bennadji Z, Ferreira F, Olivaro C (2019) Structural characterization of new immunoadjuvant saponins from leaves and the first study of saponins from the bark of *Quillaja brasiliensis* by liquid chromatography electrospray ionization ion trap mass spectrometry. Phytochem Anal 30:644–652

Yendo ACA, de Costa F, Cibulski SP, Teixeira TF, Colling LC, Mastrogiovanni M, Soulé S, Roehe PM, Gosmann G, Ferreira FA, Fett-Neto AG (2016) A rabies vaccine adjuvanted with saponins from leaves of the soap tree (*Quillaja brasiliensis*) induces specific immune responses and protects against lethal challenge. Vaccine 34:2305–2311

Yendo ACA, de Costa F, Kauffmann C, Fleck JD, Gosmann G, Fett-Neto AG (2017). Purification of an immunoadjuvant saponin fraction from *Quillaja brasiliensis* leaves by reversed-phase silica gel chromatography. In Fox C (ed.): Vaccine adjuvants: methods and protocols, methods in molecular biology, vol 1494. Ed. Humana Press, New York, USA, pp 87–93

Quillaja saponaria Molina

Carla Delporte, Maité Rodríguez-Díaz, and Bruce K. Cassels

Quillaja saponaria Mol. (Photo: Carla Delporte)

C. Delporte (✉)
Departamento de Química Farmacológica y Toxicológica, Facultad de Ciencias Químicas y Farmacéuticas, Universidad de Chile, Santiago, Chile
e-mail: cdelpor@uchile.cl

M. Rodríguez-Díaz
Escuela de Química y Farmacia, Universidad Andrés Bello, Santiago, Chile

B. K. Cassels
Departamento de Química, Facultad de Ciencias, Santiago, Chile
e-mail: bcassels@uchile.cl

Á. Máthé, A. Bandoni (eds.), *Medicinal and Aromatic Plants of South America Vol. 2*, Medicinal and Aromatic Plants of the World 7,
https://doi.org/10.1007/978-3-030-62818-5_36

Abstract *Quillaja saponaria* Mol., Quillajaceae is an endemic Chilean tree known by the vernacular name: "quillay". Infusions of its leaves and bark are employed traditionally to combat cough and bronchitis. The bark decoction is used to treat chronic skin lesions. Also, the Mapuche people use the bark as an analgesic and as a detergent. The whole aerial parts are one of the main industrial sources of triterpene saponins. Their structures and biosynthesis are quite complex. These saponins are responsible for the biological properties for which they are used in the food, cosmetic and pharmaceutical industries for their surfactant properties. Due to their powerful immunoadjuvant properties, *Quillaja* saponins are used to manufacture vaccines for human and veterinary use. Together with antigens, cholesterol and phospholipids, they form immunostimulant complexes (ISCOMs) which have proved to be an effective way of presenting antigens to the immune system. Additionally, the analgesic and anti-inflammatory properties of the main aglycone, quillaic acid, have been demonstrated in preclinical models.

Keywords "Quillay" · Triterpene saponins · Immunoadjuvant saponins · Surfactant saponins · QS-21 · Quillaic acid

1 Introduction

Quillaja saponaria Mol., or "quillay" (Quillajaceae), is one of the most important tree species of Central Chile. In traditional medicine, the bark, leaves and flowers of *Q. saponaria* are used as anti-inflammatory agents. The following monograph presents updated scientific information, referring to chemical and pharmacological studies of its main secondary metabolites (triterpenic saponins), that support its use in: (i) the food, cosmetic and pharmaceutical industries, due to its physicochemical properties, (ii) veterinary and human vaccine production due to their powerful immunoadjuvant properties, among other uses referred in this compilation. It includes the results of recent preclinical studies of its main aglycone, quillaic acid, as well of some of its derivatives that present analgesic and anti-inflammatory activities of similar potency to synthetic non-steroidal anti-inflammatory drugs.

2 Taxonomic Characteristics

The genus *Quillaja* was created by Molina (1782) with *Q. saponaria* as the type species and the only one recognized until a century ago. Originally, this genus was classified in the family Rosaceae, but later molecular and morphological phylogenetic studies confirmed its status in the Quillajaceae D. Don, of the order Fabales

Bromhead and superorder Rosanae Takhtajan (Bello et al. 2012; Tropicos 2019). *Quillaja* comprises the species *Q. saponaria* Mol. and *Q. brasiliensis* (A. St.-Hil. & Tul.) Mart. (Luebert 2013). Its common name in Spanish is "quillay", or "küllay" in the Mapuche language. It has also been known by other common names in Spanish, such as "jabón de palo" and "palo jabón" (tree soap and soap-tree) (Muñoz-Schick et al. 2012). In English it may be referred to as soapbark or Panama bark, in German as "Seifenrinde and "Panamarinde", and in French as "bois de Panama", a reference to its export via Panama since Spanish colonial times.

Synonyms *Quillaja molinae* DC., *Quillaja poeppigii* Walp., *Smegmadermos emarginatus* Ruiz & Pav., *Smegmaria emarginata* (Ruiz & Pav.) Willd. (Rodríguez et al. 2018).

3 Crude Drug Used

In traditional medicine, bark, leaves and flowers of *Q. saponaria* are used (Gupta 1995). Other, non-medical uses of considerable commercial importance are: as a source of surfactant, emulsifying and foaming agents, mainly in the food and cosmetic industry (San Martín and Briones 1999).

4 Major Chemical Constituents and Bioactive Compounds

The earliest isolation of the mixed quillaja saponins dates back to 1828 (Henry and Boutron-Chaillard 1828, cited by Wisniak 2016), and since then the bark of *Q. saponaria* became one of the main sources of triterpene saponins worldwide (van Setten et al. 1995; San Martín and Briones 2000), although in recent years the whole aerial parts of the tree have come into use (San Martín 2002). At least 100 different saponins have been tentatively identified in the aqueous or alcoholic extracts of "quillay" bark and wood, with quillaic acid (3β,16α-dihydroxy-23-oxoolean-12--en-28-oic acid) as the main sapogenin. These saponins are bidesmosides, glycosylated at the C-3 and C-28 positions of the aglycone with different oligosaccharide and aliphatic acid substituents (Higuchi et al. 1987, 1988; Kensil et al. 1991; Guo et al. 1998; Guo and Kenne 2000; Nyberg et al. 2000; van Setten et al. 1995, 2000; Nord and Kenne 1999; Kite et al. 2004). The following figure shows a generic quillaja-saponin with the aglycone and the aliphatic acid dimer in red.

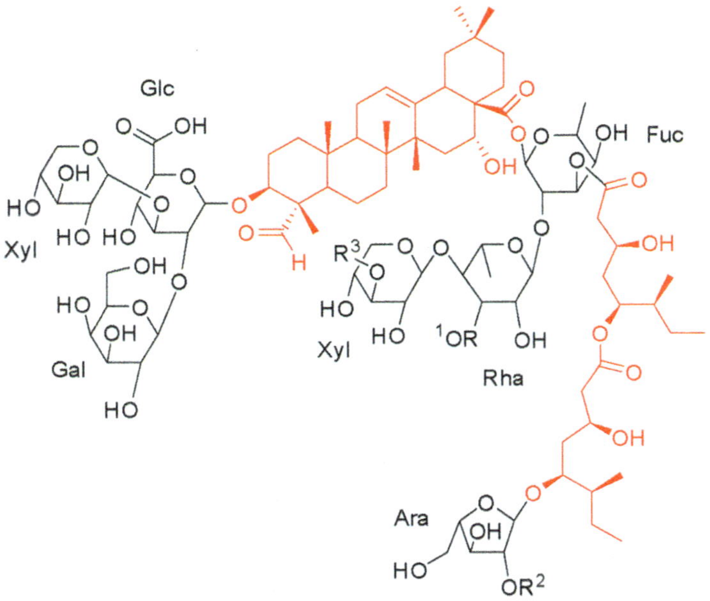

The acyl side chain bound at O-3 or O-4 of the fucose residue linked to the C-28 carboxyl group seems to play an important role in some biological activities (Fleck et al. 2019). In addition, saponins with a different aglycone (phytolaccagenic acid) and a novel fatty acyl group [(S)-2-methylbutanoyl] were reported by Nord and Kenne (2000). The amphiphilic structure of saponins, due to their lipophilic aglycones and hydrophilic saccharide side chains, is responsible for their detergent, foaming and chemical properties.

Quil-A, a fraction of the saponin mixture with a high capacity to stimulate both humoral and cellular immunity, leading to several preclinical studies (Pham et al. 2006), consists of four different saponins (QS-7, QS-17, QS-18 and QS-21), whose structures were elucidated and characterized (van Setten et al. 1995). QS-21 has been the focus of vaccine development research and is now beginning to be used in vaccines against herpes zoster and malaria (Agenus 2019). This saponin stimulates a strong Th1 response through the production of cytokines IL-2 and INF-γ, and antigen-specific antibody responses of both IgG1 and IgG2a, which makes it ideal for use in vaccines against intracellular pathogens (Garçon and van Mechelen 2011). The adjuvant actions of these saponins are due to their ability to associate with cholesterol and phospholipids, forming spherical colloidal structures called ISCOMs. Many antigen formulations include an adjuvant or immunostimulatory complex (ISCOM), which in itself is not necessarily immunogenic, but works in conjunction with the antigen to improve and/or prolong the immune response (Kim et al. 2006; Kensil et al. 1991; Pham et al. 2006).

Aqueous extracts of *Quillaja* bark also contain polyphenols, tannins, proteins, starch, monosaccharides, calcium oxalate and inorganic salts (San Martín and Briones 2000).

5 Morphological Description

Quillaja saponaria (Photos 1, 2 and 3) is an evergreen tree that can grow more than 20 m tall, and its trunk's diameter can reach 1 m. Its bright green, hairless leaves are coriaceous, simple, alternate, oblong and short-petioled, either smooth-edged or slightly dentate. Its greenish-white flowers are hermaphroditic, flattened and star shaped, either solitary or arranged as corymbose racemes growing from the leaf axils. The calyx consists of five sepals and the corolla of five petals. The androecium is made up of 10 stamens, arranged in two series of free filaments. The pistil consists of five ovaries, each with a filiform style and stigma. When the fruits mature, the five carpels, which are originally fused to the base, give rise to 5 dehiscent woody follicles arranged in a star shape, containing numerous winged seeds (Rodríguez et al. 1983; Hoffmann et al. 1992; Montenegro 2002). In young trees the bark is light brown and smooth (Photo 1), and as it matures it adopts a darker greyish color with longitudinal fissures. Under the microscope, it shows lignified fibers, fragments of brown amorphous masses, spherical starch granules, scanty suberized cells, sieve tubes with sieve areas, large sclereids and numerous hexagonal calcium oxalate crystals (Cañigueral et al. 1998).

Photo 1 *Q. saponaria* bark. (Photo: C. Delporte)

Photo 2 *Q. saponaria* leaves. (Photo: C. Delporte)

Photo 3 *Q. saponaria* flowers. (Photo: B.K. Cassels)

6 Geographical Distribution

According to Benedetti et al. (2000) (in Correa and Martínez 2013), *Q. saponaria* is one of the most important tree species of the Chilean evergreen sclerophyllous forest type. It grows in Central Chile, between 30°30′ and 38° S, from the semiarid Ovalle (IV Region of Coquimbo) to Collipulli (IX Region of Araucanía), in the transition zone to the southern deciduous forests. It is important to highlight the broad environmental spectrum under which this species develops, from the Western slopes of the coastal range to the north and south sides of the foothills of the Andes, with minimum mean temperatures ranging from −3.2 °C to 9.4 °C and maximum mean temperatures between 16.5 °C and 31.3 °C (Gotor 2008, in Correa and Martínez 2013), but extremes of −5.7 °C and 40.7 °C. According to Luebert (2013), there is no duly documented evidence that *Q. saponaria* can be found in Peru or Bolivia, outside Chile, refuting authors who may have based their claims on the fact that Ruiz and Pavón's 1777–1788 botanical expedition visited all three countries, at that time parts of the Spanish Viceroyalty of Peru. *Quillaja* forests cover about 320,000 hectares, with an estimated 4.5 tons of biomass production per 0.5 hectares (Magedans et al. 2019).

7 Ecological Requirements

Q. saponaria grows best in Mediterranean climates, with yearly average temperatures around 14 °C and rainfall concentrated in the winter months, ranging from 150 to 1500 mm (Correa and Martínez 2013). It grows mainly in sunny flat areas and hillsides which can be located up to 1500 m.a.s.l., often in places with sparse vegetation (Hoffmann et al. 1992; Gupta 1995). Its root system is well developed, both in depth and horizontally, a quality that allows it to efficiently capture nutrients and water from the soil. Due to this characteristic, the species has been used for soil stabilization, afforestation and recovery of degraded land, due to its high capacity to develop in poor and eroded soils (Valenzuela 2007; Correa and Martínez 2013). It forms communities with other hard-leaved species such as "litre" (*Lithraea caustica*), "peumo" (*Cryptocarya alba*), "maitén" (*Maytenus boaria*), "boldo" (*Peumus boldus*) and "huingán" (*Schinus polygamus*), among others. Where it grows further south, it cohabits with *Nothofagus* spp. in the mesomorphic vegetational zone of Chile (Vita 1974; Novoa-Cortez and Contreras-Leiva 2010; Correa and Martínez 2013). It is an intolerant species, which in conditions of higher humidity is displaced by shade-tolerant species, such as *P. boldus* and *C. alba*. Studies have been conducted to determine the "quillay's" water and fertilizer requirements, specifically nitrogen, phosphorus, potassium and boron for optimum development. Valenzuela (2007) demonstrated that the survival of *Q. saponaria* was favored mainly by irrigation, since all treatments with fertilizers plus irrigation resulted in higher percentages of survival of young plants in the field. Due to overharvesting, this species has been declared vulnerable. Therefore, to fell it or otherwise exploit it, authorization must be requested from the Chilean Agriculture and Livestock Service (SAG, for its Spanish acronym), which grants permits only if its subjects are isolated trees that do not constitute a forest (Mintierras 1955).

8 Traditional Use (Part(s) Used) and Common Knowledge

In ethnomedicine, the infusions of *Q. saponaria* bark and leaves were recommended to treat respiratory diseases (as an expectorant) and for skin conditions. The vaginal injection was used to counter bleeding and leukorrhea. In addition, "quillay" bark is still believed to treat baldness, as well as to stimulate the gastric secretion and act as a diuretic. Its flowers, prepared in the form of baths, have been used against chronic rheumatism and as an infusion to rub limbs and diseased areas. The Mapuche use the bark for toothache and as a detergent (Zin and Weiss 1981; Hoffmann et al. 1992; Minsalud 2010).

An early European mention of "quillay" is that of Frézier (1716), who reported that "its bark foams in water, like soap, making it better (than soap) for washing woolen goods but not for linen, which it stains yellow. All the Indians use it to wash their hair, and clean their head, instead of a comb; it is believed that this is what makes it (their hair) black." Molina, the author of the scientific name *Quillaja saponaria*, says that "what makes this tree precious for Chileans is its bark which, pounded and dissolved in water, makes a great foam like the most perfect soap, removes stains well, defats wool, and cleans in the best way all sorts of fabric and cloth. For this reason, the Peruvians take away a large amount of this bark. Molina (1782) claimed that "its name comes from the verb *quillcan* which means "to wash", possibly his interpretation of what nearly 250 years later might be "kücha" (to wash) in the Mapuche language (modern spelling). In this language *Q. saponaria* is called "küllay"; "to wash (one hair) with quillaja" is "küllaytuw", and a possible cognate is "külchaf" (to get wet).

9 Modern Medicine Based on Its Traditional Medicinal Uses

Quillaja saponins have aroused scientific interest due to their cytotoxic, antiviral and immune adjuvant activities (Palatnik de Sousa et al. 2004; Roner et al. 2007; Fleck et al. 2019). Studies have also shown that "quillay" extracts rich in saponins may be larvicidal, nematocidal, fungicidal and hepatoprotective (San Martín and Briones 2000; Pelah et al. 2002). The less toxic fractions of the saponins of this species have been used experimentally in the control and prevention of infectious diseases, cancer, autoimmune disorders, treatment of AIDS and pneumococcal infection, malaria control and tuberculosis treatment (Kensil et al. 1991). Certain fractions of aqueous bark extracts of *Q. saponaria* are more significant immune response adjuvants. These associations can be made with different types of antigens, obtaining increased *in vivo* antibody titers. For example, QS-21 stimulated cytotoxic T-lymphocyte production, induced Th1 cytokines interleukin-2 and interferon-gamma, and antibodies of the IgG2a isotype (Kensil et al. 1991; Pham et al. 2006). Furthermore, clinical studies demonstrated that QS-21 significantly outperformed other classes of adjuvants including glucan formulations, peptidoglycans, bacterial nucleosides and bacterial lipopolysaccharides, in increasing antibody responses as well as T-cell responses against target antigens (Fleck et al. 2019). Kim et al. (2006) demonstrated that QS-21A (a mixture of QS-21A$_{apiose}$ and QS-21A$_{xylose}$), obtained by synthesis, is one of the most promising immunoadjuvants due to its potency and safety. QS-21A forms conjugates with antigens present in certain types of cancer, HIV glycoproteins and protozoan proteins of malaria, amplifying the immune response mediated by T cells and B cells in microgram doses. With respect to the structural requirements for the immunoadjuvant activity of quillaja saponins, the aldehyde group at C-23 of the aglycone is essential for Th1 stimulating activity, presumably because it could form Schiff's bases with the amino groups of certain T cell receptors; while the hydrophobic acyl side chain at C-28 of *Q. saponaria* saponins might stimulate the cytotoxic response of $CD8^+$ T cells, but also play an important role in the toxicity of certain saponin fractions (Fleck et al. 2019).

Recent studies have focused on the activities of quillaic acid and its derivatives against arachidonic acid (AA) or phorbol ester (TPA) induced inflammation. Quillaic acid exhibited strong topical anti-inflammatory activity in both models, and some of the derivatives exhibited potent activities compared to synthetic non-steroidal anti-inflammatory drugs. The structural modifications performed, and the biological results suggest that both the aldehyde group at C-23 and the carboxyl group at C-28 are relevant to the anti-inflammatory activity in these models (Zhang and Popovich 2009; Rodríguez-Díaz et al. 2011). Also, Arrau et al. (2011, 2019) demonstrated *in vivo* that *Q. saponaria* saponins, quillaic acid, its methyl ester, and an oxidized derivative of the latter, elicit dose-dependent antinociceptive effects in two murine thermal models. The results obtained with the saponin mixture validate the folk use of this species as an analgesic.

The ground inner bark and/or leaves and wood of *Q. saponaria* (and apparently, also *Q. brasiliensis*) are known in commerce as "quillaia" and the aqueous extract, registered as a food additive (E 999) is referred to as "quillaia extract". This product is used as a foaming agent and emulsifier and is also included as a surfactant in cosmetics. Quite recently the European Food Safety Authority's Panel on Food Additives and Nutrient Sources Added to Food provided a scientific opinion on E 999 when used as a food additive and an evaluation of the safety of its proposed use in flavorings (EFSA 2019). The interfacial effects of *Q. saponaria* saponins and their complexes with other surfactants have been reviewed recently (Reichert et al. 2019).

Quillaja bark exports fell from 420,869 kg in 2002 to 159,837 kg in 2018 as a consequence of increased shipping of spray-dried aqueous extracts that rose from 182,302 kg to 618,491 kg in the same period, about half of the total to the USA

(Aduana 2019). It is noteworthy that since the original proposal twenty years ago (San Martín and Briones 1999) these products are obtained to an increasing extent from all the above-ground parts of farmed saplings instead of adult trees which may be killed in the process (Desert King International 2019).

10 Conclusions

Vaccine adjuvation is the main biological property studied for quillaja saponins. These have shown a potent immunogenic activity and the ability to modulate immunity towards a cell-mediated response or to enhance antibody production. The tree synthesizes a complex mixture of triterpenoid saponins with closely related structures, and their separation and purification are expensive. Laboratory synthesis of some of these secondary metabolites have been found to furnish pure compounds, as well as bioactive analogs through structural modifications designed by medicinal chemistry approaches. These new molecules are still waiting to undergo detailed mechanistic and structure-activity relationship studies to find compounds showing improved potency, selectivity and pharmacokinetic profiles, as well as reduced toxicity. It is important to mention that toxicity assessments of quillaja saponins rely mostly on preclinical models, and clinical studies are still lacking.

Sustainable production of *Q. saponaria* biomass is critical for the large scale use of its saponins. More management and cultivation studies are needed to improve the economic exploitation of this valuable resource. Considering the wide geographic distribution of *Q. saponaria* in Chile, each zone requires specific management and cultivation studies. The improvement of the performance of quillaja saponins taking advantage of biological factors could presumably be achieved by manipulating the plant's defense responses using abiotic factors and signaling molecules. It should also be borne in mind that at present the waste from industrial quillaja saponins extraction processes is discarded. Taking into account that practically every aerial part of the tree is used in the extraction of saponins, farther utilization of these wastes should be studied both chemically and pharmacologically. Finally, it should be mentioned that quillaic-acid is a richly functionalized molecule, whose biological activities and chemical manipulation have only seldom been studied. This species could be a valuable source of new derivatives with useful biological activities.

References

Aduana (2019) https://www.aduana.cl/aduana/site/edic/base/port/estadisticas.html. (Last access, August 2019)

Agenus (2019) https://agenusbio.com/wp-content/uploads/2019/05/QS-21-Newsletter-Final.pdf. (Last access, November 2019)

Arrau S, Delporte C, Cartagena C, Rodríguez-Díaz M, González P, Silva X, Cassels BK, Miranda HF (2011) Antinociceptive activity of *Quillaja saponaria* Mol. saponin extract, quillaic acid and derivatives in mice. J Ethnopharmacol 133(1):164–167

Arrau S, Rodríguez-Díaz M, Cassels BK, Valenzuela-Barra G, Delporte C, Barriga A, Miranda HF (2019) Antihyperalgesic activity of quillaic acid obtained from *Quillaja saponaria* Mol. Curr Top Med Chem 133(1):164–167

Bello MA, Rudall PJ, Hawkins JA (2012) Combined phylogenetic analyses reveal interfamilial relationships and patterns of floral evolution in the eudicot order Fabales. Cladistics 28:393–421

Cañigueral S, Vila R, Wichtl M (1998) Plantas medicinales y drogas vegetales para infusión y tisana. Un manual de bases científicas para farmacéuticos y médicos, 1ª Ed. OEMF International, Milan, Italia, 606 pp

Correa C, Martínez A (2013) Información Tecnológica de Productos Forestales no Madereros del Bosque Nativo en Chile. Proyecto CONAF – INFOR, Santiago. http://www.pfnm.cl/paqtecnologicos/quillay/informacion_quillay.pdf. (Last access, October 2019)

Desert King International (2019). http://desertking.com/quillaja-management/. (Last access, November 2019)

EFSA (2019) Panel on Food Additives and Flavourings (FAF). Re-evaluation of Quillaia extract (E 999) as a food additive and safety of the proposed extension of use. EFSA J 17(3):5622. 50 pp

Fleck JD, Andresa B, Pereira da Silva F, Artur E, Olivaro C, Ferreira F, Gasparin S (2019) Saponins from *Quillaja saponaria* and *Quillaja brasiliensis*: particular chemical characteristics and biological activities. Molecules 24(1):171

Frézier AF (1716) Relation du voyage de la mer du Sud, aux côtes du Chili, du Pérou et de Brésil, fait pendant les années 1712, 1713, et 1714. Vol I, p 206. (1716 or 2nd ed. 1732?), cited by Molina (1782, pp 185–186)

Garçon N, van Mechelen M (2011) Recent clinical experience with vaccines using MPL- and QS-21-containing adjuvant systems. Expert Rev Vaccines 10(4):471–486

Guo SJ, Kenne L (2000) Characterization of some O-acetylated saponins from Quillaja saponaria Molina. Phytochem 54:615 –623

Guo SJ, Kenne L, Guo SJ, Kenne L, Lundgren LN, Rönnberg B, Sundquist BG (1998) Triterpenoid saponins from *Quillaja saponaria*. Mol Phytochem 48:175–180

Gupta M (ed) (1995) Plantas medicinales iberoamericanas. Ed. CYTED y Andrés Bello, Panamá. 617 pp

Higuchi R, Tokimitsu Y, Fukioka T, Komori T, Kawasaki T, Oakenful D (1987) Structure of desacylsaponins obtained from the bark of *Quillaja saponaria*. Phytochemistry 26(1):229–235

Higuchi R, Tokimitsu Y, Komori T (1988) An acylated triterpenoid saponin from *Quillaja saponaria*. Phytochemistry 27:1165–1168

Hoffmann A, Farga C, Lastra J, Veghazi E (1992) Plantas Medicinales de uso común en Chile. 2nd ed. Ed. Fundación Claudio Gay, Santiago, Chile, 271 pp

Kensil CR, Patel U, Lennick M, Marciani D (1991) Separation and characterization of saponins with adjuvant activity from *Quillaja saponaria* Molina cortex. J Immunol 146:431–437

Kim Y, Wang P, Navarro-Villalobos M, Rohde BD (2006) Synthetic studies of complex immunostimulants from *Quillaja saponaria*: Synthesis of the potent clinical immunoadjuvant QS-21Aapi. J Am Chem Soc 128(36):11906–11915

Kite GC, Howes M-JR, Simmonds MSJ (2004) Metabolomic analysis of saponins in crude extracts of *Quillaja saponaria* by liquid chromatography/mass spectrometry for product authentication. Rapid Commun Mass Spectrom 18:2859–2870

Luebert F (2013) Taxonomy and distribution of the genus *Quillaja* Molina (Quillajaceae). Feddes Rep 124:57–162

Magedans YV, Yendo AC, de Costa F, Gosmann G, Fett-Neto AG (2019) Foamy matters: an update on *Quillaja* saponins and their use as immunoadjuvants. Future Med Chem. https://doi.org/10.4155/fmc-2018-0438

Minsalud (2010) Medicamentos herbarios tradicionales, Ministerio de Salud, Gobierno de Chile. Santiago. 231 pp. https://www.minsal.cl/sites/default/files/files/Libro%20MHT%202010.pdf. (Last access, October 2019)

Mintierras (1955) Reglamenta explotación de quillay y otras especies forestales. Ministerio de Tierras y Colonización, Gobierno de Chile. Decreto 366, 26-02-55. Santiago, Chile

Molina JI (1782) Saggio sulla storia naturale del Chili. Ed. Stamperia de S. Tomasso D'Aquino, Bologna, Italy, 568 pp

Montenegro G (2002) Chile. Nuestra flora útil: guía de plantas de uso apícola, en medicina folklórica, artesanal y ornamental. Ed Universidad Católica de Chile, Santiago, Chile, 267 pp

Muñoz-Schick M, Moreira-Muñoz A, Moreira S (2012) Origen del nombre de los géneros de plantas vasculares nativas de Chile y su representatividad en Chile y el mundo. Gayana Bot 69(2):309–359

Nord LI, Kenne L (1999) Separation and structural analysis of saponins in a bark extract from *Quillaja saponaria* Molina. Carbohydr Res 320:70–81

Nord LI, Kenne L (2000) Novel acetylated triterpenoid saponins in a chromatographic fraction from *Quillaja saponaria* Molina. Carbohydr Res 329:817–829

Novoa-Cortez FF, Contreras-Leiva M (eds) (2010) Biodiversidad: Flora y Vegetación en Minera Los Pelambres. Centro de Ecología Aplicada, 296 pp

Nyberg NT, Kenne L, Rönnberg B, Sundquist B (2000) Separation and structural analysis of some saponins from Quillaja saponaria. Carbohydr Res 323:87 –97

Palatnik de Sousa CB, Santos WR, Casas CP, Paraguai de Souza E, Tinoco L, Da Silva BP, Palatnik M, Parente J (2004) Protective vaccination against murine visceral leishmaniasis using aldehyde containing *Quillaja saponaria* sapogenins. Vaccine 22:2470–2479

Pelah D, Abramovich Z, Markus A, Wiesman Z (2002) The use of commercial saponin from *Quillaja saponaria* bark as a natural larvicidal agent against *Aedes* and *Culex pipiens*. J Ethnopharmacol 81:407–409

Pham HL, Ross BP, McGeary RP, Shaw PN, Hewavitharana AK, Davies NM (2006) Saponins from *Quillaja saponaria* Molina: isolation, characterization and ability to form immunostimulatory complexes (ISCOMs). Curr Drug Deliv 3:389–397

Reichert CL, Salminen H, Weiss J (2019) *Quillaja* saponin characteristics and functional properties. Annu Rev Food Sci Technol 10:43–73

Rodríguez R, Matthei O, Quezada M (1983) Flora Arbórea de Chile. Ed. Universidad de Concepción, Concepción, Chile, 408 pp

Rodríguez R, Marticorena C, Alarcón D, Baeza C, Cavieres L, Finot V, Fuentes N, Kiessling A, Mihoc M, Pauchard A, Ruiz E, Sánchez P, Marticorena A (2018) Catálogo de las plantas vasculares de Chile. Gayana Bot 75(1):1–430

Rodríguez-Díaz M, Delporte C, Cartagena C, Cassels BK, González P, Silva X, Fredy Leóne F, Wessjohann LA (2011) Topical anti-inflammatory activity of quillaic acid from *Quillaja saponaria* Mol. and some derivatives. J Pharm Pharmacol 63(5):718–724

Roner MR, Sprayberry J, Spinks M, Dhanji S (2007) Antiviral activity obtained from aqueous extracts of the Chilean soapbark tree (*Quillaja saponaria* Molina). J Gen Virol 88:275–285

San Martín R (2002) Proceso de producción de extracto de quillay de alta pureza basado en el uso de toda la biomasa proveniente de la madera obtenida del raleo y poda del árbol, así como plantaciones de quillay. Patent Application CL 200202572

San Martín R, Briones R (1999) Industrial uses and sustainable supply of *Quillaja saponaria* (Rosaceae) saponins. Econ Bot 53(3):302–311

San Martín R, Briones R (2000) Quality control of commercial quillaja (*Quillaja saponaria* Molina) extracts by reverse phase HPLC. J Sci Food Agric 80:2063–2068

Tropicos. Missouri Botanical Garden (2019). http://www.tropicos.org/Name/40021850. (Last access, October 2019)

Valenzuela L (2007) Evaluación de un ensayo de riego y fertilización de quillay (*Quillaja saponaria* Mol.), en la comuna de San Pedro, Provincia de Melipilla, Región Metropolitana. Santiago: Universidad de Chile. Memoria para optar al Título Profesional de Ingeniero Forestal.

Facultad de Ciencias Forestales. Universidad de Chile, Santiago. 74 pp. www.cybertesis.cl/tesis/uchile/2007/valenzuela_l/sources/valenzuela_l.pdf. (Last access, November 2019)

van Setten DC, van de Werken G, Zomer G, Kersten GF (1995) Glycosyl compositions and structural characteristics of the potential immuno-adjuvant active saponins in the *Quillaja saponaria* Molina extract Quil A. Rapid Commun Mass Spectrom 9:660–666

van Setten DC, Zomer G, van de Werken G, Wiertz EJH, Leeflang BR, Kamerling JP (2000) Ion trap multiple-stage tandem mass spectrometry as a pre-NMR tool in the structure elucidation of saponins. Phytochem Anal 11:190–198

Vita A (1974) Algunos antecedentes para la silvicultura del Raulí (*Nothofagus alpina* (Poepp. et Endl.) Oerst. Algunos antecedentes para la silvicultura del Quillay (*Quillaja saponaria* Mol.). Universidad de Chile. Facultad de Ciencias Forestales. Boletín Técnico N° 28. Santiago, Chile, 31 pp

Wisniak J (2016) Antoine François Boutron-Chaillard. Revista CENIC Ciencias Biológicas 47(1):58–67

Zhang W, Popovich DG (2009) Chemical and biological characterization of oleanane triterpenoids from soy. Molecules 14:2959–2975

Zin J, Weiss C (1981) La Salud por Medio de las Plantas Medicinales. Ed. Salesiana, Santiago, Chile, 387 pp

Smallanthus sonchifolius (Poepp.) H. Rob

Jerónimo Ulloa, Flavia Redko, Cesar Catalán, and Liliana Muschietti

Smallanthus sonchifolius. (Photos: L. Muschietti and C. Catalán)

Abstract *Smallanthus sonchifolius* (Poepp.) H. Rob., (family Asteraceae) commonly known as "yacón" or "yakon", is an herbaceous perennial species native to South America. Its tuberous roots have a sweet taste and are used as traditional food and eaten either raw alone or in fruit salads. They can be also boiled, baked or used

J. Ulloa · F. Redko · L. Muschietti (✉)
Facultad de Farmacia y Bioquímica, Departamento de Farmacología, Cátedra de Farmacognosia, Universidad de Buenos Aires, Ciudad Autónoma de Buenos Aires, Argentina
e-mail: lmusch@ffyb.uba.ar

C. Catalán
Instituto de Química Orgánica, Facultad de Bioquímica, Química y Farmacia, Universidad Nacional de Tucumán (UNT), San Miguel de Tucumán, Argentina

Á. Máthé, A. Bandoni (eds.), *Medicinal and Aromatic Plants of South America Vol. 2*, Medicinal and Aromatic Plants of the World 7,
https://doi.org/10.1007/978-3-030-62818-5_37

to prepare beverages, syrup or juice. The young stems are used as a vegetable like celery. "Yacón"s roots store large amounts of fructo-oligosaccharides that are not metabolized in the human digestive tract and hence their consumption does not enhance the level of glucose in the blood. "yacón" is traditionally used for the treatment of diabetes in folk medicine. The antidiabetic properties are attributed to the leaves which are dried and used to prepare infusions. Studies have reported that extracts of leaves reduce glycemia in the plasma of diabetic rats and some constituents of "yacón" leaves inhibit the α-glucosidase enzyme involved in diabetes. "Yacón" display other interesting properties such as antifungal, antibacterial, anti-inflammatory, antioxidant, antiparasitic and cytotoxic activities in different cancer cells. The most frequently investigated "yacón" secondary metabolites are sesquiterpene lactones (STLs) of the melampolide type, being enhydrin, uvedalin, sonchifolin, and polimatin B the main STLs identified in "yacón" leaves. Flavonoids, phenolic acids, monoterpenes and diterpenes have also been reported.

Keywords "Yacón" · Asteraceae · Sesquiterpene lactones · Antidiabetic activity

1 Introduction

Smallanthus sonchifolius is a neglected and long ignored Andean crop, possibly due to its low nutritional value, despite the sweet and watery flavor of its edible tuberous roots. However, in recent decades, interest in this crop has strongly resurfaced due to its multiple beneficial effects on human health. Roots contain high amounts of fructo-oligosaccharides that have been associated with the better control of diabetes, cardiovascular diseases, obesity and cancer prevention. The hypoglycemic effect of infusions and leaf extracts has been demonstrated. In addition to flavonoids, kaurenoic acid derivatives, acyclic diterpenoids, and phenolic acids, the leaves are rich in melampolide-type sesquiterpene lactones that possess strong pharmacological effects.

Consequently, "yacón" appears to be an attractive source of prebiotics (fructooligosaccharides) and food to prevent diabetic and control obesity.

2 Taxonomic Characteristics

Smallanthus (L.) Mack., the "yacón" genus, comprises 23 species ranging from the Southeastern United States to central Argentina. Species of *Smallanthus* are herbs, shrubs or small trees with tuberous roots, opposite, trinerved leaves, paleate receptacle, ray corollas externally hairy at the base, pappus lacking, and marginal cypselae partially embraced by the inner phyllaries of the involucre (Vitali et al. 2015).

S. sonchifolius (Poepp.) H. Rob. and related species were originally classified under the genus *Polymnia* (Wells 1965; Asami et al. 1989; Ohyama et al. 1990; Grau and Rea 1997), although the genus *Smallanthus*, rediscovered by Robinson (1978) had already been proposed in 1933. The new classification, *S. sonchifolius* (Poepp.) H. Roeb., is currently preferred. *Polymnia sonchifolia* Poepp. & Endl. and *Polymnia edulis* Wedd. are considered synonyms (Grau and Rea 1997; Yan et al. 1999). The etymology of *sonchifolius* derives from *Sonchus*, a genus of Cichorieae (Asteraceae), and "folium", a Latin term meaning "leaf", by the general shape of the leaves in members of this species that resembles those of some species of *Sonchus* L (Vitali et al. 2015).

Common Names "arboloco" (Colombia); "aricoma" (Peru and Bolivia; Aymara name); "jiquima" and "jiquimilla" (Venezuela and Colombia after the similarity of the roots to those of the legume "jícama," *Pachyrrhizus* sp.); "yacón", "llacón", "llacjon", "llag'on" (Peru, Bolivia, and Argentina; from the Quechua yakku = insipid, and unu = water, alluding to the watery roots, which are tasteless when just harvested) (Zardini 1991), "poire de terre" (France) and "yacon strawberry" (England) (Ojansivu et al. 2011).

It is possible to differentiate some thirty different varieties of "yacón" by the color of the stem and the storage roots, shape of the leaves, sugar content, etc. (Manrique et al. 2005; Dostert et al. 2009). Some wild species closely related to "yacón" such as *S. riparius* (H.B.K.) H. Rob. and *S. siegesbeckius* (DC.) H. Rob. have similar appearance and morphology. However, since "yacón" reproduces exclusively vegetatively and is not found in wild populations, there is no risk of confusing it with the other species.

3 Crude Drug Used

"Yacón" roots have a sweet flavor and crunchiness and are commonly eaten raw. In many countries, there is a wide variety of products derived from "yacón" roots (flour, slices, juices, purees and sweeteners). In Japan, the syrup is a particularly popular dietary supplement for people suffering from diabetes (Ojansivu et al. 2011). Dried leaves are consumed as tea (Delgado et al. 2013).

4 Major Chemical Constituents and Bioactive Compounds

Smallanthus sonchifolius is a neglected and long ignored Andean crop. Possibly, this is due to its low nutritional value, although its edible tuberous roots have a sweet and watery flavor. In recent decades, interest in this crop has strongly resurfaced, which is mainly due to its multiple beneficial effects on human health.

4.1 Roots

A major portion of tuberous root biomass is composed of water that usually exceeds 70% of fresh weight. Due to a high-water content, the root energy value is low.

Fructo-oligosaccharides (FOS) 70–80% of roots dry matter is composed of saccharides, mainly FOS. The underground storage organs of "yacon" accumulate over 60% (on dry mass basis) of inulin type β (2→1) fructans, mainly oligomers (GF_2–GF_{16}). The fructo-oligosaccharides represent mainly oligosaccharides from trisaccharide to decasaccharide with terminal sucrose (inulin type fructo-oligosaccharides). "Yacón" fructans are of relatively low molecular mass and "yacón" has a significant fructose (3–22% of root dry matter) and glucose (2–5% of root dry matter) content. Sugar content in roots may vary according to location, farming, the growing season, harvest time and the post-harvest temperature (Lachman et al., 2003; Pedreschi et al. 2003; Valentová and Ulrichová 2003; Valentová et al. 2006; Douglas et al. 2007; Sumiyanto et al. 2012; Delgado et al. 2013). In Brazil, "yacón" root carbohydrates are composed of 34–55% FOS, 7–9% glucose, 13–14% fructose and 10–13% sucrose. However, in Peru, the carbohydrate composition of "yacón" roots was 40–70% FOS, 5–15% sucrose, 5–15% fructose and less than 5% glucose (Delgado et al. 2013).

Hydroxycinnamic Acids Roots show antioxidant activity and one of the major antioxidants were identified as chlorogenic acid. Five different caffeic acid derivates were determined in the roots of "yacón" as major water-soluble phenolic compounds. Two of these were chlorogenic acids: 3-caffeoylquinic acid and 3,5-dicaffeoylquinic acid. All these compounds have antioxidant properties. Ferulic acid was also identified as a constituent of "yacón" tubers (Takenaka et al. 2003; Valentová and Ulrichová 2003; Ojansivu et al. 2011; Delgado et al. 2013).

Vitamins, Minerals and Aminoacids "yacón" contains small amounts of vitamins and minerals and does not contain starch. The most abundant minerals in "yacón" are calcium and potassium. The juice has been reported to be rich in free essential amino acids (Delgado et al. 2013).

Other Compounds 4′-Hydroxyacetophenone related antifungal phytoalexins [4′-hydroxy-3′-(3-methylbutanoyl)-acetophenone; 4′-hydroxy-3′-(3-methylbutenyl)-acetophenone and 5-acetyl-2-(1-hydroxy-1-methylethyl)-benzofurane] have been isolated from "yacón" tuber acetone extracts (Valentová and Ulrichová 2003).The tuberous roots contain 0.3–3.7% protein.

4.2 Leaves

In leaves, different classes of secondary metabolites respond differently to the effect developmental stage and environmental regulation. This was evaluated by Padilla-González et al. (2019), who observed that adult "yacón" plants are metabolically richer and more diverse than their young counterparts. On the other hand, the early stages of plant development are characterized by rather poor secondary metabolism.

Phenolic Compounds Several different phenolic acids have been identified from extracts from "yacón" leaves: protocatechuic, chlorogenic, caffeic and ferulic acids (Valentová and Ulrichová 2003; Genta et al. 2010; Milella et al. 2011; Sumiyanto et al. 2012; Delgado et al. 2013; Russo et al. 2015). Flavonoids such as rutin, myricetin, kaempferol and quercetin are also present. Gallic acid and rutin are the most abundant in the leaves (de Andrade et al. 2014; Russo et al. 2015; Lock et al. 2016). The biosynthesis of flavonoids and chlorogenic acids in young leaves is associated with high levels of solar radiation. These compounds may act as protection of internal tissues against damage induced by UV rays, especially free radicals and other reactive oxygen species generated by radiation (Padilla-González et al. 2019). The polyphenolic content of different yacon leaf methanol extracts as well as the flavonoid and tannin content were determined by Russo et al. (2015).

The polyphenolic content of different "yacón" leaf methanol extracts as well as the flavonoid and tannin content were determined by Russo et al. (2015).

Essential Oil β-Pinene, caryophyllene and γ-cadinene were the main components identified in the essential oil from dried leaves of "yacón" using a headspace solid-phase microextraction (HS-SPME) procedure (Adam et al. 2005; Ojansivu et al. 2011). On the other hand, the chemical constituents of "yacón" essential oil grown in China were analyzed by gas chromatography/mass spectrometry (GC/MS). Twenty-one components were identified in the volatile oil from leaves (yield of 0.1%), most of them being sesquiterpenes (55.7%); the main components were β-phellandrene (26.3%), β-cubebene (17.7%), β-caryophyllene (14.0%) and β-bourbonene (10.2%) (Li et al. 2009).

Diterpenoids *ent*-kaurenoic acid and related diterpenoid substances have been found such as smallanthaditerpenic acids A, B, C and D (Valentová and Ulrichová 2003; Dou et al. 2008; Raga et al. 2010; Xiang et al. 2010; Ojansivu et al. 2011; Padla et al. 2012), the acyclic diterpenoids, smaditerpenic acid E and F (Mercado et al. 2010) and *ent*-kaurane-3β,16β,17,19-tetrol and *ent*-kaurane-16β,17,18,19-tetrol (Dou et al. 2010).

Sesquiterpene Lactones Phytochemical studies of "yacón" leaves showed the presence of several melampolide-type sesquiterpene lactones (STLs) such as sonchifolin, uvedalin, enhydrin, polymatin A, B and C, 8β-epoxyangeloyloxy-9α-acetyloxy-14-oxo-1(10)*E*,4*E*,11(13)-germacratrien-6α,12-olide, fluctuanin, fluctuadin, 8-ß-angeloyloxy-9-α-hydroxy-14-oxo-acanthospermolide [smallanthin]

(Valentová and Ulrichová 2003; Schorr et al. 2007; Hong et al. 2008; Choi et al. 2010; Genta et al. 2010; Mercado et al. 2010; Siriwan et al. 2011a, b; Serra-Barcellona et al. 2012, 2017; Frank et al. 2013; De Ford et al. 2015; de Andrade et al. 2017; Ulloa 2017; Ulloa et al. 2017; Yuan et al. 2017).

The dimers enhydrofolin and uvedafolin have also been identified in "yacón's" leaves. Dimeric sesquiterpene lactones are rare in Asteraceae, especially in melampolides (Schorr et al. 2007; Kitai et al. 2015). It is noticeable to see how the composition in sesquiterpene lactones from the leaves of *S. sonchifolius* varies with accessions of different geographic origins (Mercado et al. 2014). The content of STLs is apparently rather stable (with a slight increase) over time. The biosynthesis of STLs in Asteraceae is strongly related to the leaf developmental stage, with higher expression in leaf primordia and young leaves and subsequent accumulation in capitate glandular trichomes. A possible protective role for these molecules in the aerial parts of "yacón" is supported by the fact that the presence of STLs throughout the development of "yacón" is relatively constant. Moreover, their compartmentalization in capitate glandular trichomes responds to their highly nonspecific toxicity that requires compartmentalization to prevent autotoxicity and to protect the plant against herbivores (Mercado et al. 2012; Padilla-González et al. 2019).

4.3 Flowers

In flower extracts, myricetin and gallic acid were the phenolic compounds found in the highest amounts (de Andrade et al. 2014).

5 Morphological Description

"Yacón", native of the Andes and closely related to the sunflower, is a vigorous, herbaceous perennial plant, 1.5–3 m tall, heterophyllous and stems pubescent. The plant produces large tuberous roots like sweet potatoes in appearance, but they have a much sweeter taste and crunchy flesh. The plants are extremely hardy and can grow under hot or cold conditions. "yacón" has large opposite sagittate leaves, petiolate, acute at the apex, margins irregularly dentate, mucronate, trinerved, acrodromous, petioles winged, wings uniform in width, entire, auriculate at the base, auricles discontinuous with the opposite leaf, pubescent on both faces. Lower leaves 17–30 cm long, lamina 10–13 cm wide, subtriangular, truncate at the base, petioles 6–13 cm long; upper leaves 9–25 cm long, lamina 4.5–11 cm wide, ovate or elliptic, obtuse at the base, petioles 1.5–4 cm long (Vitali et al. 2015). Each stem produces 13–16 pairs of leaves before flowering and, as the plant approaches harvest, the leaves reduce their number and size. "Yacón" has the typical inflorescence grouping of flowers in also called *capitulum*. It is composed of two types of flowers: the females that are located around the chapter, with an intense yellow or pale orange color and in numbers from 12 to 16 and the male or tubular that are close together, in greater numbers and occupy the center of the capitulum. Each male flower is formed by a bunch of stamens. Only small quantities of seeds with low germination power are produced (15–25% at best). The underground system is heterogeneous in nature, distinguished by having two kinds of roots, consisting of rhizophores, and adventitious thin and tuberous roots. The rhizophores and roots have secretory ducts of lipid in the inner cortical layers, which is derived from the meristematic endoderm (Machado et al. 2004). The tuberous, thickened, fusiform or ovate roots store sugars in the form of fructo-oligosaccharides, a special type of sugars with attributes beneficial to human health. They are the organs of economic interest, which can be found in varieties that externally have cream, whitish or purple color Depending on the cultivar, they have a different pulp color, they can be white, orange, cream or purple. The number of roots per plant varies from 3 to 35 with an average of 12. It also produces some thin, fibrous, non-thickened roots, whose function is fixation and absorption (Valderrama Cabrera 2005). Tuberous root crops, in which tuberous roots are formed after cessation of stem growth, seem to have a mechanism similar to a potato. On the other hand, the similarities with potato seem to be low in crops where tuberous roots thicken from the base of the stem (Lachman et al. 2003).

6 Geographical Distribution

Smallanthus sonchifolius ("Yacón") is native to the central highlands of the Peruvian Andean region and is cultivated from Southern Colombia to Northwestern Argentina. The plant grows in warm and temperate Andean valleys but can be found at altitudes up to 3200 m.a.s.l. (Ojansivu et al. 2011). Usually, only a few "yacón" plants are cultivated for family consumption. Beyond home use, "yacón" is grown as a cash crop to be marketed at the local level.

In Argentina, "yacón" is present only in a few localities. in the Salta and Jujuy provinces. In Bolivia, its cultivation is very common in most Andean departments. In La Paz, it is most likely the crop with the largest cultivated area and the largest germplasm diversity.

It is grown in many localities throughout the Peruvian Sierra. The largest germplasm diversity is found in southeastern Peru. Another region of diversity and widespread cultivation is in northern Peru, particularly the province of Cajamarca.

Within Ecuador, "yacón" is predominantly grown in the southern provinces of Loja, Azuay and Cañar, in the central highland's provinces, such as in Bolivar and Chimborazo and in the north of the country, namely in Pichincha, Imbabura and Carchi. Wells (1965) reported the presence of "yacón's" cultivation in Cauca, Colombia.

A strip stretching along the eastern Andean slopes, from the Apurimac river basin (12° S) in Peru to the La Paz river basin (17° S) in Bolivia encloses the area richest in "yacón" germplasm. This is also an area where at least three wild species taxonomically very close to *S. sonchifolius* species occur spontaneously (*S. macroscyphus*, *S. riparius* and *S. siegesbeckius*). Thus, this area seems to be the most likely 'origin centre' of the species (Grau and Rea 1997).

In recent years, the cultivation of this species has been expanded to several countries such as Italy, France, Germany, USA (several states), Russia and Japan (Schorr et al. 2007). In New Zealand, the crop has reached the supermarkets as a specialty vegetable (Grau 1993). In China and Brazil there are relatively large experimental fields of "yacón" cultivation and some derived products (tea, syrup, fruit) are marketed albeit in a limited way.

7 Ecological Requirements

In its natural range, *S. sonchifolius* is commonly found in the mountain forests, inter-Andean valleys, coastal areas, and disturbed habitats such as river banks, roadsides, and close to cultivation fields generally taking advantage of vegetation gaps, from sea level to above 2000 m. Flowering from April to July (Vitali et al. 2015).

Temperature *S. sonchifolius* develops well in the mountains and in the inter-Andean valleys; with annual average temperatures of 14–20 °C. Temperatures below 10 °C retard their growth and lengthen the vegetative period, reducing yields. If the temperature exceeds 26 °C, and the soil moisture is insufficient, the plant becomes excessively stressed which affects its normal development. It is very susceptible to frost, but this limitation is compensated by an excellent regrowth capacity. The crop descends to the coast, without major problem.

Humidity The crop normally develops in a range of 550–1000 mm of annual rainfall. However, it is important that in the first five months after sowing, a consistent and frequent water supply is not lacking. After flowering, the soil must maintain enough humidity to promote tuberization and a good plant development.

Light In general, the crop must receive at least nine hours of light. It grows well under the shade of fruit trees and other shrubs, and in full sun. Similarly, it develops well associated with corn, vegetables and other crops.

Soil "Yacón" thrives best in loose soils with a sandy loam or sandy clay loam texture and in soils recently incorporated into agriculture with a good supply of organic matter. They must be deep, porous, with good drainage and good water retention. The ideal pH is between 6 and 7.5, although it also tolerates mildly acidic soils. Preferably avoid soils with high salinity (Valderrama Cabrera 2005).

8 Traditional Use (Part(s) Used) and Common Knowledge

"Yacón's" history goes far back beyond the Inca's civilization. The first representations on textiles and ceramics have been found in an archaeological deposit of Nazca and date back to 500–1200 A.D. (Valentová and Ulrichová 2003). The first written record of the plant is from 1615 in a list of 55 native crops cultivated by the Andeans. One other early allusion is from 1653, describing "yacón"'s use as a fruit and its capacity to withstand several days of transport by sea (Ojansivu et al. 2011). The first available records of the use of "yacón" in Argentina are archaeological roots recovered at Pampa Grande, Serranía Las Pirguas, Guachipas, Province of Salta (collection of Museo de Ciencias Naturales de La Plata, Argentina) associated with the Candelaria Culture, which developed between 1 and 1000 A.D. in northwestern Argentina (Zardini 1991).

Its use as a traditional food (fruit) in the Andean region has been reported. It is usually peeled and eaten raw as such or in a fruit salad (Ojansivu et al. 2011). Young stems are used as a vegetable like celery (Valentová and Ulrichová 2003). Sometimes, "yacón" is kept in the sun for several days to increase the sweetness of the roots. It can be also boiled or baked or used to prepare a refreshing drink. Several supermarkets in Peru offer syrup, juice and jam prepared from "yacón" root, while dried ground leaves are sold to make tea (Delgado et al. 2013). In New Zealand, it is available in supermarkets as a specialty vegetable. In Japan and Brazil, there are many processed "yacón" products on the market, for example, air-dried tuber slices, "yacón" flour and "yacón" syrup. "Yacón" tuberous roots serve as a source of raw material for the production of sweet pastries, fermented vegetables and ethanol; they can be used as "chips" in dehydrated form. Another product is clarified "yacón" juice obtained after treating with active carbon and filtering (Lachman et al. 2003). "Yacón" is also used as an offering during Andean religious festivities.

Besides as a food, it is also recognized as a medicinal plant. In Bolivia and other countries where this crop is currently produced, it is commonly consumed by people suffering from diabetes or from various digestive or renal disorders (Serra-Barcellona et al. 2012; Sugahara et al. 2015). Its tuberous roots have a sweet taste and because the human body is not able to metabolize the fructo-oligosaccharides, "yacón" does not put on body weight. "Yacón" sweetness is caused by fructose, which is 70% sweeter than table sugar and does not stimulate insulin production and does not bring a glycaemic reaction. It has been consumed commonly by diabetics and persons suffering from digestive disorders. Fructo-oligosaccharides are the products recognized and used as food ingredients and prebiotics (Campos et al. 2012). The strong demand is not only due to the sweet taste of "yacón" that makes it pleasant to eat but to its active components that have a positive effect on the digestive system and activity against cancer and diabetes. Antidiabetic medicinal properties were attributed mainly to "yacón" leaves (Serra-Barcellona et al. 2012). Dried leaves are used to prepare a medicinal infusion or mixed with common tea leaves in Japan (Lachman et al. 2003). Other traditional uses are as refreshments during field-work, occasionally for skin rejuvenation and to relieve intestinal, hepatic and renal disorders (Ojansivu et al. 2011).

9 Modern Medicine Based on Its Traditional Medicine Uses

The use of "yacón" as food, also in view of its health effects and potential dietotherapy applications, have been reviewed by Yan et al. (2019). Tubers have been used as a functional food for its prebiotic properties given its fructo-oligosaccharide content and evaluated both *in vivo* and *in vitro* (Pedreschi et al. 2003; Bibas Bonet et al. 2010; Campos et al. 2012; Utami et al. 2013; Fabersani et al. 2018). The antioxidant activity of the roots may be mainly due to the presence of chlorogenic acid and L-tryptophan and other phenolic compounds (Ojansivu et al. 2011). The diet for rats supplemented with "yacón" root flour produced an improvement in oxidative parameters that indicate its antioxidant activity *in vivo* (Habib et al. 2015). This supplementation also showed positive effects on mineral intestinal absorption, bone mass and biomechanical properties (Lobo et al. 2007). A study of the subchronic oral consumption of dried "yacón" root flour as a diet supplement in normal rats showed that it did not produce any negative response, toxicity or adverse nutritional effect (Genta et al. 2005).

Leaves of "yacón" have been traditionally used in the Andes to treat diabetes and digestive diseases (Yan et al. 2019). Several studies have evaluated the antidiabetic effect of different "yacón" leaf extracts and constituents. A decoction significantly decreased glycemic levels in diabetic rats and improved insulin production. In addition, diabetic-dependent alterations in urinary parameters were attenuated by this decoction. This protective effect against renal damage in diabetic nephropathy is presumably mediated by TGF-b/Smads signals (Honoré et al. 2012). Aqueous extracts of "yacón" leaves extract demonstrated immediate hypoglycaemic activity. Moreover, *ent*-kaurenoic acid, isolated from the dichloromethane extract of "yacón" leaves, intraperitoneally administered to normoglycaemic mice produced a hypoglycemic effect that was not observed when orally administered (Raga et al. 2010). According to Genta et al. (2010), enhydrin, the major STL of "yacón" leaves, is involved in the *in vivo* hypoglycemic activity, since it proved effective to reduce post-prandial glucose levels and was useful in the treatment of diabetic animals. Caffeic acid derivatives present in the n-butanol extract are also in part responsible of the hypoglycemic effect of "yacón" leaves (Genta et al. 2010). The use of 10% decoction and enhydrin is safe in rat at doses in which their hypoglycemic effect is demonstrated (Serra-Barcellona et al. 2012). Another report on its antidiabetic activity has shown that both a 10% decoction of the leaves and enhydrin inhibit α-glucosidase. *In vivo* studies showed that both treatments yielded a reduction in glycemic levels compared to diabetes controls (Serra-Barcellona et al. 2017). The ethanolic extract of the leaves alone or in combination with "maca" (*Lepidium meyenii*) extract was shown to have beneficial effects on a murine diabetes model as seen by a decrease in blood glucose level after treatment. "Yacón" extract presented a high polyphenolic content associated with its antioxidant activity (Gonzales et al. 2013). Some studies in humans' beings have been reported for both the leaves (infusion) and the tuberous roots (Mayta et al. 2004; Gordillo et al. 2012; Lock et al. 2016). Smallanthaditerpenic acids A, B, C and D, kaurene derivatives isolated from leaves, presented *in vitro* α-glucosidase inhibitory activity (Xiang et al. 2010). It was seen that inhibition of α-amylase by "yacón" leaf methanol extracts was stronger than that of α-glucosidase, both carbohydrates hydrolyzing enzymes of the digestive tract (Russo et al. 2015). In a study of the effects on dysmetabolism and cardiomyopathy in diabetic rats produced by treatment with "yacón" leaf extract, it was observed that treated diabetic rats showed a decrease in glycemia, an increase in insulin levels and a reduction in triglyceride and fatty acid serum levels. This was accompanied by amelioration of the pancreatic islet injury, as well as an increase in antioxidant enzyme activity and a decrease in the fibrosis and cellular disorganization in cardiac tissue (dos Santos et al. 2018).

Ent-kaurenoic acid from "yacón" exhibited low antimicrobial activities against *Escherichia coli, Pseudomonas aeruginosa, Staphylococcus aureus, Candida albicans* and *Trichophyton mentagrophytes*, and it was found to be inactive against *Aspergillus niger*. At different concentrations, they possess specific antibacterial activity, being active only against Gram positive organisms (*Staphylococcus aureus, Staphylococcus epidermidis* and *Bacillus subtilis*) (Padla et al. 2012).

Fluctuanin, uvedalin and enhydrin exhibited strong antibacterial activity against *Bacillus subtilis* but very low antifungal activity against *Pyricularia oryzae* (Inoue et al. 1995; Lin et al. 2003; Genta et al. 2010). Enhydrin, polymatin B, *allo*-schkuhriolide were tested for antimicrobial activity, but the bioactivity against *Staphylococcus aureus* was only observed with enhydrin compared to polymatin B, and *allo*-schkuhriolide (Choi et al. 2010). Furthermore, a mixture of enhydrin and uvedalin displays antibacterial properties against *S. aureus* and apparent lack of activity against the Gram-negative bacterial strains (de Andrade et al. 2017).

S. sonchifolius leaves also have antiparasitic activity (Frank et al. 2013; Ulloa et al. 2017). The sesquiterpene lactones enhydrin, uvedalin, and polymatin B were isolated by bioassay-guided fractionation technique. These STLs, as well as the dichloromethane extract, showed *in vitro* antiparasitic activity on different life stages of *Trypanosoma cruzi* (RA strain), the causal agent of Chagas disease (Frank et al. 2013). In addition, these compounds showed anti-*T. cruzi* (Dm28C strain) and anti-*Leishmaniamexicana* activity. Ultrastructural alterations, such as an intense vacuolization and mitochondrial swelling were observed in both *L. mexicana* promastigotes and *T. cruzi* epimastigotes. In the *in vivo* study, enhydrin and uvedalin displayed a significant decrease in circulating parasites and no signs of hepatotoxicity were detected (Ulloa et al. 2017).

Enhydrin and uvedalin also display anti-inflammatory activity (Schorr et al. 2007; Genta et al. 2010). Hong et al. (2008) reported the inhibitory activity of LPS-induced nitric oxide production of various sesquiterpene lactones isolated from the leaves. Among them, the activity of enhydrin, uvedalin, polymatin A and B, sonchifolin, fluctuanin and *allo*-schkuhriolide were reported. Enhydrin, uvedalin, fluctuanin and polymatin B, which have an acetoxy group at the C-9 position, exhibited strong inhibition of NO production (Hong et al. 2008). "Yacón" aqueous and acetone leaf extracts exhibited antiedematogenic activity *in vivo*. The acetone leaf rinse extract furnished the best results regarding neutrophil migration inhibition, and cytokine inhibition (NO, TNF-α and PGE2). Results would indicate that the anti-inflammatory activity is due to the presence of both sesquiterpene lactones and chlorogenic acid derivatives, although sesquiterpene lactones seem to have more pronounced effects (Oliveira et al. 2013). "Yacón" leaf methanolic extracts, as well as aqueous extracts, showed antioxidant and radical scavenging activities *in vitro* related to their polyphenolic content (de Andrade et al. 2014; Russo et al. 2015).

A methanol extract of *S. sonchifolius* leaf was tested against a human hepatocellular carcinoma cell line (HepG2) and an inhibitory effect on proliferation and cell migration was observed. The extract also induced cell cycle arrest and necrosis. Active components are presumably melampolide-type STLs (Myint et al. 2019). The cytotoxicity of enhydrin, uvedalin and sonchifolin in cervical cancer cells was evaluated and it was found that these STLs inhibited cell proliferation and induced apoptosis in both a dose- and time-dependent manner. A significant cell death induction was supported by morphological studies (Siriwan et al. 2011a). Uvedalin, enhydrin and sonchifolin have shown chemopreventive activity on 12-*O*-tetradecanoylphorbol 13-acetate (TPA)-induced Raji cell deformation (Siriwan et al. 2011b). Chlorodalin, an uvedalin chloride isolated from an ethanolic extract of the leaves, was evaluated for its antitumor activity against human tumor cell lines along with four other known compounds (Yuan et al. 2017). *In vitro* cytotoxicity assays showed that uvedalin, enhydrin, polymatin B, sonchifolin, fluctuanin and 8-ß-angeloyloxy-9-α-hydroxy-14-oxo-acanthospermolide [smallanthin] display poor cytotoxic effects on peripheral blood mononuclear cells (PBMC) of healthy human subjects, whereas strong cytotoxicity was observed in leukemia and pancreas cancer cells. As for the mechanism of action of polymatin B, oxidative stress seems to be involved. Reactive oxygen species formation induced different effects on the different cancer cell lines: apoptosis, necroptosis, or neither of these. Additionally, cells also died partly by necrosis (De Ford et al. 2015). Besides, endophytic fungi associated with the plant were evaluated in order to characterize new cytotoxic agents. Thirty two fungal strains were isolated while *Nigrospora sphaerica* and *Phoma betae* yielded potential cytotoxic compounds (Gallo et al. 2009).

The inhibitory activity on acetylcholinesterase and butyrylcholinesterase (targets for the treatment of neurodegenerative disorders such as Alzheimer's disease) was found to be weak (Russo et al. 2015). In addition, the neuroprotective effects of an ethanolic "yacón" leaf extract (YLE) were evaluated on a lipo-polysaccharide (LPS)-induced neuroinflammation model *in vitro* and *in vivo* (Baek et al. 2018). "Yacón's" consumption has provided well-known benefits such as immune stimulatory effect (Vaz-Tostes et al. 2014), protection against colon cancer (Moura et al. 2012), acting in the reduction of serum lipids (Habib et al. 2011) and intestinal regularization (Lobo et al. 2011). The cosmetic potential of both root and leaves extracts has been tested on different molecular targets in relation to skin aging (Duque et al. 2017).

10 Conclusions

S. sonchifolius is a species mainly used as a functional food (roots) for its prebiotic and antidiabetic properties. In recent years, there has been a growing interest in the pharmacological properties of "yacón". For this reason, cultivation has expanded to several countries such as Italy, France, Germany, USA, Czech Republic, Russia and Japan, where several commercial products from "yacón" are offered in markets and online. Its use as a hypoglycemic agent has been validated by numerous assays. In addition, other biological activities, such as antifungal, antibacterial, anti-inflammatory, antioxidant, antiparasitic and antitumoral have also been reported for extracts and pure compounds isolated from them. "Yacón" is a rich source of STLs of the melampolide type, many of which have shown promising anti-inflammatory and cytotoxic activities in different cancer cell lines.

References

Adam M, Juklova M, Bajer T, Eisner A, Ventura K (2005) Comparison of three different solid-phase microextraction fibres for analysis of essential oils in yacon (*Smallanthus sonchifolius*) leaves. J Chromatogr A 1084:2–6

Asami T, Kubota M, Minamisawa K, Tsukihashi T (1989) Chemical composition of yacon, a new crop from the Andean Highlands. Jpn J Soil Sci Plant Nutr 60:122–126

Baek S, Choi NH, Lee KP, Jhun H, Kim J (2018) *Smallanthus sonchifolius* leaf attenuates neuroinflammation. J Exerc Nutr Biochem 22(2):31–35. https://doi.org/10.20463/jenb.2018.0014

Bibas Bonet ME, Meson O, de Moreno de LeBlanc A, Dogi CA, Chaves S, Kortsarz A, Grau A, Perdigón G (2010) Prebiotic effect of yacon (*Smallanthus sonchifolius*) on intestinal mucosa using a mouse model. Food Agric Immunol 21(2):175–189. https://doi.org/10.1080/09540100903563589

Campos D, Betalleluz-Pallardel I, Chirinos R, Aguilar-Galvez A, Noratto G, Pedreschi R (2012) Prebiotic effects of yacon (*Smallanthus sonchifolius* Poepp. & Endl), a source of fructo-oligosaccharides and phenolic compounds with antioxidant activity. Food Chem 135(3):1592–1599

Choi JG, Kang OH, Lee YS, Oh YC, Chae HS, Obiang-Obounou B, Park SC, Shin DW, Hwang BY, Kwon DY (2010) Antimicrobial activity of the constituents of *Smallanthus sonchifolius* leaves against methicillin-resistant *Staphylococcus aureus*. Eur Rev Med Pharmacol Sci 14(12):1005–1009

de Andrade EF, de Souza LR, Ellendersen LN, Masson ML (2014) Phenolic profile and antioxidant activity of extracts of leaves and flowers of yacon (*Smallanthus sonchifolius*). Ind Crop Prod 62:499–506

de Andrade EF, Carpiné D, Dagostin JLA, Barison A, Rüdiger AL, de Muñiz GIB, Masson ML (2017) Identification and antimicrobial activity of the sesquiterpene lactone mixture extracted from *Smallanthus sonchifolius* dried leaves. Eur Food Res Technol 243(12):2155–2161. https://doi.org/10.1007/s00217-017-2918-y

De Ford C, Ulloa JL, Catalán CAN, Grau A, Martino VS, Muschietti LV, Merfort I (2015) The sesquiterpene lactone polymatin B from *Smallanthus sonchifolius* induces different cell death mechanisms in three cancer cell lines. Phytochemistry 117:332–339

Delgado GT, Tamashiro WM, Marostica MR Jr, Pastore GM (2013) Yacon (*Smallanthus sonchifolius*): a functional food. Plant Foods Hum Nutr 68:222–228

Dos Santos KC, Cury SS, Ferraz APCR, Corrente JE, Gonçalves BM, de Araújo Machado LH, Carvalho RF, de Melo Stevanato Nakamune AC, Fabro AT, Freire PP, Corrêa CR (2018) Recovery of cardiac remodeling and dysmetabolism by pancreatic islet injury improvement in diabetic rats after yacon leaf extract treatment. Oxidative Med Cell Longev:1–9. https://doi.org/10.1155/2018/1821359

Dostert N, Roque J, Cano A, La Torre MI, Weigend M (2009) Factsheet: Datos botánicos de Yacón. Proyecto Perúbiodiverso – PBD. Museo de Historia Natural, Universidad Nacional Mayor de San Marcos, Lima, Peru, 15 pp

Dou DQ, Tian F, Qiu YK, Kang TG, Dong F (2008) Structure elucidation and complete NMR spectral assignments of four new diterpenoids from *Smallanthus sonchifolius*. Magn Reson Chem 46(8):775–779

Dou DQ, Tian F, Qiu YK, Xiang Z, Xu BX, Kang TG, Dong F (2010) Studies on chemical constituents of the leaves of *Smallanthus sonchifolius* (yacon): Structures of two new diterpenes. Nat Prod Res 24(1):40–47

Douglas JA, Follett JM, Douglas MH, Deo B, Scheffer JJC, Littler RA, Manley-Harris M (2007) Effect of environment and time of planting on the production and quality of yacon (*Smallanthus sonchifolius*) storage roots. N Z J Crop Hortic Sci 35(1):107–116. https://doi.org/10.1080/01140670709510174

Duque L, Bravo K, Osorio E (2017) A holistic anti-aging approach applied in selected cultivated medicinal plants: a view of photoprotection of the skin by different mechanisms. Ind Crop Prod 97:431–439

Fabersani E, Grande MV, Coll Aráoz MV, Zannier ML, Sánchez SS, Grau A, Oliszewski R, Honoré SM (2018) Metabolic effects of goat milk yogurt supplemented with yacon flour in rats on high-fat diet. J Funct Foods 49:447–457

Frank FM, Ulloa J, Cazorla SI, Maravilla G, Malchiodi EL, Grau A, Martino VS, Catalán CAN, Muschietti LV (2013) Trypanocidal activity of *Smallanthus sonchifolius*: Identification of active sesquiterpene lactones by bioassay-guided fractionation. Evid Based Complement Alternat Med, Article ID 627898:1–8

Gallo MBC, Chagas FO, Almeida MO, Macedo CC, Cavalcanti BC, Barros FWA, de Moraes MO, Costa-Lotufo LV, Pessoa C, Bastos JK, Pupo MT (2009) Endophytic fungi found in association with *Smallanthus sonchifolius* (Asteraceae) as resourceful producers of cytotoxic bioactive natural products. J Basic Microbiol 49(2):142–151

Genta SB, Cabrera WM, Grau A, Sánchez SS (2005) Subchronic 4-month oral toxicity study of dried *Smallanthus sonchifolius* (yacon) roots as a diet supplement in rats. Food Chem Toxicol 43(11):1657–1665

Genta SB, Cabrera WM, Mercado MI, Grau A, Catalán CAN, Sánchez SS (2010) Hypoglycemic activity of leaf organic extracts from *Smallanthus sonchifolius*: constituents of the most active fractions. Chem-Biol Interact 185(2):143–152

Gonzales GF, Gonzales-Castañeda C, Gasco M (2013) A mixture of extracts from Peruvian plants (black maca and yacon) improves sperm count and reduced glycemia in mice with streptozotocin-induced diabetes. Toxicol Mech Methods 23(7):509–518

Grau A (1993) Yacon: a highly productive root crop. New Plants Dossier No. 9. Crop & Food Research Institute Internal Report, Biodiversity Programme, Invermay, New Zeland

Grau A, Rea J (1997) Yacon. *Smallanthus sonchifolius* (Poep. & Endl.) H. Robinson. In: Hermann M, Heller J (eds) Andean roots and tubers: Ahipa, arracacha, maca and yacon. Ed. IPGRI, Rome, Italy, pp 199–242

Gordillo GC, Negrón LP, Zúñiga TH, Flores E, Moreyra R, Fuertes C, Guerra GA, Apesteguía A, Quintana AM (2012) Efecto hipoglicemiante del extracto acuoso de las hojas de *Smallanthus sonchifolius* (Yacón) en pacientes con diabetes mellitus tipo 2. Ciencia Invest 15(1):42–47

Habib NC, Honoré SM, Genta SB, Sánchez SS (2011) Hypolipidemic effect of *Smallanthus sonchifolius* (yacon) roots on diabetic rats: biochemical approach. Chem Biol Interact 194:31–39

Habib NC, Serra-Barcellona C, Honoré SM, Genta SB, Sánchez SS (2015) Yacon roots (*Smallanthus sonchifolius*) improve oxidative stress in diabetic rats. Pharm Biol 53(8):1183–1193

Hong SS, Lee SA, Han XH, Lee MH, Hwang JS, Park JS, Oh KW, Han K, Lee MK, Lee H, Kim W, Lee D, Hwang BY (2008) Melampolides from the leaves of *Smallanthus sonchifolius* and their inhibitory activity of LPS-induced nitric oxide production. Chem Pharm Bull 56(2):199–202

Honoré SM, Cabrera WM, Genta SB, Sánchez SS (2012) Protective effect of yacon leaves decoction against early nephropathy in experimental diabetic rats. Food Chem Toxicol 50(5):1704–1715

Inoue A, Tamogami S, Kato H, Nakazato Y, Akiyama M, Kodama O, Akatsuba T, Hashidoko Y (1995) Antifungal melampolides from leaf extracts of *Smallanthus sonchifolius*. Phytochemistry 39(4):845–848

Kitai Y, Hayashi K, Otsuka M, Nishiwaki H, Senoo T, Ishii T, Sakane G, Sugiura M, Tamura H (2015) New sesquiterpene lactone dimer, uvedafolin, extracted from eight yacon leaf varieties (*Smallanthus sonchifolius*): Cytotoxicity in HeLa, HL-60, and murine B16-F10 melanoma cell lines. J Agric Food Chem 63(50):10856–10861

Lachman J, Fernández EC, Orsák M (2003) Yacon [*Smallanthus sonchifolius* (Poepp. et Endl.) H. Robinson] chemical composition and use – a review. Plant Soil Environ 49(6):283–290

Li J, Liu J, Lan H, Zheng M, Rong T (2009) GC-MS analysis of the chemical constituents of the essential oil from the leaves of yacon (*Smallanthus sonchifolius*). Front Agric China 3(1):40–42. https://doi.org/10.1007/s11703-009-0008-z

Lin F, Hasegawa M, Kodama O (2003) Purification and identification of antimicrobial sesquiterpene lactones from yacon (*Smallanthus sonchifolius*) leaves. Biosci Biotechnol Biochem 67(10):2154–2159

Lobo AR, Colli C, Alvares EP, Filisetti TMCC (2007) Effects of fructans-containing yacon (*Smallanthus sonchifolius* Poepp and Endl.) flour on caecum mucosal morphometry, calcium and magnesium balance, and bone calcium retention in growing rats. Br J Nutr 97(4):776–785

Lobo AR, Cocato ML, Borelli P, Gaievski EHS, Crisma AR, Nakajima K, Nakano EY, Colli C (2011) Iron bioavailability from ferric pyrophosphate in rats fed with fructan containing yacon (*Smallanthus sonchifolius*) flour. Food Chem 126:885–891

Lock O, Pérez E, Villar M, Flores D, Rojas R (2016) Bioactive compounds from plants used in Peruvian traditional medicine. Nat Prod Commun 11(3):315–337

Machado SR, Oliveira DMT, Dip MR, Menezes NL (2004) Morfoanatomia do sistema subterrâneo de *Smallanthus sonchifolius* (Poepp. & Endl.) H. Robinson (Asteraceae). Rev Bras Bot 27(1):115–123

Manrique I, Párraga A, Hermann M (2005). Yacon syrup: principles and processing. Series: Conservación y uso de la biodiversidad de raíces y tubérculos andinos: Una década de investigación para el desarrollo (1993–2003). No. 8B. International Potato Center, Universidad Nacional Daniel Alcides Carrión, Erbacher Foundation, Swiss Agency for Development and Cooperation, Lima, Peru, 31 pp

Mayta P, Payano J, Peláez J, Pérez M, Pichardo L, Puycan L (2004) Reducción de la respuesta glicémica posprandial post-ingesta de raíz fresca de yacón en sujetos sanos. CIMEL 9(1):7–11

Mercado MI, Coll Aráoz MV, Grau A, Catalán CAN (2010) New acyclic diterpenic acids from yacon (*Smallanthus sonchifolius*) leaves. Nat Prod Commun 5(11):1721–1726

Mercado MI, Coll Aráoz MI, Coll MV, Ruiz AI, Grau A, Ponessa GI (2012) Análisis estructural, ultraestructural y desarrollo ontogenético de los tricomas glandulares de *Smallanthus sonchifolius* (Asteraceae), «yacón». Lilloa 49(1):4051

Mercado MI, Coll Aráoz MV, Manrique I, Grau A, Catalán CAN (2014) Variability in sesquiterpene lactones from the leaves of yacon (*Smallanthus sonchifolius*) accessions of different geographic origin. Genet Resour Crop Evol 61(6):1209–1217. https://doi.org/10.1007/s10722-014-0103-8

Milella L, Martelli G, Salava J, Fernández E, Ovesná J, Greco I (2011) Total phenolic content, RAPDs, AFLPs and morphological traits for the analysis of variability in *Smallanthus sonchifolius*. Genet Resour Crop Evol 58(4):545–551. https://doi.org/10.1007/s10722-010-9597-x

Moura NA, Caetano BFR, Sivieri K, Urbano LH, Cabello C, Rodrigues MAM, Barbisan LF (2012) Protective effects of yacon (*Smallanthus sonchifolius*) intake on experimental colon carcinogenesis. Food Chem Toxicol 50:2902–2910

Myint PP, Dao TTP, Kim YS (2019) Anticancer activity of *Smallanthus sonchifolius* methanol extract against human hepatocellular carcinoma cells. Molecules 24(17):3054

Ohyama T, Ito O, Yasuyoshi S, Ikarashi T, Minamisawa K, Kubota M, Tsukihashi T, Asami T (1990) Composition of storage carbohydrates in tubers of yacon (*Polymnia sonchifolia*). Soil Sci Plant Nutr 36:167–171. https://doi.org/10.1080/00380768.1990.10415724

Ojansivu I, Ferreira CL, Salminen S (2011) Yacon, a new source of prebiotic oligosaccharides with a history of safe use. Trends Food Sci Technol 22:40–46. https://doi.org/10.1016/j.tifs.2010.11.005

Oliveira RB, Chagas-Paula DA, Secatto A, Gasparoto TH, Faccioli LH, Campanelli AP, da Costa FB (2013) Topical anti-inflammatory activity of yacon leaf extracts. Braz J Pharmacogn 23(3):497–505

Padilla-González GF, Frey M, Gómez-Zeledón J, Da Costa FB, Spring O (2019) Metabolomic and gene expression approaches reveal the developmental and environmental regulation of the secondary metabolism of yacon (*Smallanthus sonchifolius*, Asteraceae). Sci Rep 9(1):1–15. https://doi.org/10.1038/s41598-019-49246-2

Padla EP, Solis LT, Ragasa CY (2012) Antibacterial and antifungal properties of *ent*-kaurenoic. acid from *Smallanthus sonchifolius*. Chin J Nat Med 10(6):408–414. https://doi.org/10.1016/S1875-5364(12)60080-6

Pedreschi R, Campos D, Noratto G, Chirinos R, Cisneros-Zevallos L (2003) Andean yacon root (*Smallanthus sonchifolius* Poepp. Endl) fructo-oligosaccharides as a potential novel source of prebiotics. J Agric Food Chem 51(18):5278–5284

Raga DD, Alimboyoguen AB, Del Fierro RS, Ragasa CY (2010) Hypoglycaemic effects of tea extracts and *ent*-kaurenoic acid from *Smallanthus sonchifolius*. Nat Prod Res 24(18):1771–1782

Robinson H (1978) Studies in the Heliantheae (Asteraceae). XII. Re-establishment of the genus *Smallanthus*. Phytologia 39:47–53

Russo D, Valentão P, Andrade PB, Fernandez EC, Milella L (2015) Evaluation of antioxidant, antidiabetic and anticholinesterase activities of *Smallanthus sonchifolius* landraces and correlation with their phytochemical profiles. Int J Mol Sci 16(8):17696–17718. https://doi.org/10.3390/ijms160817696

Schorr K, Merfort I, Da Costa FBA (2007) Novel dimeric melampolide and further terpenoids from *Smallanthus sonchifolius* (Asteraceae) and the inhibition of the transcription factor NF-κB. Nat Prod Commun 2:367–374

Serra-Barcellona C, Cabrera WM, Honoré SM, Mercado MI, Sánchez SS, Genta SB (2012) Safety assessment of aqueous extract from leaf *Smallanthus sonchifolius* and its main active lactone, enhydrin. J Ethnopharmacol 144:362–370

Serra-Barcellona C, Habib NC, Honoré SM, Sánchez SS, Genta SB (2017) Enhydrin regulates postprandial hyperglycemia in diabetic rats by inhibition of α-glucosidase activity. Plant Foods Hum Nutr 72(2):156–160. https://doi.org/10.1007/s11130-017-0600-y

Siriwan D, Naruse T, Tamura H (2011a) Effect of epoxides and α-methylene-γ-lactone skeleton of sesquiterpenes from yacon (*Smallanthus sonchifolius*) leaves on caspase-dependent apoptosis and NF-κB inhibition in human cervical cancer cells. Fitoterapia 82(7):1093–1101

Siriwan D, Miyawaki C, Miyamoto T, Naruse T, Okazaki K, Tamura H (2011b) Chemopreventive activity of sesquiterpene lactones (SLs) from yacon against TPA-induced Raji cells deformation. Pak J Biol Sci 14(10):605–609

Sugahara S, Ueda Y, Fukuhara K, Kamamuta Y, Matsuda Y, Murata T, Kuroda Y, Kabata K, Ono M, Igoshi K, Yasuda S (2015) Antioxidant effects of herbal tea leaves from yacon (*Smallanthus sonchifolius*) on multiple free radical and reducing power assays, especially on different superoxide anion radical generation systems. J Food Sci 80:C2420–C2429

Sumiyanto J, Dayan FE, Cerdeira AL, Wang Y-H, Khan IA, Moraes RM (2012) Oligofructans content and yield of yacon (*Smallanthus sonchifolius*) cultivated in Mississippi. Sci Hortic 148:83–88

Takenaka M, Yan X, Ono H, Yoshida M, Nagata T, Nakanishi T (2003) Caffeic acid derivatives in the roots of yacón (*Smallanthus sonchifolius*). J Agric Food Chem 51(3):793–796

Ulloa J (2017) Actividad antiparasitaria y citotóxica de especies argentinas de la familia Asteraceae: identificación de compuestos bioactivos. Thesis. Facultad de Farmacia y Bioquímica, Universidad de Buenos Aires, Buenos Aires, Argentina, 284 pp

Ulloa JL, Spina R, Casasco A, Petray PB, Martino V, Sosa MA, Frank FM, Muschietti LV (2017) Germacranolide-type sesquiterpene lactones from *Smallanthus sonchifolius* with promising activity against *Leishmania mexicana* and *Trypanosoma cruzi*. Parasit Vectors 10(1):567. https://doi.org/10.1186/s13071-017-2509-6

Utami NWA, Sone T, Tanaka M, Nakatsu CH, Saito A, Asano K (2013) Comparison of yacon (*Smallanthus sonchifolius*) tuber with commercialized fructo-oligosaccharides (FOS) in terms of physiology, fermentation products and intestinal microbial communities in rats. Biosci Microbiota Food Health 32(4):167–178. https://doi.org/10.12938/bmfh.32.167

Valderrama Cabrera M (2005) Manual del cultivo de yacón. PYMAGROS (Productores y Mercados del Agro de la Sierra). Ed. Convenio COSUDE-MINAG. Cajamarca, Peru, 47 pp

Valentová K, Ulrichová J (2003) *Smallanthus sonchifolius* and *Lepidium meyenii* prospective Andean crops for the prevention of chronic diseases. Biomed Papers 147(2):119–130

Valentová K, Lebeda A, Doležalová I, Jirovský D, Simonovska B, Vovk I, Kosina P, Gasmanová N, Dziechciarková M, Ulrichová J (2006) The biological and chemical variability of yacon. J Agric Food Chem 54(4):1347–1352

Vaz-Tostes MG, Viana ML, Grancieri M, Luz TCS, Paula H, Pedrosa RG, Costa NMB (2014) Yacon effects in immune response and nutritional status of iron and zinc in preschool children. Nutrition 30:666–672

Vitali MS, Sancho G, Katinas L (2015) A revision of *Smallanthus* (Asteraceae, Millerieae), the yacon genus. Phytotaxa 214(1):1–84. https://doi.org/10.11646/phytotaxa.214.1.1

Wells JR (1965) A taxonomic study of *Polymnia* (Compositae). Brittonia 17:144–159

Xiang Z, He F, Kang TG, Dou DQ, Gai K, Shi YY, Kim YH, Dong F (2010) Anti-diabetes constituents in leaves of *Smallanthus sonchifolius*. Nat Prod Commun 5(1):95–98

Yan X, Suzuki M, Ahnishi-Kameyama M, Sada Y, Nakanishi T, Nagata T (1999) Extraction and identification of antioxidants in the roots of yacon (*Smallanthus sonchifolius*). J Agric Food Chem 47:4711–4713

Yan MR, Welch R, Rush EC, Xiang X, Wang X (2019) A sustainable wholesome foodstuff; health effects and potential dietotherapy applications of yacon. Nutrients 11(11):2632. https://doi.org/10.3390/nu11112632

Yuan Y, Win Aung KK, Ran X, Wang XT, Dou DQ, Dong F (2017) A new sesquiterpene lactone from "yacón" leaves. Nat Prod Res 31(1):43–49

Zardini E (1991) Ethnobotanical notes on "yacon" *Polymnia sonchifolia* (Asteraceae). Econ Bot 45(1):72–85

Smilax campestris Griseb.

Manuel Minteguiaga and Fernando Ferreira

Smilax campestris Griseb. (Photo: Rosângela G. Rolim. With permission by the author.)

M. Minteguiaga (✉)
Espacio de Ciencia y Tecnología Química, Centro Universitario Regional Noreste/Sede Tacuarembó, Universidad de la República, Tacuarembó, Uruguay

Laboratorio de Biotecnología de Aromas, Departamento de Química Orgánica, Facultad de Química, Universidad de la República, Montevideo, Uruguay
e-mail: manuel.minteguiaga@cut.edu.uy

F. Ferreira
Espacio de Ciencia y Tecnología Química, Centro Universitario Regional Noreste/Sede Tacuarembó, Universidad de la República, Tacuarembó, Uruguay

Laboratorio de Carbohidratos y Glicoconjugados, Departamento de Química Orgánica, Facultad de Química, Universidad de la República, Montevideo, Uruguay

Á. Máthé, A. Bandoni (eds.), *Medicinal and Aromatic Plants of South America Vol. 2*, Medicinal and Aromatic Plants of the World 7,
https://doi.org/10.1007/978-3-030-62818-5_38

Abstract *Smilax campestris* Griseb. (Smilacaceae) ("zarzaparrilla") is a medicinal plant species widely used by traditional communities in South America. In general, the plant parts used by popular medicine are the rhizophore (underground stem) and roots, in the form of infusions or decoctions. Their main indications are: antirheumatic, antisyphilitic, diuretic, diaphoretic, anti-inflammatory, antiseptic, and to treat skin conditions (i.e. psoriasis) and gout; among others. There is lacking scientific evidence regarding the phytochemistry of such aqueous extracts, as well as their biological and pharmacological properties. Both aerial and underground organs have been described to contain polyphenolic acids, flavonoids (aglycones, glycosides and proanthocyanidin oligomers/polymers) and saponins; but no systematic research on the isolation of other families of natural products has been conducted, yet. The production of secondary metabolites (in particular flavonoids) is highly dependent on both environmental conditions (solar radiation, herbivores, etc.) of habitats and the phenological stage of the plant itself. The scientific pharmacological *in vitro* tests performed on *S. campestris* extracts have confirmed their antioxidant and anti-inflammatory activities, as well as the lack of acute toxicity. However, they have failed to demonstrate diuretic effects in *in vivo* models; results on the antimicrobial potential are controversial. Further research is needed to evaluate the other ethnomedicinal properties attributed to the vegetal drug. This chapter presents a critical review of the special literature on *S. campestris* including the most updated information available in journals, books and the internet.

Keywords "Zarzaparrilla" · Flavonoids · Antioxidant · Anti-inflammatory

1 Introduction

Smilax campestris Griseb. has been recognized as medicinal plant in South America, from the early times of indigenous traditions, Portuguese and Spanish conquerors' practices, until recent research. Although, this is the most widely distributed *Smilax* spp. in Southern Brazil, Argentina, and Uruguay (The Pampas Region), only a few studies have been conducted in order to describe its medicinally active substances. The aim of this work is to review the findings of the most important studies related to this plant species.

2 Taxonomic Characteristics

Smilax L. (Smilacaceae; Monocotyledoneae) is a genus with more than 300 subwoody species, distributed in the temperate and tropical warm and subtropical regions of the globe: mainly in Asia, Oceania and the Americas; and with few species in Africa and Europe (Conran 1998; Guaglianone and Gattuso 2006; Salas-Coronado et al. 2017).

The generic name was given after the Greek myth of the love affair between Crocus and the nymph Smilax (Benigni et al. 1962).

The American *Smilax* spp. have been early recognized by indigenous populations as medicinal plants, and adopted by Spanish and Portuguese people as they traded the vegetal drug to Europe between the XVI and XIX centuries (Benigni et al. 1962; Medeiros et al. 2007; Montenegro 2007; Scarpa and Rosso 2014). These medicinal species were highly appreciated for the treatment of syphilis, reinforced by the fact of the massive epidemics in Europe and America by the beginning of the XVI century (Benigni et al. 1962; Medeiros et al. 2007).

In Argentina, eight *Smilax* spp. grow, while in Brazil 32 species occur. The species cited in Uruguay are *S. campestris* Griseb., *S. brasiliensis* Spreng. and *S. cognata* Kunth (synonym *S. montevidensis* Hort. ex Steud.) (Gibert 1873; Paccard 1905; Arrillaga de Maffei 1969; Guaglianone and Gattuso 1991; Rugna et al. 2003a; Alonso-Paz et al. 2008; Soares 2012; Martins et al. 2013). However, in Southern Brazil, Argentina, and Uruguay (The Pampas Region), *S. campestris* is the most abundant species of the *Smilax* L. genus (Guaglianone and Gattuso 1991; Salaverry et al. 2020).

Synonyms *Smilax campestris* var. *marginulata* (Mart. ex Griseb.) A. DC.; *Smilax campestris* var. *marginulatum* A. DC.; *Smilax campestris* var. *rubiginosa* (Griseb.) A. DC.; *Smilax campestris* var. *spruceana* A. DC. (The Plant List 2019).

3 Crude Drug Used

The plant organs used for medicinal purposes are the rhizophores (underground stems) and roots of *S. campestris* (Photo 1), mainly in the form of aqueous infusions or decoctions (Arrillaga de Maffei 1969; Gattuso 1995; Freire et al. 1997; Gorzalczany et al. 1999; Gupta 2006; Guaglianone and Gattuso 2006; Montenegro, 2007; Medeiros et al. 2007; Alonso-Paz et al. 2008; Barboza et al. 2009; Acosta et al. 2013; Sella et al. 2013; Soares et al. 2014; Salas-Coronado et al. 2017; Colares et al. 2017). In Uruguay, it is consumed together with the popular *Ilex paraguariensis* A. St. Hil. infusion named "mate" (5% of underground organs in water; Alonso-Paz et al. 2008). Occasionally, the aerial parts might also be employed as infusions or decoctions, in particular in Southern Brazil (Coelho de Souza et al. 2004).

After the collection, the plant material (underground stems and roots) is washed with water to eliminate remains of earth, dried, cut and then stored in plastic bags or like a bunch (Martins 2009; Soares 2012). The plant material collected is from both male and female plants, since *S. campestris* is a dioecious species.

The plant material employed in traditional medicine in Bolivia, Brazil, Argentina, Paraguay and Uruguay is sourced from wild populations. This leads to extractivism jeopardizing the future of germplasm resources (Soares 2012; Soares et al. 2014).

Photo 1: Rhizophore and roots of *S. campestris* used in traditional medicine, in South America. (Photo: Aline R. Martins)

Overexploitation is a major problem, since the most frequently consumed plant parts are the underground organs, which means that no new plants can regenerate once the material is harvested (Soares 2012; Soares et al. 2014). Another important aspect to consider is that the seed germination rate and seedling development rate are too slow (Martins et al. 2012; Soares et al. 2014).

In Brazil (São Paulo State), the vegetal drug is marketed without correct taxonomic identification and with high level of foreign organic matter contamination (Medeiros et al. 2007; Martins et al. 2014; Soares 2012; Soares et al. 2014). In the market, "salsaparilha" materials are most frequently classified and named according to the geographical origin rather than according to botanical considerations, such "salsaparilha do Pará" (putatively, *S. papyraceae* and *S. syphilitica*), "salsaparilha de Minas Gerais" (*S. officinalis*), and "salsaparilha do México" (*S. medica*); among others (Benigni et al. 1962; Soares 2012).

S. campestris was included in the Second Edition (1921) of the Argentinean Pharmacopeia [together with other alien *Smilax* species (e.g.: *S. medica*, *S. ornata*, *S. papyraceae* and *S. officinalis*)], also popularly known as "zarzaparrilla". Remarkably, it has been excluded from the Third (1943) and following editions (Gattuso 1995; Rugna et al. 2003a, b). Similarly, in the First Edition (1929) of the Brazilian Pharmacopeia monographs of four "salsaparilha" species (*S. medica*, *S. syphilitica*, *S. papyraceae* and *S. officinalis*) were included: three native and one alien species. *S. campestris* was cited as eventual adulterant. In the Second (1959) and following editions these species have been excluded (Martins 2009; Soares 2012; De Lima 2016). Remarkably, currently, *S. campestris* is included in the Argentinean Food Codex and MERCOSUR Technical Regulation as a safe natural additive (flavoring agent) in drinks and foods (MERCOSUR 2006; Código Alimentario Argentino 2019), despite lacking scientific information on this subject.

4 Major Chemical Constituents and Bioactive Compounds

The original phytochemical screening of *S. campestris* was performed by Bandoni et al. (1976) with classical chemical reactions over the extract of the aerial parts. They found positive results for the following families of secondary metabolites: phenols, flavonoids (trace levels), leucoanthocyanidins, tannins, and water insoluble steroids and alkaloids (trace levels). Since then, other authors have conducted additional phytochemical screenings from the aerial parts, with positive results for: phenols and flavonoids (Ferreira and Vargas 1999; Rugna et al. 2003b; Morais et al. 2014), sterols/triterpenes and saponins (Ferreira and Vargas 1999; Rugna et al. 2003b; Morais et al. 2014), indole-derived and other alkaloids (Rugna et al. 2003b; Morais et al. 2014), and coumarins (Rugna et al. 2003b; Morais et al. 2014).

Rugna et al. (2003b) identified in the infusions from rhizomes (rizophores): phenols, flavonoids, indole-derived alkaloids, and steroids. Furthermore, from these underground organs there were identified by histochemical and microscopical analysis: starch, calcium oxalate crystals (raphides), and lipids (Martins 2009; Martins et al. 2010; Soares 2012; Sella et al. 2013; Colares et al. 2017).

Polyphenols (acids and flavonoids) is the main group of secondary metabolites identified and isolated from the underground and aerial organs of *Smilax* spp., being not *S. campestris* an exception (Salas-Coronado et al. 2017). However, the polyphenolic profiles of this species' leaves, roots and rhizophores are qualitatively similar (with minor quantitative differences) when compared different populations, and even, when the comparison is performed against other *Smilax* spp. (Martins 2009; Soares 2012). Thus, phytochemical determinations are not enough to differentiate closely related species (Martins 2009, Soares 2012).

Phenolic acids [chlorogenic, *p*-coumaric, ferulic, and (*E*)-cinnamic] have been identified in the aerial and underground organs of *S. campestris* (Martins 2009). According to this author, these compounds might contribute with the antirheumatic, antioxidant, and anti-inflammatory properties described for the vegetal drug.

Regarding flavonoids, all the alcoholic and aqueous extracts from aerial parts of Argentinean and Brazilian populations exhibited the presence of the flavonols quercetin and rutin (Rugna et al. 1999, 2002, 2004, 2005, 2007a; Martins 2009; Morais et al. 2014). The aerial organs also contained other flavonol glycosides [quercetin-3-*O*-β-D-glucoside, quercetin-3-*O*-α-L-rhamnoglucoside, quercetin-3-*O*-rhamnogalactoglucoside, kaempferol, kaempferol-3-*O*-glucoside, kaempferol-3-*O*-rhamnoglucoside, isorhamnetin, isorhamnetin-3-*O*- rhamnoglucoside], the flavanol (+)-catechin and epicatechin; and the proanthocyanidins procyanidin and propelargonidin (oligomeric/polymeric flavonoids; condensed tannins) (Rugna et al. 1999, 2002, 2004, 2005, 2007a, b, 2008, 2010, 2011). The composition of the underground organs is like the aerial parts (Rugna et al. 2005, 2007b; Martins 2009). Recently, Salaverry et al. (2020) performed an HPLC-MS/MS screening of the rhizophore aqueous infusion confirming the presence of several of the named compounds (quercetin and catechin derivatives), as well as quercetin-3-*O*-galactoside, quercetin-3-*O*-rhamnoside and others mono- and di-glycosylated flavonoids preliminary identified. These flavonoids could have a key role in the anti-inflammatory activity of rhizophore aqueous extracts (Salaverry et al. 2020).

The reported flavonoid anthocyanins are found mainly in the *S. campestris* fruits, such as: cyanidin, pelargonidin, cyanidin-3-*O*-glucoside and cyanidin-3-*O*-rhamnoglucoside; which surely are associated with their dark blue and violet colors (Rugna et al. 2004, 2005). Their precursors, proanthocyanidins (condensed tannins such procyanidin and propelargonidin), are more concentrated in the leaves but are also present in all the plant organs at different concentration levels (Rugna et al. 2005, 2007b).

The differences in flavonoid patterns in specimens from different phytogeographic regions might indicate the existence of chemical races or chemotypes, which lead to the possibility to employ them as chemo-markers, in special quercetin and rutin that are biosynthesized all along the year (Rugna et al. 1999, 2013). In addition, the concentration/identity of flavonoids in aerial or underground parts is highly dependent on the internal and external factors to the plants such as: dioecism (Rugna et al. 2002), phenological stage (Rugna et al. 2002, 2007b, 2008, 2013; Salaverry et al., 2020), different plant organs (Rugna et al. 2004), environmental temperature patterns (Rugna et al. 2013), incidence of the solar radiation (Rugna et al. 2007a), and biotic interactions (i.e. with the *Agraulis vanillae* L. larvae) (Rugna et al. 2011). All these metabolic changes could be part of complex reaction mechanisms of fitness, protection and adaptation to the environment/ecosystems where *S. campestris* grows and develops.

In experiments with cultivated plant specimens of *S. campestris*, Rugna et al. (2010) found that the level of polyphenols (flavonoids and tannins) is modulated by the type and concentration of applied phytohormones [benzylaminopurin (BAP) and indoleacetic acid (IAA)]. High levels of BAP diminish the concentration of polyphenols independently of the concentration of IAA in the treatment, but with BAP constant, the polyphenols increase when IAA increases (Rugna et al. 2010).

In general, all the *Smilax* spp. give positive results in the foam test for saponins (Soares 2012). From *S. campestris* aerial parts, it was isolated a steroidal saponin (smilaxsaponin) (Domínguez 1928). Recently, it was reported the presence of sarsapogenin in *S. campestris* rhizophore aqueous extract, as well as other not identified related-glycosides (Salaverry et al. 2020); but it is also probable the occurrence of smilagenin and other saponins such in several *Smilax* spp. (Benigni et al. 1962; Bernardo et al. 1996). These saponins might be responsible for the antimicrobial action of the rhizophore extracts (Colares et al. 2017).

5 Morphological Description

Originally described in 1842 by the German botanist August Grisebach, the plant species are twining-climbing perennial vines that grow until 2–4 m high, with lignified rhizophores (underground stems) provided of storage function (Grisebach 1842; Guaglianone and Gattuso 1991; Martins et al. 2010). *S. campestris* presents a brownish underground system (10–15 cm depth; Photo 1) composed by a 5–10 cm length branching stems with swollen nodes (named knottys); those nodes can produce subsequent ramifications in the underground stem and with them, new aerial stems and adventitious roots (Gattuso 1995; Martins 2009; Martins et al. 2010; Álvarez 2019). The morphoanatomy of this complex subterraneous system is described in the literature (Gattuso 1995; Martins 2009; Martins et al. 2010; Colares et al. 2017), and the current trend is to name it as rhizophore instead rhizome, such some other authors stated previously. The stems have stingers in the form of triangular-curved thorns. Morphologically, the leaves are disposed alternately along the stem with variable sizes and shapes: ovate, lanceolate-ovate, obtuse, elliptic or rarely cordate. In appearance to the touch are leathery, with entire margin, obtuse/acute apiculated apex and rounded base. In general, the leaves have 3–5 main parallel veins (basal acrodrome venation), from which emerge straight secondary veins (Grisebach 1842; Guaglianone and Gattuso 1991; Martins et al. 2010; Álvarez 2019). In some ecotypes, it is possible to find spiniform emergences in the veins and leaf margins (mainly in the abaxial face), which are associated to environmental induction (Martins et al. 2013; Zanette-Silva et al. 2013; Durán et al. 2013; Álvarez 2019). The petioles are short with two lignified tendrils at their bases, which are employed for twinning with the surrounding vegetation.

The high variation in the leaf morphology of *S. campestris* (with different morphotypes in the area of occurrence due to the ecological factors) makes sometimes difficult the species identification in the genus *Smilax*, even more, if it is considered that it presents dioecism (Guaglianone and Gattuso 1991; Rugna et al. 1999, 2002; Guaglianone and Gattuso 2006; Martins et al. 2013; Martins et al. 2014). Furthermore, the unequivocal determination to species level is also difficult if considered the underground organs (Martins 2009; Martins et al. 2010; Sella et al. 2013). The microanatomy of the aerial organs also has been studied by several authors in different regions of occurrence (Guaglianone and Gattuso, 1991; Martins 2009; Martins et al. 2013, 2014; Sella et al. 2013; Cabral et al. 2018; Arriola et al. 2018). The inflorescences of *S. campestris* are arranged in umbels (30–45 flowers each; white, pink or purple-red-colored) and protected by ovate-triangulate bracts; the staminate flowers (male-flowers) contained six stamens while pistillate flowers (female-flowers) have three carpels and three stigmas with superior ovary (hypogynous) (Guaglianone and Gattuso 1991, 2006; Gattuso 1995; Rosa and Ferreira 1999; Rugna et al. 2005; Alonso-Paz et al. 2008; Martins et al. 2013; Álvarez 2019). Blooming: typically, from the end of winter to spring (July-September) (De Lima 2016; Álvarez 2019). The fruits are black/violet berries that contain 1–3 seeds with typical caruncles (Guaglianone and Gattuso 1991, 2006; Gattuso 1995; Rosa and Ferreira 1999; Rugna et al. 2005; Alonso-Paz et al. 2008; Martins et al. 2013; Álvarez 2019).

6 Geographical Distribution

According to the most of literature reports, *S. campestris* is distributed extensively through the South American continent: Southern to Northeastern Brazil, Bolivia, Paraguay, Uruguay and Center and Northeastern Argentina (Guaglianone and Gattuso 1991; Gattuso 1995; Freire et al. 1997; Gupta, 2006; Alonso-Paz et al. 2008). Nevertheless, it is also cited in the literature, as a native species to Mesoamerica and the Caribe region (Molares et al. 2012; Alonso-Castro et al. 2015). This could, however, be due to a possible misidentification. *S. campestris* has been described to occur from sea-level to 1800 m altitude (Durán et al. 2013; Álvarez 2019).

7 Ecological Requirements

S. campestris grows in the edges of tropical and subtropical riparian forests or scrublands, both in highlands and in lowlands (Guaglianone and Gattuso 1991, 2006; Rugna et al. 2004, 2013; Martins et al. 2012; Álvarez 2019). In some cases, it grows over the dunes in "restinga" forests (near the sea) in Southern Brazil, where it is adapted to high radiation levels, salinity and water scarcity (Zanette-Silva et al. 2013; Romeu et al. 2013; Arriola et al. 2018; Cabral et al. 2018).

S. campestris phenotypes are partially determined by the ecological traits that exist in the multiple ecosystems where the plant grown, demonstrating a wide plasticity (Martins 2009; Rugna et al. 2013; Cabral et al. 2018). Thus, field studies have found that the presence of thorns is correlated with the vegetal density of the environment (both in the canopy and in the soil): the higher the density, the more thorns are developed in the stems and leaves (Zanette-Silva et al. 2013). In another study, Cabral et al. (2018) verified that in different "restinga" environments the morphology of the leaves varies broadly, similarly to the morphological parameters related with the leaf development, with plant individuals growing near to the sea (xerophiles). Characteristically, these plants produce less biomass, which is surely due to the water stress and salinity of the environment.

In the natural environments, the plants are attacked by herbivores and galling insects, thus, the proliferation of thorns might be correlated with these complex biotic interactions (Romeu et al. 2013; Arriola et al. 2018). Moreover, *S. campestris* is considered as a super-host of galling herbivores because at least 4 of them produce galls in the stem and the leaves, being considered as an important phytosociological community member of the riparian forest (Arriola et al. 2018).

Some experiments done at laboratory conditions demonstrated that the total absence of light improves the germination of *S. campestris* seeds (compared to illumination conditions), obtaining the best performance (91%) at 35 °C in the darkness after 51 days from sowing (Rosa and Ferreira 1999). Despite it, Martins et al. (2012) did not find any differences in germination by the influence of the light conditions, while at 30 °C there were produced significantly more seedlings than at 20–30 °C regime. These results suggest that *S. campestris* seeds require high temperatures to germinate adequately, typical behavior of tropical forest plant species (Rosa and Ferreira 1999; Martins 2009; Martins et al. 2012).

Some alterations in the ecosystem by the introduction of alien species benefit the growth and survival of *S. campestris*, i.e. when *Pinus taeda* is a dominant species the frequency of the vines augmented by a factor of 9. This may be explained because *S. campestris* requires advanced woody environments by its climbing habit (Fischer et al. 2014).

8 Traditional Use (Part(s) Used) and Common Knowledge

Smilax campestris Griseb. is popularly known in Argentina, Bolivia, Paraguay and Uruguay as "zarzaparrilla", "zarsaparrilla", "salsaparrilla", "zarzaparrilla blanca", "zarzaparrilla negra", "zarza blanca", "zarza espinosa", "zarzamora", "yapecanga", "morenita brava", "sacha-mora", "verdenaso" and "raíz de la China" (Guaglianone and Gattuso 1991; Gattuso 1995; Gaglianone and Gattuso 2006; Montenegro 2007; Barboza et al. 2009; Colares et al. 2017). In Brazil, is popularly recognized as: "salsaparrilha", "sarsaparrilha", "sarza", "salsa parrilha", "salsa americana", "japicanga", "japecanga", "japecanga-do-campo", "japecanga-verdadeira", "japecanga-mineira", "nhupicanga", "japeganga", "dente de cachorro", "legação"; among others (Medeiros et al. 2007; Romeu et al. 2013).

In English, it is named "greenbrier", shared with other members of *Smilax* genus (Martins et al. 2012). South American indigenous populations denominated this species as: "*yuá'pecá*" (Guarani tradition; Mereles et al. 1994); "kajahuatana" (Quechua tradition; Gupta 2006) and "nallin lhue" ("tararira tooth" in Toba tradition; Martínez and Cúneo 2009; Scarpa and Rosso 2014). This latter derived from nallin: "wolf fish" and *lhue*: teeth, that refers to the sharp stingers of the plant (Martínez and Cúneo 2009).

The popular name "zarzaparrilla" and its derived ones are applied at least to 19 *Smilax* spp., and even other species are known with the same common name, such *Herreria montevidensis* Klotzsch ex Griseb., *Herreria sarsaparrilha* Mart. (Herreriaceae) and *Muehlenbeckia sagittifolia* (Ortega) Meisn. (Polygonaceae), which could be found as adulterants in vegetal drug commercial packages (Arrillaga de Maffei 1969; Medeiros et al. 2007; Martins 2009; Soares 2012; Sella et al. 2013; Martins et al. 2014; Colares et al. 2017).

In the ethnobotanical traditions of South American countries, the therapeutic properties attributed to rhizophore and roots aqueous extracts (infusions and decoctions) are: antirheumatic, antisyphilitic, diuretic, diaphoretic, anti-inflammatory, weak laxative, to treat skin conditions (i.e. psoriasis), as antipruritic and antiseptic (antifungal), to treat the gout and as blood cleanser including to reduce its cholesterol level (Benigni et al. 1962; Arrillaga de Maffei 1969; Gattuso 1995; Freire et al. 1997; Gorzalczany et al. 1999; Gupta 2006; Guaglianone and Gattuso 2006;

Montenegro 2007; Medeiros et al. 2007; Alonso-Paz et al. 2008; Barboza et al. 2009; Martins 2009; Acosta et al. 2013; Sella et al. 2013; Soares 2012; Soares et al. 2014; De Lima, 2016; Salas-Coronado et al. 2017; Colares et al. 2017; Álvarez 2019; Salaverry et al. 2020). In Mesoamerica and the Caribe, these infusions are indicated against obesity (Molares et al. 2012; Alonso-Castro 2015), but we have some concerns about the correct plant identity because *S. campestris* is in general considered in the literature as a South American native species (Guaglianone and Gattuso 1991; Gattuso 1995; Freire et al. 1997; Gupta 2006; Alonso-Paz et al. 2008).

The infusions made of aerial parts (generally leaves) are, in general, cited for anti-inflammatory, antisyphilitic, antirheumatic, diuretic, tonic and digestive properties being consumed as refreshing beverages (Gattuso 1995; Rosa and Ferreira 1999; Rugna et al. 2002, 2005; Alonso-Paz et al. 2008; Barboza et al. 2009; Barneche et al. 2010; Morais et al. 2014; Álvarez 2019; Salaverry et al. 2020). Remarkably, in Southern Brazil, the ethnobotanical use of aerial part infusions is more related to inflammatory processes and microbial infections: for diarrhea, wounds heal, and vaginal fungus (the latter especially from flower extracts) (Coelho de Souza et al. 2004).

In Center Argentina (Córdoba Province), *S. campestris* fruits (known locally as "zarzamora") are eaten by the rural population because their sweet purple pulp, and thus, this species is considered as an edible plant (Arias Toledo et al. 2007). In Northeastern Argentina (Misiones Province), those fruits are employed for staining purposes (blue and violet colors) (Scarpa 2017).

9 Modern Medicine Based on Its Traditional Medicine Uses

The 50% MeOH rhizophore extract was tested for antioxidant potential through TRAP *in vitro* assay, demonstrating that the activity level varies according to the seasonality, which may be associated with the dynamic in polyphenol content (especially proanthocyanidins) (Rugna et al. 2003a). In another work, Morais et al. (2014) assessed for antioxidant activity through DPPH methodology finding the best results for EtOH leaves extract and its butanol fraction with IC_{50} values of 2.06 and 13.61μg/ml (respectively), even better than the value for BHT positive control (16.36μg/ml); these results were correlated with a higher content of polyphenols in such extracts. Similar results were obtained by De Lima (2016) with the same methodology (IC_{50} = 13.1μg/ml for leaves EtOH extract), confirming the antioxidant activity of *S. campestris* alcoholic extracts.

Recently, it was evaluated the *in vitro* anti-inflammatory activity from the rhizophore aqueous extracts from *S. campestris* collected in Argentina, employing the model of human THP-1 macrophages activated with bacterial lipopolysaccharide (LPS) (Salaverry et al., 2020). With this model, the extracts modulated the macrophage response diminishing the released levels of pro-inflammatory cytokines (TNF-α, IL-1β and IL-6) and chemokines (IL-8 and MCP-1) when inflammation was stimulated by LPS. Moreover, it was assessed and confirmed the inhibition of metalloprotease activity (gelatinase), the expression of NF-κB inflammatory transcription factor in the nuclear fraction, and the production of superoxide anion responsible for oxidative stress (Salaverry et al. 2020). On the other hand, the authors corroborated that the extracts were not cytotoxic against macrophages at basal or under simulated inflammation conditions. All this information highlights the immunomodulatory effects of the *S. campestris* rhizophore extracts without exerting cytotoxic effects, especially at the higher dose assessed by the authors (10μg/ml) (Salaverry et al. 2020).

Regarding antimicrobial tests, some organic fractions of leaves EtOH extracts have demonstrated activity against several drug-resistants *Candida* spp. and *Cryptococcus* spp. yeasts, i.e. the most active was the ethyl acetate fraction against *C. krusei* with a MIC of 62.5 μg/ml, but this value was higher than the positive controls fluconazole and amphotericin B (Morais et al. 2014). Battista et al. (2007) demonstrated that 80% MeOH extracts from *S. campestris* rhizophores and leaves inhibited the growth of *C. albicans* and *Aspergillus niger* strains in diffusion tests, being the former the most active; while the aqueous or methanolic extracts of the roots, and the aqueous extracts of rhizophore and leaves were inactive. In an antimicrobial screening from the Uruguayan woody Flora, Barneche et al. (2010) found that the acetone extracts of *S. campestris* leaves were active (employing diffusion assay) against some bacteria strains of *Listeria inocua*, *Bacillus subtilis* and *Pseudomonas aeruginosa*, and the yeast *C. albicans*.

However, in a screening of medicinal plants employed in Southern Brazil, Coelho de Souza et al. (2004) did not find any level of antimicrobial activity of aerial parts MeOH extracts against several bacteria and yeasts (including *B. subtilis* and *C. albicans*) using the agar-diffusion assay. In the same line, De Lima (2016) also found total lack of antimicrobial activity from EtOH aerial parts extracts (from Western-Central Brazil), assessing for 19 different bacteria and fungi strains. Despite the authors aware of the limitation of the agar-diffusion antimicrobial screening method and the variation in the metabolic profile of plant material (Coelho de Souza et al. 2004; De Lima 2016), these results call for a systematic study about the real antimicrobial potential from *S. campestris* in all the regions where the plant species occur.

The organic macerated (1:1 CH_2Cl_2-MeOH), 70% MeOH (Soxhlet) and aqueous extracts from *S. campestris* leaves from Bolivia were subjected to *in vitro* research on dopaminergic activity to treat neurological and neuropsychiatric disorders such Parkinson's disease and schizophrenia, among others (Luedtke et al. 2002). Nevertheless, such extracts did not demonstrate significant activity and were discarded in an initial screening.

In the literature, there are only two pre-clinical studies with animal models. The diuretic properties of *S. campestris* rhizophore and roots aqueous infusions were tested through an *in vivo* classical model with Sprague-Dawley rats (Debenedetti et al. 2000). In this work, orally administering extract doses of 250–1000 mg/kg (equivalent to 3.0–12.0 g of plant material/kg), it was not detected diuretic effect after 10 h of assay; thus, not supporting such ethnobotanical indication (Debenedetti et al. 2000). Also, the oral acute toxicity of the infusion of *S. campestris* rhizophore and roots was evaluated against CF-1 mice (Gorzalczany et al. 1999). Even applying a high dose of plant extract (14.4 g/kg) there was not observed acute toxic signs in the behavior during 15 days of observation, and after the sacrifice and posterior necropsy of the animals. Considering that the normal dose of an infusion in humans is 100 times lesser, the use of such plant extracts might be considered safe, at least from the acute toxicity point of view (Gorzalczany et al. 1999). Nevertheless, more clinical research is necessary to assess for continuous use, to support the inclusion of this plant extract in Argentina and MERCOSUR food additives regulations (MERCOSUR 2006; Código Alimentario Argentino 2019).

An assessment of the mutagenic and cytotoxic properties from the aerial parts infusions was performed by Ferreira and Vargas (1999) applying the *in vitro Salmonella thyphimurium* microsome assay (Ames test). The authors observed an absence of mutagenic and cytotoxic effects for the infusion, even though at higher concentrations (250–500 mg/plate) it was evidenced a trend towards the induction of both detrimental effects (Ferreira and Vargas 1999). These results highlight the need for caution in the frequent and long-term consumption of *S. campestris* teas, although more experimental evidence is required. For other *Smilax* spp., it is reported that the excess (more than 3–5 cups/day) in the rhizophores tea drinking can cause nausea, vomiting, salivation and decrease in the blood pressure (Medeiros et al. 2007).

Another aspect regarding toxicity and which requires further attention is the fact that certain medicinal plants might result toxic owing to fungi contaminations and their mycotoxins. Thus, dried *S. campestris* (plant organ not informed) has been reported by its contamination produced by the fungus *Aspergillus alliaceus*, which eventually can synthesize ochratoxin A (a suspected carcinogen) (Rizzo et al. 2004). For preventing this problem, serious quality control procedures must be applied to ensure safety to medicinal plant consumers.

10 Conclusions

The information about the ethnobotanical uses of *S. campestris* rhizophores and leaves as medicinal infusions and decoctions in South America, is abundant. Scientific evidence is, however, still very scarce. Regarding the chemical constituents of the species, only flavonoids (present in all plant organs) have been partially characterized as bioactive compounds, even though phytochemical screenings have demonstrated the presence of coumarins, alkaloids, saponins, steroids, and tannins. There are a few studies regarding the *in vitro* bioactivities of the plant material, in particular, the antioxidant, antimicrobial and anti-inflammatory effects of rhizophores and leaf-extracts. The only pre-clinical test for the diuretic effects (performed on rats) has failed to support a relevant traditional use. No further clinical studies have been performed, to date. In view of these considerations, more scientific studies are needed to provide a rational base for the safe use of this plant species in medicine. To date, the medicinal use of *S. campestris* requires caution. There is also no sufficient information supporting the inclusion of this species in the South American Pharmacopeias.

Acknowledgments The authors wish to acknowledge to *Centro Universitario de Tacuarembó* (CUT-UdelaR). Prof. Rosângela G. Rolim (UFRGS), Prof. Aline R. Martins (UNESP) and Prof. Diego Passarella (CUT-UdelaR) are recognized for his contribution with photographs of the plant material.

References

Acosta ME, Vignale ND, Ladio AH (2013) Uso y comercialización de especies medicinales en la ciudad de San Salvador de Jujuy. Agraria 7(14):74–81

Alonso-Castro AJ, Domínguez F, Zapata-Morales J, Carranza-Álvarez C (2015) Plants used in the traditional medicine of Mesoamerica (Mexico and Central America) and the Caribbean for the treatment of obesity. J Ethnoparmacol 175:335–345

Alonso-Paz E, Bassagoda MJ, Ferreira F (2008) Yuyos. Uso racional de las plantas medicinales. 2° ed. Ed. Fin de Siglo, Montevideo, Uruguay, 222 pp

Álvarez MJ (2019) Pharmacological properties of native plants from Argentina. Ed. Springer Nature Switzerland AG, Cham, Switzerland, 255 pp

Arias Toledo B, Colantonio S, Galetto L (2007) Knowledge and use of edible and medicinal plants in two populations from the Chaco forest, Córdoba Province, Argentina. J Ethnobiol 27(2):218–232

Arrillaga de Maffei BR (1969) Plantas Medicinales. Ed. Nuestra Tierra, Montevideo, Uruguay, 60 pp

Arriola IA, Melo-Júnior JCF, Ferreira BG, Isaias RMS (2018) Galls on *Smilax campestris* Griseb. (Smilacaceae) protect the insects against restinga constraints, but do not provide enriched nutrition. Braz J Bot 41(1):145–153

Bandoni AL, Mendiondo ME, Rondina RVD, Coussio JD (1976) Survey of Argentine medicinal plants: folklore and phytochemical-screening. Econ Bot 30(2):161–185

Barboza GE, Cantero JJ, Núñez C, Pacciaroni A, Ariza Espinar L (2009) Medicinal plants: a general review and a phytochemical and ethnopharmacological screening of the native Argentine Flora. Kurtziana 34(1–2):7–365

Barneche S, Bertucci A, Haretche F, Olivaro C, Cerdeiras MP, Vázquez A (2010) Prospección química y microbiológica del bosque de galería del río Uruguay. Braz J Pharmacogn 20(6):878–885

Battista SM, Garcia GF, Rugna AZ, Wagner ML, Gurni AA (2007) Actividad antimicótica en extractos de diferentes órganos de *Smilax campestris* Griseb-Smilacaceae. Bol Latinoam Caribe Plant Med Aromáticas 6(6):330–331

Benigni R, Capra C, Cattorini PE (1962) Piante Medicinali. Chimica Farmacologia e Terapia. Volume 1. Ed. Inverni & Della Beffa, Milan, Italy, 1833 pp

Bernardo RR, Pinto AV, Parente JP (1996) Steroidal saponins from *Smilax officinalis*. Phytochemistry 43(2):465–469

Cabral RDC, de Melo-Junior JCF, Matilde-Silva M (2018) Plasticidade morfoanatômica foliar em *Smilax campestris* (Smilacaceae) em gradiente ambiental de restinga, SC, Brasil. Hoehnea 45(2):173–183

Código Alimentario Argentino (2019) Capítulo XVIII. Aditivos Alimentarios. Published on the Internet: https://www.argentina.gob.ar/anmat/codigoalimentario. (Last access, November 2019)

Coelho de Souza G, Haas APS, von Poser GL, Schapoval EES, Elisabetsky E (2004) Ethnopharmacological studies of antimicrobial remedies in the south of Brazil. J Ethnopharmacol 90:135–143

Colares MN, Alonso SMM, Spegazzini E, Hernández MP, Perrotta VG, Novoa MC, Arambarri AM (2017) Anatomy and histochemistry of the underground organs of climbing medicinal plants from the shore of Rio de la Plata (Argentina). Lilloa 54(2):123–140

Conran JG (1998) Smilacaceae. In: Kubitzki K (ed) Flowering plants-monocotyledons. the families and genera of vascular plants, vol 3. Ed. Springer, Berlin/Heidelberg, Germany, pp 417-422

De Lima TF (2016) Propriedade antimicrobiana e atividade antioxidante do extrato etanólico das folhas de *Smilax campestris* Grisebach. Thesis. Universidade Federal da Grande Dourados, Dourados, Brazil, 60 pp

Debenedetti S, Miño J, Rojo A, Acevedo C (2000) Ensayo del efecto diurético de los extractos acuosos de *Amaranthus muricatus* (Moquin) Gill. ex Hicken, *Bauhinia candicans* Benth. y *Smilax campestris* Griseb. Acta Farm Bonaer 19(1):17–20

Domínguez JA (1928) Contribución a la Materia Médica Argentina. Ed. Peuser, Buenos Aires, Argentina, 433 pp

Durán A, Gaspar S, Salce B, Romeo RA (2013) Lianas y enredaderas más destacadas del circuito turístico Jaire-Tiraxi-Tesorero-Tilquiza, Dpto. Dr. Manuel Belgrano, Jujuy (Argentina). Agraria 7(14):149–154

Ferreira ICFS, Vargas VMF (1999) Mutagenicity of medicinal plant extracts in *Salmonella*/microsome assay. Phytother Res 13:397–400

Fischer FM, Oliveira JM, Dresseno ALP, Pillar VD (2014) The role of invasive pine on changes of plant composition and functional traits in a coastal dune ecosystem. Nat Conservaçao 12(1):19–23

Freire SE, Arambarri AM, Mandrile EL, Degenhardt S (1997) Anatomía y Etnobotánica de las Especies Medicinales de Monocotiledóneas de la Estepa Pampeana: Claves para su Determinación. Acta Farm Bonaer 16(2):69–82

Gattuso SJ (1995) Exomorfología y Anatomía de *Smilax campestris* (Smilacaceae). Acta Farm Bonaer 14(3):181–190

Gibert E (1873) Enumeratio Plantarum sponte nascentum Agro Montevidensi cum Synonimis Selectis. Ed. Asociación Rural del Uruguay, Montevideo, Uruguay. 146 pp. In: facsimile: https://autores.uy/. (Last access, November 2019)

Gorzalczany S, Rojo A, Rondina R, Debenedetti S, Acevedo C (1999) Estudio de toxicidad aguda por vía oral de Plantas Medicinales Argentinas. Acta Farm Bonaer 18(3):221–224

Grisebach AHR (1842) *Smilax campestris*. In: von Martius CFP et al (eds) Flora Brasiliensis 3(1):15. In: facsimile: http://florabrasiliensis.cria.org.br/. (Last access, November 2019)

Guaglianone ER, Gattuso S (1991) Estudios taxonómicos sobre el género *Smilax* (Smilacaceae) I. Bol Soc Argent Bot 27(1-2):105–129

Guaglianone ER, Gattuso S (2006) Smilacaceae Vent. Flora del Valle de Lerma. Aportes Botánicos de Salta-Ser 7(16):1–7

Gupta MP (2006) Medicinal plant originating in the Andean high plateau and central valleys region of Bolivia, Ecuador and Perú. Ed. United Nations Industrial Development Organisation (UNIDO), 305 pp

Luedtke RR, Freeman RA, Martin MW, Bastien JW, Zalles-Asin J, Reinecke MG (2002) Pharmacological survey of medicinal plants for activity at dopamine receptor subtypes. I. Activation of D1-like receptor linked adenylyl cyclase. Pharm Biol 40(4):315–325

Martínez GJ, Cúneo P (2009) Las denominaciones vernáculas y el conocimiento toba del entorno vegetal. Rev Dialectol Tradic Pop 64(2):149–168

Martins AR (2009) Morfoanatomia, germinação e perfil químico de espécies de *Smilax* L. (Smilacaceae). Thesis. Universidade Estadual de Campinas, Campinas, Brazil, 158 pp

Martins AR, Pütz N, Soares AN, Bombo AB, Glória BA (2010) New approaches to underground systems in Brazilian *Smilax* species (Smilacaceae). J Torrey Bot Soc 137(2):220–235

Martins AR, Soares AN, Bombo AB, Fidelis A, Coelho ADL, da Glória BA (2012) Germination and seedling morphology of four South American *Smilax* (Smilacaceae). Rev Biol Trop 60(1):495–504

Martins AR, Bombo AB, Soares NA, da Glória BA (2013) Aerial stem and leaf morphoanatomy of some species of *Smilax*. Braz J Pharmacogn 23(4):576–584

Martins AR, Soares MKM, Redher VLG, Bajay MM, Villela PMS, Zucchi MI, da Glória BA (2014) Use of anatomical, chemical, and molecular genetic characteristics in the quality control of medicinal species: a case study of sarsaparilla (*Smilax* spp.). Econ Bot 68(4):410–425

Medeiros MFT, de Senna-Valle L, Andreata RHP (2007) Histórico e o uso da "salsa parrilha" (*Smilax* spp.) pelos boticários no Mosteiro de São Bento. Rev Bras Biociencias 5(S1):27–29

MERCOSUR (2006) Resolución GMC No. 10/06. Reglamento técnico MERCOSUR sobre aditivos aromatizantes/saborizantes. Published on the Internet: https://www.mercosur.int/documentos-y-normativa/normativa/. (Last access, November 2019)

Mereles F, Degen R, Ortíz M (1994) La diversidad florística del macizo Acahay, Paraguari, Paraguay. Rojasiana 2(2):129–163

Molares S, Arenas PM, Aguilar A (2012) Etnobotánica urbana de los productos vegetales adelgazantes comercializados en México DF. Bol Latinoam Caribe Plant Med Aromáticas 11(5):400–412

Montenegro P (2007) Materia Médica Misionera. 1° ed. Ed. Universitaria de la Universidad Nacional de Misiones. Posadas, Argentina. 316 pp [Re-edited from the original from 1710]

Morais MI, Pinto MEA, Araújo SG, Castro AHF, Duarte-Almeida JM, Rosa LH, Rosa CA, Johann S, Lima LARS (2014) Antioxidant and antifungal activities of *Smilax campestris* Griseb. (Smilacaceae). Nat Prod Res 28(16):1275–1279

Paccard E (1905). Lista de algunas plantas medicinales de las Repúblicas Oriental y Argentina. Ed. Barreiro y Ramos, Montevideo, Uruguay, 128 pp. In: fascimile: https://autores.uy/. (Last access, November 2019)

Rizzo I, Vedoya G, Maurutto S, Haidukowski M, Varsavsky E (2004) Assessment of toxigenic fungi on Argentinean medicinal herbs. Microbiol Res 159:113–120

Romeu B, Batilani-Filho M, Faveri SB, Hernández MIM (2013) Influência das características da vegetação na ocorrência de herbivoria por galhadores e mastigadores sobre salsaparrilha (*Smilax campestris*) em ambiente de restinga. In: Batilani-Filho M, Zanette-Silva L, Segal B, Hernández MIM (eds) Ecologia de campo: abordagens no mar, na terra e em águas continentais. Ed. PPG Ecologia, Universidade Federal de Santa Catarina, Brazil, pp 230–240

Rosa SGT, Ferreira AG (1999) Germination of medicinal plant: *Smilax campestris* Griseb. (Salsaparrilha). Acta Hortic 502:105–111

Rugna AZ, Gurni AA, Wagner ML (1999) Progress in studies of flavonols from *Smilax campestris* Griseb. Smilacaceae. Acta Hortic 501:191–194

Rugna AZ, Gurni A, Wagner ML (2002) Estudio Variacional de Flavonoles en Ejemplares Masculinos y Femeninos de *Smilax campestris* Griseb. Smilacaceae. Acta Farm Bonaer 21(2):119–121

Rugna AZ, Polo J, Evelson P, Gurni AA, Llesuy S, Wagner ML (2003a) Antioxidant activity in rhizomes from *Smilax campestris* Griseb. Smilacaceae. Mol Med Chem 1:21–25

Rugna AZ, Vugin A, Gurni A, Wagner ML (2003b) Marcha fitoquímica comparativa entre las hojas y los rizomas de *Smilax campestris* Griseb. Smilacaceae. Dominguezia 19(1):25–29

Rugna AZ, Gurni AA, Wagner ML (2004) Estudio Comparativo de Flavonoides en los órganos aéreos de *Smilax campestris* Griseb. Smilacaceae. Acta Farm Bonaer 23(3):379–382

Rugna AZ, Gurni AA, Wagner ML (2005) Fitoquímica comparativa de flavonoides en los diferentes órganos de *Smilax campestris* Griseb. Smilacaceae. Dominguezia 21(1):17–23

Rugna AZ, Ricco R, Gurni AA, Wagner ML (2007a) Efectos de la Radiación Solar sobre la Producción de Polifenoles em Ejemplares Femeninos de *Smilax campestris* Griseb. Smilacaceae. Lat Am J Pharm 26(3):420–423

Rugna AZ, Gurni AA, Wagner ML (2007b) Ritmo de producción de polifenoles de *Smilax campestris* Griseb. (Smilacaceae). Bol Latinoam Caribe Plant Med Aromáticas 6(5):297–298

Rugna AZ, Ricco R, Gurni A, Wagner M (2008) Variaciones en el Contenido de los Polifenoles Foliares em *Smilax campestris* Griseb. - Smilacaceae - según su Grado de Desarrollo. Lat Am J Pharm 27(2):247–249

Rugna AZ, Romero O, Mazzeo M, Santamaría JM, Gurni AA, Wagner ML (2010) Producción de polifenoles en ejemplares de *Smilax campestris* Griseb. (Smilacaceae) cultivadas en condiciones controladas de cultivo. Dominguezia 26(2):31–36

Rugna AZ, Ricco RA, Gurni AA, Wagner ML (2011) *Smilax campestris* Griseb. –Smilacaceae-: variaciones en la producción de polifenoles en hojas sanas y atacadas por la oruga de la mariposa *Agraulis vanillae* L. –Heliconidae. Dominguezia 27(2):27–33

Rugna AZ, Gurni AA, Wagner ML (2013) Phenological variations of polyphenols in *Smilax campestris* (Smilacaceae). Turk J Bot 37:350–354

Salas-Coronado R, Hernández-Carlos B, Llaguno-Guilberto J, Santos-Sánchez NF (2017) Phenolic compounds in Genus *Smilax* (Sarsaparilla). In: Soto-Hernandez M, Palma-Tenango M, Garcia-Mateos MR (eds) Phenolic compounds – natural sources, importance and applications. Ed. IntechOpen, London, UK, pp 233–260

Salaverry LS, Parrado AC, Mangone FM, Dobrecky CB, Flor SA, Lombardo T, Sotelo AD, Saccodossi N, Rugna AZ, Blanco G, Canellada A, Rey-Roldán EB (2020) *In vitro* anti-inflammatory properties of *Smilax campestris* aqueous extract in human macrophages, and characterization of its flavonoid profile. J Ethnoparmacol 247:112282

Scarpa GF (2017) Etnobotánica histórica de grupos criollos de Argentina II: puesta en valor, adscripción cultural y análisis de los usos no medicinales presentados por el gobierno argentino en la Exposición Universal de París 1889. Bonplandia 26(2):77–102

Scarpa GR, Rosso CN (2014) La etnobotánica moqoit inédita de Raúl Martínez Crovetto I: Descripción, actualización y análisis de la nomenclatura indígena. Bol Soc Argent Bot 49(4):623–647

Sella M, Debenedetti S, Valle C, Gattuso S, Di Bernardi T, Spegazzini E (2013) Parámetros histoanatómicos para el reconocimiento de dos especies denominadas "zarzaparrilla". Memoria digital. Congreso SILAE XXII (Costa Rica 2013) Version 3.25112013:105-106. Published on the Internet https://www.silae.it/docs/eventi/sl22/Atti_XXII_SILAE_Puntarenas_v3.pdf. (Last access, November 2019)

Soares MKM (2012) Anatomia e perfil químico da salsaparrilha comercializada no estado de São Paulo. Thesis. Universidade de São Paulo, Piracicaba, Brazil, 69 pp

Soares MKM, Martins AR, Montanari-Junior I, Figueira GM, Zucchi MI, Bajay MM, da Glória BA (2014) The sarsaparilla market in the state of São Paulo (Brazil) and the challenges of cultivation. Braz J Pharmacogn 24:73–79

The Plant List (2019) Version 1.1 Published on the Internet. http://www.theplantlist.org. (Last access, November 2019)

Zanette-Silva L, Paitach RL, Faveri SB, Hernández MIM (2013) Efeito de fatores estruturais do ambiente sobre a variação fenotípica de espinhos em salsaparrilha, *Smilax campestris* (Smilacaceae). In: Batilani-Filho M, Zanette-Silva L, Segal B, Hernández MIM (eds) Ecologia de campo: abordagens no mar, na terra e em águas continentais. Universidade Federal de Santa Catarina, Brazil, PPG Ecologia, pp 222–229. ISBN 978-85-8328-029-3

Solanum crispum Ruiz & Pav.

Bruce K. Cassels

Solanum crispum growing wild in La Aurora, Curacaví, Santiago Metropolitan Region, Chile. (Photo by the author)

Abstract *Solanum crispum* is a bush that grows in moist places from semiarid to humid temperate regions in Chile. Its aerial parts were used traditionally to treat severe but not clearly identified conditions characterized by fever and it still recommended by herbalists, as a febrifuge. Different extracts of the plant exhibited antipyretic and/or anti-inflammatory activities in animal and human studies. Chemical analyses have identified several steroidal (solasodane) alkaloids, daucosterol, and the coumarin scopoletin, all of them could be possible responsible for the observed effects.

B. K. Cassels (✉)
Department of Chemistry, Faculty of Sciences, University of Chile, Santiago, Chile
e-mail: bcassels@uchile.cl

Á. Máthé, A. Bandoni (eds.), *Medicinal and Aromatic Plants of South America Vol. 2*, Medicinal and Aromatic Plants of the World 7,
https://doi.org/10.1007/978-3-030-62818-5_39

Keywords Solanaceae · "Natre" · "Natri" · Tomatillo · Medicinal · Febrifuge · Ornamental

1 Introduction

Solanum crispum is a flowering plant species, in the family Solanaceae. It is native to Chile, while it is often listed as also growing in Peru. This species is not listed in a recent annotated checklist of Peruvian *Solanum* species (Särkinen et al. 2015). This is a widespread error that probably exists since the time of its collection and description by Ruiz and Pavón (late eighteenth century) when Chile was a dependency of the Viceroyalty of Peru. The large *S. crispum* shrub has flowers resembling those of the closely related potato. For this reason, it is commonly mentioned in the horticultural literature as "Chilean potato tree" or "Chilean potato vine", although it is neither a tree nor a vine. A popular cultivar in the British Isles is called "glasnevin", after the location of the Irish National Botanic Gardens, where it was selected. In autumn the plant produces small poisonous berries.

2 Taxonomic Characteristics

Solanum crispum Ruiz & Pav. (Solanaceae) is endemic to Chile and is often mentioned in the literature as *S. congestiflorum* Dunal, *S. gayanum* (J. Rémy) Phil. f., *S. ligustrinum* Lodd., and *S. tomatillo* (J. Rémy) Phil. f. (The Plant List 2019). Its native names are "natri" or "natre", "tomatillo" (little tomato) and the probably obsolete "hierba del chavalongo" in reference to its historic use. Based largely on its morphology, it is classified as a member of the "*nitidum* group" of the Dulcamaroid clade of *Solanum* L. (Knapp 2013).

3 Crude Drug Used

Leaves, sometimes including small twigs and (rarely) the stem bark are used in infusions. These are very bitter and have given rise to the popular Chilean expression "bitter as natri". They are taken orally, at least in the dominant "europeanized" culture. However, as late as the second half of the nineteenth century, rectal administration was sometimes practiced, and it also seems likely that it is still used in this way by some of the Mapuche people of Central-Southern Chile and Argentina. This route deserves study, considering its avoidance of possibly extensive modification of the plant constituents by the gut flora and first-pass metabolism in the liver of the absorbed substances.

4 Major Chemical Constituents and Bioactive Compounds

The earliest modern attempt to isolate a secondary metabolite from *S. crispum* dates back to 1953 (Bianchi et al. 1953), but prior to 1889 the preparation of "natrina" and crystalline salts of this substance were described, and the products were tested on patients (Murillo 1889). The persistent foam formed in aqueous *S. crispum* preparations (and even in the urine of patients treated with these products: Murillo 1889) points to the presence of saponins. In spite of several subsequent papers, however, the only compounds that may be considered to be correctly characterized, obtained directly from the plant (*e.g.* without previous hydrolysis), and identified beyond reasonable doubt are the coumarin scopoletin and the widespread phytosterol glycoside daucosterol or β-sitosterol 3-*O*-β-D-glucoside (Delporte et al. 1998).

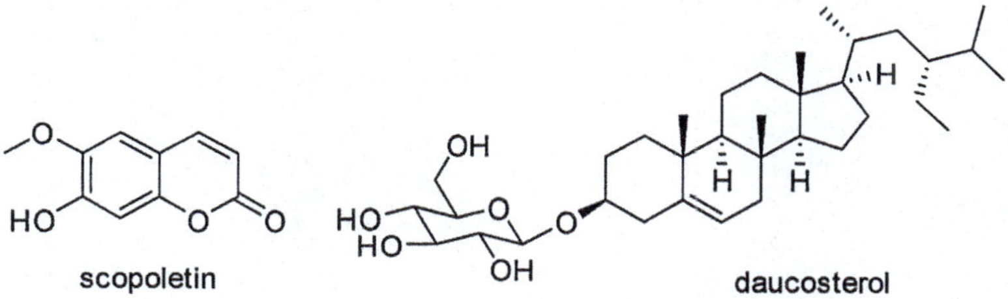

In the first chemical publications the plant was called *S. tomatillo*, and it was shown that the main aglycone obtained upon acid hydrolysis of an extract of the plant material was solasodine, together with much smaller amounts of tomatillidine and its 5α,6-dihydro derivative (Bianchi et al. 1960, 1965).

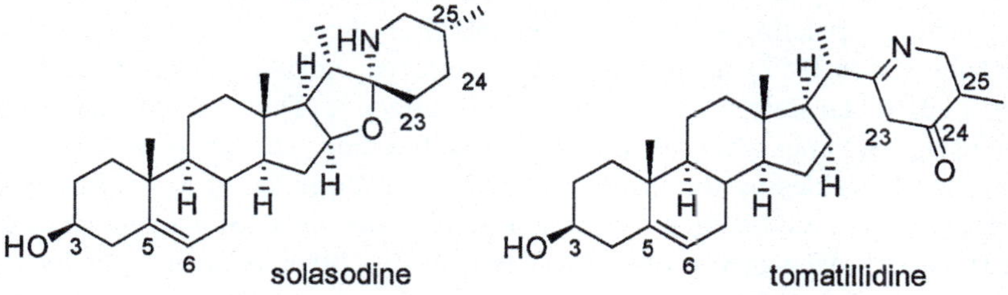

It is not unreasonable to suspect that, barring errors in the structure determinations, the two latter compounds are isolation artefacts. One of the glycosidic alkaloids was supposedly identified as solanine (Bianchi et al. 1961), which might be somewhat dubious as the aglycone solanidine has a different modified steroidal backbone (solanidane) from the solasodane structure isolated from the plant on several occasions. A more recent paper by the same author, on the plant called on that occasion *S. congestiflorum*, somewhat surprisingly does not mention solasodine but instead four new analogs: solacongestidine, solafloridine 23- and

24-oxosolacongestidine, the latter supposedly the C-25 epimer of 5α,6-dihydrotomatillidine (Sato et al. 1969).

The apparent absence of solasodine from the leaves and twigs of "*S. congestiflorum*", reported seasonal differences in the composition of "*S. tomatillo*" and the likely interindividual and interpopulational variation of *S. crispum* point to the importance of performing new studies, using methodologies that differ profoundly from those employed half a century ago.

Hydrolysis of a non-alkaloidal glycoside fraction of what at the time was designated as *S. tomatillo* led to the isolation of the steroidal genin diosgenin (an analog of solasodine with the NH group replaced by an O), and its presumed dehydration artefact 25*R*-spirosta-3,5-diene (Morales et al. 1974):

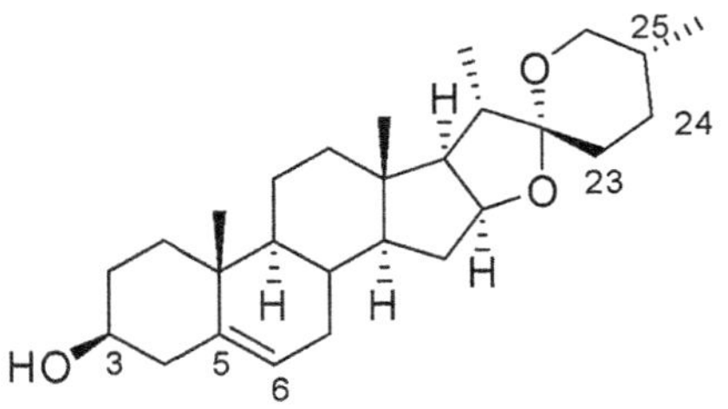

diosgenin

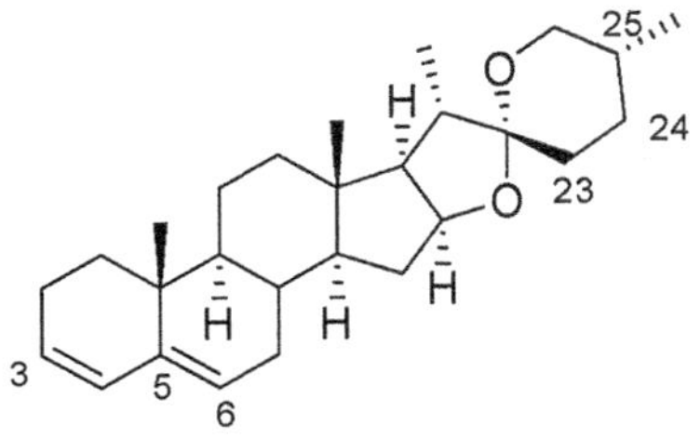

25*R*-spirosta-3,5-diene

When scopoletin and daucosterol were isolated from *S. crispum* (Delporte et al. 1998; called *S. ligustrinum* in that paper), it was shown that aqueous extracts (such as the decoction used traditionally), and also extracts prepared with different organic solvents, presented antipyretic and anti-inflammatory activities after oral administration in guinea pigs. The dichloromethane extract and a chromatographic fraction enriched in scopoletin, which is a known inhibitor of prostaglandin synthase (Farah and Samuelsson 1992), also exhibited these effects. Oral administration of the mixture of alkaloidal steroid glycosides was also effective. There are no studies on the individual components of such mixtures, but solasodine, which may well be released from the glycosides by the intestinal biota, has been reported to present an anti-inflammatory action like that of cortisone at the same dose (Steinegger and Hänsel 1988). Finally, daucosterol was anti-inflammatory at 30 mg/kg. This glucoside may also undergo metabolic hydrolysis to afford β-sitosterol, which shows antipyretic and anti-inflammatory activities at high doses (Gupta et al. 1980). These results support the traditional use of aqueous extracts of *S. crispum* for fever and, possibly, inflammatory conditions. One may suggest that the *in vivo* effects are due to the additive or synergistic action of the steroid alkaloids, scopoletin, and daucosterol.

A few years ago, a decoction of *S. crispum* leaves was analyzed by HPLC-MS, revealing the presence of many phenolic compounds (acids and flavonoids) which were only tentatively identified. In addition, seven steroidal alkaloids, dehydrospirosolane (like solasodine) and/or solanidane (like solanine) were noted, but only the molecular ion (M + 1) of the genin was observed. The content of phenolics was thought to explain the protective effect of the extract on human erythrocytes and biological membrane models against damage by hypochlorous acid. The decoction produced concentration-dependent changes in the red cell morphology, and it was determined that this effect was mainly due to the interaction of some of its components with the outer membrane (Suwalsky et al. 2016). An earlier paper by the same team had demonstrated similar effects for solasodine, both in erythrocytes and in synthetic phospholipid bilayers, suggesting that this and other steroids increased membrane fluidity (Manrique-Moreno et al. 2014). Also, *S. crispum* inhibited calcium release from the endoplasmic reticulum, and its hydroalcoholic extract has an *in vitro* vasodilatory effect presumably mediated by nitric oxide (NO•) (Vinet et al. 2016). Although these studies are interesting in themselves, *in vivo* work is necessary to determine if the observed effects are of any relevance from the therapeutic or toxicologic point of view.

5 Morphological Description

Solanum crispum usually grows in its native habitats as a scraggly bush, about 1–2 m tall, although with abundant watering and adequate support it may climb more than 3 m, and pruning can promote a denser habit. The ovate leaves, are about 3–7 cm long and 1–3 cm wide with wavy margins, often slightly folded upwards along the central vein and curved downwards lengthwise sometimes showing a finely haired, grayish lower surface. The pentamerous flowers grow in showy clusters that bloom all summer, with light blue to violet-colored corollas about 2 cm in diameter and prominent yellow stamens. The fruit is a round berry somewhat less than 1 cm in diameter, orange to bright red when ripe. All parts of the plant are considered poisonous.

6 Geographical Distribution

Solanum crispum grows spontaneously in Chile from the Aconcagua valley (about 33° S) to the main island of the Chiloé archipelago (about 42° S), from close to sea level to above 2500 m.a.s.l. It has also been collected around populated areas in Neuquén Province (Argentina, about 39° S), where it is believed to have been introduced from Chile as a valued medicinal plant (Knapp 2013). Mention of its occurrence in Mendoza Province (Argentina, about 33° S) is almost certainly in error, as "Santa Rosa de los Andes to Uspallato (*sic*) Pass (Moseley s.n.)" refers to the Chilean side of the Andes range. It is grown as a popular ornamental, mainly in Europe where it is known as the Chilean potato tree (or bush, or vine). The Glasnevin cultivar is mentioned most (Gardenia 2019; British Horticultural Society 2020).

7 Ecological Requirements

In the semi-arid to Mediterranean-type climate of the northern half of its range it prefers relatively shaded spots with adequate moisture, close to the Chilean sclerophyllous forest, although it also grows in open spaces. Further south it is associated with the humid temperate *Nothofagus* forest margins.

As a cultivated plant, in gardens it tolerates light frosts, neutral to slightly alkaline, well-drained soils are recommended, and the plant grows best in full sun ensuring sufficient moisture in desert or semi-desert conditions (Gardenia 2019). It is easily propagated by cuttings as it is fast-growing and, if provided with adequate support, it is said to attain a height of eight meters in 10 years (British Horticultural Society 2020).

8 Traditional Use (Part(s) Used) and Common Knowledge

One of the oldest mentions of the use of *S. crispum*, at the end of the eighteenth century, is as a febrifuge: "The inhabitants make frequent use of it, boiled or infused, for the inflammatory fevers called *Congo* and *Chavalongo*" (Ruiz and Pavón 1799), also cited by Murillo (1889). "Chavalongo" has been translated as "fever of the head" (Febrés 1765). Some authors have associated this condition with typhoid fever, but it seems likely that its meaning referred broadly to the Old-World febrile diseases that decimated the native populations from the mid-sixteenth to the eighteenth centuries (Martínez-Sanz 2005). Citing extensive experience in hospitals and in private practice, Murillo (1889) mentioned his earlier use of "natri" to treat typhoid fever and confirmed the antipyretic effect of the leaves and/or twigs and bark and its pharmaceutical preparations "which are inferior … to quinine and its salts …; but one would seek in vain a medicament more innocent, easier to administer, (and) presenting less drawbacks in its use". It is still sold on the streets and in herbalists' shops in Chile, for use against fevers of any origin (Muñoz et al. 1981, 2001). In the Mapuche cultural area it is imported and consumed as far south as 43° S latitude in Chubut Province, Argentina (Molares and Ladio 2014).

9 Modern Medicine Based on Its Traditional Medicinal Uses

Solanum crispum is not generally used in modern medicine. Murillo (1889), acknowledging its limited effectiveness and short duration of action, concludes "Let us leave the "natri", administered orally or in repeated enemas, as the humble but effective servant of the peasant who lives far from the feverish activity that we call civilization". However, its anti-inflammatory and antipyretic actions are well documented in animal studies (Delporte et al. 1998), and its vasorelaxant and antioxidant activity and its antagonism of calcium release from the endoplasmic reticulum have been assayed *ex vivo* and *in vitro* (Suwalsky et al. 2016; Vinet et al. 2016), so hopefully future pharmacological and pharmaceutical research on this plant will lead to a revival of interest in its use.

10 Conclusions

The antipyretic and anti-inflammatory effects of the intensely bitter *S. crispum* do not seem to be particularly attractive for most present-day urban citizens, given the abundance of efficacious, inexpensive and easily administered synthetic products. Nevertheless, its components deserve to be studied with up-to-date methodologies to confirm their structures and determine their mechanism(s) of action which may hold some surprises in store. The saponins and, in particular the solasodine (and tomatillidine?) glycosides, should be examined with regard to their local effects on the intestinal epithelium, the intestinal biota, their metabolism by the latter and their pharmacokinetics. It must be pointed out that solasodine and some of its derivatives exhibit potential anticancer, antifungal and anti-trypanosomal activities that certainly merit further investigation.

References

Bianchi E, Albertz A, Castillo R (1953) Natrine, the alkaloid of *Solanum crispum* (Natri). Ann Chim 43:208–222

Bianchi E, Díaz F, Garbarino J (1960) Ricerche sugli alcaloidi del genere *Solanum*. Nota I: Gli alcaloidi del *Solanum tomatillo* (Phil.). Gazz Chim Ital 90:894–902

Bianchi E, Díaz F, Charlín M, Garbarino J (1961) Ricerche sugli alcaloidi del genere *Solanum*. Nota II: Glucoalcaloidi della solasodina isolati dal *Solanum tomatillo* (Phil.). Rendiconti Accad Naz Sci XL 11:465–469

Bianchi E, Djerassi C, Budzikiewicz H, Sato Y (1965) Structure of tomatillidine. J Organomet Chem 30(3):754–760

Delporte C, Backhouse N, Negrete R, Salinas P, Rivas P, Cassels BK, San Feliciano A (1998) Antipyretic, hypothermic and antiinflammatory activities and metabolites from *Solanum ligustrinum* Lood. Phytother Res 12(2):118–122

Farah HM, Samuelsson G (1992) Pharmacologically active phenylpropanoids from *Senra incana*. Planta Med 58:14–18

Febrés A. Arte de la lengua general del reyno de Chile. Lima (1765). https://archive.org/details/artedelalenguage00febr/page/n6. (Last access, November 2019), 682 pp

Gardenia, Creating Gardens (2019). https://www.gardenia.net/plant/Solanum-crispum-Glasnevin. (Last access, November 2019)

Gupta MP, Nath R, Srivastava N, Shanker K, Kishor K, Bhargava KP (1980) Anti inflammatory and antipyretic activities of β-sitosterol. Planta Med 39:157–163

Knapp S (2013) A revision of the Dulcamaroid clade of *Solanum* (Solanaceae). PhytoKeys 22:1–432

Manrique-Moreno M, Londoño-Londoño J, Jemioła-Rzemińska M, Strzałka K, Villena F, Avello M, Suwalsky M (2014) Structural effects of the *Solanum* steroids solasodine, diosgenin and

solanine on human erythrocytes and molecular models of eukaryotic membranes. Biochim Biophys Acta 1838:266–277
Martínez-Sanz P (2005) La viruela y Fray Chaparro. Ars Med (Santiago) 10:119–127
Molares S, Ladio A (2014) Medicinal plants in the cultural landscape of a Mapuche-Tehuelche community in arid Argentine Patagonia: an eco-sensorial approach. J Ethnobiol Ethnomed 10:61
Morales R, Urzúa A, Cassels BK (1974) Aislamiento de diosgenina de *Solanum tomatillo*. Contrib Cient Tecnol (UTE) 14:57–63
Muñoz M, Barrera E, Meza I (1981) El uso medicinal y alimenticio de plantas nativas y naturalizadas en Chile. Museo Nacional de Historia Natural, Santiago, Chile. Publicación Ocasional N° 33:3–91
Muñoz O, Montes C, Wilkomirsky T (2001) Plantas medicinales de uso en Chile: química y farmacología. Ed. Universitaria, Santiago, Chile, 330 pp
Murillo A (1889) Plantes médicinales du Chili. Exposition Universelle de Paris 1889, Section Chilienne, Paris, France, pp 142–149
Royal Horticultural Society (2020). https://www.rhs.org.uk/plants/97176/Solanum-crispum-Glasnevin/Details. (Last access, April 2020)
Ruiz H, Pavon J (1799) Flora Peruviana, et Chilensis. Ed. de Sancha G. Madrid. Volume II. p 31 and figure 158a https://gallica.bnf.fr/ark:/12148/bpt6k96491h/f34.image. (Last access, November 2019)
Särkinen T, Baden M, Gonzáles P, Cueva M, Giacomin LL, Spooner DM, Simon R, Juárez H, Nina P, Molina J, Knapp S (2015) Annotated list of *Solanum* L. (Solanaceae) for Peru. Rev Peru Biol 22(1):3–62
Sato Y, Sato Y, Kaneko H, Bianchi E, Kataoka H (1969) Alkaloids from *Solanum congestiflorum*. J Organomet Chem 34(6):1577–1582
Steinegger E, Hänsel R (1988) Lehrbuch der Pharmakognosie und Phytopharmazie. Ed. Springer, Berlin/Heidelberg, Germany, p 804
Suwalsky M, Ramírez P, Avello M, Villena F, Gallardo MJ, Barriga A, Manrique-Moreno M (2016) Morphological effects and antioxidant capacity of *Solanum crispum* (Natre) in vitro assayed on human erythrocytes. J Membr Biol 249(3):349–361
The Plant List (2019) Version 1.1 published on the Internet. http://www.theplantlist.org. (Last access, November 2019)
Vinet R, Álvarez R, Knox M, Guzmán L, Martínez JL, Flores E (2016) Vasodilatory properties of *Solanum crispum* Ruiz & Pav. a South American native plant. Bol Latinoam Caribe Plant Med Aromát 15(2):94–98

Solanum sisymbriifolium Lam.

Lucía Soledad Di Ciaccio, Julián Guariniello, Jésica Ianicelli, and Alejandro Escandón

Solanum sisymbriifolium Lam. Estación Experimental Agropecuaria Área Metropolitana de Buenos Aires (EEA AMBA), Udaondo, Buenos Aires, Argentina. (Photo: Di Ciaccio, L.S)

L. S. Di Ciaccio (✉)
Instituto de Genética "Ewald A. Favret", Instituto Nacional de Tecnología Agropecuaria (INTA), Buenos Aires, Argentina

Agencia Nacional de Promoción Científica y Tecnológica (ANPCyT), Buenos Aires, Argentina
e-mail: diciaccio.lucia@inta.gob.ar

J. Guariniello · A. Escandón
Instituto de Genética "Ewald A. Favret", Instituto Nacional de Tecnología Agropecuaria (INTA), Buenos Aires, Argentina

J. Ianicelli
Instituto de Genética "Ewald A. Favret", Instituto Nacional de Tecnología Agropecuaria (INTA), Buenos Aires, Argentina

Consejo Nacional de Investigaciones Científicas y Tecnológicas (CONICET), Buenos Aires, Argentina

Á. Máthé, A. Bandoni (eds.), *Medicinal and Aromatic Plants of South America Vol. 2*, Medicinal and Aromatic Plants of the World 7,
https://doi.org/10.1007/978-3-030-62818-5_40

Abstract *Solanum sisymbriifolium* (Lam.) is a shrub that belongs to the Solanaceae family. It is native to South America. It grows in ruderal locations and wild, along roadsides and in waste places, landfills and plowed fields. It is naturally pollinated by insects and widely propagated by seeds, although vegetative propagation by rhizomes is also possible. Due to its vast dispersion and rapid growth, it is also considered an invasive weed. In folk medicine, its roots and the entire above-ground plant are mostly used as a diuretic, analgesic, contraceptive, antisyphilitic and hepatoprotective. According to preclinical studies a variety of pharmacological activities can be related to its empirical use, especially, the molluscicidal, antihypertensive, anticonvulsant, antimicrobial and antioxidant effects. The plant is also used in agriculture as a trap-crop in nematode control. *S. sisymbriifolium* is also used as a food source. The cultivation of sterile plants or the *in vitro* production of metabolites seem to reveal opportunities for the exploitation of this promising species.

Keywords Solanaceae · "Espina colorada" · Glycoalkaloids · Steroidal saponins · Antihypertensive

1 Introduction

The Solanaceae family includes several species known for their food importance and for its capacity for the synthesis of steroidal alkaloids (Dinan et al. 2001). In this sense, Nohara et al. (2007) demonstrated the presence of a considerable amount of solasodine, spirosolane, solanidane, spirostane and furostane glycosides in different species of the genus *Solanum*.

The study of *Solanum sisymbriifolium* is of great interest, since it is widely employed in popular medicine in the Southern Cone. In addition, it contains solasonine (a glycoalkaloid of steroid nature) in its leaves and roots (Ferro et al. 2005). The product of its hydrolysis is the alkaloid solasodine, which can be transformed, through simple chemical reactions, into various steroidal hormones of pharmacological interest, for its application in health and in the agricultural sector.

2 Taxonomic Characteristics

The Solanaceae family is one of the largest families, with about 102 genera and 2460 species (Stevens 2001). It is widely distributed in tropical, subtropical and temperate regions around the world. However, they are concentrated in Australia and Central and South America, where at least 40 genera are endemic (Heywood 1985). In Argentina, the family is represented by about 41 genera and 528 species in different biomes (Instituto de Botánica Darwinion 2019). *Solanum sisymbriifolium* (Lam.) was described by Lamarck and published in *Tableau encyclopédique et méthodique des trois règnes de la nature* in 1794. It belongs to the Equisetopsida class, Asterales order, Solanaceae family and *Solanum* genus (Tropicos 2019).

Synonyms *Solanum sisymbriifolium* Lam. var. *macrocarpum*, *Solanum balbisii* Dunal var. *bipinnata*, *Solanum balbisii* Dunal var. *purpureum*, *Solanum sisymbriifolium* Lam. var. *brevilobum*, *Solanum decurrens* Balb., *Solanum balbisii* Dunal, *Solanum bipinnatifidum* Larrañaga, *Solanum sisymbriifolium* Lam. f. *albiflorum*, *Solanum sisymbriifolium* Lam. *f. lilacinum*, *Solanum sisymbriifolium* Lam. var. *heracleifolium*, *Solanum subviscidum* Schrank, *Solanum balbisii* Dunal var. *oligospermum*, *Solanum sisymbriifolium* Lam. var. *oligospermum*, *Solanum sisymbriifolium* Lam. var. *acutilobum purpureiflorum*. (Instituto de Botánica Darwinion 2019).

3 Crude Drug Used

In ethnomedicine and modern medicine the use of roots and whole above-ground plant parts (leaves, flowers and fruits) has been reported, as a crude drug (More 2017).

4 Major Chemical Constituents and Bioactive Compounds

Chemical screenings showed the presence of alkaloids, phenols, terpenoids (steroids) and glycosides, as the major phytoconstituents on leaves (More 2017). The fruits contain solamargine and β-solamarine (glycoalkaloids) (Bagalwa et al. 2010), sisymbriifolin (a neolignan) and carpesterol (a rare C_{30} sterol) (Chakravarty et al. 1996). The aglycone, solasodine, can be obtained from the leaves and fruits (Chauhan et al. 2011). The chemical constituents found in roots were cuscohygrine (Evans and Somanabandhu 1980), solacaproine (Maldoni 1984), and isonuatigenin-3-*O*-β-solatriose (Ferro et al. 2005). The essential oil of the flowers has a high content of aldehydes and aliphatic hydrocarbons and the essential oil of the fruits has a high amount of fatty acids and their derivatives. Heptadecane and octadeca-9,12,15-trienal were the main compounds in flowers, whereas the fruits contained hexadecanoic acid and ambrettolide (17-oxacycloheptadec-6-en-1-one. It would be valuable to confirm the identity and content of the latter compound, given its appreciation in the perfume industry) (Pasdaran et al. 2017). Seeds have 20.6% crude oil and contain linoleic acid, oleic acid and palmitic acid as major fatty acids (Nolasco et al. 2001).

5 Morphological Description

S. sisymbriifolium is a perennial shrub (Fig. 1). Stems are 70–200 cm in length, erect, branched, up to 4 cm diameter at the woody base, viscous villous with sparsely stellate, glandular, eglandular and simple hairs; prickles dense, yellow or orange yellow, subulate, up to 17 mm long, very sharp. Leaves are alternate, with leaf-blades ovate-oblong, 10–15 cm in length and 5–10 cm in width, deeply pinnatisect or pinnatifid, margin sinuate, and lobes rounded, with many prickles along main veins on both surfaces. Inflorescences are racemose, 6–14 flowered; peduncles up to 20 cm long, un-branched, glandular and simple pilose, densely spiny; pedicels slender, glandular-pilose, and slightly spinulose, up to 1.5 cm; calyx cup shaped, green, united at base, with prickles in hermaphrodite flowers; corolla whitish or lilacs, bluish and pale violet, rotate, 1.8 × 3.0 cm in diameter, with broadly triangular lobes, stellate-pilose outside; stamens equal; filaments slender, glabrous, 0.2 cm long; anthers 0.7–0.8 cm long; ovary ovoid, 0.2 cm, superior, glabrous; style 1–1.2 cm, long. Fruiting pedicels deflexed, densely glandular-pilose and viscid; fruiting calyx enlarged 1.1–1.3 × 0.4–0.6 cm, enveloping most berry, ultimately flaring wide and exposing it (D'Arcy 1974).

6 Geographical Distribution

It is native to the Southern Cone (Brazil, Argentina, Paraguay, Uruguay, Chile, Bolivia and Peru) and naturalized to North America (Canada, Mexico, USA), Europe (Spain, the Netherlands), Asia (India, China, Taiwan), Africa (South Africa, Congo, Swaziland), and Australasia (Australia, New Zealand) (Germplasm Resources Information Network 2019; Flora Argentina 2019).

Fig. 1 Crude drug used of *S. sisymbriifolium*. (**a**) Leaves. (**b**) Flowers. (**c**) Roots. (**d**) Ripe Fruits. (Photos: Di Ciaccio, L.S; Guariniello, J. Gobernador Guillermo Udaondo 1695, Estación Experimental Agropecuaria Área Metropolitana de Buenos Aires (EEA AMBA), Udaondo, Buenos Aires, Argentina. September/November, 2019)

7 Ecological Requirements

Solanum sisymbriifolium is found along roadsides and in waste places, landfills, and plowed fields, as well as non-native ranges (Hill and Hulley 1995). This is a fast-growing species that adapts to all types of soils (mainly fertile soils) and is shade tolerant. (Global Invasive Species Database 2015). It reproduces both sexually by entomophilic pollination (carpenter bees) and by rhizomes (D'arcy 1974; Hill and Hulley 1995; Bean 2012). The seeds are highly viable and present zoochory (Timmermans 2005).

Since this species tends to be invasive, its introduction to cultivation in a new region should be considered before implementation (Global Invasive Species Database 2015)

8 Traditional Use (Part(s) Used) and Common Knowledge

Common names: sticky nightshade, litchi tomato, red buffalo-bur, fire and ice, "morelle de Balbis", "tutiá", "revienta caballos" (busts horses), "espina colorada" (red thorn), "pocotillo", "cardo" (thistle), "comida de víbora" (viper food), "tomatillo del campo", "guindilla de campo", "joá-das-taperas", "togué' hâlá" (red fruit) (Scarpa and Rosso 2014; Germplasm Resources Information Network 2019; Flora Argentina 2019).

S. sisymbriifolium is a medicinal plant widely used by the native people of South America and it is also important in Ayurvedic medicine (Prajapati et al. 2013). Roots are used in the treatment of hypertensive diseases in Paraguay, and as diuretic, analgesic, contraceptive, antisyphilitic and hepatoprotective in Argentine. Added to the "tereré" (the cold "mate", infusion) of the Guaraní people, it is used as a muscle relaxant and anti-inflammatory for waist pain, and as an antispasmodic for intestinal soreness (Rondina et al. 2008).

In Argentina, the aerial parts are used to treat diarrhea and infections of respiratory and urinary tracts. Flowers are used as analgesic (India), leaves as febrifuge (Peru) and as diuretic (Brazil). The plant is also used as emmenagogue and for fertility regulation (More 2017). The fruits of *S. sisymbriifolium* are consumed, raw or boiled, by the Chorote people from north-west Argentina and south-west Paraguay (Kinupp 2007; Rapoport et al. 2009).

9 Modern Medicine Based on Its Traditional Medicine Uses

The hydroalcoholic crude extract of its roots and its butanolic fraction (which contains the steroidal saponin nuatigenin-3-*O*-β-chacotriose) was detected to show antihypertensive action, in anesthetized normotensive rats (Ibarrola et al. 2011).

Remarkably, the alcoholic extract of leaves has antinociceptive, antidiarrheal, neuropharmacological and cytotoxic activity (Apu et al. 2013).

In addition, the aerial parts (specifically, the solamargine contained in fruits) have significant molluscicidal activity, that could be used to control snails, hosts of the parasitic agent *Schistosoma mansoni* (Bagalwa et al. 2010).

Chauhan et al. (2011) reported that solasodine (steroidal alkaloid) isolated from fruits has anticonvulsant activity in rodents. In this context, it is important to remark that this alkaloid -isolated from other *Solanum* species- has been studied as a precursor for the synthesis of corticosteroids and oral contraceptives (progesterone). Furthermore, it has shown anticancer activity, an interesting insecticide property and an important anti-accelerator cardiac action (Kittipongpatana et al. 1999; Kreft et al. 2000; Jacob and Malpathak 2005; Marzouk et al. 2005).

Finally, Pasdaran et al. (2017) concluded that the essential oils of litchi tomato flowers have considerable antioxidant and antibacterial activity against *Staphylococcus aureus*.

On the other hand, despite being considered a weed (Global Invasive Species Database 2015), *S. sisymbriifolium* has a relevant application in agriculture. The species has also been studied for use as a nematode trap plant in potato crops (Timmermans 2005), as a rootstock for commercial varieties of tomato (Ibrahim et al. 2001; Mitidieri et al. 2015) and as biofumigant in banana crop (Pestana et al. 2009); so, it could be included as a tool for integrated pest management of crops. In addition, it was included in breeding programs for the transfer of pest resistance genes, through the generation of tomato hybrids (Piosik et al. 2019).

10 Conclusions

S. sisymbriifolium is a promising species for the human and veterinary pharmaceutical industry. It is suitable for the development of drugs, such as steroid hormones (progesterone). Importantly, it also has a potential for both agriculture and the agrochemical industry. From ethnopharmacological and pharmacological studies on *S. sisymbriifolium*, various biological and medicinal activities have been reported: these include nematocidal, molluscicidal, antimicrobial, anticonvulsant, antihypertensive, antitumor and antioxidant effects. The selection and cultivation of sterile plants or the *in vitro* production of metabolites could speed up the exploitation of this species, in view of its dispersion control.

References

Apu AS, Bhuyan SH, Matin M, Hossain F, Khatun F, Jamaluddin ATM (2013) Analgesic, neuropharmacological, antidiarrheal, and cytotoxic activities of the extract of *Solanum sisymbriifolium* (Lam.) leaves. Avicenna J Phytomed 3(4):302–312

Bagalwa JJM, Voutquenne-Nazabadioko L, Sayagh C, Bashwira AS (2010) Evaluation of the biological activity of the molluscicidal fraction of *Solanum sisymbriifolium* against non target organisms. Fitoterapia 81:767–771

Bean AR (2012 onwards) *Solanum* species of eastern and northern Australia. Version: 6th September 2018. https://www.delta-intkey.com/solanum/www/sisymbri.htm. (Last access, May 2020)

Chakravarty AK, Mukhopadhyay S, Saha S, Pakrashi SC (1996) A neolignan and sterols in fruits of *Solanum sisymbriifolium*. Phytochemistry 41:935–939

Chauhan K, Sheth N, Ranpariya V, Parmar S (2011) Anticonvulsant activity of solasodine isolated from *Solanum sisymbriifolium* fruits in rodents. Pharm Biol 49(2):194–199

D'Arcy WG (1974) Solanum and its close relatives in Florida. Ann Mo Bot Gard 61(3):819–867

Dinan L, Harmatha J, Lafont R (2001) Chromatographic procedures for the isolation of plant steroids. J Chromatogr A 935(1–2):105–123

Evans WC, Somanabandhu A (1980) Nitrogen-containing non-steroidal secondary metabolites of *Solanum*, *Cyphomandra*, *Lycianthes* and *Margaranthus*. Phytochemistry 19:2351–2356

Ferro EA, Alvarenga NL, Ibarrola DA, Hellión-Ibarrola MC, Ravelo AG (2005) A new steroidal saponin from *Solanum sisymbriifolium* roots. Fitoterapia 76(6):577–579

Flora Argentina y del Cono Sur (2019) www.floraargentina.edu.ar. (Last access, September 2019)

Germplasm Resources Information Network (2019). www.ars-grin.gov. (Last access, September 2019)

Global Invasive Species Database (2015) Species profile *Solanum sisymbriifolium*. http://www.iucngisd.org/gisd/species.php?sc=1216. Last access, November 2019

Heywood VH (1985) Las plantas con flores. Ed. Reverté, Madrid, Spain, pp 39–40

Hill MP, Hulley PE (1995) Biology and host range of *Gratiana spadicea* (Klug, 1829) (Coleoptera: Chrysomelidae: Cassidinae), a potential biological control agent for the weed *Solanum sisymbriifolium* Lamarck (Solanaceae) in South Africa. Biol Control 5(3):345–352

Ibarrola DA, Hellion-Ibarrola MC, Montalbetti Y, Heinichen O, Campuzano M, Kennedy ML, Alvarenga N, Ferro EA, Dölzvargas JH, Momose Y (2011) Antihypertensive effect of nuatigenin-3-O-β-chacotriose from *Solanum sisymbriifolium* Lam. (Solanaceae) (*nuatî pytâ*) in experimentally hypertensive (ARH+DOCA) rats under chronic administration. Phytomedicine 18(8–9):634–640

Ibrahim M, Munira MK, Kabir MS, Islam AKMS, Miah MMU (2001) Seed germination and graf compatibility of wild *Solanum* as rootstock of tomate. J Biol Sci 1(8):701–703

Instituto de Botánica Darwinion (2019) Cono Sur, Catálogo de Plantas Vasculares. www.darwin.edu.ar. (Last access, September 2019)

Jacob A, Malpathak N (2005) Plantlet regeneration enhances solasodine productivity in hairy root cultures of *Solanum khasianum* Clarke. In Vitro Cell Dev Biol Plant 41:291–295

Kinupp VF (2007) Plantas alimentícias não-convencionais da Região Metropolitana de Porto Alegre, RS. Thesis, Universidade Federal do Rio Grande do Sul. Faculdade de Agronomia. Programa de Pós-Graduação em Fitotecnia, Rio Grande do Sul, 562 pp

Kittipongpatana N, Porter JR, Hock RS (1999) An improved high performance liquid chromatographic method for the quantification of solasodine. Phytochem Anal 10:26–31

Kreft S, Zel J, Pukl M, Umek A, Strukelj B (2000) Non-aqueous capillary electrophoresis for the simultaneous analysis of solasodine and solasonine. Phytochem Anal 11:37–40

Maldoni BE (1984) Alkaloids of *Solanum sisymbriifolium*. An Asoc Quim Argent 72:265–267

Marzouk Z, Cheria J, Bakhrouf A, Guedira K, Boukef K (2005) Influence of vegetation stage on the solasodine content in four natural *Solanum sodomeum* L. populations of Tunisia. J Food Agric Environ 3:243–245

Mitidieri MS, Piris E, Brambilla V, Barbieri M, Cap G, González J, Del Pardo K, Ciapone M, Celié R, Arpía E, Paunero I, Peralta R, Verón R, Sánchez F (2015) Evaluación de parámetros de rendimiento y sanidad de dos híbridos comerciales de tomate (*Solanum lycopersicum* L.) injertados sobre *Solanum sisymbriifolium* (Lam.), en un invernadero con suelo biosolarizado. Hortic Argent 34(84):5–17

More GK (2017) A review of the ethnopharmacology, phytochemistry and pharmacological relevance of the South African weed *Solanum sisymbriifolium* Lam. (Solanaceae). Environ Dev Sustain 21(1):37–50

Nohara T, Ikeda T, Fujiwara Y, Matsushita S, Noguchi E, Yoshimitsu H, Ono M (2007) Physiological functions of solanaceous and tomato steroidal glycosides. J Nat Med 61(1):1–13

Nolasco SM, Quiroga OE, Pereyra A, Wiese B (2001) Caracterización química del aceite y harina residual de *Solanum sisymbriifolium* Lam. Grasas Aceites 52(2):123–126

Pasdaran A, Pasdaran A, Mamedov N (2017) Antibacterial and antioxidant activities of the volatile composition of the flower and fruit of *Solanum sisymbriifolium* (Litchi tomato). Pharm Sci 23:66–71

Pestana M, Gouveia M, Abrantes I (2009) Efeitos de *Solanum sisymbriifolium* e *S. nigrum* sobre o nemátode-das-lesões-radiculares, *Pratylenchus goodeyi*, parasita da bananeira. Rev Ciênc Agron 32(2):173–181

Piosik L, Ruta-Piosik M, Zenkteler M, Zenkteler E (2019) Development of interspecific hybrids between *Solanum lycopersicum* L. and *S. sisymbriifolium* Lam. via embryo calli. Euphytica 215:31

Prajapati RP, Karkare VP, Kalaria MV, Parmar SK, Sheth NR (2013) Pharmacognostic and phytochemical evaluation of the *Solanum sisymbriifolium* leaf. Afr J Biotechnol 12(42):6133–6139

Rapoport EH, Marzocca A, Drausal BS (2009) Malezas comestibles del cono sur y otras partes del planeta. Ed. INTA, Buenos Aires, Argentina, p 216

Rondina RV, Bandoni AL, Coussio JD (2008) Especies medicinales argentinas con potencial actividad analgésica. Dominguezia 24(1):47–69

Scarpa GR, Rosso CN (2014) La etnobotánica moqoit inédita de Raúl Martínez Crovetto I: Descripción, actualización y análisis de la nomenclatura indígena. Bol Soc Argent Bot 49(4):623–647

Stevens PF (2001) Angiosperm phylogeny website. Version 14, July 2017. http://www.mobot.org/MOBOT/research/APweb/. (Last access, November 2019)

Timmermans BGH (2005) *Solanum sisymbriifolium* (Lam.): a trap crop for potato cyst nematodes. Thesis, Wageningen University, the C.T. de Wit Graduate School for Production Ecology and Resource Conservation (PE&RC), Wageningen, Netherland, 135 pp

Tropicos (2019) Missouri botanical garden. http://www.tropicos.org. (Last access, November 2019)

Xenophyllum poposum (Philippi) V.A. Funk

César A. N. Catalán and Diego A. Sampietro

Xenophyllum poposum (Philippi) V.A. Funk. (Photos: Alfredo Grau.) Argentina, Jujuy Province, Rinconada Dept., 5121 m.a.s.l., 22° 29,340′ S, 66° 47,300′ W, west slope of the Salle volcano on stony-sandy soil

Abstract *Xenophyllum poposum* (Philippi) V.A. Funk (syn. *Werneria poposa* Philippi) is a creeping sub-shrub, 3–5 cm tall with an unpleasant smell, commonly known as "poposa" or "pupusa". It is widely used in Andean traditional medicine to treat hypertension, altitude sickness (mountain sickness), hepatic diseases and digestive disorders. Mono- and sesquiterpenoids accompanied by benzofuran

C. A. N. Catalán (✉)
Instituto de Química Orgánica, Facultad de Bioquímica, Química y Farmacia, Universidad Nacional de Tucumán, S. M. de Tucumán, Argentina
e-mail: ccatalan@fbqf.unt.edu.ar; cancatalan@gmail.com

D. A. Sampietro
Facultad de Bioquímica, Química y Farmacia, Laboratorio de Biología de Agentes Bioactivos y Fitopatógenos (LABIFITO), Universidad Nacional de Tucumán, S. M. de Tucumán, Argentina

Á. Máthé, A. Bandoni (eds.), *Medicinal and Aromatic Plants of South America Vol. 2*, Medicinal and Aromatic Plants of the World 7,
https://doi.org/10.1007/978-3-030-62818-5_41

derivatives, such as tremetone and 6-hydroxytremetone, have been identified in its essential oil while *p*-hydroxyacetophenone and benzofuran derivatives, flavonoids, coumarins and diterpenoids, were isolated from organic extracts of aerial parts. 4-Hydroxy-3-(isopenten-2-yl)-acetophenone (**1**) is by far the dominant component of the methanol extract along with minor amounts of 6-hydroxytremetone (**2**); both compounds showing a hypotensive effect on rats and mice. Two components of its essential oil, namely, T-cadinol (10-*epi*-α-cadinol) and α-cadinol possess significant anti-inflammatory activity and inhibit nitric oxide production. All available information seems to support the uses of *X. poposum* in traditional medicine. Due to its numerous bioactive metabolites, strong effects should be expected after the consumption of this herb. Consequently, the dosage (amount consumed) should be carefully controlled to avoid undesirable effects. Further in-depth investigations are needed to chemically characterize the water soluble components of this plant and to identify possible chemotypes, as well as define quality parameters.

Keywords *p*-Hydroxyacetophenone derivatives · 6-Hydroxytremetone · Essential oil · Antifungal · Antihypertensive · Anti-inflammatory

1 Introduction

Xenophyllum poposum, commonly known as "poposa" or "pupusa", is a wild medicinal plant widely used in Andean folk medicine to treat hypertension, altitude sickness ("apunamiento"), digestive disorders and chills. It is used also to season soups and broths. It is a creeping sub-shrub that grows in the high mountains of Southern Peru to Northwest Argentina and Northern Chile between 3200 and 5300 m.a.s.l. This plant along with "chachacoma" (*Senecio* spp.) and "copa-copa" (*Artemisia copa* Phil.) are probably the most common Asteraceae in the folk pharmacopoeia of Northwestern Argentina. The hypotensive activity of "poposa" has been demonstrated. The antifungal compound 4-hydroxy-3-(3-methyl-2-butenyl)-acetophenone (4-HMBA) is by far the main component of this species which together with 6-hydroxytremetone are the active principles responsible for the vasodilatation effect. Other identified bioactive metabolites were diterpenoids, benzofuran derivatives, flavonoids and coumarins; the aerial parts yield an essential oil consisting of mono- and sesquiterpenoids along with benzofuran derivatives. Very little is known on the water-soluble metabolites. This species possesses a plethora of bioactive metabolites, therefore additional studies on its pharmacological effects are urgently needed.

2 Taxonomic Characteristics

Xenophyllum is a genus of herbaceous plants belonging to the Asteraceae family, subfamily Asteroideae, tribe Senecioneae, subtribe Senecioninae, with 21 species recognized. This genus was extracted from *Werneria sensu lato* and is characterized by mat- or hummock-forming perennials that grow at high elevations in the Andes from Colombia to northern Argentina and Northern Chile (Funk 1997). The name *Xenophyllum* derives from the unusual forms of the leaves (*xeno* = strange; *phyllum* = leaf). *X. poposum* (Philippi) V.A. Funk is a sub-shrub with an unpleasant smell that grows in high mountains from Southern Peru and Western Bolivia to Northern Chile and Northwestern Argentina. Synonyms: *Werneria poposa* Phil., *Werneria lorentziana* Hieronymus (Cabrera 1978). ***Common names:*** "poposa", "popusa", "pupusa", "pupusa de la ciénaga", "akhana", "fosfosa" (Giberti 1983; Barboza et al. 2009); "urqu pura pura" (Aymara language) (Alonso and Desmarchelier 2006).

3 Crude Drug Used

In traditional medicine, the aerial parts (twigs and leaves) of *Xenophyllum poposum* are used (Giberti 1983; Alonso and Desmarchelier 2006; Barboza et al. 2009; Benites et al. 2011; González et al. 2012; Lock and Rojas 2019).

Infusions prepared from a handful of aerial parts (twigs and leaves) in one liter of water are used as a digestive. Drink a cup after lunch and dinner (Alonso and Desmarchelier 2006). Three cups of the infusion per day are advised against the altitude sickness (Alonso and Desmarchelier 2006). In Bolivia, the root infusion is used as a febrifuge; it is also used against pulmonary and stomach hemorrhages. The powdered plant material, mixed with chicken fat, is also used externally. The ointment thus obtained is used topically to treat rheumatic pain and relieve bumps and sprains (Meyer 1943; Zardini 1984; Abella et al. 2000). In addition, mixed with *Artemisia copa* ("copa-copa"), an infusion is prepared to relieve rheumatic pain (Pérez de Nucci 1988; Pochettino and Martínez 1998; Alonso and Desmarchelier 2006).

4 Major Chemical Constituents and Bioactive Compounds

X. poposum has been chemically investigated by several authors (Ponce and Gros 1991, 1995; Ponce 1994; Córdova et al. 1998; Lock Sing de Ugaz 1998; Abella et al. 2000; Benites et al. 2011, 2012; González et al. 2012; Mercado et al. 2015). The following type of compounds were identified: diterpenoids (Ponce and Gros 1991; Córdova et al. 1998; Lock Sing de Ugaz 1998), *p*-hydroxyacetophenone and benzofuran derivatives (Ponce and Gros 1991; Lock Sing de Ugaz 1998; Benites et al. 2011; González et al. 2012; Mercado et al. 2015), flavonoids and coumarins (Ponce and Gros 1995; Lock Sing de Ugaz 1998), and, in the essential oil, mono- and sesquiterpenoids accompanied by benzofuran derivatives (tremetone, dehydrotremetone and 6-hydroxytremetone) (Abella et al. 2000; Benites et al. 2011; González et al. 2012; Mercado et al. 2015). Reviews on the chemistry (Lock Sing de Ugaz 1998; Lock 2006) and biological activities of *Werneria* and *Xenophyllum* (Lock and Rojas 2019) have been published. The secondary metabolites of *X. poposum* are produced by secretory tissues in the schizogenous ducts that go through the leaves and stems (Abella et al. 2000; Mercado et al. 2015) and in secretory cavities of the leaves (leaf glands).

The essential oil of *X. poposum* has been obtained by steam water distillation with yields of 0.5% (aerial parts) up to 2.2% (leaves) based on dry plant material (Abella et al. 2000; Benites et al. 2011; González et al. 2012; Mercado et al. 2015). A collection of *X. poposum* coming from an undisclosed location of the Argentine Puna (probably Jujuy Province), showed an essential oil dominated by monoterpenes (*ca.* 75%) with α-pinene (5.5%), β-pinene (21.8%), α-terpinene (5.2%), β-phellandrene (4.8%), isopulegol (4.8%), terpinen-4-ol (5.3%), α-terpineol (2.1%), citronellol (4.6%), neral (2.6%) and geranial (2.8%) as major components accompanied by smaller amounts of the sesquiterpenes γ-muurolene (0.3%), α-muurolene (0.1%), γ-cadinene (0.8%), δ-cadinene (2.9%) and β-eudesmol (1.9%) (Abella et al. 2000); notably, the presence of tremetone analogs was not reported in this oil. A collection from Tarapaca, Chile, yielded an essential oil with a qualitatively similar composition but with similar amounts of monoterpenoids (*ca.* 35%) and sesquiterpenoids (*ca.* 34%) accompanied by significant amounts of benzofurans, namely, tremetone (9.5%) and 6-hydroxytremetone (8.3%) (Benites et al. 2011). On the other hand, collections from the Provinces of Catamarca and Tucumán in northwestern Argentina gave essential oils dominated by sesquiterpenoids (>70%) with very low content of monoterpenoids (<5%) along with significant amounts of 6-hydroxytremetone (14.7% and 11.8% resp.) and tremetone (1.7% and 0.6% resp.) (González et al. 2012; Mercado et al. 2015). Major sesquiterpenoid components were δ-cadinene (16.5% and 18.8% respectively), 10-*epi*-α-cadinol (T-cadinol) (12.0% and 10.0% resp.), *a*-cadinol (8.8% and 5.7%); *γ*-cadinene (7.5% and 9.1% resp.), 1-*epi*-cubenol (4.2% and 1.8% resp.) and *a*-muurolene (3.0% and 3.3% resp.). Later collections of the Catamarca population at "Cerro Pabellón", Andalgalá Department, in December 2005, March 2009 and March 2011, yielded volatile oils with chemical composition essentially identical to the first collection (March 2003) (González et al. 2012).

Variations in the essential oil composition of the different collections of "poposa" may be due to the existence of chemotypes, phenological stage and/or environmental conditions. However, the stability of the chemical profile of the oils from the same plant stock during several years as with the collections of Catamarca and Tucumán provinces (González et al. 2012; Mercado et al. 2015), strongly supports that they represent a true chemotype (Medina-Holgín et al. 2007; Marcial et al. 2016), different from the Chilean collection (Benites et al. 2011). Most of the essential oil components are bioactive. Thus, 6-hydroxytremetone (**2**) showed significant antifungal activity against *Aspergillus fumigatus* (González et al. 2012), anticancer activity (Liu et al. 2010); plant growth inhibitory (Céspedes et al. 2002), allergenic (Hausen and Helmke 1995), phytotoxic (et al. 2011a, b) and anti-HIV activities (Piacente et al. 1994). Tremetone, a minor component in the essential oil and methanol extract of *X. poposum* exhibited a morphine-like analgesic effect (Benites et al. 2012). However, it should be noted that upon microsomal activation, tremetone is metabolized to a product toxic to mammalians (Beier et al. 1987, 1993). T-cadinol (10-*epi*-α-cadinol) inhibits induced intestinal hypersecretion in mice and contractions of the isolated guinea pig ileum (Cleason et al. 1991a); T-cadinol and α-cadinol possess significant anti-inflammatory activity and inhibit nitric oxide production induced by lipopolysaccharides (Tung et al. 2011). Thus, *X. poposum* contains a plethora of powerful bioactive metabolites, the most relevant being the inhibitory effect of T-cadinol on intestinal hypersecretion and ileum contractions (Cleason et al. 1991a), as a calcium antagonist (Cleason et al. 1991b) and as anti-inflammatory (Tung et al. 2011); and the hypotensive and vasodilatation properties of 4-hydroxy-3-(3-methyl-2-butenyl)-acetophenone (4-HMBA) and 6-hydroxytremetone (Córdova et al. 1998; Palacios et al. 2018; Cifuentes et al. 2018). Due to the presence of so many bioactive metabolites, strong effects should be expected after the consumption of this herb, which undoubtedly deserves to be investigated in depth to accurately characterize this plant, define quality parameters and chemotypes.

5 Morphological Description

Xenophyllum poposum is a creeping sub-shrub 3–5 cm high, very branchy that gives off an unpleasant smell when it is crushed. Densely leafy twigs. Spiraled leaves, imbricated along the stem, fleshy, linear, rounded at the apex and widened at the base in a membranous sheath with glabrous lamina and woolly sheath inside, 5–8 mm long and 1 mm wide. Capitulum: apical, solitary, sessile, radial. Flared involucre, 5–8 mm high by 5–6 mm in diameter; phyllaries: 8–10, oblong, obtuse, welded halfway, glabrous. Dimorphic tubulose hermaphroditic flowers. Ray flowers: few, white ligule 6–7 mm long. Disc flowers: yellow or lilac, hermaphrodites, with tubular corolla. Cylindrical, glabrous ribbed achene; white-yellowish pappus (see photos) (Cabrera 1978).

X. poposum is frequently confused with *X. incisum* (Philippi) V.A. Funk (syn. *Werneria incisa* Philippi), a closely related species that grows in very humid places near streams and on the banks of Andean lagoons, which differs by having trilobed leaves at the apex. Common names for *X. incisum* are "pupusa de agua", "pupusa del cerro" and also "poposa" or "popusa". Both species have essentially the same use in traditional medicine (Giberti 1983; De Marchese et al. 2007).

6 Geographical Distribution

Xenophyllum poposum grows in high mountains of Southern Peru, Bolivia, Northern Chile and Northwestern Argentina. In Peru, it grows from Lima to Tacna between 3900 and 5140 m in places with little vegetation among rocky areas. In Bolivia, it grows between 3500 and 5200 meters above sea level in the highlands of the Andean region in the western part of the country (departments of La Paz, Cochabamba, Potosí and Oruro). In Chile, *X. poposum* grows in the Tarapacá Region and in the Antofagasta Region, between 3200 and 5300 m.a.s.l., where it is traditionally used as a seasoning in soups and broths. In Argentina, it is found in the north-western region of the country (provinces of Jujuy, Catamarca, La Rioja, Tucumán and Salta) between 3500 and 5300 meters above sea level, where it is consumed as an infusion against colic, digestive diseases, liver disorders, as an anti-inflammatory, colds, strong cough, altitude sickness ("soroche" or "apunamiento") and for the treatment of hypertension (Giberti 1983; Alonso and Desmarchelier 2006; Barboza et al. 2009).

7 Ecological Requirements

X. poposum is an endangered species that grows wild only. It blooms in autumn and spring. Its reproduction is through seed and by the division of bushes. *X. poposum* grows in sandy-rocky soils. Its habitats are the rocky slopes of the high Andean mountains and surrounding areas. It is strongly aromatic when crushed, giving off a foul odor, very characteristic.

8 Traditional Use (Part(s) Used) and Common Knowledge

In Chile, the aerial parts are traditionally used as a seasoning in soups and broths. usions of the aerial parts (stem, leaves and sometimes also flowers), (Photo 1) are used for insomnia, nervous states, altitude sickness (mountain sickness), stomach pain and headaches (Barrera et al. 1981). In Bolivia, the root infusion is used as a febrifuge; it is also used against pulmonary and stomach hemorrhages. In most of its natural range, infusions of aerial parts are consumed to treat diarrhea and gastro-intestinal bloating, for mountain sickness, against abdominal pain, flu, indigestion, intestinal inflammation, rheumatism and to stabilize arterial pressure (Giberti 1983; Abella et al. 2000; Barboza et al. 2009). Infusions are prepared using a handful of aerial parts (twigs and leaves) in one liter of water; drink up to three cups a day. In Northern Argentina, infusions of the aerial parts are used in the treatment of hyper-tension (Barboza et al. 2009) and as digestive, cough suppressant, and for renal colic, cold and pneumonia (Alonso and Desmarchelier 2006). An ointment made of the powdered plant mixed with chicken fat is externally used for topical treatment

Photo 1 *Xenophyllum poposum* (Philippi) V.A. Funk. Twigs with leaves and flowers. (Photo: Alfredo Grau)

of rheumatic pain and relief of bumps and sprains (Meyer 1943; Zardini 1984; Abella et al. 2000). Mixed with *Artemisia copa* ("copa-copa"), an infusion is prepared to relieve rheumatic pain (rez de Nucci 1988; Pochettino and Martínez 1998; Alonso and Desmarchelier 2006). *X. poposum* is also used to treat chills, as a food condiment for pneumonia convalescents and added in small quantity to the "yerba mate" (*Ilex paraguariensis* A. St. Hil.) infusion. (Giberti 1983; Alonso and Desmarchelier 2006). In Northwestern Argentina, it is also used as a condiment to flavor foods (Scarpa and Arenas 1996). To be used as a digestive, an infusion is prepared with a handful of aerial parts (twigs and leaves) in one liter of water. Drink a cup after lunch and dinner. An alternative is the use of two teaspoons of plant per cup (Alonso and Desmarchelier 2006). Three cups of the infusion per day are advised against the altitude sickness (Alonso and Desmarchelier 2006).

9 Modern Medicine Based on Its Traditional Medicine Uses

The hypotensive activity of "poposa" has been demonstrated in rats and mice. A hydroalcoholic extract of *X. poposum*, administered orally, significantly reduced blood pressure and heart rate in both normotensive and hypertensive rats due to an increase in vasodilatation and a decrease in heart rate (Córdova et al. 1998; Palacios et al. 2018; Cifuentes et al. 2018). Similar results were obtained with mice (Cifuentes et al. 2018). The most active metabolites that caused vasodilatation were 4-hydroxy-3-(3-methyl-2-butenyl)-acetophenone (4-HMBA) (**1**) and 5-acetyl-6-hydroxy-2-isopropenyl-2,3-dihydrobenzofuran (6-hydroxytremetone; dihydroeuparin) (**2**). Vasodilation was dependent on the vascular endothelium and was inhibited by N(G)-nitro-L-arginine methyl ester (L-NAME), an endothelial nitric oxide synthase inhibitor (Palacios et al. 2018). These results are consistent with the use of "poposa" in Andean traditional medicine for the treatment of hypertension and altitude sickness.

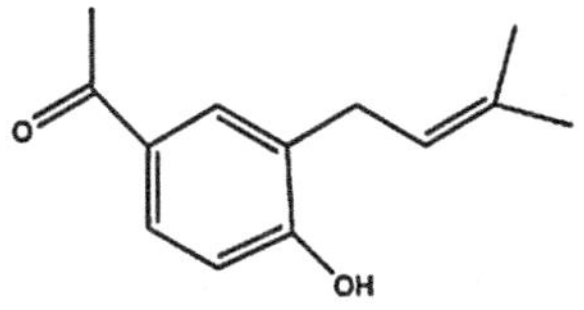

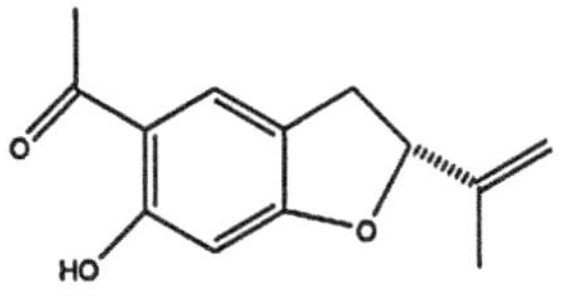

1 **2**

A patent has been registered on the semi-synthetic oximes prepared from compounds **1** and **2** as hypotensive and vasodilator agents (Paredes Poblete et al. 2016). The prenylated *p*-hydroxyacetophenone derivative **1** proved to be an effective antifungal agent possessing also a moderate antibacterial activity (Tomás-Barberán et al. 1990). It shows antifungal activity against *Candida albicans* and a synergistic effect with fluconazole, inhibiting filamentation and thickening of the cell wall (Soberon et al. 2015). Structural and vibrational properties of compound **1** were studied by infrared and Raman spectroscopies (Lizarraga et al. 2012). 6-Hydroxytremetone (**2**) displays potent anticancer activity against HL-60 human leukemia and HeLa cell lines cells (Liu et al. 2010); plant growth inhibitory (Céspedes et al. 2002), allergenic (Hausen and Helmke 1995), phytotoxic (Romano et al. 2011a, b) and anti-HIV activities (Piacente et al. 1994). Compound **2** also showed significant antifungal activity against *Aspergillus fumigatus* but weak activity against the fungi *Candida albicans, Cryptococcus neoformans and Trichophyton rubrum* and weak activity against the bacteria *Acinetobacter baumannii, Escherichia coli* ATCC 25922, *Salmonella typhimurium* and *Staphylococcus aureus* MRSA (González et al. 2012). The natural enantiomer isolated from *X. poposum* was characterized as the 2*R*-(-)-form and its crystalline structure was determined by X-rays (Romano et al. 2008); also, the structural and vibrational properties of 2*R*-(-)-6-hydroxytremetone and a complete assignment of the observed bands in the IR spectrum was performed (Romano et al. 2011a, b). 6-Hydroxytremetone (**2**) also showed *in vitro* antifungal activity on the phytophatogenic fungus *Fusarium semitectum* (González et al. 2007).

The essential oil, hexane, dichloromethane (DCM) and methanol extracts of "poposa" displayed antibacterial activity *in vitro* against *Staphylococcus aureus, Bacillus subtilis* and *Saccharomyces cerevisiae* while topical applications of DCM and hexane extracts exhibited anti-inflammatory and analgesic activity. The topical application of hexane and DCM extracts significantly decreased mouse ear edema induced by TPA and showed good analgesic activity in the acetic acid-induced writhing test (Erazo et al. 2006).

10 Conclusions

Experimental data support the uses of *Xenophyllum poposum* ("poposa") in traditional medicine. Particularly relevant is the hypotensive effect due to increased vasodilatation produced by its main metabolites 4-hydroxy-3-(3-methyl-2-butenyl)-acetophenone (**1**) and 5-acetyl-6-hydroxy-2-isopropenyl-2,3-dihydrobenzofuran (6-hydroxytremetone; dihydroeuparin) (**2**) (Córdova et al. 1998; Palacios et al. 2018; Cifuentes et al. 2018). It should be noted here that a phytochemical screening of *X. poposum* from Peru (as *Werneria poposa*) tested positive for alkaloids (Lock Sing de Ugaz 1998) but no later work was published on the presence of alkaloids in this herb. Consequently, because the genus *Xenophyllum* belongs to the Senecioneae tribe of the Asteraceae family, the presence of hepatotoxic pyrrolizidine alkaloids should be urgently clarified. In addition, almost nothing is known about the water-soluble compounds of "poposa", therefore further studies on its metabolites, such as flavonoid glycosides, cardiac glycosides, chlorogenic acid and caffeoylquinic acids, are to be expected.

Acknowledgments The author thanks Dr. Alfredo Grau, Faculty of Natural Sciences and Miguel Lillo Institute, National University of Tucumán, for the photos of *Xenophyllum poposum*.

References

Abella L, Cortella AR, Velasco-Negueruela A, Pérez-Alonso M (2000) Ethnobotany, volatile oils and secretion tissues of *Werneria poposa* from Argentina. Pharm Biol 38(3):197–203

Alonso J, Desmarchelier C (2006) Plantas medicinales autóctonas de la Argentina. Ed. Fitociencia, Buenos Aires, Argentina, pp 383–385

Barboza GE, Cantero JJ, Núñez C, Pacciaroni A, Ariza Espinar L (2009) Medicinal plants: a general review and a phytochemical and ethnopharmacological screening of the native Argentine Flora. Kurtziana 34(1–2):7–365

Barrera E, Meza I, Muñoz M (1981) El uso medicinal y alimenticio de plantas nativas y naturalizadas en Chile. Ed. Museo Nacional de Historia Natural, Santiago de Chile, p 34

Beier RC, Norman JO, Irvin RT, Witzel DA (1987) Microsomal activation of constituents of 16-white snakeroot (*Eupatorium rugosum* Houtt) to form toxic products. Am J Vet Res 48:583–585

Beier RC, Norman JO, Reagor MC, Rees MS, Mundy BP (1993) Isolation of the major component in white snakeroot that is toxic after microsomal activation: possible explanation of sporadic toxicity of white snakeroot plants and extracts. Nat Toxins 1:286–293

Benites J, López J, Rojo LE, Díaz P, Rojas M, Fuentes R, Moiteiro C, Venancio F (2011) Chemical composition of the essential oil of *Xenophyllum poposum*. Chem Nat Compd 46(6):988–989

Benites J, Gutiérrez E, López J, Rojas M, Rojo L, Costa M do C, Vinardell MP, Calderón PB (2012) Evaluation of analgesic activities of tremetone derivatives isolated from the Chilean altiplano medicine *Parastrephia lepidophylla*. Nat Prod Commun 7(5):611–614

Cabrera AL (1978) Flora de la Provincia de Jujuy. República Argentina. Parte X. Compositae. Colección Científica del INTA, Buenos Aires, Argentina, p 474

Céspedes CL, Uchoa A, Salazar JR, Perich F, Pardo F (2002) Plant growth inhibitory activity of *p*-hydroxyacetophenones and tremetones from Chilean endemic *Baccharis* species and some analogous: a comparative study. J Agric Food Chem 50:2283–2292

Cifuentes F, Palacios J, Kuzmicic J, Carvajal L, Muñoz F, Quispe C, Nwokocha CR, Morales G, Norambuena-Soto I, Chiong M, Paredes A (2018) Vasodilator and hypotensive effects of pure compounds and hydroalcoholic extract of *Xenophyllum poposum* (Phil.) V.A Funk (Compositae) on rats. Phytomedicine 50:99–108

Claeson P, Andersson R, Samuelsson G (1991a) T-cadinol: a pharmacologically active constituent of scented myrrh: introductory pharmacological characterization and high field ^{1}H- and ^{13}C-NMR data. Planta Med 57:352–356

Claeson P, Zygmunt P, Hogestatt ED (1991b) Calcium antagonistic properties of the sesquiterpene ℘-cadinol: a comparison with nimodipine in the isolated rat aorta. Pharmacol Toxicol 69:173–177

Córdova A, Lock O, Jurupe H (1998) Estudio químico farmacológico de la *Werneria poposa* Philippi. Rev Soc Quim Peru 64:264–272

De Marchese MJA, De Heluani CS, Catalán CAN, Griffin CA, Vaughn JB, Herz W (2007) Incisol, an alcohol with a novel sesquiterpene skeleton from *Xenophyllum incisum*. Biochem Syst Ecol 35:169–175

Erazo S, Gogoux F, Delporte C, Backhouse N, Negrete R, Zaldívar M, Belmonte E, Cayunao C, Bachiller I, Bandoni A, Di Leo Lira P, van Baren C, Gianninoto H, García R (2006) Actividades biológicas de *Xenophyllum poposum* Phil. (Asteraceae), planta del altiplano chileno. Rev Fitotec Mex 6:165–166

Funk VA (1997) *Xenophyllum*, a new Andean genus extracted from *Werneria s.l.* (Compositae: Senecioneae). Novon 7:235–241

Giberti GC (1983) Herbal Folk Medicine in Northwestern Argentina: Compositae. J Ethnopharmacol 7(3):321–341

González AM, Romano E, Tracanna MI, Rachi A, Medvescigh J, Popich S, Benavente A (2007) Producto natural como alternativa de control de enfermedades fúngicas en olivo. Bol Latinoam Caribe Plant Med Aromát 6(6):407–408

González AM, Tracanna MI, Amani SM, Schuff C, Poch MJ, Bach H, Catalán CAN (2012) Chemical composition, antimicrobial and antioxidant properties of the volatile oil and methanol extract of *Xenophyllum poposum*. Nat Prod Commun 7(12):1663–1666

Hausen BM, Helmke B (1995) Butenylbithiophene, alpha-terthienyl and hydroxytremetone as contact allergens in cultivars of marigold (*Tagetes* sp.). Contact Dermatitis 33:33–37

Liu Q, Shen L, Wang T, Chen J, Qi W, Gao K (2010) Novel modified furanoeremophilane-type sesquiterpenes and benzofuran derivatives from *Ligularia veitchiana*. Food Chem 122:55–59

Lizarraga E, Romano E, Rudyk R, Catalán CAN, Brandan SA (2012) Structural study, coordinated normal analysis and vibrational spectra of 4-hydroxy-3-(3-methyl-2-butenyl)-acetophenone. Spectrochim Acta A 97:202–208

Lock O, Rojas R (2019) Phytochemistry and biological activities of *Werneria* and *Xenophyllum* species. Bol Latinoam Caribe Plant Med Aromát (BLACPMA) 18(3):223–238

Lock Sing OR (2006) Diversidad química en el género *Werneria*. Rev Soc Quim Peru 72:32–43

Lock Sing de Ugaz O (1998) Avances en el estudio del género *Werneria* y sus metabolitos secundarios. Rev Quím PUCP 12(1):69–85

Marcial G, de Lampasona MP, Vega MI, Lizarraga E, Viturro CI, Slanis A, Juárez MA, Elechosa MA, Catalán CAN (2016) Intraspecific variation in essential oil composition of the medicinal plant *Lippia integrifolia* (Verbenaceae). Evidence for five chemotypes. Phytochemistry 122:203–212

Medina-Holguín AL, Micheletto S, Holguín FO, Rodriguez J, O'Conell MA, Martin C (2007) Environmental influences on essential oils in roots of *Anemopsis californica*. HortScience 4:1578–1583

Mercado MI, Lizarraga E, Catalán CAN, Ponessa GI (2015) Composición del aceite esencial de *Xenophyllum poposum* (Asteraceae) de las provincias de Catamarca y Tucumán. Estructuras foliares y caulinares asociadas a su síntesis. Lilloa 52(Suppl):56–61

Meyer T (1943) Notas sobre un viaje etnobotánico al departamento Santa Victoria (Salta). Rev Geogr Am 10:121–132

Palacios J, Quispe C, Paredes A, Cifuentes F (2018) Uso potencial de *Xenophyllum poposum* (Phil.) V.A Funk (Compositae) en el tratamiento del mal de altura e hipertensión. Med Plant Commun 1:24

Paredes Poblete AG, Morales Borcosque GS, Cifuentes Jorquera FE (2016) US Pat. Appl. Publ. (2016), US 20160130219 A1 20160512

Pérez de Nucci A (1988) La medicina tradicional del noroeste Argentino. Historia y presente. Ed. del Sol, Buenos Aires, Argentina, 158 pp

Piacente S, Aquino R, de Tommasi N, Pizza C, de Ugaz OL, Orellana HC, Mahamood N (1994) Constituents of *Werneria ciliolata* and their *in vitro* anti-HIV activity. Phytochemistry 36:991–996

Pochettino M, Martínez M (1998) Estudio etnobotánico en el Departamento de Molinos, Salta. In: Amat AG (ed) Farmacobotánica y Farmacognosia en Argentina (1980–1998). Ed. Científicas Americanas, La Plata, Argentina, pp 55–86

Ponce MA (1994) Estudio fitoquímico de *Werneria poposa* Philippi. Thesis, Facultad de Ciencias Exactas y Naturales, Universidad de Buenos Aires, Argentina, 206 pp

Ponce MA, Gros GE (1991) Prenylated aromatic ketones from *Werneria poposa*. J Argent Chem Soc 79:197–200

Ponce MA, Gros GE (1995) A flavonoid and coumarins from *Werneria poposa*. J Argent Chem Soc 83:93–95

Romano E, Raschi AB, Benavente AM, González AM, Piro OE (2008) Characterization and crystal structure of 2*R*-(-)-6-hydroxytremetone from *Xenophyllum poposum* (Phil.) V.A. Funk. Nat Prod Res 22(2):124–135

Romano E, Raschi AB, Benavente A, Brandan SA (2011a) Structural analysis, vibrational spectra and coordinated normal of 2*R*-(-)-6-hydroxytremetone. Spectrochim Acta A Mol Biomol Spectrosc 84(1):111–116

Romano E, Raschi AB, González AM, Jaime G, Fortuna MA, Hernández LR, Bach H, Benavente AM (2011b) Phytotoxic activities of (2*R*)-6-hydroxytremetone. Plant Physiol Biochem 49(6):671–675

Scarpa G, Arenas P (1996) Especias y colorantes en la cocina tradicional de la puna jujeña. Candollea 51:483–514

Soberón JR, Lizarraga EF, Sgariglia MA, Carrasco Juárez MB, Sampietro DA, Ben Altabef A, Catalán CAN, Vattuone MA (2015) Antifungal activity of 4-hydroxy-3-(3-methyl-2-butenyl)-acetophenone against *Candida albicans*. Evidence for the antifungal mode of action. Antonie van Leeuwenhoek J Microbiol 108:1047–1057

Tomás-Barberán F, Iniesta-San Martín E, Tomás-Lorente F, Rumbero A (1990) Antimicrobial phenolic compounds from three Spanish *Helichrysum* species. Phytochemistry 29:1093–1095

Tung YT, Huang CC, Ho ST, Kuo YH, Lin CC, Lin CT, Wu JH (2011) Bioactive phytochemicals of leaf essential oils of *Cinnamomum osmophloeum* prevent lipopolysaccharide/D-galactosamine (LPS/D-GalN)-induced acute hepatitis in mice. J Agric Food Chem 59:8117–8123

Zardini EM (1984) Etnobotánica de Compuestas argentinas con especial referencia a su uso farmacológico. Acta Farm Bonaer 3(1):77–99

Zuccagnia punctata Cav.

María Inés Isla, María Alejandra Moreno, María de los Ángeles Álvarez, and Iris Catiana Zampini

Zuccagnia punctata Cav. (Photos: by Dra. María Inés Isla, in Amaicha del Valle, Tucumán, Argentina.)

Abstract *Zuccagnia punctata* Cav. (Fabaceae, Caesalpiniaceae) is an argentine medicinal plant species, belonging to a monotypic genus representatives of which grow in the arid and semiarid Argentine regions, in the biogeographic Province of Monte. The taxonomic characteristics, traditional use, phytochemistry and pharmacological activity of *Z. punctata* aerial parts (leaves and stems) and flowers are described. A wide range of traditional uses have been described and scientifically validated by *in vitro* and *in vivo* assays. These include the antibacterial, antifungal, anti-inflammatory, antioxidant, antitumoral, antihypertensive activities, as well as the effect in the prevention of cardiovascular disease related with hypercholesterolemia and endothelial dysfunction, between others. Compounds

M. I. Isla (✉) · M. A. Moreno · M. de los Ángeles Álvarez · I. C. Zampini
Instituto de Bioprospección y Fisiología Vegetal, (INBIOFIV, CONICET-Facultad de Ciencias Naturales e IML, Universidad Nacional de Tucumán), San Miguel de Tucumán, Argentina
e-mail: misla@csnat.unt.edu.ar

Á. Máthé, A. Bandoni (eds.), *Medicinal and Aromatic Plants of South America Vol. 2*, Medicinal and Aromatic Plants of the World 7,
https://doi.org/10.1007/978-3-030-62818-5_42

from different chemical groups (phenolic compounds and essential oils) have also been isolated and identified in *Z. punctata*. The main bioactive phytochemicals in above-ground organs (leaves, stems and flowers) have been identified as 2′,4′-dihydroxy-3′-methoxychalcone and 2′,4′-dihydroxychalcone and proposed as bioactive chemical markers. No genotoxic effect of *Z. punctata* extracts has been demonstrated. Oral administration of *Z. punctata* extract and some of their isolated compounds, were not toxic in rabbits and mice models. Although no human clinical trials have been carried out so far, *Z. punctata* is a species widely used by local populations of the Monte ecoregion for many years, without manifestations of obvious toxicity. Therefore, extracts of aerial parts of *Z. punctata* could be used in herbal medicines as standardized extracts indicating the amount of major bioactive compounds (chalcones). In addition, it could be included in several phytotherapeutic preparations such as capsules, creams, gels, microencapsulates.

Keywords "Jarilla poposa" · Argentine medicinal plants · Anti-inflammatory · Antibiotic · Antioxidant · Antifungal · Antitumoral

1 Introduction

This chapter integrates the traditional knowledge and the results of the scientific investigations into *Zuccagnia punctata*, a promising native medicinal plant species of Argentina. The data, the analysis and discussion presented in this review, do not only contribute to the technological development in the field of medicinal products using this plant, but also dimension the importance of conservation and sustainable use of this species.

2 Taxonomic Characteristics

Zuccagnia punctata Cav. (Fabaceae, Caesalpiniaceae), commonly known as "jarilla pispito", "jarilla macho", "jarilla poposa", "pus pus" and "lata" belongs to a monotypic genus comprised of this argentine endemic species, characteristic of xerophytic plants from the biogeographic Province of Monte (Cabrera and Willink 1973; Cabrera 1976; Zuloaga and Morrone 1999; Ulibarri 2005). Etymologically, *Zuccagnia* is dedicated to the Italian physicist and botanist Attilio Zuccagni (1754–1807) and the common name "jarilla", has an Arabic etymology, as the diminutive of the word "jara" or "sá'ra" which means land full of vegetation.

3 Crude Drug Used

Parts of the plant used: the crude drug used includes leaves, branches, and flowers of the plant, which have only been exposed to the collection and drying process (Photo 1).

4 Major Chemical Constituents and Bioactive Compounds

Ethanolic and methanolic extracts of *Zuccagnia punctata* aerial parts have been found to contain high amounts of phenolic compounds. Chalcones (2′,4′-dihydroxy-3′-methoxychalcone (DHMC), 2′,4′-dihydroxychalcone (DHC), 2′,4′-dihydroxy-dihydrochalcone, and trihydroxy dihydrochalcone hexoside); flavones such as 3,5,7-trihydroxyflavone (galangin), 3,5-dihydroxy-7-methoxy flavone (izalpinin), 3,5,4′-trihydroxy-7-methoxy flavone (rhamnocitrin), 3-hydroxy-7,8-dimethoxyflavone, 3,7-dihydroxyflavone (3,7-DHF); flavanones such as 7-hydroxy flavanone (7-HF), 5,7-dihydroxyflavanone (pinocembrin), 5-hydroxy-7-methoxy flavanone (pinostrobin), and 7-hydroxy-8-methoxyflavanone; caffeic acid derivates such as 1-methyl-3-(4′-hydroxyphenyl)-propyl caffeate and 1-methyl-3-(3′,4′-dihydroxyphenyl)-propyl caffeate; quinic acid; galloyl quinic acid, and catechin derivates were identified (Pederiva et al. 1975; Pederiva and Giordano 1984; Ortega et al. 2000; Svetaz et al. 2004; Morán Vieyra et al. 2009; Agüero et al. 2010; Chieli

Photo 1 *Zuccagnia punctata* Cav. Photo: Maria de los Angeles Álvarez, in Amaicha del Valle, Tucumán, Argentina

et al. 2012; Nuño et al. 2014, 2018; Moreno et al. 2018a; Zampini et al. 2005, 2008, 2012).

Hydroalcoholic extracts from fresh and dried flowers contain two chalcones as major compounds (2′,4′-dihydroxychalcone and 2′,4′-dihydroxy-3′-methoxychalcone) (Moreno et al. 2015a).

In infusions of aerial parts of *Z. punctata* were identified by HPLC-DAD-ESI-MS/MS the flavonoids quercetin glycoside, naringenin rutinoside, galangin dirhamnoside, 2′,4′-dihydroxy-dihydrochalcone and trihydroxy-dihydrochalcone hexoside, as well as proanthocyanidins and hydrolyzable tannins (Carabajal et al. 2020a), which agreed with the results of previous phytochemical research on hydroalcoholic extracts (ethanol 60°) of this plant from the same zone (Moreno et al. 2018a). Pinocembrin, dihydroxyflavanone, 3,7,8-trihydroxy-dihydroflavanone, galangin, 7-hydroxy-8-methoxyflavanone, pinobanksin-5-methyl ether, 7-HF, 3,7-dihydroxy-8-methoxyflavone, and 3,7-DHF were also detected in the infusions (Carabajal et al. 2020a). Other chalcone present in infusion was 2′,4′,4-trihydroxy-6′-methoxychalcone. The chalcone DHC was the main constituents of infusion, ethanolic, methanolic, and dichloromethane (DCM) extracts and could be considered as a chemical marker of *Z. punctata* (Svetaz et al. 2004; Zampini et al. 2005; Moreno et al. 2018a; Nuño et al. 2014, 2018; Carabajal et al. 2020a). Moreno et al. (2015b) have reported the content of total polyphenolic compounds on the foliar surface of *Z. punctata* (177 ± 13µg GAE/cm^2) and the presence of two major constituents, DHC and DHMC. The content of both compounds was 99.25µg DHC/cm^2 of leaf and 73.38µg DHMC/cm^2 of leaf. Histochemical analysis revealed on the foliar surface a high accumulation of chalcones. Consequently, the authors have proposed that the chalcones would act as defense mechanisms against UV radiation for the protection of photosynthetic tissues. The chalcones have received attention due to the interesting biological properties such as antioxidant, cytotoxic, anticancer, antimicrobial, antiprotozoal, antiulcer, antihistaminic and anti-inflammatory activities, for treatment of cancer, and cardiovascular disorders (Batovska and Todorova 2010). Other authors reported that chalcones are potent α-glucosidase, α-amylase and lipase inhibitors and cardiovascular agents (Mahapatra et al. 2015; Mahapatra and Bharti 2016; Cai et al. 2017; Bale et al. 2018). Butassi et al. (2015) have described the variability of chalcone content in four batches of *Z. punctata* dry extracts prepared with samples collected in the same place during different months of 1 year using DCM as extraction solvent, indicating a content of 9.64–16.54 mg chalcones/g extract. These results show the need to standardize the extracts by quantifying the chemical quality markers before using them.

5 Morphological Description

Z. punctata is a glutinous and aromatic shrub of 1–2.5 m with pseudo-paripinnate resinous leaves of 3–5 cm with subopposite leaflets (5–13 pares), nanophyll with acuminate apex, rounded base and entire margin (Lersten and Curtis 1996; Mercado et al. 2013). The plants bloom from August to March and fructifies from November to April (Ulibarri 2005). The yellow flowers, born in erect racemes, have a funnel-shaped calyx, with 5 sepals, corolla with 5 free petals, and 10 free stamens (Burkart et al. 1999; Zuloaga and Morrone 1999; Ulibarri 2005). The fruits are reddish brown, capsuliforms, indehiscent, leathery, ovoid-acute, oblique, compressed and estipitade (Burkart et al. 1999; Zuloaga and Morrone 1999; Ulibarri 2005).

Both epidermal surfaces of the leaves and the rachis present epidermal cells with straight anticlinal walls, thick cuticle, cyclocytic stoma, sunken capped glandular trichomes located in crypts and unicellular non-glandular trichomes arranged at the margins (Lersten and Curtis 1996; Álvarez et al. 2012; Mercado et al. 2013; Moreno et al. 2015b). In section, leaflets are iso-lateral and amphi-stomatic. The middle vein presents a collateral vascular bundle with sclerenchymatous layers at the phloem pole. Idioblasts containing druses in the mesophyll are abundant (Mercado et al. 2013). Moreno et al. (2015b) have also reported the content of cuticular wax in *Z. punctata* leaves (88μg/cm^2).

6 Geographical Distribution

Zuccagnia punctata is is widely distributed in the arid and semiarid areas of Western Argentina (Jujuy, Salta, Tucumán, Catamarca, La Rioja, San Juan, Mendoza and San Luis Provinces) from 700 to 2700 m.a.s.l. (Cabrera and Willink 1973; Cabrera 1976; Burkart et al. 1999; Zuloaga and Morrone 1999; Ulibarri 2005). It has a true resemblance to the true "jarillas" species of the genus *Larrea* (Zygophyllaceae), with which it coexists. *Z. punctata* inhabits the same environment as *Larrea cuneifolia* Cav. and *Larrea divaricata* Cav., and form a natural arid community named "jarillal" (Ulibarri 2005). Other plant species growing in this eco-region are *Bulnesia retama* (Gillies ex Hook. & Arn.) Griseb., *Monttea aphylla* (Miers) Benth. & Hook., *Gochnatia glutinosa* (D. Don) Hook. & Arn, *Plectrocarpa rougesii* Descole, O'Donell & Lourteig, *Mimosa ephedroides* (Gillies ex Hook. & Arn.) Benth., *Bougainvillea spinosa* (Cav.) Heimerl, *Prosopis alba* Griseb. and *Prosopis nigra* Griseb.

This species has been included in the preliminary red list of endangered plants, in category 3 (Norma RE-84-2010-SADS).

Recently, a micropropagation method has been reported for *Z. punctata* by Álvarez et al. (2019). *In vitro* cultures were established by culturing seedling explants on Murashige-Skoog medium. By using this method regenerated plants could be obtained in 10 weeks.

7 Ecological Requirements

The native populations of *Z. punctata* grow on poor quality soils, with limited water supply. They are found in sandy, rocky soils, with very low fertility values (low organic matter). The climate in the Monte region (Northwestern Argentina), where this species is found, is characterized by its extreme aridity, a wide daily temperature fluctuation, rains concentrated in summer (reached 200 mm per year), strong annual sunshine, frequent dry winds and low atmospheric humidity (Paoli et al. 2002; Lozano 2011).

8 Traditional Use (Part(s) Used) and Common Knowledge

The infusion and decoction in water, as well as maceration in ethanol of aerial parts of *Z. punctata*, with and without flowers or fruits, have been used extensively as a traditional medicine in Argentina as foot antiseptic and rubefacient, and against bacterial and fungal infections, asthma, arthritis, rheumatism, inflammations, and tumors (Ratera and Ratera 1980; Toursarkissian 1980; Del Vitto et al. 1997; Ortega et al. 2000; Carabajal et al. 2020a). The stems are used in construction of houses roofs (Ulibarri 2005; Carabajal et al. 2020a), also it is used for dye wool of leaden color (Burkart 1952; Ratera and Ratera 1980; Toursarkissian 1980; Barboza et al. 2009) and the resin of leaves were used in the ritual of mummification and preservation of dead ancestors (Diario Chilecito 2019). This plant is widely used in traditional medicine and no toxic effects on health were reported. Recently, *Z. punctata* was cited as a plant species of economic interest in Argentina (Cantero et al. 2019).

9 Modern Medicine Based on Its Traditional Medicine Uses

9.1 Antibacterial Activity

Zuccagnia punctata is a plant species used in argentine traditional medicine, principally as an antiseptic and a vulnerary (Ratera and Ratera 1980; Toursarkissian 1980; Ortega et al. 2000; Carabajal et al. 2020a). The antibacterial activity of hydroalcoholic extracts from aerial parts of *Z. punctata* against antibiotic-multiresistant Gram negative bacteria isolated from human infections (*Acinetobacter baumannii, Enterobacter cloacae, Escherichia coli, Klebsiella pneumoniae, Morganella morganii, Proteus mirabilis, Pseudomonas aeruginosa, Serratia marcescens, Stenotrophomonas maltophilia*) was demonstrated (MIC values ranging from 25 to 200μg/ml). MBC values were identical or two-fold higher than the corresponding MIC values (Zampini et al. 2005). The dihydroxychalcone (DHC) isolated from

Z. punctata exhibited strong antibacterial activity against *A. baumannii, E. cloacae, M. morganii, P. mirabilis, P. aeruginosa, S. marcescens, S. maltophilia* (Zampini et al. 2005). *Z. punctata* extract, and three flavonoids isolated from them (DHC, 7-HF and 3,7-DHF), were active against four different serotypes of *Streptococcus pneumoniae* (Gram positive bacteria) with MIC values between 50 and 500µg/ml (Zampini et al. 2012). The antibacterial activity of *Z. punctata* extract, DHC, 3,7-DHF and 7-HF were also examined by using an *S. pneumoniae* infection model in mice. The natural products were orally administered to infected mice. The treatment with *Z. punctata* extract (1 mg/mice) and 7-HF (1 mg/mice) significantly reduced the number of viable *S. pneumoniae* in lung ($p < 0.01$) while DHC and 3,7-DHF had no effect in the used mice model. These results suggest that the antibacterial activity of the extract on *S. pneumonia, in vivo*, might be related, at least in part, to the action of 7-HF present in the plant (Zampini et al. 2012). *Z. punctata* extract and 7-HF could be considered as antibacterial natural drugs against *S. pneumoniae* (Zampini et al. 2012). The possible hepatotoxic and nephrotoxic effects of *Z. punctata* extract (1 mg/mice) were analyzed in mice with and without streptococcal infections. In both cases, the activities of alanine transaminase (ALT) and aspartate transaminase (AST) enzymes and the levels of creatinine and urea in blood were not changed by *Z. punctata* extract as compared to the control values. Therefore, intake once or twice a day of 1 mg of plant extract for 7 days did not result in toxicity (Zampini et al. 2012).

Nuño et al. (2018), demonstrated that non-polar extracts obtained from the aerial parts of *Z. punctata* affected the growth of several antibiotic-resistant clinical isolates of *Staphylococcus aureus*. The MIC values for *Z. punctata* extract were between 125 and 500µg/ml. For the isolated compounds (DHC, DHMC, DHF, HF) the MIC values were between 25 and 400µg/ml and the MBC values were between 50 and > 400µg/ml against all strains assayed. The order of reactivity for isolated compounds on *S. aureus* was DHC > DHF=HF > DHMC. Previously, other authors reported that OH groups of chalcones induce and enhance the antimicrobial activity (Hatano et al. 2000). DHC produces visible cellular aggregation, blebs on the cellular surface and cytoplasmic coagulation of *S. aureus* (Nuño et al. 2018). The *Z. punctata* extract and the isolated compounds (principally DHC) also inhibit biofilm formation and reduce the metabolism of *S. aureus* cells (Nuño et al. 2018).

The flower extracts also were active on methicillin-resistant *Staphylococcus aureus* (MIC = 250µg GAE/ml) and *Enterococcus faecalis* (MIC = 500µg GAE/ml) (Moreno et al. 2015a). These results indicate that the aerial parts of *Z. punctata* have great possibilities to obtain broad-spectrum antibiotics.

9.2 *Antimycotic Activity*

The native people of Northwestern Argentina have reported the use of *Z. punctata* as antifungal (Ratera and Ratera 1980; Toursarkissian 1980; Ortega et al. 2000; Carabajal et al. 2020a). The ethnopharmacological use of this plant species was confirmed by several scientific reports. Thus, the effect on yeasts such as *Candida albicans, Saccharomyces cerevisiae,* and *Cryptococcus neoforms* was demonstrated. The MIC values were between 62.5 and 250μg/ml (Svetaz et al. 2007; Agüero et al. 2010; Nuño et al. 2014; Moreno et al. 2018b). The antifungal activity was attributed to DHC (Svetaz et al. 2007; Agüero et al. 2010; Nuño et al. 2014; Moreno et al. 2018a, b). The antimicrobial activity and particularly the antifungal action of chalcones have been largely attributed to the reactive enone moiety (Lahtchev et al. 2008).

Butassi et al. (2015) have described the synergistic effect between *Z. punctata* DCM extract and *Larrea nitida* DCM extract on the *C. albicans* and *C. glabrata* growth. Nuño et al. (2014) showed that *Z. punctata* DCM extracts and the isolated chalcones were effective not only as inhibitors of *Candida* growth but as inhibitors upon biofilm formation as well as on preformed *Candida* biofilm and yeast germ tube formation in doses lower than MIC values. Chalcones concentration necessary to produce the 50% of inhibition of yeast germ tube formation (12.5μg/ml for DHC and 30μg/ml for DHMC) were like the concentration necessary to produce the 50 and 87% biofilm inhibition of *C. albicans*, respectively (Nuño et al. 2014). Furthermore, these compounds can inhibit exoenzymes (phospholipase and hemolysin), which are responsible for the invasion mechanisms of the *Candida* strains (Nuño et al. 2014). All these effects could moderate colonization, thus suppressing the pathogen invasive potential. *Z. punctata* flowers ethanolic extracts were active on six *Candida* species with MIC values between 30 and 120μg GAE/ml (Moreno et al. 2015a).

Z. punctata aerial parts were also active on dermatophytes fungi isolated from skin infection (*Microsporum gypseum, Tricophyton rubrum,* and *T. mentagrophytes*) (MIC values 8–16μg/ml) (Agüero et al. 2010). The antifungal activity on dermatophytes was attributed to DHC (MIC = 4–32μg/ml) and DHMC (MIC = 8μg/ml) (Agüero et al. 2010). The essential oils obtained from *Z. punctata* also showed antifungal activity against the dermatophytes *M. gypseum, T. rubrum* and *T. mentagrophytes* with MIC values between 15.6 and 125μg/ml and were not active against the yeasts *C. albicans, C. tropicalis, S. cerevisiae* or *C. neoformans* (Álvarez et al. 2012). The major constituents of the oil, linalool and (−)-5,6-dehydrocamphor were also tested against the same fungi strains. Linalool showed moderate activity against dermatophytes (MIC = 125–250μg/ml) being inactive against the yeasts *C. albicans, C. tropicalis, C. neoformans*. The other major constituent, (−)-5,6-dehydrocamphor, was also inactive up to 250μg/ml. According to these results, the main components of essential oils from *Z. punctata* would be not responsible for its antidermatophytes activity. The activity of the essential oils was attributed to minor components such as thymol and carvacrol (MIC values of 15.6–31.2μg/ml), (Álvarez et al. 2012). According to Tangarife-Castaño et al. (2011) classification system for antifungal activity in plant derivatives, the polyphenolic extracts and essential oils of *Z. punctata* could be considered a strong natural antifungal.

9.3 Antioxidant Activity

The antioxidant capacity of hydroalcoholic extracts of *Z. punctata* aerial parts and compounds isolated from them, such as 7-HF, DHC and 3,7-DHF, was demonstrated (Ávila et al. 2001; Morán Vieyra et al. 2009; Moreno et al. 2018a). The antioxidant capacity of *Z. punctata* aerial parts infusions was also evaluated on the ABTS radical cation assay (SC_{50} = 17.1μg/ml, equivalent to 9.5μg GAE/ml), (Carabajal et al. 2017). The authors showed antioxidant capacity in teas of *Z. punctata* and mixtures of *Z. punctata* with *Larrea cuneifolia*, *Larrea divaricata* and *Hovenia dulcis*. Infusions of *Z. punctata* and mix with *L. cuneifolia* and *L. divaricata* showed also hydroxyl radical, H_2O_2 and superoxide anion scavenging capacity and protection on lipid oxidation (Carabajal et al. 2020a, b). The antioxidant potential of "jarilla" infusions was similar to that exhibited by black tea (*Camelia sinensis*) and even higher than that of mint (*Mentha piperita*) and melissa (*Melissa officinalis*) (Carabajal et al. 2020b). When *Z. punctata* infusions were subjected to a gastro-duodenal digestion in an *in vitro* process, a reduction in phenolic compounds content and in antioxidant capacity was observed (Carabajal et al. 2020b). The polyphenolic extracts of flowers exhibited also antioxidant activity (free radical scavenging and inhibitory activity on lipoperoxidation) (Moreno et al. 2015a).

9.4 Anti-inflammatory Activity

Z. punctata is used in popular medicine as rubefacient and anti-inflammatory. The inhibitory effect of ethanolic extract of aerial parts and isolated flavonoids from *Z. punctata* aerial parts on pro-inflammatory enzymes such as cyclooxigenase-2 (COX-2), lipooxygenase (LOX) and phospholipase A-2 ($sPLA_2$) was demonstrated (Alberto et al. 2007; Nuño et al. 2014). These enzymes are key in the synthesis of prostanoids and eicosanoids from poly-unsaturated fatty acids, involved in various inflammatory and allergic disorders. For the $sPLA_2$, DHC was the most active compound follows by *Z. punctata* ethanolic extract and DHMC, 7-HF and 3,7-DHF (Nuño et al. 2014). The pure compounds were more potent than commercial anti-inflammatory drugs, such as naproxen. The compounds 3,7-DHF and DHMC showed the highest activity on LOX with IC_{50} values of 57.8 and 63.4μg/ml, respectively. DHC showed the highest COX-2 inhibitory activity with IC_{50} value of 1.72μg/ml followed by DHF, HF and *Z. punctata* extract. The DHMC was not active against COX-2 at the tested concentrations (10–150μg/ml). The DHC showed similar activity to the synthetic anti-inflammatory drugs assayed (indomethacin and nimesulide), (Nuño et al. 2014). The polyphenolic extracts of flowers of *Z. punctata* exhibited also anti-inflammatory (inhibition of LOX and COX enzymes) activity (Moreno et al. 2015a, b).

Moreover, "jarilla" aerial parts infusions exhibited notable LOX inhibition (Carabajal et al. 2020a). Concerning the activity against the other pro-inflammatory enzyme, hyaluronidase, the infusions did not demonstrate a good inhibitory action. The inhibitory ability on pro-inflammatory enzymes and its antioxidant capacity support the potential use of the *Z. punctata* as phytomedicine that could be used to prevent the development of chronic inflammatory pathologies. Studies previous showed an effect on chronic inflammatory diseases such as peptic ulcer and gout. The principal factors causing peptic ulcer are inadequate dietetic habits, prolonged use of non-steroidal anti-inflammatory drugs, stress and infection by *Helicobacter pylori*, in addition to other factors of genetic origin. Cytoprotective effects of chalcones from *Z. punctata* (Ortega et al. 2000; De la Rocha et al. 2003) and antiulcer activity of the *Z. punctata* extract has been reported (Falcao et al. 2008). The methanolic extract and aqueous extract (infusion) of *Z. punctata*, as well as DHC and DHMC isolated from them, were able to inhibit 84, 43, 63 and 25% of necrotizing gastric damage, respectively (100 mg/kg). Furthermore, xanthine oxidase (XO) catalyzes the oxidation of xanthine and hypoxanthine into uric acid. Its overproduction and/or excretion failure from human plasma (called hyperuricemia) induce an inflammatory process, an underlying cause of "gout", but also of hypertension or diabetes. Despite *Z. punctata* infusions showed XO inhibitory activity. In XO reaction, molecular oxygen acts as an electron acceptor, also generating O_2^- radical anion and H_2O_2 as products. It is interesting to note that *Z. punctata* showed O_2^- radical anion scavenging capacity and inhibited its synthesis by the enzymatic pathway.

9.5 Nematicidal Activity

Due to the appearance of drug- resistant worms, new therapeutic drugs with good efficacy and low toxicity are urgently needed. The flavonoids 7-HF, 3,7-DHF and DHC, isolated from aerial parts of *Z. punctata*, were assayed on the free-living nematode *Caenorhabditis elegans*. Only DHC showed an anthelmintic effect and alteration of egg hatching and larval development processes of *C. elegans*. At 17μg/ml, DHC was able to kill 50% of adult nematodes (D'Almeida et al. 2015). Therefore, DHC could be proposed as a potential anthelmintic drug.

9.6 Antihypertensive Activity

Recently, *in vitro* studies showed that *Z. punctata* extract and its major flavonoids have a vasorelaxant effect, sensitize acetylcholine-response, reduce phenylephrine vasoconstriction and antagonize angiotensin II-contractile response in arteries from hypercholesterolemic rabbits (Roco et al. 2017). Oral administration of *Z. punctata* extract on a rabbit model of hypercholesterolemia, reduced mean blood pressure, total cholesterol, triglycerides by about 65 and 50% respectively, reduced oxidative stress parameters (thiobarbituric acid reactive substances), aortic intima/media ratio and increased glutathione reduced/oxidized index in rabbits fed food enriched with 1% cholesterol (HD rabbits). No differences were found in aspartate aminotransferase, alanine aminotransferase, bilirubin, creatinine. Acetylcholine relaxation was normalized and contractile response to norepinephrine and angiotensin II was reduced in HD rabbits orally administrated with *Z. punctata* extract. The effects of *Z. punctata* extract and rosuvastatin were similar in almost all clinical and biochemical parameters except for TC (Roco et al. 2018). Oral administration of flavonoid-rich *Z. punctata* extract at doses of 2.5 mg GAE/day was not toxic as there was no sedation, itching and mortality seen; no differences existed on weight gain, organ size (heart and kidney), liver and kidney function or hematological parameters in experimental rabbits. These results are in harmony with previous *in vivo* studies using a mice model (Zampini et al. 2012).

Oral administration of *Z. punctata* extract as a natural product in the prevention of cardiovascular disease related to hypercholesterolemia and endothelial dysfunction is very promising.

9.7 Inhibition of Drug Resistance Mechanisms

One of the major drug resistance mechanisms in cells involves decreased uptake of water-soluble drugs, which require transporter to enter cells. This decrease can be attributed to the upregulation of drug transporters, such as P-glycoprotein (P-gp). These transporters can export a range of anticancer drugs from the cells, thus lowering the concentration to below the level required to provide a cytotoxic effect. *Z. punctata* hydroalcoholic extracts and its flavonoids 3,7-DHF and DHC, were able to modulate activity and expression of the membrane ABC transporters, in particular P-gp, the best-known membrane efflux pump involved in drug resistance. P-gp is physiologically expressed in the apical membrane of intestine, liver and kidney cells (Álvarez et al. 2010). The studies were realized in a human proximal tubule cell line (HK-2), and the results suggested an impact of *Z. punctata* extract and some of its flavonoid components on the pharmacokinetics of drugs, which are P-gp substrates, as well as a potential role on multidrug resistance modulation (Chieli et al. 2012).

9.8 Genotoxic/Antigenotoxic Effect

The hydroalcoholic extract of *Z. punctata* and DHC were investigated for genotoxicity/antigenotoxicity on human hepatoma HepG2 cells. The results indicated that *Z. punctata* extract, under the experimental conditions tested, neither affected cell viability nor induced DNA damage, and protected HepG2 cells against DNA damage induced by the direct-acting genotoxic compound (4-nitroquinoline-N-oxide). The antigenotoxic effect observed could be explained for the presence of DHC in this plant species (Zampini et al. 2012). This genotoxicological study suggests that the antigenotoxic properties of *Z. punctata* are of great pharmacological importance and might be beneficial for cancer prevention.

10 Conclusions

In this chapter, the broad medicinal potential of *Zuccagnia punctata* aerial parts extracts (antibiotics, antimycotic, anti-inflammatory, nematicide, antigenotoxic, antihypertensive, antioxidant effect) as well as their standardization is surveyed from the chemical point of view. Recent results in the determination of main bioactive constituents is also summarized. In accordance with the Argentine Regulations by the "Administración Nacional de Medicamentos, Alimentos y Tecnología Médica" (ANMAT 2015), based on its long history of medicinal use in Argentina by different communities, *Z. punctata* could be considered as an herbal medicine of traditional use (Art. 2). Considering the EMA guidelines (2010), the dry extracts of aerial parts from *Z. punctata* could be used in herbal medicinal products, as standardized extracts, declaring the quantity of main bioactive compounds or marker compounds (chalcones). Alternative uses include quantitative extracts (dry extract), dry or liquid extracts or incorporation into other herbal preparations or pharmaceutical preparations (gel, creams, microcapsules, capsules, tablets). While progress has been made in exploring the pharmacological properties of this species, still much research is needed, especially in the domain human clinical trials. Similarly, more studies are necessary to develop the good practices of production, including the harvesting in wild growing populations and the selection of best germplasm for cultivation trials to promote their sustainable use.

Acknowledgments The authors acknowledge the cooperation of the inhabitants of the studied areas and the financial support from the SCAIT-UNT (PIUNT 2018 G-637), the Agencia Nacional de Promoción Científica y Técnica (ANPCyT, PICT 2014-3136 and PICT 2017-4436), the Consejo Nacional de Investigaciones Científicas y Técnicas (CONICET, PUE -0011, Argentina).

References

Agüero MB, González M, Lima B, Svetaz L, Sánchez M, Zacchino S, Feresin G, Schmeda-Hirschmann G, Palermo J, Wunderlin D, Tapia A (2010) Argentinean propolis from *Zuccagnia punctata* Cav. (Caesalpinieae) exudates: phytochemical characterization and antifungal activity. J Agric Food Chem 58:194–201

Alberto MR, Nieva Moreno MI, Zampini IC, Isla MI (2007) Anti-inflammatory activity of structurally related natural flavonoids. BLACPMA 6:308–309

Álvarez AI, Real R, Pérez M, Mendoza G, Prieto JG, Merino G (2010) Modulation of the activity of ABC transporters (P-glycoprotein, MRP2, BCRP by flavonoids and drug response). J Pharm Sci 99:598–617

Álvarez SL, Cortadi A, Juárez MA, Petenatti E, Tomi F, Casanova J, van Baren CM, Zacchino S, Vila R (2012) (−)-5,6-dehydrocamphor from the antifungal essential oil of *Zuccagnia punctata*. Phytochem Lett 5:194–199

Álvarez M, Enrico R., Zampini IC, Isla MI (2019) Micropropagación de una especie endémica de Argentina, *Zuccagnia punctata*. XIV Jornadas Internas de Investigación, Docencia y Extensión de la Facultad de Ciencias Naturales e Instituto Miguel Lillo. 3–4 December 2019, San Miguel de Tucumán, Tucumán, Argentina, p 10

ANMAT (2015) (Administración Nacional de Medicamentos, Alimentos y Tecnología Médica). Resolución N° 5418

Ávila V, Bertolotti SG, Criado S, Pappano N, Debattista N, García NA (2001) Antioxidant properties of natural flavonoids: quenching and generation of singlet molecular oxygen. J Food Sci Technol 36:25–33

Bale AT, Khan KM, Salar U, Chigurupati S, Fasina T, Ali F, Perveen S (2018) Chalcones and bis-chalcones: as potential α-amylase inhibitors; synthesis, *in vitro* screening, and molecular modeling studies. Bioorg Chem 79:179–189

Barboza GE, Cantero JJ, Nuñez C, Pacciaroni A, Ariza Espinar L (2009) Medicinal plants: a general review and a phytochemical and ethnopharmacological screening of the native Argentine Flora. Kurtziana 32:7–365

Batovska DI, Todorova IT (2010) Trends in utilization of the pharmacological potential of chalcones. Curr Clin Pharmacol 5:1–29

Burkart A (1952) Las Leguminosas argentinas, silvestres y cultivadas, 2nd edn. Ed. ACME, Buenos Aires, Argentina, 590 pp

Burkart R, Bárbaro NO, Sánchez RO, Gómez DA (1999) Eco-regiones de la Argentina. Ed. Secretaría de Recursos Naturales y Desarrollo Sustentable, Buenos Aires, Argentina, 42 pp

Butassi E, Svetaz LA, Ivancovich JJ, Feresin GE, Tapia A, Zacchino SA (2015) Synergistic mutual potentiation of antifungal activity of *Zuccagnia punctata* Cav. and *Larrea nitida* Cav. extracts in clinical isolates of *Candida albicans* and *Candida glabrata*. Phytomedicine 22:666–578

Cabrera AL (1976) Regiones Fitogeográficas Argentinas. In: Kugler WF (ed) Enciclopedia Argentina de Agricultura y Jardinería. Segunda edición. Ed. ACME, Buenos Aires, Argentina, 85 pp

Cabrera AL, Willink A (1973) Biogeografía de América Latina. Ed. O. E. A. Serie de Biología, Washington, DC, Argentina, 120 pp

Cai CY, Rao L, Rao Y, Guo JX, Xiao ZZ, Cao JY, Wang B (2017) Analogues of xanthone-chalcones and bis-chalcones as α-glucosidase inhibitors and anti-diabetes candidates. Eur J Med Chem 130:51–59

Cantero JC, Nuñez CO, Bernardello G, Amuchástegui A, Mulko J, Brandolin P, Palchetti MV, Iparraguirre J, Virginil N, Ariza Espinar L (2019) Las plantas de importancia económica en Argentina. 1st Ed, UniRío Editora, Universidad Nacional de Río Cuarto, Río Cuarto, Argentina, 937 pp. https://botanicaargentina.org.ar/wp-content/uploads/2019/06/978-987-688-332-0-baja-resoluci%C3%B3n.pdf. (Last access, December 2019)

Carabajal MPA, Isla MI, Zampini IC (2017) Evaluation of antioxidant and antimutagenic activity of herbal teas from native plants used in traditional medicine in Argentina. S Afr J Bot 110:258–265

Carabajal MPA, Perea MC, Isla MI, Zampini IC (2020a) The use of Jarilla native plants in a Diaguita-Calchaquí indigenous community from northwestern Argentina: an ethnobotanical, phytochemical and biological approach. J Ethnopharmacol 247, 112258

Carabajal MPA, Isla MI, Borsarelli CD, Zampini IC (2020b) Influence of *in vitro* gastro-duodenal digestion on the antioxidant activity of single and mixed three "Jarilla" species infusions. J Herb Med 19:100296

Chieli E, Romiti N, Zampini IC, Garrido G, Isla MI (2012) Effects of *Zuccagnia punctata* extracts and their flavonoids on the function and expression of ABCB1/P-glycoprotein multidrug transporter. J Ethnopharmacol 144:797–801

D'Almeida RE, Alberto MR, Morgan PG, Sedensky MM, Isla MI (2015) Effect of structurally related flavonoids from *Zuccagnia punctata* Cav. on *Caenorhabditis elegans*. Acta Parasitol 60:164–172

De la Rocha N, María AOM, Gianello JC, Pelzer L (2003) Cytoprotective effects of chalcones from *Zuccagnia punctata* and melatonin on gastroduodenal tract in rats. Pharmacol Res 48:97–99

Del Vitto LA, Petenatti EM, Petenatti ME (1997) Recursos herbolarios de San Luis (República Argentina) Primera parte: Plantas Nativas. Herbal resources of San Luis (Argentina). First part: native plants. Multequina 6:49–66

Diario Chilecito.com (2019) Available in http://www.diariochilecito.com.ar/articulo/4796.html. (Last access, December 2019)

EMA (2010) Guideline on declaration of herbal substances and herbal preparations in herbal medicinal products/traditional herbal medicinal products. Available at http://www.ema.europa.eu/docs/en_GB/document_library/Scientific_guideline/2009/09/WC500003272.pdf. (Last access, December 2019)

Falcao HS, Mariath IR, Diniz MFFM, Batista LM, Barbosa-Filho JM (2008) Plants of the American continent with antiulcer activity. Phytomedicine 15:132–146

Hatano T, Shintani Y, Aga Y, Shiota S, Tsuchiya T, Yoshida T (2000) Phenolic constituents of licorice. VIII. Structures of glycophenone and glycoisoflavanone, and effects of licorice phenolics on methicillin-resistant *Staphylococcus aureus*. Chem Pharm Bull 48:1286–1292

Lahtchev KL, Batovska DI, Parushev SP, Ubiyvovk VM, Sibirny AA (2008) Antifungal activity of chalcones: a mechanistic study using various yeast strains. Eur J Med Chem 43:2220–2228

Lersten NR, Curtis JD (1996) Survey of leaf anatomy, especially secretory structures, of tribe Caesalpinieae (Leguminosae, Caesalpinioideae). Pl Syst Evol 200:1–39

Lozano L (2011) Ecofisiología de la tuna (*Opuntia ficus-indica* (L.) Mill.). Horticultura Argentina 30(72):37–52

Mahapatra DK, Bharti SK (2016) Therapeutic potential of chalcones as cardiovascular agents. Life Sci 148:154–172

Mahapatra DK, Asati V, Bharti SK (2015) Chalcones and their therapeutic targets for the management of diabetes: structural and pharmacological perspectives. Eur J Med Chem 92:839–865

Mercado MI, Ruiz AI, Zampini IC, Nuño G, Cuello S, Isla MI, Ponessa GI (2013) Arquitectura y morfoanatomía foliar y caulinar de *Zuccagnia punctata* (Fabaceae). Histolocalización de Compuestos Bioactivos. Lilloa 50:58–68

Morán Vieyra F, Boggetti H, Zampini I, Ordoñez R, Isla M, Álvarez R, De Rosso V, Mercadante A, Borsarelli C (2009) Singlet oxygen quenching and radical scavenging capacities of structurally related flavonoids present in *Zuccagnia punctata* Cav. Free Radic Res 43:553–564

Moreno A, Nuño G, Cuello S, Sayago J, Alberto MR, Zampini C, Isla MI (2015a) Anti-inflammatory, antioxidant and antimicrobial activity characterization and toxicity studies of flowers of "Jarilla", a medicinal shrub from Argentina. Nat Prod Commun 6:991–994

Moreno MA, Nuño G, Cuello S, Zampini IC, Mercado M, Ponessa G, Sayago JE, Isla MI (2015b) Histochemical localization and characterization of chalcones in foliar surface of *Zuccagnia punctata* Cav. Insight into their physiological role. Phytochem Lett 13:134–140

Moreno MA, Gómez-Mascaraque L, Arias M, Zampini IC, Sayago JE, Pino Ramos LL, Schmeda-Hirschmann G, López-Rubio A, Isla MI (2018a) Electrosprayed chitosan microcapsules as delivery vehicles for vaginal phytoformulations. Carbohydr Polym 201:425–437

Moreno MA, Córdoba S, Zampini IC, Mercado MI, Ponessa G, Sayago JE, Pino Ramos LL, Schmeda-Hirschmann G, Isla MI (2018b) Argentinean *Larrea* dry extracts with potential use in vaginal candidiasis. Nat Prod Commun 13(2):171–174

Norma RE-84-2010-Secretaria de Agricultura y Desarrollo sustentable de la Nación, Argentina. http://www.loa.org.ar/legNormaDetalle.aspx?id=11091. (Last access, December 2019)

Nuño G, Alberto M, Zampini I, Cuello S, Ordoñez R, Sayago J, Baroni V, Wunderlin D, Isla MI (2014) The effect of *Zuccagnia punctata* Cav, an Argentina medicinal plant, on virulence factors from *Candida* species. Nat Prod Commun 9:933–936

Nuño G, Alberto MR, Arena ME, Zampini IC, Isla MI (2018) Effect of *Zuccagnia punctata* Cav. (Fabaceae) extract on pro-inflammatory enzymes and on planktonic cells and biofilm from *Staphylococcus aureus*. Toxicity studies. Saudi J Biol Sci 25:1713–1719

Ortega CA, María AOM, Gianello JC (2000) Chemical components and biological activity of *Bidens subalternans*, *B. aurea* (Astereaceae) and *Zuccagnia punctata* (Fabaceae). Molecules 5:465–470

Paoli H, Bianchi AR, Yáñez CE, Volante JN, Fernández DR, Mattalía MC, Noé YE (2002) Recursos Hídricos de la Puna, Valles y Bolsones áridos del Noroeste Argentino. Convenio INTA EEA Salta-CIED, Argentina, pp 51–54

Pederiva R, Giordano O (1984) 3,7-Dihydroxy-8-methoxyflavone from *Zuccagnia punctata*. Phytochemistry 23:1340–1341

Pederiva R, Kavka J, D'Arcangelo AT (1975) Chalconas y flavanonas aisladas de *Larrea nitida* Cav. An Asoc Quim Argent 63:85–90

Ratera EL, Ratera MO (1980) Plantas de la flora argentina empleadas en medicina popular. Ed. Hemisferio Sur. Buenos Aires, Argentina, 189 pp

Roco J, Alarcón G, Sierra L, Zampini C, Isla MI, Jérez S (2017) Beneficial effects of hydroalcoholic extract and flavonoids from *Zuccagnia punctata* in a rabbit model of vascular dysfunction induced by high cholesterol diet. Med Chem Res 26:2336–2344

Roco J, Alarcón G, Medina M, Zampini C, Isla MI, Jérez S (2018) Oral administration of *Zuccagnia punctata* extract improves lipid profile, reduces oxidative stress and prevents vascular dysfunction in hypercholesterolemic rabbits. Phytomedicine 48:104–111

Svetaz L, Tapia A, López S, Furlán R, Petenatti E, Pioli R, Schmeda-Hirschmann G, Zacchino S (2004) Antifungal chalcones and new caffeic acid esters from *Zuccagnia punctata* acting against soybean infecting fungi. J Agric Food Chem 52:3297–3300

Svetaz L, Agüero MB, Álvarez S, Luna L, Feresin G, Derita M, Tapia A, Zacchino S (2007) Antifungal activity of chalcones from *Zuccagnia punctata* Cav. acting against clinically important fungi and studies of mechanism of action. Planta Med 73:1074–1080

Tangarife-Castaño V, Correa-Royero J, Zapata-Londono B, Durán C, Stanshenko E, Mesa-Arango AC (2011) Anti-*Candida albicans* activity, cytotoxicity and interaction with antifungal drugs of essential oils and extracts from aromatic and medicinal plants. Infection 15:160–167

Toursarkissian M (1980) Plantas medicinales de la Argentina; sus nombres botánicos, vulgares, usos y distribución geográfica. Ed. Hemisferio Sur, Buenos Aires, Argentina, 178 pp

Ulibarri EA (2005) *Zuccagnia punctata* (Leguminosae): ¿nuevo o viejo endemismo argentino? Darwiniana 43:212–215

Zampini IC, Vattuone MA, Isla MI (2005) Antibacterial activity of *Zuccagnia punctata* Cav. ethanolic extracts. J Ethnopharmacol 102:450–456

Zampini IC, Villarini M, Moretti M, Dominici L, Isla MI (2008) Evaluation of genotoxic and antigenotoxic effects of hydroalcoholic extracts of *Zuccagnia punctata* Cav. J Ethnopharmacol 115:330–335

Zampini IC, Villena J, Salva S, Herrera M, Isla MI, Álvarez S (2012) Potentiality of standardized extract and isolated flavonoids from *Zuccagnia punctata* for the treatment of respiratory infections by *Streptococcus pneumoniae*: *In vitro* and *in vivo* studies. J Ethnopharmacol 140:287–292

Zuloaga FO, Morrone O (1999) Catálogo de las Plantas Vasculares de la República Argentina. II. Fabaceae-Zygophyllaceae (Dicotyledoneae). Monography Systematic Botany. Missouri Botany Garden, USA, p 74

The manufacturer's authorised representative in the EU is Springer Nature Customer Service Centre GmbH, Europaplatz 3, 69115 Heidelberg, Germany. If you have any concerns regarding our products, please contact ProductSafety@springernature.com

Printed and bound by CPI Group (UK) Ltd, Croydon, CR0 4YY
30/09/2026
02225445-0016